# Reagents for Organic Synthesis

# Reagents for Organic Synthesis

VOLUME 6

**Mary Fieser**

**Louis F. Fieser**

A WILEY-INTERSCIENCE PUBLICATION
JOHN WILEY & SONS
NEW YORK · LONDON · SYDNEY · TORONTO

Library of Congress Catalog Card Number: 66–27894

ISBN 0-471-25873-3

Printed in the United States of America.

10 9 8 7 6 5 4 3 2 1

# PREFACE

This volume covers literature for the most part from August 1974 through December 1975. It includes references to about 400 reagents previously discussed in this series and to about an equal number of reagents that are included for the first time. Of course we realize that few of these newer reagents will become as valuable as, say, Grignard reagents, but we have tried to include reagents that open new vistas in organic synthesis. For example, we have included for the first time a reagent based on the lanthanide ytterbium, because, with the exception of cerium, this area has been rather neglected by organic chemists.

We are again indebted to colleagues who have furnished suggestions or additional information. We are particularly grateful to the following chemists who have again been most helpful in the preparation of the book: Professors John A. Secrist and Robert H. Wollenberg, Dr. William Moberg, Dr. Mark A. Wuonola, Dr. James V. Heck, Dr. Oljan Repič and Dr. Rick Danheiser. We acknowledge with gratitude the help for the first time of Dr. Ving Lee, Dr. May Lee, Dr. David Hesson, David Wenkert, Homer Pearce, Peter Ulrich, William Roush, Howard Simmons, III, Donald W. Landry, John Lechleiter, Stephen Kamin, Janice Smith, and Dale Boger. The photograph on the dust cover was taken by Mary Hanlon-Wollenberg.

We thank the Chemistry Department of Harvard University for use of an office and for limited access to the library.

MARY FIESER
LOUIS F. FIESER

*Belmont, Massachusetts*
*November 1976*

*v*

# CONTENTS

Reagents    1

Index of Reagents according to types    677

Author index    693

Subject index    727

# Reagents for Organic Synthesis

# A

**Acetic acid–Acetic anhydride.**

*Isomerization of vinylcarbinols to allylic alcohols.*[1] Treatment of a tertiary vinylcarbinol, available by the reaction of a ketone with a vinylmagnesium halide, with acetic acid–acetic anhydride and a trace of *p*-toluenesulfonic acid at ~20° (cooling) for several hours results in isomerization to the acetate of the isomeric primary allylic alcohol. An example is the preparation of 3,3-dimethylallyl alcohol from 2-methyl-3-butene-2-ol.

$$(CH_3)_2\overset{\underset{|}{OH}}{C}-CH=CH_2 \xrightarrow[\text{TsOH}]{\text{HOAc, Ac}_2\text{O}} (CH_3)_2C=CHCH_2OAc \xrightarrow[60-70\%]{\text{KOH, H}_2\text{O}} (CH_3)_2C=CHCH_2OH$$

[1] J. H. Babler and D. O. Olsen, *Tetrahedron Letters*, 351 (1974); J. H. Babler, D. O. Olsen, and M. Turner, *Org. Syn.*, submitted (1975).

**Acetic anhydride, 1, 3; 2, 7–10; 5, 3–4.**

*α-Methylenelactam rearrangement.* Ferles[1] and Rueppel and Rapoport[2] have reported that treatment of 1-methylnipecotic acid (1) with refluxing acetic anhydride for several hours and then with potassium carbonate at 0° leads to 1-methyl-3-methylene-2-piperidone (2) in high yield. Actually, before their work this rearrangement had been carried out with lysergic acid and derivatives,[3] but

(1)                    (2)

had attracted little notice. Rapoport *et al.*[4] have recently shown that this rearrangement is a general reaction for five- and six-membered cyclic β-amino acids and have established the mechanism shown in equation I. The rearrangement involves formation of a mixed anhydride, β-elimination, and, finally, recyclization.

Substituents on nitrogen and at the α- and α'-positions to nitrogen have little effect on the reaction. The high selectivity of the rearrangement is shown by the fact that the four isomers of the acid (3) yield exclusively the same unsaturated lactam (4), which has the thermodynamically more stable *trans*-configuration with respect to the carbonyl group.[5]

(3, four isomers)            (4)

The rearrangement has been used by Rapoport *et al.*[6] to obtain the fused pyridone–lactone DE ring system in a total synthesis of *dl*-camptothecin (5).

(5)

[1] M. Ferles, *Czech. Commun.*, **29**, 2323 (1964).

[2] M. L. Rueppel and H. Rapoport, *Am. Soc.*, **94**, 3877 (1972).

[3] W. A. Jacobs and L. C. Craig, *ibid.*, **60**, 1701 (1938); A. Stoll, A. Hofmann, and F. Troxler, *Helv.*, **32**, 506 (1949).

[4] D. L. Lee, C. J. Morrow, and H. Rapoport, *J. Org.*, **39**, 893 (1974).

[5] D. Thielke, J. Wegener, and E. Winterfeldt, *Angew. Chem. internat. Ed.*, **13**, 602 (1974).

[6] J. T. Plattner, R. D. Gless, and H. Rapoport, *Am. Soc.*, **94**, 8613 (1972); C. Tang and H. Rapoport, *ibid.*, **94**, 8615 (1972).

## Acetic anhydride–Boron trifluoride etherate.

*Regeneration of* $\Delta^5$*-steroids from* $3\alpha,5$*-cyclosteroids.* The 5,6-double bond of steroids is often protected by conversion to the *i*-steroid (1). The $3\beta$-ol-$\Delta^5$-steroid was regenerated in recent research by reflux in acetic acid with

fused zinc acetate or by treatment with a large quantity of alumina. Indian chemists[1] report that the conversion of (1) to (2) can be carried out in quantitative yield by treatment with acetic anhydride and $BF_3 \cdot$ etherate at $0°$ for about 15 min. Simultaneous attack of the reagents on the cyclopropane ring is considered to be involved.

[1] C. R. Narayanan, S. R. Prakash, and B. A. Nagasampagi, *Chem. Ind.*, 966 (1974).

## Acetic anhydride–Methanesulfonic acid.

*Pummerer rearrangement* (5, 3–4). 2-Phenylsulfonyl ketones (1)[1] can be converted into (2) by a Pummerer rearrangement using acetic anhydride and a catalytic amount of methanesulfonic acid in $CH_2Cl_2$ at $20°$.[2] The rearrangement provides a general synthesis of $\alpha$-phenylthio-$\alpha,\beta$-unsaturated ketones (2).

[1] H. J. Monteiro and J. P. de Souza, *Tetrahedron Letters*, 921 (1975).
[2] H. J. Monteiro and A. L. Gemal, *Synthesis*, 437 (1975).

## Acetic anhydride–Phosphoric acid.

*Aromatization of* $\alpha$*-tetralone oximes* (1).[1] When the oximes (1) are heated in acetic anhydride and anhydrous phosphoric acid (1, 860) for 30 min. at $80°$, $N$-(1-naphthyl)acetamides (2) are obtained in 82–93% yields. Lower yields have been obtained with acetic acid–acetic anhydride containing hydrogen chloride or hydrogen bromide.

[1]M. S. Newman and W. M. Hung, *J. Org.*, **38**, 4073 (1973).

## Acetic anhydride–Pyridine.

*Enamides.*[1] Treatment of an aldoxime with refluxing acetic anhydride and a base affords the corresponding nitrile. Ketoximes under these conditions are converted into enamides or enimides. Thus cholestanone oxime (1), when refluxed in acetic anhydride and pyridine for 10 hr., is converted into the enimide (2) in high yield. When (2) is chromatographed on alumina, it is converted quantitatively into the enamide (3).

Enamides can also be obtained in comparable yields by the reaction of ketoximes with acetic anhydride, DMF, and chromium(II) acetate or titanium(III) acetate at room temperature. Under these conditions the enamides are not further acetylated. However, Boar and Barton consider acetic anhydride in pyridine to be the reagent of choice for reductive acetylation of ketoximes.

*Corticosteroid side-chain synthesis.* The paper[1] reports a potentially useful reaction of enamides. Thus the enamide (3) is oxidized by lead tetraacetate in benzene to 2α-acetoxy-5α-cholestane-3-one (4) in 69% yield. This reaction has

been developed into a one-pot synthesis of the corticosteroid side chain from 20-keto steroids as formulated in equation II[2]:

*13*-epi-*Steroids.*[3] Treatment of the oxime of 3β-acetoxyandrostene-5-one-17 (1) with acetic anhydride and pyridine (prolonged reflux) gives a mixture of (2) and (3). If the reaction mixture is chromatographed on alumina, (3) is ob-

tained as the only product (85% yield). These products are 13-*epi*-steroids, previously obtained less conveniently by irradiation of 17-ketosteroids with ultraviolet light.[4]

This reaction, which involves ring opening and reclosing, provides evidence for a radical mechanism for the acetic anhydride–pyridine reaction.

Reduction of (1) with chromous acetate in the presence of acetic anhydride gives the normal enamide with no epimerization of the 13-methyl group.

[1] R. B. Boar, J. F. McGhie, M. Robinson, D. H. R. Barton, D. C. Horwell, and R. V. Stick, *J.C.S. Perkin I,* 1237 (1975).
[2] R. B. Boar, J. F. McGhie, M. Robinson, and D. H. R. Barton, *J.C.S. Perkin I,* 1242 (1975).
[3] R. B. Boar, F. K. Jetuah, J. F. McGhie, M. S. Robinson, and D. H. R. Barton, *J.C.S. Chem. Comm.,* 748 (1975).
[4] L. F. Fieser and M. Fieser, *Steroids,* Rheinhold, New York, 1959, p. 464.

**Acetic anhydride–Sodium acetate.**

*Pummerer rearrangement* (5, 3–4). The Pummerer rearrangement of β-hydroxy sulfoxides to derivatives of α-hydroxy aldehydes has been extended to the rearrangement of β-keto sulfoxides.[1] Thus rearrangement of (1) with acetic anhydride–sodium acetate in toluene under reflux gives the S-aryl thioester (2) in 74% yield. The ester is hydrolyzed by base to mandelic acid (3). In the absence of sodium acetate the normal product of the Pummerer rearrangement

$$C_6H_5\underset{\underset{O}{\parallel}}{C}CH_2\underset{\underset{O}{\parallel}}{S}C_6H_5 \xrightarrow[74\%]{\underset{NaOAc}{Ac_2O}} C_6H_5\underset{\underset{OAc}{|}}{C}H\underset{\underset{O}{\parallel}}{C}SC_6H_5 \xrightarrow{NaOH} C_6H_5\underset{\underset{OH}{|}}{C}HCOOH$$

$$(1) \qquad\qquad\qquad\qquad (2) \qquad\qquad\qquad\qquad (3)$$

$$\downarrow Ac_2O$$

$$C_6H_5\underset{\underset{O}{\parallel}}{C}\overset{\overset{OAc}{|}}{C}HSC_6H_5 \xrightarrow{NaOAc}$$

$$(4)$$

(4) is formed. This is converted into (2) by sodium acetate or pyridine. There-fore in the modified reaction an intramolecular oxidation–reduction is involved.

Note that β-keto sulfoxides are readily available by the reaction of esters with dimsylsodium (**1**, 310–311). The reaction thus constitutes a method for prepara-tion of one-carbon homologated α-hydroxy acids from esters.

$$RCOOR' \xrightarrow{CH_3SOC\overset{-}{H}_2Na} R\underset{\underset{O}{\parallel}}{C}CH_2\overset{\overset{O}{\parallel}}{S}CH_3 \xrightarrow[NaOAc]{Ac_2O} R\underset{\underset{OAc}{|}}{C}H\overset{\overset{O}{\parallel}}{C}SCH_3 \longrightarrow R\underset{\underset{OH}{|}}{C}HCOOH$$

[1]S. Iriuchijima, K. Maniwa, and G. Tsuchihashi, *Am. Soc.*, **97**, 596 (1975).

### Acetic anhydride–Triethylamine, 5, 4.

*α-Aryl-γ-benzylidene-Δ$^{α,β}$-butenolides.*[1] These substances can be obtained in 40–85% yield by heating phenylpropargyl aldehydes or ketones with arylacetic acids in the presence of acetic anhydride and triethylamine at 150° for 18 hr. followed by treatment with acid.

$$C_6H_5C\equiv C\overset{\overset{O}{\parallel}}{C}R \quad + \quad ArCH_2COOH \xrightarrow{Ac_2O,\ N(C_2H_5)_3} \begin{bmatrix} C_6H_5C\equiv C - \overset{\overset{O}{\parallel}}{C}R \\ \underset{HOOC}{\diagdown}\underset{Ar}{C}\diagup \end{bmatrix} \xrightarrow{H^+} C_6H_5CH= $$

$$(R = H, CH_3, C_6H_5)$$

[1]Y. S. Rao and R. Filler, *Tetrahedron Letters*, 1457 (1975).

### Acetoinenediolcyclopyrophosphate [Di-(4,5-dimethyl-2-oxo-1,3,2-dioxaphos-pholenyl)oxide] (1). Mol. wt. 282.13, m.p. 84–86°.

*Preparation:*

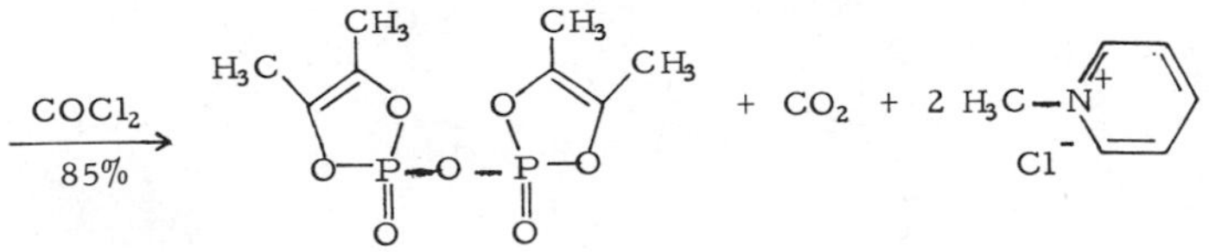

$$\xrightarrow[85\%]{COCl_2} \quad + \quad CO_2 \quad + \quad 2\ H_3C-N^+ \quad Cl^-$$

(1)

*Phosphodiesters.*[1] The pyrophosphate (1) reacts with alcohols or phenols in the presence of 1 eq. of a base (2,4,6-collidine) to give (2) and the collidinium salt (3). The product (2) can react with another alcohol in the absence of a base to give dialkylacetoinyl phosphates (4) in yields of about 95% (4, 536). The

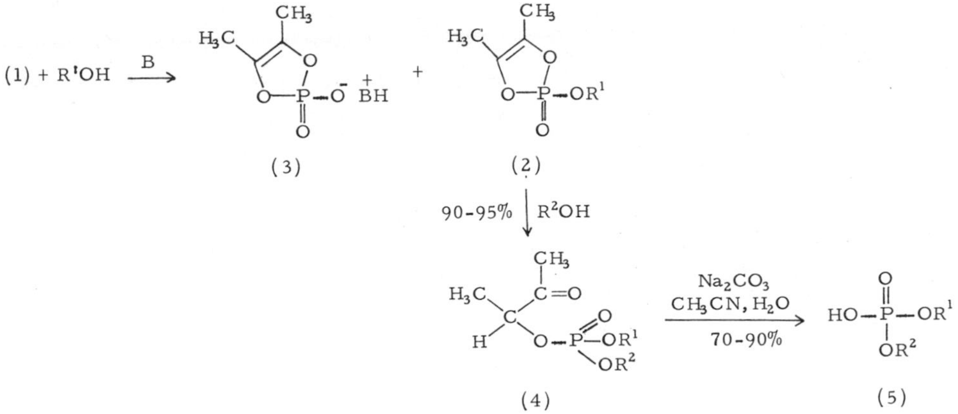

$$(1) + R'OH \xrightarrow{B}$$

(3)          (2)

$$90\text{-}95\% \downarrow R^2OH$$

$$(4) \xrightarrow[70\text{-}90\%]{\substack{Na_2CO_3 \\ CH_3CN,\ H_2O}} (5)$$

(4)          (5)

acetoinyl blocking group can be removed from (4) by sodium carbonate in water–acetonitrile or by triethylamine in aqueous pyridine.

If one of the two alcohols is a primary, secondary diol the reagent reacts selectively with the primary hydroxyl group.

*4,5-Dimethyl-2-chloro-2-oxo-1,3,2-dioxaphosphole* (2). This new reagent for phosphorylative coupling of two different alcohols can be prepared from (1) in 82% overall yield as formulated. This new reagent undergoes displacement

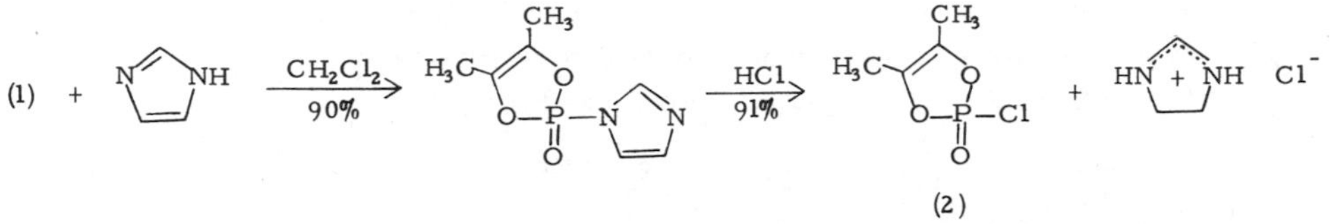

$$(1) + \quad \xrightarrow[90\%]{CH_2Cl_2} \quad \xrightarrow[91\%]{HCl} \quad + \quad HN^+\ NH\ \ Cl^-$$

(2)

with alcohols in the presence of a base (triethylamine or pyridine) in yields of 90–95%. It has been used for the synthesis of a pyrophosphate (3).[2]

(2) + R'OH + N(C$_2$H$_5$)$_3$ $\xrightarrow{90-95\%}$ [structure: 1,2-dimethylethylenedioxy phosphate with $CH_3$, $H_3C$, O, O–P–OR', ‖O] + $H\overset{+}{N}(C_2H_5)_3$ Cl$^-$

(2) + [$H_3CO$, $H_3CO$, PO$^-$, ‖O] [$H_3C–\overset{+}{N}$ pyridinium] $\longrightarrow$ [structure with $CH_3$, $CH_3$, $H_3CO$, $H_3CO$, P–O–P–O, O]

(3)

*One-flask phosphorylative coupling to unsymmetrical dialkyl phosphates* (4). Ramirez and Marecek[3] have developed a convenient method for the preparation of dialkylacetoinyl phosphates (3) without isolation of intermediates. The reagent is the *p*-nitrophenyl(1,2-dimethylethylenedioxy)phosphate (2), prepared by the reaction of *p*-nitrophenol with (1).[4] The reaction is carried out in $CH_2Cl_2$ by dropwise addition of R$^1$OH and triethylamine (1 eq. of each) to

[structure: $H_3C$, $CH_3$, O, O–P–OC$_6$H$_4$·NO$_2$, ‖O] + R'OH $\xrightarrow[N(C_2H_5)_3]{CH_2Cl_2}$ [ [structure: $H_3C$, $CH_3$, O, O–POR', ‖O] + p-NO$_2$C$_6$H$_4$O$^-$ $\overset{+}{H}N(C_2H_5)_3$ ]

(2)                                                                 (a)

91–93% $\downarrow$ R$^2$OH

[structure: $H_3C$, $CH_3$, C=O, C, H, O–P–OR', OR$^2$] $\xrightarrow{Na_2CO_3}$ HO–P–OR$^1$, ‖O, OR$^2$

(3)                              (4)

(2); after 15–30 min. at 25°, R$^2$OH (1 eq.) is added, and the reaction is allowed to proceed for 1–2 hr. at 25°. The final step involves the usual hydrolysis of the triester (3). This one-flask procedure is possible because the phenolic salt formed in the first step catalyses the reactions involving both R$^1$OH and R$^2$OH. Aryl phosphates are not formed because electron-withdrawing substituents decrease the reactivity of phenols toward both (2) and the intermediate (a).

[1]F. Ramirez, J. F. Marecek, and I. Ugi, *Synthesis*, 99 (1975); *Idem, Am. Soc.*, 97, 3809 (1975).

[2]F. Ramirez, H. Okazaki, and J. F. Marecek, *Synthesis*, 637 (1975).

[3]F. Ramirez and J. F. Marecek, *J. Org.*, 40, 2849 (1975).

[4]The corresponding phosphate in which the aryl group is $C_6F_5$ was also used with comparable results.

**Acetone, $CH_3COCH_3$.**

Ketones can be recovered in acceptable yields from hydrazones and oximes by exchange with acetone in a sealed tube at 50–80° (*cf.* **Pyruvic acid, 1,** 974).[1]

[1] S. R. Maynez, L. Pelavin, and G. Erker, *J. Org.*, **40,** 3302 (1975).

**3β-Acetoxy-17β-chloroformylandrostene-5 (1).**

(1)

This sterol[1] has been used to resolve squalene-2,3-diol by way of the diastereoisomeric esters.[2]

[1] J. Staunton and E. J. Eisenbraun, *Org. Syn.*, **42,** 4 (1962).
[2] R. B. Boar and K. Damps, *Tetrahedron Letters,* 3731 (1974).

**Acetyl bromide, $CH_3COBr$.** Mol. wt. 122.95, b.p. 75–77°. Suppliers: Aldrich, Eastman, others.

*Selective O-acetylation with AcBr and $CF_3COOH$.* Hydroxyl groups can be selectively acetylated in the presence of primary and secondary amino groups with acetyl bromide and trifluoroacetic acid.[1]

This system was used successfully to obtain diacetylapomorphine (1) in 70% yield. Acetylation of apomorphine with acetic anhydride catalyzed by a base

(1)                    (2)

(NaOAc, pyridine, triethylamine) or with acetic anhydride and acetic acid results mainly or exclusively in formation of the ring-opened product (2).[2] Success with the new system is attributed to the weak basicity of trifluoroacetate anion.

[1] R. J. Borgman, J. J. McPhillips, R. E. Stitzel, and I. J. Goodman, *J. Medicin. Chem.,* **16,** 663 (1973); R. J. Borgman, M. R. Baylor, J. J. McPhillips, and R. E. Stitzel, *ibid.,* **17,** 427 (1974).
[2] R. J. Borgman, R. V. Smith, and J. E. Keiser, *Synthesis,* 249 (1975).

**Acetyl p-toluenesulfonate, 2,** 14–15; **4,** 8.

*Cyclization.*[1] The dione (1) is converted into the estrane (2) by acetyl p-toluenesulfonate in acetic anhydride. Cyclization of (1) by the usual reagents is accompanied by dehydration. Thus cyclization of (3), which lacks a $C_{11}$-keto group, gives (4).

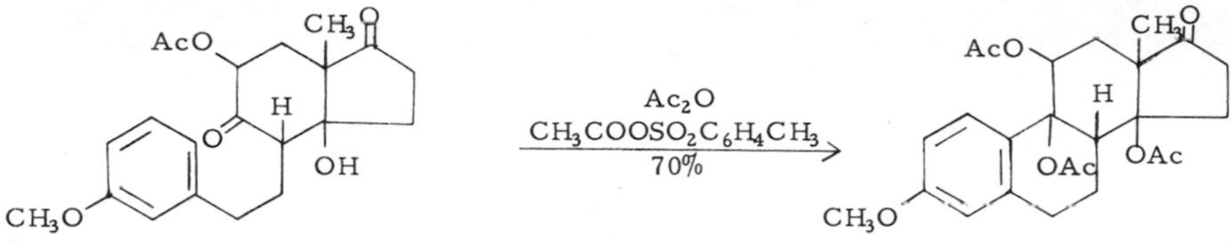

(1)

$$\xrightarrow[\text{70\%}]{\underset{CH_3COOSO_2C_6H_4CH_3}{Ac_2O}}$$

(2)

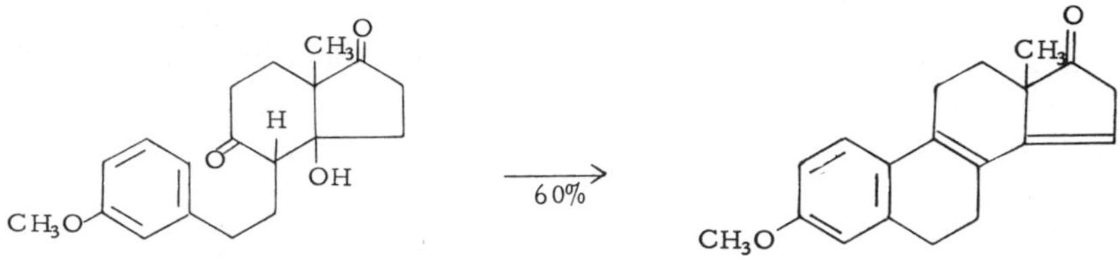

(3)

$$\xrightarrow{60\%}$$

(4)

[1] A. R. Daniewski, *J. Org.,* **40,** 3124 (1975); A. R. Daniewski and M. Koćor, *ibid.,* **40,** 3136 (1975).

**Adogen 464.** This is a methyltrialkylammonium chloride, in which the alkyl groups are a mixture of $C_8$–$C_{10}$ straight chains. Adogen is a trade name registered by Ashland Chemical Co. Supplier: Aldrich.

*Methylenation of catechols.* Bashall and Collins[1] have converted various o-dihydroxyphenols into the methylenedioxy derivative by addition of a solution of the phenol in aqueous NaOH to a mixture of dibromomethane, water, and this phase-transfer catalyst. Reported yields are 76–86%. Classical methods require anhydrous conditions and aprotic solvents.

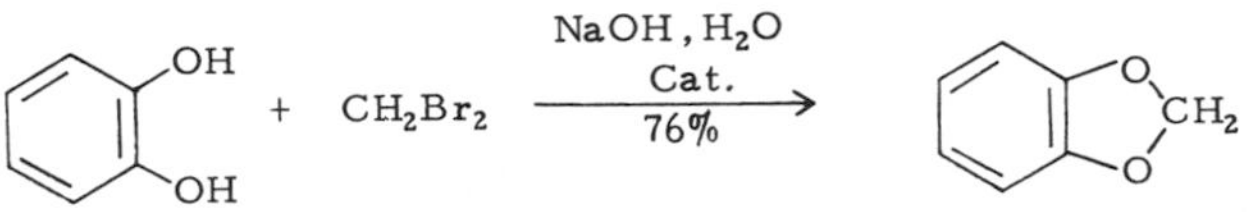

[1] A. P. Bashall and J. F. Collins, *Tetrahedron Letters,* 3489 (1975).

**β-Alanine, 1,** 16.

*Knoevenagel condensation.* Use of β-alanine as catalyst markedly improves the yield in the condensation of ethyl pyruvate and malononitrile. Potassium sulfide is much less effective.[1]

$$\underset{C_2H_5OOC}{\overset{CH_3}{>}}C=O + H_2C(CN)_2 \quad \xrightarrow[80-83\%]{\substack{20-40^\circ \\ 1-2\ hr. \\ Cat.}} \quad \underset{C_2H_5OOC}{\overset{CH_3}{>}}C=C(CN)_2$$

[1] Y. Yamada, K. Iguchi, K. Hosaka, and K. Hagiwara, *Synthesis*, 669 (1974).

**Alkyl orthotitanates,** $Ti(OR)_4$. Fluka supplies a number of these reagents: $R = n\text{-}C_3H_7, i\text{-}C_3H_7, n\text{-}C_4H_9$.

*β-Hydroxy esters.*[1] The reaction of ketene, a carbonyl compound, and an alkyl orthotitanate at room temperature leads to a β-hydroxy ester after hydrolysis. Yields are satisfactory when R is primary or secondary, but rather low when R is tertiary. Yields are also low in the case of hindered ketones.

$$CH_2=C=O + \underset{R^2}{\overset{R^1}{>}}C=O + Ti(OR)_4 \longrightarrow R^1-\underset{\underset{-Ti-}{\overset{|}{O}}}{\overset{\overset{R^2}{|}}{C}}-CH_2-C\overset{\nearrow O}{\underset{\searrow OR}{}} \quad \xrightarrow{H_3O^+} \quad R^1-\underset{\underset{OH}{|}}{\overset{\overset{R^2}{|}}{C}}CH_2COOR$$

[1] L. Vuitel and A. Jacot-Guillarmod, *Helv.*, **57**, 1703 (1974).

**Allyl alcohol, O,2-dilithio derivative,** $CH_2=C(Li)CH_2OLi$. Mol. wt. 69.94.
This reagent can be prepared by reaction of 2-bromoallyl alcohol in ether with 2.5 eq. of *t*-butyllithium at $-78$ to $0^\circ$. It reacts with carbonyl compounds to form, after hydrolysis, unsaturated diols of type (1) in 65–75% yield.[1]

$$>\!C=O + \underset{Li}{\overset{|}{CH_2}}=CCH_2OLi \quad \xrightarrow{65-75\%} \quad >\!\underset{\underset{CH_2}{\overset{||}{C}-CH_2OH}}{\overset{\nearrow OH}{C}}$$

(1)

[1] E. J. Corey and G. N. Widiger, *J. Org.*, **40**, 2975 (1975).

**S-Allyl  N,N-dimethyldithiocarbamate,** $CH_2=CHCH_2S\overset{\overset{S}{||}}{C}N(CH_3)_2$  (1). Mol. wt. 161.29.
The carbamate is prepared from allyl chloride, dimethylamine, and carbon disulfide (no experimental details).[1]

*trans-Disubstituted alkenes.*[1] The lithium salt of (1), prepared with LDA in THF at $-60^\circ$, is alkylated in high yield, possibly owing to chelation of the salt. The product (2) undergoes allylic rearrangement gradually at $25^\circ$ or rapidly at 100° to give the derivative of the E-olefin (3).

The product (3) can be alkylated again at the α-position to give (4) in excellent yield. If (4) is heated, it rearranges partially to give mixtures of (4) and (5).

The process thus provides a stereoselective route to *trans*-disubstituted alkenes.

If methallyl chloride, $CH_2=C(CH_3)CH_2Cl$, is used in place of allyl chloride, trisubstituted olefins can be prepared, usually as a mixture of E- and Z-isomers (6) and (7).

(6, E)                    +                    (7, Z)

Hayashi and Midorikawa[2] have since found that the dimethyldithiocarbamate group of the products obtained above can be reductively cleaved in high yield by either lithium in ethylamine or sodium in liquid ammonia. Consequently, di- and trisubstituted olefins are readily prepared in this way. The method was used successfully for synthesis of several *trans*-alkenol sex pheromones of insects, as formulated for the synthesis of E-7-tetradecene-1-ol acetate (8) in high overall yield.

1) LiN(C$_6$H$_{11}$)($i$-C$_3$H$_7$)
2) I(CH$_2$)$_5$OTHP
$\xrightarrow{\hspace{2cm}}$
90–93%

$$CH_3(CH_2)_5 \diagdown C=C \diagup^H_{\phantom{x}} \diagdown^H_{C} (CH_2)_5OTHP$$

$$\underset{\underset{S}{\overset{\|}{}}}{S-C-N(CH_3)_2}$$

$\xrightarrow{\text{Li/C}_2\text{H}_5\text{NH}_2}$
90%

$$CH_3(CH_2)_5\diagdown C=C \diagup^H_{\phantom{x}}\diagdown (CH_2)_6OTHP$$

$\xrightarrow[85\%]{\text{CH}_3\text{COCl/HOAc}}$

$$CH_3(CH_2)_5\diagdown C=C\diagup^H_{\phantom{x}}\diagdown (CH_2)_6OCOCH_3$$

$$(8)$$

The dimethyldithiocarbamate group can also be cleaved reductively with deactivated Raney nickel. This method favors formation of internal olefins and is useful for synthesis of olefins with an isopropylidene group, $=C(CH_3)_2$. For example, (9) is reduced mainly to (10)[3]:

$$CH_2=C\diagup^{CH_3}_{\phantom{x}} \quad \xrightarrow{92\%} \quad C_6H_5CH_2\diagdown C=C\diagup^{CH_3}_{\phantom{x}}\diagdown CH_3 \quad + \quad C_6H_5CH_2CH_2C\diagup^{CH_3}_{\phantom{x}}\diagdown CH_2$$

$$\underset{CHSCN(CH_3)_2}{\underset{CH_2C_6H_5}{}}$$

(9)                    (10)         6:1        (11)

*α,β-Unsaturated aldehydes.* Japanese chemists[4] have reported two methods for conversion of an alkyl halide, RX, into (E)-α,β-unsaturated aldehydes, RCH=CHCHO, one of which is formulated. S-Allyl N,N-dimethyldithio-carbamate (1) is converted into the lithium salt, which is sulfenylated with dimethyl disulfide. The initial product (a) under conditions of work-up under-goes allylic rearrangement to S-γ-methylthioallyl dithiocarbamate (2). This is treated with LDA and then an allyl halide. The α,β-unsaturated aldehyde (4) was obtained by hydrolysis with mercuric ion in aqueous acetonitrile.

$$CH_2=CHCH_2S\overset{\overset{S}{\|}}{C}N(CH_3)_2 \quad \xrightarrow[\text{2) CH}_3\text{SSCH}_3]{\text{1) LDA}} \quad \left[ \begin{array}{c} SCH_3 \\ C \\ CH \diagdown S \\ \| \quad | \\ CH_2 \quad C \\ S \diagdown N(CH_3)_2 \end{array} \right] \xrightarrow{88\%}$$

(1)                                              (a)

$$\begin{array}{c} SCH_3 \\ CH \\ CH \quad S \\ | \quad \| \\ CH_2 \quad C \\ S \diagdown N(CH_3)_2 \end{array} \quad \xrightarrow[\text{2) RX}]{\text{1) LDA}} \quad CH_3SCH \diagup^H_{\phantom{x}} C \diagdown C \diagup^{S\overset{\|}{C}N(CH_3)_2}_{\phantom{x}} \quad \xrightarrow[\substack{70-90\% \\ \text{from (2)}}]{\substack{\text{Hg(II)} \\ \text{CH}_3\text{CN, H}_2\text{O}}} \quad R\diagdown C=C\diagup^H_{\phantom{x}}\diagdown CHO$$

(2)                              (3)                              (4)

Use of S-methallyl N,N-dimethyldithiocarbamate in place of (1) results in synthesis of aldehydes of type (5), (E)-β-alkyl-α-methylacroleins.

$$\underset{H}{\overset{R}{\phantom{x}}}C=C\underset{CHO}{\overset{CH_3}{\phantom{x}}}$$

(5)

[1] T. Hayashi, *Tetrahedron Letters*, 339 (1974).
[2] T. Hayashi and H. Midorikawa, *Synthesis*, 100 (1975),
[3] T. Nakai, H. Shiono, and M. Okawara, *Chem. Letters*, 249 (1975).
[4] *Idem, Tetrahedron Letters*, 3625 (1974).

**Allylidenetriphenylphosphorane, 5, 7–9.**

*Polyene synthesis.*[1] Allylidenetriphenylphosphorane (1)[2] reacts with methyl *cis*-3-chloropropenoate (2) to form the stabilized ylide (3), presumably by Michael addition (a) and loss of chloride ion (b), and deprotonation. The ylide reacts with aldehydes and ketones to form polyenes, (4) and (5).

$$(C_6H_5)_3P=CHCH=CH_2 + \underset{Cl}{\overset{H}{\phantom{x}}}C=C\underset{COOCH_3}{\overset{H}{\phantom{x}}} \longrightarrow \left[ (C_6H_5)_3\overset{+}{P}CH=CHCH_2\overset{Cl}{\underset{|}{C}}\overset{-}{H}CHCOOCH_3 \right.$$

(1)                (2)                                (a)

$$\xrightarrow{-Cl} (C_6H_5)_3\overset{+}{P}CH=CHCH_2CH=CHCOOCH_3 \left. \right] \xrightarrow{-H^+} (C_6H_5)_3P=CHCH=CHCH=CHCOOCH_3$$

(b)                                                    (3)

$$(CH_3)_2CH(CH=CH)_3COOCH_3 \xleftarrow[67\%]{(CH_3)_2CHCHO} (3) \xrightarrow[84\%]{C_6H_5CHO} C_6H_5(CH=CH)_3COOCH_3$$

(5, 1:1.5 cis:trans)                                      (4, all trans)

Other β-chloro-α,β-unsaturated carbonyl compounds can replace (2) as shown in the examples.

*Examples:*

The procedure permits the synthesis of compounds with three double bonds conjugated with a carbonyl group. In all cases the γ-proton of the phosphorane is replaced by the β-carbon of the unsaturated ester or ketone.

Unfortunately, this reaction does not appear to be useful for synthesis of isoprenoid polyenes. Thus the ylide (6), prepared as shown, reacts with isobutyraldehyde to give four isomers in comparable amounts. Apparently both E- and Z-isomers of (6) are formed and each isomer affords a pair of *cis, trans*-isomers in the reaction with the aldehyde.

$$(1) + CH_3CCl=CHCOOCH_3 \longrightarrow (C_6H_5)_3P=CHCH=CHC(CH_3)=CHCOOCH_3$$
$$(6)$$

[1] E. Vedejs and J. P. Bershas, *Tetrahedron Letters*, 1359 (1975).
[2] Generated from allyltriphenylphosphonium bromide with LDA in THF.

*trihapto*-**Allyltris(trimethylphosphite)cobalt(I),**  $\eta^3$-$C_3H_5$Co[P(OCH$_3$)$_3$]$_3$  (1). Mol. wt. 469.23, m.p. 154° dec.

*Preparation.*[1] This orange cobalt complex is prepared from KCo[P(OCH$_3$)$_3$]$_4$ and allyl iodide in THF (room temperature).

*Catalytic hydrogenation of arenes.* Muetterties and Hirsekorn[2] reported that this soluble complex catalyzes hydrogenation of arenes at 25°. More importantly, Muetterties *et al.*[3] find that arenes are hydrogenated with this catalyst more readily than alkenes. Thus benzene is reduced to cyclohexane at a rate three to four times as fast as cyclohexene is reduced to cyclohexane and about twice as fast as 1- or 2-hexene is reduced to hexane. Moreover, *cis*-addition of hydrogen was established in the hydrogenation of xylene and of mesitylene. The hydrogenation of benzene does not proceed through cyclohexadiene or cyclohexene. Therefore it appears that the arene is attached to the metal atom until complete reduction has occurred. Moreover, the allylcobalt complex is recovered almost quantitatively after hydrogenation. The original paper should be consulted for a possible reaction sequence.

Further evidence for *cis*-hydrogenation with this catalyst is that all *cis*-cyclohexane-$d_6$ (I) is obtained from $C_6D_6$ and only *cis*-decalin (II) is obtained from naphthalene.[4]

(I)                                        (II)

[1] E. L. Muetterties and F. J. Hirsekorn, *Am. Soc.*, **96**, 7920 (1974).
[2] *Idem, ibid.*, **96**, 4063 (1974).
[3] F. J. Hirsekorn, M. C. Rakowski, and E. L. Muetterties, *ibid.*, **97**, 237 (1975).
[4] E. L. Muetterties, M. C. Rakowski, F. J. Hirsekorn, W. D. Larson, V. J. Basus, F. A. L. Anet, *Am. Soc.*, **97**, 1266 (1975).

**Alumina, 1,** 19–20; **2,** 17; **3,** 6; **4,** 8.

*Oxidation.* When a benzene solution of 3-nitrofluorene-9-ol (1) is passed through a column of alumina, 3-nitrofluorene-9-one (2) is obtained in about

50% yield. 9-Fluorenol is not oxidized under the same conditions. Air oxidation is involved in the conversion of (1) to (2), but apparently the nitro groups are also responsible for more than half of the reaction. 2-Nitrofluorene-9-ol is also oxidized in this way to the corresponding ketone (85% yield).[1]

*Cyclobutenones.*[2] 3-Alkoxycyclobutanones, obtained by cycloaddition of ketenes to enol ethers,[3] are converted into 2-cyclobutenones by chromatography on alumina (Woelm, activity grade I) in 50–80% yield.

*Reactions on alumina surfaces.* The water present in commercially available aluminas is known to effect some undesired side reactions and has been used to some extent to effect desired transformations (*e.g.*, selective hydrolysis, **2,** 3601). Posner *et al.*[4] reasoned that replacement of the water in alumina by alcohols, thiols, or amines should result in doped-aluminas useful for organic synthesis. In the general procedure the alumina is dehydrated by heating at 400° and 0.06 torr for 24 hr. and is then stirred with the dope in an inert solvent.

Alumina impregnated with methanol was found to convert epoxides into 1,2-diol monomethyl ethers in good yield at 25°. Similar reactions were observed with $C_2H_5SH$ and $CH_3COOH$.

*Examples:*

(61%)          (30%)

74%

An even more striking reaction of this type is the reduction of aldehydes by 2-propanol on alumina. The reaction is effected in 2 hr. at 25° in yields of 65–85%. Even sensitive α,β-unsaturated aldehydes are reduced to the corresponding allylic alcohols in high yield. Thus it was possible to reduce 10-oxoundecanal to 10-oxoundecanol in 70% yield.[5]

[1] H.-L. Pan, C.-A. Cole, and T. L. Fletcher, *Synthesis*, 716 (1975).
[2] H. Mayr and R. Huisgen, *Angew. Chem. internat. Ed.*, **14**, 499 (1975).
[3] *Idem., Tetrahderon Letters*, 1349 (1975).
[4] G. H. Posner, D. Z. Rogers, C. M. Kinzig, and G. M. Gurria, *Tetrahedron Letters*, 3597 (1975).
[5] G. H. Posner and A. W. Runquist, *ibid.*, 3601 (1975).

**Aluminum bromide, 1,** 22–23; **2,** 19–21; **3,** 7; **4,** 10; **5,** 10.
*"Sludge" catalyst* (**2,** 20; **3,** 7). Details of the Princeton sludge catalyst have been published.[1] It is a heavy yellow oil that can be stored for some time under cyclohexane. Activity can be augmented or restored by addition of $AlBr_3$. The catalyst probably consists of polymerized isobutene, formed by dehydrobromination of *t*-butyl bromide. It has been used to rearrange hydrogenated Bisnor-S to diadamantane in ~70% yield.
Related catalysts have been prepared from *sec*-butyl bromide or *t*-butyl chloride and $AlCl_3$.[2]

[1] T. M. Gund, E. Osawa, V. Z. Williams, Jr., and P. v. R. Schleyer, *J. Org.*, **39**, 2979 (1974).
[2] M. Nomura, P. v. R. Schleyer, and A. A. Arz, *Am. Soc.*, **89**, 3657 (1967).

**Aluminum chloride, 1,** 24–34; **2,** 21–23; **3,** 7–9; **4,** 10–15; **5,** 10–13.
*Bromination catalyst* (**1,** 32). The *meta*-bromination of aromatic aldehydes and ketones catalyzed by $AlCl_3$ proceeds in slightly better yields and at lower temperatures (35–45°) when 1,2-dichloroethane is used as solvent.[1]
*Diels–Alder catalyst* (**4,** 10–11). Details for the synthesis of β-damascone and β-damascenone by the Diels–Alder reaction are now available.[2]
The paper includes a method for *cis*-dehydrobromination of α-bromo ketones. Thus dehydrobromination of a 80:20 mixture of (1) and (2) with lithium fluoride and lithium carbonate in DMF at 120° requires a reaction period of

70 hr., because of reluctance of the *cis-α*-bromo ketone to lose HBr. However, if a mixture of cuprous bromide and $Li_2CO_3$ is used, after 1 hr. the ratio of (1) to (2) changes to about 50:50 and some (3) has been formed. The dehydrobromination to (3) can then be completed within 4 hr. An alternative route to (3) is NBS bromination of the mixture of (1) and (2) followed by reduction with zinc and acetic acid.

*Chloromethylation of thioanisole.*[3] Chloromethylation of thioanisole under usual conditions (chloromethyl methyl ether or formaldehyde and hydrochloric acid) gives mixtures of *o*- and *p*-methylthiobenzyl chloride, which are not practicably separable. However, chloromethylation of thioanisole with methylal (1, 672) and aluminum chloride in methylene chloride gives the *para*-product in 74% yield if about 2 moles of the Lewis acid is used per mole of thioanisole. If only 1 mole of $AlCl_3$ is used, about equal *ortho*- and *para*-substitution is observed. A complex composed of thioanisole and aluminum chloride is considered to be the actual substrate.

*l-Amino-4-arylaminoanthraquinones.*[4] These quinones can be obtained readily as shown in equation I. A free amino group at $C_1$ is essential for this reaction.

(I)

$$\text{(I)} \quad + \quad C_6H_5NH_2 \quad \xrightarrow[85\%]{AlCl_3,\,20^\circ}$$

**Polystyrene–Aluminum chloride** (**4**, 13; **5**, 13). (P)—$AlCl_3$ catalyzes formation of acetals of benzaldehydes. Some catalysis is observed by the polymer alone, probably because it can entrap the water formed on acetalization.[5]

[1] R. F. Eizember and A. S. Ammons, *Org. Prep. Proc. Int.*, **6**, 251 (1974).
[2] K. S. Ayyar, R. C. Cookson, and D. A. Kagi, *J.C.S. Perkin I*, 1727 (1975).
[3] S. H. Pines, R. F. Czaja, and N. L. Abramson, *J. Org.*, **40**, 1920 (1975).
[4] G. Philip and S. V. Sunthankar, *Chem. Ind.*, 433 (1975).
[5] E. C. Blossey, L. M. Turner, and D. C. Neckers, *J. Org.*, **40**, 959 (1975).

**Aluminum hydride, 1,** 34–35, **2,** 23–24; **3,** 9–10; **5,** 13–14.
  *Benzocyclopropenes.* 7,7-Dichloro- and 7,7-difluorobenzocyclopropenes (1)

(1, X = Cl, F)

can be reduced by aluminum hydride prepared *in situ* from lithium aluminum hydride and freshly sublimed aluminum chloride in ether. The reaction is carried out at $-50$ to $20^\circ$.[1]

[1] P. Müller, *Helv.*, **57**, 704 (1974).

**Aluminum isopropoxide, 1,** 35–37; **3,** 10; **4,** 15–16; **5,** 14.
  *Reduction of* α,β-*unsaturated ketones.*[1] This reducing agent is the reagent of choice for reduction of enones of type (1) to the α,β-unsaturated alcohols (2). Usual reducing agents favor 1,4-reduction to the saturated alcohol.

(1)                    (2, 20%, each isomer)

[1] D. H. Picker, N. H. Andersen, and E. M. K. Leovey, *Syn. Commun.*, **5**, 451 (1975).

## Aluminum tricyclohexoxide, 4, 15.

*Reduction.* A recent synthesis of the carcinogenic hydrocarbon benz[a]-anthracene has been described.[1]

[1]F. U. Ahmed, T. Rangarajan, E. J. Eisenbraun, G. W. Keen, and M. C. Hamming, *Org. Prep. Proc. Int.*, **7**, 267 (1975).

## 2-Amino-2-methyl-1-propanol, 1, 37; 3, 14–15.

*Protection of —COOH* (3, 14–15). Full experimental details have been published for the protection of carboxyl groups as the 4,4-dimethyl-$\Delta^2$-oxazoline derivatives. The derivatives are stable to Grignard and hydride reagents. This mode of protection was used in the synthesis of various substituted benzoic acids from the corresponding bromo acids.[1]

Another interesting use of these oxazolines is based on the observation that the 2-phenyloxazolines are metalated very selectively by *n*-butyllithium at $-45°$ at the *ortho*-position. This reaction was used for preparation of polydeuteriobenzoic acids, for example, of (3) from (1). Electrophiles other than $D_2O$ can also be used.[2]

[1]A. I. Meyers, D. L. Temple, D. Haidukewych, and E. D. Mihelich, *J. Org.*, **39**, 2787 (1974).
[2]A. I. Meyers and E. D. Mihelich, *ibid.*, **40**, 3158 (1975).

## Ammonium persulfate (Ammonium peroxydisulfate), 1, 952–954; 2, 348; 3, 238–239; 5, 15–16.

*δ-Phenyl-γ-butyrolactone.*[1] Oxidation of γ-phenylbutyric acid (1) by ammonium persulfate in water at 90° gives δ-phenyl-γ-butyrolactone (2) in 91% yield based on acid consumed. The oxidation is believed to involve formation of the radical $C_6H_5\dot{C}H(CH_2)_2COOH$ and oxidation to $C_6H_5\overset{+}{C}H(CH_2)_2COOH$,

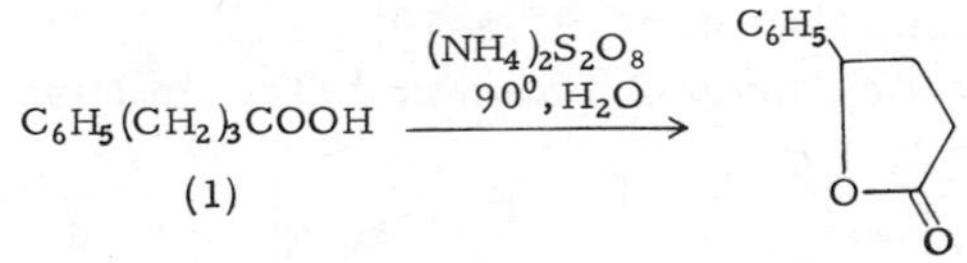

$$C_6H_5(CH_2)_3COOH \xrightarrow[90^0, H_2O]{(NH_4)_2S_2O_8}$$

(1)

(2)

which then undergoes ring closure with loss of a proton.

[1] A. Clerici, F. Minisci, and O. Porta, *Tetrahedron Letters*, 4183 (1974).

## Ammonium polysulfide, 1, 1120–1121.

*Willgerodt reaction* (1, 705–706; 1120–1121). This reaction and the Kindler modification have been reviewed in depth (172 references).[1]

[1] E. V. Brown, *Synthesis*, 358 (1975).

## Aniline, $C_6H_5NH_2$, 1, 41.

*Decarboxylative Michael reaction.*[1] Addition of a catalytic amount of aniline to a mixture of acetoacetic acid (1) and cyclohexenone (2) in ether at room temperature leads to evolution of carbon dioxide and formation of 3-acetonylcyclohexane-1-one (3). The reaction of (4) with (2) under the same

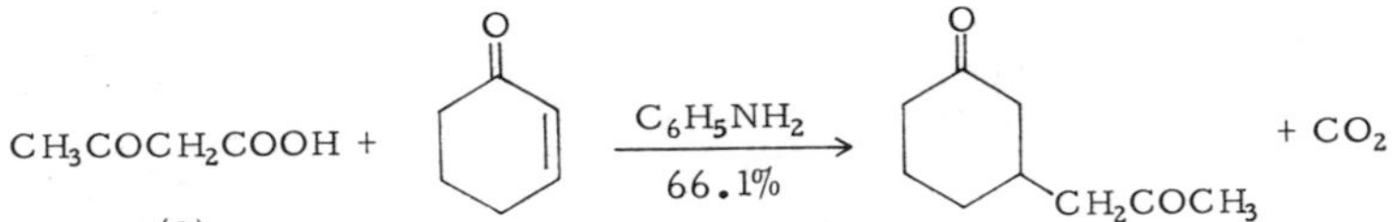

$$CH_3COCH_2COOH + \qquad \xrightarrow[66.1\%]{C_6H_5NH_2} \qquad + CO_2$$

(1)          (2)                    (3)

$$\xrightarrow[64.8\%]{C_6H_5NH_2} \qquad + CO_2$$

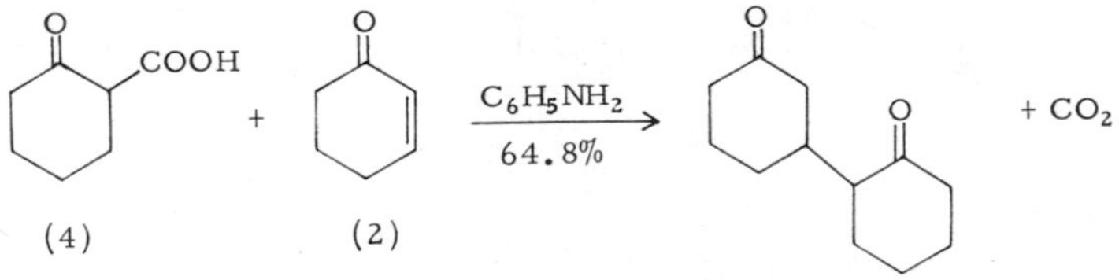

(4)          (2)                    (5)

conditions gives (5). This route to 1,5-diketones is believed to involve decarboxylation of the $\beta$-keto acid followed by a Michael addition of the resulting carbanion to the $\alpha,\beta$-unsaturated ketone.

[1] M. Yasuda, *Chem. Letters*, 89 (1975).

## Antimony(V) chloride, 1, 42; 4, 21; 5, 18–19.

*4-Chloro-4-methylcyclohexa-2,5-dienone.*[1] The reaction of *p*-cresol (1,

0.083 mole) with antimony(V) chloride (0.47 mole) in methylene chloride at
$-50°$ results in formation of the cyclohexadienone derivative (2) in high yield.
The temperature and the amount of $SbCl_5$ are critical in this reaction. The
product (2) can be converted into various dienones (3) by solvolysis assisted
by Ag(I) ion.

*Intercalated reagent.* Mélin and Herold[2] have reported the preparation of
$C_{24}SbCl_5$ as black platelets with considerable stability to air. Kagan *et al.*[3] find
that intercalated $SbCl_5$ differs considerably in reactivity from ordinary $SbCl_5$.
It reacts with alkyl bromides, iodides, and tosylates to give the corresponding
alkyl chlorides. It does not effect aromatic chlorination. In contrast, ordinary
$SbCl_5$ reacts with alkyl bromides to give $\alpha$-chloro bromides and readily chlorinates aromatic compounds.

*Examples:*

*Chlorination of* cis,cis-*cyclooctadiene-1,5.*[4] The reaction of *cis,cis*-cyclo-octadiene-1,5 (1) with $SbCl_5$ in $CCl_4$ at $-20°$ gives a mixture of *endo*- and *exo*-2-*anti*-8-dichlorobicyclo[3.2.1]octane, (2) and (3), respectively. Chlorination of (1) with $PCl_5$, $SO_2Cl_2$, $C_6H_5ICl_2$, $CuCl_2$, and $MoCl_5$ gives mixtures of

(1)                    (2)                    (3)

*cis*- and *trans*-5,6-dichlorocyclooctenes. Note that transannular reactions of (1) usually lead to bicyclo[3.3.0]octane derivatives.

Chlorination of *cis*-cyclooctene with $SbCl_5$ involves a transannular 1,5-hydride shift; a mixture of *cis*- and *trans*-1,4-dichlorocyclooctane is obtained in 78% yield.

[1] A. Nilsson, A. Ronlán, and V. D. Parker, *Tetrahedron Letters*, 1107 (1975).
[2] J. Mélin and A. Herold, *Compt. rend.*, **269**, 877 (1969).
[3] J. Bertin, J. L. Luche, H. B. Kagan, and R. Setton, *Tetrahedron Letters*, 763 (1974).
[4] S. Uemura, A. Onoe, and M. Okano, *J.C.S. Chem. Comm.*, 210 (1975).

## Antimony(V) fluoride, 5, 9.

*Cycloaddition reactions of sulfur dioxide.*[1] Addition of cyclooctatetraene to $SbF_5$ (1 eq.) in liquid $SO_2$ at $-70°$ leads to formation of (1) and (2); the former sulfone is rapidly isomerized to (2), which has been isolated.[2] Formation of (1) represents an unusual 1,5-cycloaddition. The trivial name for (1) is

(1a)                    (1b)                    (2)

9-thiabarbaralane 9,9-dioxide. It is a fluxional molecule (Cope rearrangement).

More recently, $SbF_5$-catalyzed addition of $SO_2$ to hexamethyl-Dewarbenzene (3) has been reported.[3] When the reaction is carried out at $-95°$, the spectra indicate formation of the cation (4); addition of $CH_3OH/CH_3ONa$ converts (4) into the sulfone (5), obtained in about 80% yield from (3). Some starting material is recovered.

(3)                    (4)                    (5)

*Trimethylcarbenium fluoroantimonate,* $(CH_3)_3\overset{+}{C}SbF_6^-$. This relatively stable salt can be prepared as light-colored crystals by the reaction at $-70°$ of $(CH_3)_3CCl$ with fluoroantimonic acid $(HF \cdot SbF_5)$[4] or with antimony penta-fluoride in sulfuryl chloride fluoride, $ClSO_2F$.[5] It is stable at $-20°$, but de-composes at room temperature.

This cation alkylates *t*-butanol in the presence of a hindered base such as diisopropylethylamine (4, 529)[6] to give di-*t*-butyl ether in quantitative yield.[5] This ether cannot be prepared by the usual Williamson synthesis because po-tassium *t*-butoxide does not undergo $S_N2$ reactions. It had been obtained pre-viously by the reaction of *t*-butyl chloride with silver carbonate in 31% yield.[7]

$$(CH_3)_3\overset{+}{C} \ + \ (CH_3)_3COH \rightleftharpoons (CH_3)_3\overset{+}{C}\underset{H}{O}C(CH_3)_3 \xrightarrow{R_3N} (CH_3)_3COC(CH_3)_3 \ + \ R_3\overset{+}{N}H$$

The present synthesis is apparently the first example of an electrophilic alkylation in a basic solution with a carbocation.

[1] L. A. Paquette, U. Jacobsson, and M. Oku, *J.C.S. Chem. Comm.*, 115 (1975).
[2] J. Gasteiger and R. Huisgen, *Am. Soc.*, **94**, 6541 (1972).
[3] H. Hogeveen, H. Jorritsma, and P. W. Kwant, *Tetrahedron Letters*, 1795 (1975).
[4] G. A. Olah, J. J. Svoboda, and A. T. Ku, *Synthesis*, 492 (1973).
[5] G. A. Olah, Y. Halpern, and H. C. Lin, *ibid.,* 315 (1975).
[6] S. Hünig and M. Kiessel, *Ber.*, **91**, 380 (1958).
[7] R. West, D. L. Powell, M. K. T. Lee, and L. S. Whatley, *Am. Soc.*, **86**, 3227 (1964).

α-**Azidostyrene,** $C_6H_5\overset{\overset{N_3}{|}}{C}=CH_2$. Mol. wt. 145.16, pale yellow oil.
*Preparation*[1]:

$$C_6H_5CH=CH_2 \xrightarrow{Br_2, CCl_4} C_6H_5\overset{\overset{Br}{|}}{C}HCH_2Br \xrightarrow[15-20°]{NaN_3, DMSO} C_6H_5\overset{\overset{N_3}{|}}{C}HCH_2Br \xrightarrow[DMSO]{NaOH} C_6H_5\overset{\overset{N_3}{|}}{C}=CH_2$$

(1)

*Alkyl aryl ketones.*[2] These ketones can be prepared in good to high yield by the reaction of trialkylboranes with this azide:

$$\xrightarrow[70-95\%]{} C_6H_5\underset{\overset{\|}{O}}{C}CH_2R$$

*Examples:*

$$CH_2=CH_2 \xrightarrow[95\%]{\substack{1) \ BH_3 \\ 2) \ (1)}} C_6H_5\underset{\overset{\|}{O}}{C}CH_2CH_2CH_3$$

[1] A. G. Hortmann, D. A. Robertson, and B. K. Gillard, *J. Org.*, **37**, 322 (1972).
[2] A. Suzuki, M. Tabata, and M. Ueda, *Tetrahedron Letters*, 2195 (1975).

## Azidotris[dimethylamino] phosphonium hexafluorophosphate,

$[(CH_3)_2N]_3\overset{+}{P}N_3PF_6^-$ (1). Mol. Wt. 350.20, m.p. $>250°$, stable, nonhygroscopic.
*Preparation:*

$$[(CH_3)_2N]_3P + Br_2 \xrightarrow{\text{ether}} [(CH_3)_2N]_3\overset{+}{P}BrBr^- \xrightarrow{KPF_6, H_2O}$$

$$[(CH_3)_2N]_3\overset{+}{P}Br\ PF_6^- \xrightarrow[83\% \text{ overall}]{NaN_3, (CH_3)_2CO} (1)$$

*Peptide synthesis.*[1] The reagent effects coupling of acylamino acids with amino acid esters in DMF and triethylamine at $-10$ to $0°$ in 80–97% yields. Racemization according to the test of Young is less than 3%. The condensation involves the intermediate formation of an acyl azide, which has been isolated

$$\underset{\underset{NHCOC_6H_5}{|}}{RCHCOOH} + H_2NCH_2COOC_2H_5 \cdot HCl \xrightarrow{(1)} \underset{\underset{NHCOC_6H_5}{|}}{RCHCONHCH_2COOC_2H_5}$$

in the case of benzoic acid.

[1] B. Castro and J. R. Dormoy, *Bull. soc.*, 3359 (1973).

# B

**Benzaldehyde, $C_6H_5CHO$.**

*Polymeric benzaldehyde.* Fréchet and Pellé[1] have prepared a polymeric form of benzaldehyde by chloromethylation of a divinylbenzene—styrene copolymer followed by oxidation of the chloromethylated product with DMSO. The reagent was used to protect the 4- and 6-hydroxyl group of methyl α-D-glucopyranoside (1) by formation of the acetal (2), which can then be converted

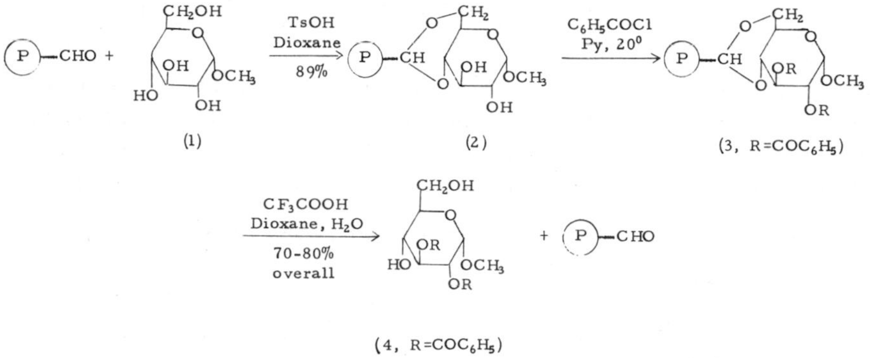

(1)                    (2)                    (3, R=COC_6H_5)

(4, R=COC_6H_5)

into various 2,3-disubstituted derivatives such as (3). Cleavage of the protective group can be effected with trifluoroacetic acid—dioxane.

[1] J. M. J. Fréchet and G. Pellé, *J.C.S. Chem. Comm.*, 225 (1975).

**Benzaldehyde ethylene dithioacetal (1).**

*Benzal derivatives of glycols.*[1] Treatment of (1) with methyl fluorosulfonate (**4**, 339–340) in $CH_2Cl_2$ gives the disulfonium salt (2). Addition of 1,2-, 1,3-, or 1,4-diols to the solution in the presence of base [$K_2CO_3$, $N(C_2H_5)_3$, Py] leads to benzal derivatives (3) in fair to good yields. Usually acidic conditions

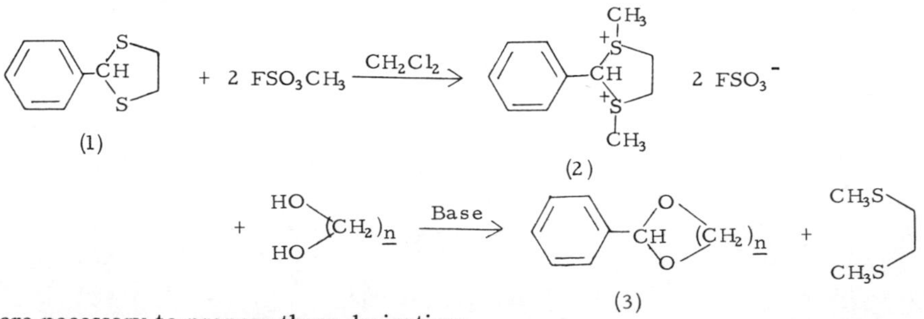

(1)                    (2)

(3)

are necessary to prepare these derivatives.

26

*Examples:*

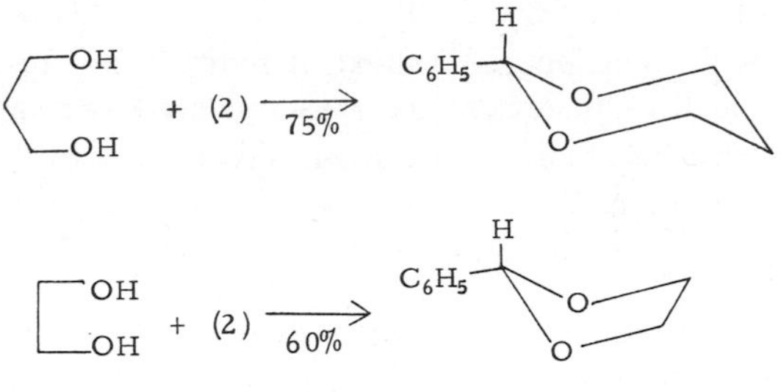

[1] R. M. Munavu and H. H. Szmant, *Tetrahedron Letters*, 4543 (1975).

**Benzamidine.** $C_6H_5C(=NH)NH_2$. Mol. wt. 120.15. Aldrich supplies the hydrochloride hydrate.

*1,3-Diazaazulenes.*[1]

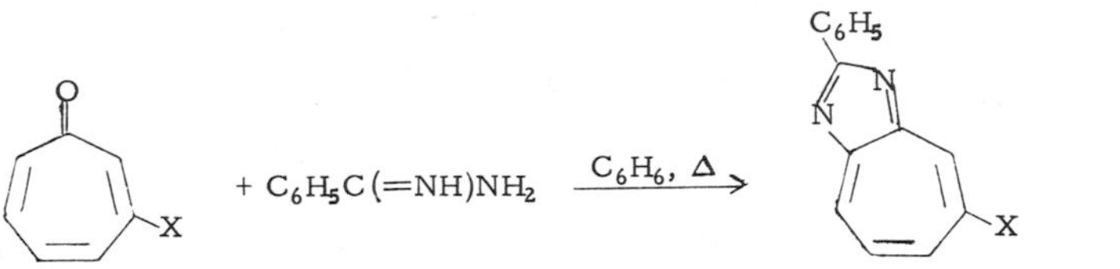

X = $OCH_3$ (40%)
X = $SCH_3$ (50%)
X = H (~6%)

[1] F. Del Cima, M. Cavazza, C. A. Veracini, and F. Pietra, *Tetrahedron Letters,* 4267 (1975).

**π-Benzenechromium  tricarbonyl,** 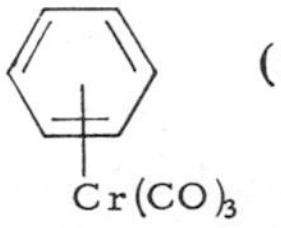 (1). Mol. wt. 214.15, m.p. 162–163° (yellow).

*Preparation.* This chromium complex is usually prepared by the reaction of benzene with chromium hexacarbonyl under high pressure.[1] Rausch[2] has recently published a simple synthesis based on a patent procedure in which equivolume amounts of benzene and 2-picoline are refluxed for an extended period under nitrogen. Carbon monoxide is evolved, and little sublimation of chromium hexacarbonyl occurs. Yields are 90–94% for reactions conducted for 96 hr. or more. The method is presumably successful because of formation of the 2-picoline complex (A), which reacts with benzene to form (1).

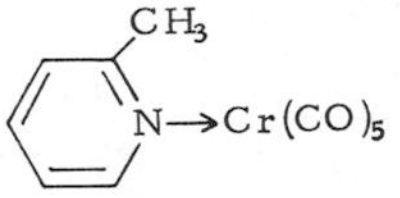

(A)

*Phenylation of carbanions* (*see also* π-**Chlorobenzenechromium tricarbonyl,** this volume). π-Benzenechromium tricarbonyl reacts readily with some carbanions (generated with LDA, potassium hexamethyldisilizane, potassium hydride) to give an intermediate that gives the free aryl derivative on oxidation with iodine. It is not clear whether the reaction involves attack of the benzene ligand or at the chromium atom. Yields are high with reactive organolithium reagents; ester enolates react satisfactorily, but ketone enolates fail to give products in useful yields.[3]

$$(1) + LiC(CH_3)_3 \longrightarrow Intermediate \xrightarrow[97\%]{I_2} C_6H_5C(CH_3)_3$$

$$(1) + KC(CH_3)_2COOC(CH_3)_3 \longrightarrow \xrightarrow[88\%]{I_2} C_6H_5C(CH_3)_2COOC(CH_3)_3$$

$$(1) + ClMgC(CH_3)_3 \longrightarrow \xrightarrow[<5\%]{I_2} C_6H_5C(CH_3)_3$$

$$(1) + LiCH_2COC_6H_5 \longrightarrow \xrightarrow[<5\%]{I_2} C_6H_5CH_2COC_6H_5$$

[1]W. Strohmeier, *Ber.,* **94,** 2490 (1961).
[2]M. D. Rausch, *J. Org.,* **39,** 1787 (1974).
[3]M. F. Semmelhack, H. T. Hall, M. Yoshifuji, and G. Clark, *Am. Soc.,* **97,** 1247 (1975).

**Benzeneselenol,** $C_6H_5SeH$. Mol. wt. 157.07, b.p. 67–68°/12 mm. Supplier: Eastman.

*Allylic alcohols.* Two laboratories[1,2] have reported a novel synthesis of allylic alcohols from two carbonyl compounds. One carbonyl compound (1) is treated with benzeneselenol (2 eq.) under acid catalysis to give a selenoacetal (2). This product can be cleaved to the carbanion (a) by *n*-butyllithium in THF at −78°. The carbanion reacts with a second aldehyde or ketone to form a β-hydroxyselenide (3). The final step involves oxidation and selenoxide fragmentation to the allylic alcohol (4). β-Hydroxyselenides have been obtained by

(1)                    (2)                    (a)

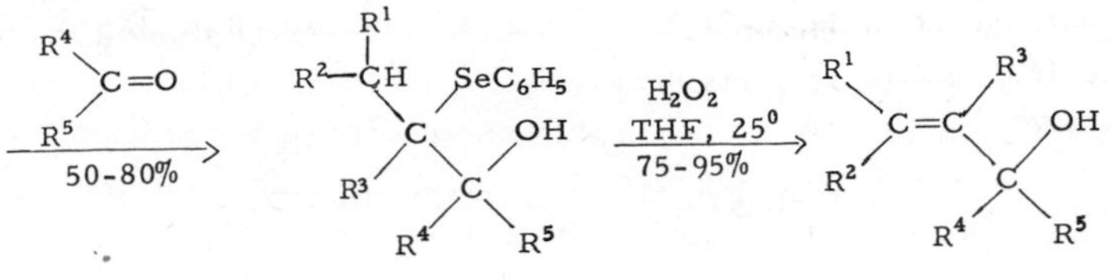

(3)                                    (4)

cleavage of epoxides with sodium selenophenolate (**5**, 272); however, this reaction can lead to two different β-hydroxyselenides from unsymmetrically substituted epoxides.

*Di- and trisubstituted epoxides.*[3] β-Hydroxyselenides (**1**), prepared by the method illustrated above, can be converted into epoxides. Alkylation with methyl iodide in the presence of silver tetrafluoroborate gives the resulting salts (**2**), which are treated with potassium *t*-butoxide in DMSO. An epoxide (**3**) and methyl phenyl selenide are formed by way of the betaine (**a**). Unfortunately, when two alkyl groups are attached to the carbon bearing the $SeC_6H_5$ group,

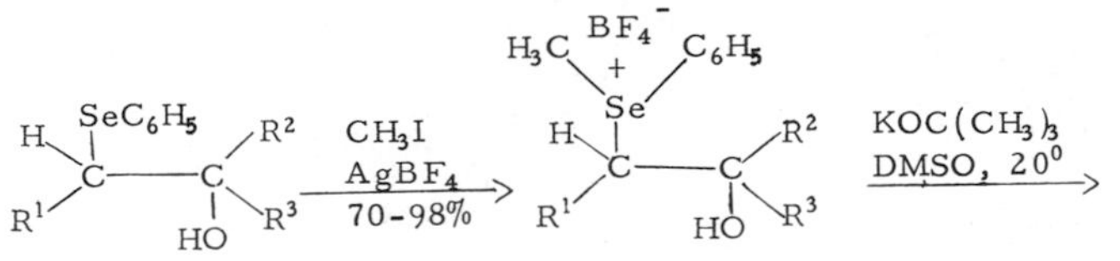

(1)                                    (2)

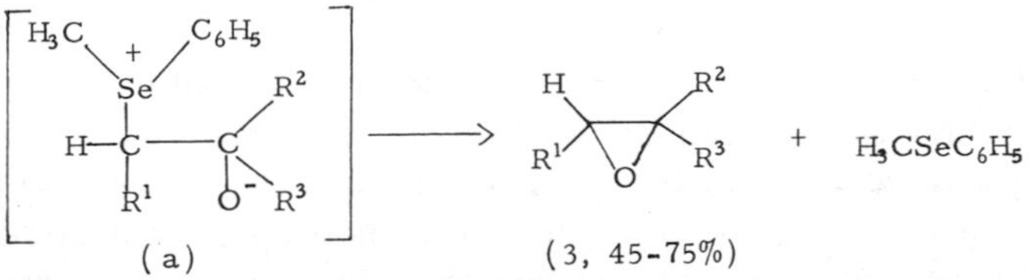

(a)                         (3, 45-75%)

the desired selenonium salts (**2**) cannot be obtained. For example, (**4**) is converted into the butanone (**5**). Thus tetrasubstituted epoxides are not accessible by this route.

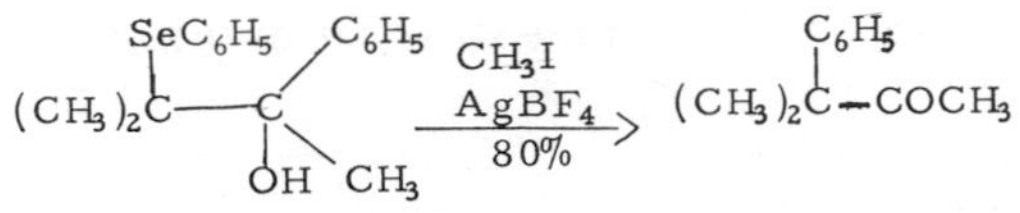

(4)                              (5)

[1] W. Dumont, P. Bayet, and A. Krief, *Angew. Chem. internat. Ed.,* **13**, 804 (1974).
[2] D. Seebach and A. K. Beck, *ibid.,* **13**, 806 (1974).
[3] W. Dumont and A. Krief, *ibid.,* **14**, 350 (1975).

**Benzenesulfenyl chloride,** $C_6H_5SCl$. Mol. wt. 145.62, b.p. 73–75°/9 mm., red, unstable to water.

   *Preparation*[1,2]:

$$C_6H_5SH + Cl_2 \xrightarrow[\text{quant.}]{CCl_4} C_6H_5SCl + HCl$$

An analogous procedure has been described for the preparation of *p*-toluene-sulfenyl chloride.[3]

   *Alkylation of allylic alcohols.* Evans and Anderson[4] have developed an ingenious method for effecting γ-alkylation of allylic alcohols that is based on the fact that sulfenate esters undergo facile [2,3] sigmatropic rearrangement to sulfoxides. Thus treatment of an allylic alcohol (1) with *n*-butyllithium

and then with benzenesulfenyl chloride gives the sulfoxide (2) by way of (a). The sulfoxide on treatment with lithium diisopropylamide and then an alkyl iodide is alkylated predominantly at the position α to the sulfur to give (3). On treatment of (3) with a thiophile, particularly trimethyl phosphite[5], the alkylated allylic alcohol (4) is obtained. The process is applicable to substituted allylic alcohols, both acyclic and cyclic.

   The method has two shortcomings. Fairly active halides are required for alkylation of the sulfoxides, and alkylation can occur both α and γ to sulfur, the relative ratio being dependent on the structure of the sulfoxide.[6]

   *Examples:*

*1,3-Alkylative carbonyl transposition.*[7] The [2,3] sigmatropic rearrangement of sulfenate esters to sulfoxides was used in the first step of a method for 1,3-transposition of an $\alpha,\beta$-unsaturated ketone with concomitant alkylation at the site of the original carbonyl group. The sequence is illustrated for the case of ethyl vinyl ketone (1). Reaction of (1) with *n*-butyllithium and then with benzenesulfenyl chloride gives the allylic sulfoxide (2). Reaction of (2) with LDA and then with diphenyl disulfide results in sulfenylation with *in situ* rearrangement and desulfenylation to yield the hydroxy enol thioester (3) directly. The $\alpha,\beta$-unsaturated aldehyde (4) is obtained on hydrolysis of (3) with mercuric chloride.

A similar transposition can be applied to saturated ketones. For example, a ketone is treated with a vinyl organometallic reagent to generate a derivative of an allylic alcohol. Reaction with benzenesulfenyl chloride then gives an allylic sulfoxide by means of a [2,3] sigmatropic rearrangement. The remaining steps are the same as those used above. The sequence for estrone methyl ether (5) is given below. The overall result is to effect the equivalent of an aldol condensation of estrone with propionaldehyde.

(5)                              (a)

(6)                              (7)

[1]H. Lecher and F. Holschneider, *Ber.*, **57**, 755 (1924).
[2]E. Kühle, *Synthesis*, 561 (1970).
[3]F. Kurzer and J. R. Powell, *Org. Syn., Coll. Vol.*, **4**, 934 (1963).
[4]D. A. Evans and G. C. Anderson, *Accts. Chem. Res.*, **7**, 147 (1974).
[5]The trimethyl phosphite is purified by distillation from sodium.
[6]D. A. Evans, G. C. Andrews, T. T. Fujimoto, and D. Wells, *Tetrahedron Letters*, 1385, 1389 (1973).
[7]B. M. Trost and J. L. Stanton, *Am. Soc.*, **97**, 4018 (1975).

**5-Benzisoxazolemethylene chloroformate,** (1). Mol. wt. 211.6.

The chloroformate is prepared by reaction in $CH_2Cl_2$ of 5-hydroxymethyl-benzisoxazole with excess phosgene. It is stored in $CH_2Cl_2$.

*Protection of amino groups.* Kemp and Hoyng[1] have used the 5-benzis-oxazolylmethyleneoxycarbonyl (Bic) group for protection of amino groups in peptide synthesis. The group is compatible with most operations of synthesis, including mildly basic conditions. However, it is isomerized to (2) by 2 eq. of triethylamine in $CH_3CN$ or DMF at 25°. The product is hydrolyzed by acid to release the amino groups. The group is also cleaved by hydrogenolysis and by HBr in acetic acid.

(1)                    (2)                    (3)

[1]D. S. Kemp and C. F. Hoyng, *Tetrahedron Letters*, 4625 (1975).

**Benzocyclopropene, 4, 402.**

*Diels–Alder reaction.*[1] The first step in a synthesis of 2-hydroxy-4,10-methano [11] annulenone (7), a $10\pi$-troponoid, is the reaction of benzocyclopropene (1) with the *o*-benzoquinone (2). As expected, (7) resembles $\alpha$-tropolone

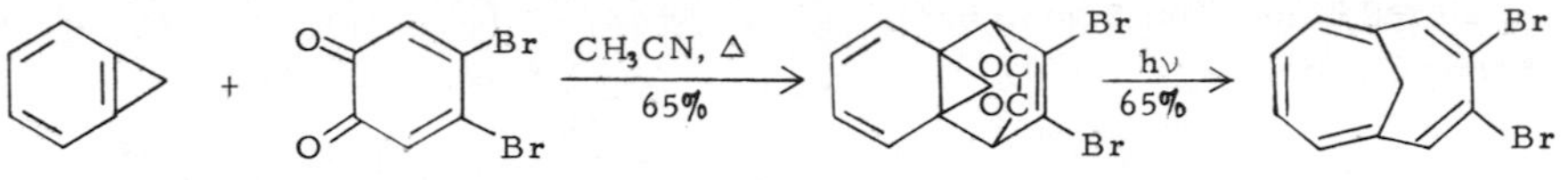

(1)           (2)                        (3)                    (4)

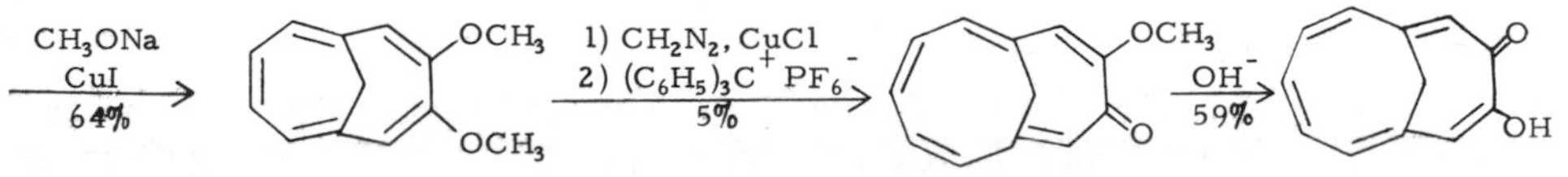

(5)                              (6)                (7)

in various properties.

[1] E. Vogel, J. Ippen, and V. Buch, *Angew. Chem. internat. Ed.*, **14**, 566 (1975).

**1,3,2-Benzodioxaborole (Catecholborane), 4, 25; 69–70; 5, 100–101.**

*Hydroboration.* Brown and Gupta[1] have reported results of an extensive study of the hydroboration of alkenes and alkynes (equations I and II). The

(I)

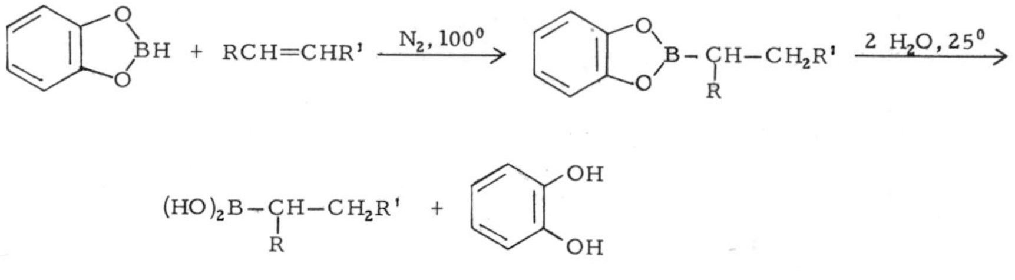

(II)

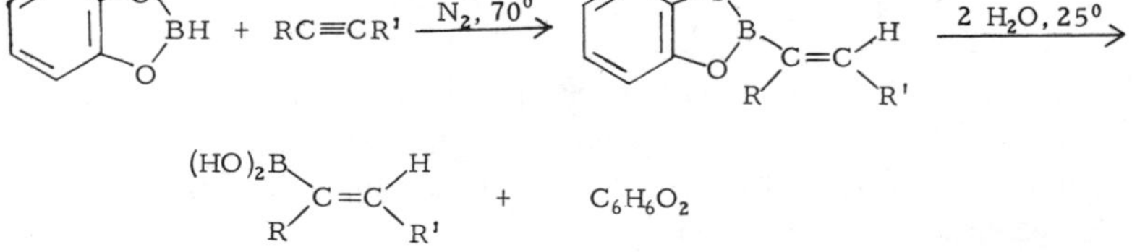

reaction is stereospecific (*cis*-addition) in the case of alkynes. It is also regioselective; thus the boron atom becomes attached preferentially to the less hindered carbon atom of the double or triple bond. However, electronic effects can overcome, in part, steric effects and can result in formation of both possible products of hydroboration. This hydroboration is particularly useful for synthesis of alkane- and alkeneboronic acids. The by-product catechol is highly soluble in water and easily removed.

[1]H. C. Brown and S. K. Gupta, *Am. Soc.,* **97,** 5249 (1975).

**S-Benzoic O,O-diethyl phosphorodithioic anhydride** (1). Mol. wt. 290.34, viscous red liquid. Unstable to light.
    *Preparation:*

$$(C_2H_5O)_2P \overset{S}{\underset{SH}{\diagup}} + C_6H_5COCl \xrightarrow{Na_2CO_3} (C_2H_5O)_2P \overset{SCOC_6H_5}{\underset{S}{\diagup}}$$

$$(1)$$

*Selective benzoylation of amines.*[1] The reagent benzoylates amino groups at room temperature. Hydroxyl groups do not react even at temperatures of 80–100°.[1]

$$(1) + RNH_2 \longrightarrow RNHCOC_6H_5 + (C_2H_5O)_2P \overset{S}{\underset{SH}{\diagup}}$$

[1]P. G. Nair and C. P. Joshua, *Chem. Ind.,* 704 (1974).

**Benzoin,** $C_6H_5CH(OH)COC_6H_5$. Mol. wt. 212.25, m.p. 134–136°. Suppliers: Aldrich, Eastman, others.
    *Phenyl alkyl ketones.* Japanese chemists[1] have used benzoin as an equivalent to the benzoyl anion $(C_6H_5CO^-)$ in a synthesis of ketones. Thus benzoin is

$$C_6H_5\overset{O}{\overset{\|}{C}}-\overset{OH}{\overset{|}{C}}HC_6H_5 \xrightarrow[50-100\%]{\underset{DMSO}{RX, NaOH}} C_6H_5\overset{O}{\overset{\|}{C}}-\overset{OH}{\overset{|}{\underset{R}{C}}}-C_6H_5 \xrightarrow[80-100\%]{\underset{CH_3OH}{NaBH_4}} C_6H_5\overset{OH}{\overset{|}{C}}H-\overset{OH}{\overset{|}{\underset{R}{C}}}-C_6H_5$$

$$\xrightarrow[40-95\%]{\underset{CH_3OH-H_2O}{NaIO_4}} R\overset{O}{\overset{\|}{C}}C_6H_5 + C_6H_5CHO$$

C-alkylated by an alkyl chloride or bromide in DMSO in the presence of sodium hydroxide.[2] The product is reduced to the diol, which is then oxidized by sodium metaperiodate to the alkyl phenyl ketone and benzaldehyde (which can be reconverted into benzoin).

[1]Y. Ueno and M. Okawara, *Synthesis,* 268 (1975).
[2]H.-G. Heine, *Ann.,* **735,** 56 (1970).

**Benzotriazolyl-N-hydroxytris(dimethylamino)phosphonium    hexafluorophosphate** (1). Mol. wt. 442.50; stable, nonhygroscropic; the perchlorate salt is explosive. Supplier: Fluka.

*Preparation from hydroxybenzotriazole:*

(1)

*Peptide synthesis.*[1] This reagent effects coupling of N-protected amino acids with amino acid esters in the presence of 2 eq. of triethylamine. The reaction is conducted at room temperature in $CH_3CN$, $CH_2Cl_2$, or $DMF-CH_3CN$. The reagent is converted into hydroxybenzotriazole. Yields of dipeptides are in the range of 80–95%. Racemization appears to be negligible.

[1] B. Castro, J. R. Dormoy, G. Evin, and C. Selve, *Tetrahedron Letters,* 1219 (1975).

**Benzoyl cyanide,** $C_6H_5COCN$. Mol. wt. 131.13, m.p. 30–32°, b.p. 206°. Supplier: Aldrich.

*Preparation.*[1]

*Selective benzoylations.*[2] The reagent has been used for selective benzoylation of carbohydrates. It is comparable to N-benzoylimidazole.[3]

[1] K. E. Koenig and W. P. Weber, *Tetrahedron Letters,* 2275 (1974).
[2] S. A. Abbas and A. H. Haines, *Carbohydrate Res.,* **39,** 358 (1975); S. A. Abbas, A. H. Haines, and A. G. Wells, *J.C.S. Perkin I,* 1351 (1976).
[3] F. A. Carey and K. O. Hodgson, *Carbohydrate Res.,* **12,** 463 (1970).

**N-Benzoylperoxycarbamic acid** (1). Mol. wt. 181.14, m.p. 98–105° dec.

*Preparation.*[1] The carbamic acid (1) is prepared by the reaction of benzoyl

(1)

isocyanate[2] and hydrogen peroxide under anhydrous conditions.

*Epoxidation.* Höft and Ganschow[1] report that this reagent converts olefins into epoxides, Schiff bases into oxaziridines, and tertiary amines into N-oxides. Epoxidation can be performed more simply by addition of benzoyl isocyanate to a solution of the alkene in THF containing excess anhydrous $H_2O_2$ and a trace of a radical inhibitor.[3] Under these conditions phenanthrene is converted at 25° into biphenyl-2,2'-dicarboxylic acid.

1,1-Carbonyldi-1,2,4-triazole (2, 61) can be used in place of benzoyl isocyanate. In this case the epoxidation reagent is believed to have structure (2). This reagent epoxidizes olefins much more rapidly than perbenzoic acid. The hydro-

$$(2)$$

gen bond present in (1) and (2) is believed to be responsible for the enhanced reactivity.

[1] V. E. Höft and S. Ganschow, *J. prakt. Chem.*, **314**, 145 (1972).
[2] A. J. Speziale and L. R. Smith, *J. Org.*, **28**, 1805 (1963).
[3] J. Rebek, Jr., S. F. Wolf, and A. B. Mossman, *J.C.S. Chem. Comm.*, 711 (1974).

**N-Benzoyl-N'-triflylhydrazine,** $C_6H_5CONHNHSO_2CF_3$ (1). Mol. wt. 268.22, m.p. 159–160°.

This hydrazine is prepared in 93% yield by reaction of benzoylhydrazine with triflic anhydride in $CH_2Cl_2$ at $-78° \rightarrow 20°$.

*Conversion of alkyl halides to hydrazones.*[1] Primary and activated secondary alkyl halides react with this triflyl hydrazine and $K_2CO_3$ (2 eq.) in acetonitrile to form acyl hydrazones of aldehydes and ketones, respectively. The reaction involves alkylation followed by elimination of $HSO_2CF_3$. Yields are high (80–90%) in the case of aldehyde derivatives, but they are lower in the case of ketone derivatives.

[1] J. B. Hendrickson and D. D. Sternbach, *J. Org.*, **40**, 3450 (1975).

**1-Benzyl-1,4-dihydronicotinamide,** (1). Mol. wt. 214.27, m.p. 120–122° dec.

*Preparation.*[1]

*Reduction of α-diketones and α-keto esters.* Nicotinamide is a component of the coenzymes diphosphopyridine nucleotide (DPN) and triphosphopyridine nucleotide (TPN), which are involved in various biological hydrogen-transfer reactions. Several laboratories have used (1) as a model of 1,4-dihydronicotin-amide. Japanese chemists[2] have been able to effect reductions of α-diketones and α-keto alcohols with (1), but only in the presence of magnesium cations. The reaction is slow in the dark and is accelerated by irradiation. For example, benzil can be reduced in this system to benzoin in yields of 75%; benzoin is reduced further to a slight extent.

$$C_6H_5CO-COC_6H_5 \xrightarrow[\text{h}\nu]{(1),\ Mg^{++}} C_6H_5\underset{OH}{C}H-COC_6H_5 \ + \ C_6H_5\underset{OH}{C}H-\underset{OH}{C}HC_6H_5$$

$$(75\%) \qquad\qquad\qquad (1\%)$$

*Reduction of α-keto esters.* The reduction of α-keto esters by this substance has also been studied.[3] Reduction of two α-keto acids to α-hydroxy acids has been effected in high yield, but only in the presence of $Mg^{2+}$ or $Zn^{2+}$ (as in enzymatic reactions of DPNH). Reduction involves direct transfer of hydrogen from the model compound. By use of a chiral derivative of 1-benzyl-1,4-di-hydronicotinamide reduction of ethyl benzoylformate to ethyl (R)-(-)-man-delate has been effected with an optical purity of 19%.

[1]P. Karrer and F. J. Stare, *Helv.*, **20**, 418 (1937); D. Mauzerall and F. H. Westheimer, *Am. Soc.*, **77**, 2261 (1955).
[2]Y. Ohnishi, M. Kagami, and A. Ohno, *Tetrahedron Letters*, 2437 (1975).
[3]*Idem, Am. Soc.*, **97**, 4766 (1975).

**Benzyldimethylanilinium hydroxide,** $C_6H_5CH_2\overset{CH_3}{\underset{CH_3}{N}}{}^{+}-C_6H_5$ $OH^-$.   Mol. wt. 229.31.

A mixture of equimolar amounts of N,N-dimethylaniline and benzyl chloride on standing deposits crystals of benzyldimethylanilinium chloride, which is converted into the hydroxide by silver oxide in methanol.

*Benzyl esters.*[1] Preparation of benzyl esters by Fischer esterification pro-ceeds with difficulty. An alternative method involves decomposition of benzyl-dimethylanilinium salts of carboxylic acids in a refluxing inert solvent.

$$RC\underset{O^-}{\overset{O}{\diagup}} \quad C_6H_5CH_2\overset{CH_3}{\underset{CH_3}{N}}{}^{+}-C_6H_5 \quad \xrightarrow[65-85\%]{\overset{C_6H_5CH_3}{\text{reflux}}} \quad RC\underset{OCH_2C_6H_5}{\overset{O}{\diagup}} \ + \ C_6H_5N(CH_3)_2$$

N-Protected amino acids can be converted into benzyl esters by this method.[2] The *t*-BOC group can be used for protection and can be removed by treatment with formic acid after the esterification step. The amino group can also be protected as a ketimine salt.

[1]K. Williams and B. Halpern, *Synthesis,* 727 (1974).
[2]*Idem, Australian J. Chem.*, **28**, 2065 (1975).

## Benzyl-*n*-butyldimethylammonium chloride, polymeric,

$$\widehat{P} - C_6H_5CH_2\overset{+}{N}(CH_3)_2(\underline{n}\text{-}C_4H_9)\ Cl^- \quad (1).$$

This polymer is prepared[1] from chloromethylated polystyrene (Bio-Rad Laboratories).

*Triphase catalysis.* Regen[2] has shown that this insoluble resin (1) can serve as a catalyst for certain aqueous phase–organic phase reactions. When the polymer is suspended in a heterogeneous mixture of 1-bromooctane and aqueous sodium cyanide, which is then heated to 110° for 4 hr., 1-cyanooctane is formed in 92% yield. The alkyl halide is unchanged in the absence of (1). The resin also catalyzes the generation of dichlorocarbene from chloroform solutions placed over aqueous sodium hydroxide. If this mixture with (1) and α-methylstyrene is heated for 40 hr. at 50°, the adduct (2) is obtained in 99% yield.

$$\overset{H_3C}{\underset{C_6H_5}{\diagdown}}C=CH_2 \xrightarrow[99\%]{:CCl_2} \overset{H_3C}{\underset{C_6H_5}{\bigvee}}\underset{Cl\ \ Cl}{\triangle}$$

$$(2)$$

Catalysis of two-phase reactions with an insoluble catalyst has the advantage that emulsions are not formed and that the catalyst can be removed from the product by filtration.

[1]S. L. Regen and D. P. Lee, *Am. Soc.*, **96**, 294 (1974).
[2]S. L. Regen, *ibid.*, **97**, 5956 (1975).

## 3-Benzyl-5-(2-hydroxyethyl)-4-methyl-1,3-thiazolium chloride,

$$\overset{H_3C}{\underset{HOCH_2CH_2}{\diagup}}\underset{S}{\left[\overset{CH_2C_6H_5}{\underset{N^+}{\diagup}}Cl^-\right]}$$

Mol. wt. 239.76, m.p. 140.5°.

$$(1)$$

This thiazolium salt (1) is prepared[1] by reaction of 5-(2-hydroxyethyl)-4-methyl-1,3-thiazole (Fluka) with benzyl chloride in $CH_3CN$ (82% yield).

*1,4-Diketones; 3-methylcyclopentenones.*[2] Thiazolium salts such as (1)[3] catalyze the addition of aldehydes to methyl vinyl ketone to give 1,4-diketones in yields of 50–70% (equation I). The same conditions can be used to convert

$$(\text{I})\quad RCHO + CH_2=CHCCH_3 \xrightarrow[50-70\%]{Cat., N(C_2H_5)_3} RCCH_2CH_2CCH_3$$

(II)  $CH_3(CH_2)_2CHO \xrightarrow[\sim 75\%]{Cat., N(C_2H_5)_3} CH_3(CH_2)_2\underset{O}{\overset{}{C}}-\underset{OH}{\overset{}{CH}}(CH_2)_2CH_3$

aldehydes into acyloins (equation II).

This addition reaction provides a route to 3-methylcyclopentenones, since 1,4-diketones are cyclized by base to cyclopentenones. The reaction has been used in a synthesis of cis-jasmone (2).

(2)

[1] H. Stetter and H. Kuhlmann, *Synthesis*, 379 (1975).
[2] *Idem, Tetrahedron Letters*, 4505 (1974).
[3] This particular salt was used because the thiazole is available commercially.

**Benzylideneacetone(tricarbonyl)iron,** $\underset{O}{\overset{\parallel}{CH_3C}}CH=CHC_6H_5Fe(CO)_3.$    Mol wt. 286.06, orange-red crystals.

This complex is prepared from benzylideneacetone and diiron nonacarbonyl (45% yield) or iron pentacarbonyl ($h\nu$).

Ergosteryl acetate tricarbonyliron can be obtained in >70% yield by use of this reagent. Use of dodecacarbonyliron gives the complex in low yield. The tricarbonyliron complex is useful for protection of the diene system during reactions of the isolated double bond.[1]

[1] G. Evans, B. F. G. Johnson, and J. Lewis, *J. Organometal. Chem.*, **102**, 507 (1975).

**(−)-N-Benzyl-N-methylephedrinium bromide,**

$\begin{array}{l} \overset{CH_3}{\underset{|}{H\blacktriangleright C}}\overset{Br^-}{\underset{}{N(CH_3)_2CH_2C_6H_5}} \\ \underset{|}{} \\ \underset{C_6H_5}{H\blacktriangleright C\blacktriangleleft OH} \end{array}$    Mol. (1)

wt. 348.29, m.p. 222°, $\alpha_D -5.1°$.

*Asymmetric alkylation.* Alkylation of the β-keto ester (1a) or the β-diketone (1b) with a primary alkyl bromide under phase-transfer conditions with this chiral catalyst leads to an optically active product (2). However, optical yields are not high (~5–6% in $CHCl_3$ as the organic solvent); products had no activity when $CHCl_3$ was replaced with butane. Lower degrees of inductions were also observed when N,N-dimethylephedrinium bromide (this volume) was used as catalyst.[1]

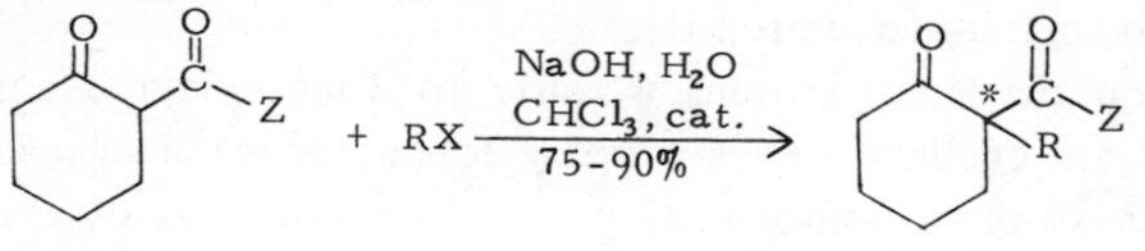

(1a, $Z=OCH_3$, 1b, $Z=CH_3$) (2)

[1] J.-C. Fiaud, *Tetrahedron Letters*, 3495 (1975).

**O-Benzylmonoperoxycarbonic acid,** $C_6H_5CH_2OC(=O)OOH$. Mol. wt. 168.14.

This peracid is prepared in 65% yield by treatment of dibenzyl peroxydicarbonate, $(C_6H_5CH_2OC(=O)O)_2$, with alkaline hydrogen peroxide. It decomposes only slowly at low temperatures. This acid is intermediate in reactivity between perbenzoic acid and *m*-chloroperbenzoic acid in epoxidation of *trans*-stilbene. The by-products are benzyl alcohol and carbon dioxide, so that the reaction proceeds under neutral conditions.[1]

[1] R. M. Coates and J. W. Williams, *J. Org.*, **39**, 3054 (1974).

**Benzylsodium,** $C_6H_5CH_2Na$. Mol. wt. 114.13.

*Preparation*[1,2]:

$$C_6H_5Cl + 2\ Na \longrightarrow C_6H_5Na + NaCl$$

$$C_6H_5Na + C_6H_5CH_3 \longrightarrow C_6H_5CH_2Na + C_6H_6$$

***Alcohol synthesis.***[2] Alcohols can be prepared by the reaction of benzylsodium (large excess) with aldehydes and ketones.

***Examples:***

$$C_6H_5CH_2Na + C_6H_5CHO \xrightarrow[42\%]{1)\ C_6H_5CH_3,\ <40^0\quad 2)\ H_2O} C_6H_5CH_2CH(C_6H_5)OH$$

$$C_6H_5CH_2Na + C_6H_5COCH_3 \xrightarrow[69\%]{} C_6H_5CH_2C(C_6H_5)(CH_3)-OH$$

$$C_6H_5CH_2Na + C_6H_5COC_6H_5 \xrightarrow[91\%]{} C_6H_5CH_2C(C_6H_5)_2-OH$$

[1] H. Gilman, H. A. Pacevitz, and O. Baine, *Am. Soc.*, **62**, 1514 (1940).
[2] W. T. Smith, Jr., and D. L. Nickel, *Org. Prep. Proc. Int.*, **7**, 277 (1975).

**Benzyl tri-*n*-butylammonium chloride,** $[(C_4H_9)_3\overset{+}{N}CH_2C_6H_5]Cl^-$. Mol. wt. 311.93, m.p. 162–164°.

This salt can be prepared in 86% yield by quaternization of tri-*n*-butylamine with benzyl chloride in acetonitrile (1 week reflux).[1]

*Ion-pair extraction.*[1] Iodides are extracted into methylene chloride or similar solvents much more readily than the corresponding chlorides. This ion-exchange method is used to convert trimethyloxosulfonium iodide into the chloride. Thus on distribution of this iodide and an equimolecular amount of the quaternary chloride between methylene chloride and water, the organic phase contains benzyl tri-*n*-butylammonium iodide and the water phase contains trimethyloxosulfonium chloride. The ion exchange is practically complete. Note that trimethyloxosulfonium chloride is preferred to the iodide for generation of dimethyloxosulfonium methylide because the chloride is more soluble in THF (**1**, 315).[1]

[1]A. Brändström and B. Lamm, *Acta Chem. Scand.*, **28 B**, 590 (1974).

**Benzyltriethylammonium chloride, 1,** 53; **3,** 19; **4,** 27–31; **5,** 26–28.

*Halomethyl aryl sulfones.* Mąkosza *et al.*[1] report that chloro- and bromomethyl aryl sulfones are converted into carbanions in 50% aqueous sodium hydroxide in the presence of this quaternary ammonium salt. The carbanions are readily alkylated and condense with carbonyl compounds to form oxiranes.

*Examples:*

$$\underset{\phantom{x}}{p\text{-}CH_3C_6H_4SO_2CH_2Br + C_2H_5Br} \xrightarrow[67\%]{\underset{50\% \text{ aq. NaOH}}{\text{Cat.}}} \underset{\phantom{x}}{p\text{-}CH_3C_6H_4SO_2\overset{\overset{\displaystyle Br}{|}}{C}HC_2H_5}$$

$$C_6H_5SO_2CHCl_2 + C_6H_5CH_2Cl \xrightarrow[84\%]{\underset{50\% \text{ aq. NaOH}}{\text{Cat.}}} C_6H_5SO_2\underset{\underset{\displaystyle CH_2C_6H_5}{|}}{C}Cl_2$$

$$p\text{-}CH_3C_6H_4SO_2CH_2Cl + (C_6H_5)_2C{=}O \xrightarrow[90\%]{\underset{50\% \text{ aq. NaOH}}{\text{Cat.}}} p\text{-}CH_3C_6H_4SO_2CH\overset{\displaystyle\diagdown}{\underset{\displaystyle O}{\diagdown}}C(C_6H_5)_2$$

*Dichlorocarbene.* Sasaki *et al.*[2] have reported that dichlorocarbene generated by phase-tranfer catalysis reacts more efficiently than dichlorocarbene generated in other ways with strained unsaturated polycyclic hydrocarbons. Thus they were able to prepare adducts of basketene (1) and snoutene (2) in the yields indicated.

(**1**, 45%)                    (**2**, 60%)

*Phosphorylation of amines.* Zwierzak[3] has improved a procedure of Todd et al.[4] for phosphorylation of amines by carrying out the reaction in a two-phase system with this catalyst. A dialkyl phosphonate and a tetrahalomethane ($CCl_4$, $CBr_4$) are the source of a dialkylphosphorohalidate, the actual phosphorylation reagent. The amine and this combination are allowed to react at 0–5° in $CH_2Cl_2$–20% aqueous NaOH with benzyltriethylammonium chloride as the phase-transfer catalyst. Yields are generally in the range 70–90%.

*1,1-Dihalospiropentanes.*[5] Methylenecyclopropanes react readily with dichloro- and dibromocarbene, generated by the method of Makosza, to give 1,1-dihalospiropentanes in fair to good yield. The reaction requires about 3 hr.

at 40–45°. Note that the same reaction with triphenylethylene requires 60 hr. The enhanced reactivity of the substituted methylenecyclopropanes is ascribed to strain in this system.

*Cyclopropane-1,1-dicarboxylic acid.* This compound can be obtained in high yield by double alkylation of diethyl malonate with 1,2-dibromoethane under phase-transfer conditions. Malonic acid cannot be alkylated in this way. 1-Cyanocyclopropanecarboxylic acid can be prepared by reaction of the dibromide with ethyl cyanoacetate (86% yield).[6]

*α,β-Unsaturated sulfones.* α,β-Unsaturated sulfones (3) can be prepared by base-induced condensation of sulfones (1) with aromatic aldehydes in a two-phase system with benzyltriethylammonium chloride as phase-transfer catalyst. Yields are generally in the range 50–85%. Aliphatic aldehydes under these conditions undergo self-condensation.[7]

$$C_6H_5\overset{O}{\underset{O}{\overset{\|}{\underset{\|}{S}}}}CH_2R \;+\; ArCHO \;\xrightarrow[\substack{NaOH,\ H_2O/CH_2Cl_2 \\ C_6H_5CH_2(C_2H_5)_3N^+Cl^- \\ \\ 25\text{-}98\%}]{} \; C_6H_5\overset{O}{\underset{O}{\overset{\|}{\underset{\|}{S}}}}-\overset{R}{\overset{\|}{C}}=CHAr \;+\; H_2O$$

(1, R=H, CH₃,  (2)                     (3)

C₆H₅)

*Synthesis of mandelic acid.*[8] Mandelic acids are obtained in 75–85% yield by reaction of benzaldehydes with chloroform and sodium hydroxide in the presence of this phase-transfer catalyst.

$$ArCHO \;\xrightarrow{:CCl_2}\; Ar\overset{Cl}{\underset{H}{\overset{\times}{C}}}\!\!-\!O \;\longrightarrow\; Ar\overset{Cl}{\overset{\|}{C}}HCOCl \;\xrightarrow{NaOH}\; Ar\overset{OH}{\overset{\|}{C}}HCOOH$$

*N-Alkylation of N-acylanilines.* These substances can be alkylated in aqueous NaOH–benzene with use of this phase-transfer catalyst.[9]

$$Ar-\overset{H}{N}-C\!\!\diagup^{O}_{\diagdown R} \;\underset{}{\overset{OH^-,\,C_6H_6,\,TEBA}{\rightleftarrows}}\; ArN\!-\!C\!\!\diagup^{O}_{\diagdown R} \;\xrightarrow[80\text{-}95\%]{R'X}\; Ar-\overset{R'}{N}-C\!\!\diagup^{O}_{\diagdown R}$$

[1] A. Jónczyk, K. Bánko, and M. Mąkosza, *J. Org.*, **40**, 266 (1975).
[2] T. Sasaki, K. Kanematsu, and N. Okamura, *ibid.*, **40**, 3322 (1975).
[3] A. Zwierzak, *Synthesis*, 507 (1975).
[4] F. R. Atherton, H. T. Openshaw, and A. R. Todd, *J. Chem. Soc.*, 660, 674 (1945).
[5] E. Dunkelblum and B. Singer, *Synthesis*, 323 (1975).
[6] R. K. Singh and S. Danishefsky, *J. Org.*, **40**, 2969 (1975).
[7] G. Cardillo, D. Savoia, and A. Umani-Ronchi, *Synthesis*, 452 (1975).
[8] A. Merz, *ibid.*, 724 (1974).
[9] R. Brehme, *ibid.*, 113 (1976).

**N-Benzyltriflamide, 5, 29.**

*Synthesis of primary amines.* Hendrickson and co-workers[1] have prepared two derivatives, (1) and (2), of N-benzyltriflamide that can be used for a one-flask Gabriel synthesis of primary amines. These reagents undergo alkylation

C₆H₅CHNHSO₂CF₃
|
COOCH₃

(1, m.p. 71-73°)

(2, m.p. 183-184°)

and subsequent elimination of $CF_3S(O)OH$ under the same basic conditions ($K_2CO_3$ in refluxing acetonitrile).

$$R\text{--}Br \;+\; (1) \xrightarrow[\text{CH}_3\text{CN, }\Delta]{\text{K}_2\text{CO}_3} \left[ \begin{array}{l} R\text{--}N\text{--}SO_2CF_3 \\ \quad\quad | \\ C_6H_5CHCOOCH_3 \end{array} \xrightarrow[-CF_3SO_2H]{K_2CO_3} \begin{array}{l} R\text{--}N \\ \quad\quad \| \\ C_6H_5\overset{}{C}COOCH_3 \end{array} \right]$$

$$\xrightarrow[55\text{-}90\%]{H_3O^+} \; R\text{--}NH_2 \;+\; \begin{array}{c} C_6H_5 \\ \phantom{} \\ CH_3OOC \end{array}\!\!\!\!>\!\!C\!=\!O$$

[1] J. B. Henrickson, R. Bergeron, and D. D. Sternbach, *Tetrahedron*, **31**, 2517 (1975).

## Benzyl trifluoromethanesulfonate (Benzyl triflate), $C_6H_5CH_2OSO_2CF_3$. Mol. wt. 240.20.

Lemieux and Kondo[1] have prepared this ester by the reaction of benzyl alcohol, 2,4,6-trimethylpyridine or pyridine in $CH_2Cl_2$, and triflic anhydride at $-60°$. It was shown to be a very powerful benzylating reagent, particularly for preparation of 2-O-benzyl derivatives of sugars.

[1] R. U. Lemieux and T. Kondo, *Carbohydrate Res.*, **35**, C 4 (1974).

## Benzyltrimethylammonium fluoride, $C_6H_5CH_2\overset{+}{N}(CH_3)_3$ $F^-$. Mol. wt. 169.25; hygroscopic (manipulate in a drybox or under argon).

Triton B (40% methanolic solution) is neutralized with 47% aqueous HF; the fluoride is dried at $100°$ at $<1$ mm. for 24 hr., pulverized, and dried again. It is stored over $P_2O_5$.

*Regiospecific alkylation of ketones.*[1] This salt converts trimethylsilyl enol ethers into quaternary ammonium enolates that undergo regiospecific monoalkylation. This method is especially useful for alkylation at the less substituted position of a cyclohexanone; lithium enolates are more useful for alkylation at

(80%)        (9%)

more highly substituted positions.

[1] I. Kuwajima and E. Nakamura, *Am. Soc.*, **97**, 3257 (1975).

**Bis(acetylacetonate)palladium(II), Pd(acac)$_2$ ,**

(1)

Mol.

wt. 304.92, m.p. 205° dec. Suppliers: ROC/RIC, Strem.

This organometallic reagent in conjunction with pyridine has been used as a homogeneous catalyst for the hydrogenation of nitrobenzene to aniline. Petroleum ether (b.p. 100–120°) is the most efficient solvent. Yield, 90%.[1]

[1]M. C. Datta, C. R. Saha, and D. Sen, *Chem. Ind.*, 1057 (1975).

**Bis(acrylonitrile)nickel(0), 3, 20.**

*Isomerization of quadricyclane.*[1] Quadricyclane (1) is isomerized mainly to norbornadiene (2) in the presence of a nickel(0) catalyst: bis(acrylonitrile)-nickel(0) or bis(1,5-cyclooctadiene)nickel(0). Evidence that a complex such as (a) is involved in the product-determining step is presented. For example, isomerization in the presence of acrylonitrile results in *exo-* and *endo-*isomers of the adduct (3).

[1]R. Noyori, I. Umeda, H. Kawauchi, and H. Takaya, *Am. Soc.*, **97**, 812 (1975).

**Bis(benzonitrile)palladium(II) chloride, $(C_6H_5CN)_2PdCl_2$, 5, 31–32.** Mol. wt. 383.85. Suppliers: ROC/RIC, Strem.

*π-Allylpalladium chloride complexes* (4, 369; 5, 500–501). Steroidal olefins are converted efficiently into π-allylpalladium complexes by treatment with bis(benzonitrile)palladium(II) chloride in refluxing chloroform (24–48 hr.).[1] π-Allyl complexes are also formed by use of disodium tetrachloropalladate, $Na_2PdCl_4$, in acetic acid–acetic anhydride containing sodium acetate (50°,

96 hr.); however, this method is less selective than the former method. For example, $\Delta^5$-cholestene (1) is converted by the first procedure only into (2), and by the second method into a mixture of (2), (3), and (4). This is the first example of isolation of diastereoisomeric $\pi$-allylpalladium complexes. Steric

effects are evidently important in these reactions.

The same paper reports that these $\pi$-allylpalladium complexes can be reduced by lithium aluminum hydride. Both (2) and (3) are reduced to $\Delta^4$-cholestene (30%) and $\Delta^5$-cholestene (70%); complex (4) is reduced quantitatively to $\Delta^5$-cholestene.

Oxidation of these $\pi$-allylpalladium complexes proceeds regio- and stereoselectively to allylic alcohols.[2] Thus treatment of (2) in pyridine (1 eq.) with $m$-chloroperbenzoic acid in petroleum ether gives $4\alpha$-hydroxy-$\Delta^5$-cholestene; the $\beta$-isomer (3) is oxidized to $6\beta$-hydroxy-$\Delta^4$-cholestene. The oxidation of (4) to $7\alpha$-hydroxy-$\Delta^5$-cholestene is even more selective. Oxidation in the absence

(4)  $\xrightarrow[\substack{82\%}]{\substack{\text{ClC}_6\text{H}_4\text{COOOH} \\ \text{Py}}}$

of pyridine is less selective, and ketones are also formed. Thus peracid oxidation of (2) in light petroleum alone gives $\Delta^5$-cholestene-4-one (16%), $\Delta^4$-cholestene-6-one (4%), $\Delta^5$-cholestene-4α-ol (5%), $\Delta^5$-cholestene-4β-ol (3%), $\Delta^4$-cholestene-6β-ol (12%), and $\Delta^4$-cholestene-6α-ol (12%).

*Indenones.* Oxidation of enynes of type (1) with this palladium complex leads to indenones (2) in 15–54% yield.[3]

(1)                                                              (2)

[1] D. N. Jones and S. D. Knox, *J.C.S. Chem. Comm.*, 165 (1975).
[2] *Idem, ibid.*, 166 (1975).
[3] W. Münzenmaier and H. Straub, *Synthesis*, 49 (1976).

**Bis($\eta^5$-cyclopentadienyl)niobium trihydride,** $(\eta^5\text{-C}_5\text{H}_5)_2\text{NbH}_3$ (1). Mol. wt. 226.11, yellow crystals.

*Preparation.* The trihydride can be prepared[1] by reduction of $(\text{C}_5\text{H}_5)_2\text{NbCl}_2$[2] in toluene with sodium bis-(2-methyoxyethoxy)aluminum hydride (**3**, 260–261; **4**, 441–442; **5**, 596) until solution is complete. The reagent (1) is obtained in 55% yield after hydrolysis with aqueous NaOH. An earlier preparation required high-pressure hydrogenation.[3]

*Complexes with acetylenes.*[1] This niobium(III) reagent readily forms complexes, (2a) and (2b), with acetylenes. When R and R′ are different, the bulkier alkyl group is oriented preferentially toward the hydride ligand. Treatment of the complexes (2) with acid yields, nearly quantitatively, the corresponding

(2a)                          (2b)

(3)

*cis*-olefin (3). Treatment of (2) with methyl fluorosulfonate liberates the original acetylene and methane; presumably, methylation occurs at the niobium. When heated with CO under pressure the alkenyl(carbonyl) complexes (4) are formed, in which the niobium atom is attached preferentially to the vinylic carbon atom with the smaller substituent. This insertion occurs with *cis* C=C stereochemistry. These complexes (4) react slowly with methyl fluorosulfonate, but methylated olefins, (6) and (7), are formed eventually; these result from methylation not at the metal, but at the $\beta$-vinylic carbon atom.

Schwartz has reported preparation of tantalum complexes related to (2).[4]

(2a), (2b) $\xrightarrow{\text{CO}}$     (4a)     +     (4b)

$\downarrow$ CH$_3$OSO$_2$F

(5)     (6)     (7)

[1] J. A. Labinger and J. Schwartz, *Am. Soc.*, **97**, 1596 (1975).
[2] C. R. Lucas and M. L. H. Green, *J.C.S. Chem. Comm.*, 1005 (1972).
[3] F. N. Tebbe and G. W. Parshall, *Am. Soc.*, **93**, 3793 (1971).
[4] J. A. Labinger, J. Schwartz, and J. M. Townsend, *ibid.*, **96**, 4009 (1974).

**Bis(cyclopentadienyl)titanium dichloride (Titanocene dichloride),** $(\eta^5\text{-}C_5H_5)_2$-TiCl$_2$. Mol. wt. 249.00, m.p. 289–291°, red, air stable. Suppliers: Alfa, ROC/RIC, Strem.

*Terpenoids.*[1] In the presence of this Ti(IV) compound as catalyst, isoprene undergoes a regioselective insertion into allyl–magnesium bonds as exemplified for the reaction of crotylmagnesium chloride (1) in equation I. Evidently, (1) reacts in the isomeric form (1b). The products (2) can undergo usual Grignard reactions.

(I)     (1a) $\rightleftharpoons$ (1b) $+ \ CH_2=C-CH=CH_2 \xrightarrow{\substack{Cp_2TiCl_2 \\ THF, 50-60°}}$

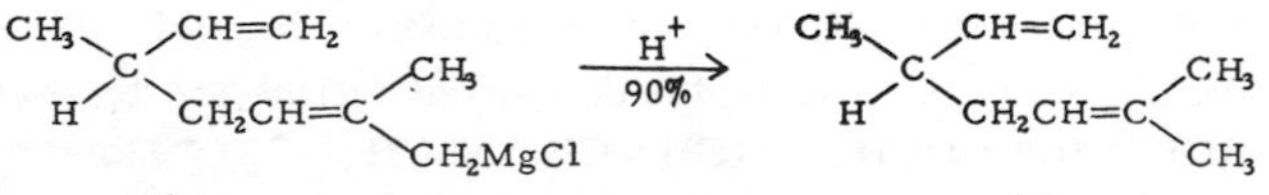

(2)                                              (3)

This insertion reaction has been applied to the syntheses of various natural terpenoids and related products. An example is the synthesis of geranylacetone. Note that *trans*-isomers are formed almost exclusively.

(II)

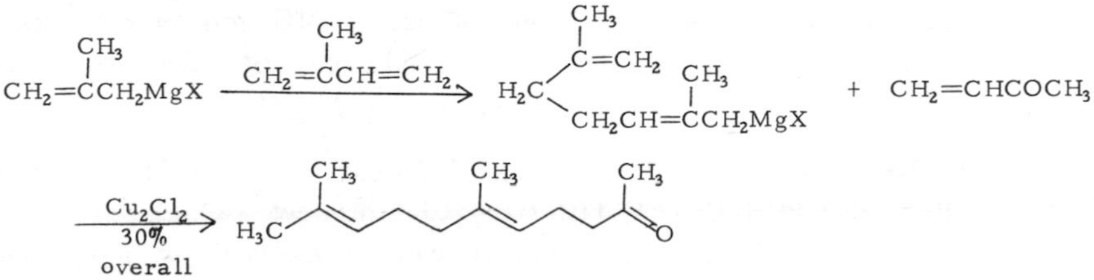

[1]S. Akutagawa and S. Otsuka, *Am. Soc.*, **97**, 6870 (1975).

**Bis(dimethylaluminum)1,3-propanedithiolate**, $(CH_3)_2 AlS(CH_2)_3 SAl(CH_3)_2$ (1). Mol. wt. 217.28.

The reagent is prepared[1] by the reaction at $0°$ of trimethylaluminum in toluene–$CH_2 Cl_2$ with 1,3-propanedithiol. The reagent is used in solution as prepared.

*Ketene thioacetals.*[1] This reagent reacts with various methyl esters to form ketene acetals (2) in 50–85% yield. The products (2) are useful for synthesis

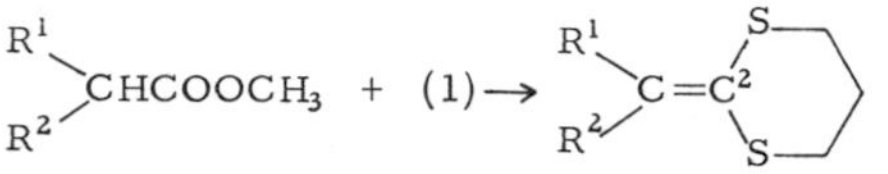

(2)

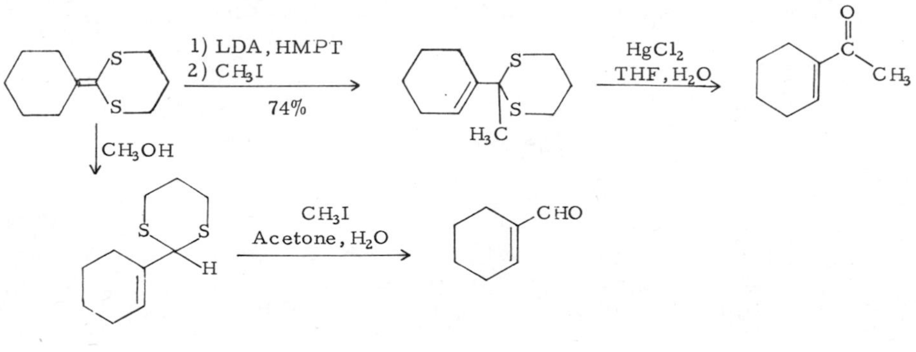

since they can be metalated by LDA in the presence of HMPT to give allyl anions, which react with various electrophiles predominately at $C_2$ in the dithiane ring. Two important transformations that can be realized with ketene thioacetals are conversion of a methyl ester into an $\alpha,\beta$-unsaturated ketone and conversion into an $\alpha,\beta$-unsaturated aldehyde. Since such aldehydes are readily oxidized to acids, the dehydrogenation of esters is possible: $RCH_2CH_2COOCH_3 \rightarrow RCH{=}CHCOOCH_3$.

*Compare* **bis(dimethylaluminum)1,2-ethanedithiolate, 5, 35–36.**

[1] E. J. Corey and A. P. Kozikowski, *Tetrahedron Letters*, 925 (1975).

## 1,8-Bis(dimethylamino)naphthalene, 3, 22; 4, 35.

*Debromination.*[1] Ho and Wong have reported that some *vic*-dibromides are debrominated by this very hindered base (DMF, reflux) in surprisingly high yields. On the other hand, indene dibromide does not undergo this reaction.

[1] T.-L. Ho and C. M. Wong, *Syn. Commun.*, **5**, 87 (1975).

## [1,3-Bis(diphenylphosphino)propane]nickel(II)          chloride,          Ni(dppp)Cl$_2$, Ni[(C$_6$H$_5$)$_2$PCH$_2$CH$_2$CH$_2$P(C$_6$H$_5$)$_2$]Cl$_2$. Mol. wt. 542.03.

*Preparation.*[1] The red complex can be prepared in almost quantitative yield by the reaction of nickel(II) chloride hexahydrate with 1,3-bis(diphenylphosphine)propane (Strem) in isopropanol–methanol. The mixture is heated for 30 min. and then allowed to cool to room temperature, when the complex separates.

*Cross-coupling of Grignard reagents with halides.*[2] This phosphine–nickel complex catalyzes the cross-coupling of alkyl, alkenyl, aryl, and hetaryl Grignard reagents with aryl, hetaryl, and alkenyl halides. Alkyl halides usually do not undergo this reaction. *Caution:* The coupling reaction is usually exothermic; the Grignard reagent should be added to a cold solution of the catalyst and the organic halide in ether. After the exothermic reaction subsides the mixture is usually refluxed for 3–20 hr.

*Examples:*

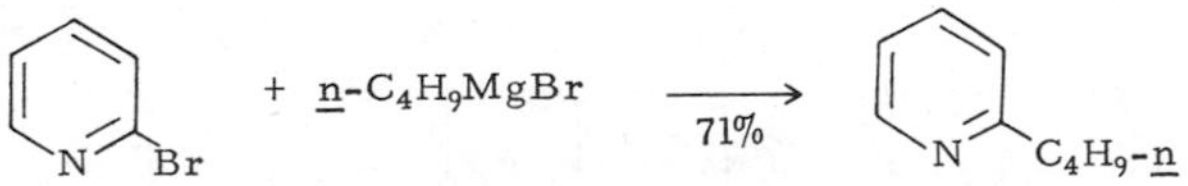

*Metacyclophanes.* Japanese chemists[3] have effected a one-step synthesis of [n] metacyclophanes and [n] (2,6)-pyridinophanes (2a and 2b) by coupling of 1,3-dichlorobenzene (1a) or of 2,6-dichloropyridine (1b) with di-Grignard reagents catalyzed by this nickel(II)—phosphine complex. Highest yields (33% in the case of 2b) were obtained with n = 10. Yields of pyridinophanes were

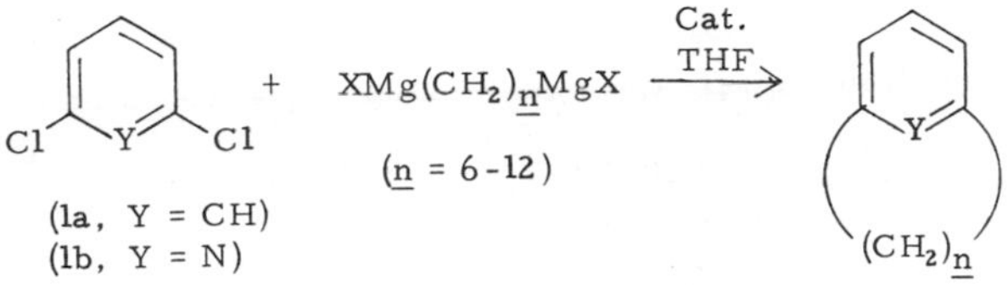

(1a, Y = CH)
(1b, Y = N)

($n$ = 6 -12)

(2)

somewhat higher than those of metacyclophanes.

This reaction was used to synthesize racemic muscopyridine (3) in 20% yield from 2,6-dichloropyridine and the di-Grignard reagent from 2-methyl-1,10-dibromodecane.

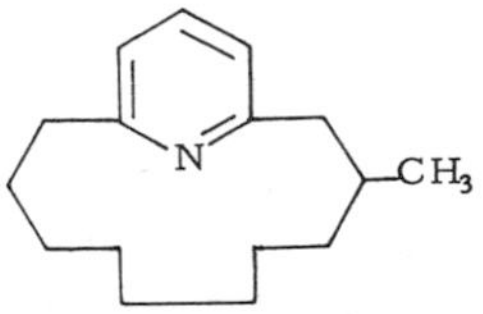

(3)

[1] G. R. Van Hecke and W. D. Horrocks, Jr., *Inorg. Chem.*, **5**, 1968 (1966).

[2] K. Tamao, K. Sumitani, and M. Kumada, *Am. Soc.*, **94**, 4374 (1972); K. Tamao, Y. Kiso, K. Sumitani, and M. Kumada, *ibid.*, **94**, 9268 (1972); Y. Kiso, K. Tamao, N. Miyake, K. Yamamoto, and M. Kumada, *Tetrahedron Letters*, 3 (1974); M. Kumada, K. Tamao, and K. Sumitani, *Org. Syn.*, submitted (1975).

[3] K. Tamao, S. Kodama, T. Nakatsuka, Y. Kiso, and M. Kumada, *Am. Soc.*, **97**, 4405 (1975).

## Bis-(2,2-dipyridyl)-silver(II) peroxydisulfate, $Ag(dipy)_2S_2O_8$. Mol. wt. 684.44.

The reagent is prepared in the same way as the corresponding pyridine complex.[1]

*Oxidative acetoxylation of arenes.*[2] This Ag(II) salt oxidizes arenes in acetic acid containing sodium acetate to acetoxyarenes usually in high yield. The

$$o:m:p = 68:1:31$$

$$95:5$$

isomer distribution is similar to that observed in anodic acetoxylation.

This reaction can also be conducted with silver(I) acetate in the presence of 2,2'-dipyridine and potassium peroxydisulfate.

[1] W. G. Palmer, *Experimental Inorganic Chemistry*, Cambridge University Press, Cambridge, 1965, p. 158.
[2] K. Nyberg, and L.-G. Wistrand, *Acta Chem. Scand.*, **B29**, 629 (1975),

**Bis(4-methylpiperazinyl)aluminum hydride,** (1). Mol. wt. 226.29.

The reagent is prepared by the reaction of aluminum hydride in THF with 2 moles of the amine.

$$C_6H_5COOH \xrightarrow[\substack{\text{1) (1)} \\ \text{2) } H_2O \\ 86\%}]{} C_6H_5CHO$$

$$n\text{-}C_5H_{11}COOH \xrightarrow[72\%]{} n\text{-}C_5H_{11}CHO$$

*Reduction of acids or esters to aldehydes.*[1] Diaminoaluminum hydrides, particularly those derived from secondary cyclic amines such as N-methyl-piperazine or morpholine, are useful for reduction of carboxylic acids or esters to aldehydes, without contamination by the corresponding alcohols. The reduction is carried out in refluxing THF (6–20 hr.). In general, 2 moles of reagent are used for each mole of substrate. The reduction is applicable to both aromatic and aliphatic esters. Yields are in the range of 50–85%.

*Reduction of amides to aldehydes.*[2] Several types of amides, including N,N-dialkylamides and N-acylmorpholines, are reduced by excess reagent to aldehydes in 65–80% yield. One example of reduction of an N-monosubstituted

$$RCON\begin{matrix}R' \\ \\ R''\end{matrix} \xrightarrow[\text{2) H}_2\text{O}]{\text{1) (1)}} RCHO$$

amide, N-methylbenzanilide → benzaldehyde (42% yield), has been reported.

[1]M. Muraki and T. Mukaiyama, *Chem. Letters*, 1447 (1974); *idem, ibid.*, 215 (1975).
[2]*Idem, ibid.*, 875 (1975).

### 1,3-Bis(methylthio)allyllithium, 4, 38–39.

Details for the preparation of this reagent and for its use in the preparation of γ-hydroxy-α,β-unsaturated aldehydes have been published.[1]

[1]B. W. Erickson, *Org. Syn.*, 54, 19 (1974).

### Bis(methylthio)(trimethylsilyl)methyllithium, $(CH_3S)_2C\begin{matrix}Li \\ \\ Si(CH_3)_3\end{matrix}$    (1).

Mol. wt. 186.34.

This reagent is prepared from bis(methylthio)(trimethylsilyl) methane.

$$(CH_3S)_2CH_2 \xrightarrow[84\%]{\begin{array}{c}\text{1) } \underline{n}\text{-BuLi, THF, -60}^0 \\ \text{2) } (CH_3)_3SiCl\end{array}} (CH_3S)_2C\begin{matrix}H \\ \\ Si(CH_3)_3\end{matrix} \xrightarrow[\text{THF, -60}^0]{\underline{n}\text{-BuLi}} (1)$$

b.p. 67-70°/10 torr

*Ketene thioacetals.*[1] The reagent reacts with aldehydes and ketones in THF at 20° to give, after acid hydrolysis, ketene thioacetals:

$$(1) + {\small >}C{=}O \longrightarrow (CH_3S)_2C\begin{matrix}Si(CH_3)_3 \\ \\ C{-}OLi \\ |\end{matrix} \xrightarrow{H_3O^+} (CH_3S)_2C{=}C{\small <}$$

*Examples:*

$$(1) + HCHO \xrightarrow{86\%} (CH_3S)_2C{=}CH_2$$

$$(1) + C_4H_9CHO \xrightarrow{80\%} (CH_3S)_2C{=}CHC_4H_9$$

$$(1) + O{=}\hexagon \xrightarrow{80\%} (CH_3S)_2C{=}\hexagon$$

*Conjugate addition.*[2] The metalated thioacetal (1), unlike the corresponding 1,3-dithiane derivatives, undergoes Michael addition with cyclic enones; the resulting enolates (a) can be alkylated directly with alkyl halides to form α,β-disubstituted ketones (3), predominately as the *trans*-isomers. The corresponding 1,4-diketones can be obtained by hydrolysis catalyzed by mercuric chloride or

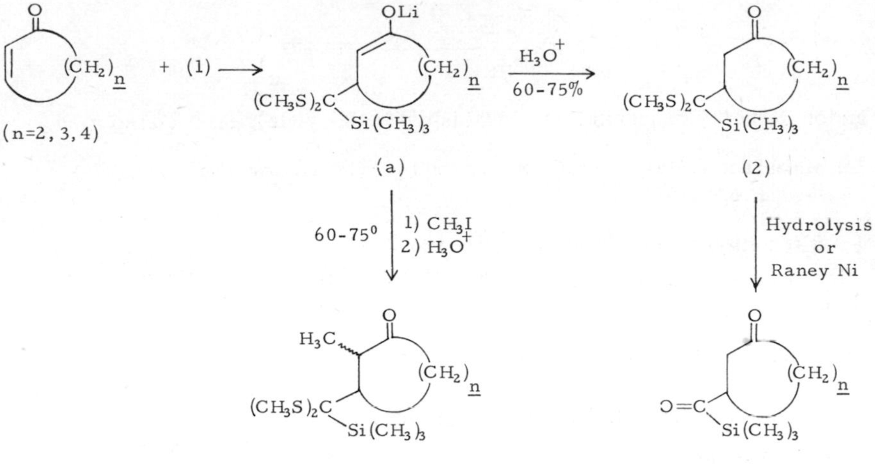

(a)

(2)

(3)

(4)

by treatment with Raney nickel. For example, the diketones (5)–(7) have been prepared in this way in the indicated yields.

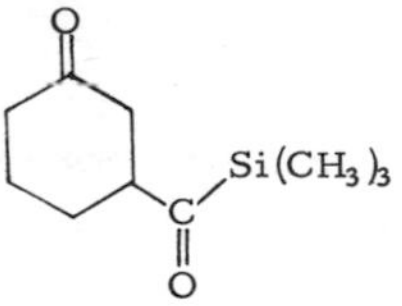

(5, 71%) 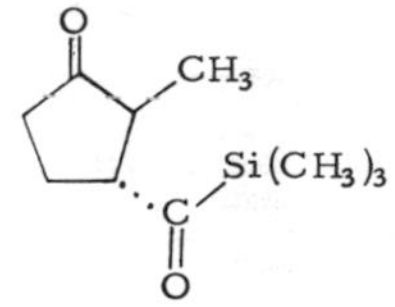          (6, 66%) 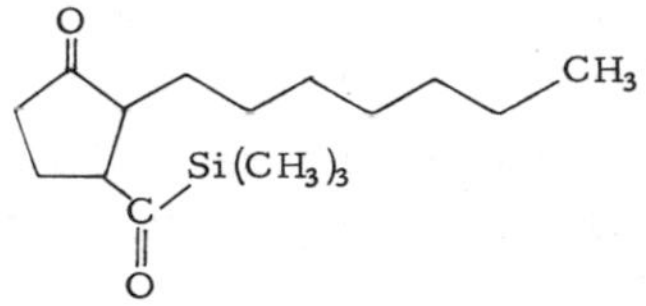          (7, 60%)

Bis(methylthio)(trimethylstannyl)methyllithium, $(CH_3S)_2C\begin{smallmatrix}Li\\\\Sn(CH_3)_3\end{smallmatrix}$ , also undergoes Michael addition to cyclic enones in comparable yields. In this case a further transformation of the intermediate (b) corresponding to (a) is possible: transmetalation to the dilithium derivative (c), which can be alkylated to give (8).

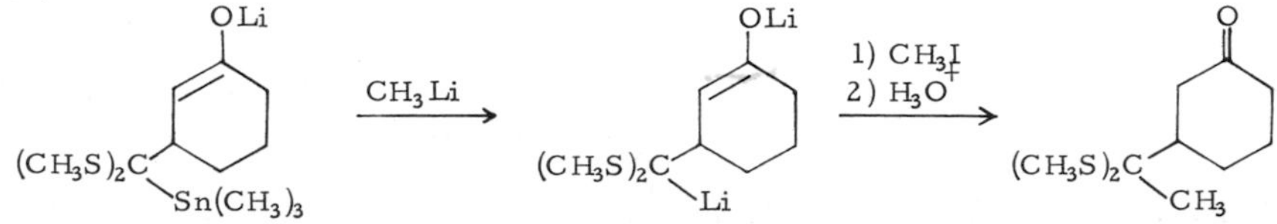

(b)          (c)          (8)

*S-Methyl thiocarboxylates.* Ketene S,S-dimethylacetals (1) can be hydrolyzed to S-methyl thiocarboxylates (2) by trifluoroacetic acid in water. Ketene thioacetals derived from 1,3-dithiane are unchanged by this procedure.[3]

$$\underset{(1)}{\underset{R^2}{\overset{R^1}{\phantom{x}}}C=C\underset{SCH_3}{\overset{SCH_3}{\phantom{x}}}} \quad \xrightarrow[65-90\%]{\underset{H_2O}{CF_3COOH}} \quad \underset{(2)}{\underset{R^2}{\overset{R^1}{\phantom{x}}}CH\!-\!C\underset{SCH_3}{\overset{O}{\phantom{x}}}}$$

[1] D. Seebach, M. Kolb, and B.-T. Gröbel, *Ber.,* **106**, 2277 (1973).
[2] D. Seebach and R. Bürstinghaus, *Angew. Chem. internat. Ed.,* **14**, 57 (1975).
[3] *Idem, Synthesis,* 461 (1975).

## Bismuth trioxide–Stannic oxide, $Bi_2O_3$–$SnO_2$.

*Dehydrodimerization of toluene to stilbene.*[1] Toluene can be oxidized to stilbene when heated at elevated temperatures in a conventional flow system over a catalyst composed of bismuth trioxide and stannic oxide. Benzene and

$$2\ C_6H_5CH_3 \xrightarrow[\Delta]{O_2,\ Cat.} C_6H_5CH_2CH_2C_6H_5 \longrightarrow C_6H_5CH=CHC_6H_5$$

dibenzyl are also formed.

[1] K. H. Liu, A. Masuda, and Y. Yamazaki, *Tetrahedron Letters,* 3007 (1975).

**1,2-Bis($\beta$-tosylethoxycarbonyl)diazene,** $TsCH_2CH_2OOC-\overset{\overset{\textstyle N-COOCH_2CH_2Ts}{\|}}{N}$ Mol. wt. 482.54, m.p. $\sim 5°$.

The reagent is prepared from $\beta$-tosylethoxycarbonyl chloride (2-*p*-tolyl-sulfonylethyl chloroformate)[2] as formulated:

$$2\ TsCH_2CH_2OCOCl + H_2NNH_2 \xrightarrow[\sim 100\%]{\underset{Dioxane}{MgO}} \underset{TsCH_2CH_2OCONH}{HNOCOCH_2CH_2Ts} \xrightarrow[40-50\%]{N_2O_4} (1)$$

*Diels–Alder reactions.*[1] The diazene (1) undergoes a Diels–Alder reaction with cyclopentadiene to give the expected adduct in >90% yield. The reaction with cyclobutadiene (liberated by oxidation of cyclobutadiene iron tricarbonyl) has been studied in detail. The adduct (2) is formed in 45% yield. Deblocking of (2) by treatment with base and then acid leads to 2,3-diazabicyclo[2.2.0]-hexene-5 (4). Oxidation of a hydrazo group is known to lead to an azo group.

The product in the case of (4) would be 2,3-diaza-Dewar benzene (5). Masamune *et al.* have presented evidence for the transient formation of (5) followed by a reverse Diels–Alder reaction to form cyclobutadiene (6). Thus mild oxidation of

(4) results in formation of the *syn*-dimer of cyclobutadiene (7). Oxidation of (4) in the presence of a quinone results in formation of an adduct such as (8).

[1] S. Masamune, N. Nakamura, and J. Spadaro, *Am. Soc.*, **97**, 918 (1975).
[2] A. T. Kader and C. J. M. Sterling, *J. Chem. Soc.*, 258 (1964).

*trans*-1,2-Bis(tri-*n*-butylstannyl)ethylene, 5, 43–44.
    *Improved preparation*[1]:

[1] E. J. Corey and R. H. Wollenberg, *J. Org.*, **40**, 3788 (1975).

**Bis(tri-*n*-butyl)tin oxide,** $(Bu_3Sn)_2O$. Mol. wt. 596.08, b.p. $180°/2$ mm. Suppliers: Alfa, ROC/RIC.
    *Oxetane synthesis.*[1] Two new syntheses of oxetane using this reagent have been reported (equations I and II). A third synthesis employs tributylethoxytin (III).

(II)   $2 \, HO(CH_2)_3Br + (Bu_3Sn)_2O \xrightarrow[-H_2O]{C_6H_6, 80^0} 2 \, Bu_3SnO(CH_2)_3Br \xrightarrow[20\%]{\Delta} 2 \; [\;\square\;]_O + 2 \, Bu_3SnBr$

(III)   $AcO(CH_2)_3Br + Bu_3SnOC_2H_5 \xrightarrow[-C_2H_5OAc]{100^0} Bu_3SnO(CH_2)_3Br \xrightarrow[40\%]{220^0} [\;\square\;]_O$

[1] J. Biggs, *Tetrahedron Letters*, 4285 (1975).

## Bis(triethylphosphine)nickel(II) bromide, $[(C_2H_5)_3P]_2NiBr_2$. Mol. wt. 454.84, air stable.

The salt is prepared by reaction of triethylphosphine (2 moles) with nickel(II) bromide.

*Coupling of aryl halides with alkylmetals* (*cf.* **4**, 163–164). Morrell and Kochi[1] have studied nickel(II) catalysis of this reaction with bis(triethylphosphine)nickel(II) bromide as the catalyst (equation I). They have proposed the

(1)   $Ar-Br + CH_3M \xrightarrow{[(C_2H_5)_3P]_2NiBr_2} ArCH_3 + MBr$

catalytic cycle formulated in scheme I.

*Scheme I.*

$[(C_2H_5)_3P]_2Ni\overset{Br}{\underset{Br}{\big\langle}} + ArBr \longrightarrow [(C_2H_5)_3P]_2Ni\overset{Ar}{\underset{Br}{\big\langle}}$

$L_2Ni\overset{Ar}{\underset{Br}{\big\langle}} + CH_3Li \xrightarrow{-LiBr} L_2Ni\overset{Ar}{\underset{CH_3}{\big\langle}}$

$L_2Ni\overset{Ar}{\underset{CH_3}{\big\langle}} \longrightarrow [L_2Ni\overset{+}{\phantom{.}}\overset{Ar}{\underset{CH_3}{\big\langle}} \cdot ArX\cdot^-]$

$L_2Ni\overset{+}{\phantom{.}}\overset{Ar}{\underset{CH_3}{\big\langle}} \longrightarrow L_2Ni^+ + ArCH_3$

$L_2Ni^+ + ArX\cdot^- \longrightarrow L_2NiArX$

[1] D. G. Morrell, and J. K. Kochi, *Am. Soc.*, **97**, 7267 (1975).

## Bis(trimethylsilyl)amidocopper, $[(CH_3)_3Si]_2NCu$. Mol. wt. 223.93.

This organocopper reagent is prepared *in situ* from hexamethyldisilazane, *n*-butyllithium, and cuprous iodide.

*Primary arylamines.* Reaction of aryl iodides with this reagent in refluxing pyridine leads to silyl-protected amines, convertible into free amines by methanolysis.[1]

$ArI + [(CH_3)_3Si]_2NCu \xrightarrow[-CuI]{} ArN[Si(CH_3)_3]_2 \xrightarrow[30-60\%]{CH_3OH} ArNH_2 + 2 \, (CH_3)_3SiOCH_3$

[1] F. D. King and D. R. M. Walton, *J.C.S. Chem. Comm.*, 256 (1974).

Bis(trimethylsilyl)formamide, $HC\overset{OSi(CH_3)_3}{\underset{NSi(CH_3)_3}{\diagup}}$ (1). Mol. wt. 189.41, b.p. 54–55°/13 mm. Supplier: Fluka.

The reagent is prepared by the reaction of formamide with 2 eq. of trimethylchlorosilane and triethylamine (71% yield).

The reagent reacts with enolizable active methylene compounds to give trimethylsilyl ethers (equation I). Nonenolizable active methylene compounds react to form aminomethylene compounds (equation II).[1]

$$(\text{I}) \quad RCOCH_2COR' + (1) \rightarrow RC\overset{OSi(CH_3)_3}{=}CHCOR' + HC\overset{O}{\underset{NHSi(CH_3)_3}{\diagup}}$$
$$(65-80\%)$$

$$(\text{II}) \quad \overset{NC}{\underset{R}{\diagdown}}CH_2 + (1) \rightarrow \overset{NC}{\underset{R}{\diagdown}}C=CHNH_2 + (CH_3)_3SiOSi(CH_3)_3$$
$$(R = CN, COOCH_3) \qquad (60-65\%)$$

[1]W. Kantlehner, W. Kugel, and H. Bredereck, *Ber.,* **105,** 2264 (1972).

**O,N-Bis[trimethylsilyl]hydroxylamine,** $(CH_3)_3SiNHOSi(CH_3)_3$ (1). Mol. wt. 177.39; b.p. 79–80°/97 torr, 78–80°/100 torr.

*Preparation*[1,2]:

$$H_2NOH \cdot HCl \xrightarrow[\underset{70\%}{2)\ (CH_3)_3SiCl,\ Pet.\ ether,\ \Delta}]{1)\ HN(C_2H_5)_2,\ THF} (1)$$

*O-Arenesulfonyl- and O-arenecarbonylhydroxylamines.*[2]

$$(1) \xrightarrow[\text{Ether}]{n\text{-BuLi}} [(CH_3)_3Si]_2NOLi \begin{cases} \xrightarrow[\sim 55\%]{ArSO_2Cl} ArSO_2ON[Si(CH_3)_3]_2 \xrightarrow[80\%]{H_3O^+} ArSO_2ONH_2 \\[2ex] \xrightarrow[\sim 55\%]{ArCOCl} ArCOON[Si(CH_3)_3]_2 \xrightarrow[80\%]{H_3O^+} ArCOONH_2 \end{cases}$$

[1]R. West, P. Boudjouk, and A. Matuszko, *Am. Soc.,* **91,** 5184 (1969).
[2]F. D. King and D. R. M. Walton, *Synthesis,* 788 (1975).

**Bis(triphenylphosphine)nickel(II) bromide,** $[P(C_6H_5)_3]_2NiBr_2$ (1). Mol. wt. 743.07.

This organometallic reagent is prepared from $NiBr_2 \cdot 6H_2O$ and triphenylphosphine in butanol.

*Linear dimerization of 1,3-butadiene.*[1] Butadiene is dimerized to (E,E)-1,3,6-octatriene in 95% yield by a catalyst system composed of $NaBH_4$ and $[P(C_6H_5)_3]_2NiBr_2$ in a ratio of 2:1. The polymer-bound analogue (2) of

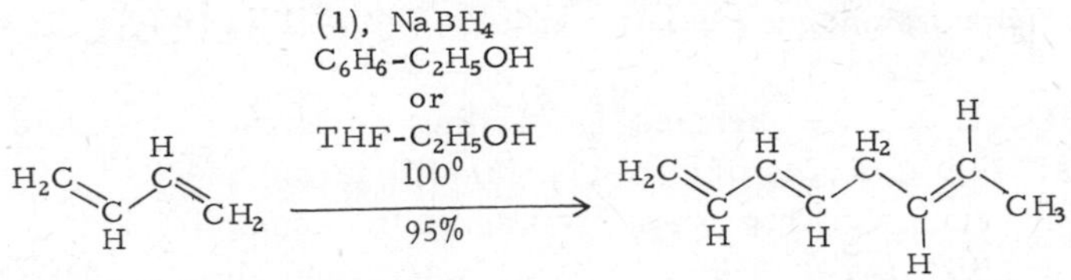

(1),  NaBH₄
C₆H₆-C₂H₅OH
or
THF-C₂H₅OH
100°
95%

(1) is equally effective and permits homogeneous catalysis and easy recovery

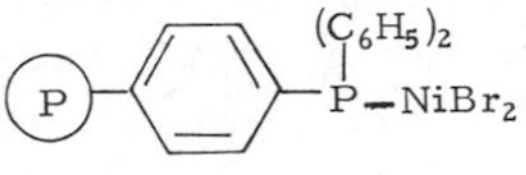

(2)

of the catalyst. Bis(triphenylphosphine)nickel dichloride is somewhat less effective than the dibromide.

[1]C. U. Pittman, Jr., and L. R. Smith, *Am. Soc.*, **97**, 341 (1975).

**Bis(triphenylphosphine)nickel(II)  chloride,  [(C₆H₅)₃P]₂NiCl₂.**     Mol. wt. 654.15.

*Biaryls.*[1]  Nickel complexes of the type NiCl₂L₂[2] catalyze the cross-coupling of sterically hindered aryl Grignard reagents with an aryl bromide to give biaryls in high yield as shown in the example. Thus 2-phenylmesitylene can be

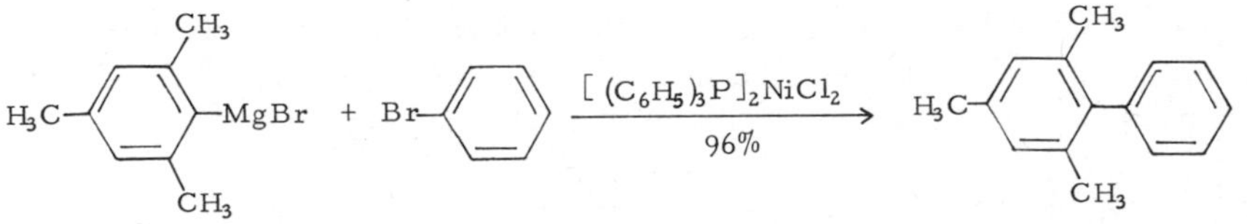

obtained in high yield as shown. However, it is obtained in only 48% yield by coupling phenylmagnesium bromide with mesityl bromide.

[1]K. Tamao, A. Minato, N. Miyake, T. Matsuda, Y. Kiso, and M. Kumada, *Chem. Letters,* 133 (1975).
[2]L = triphenylphosphine,  1,2-bis(diphenylphosphino)propane, or 1,2-bis(dimethylphosphino)ethane.

**Bis(triphenylphosphine)palladium(II)  chloride,  PdCl₂[(C₆H₅)₃P]₂.** Mol. wt. 702.17, air-stable, yellow. Suppliers: ROC/RIC, Strem.

*Carboalkylation.*[1]  Aryl and benzyl halides undergo carboalkylation with CO and an alcohol in the presence of this palladium(II) complex as catalyst. A variety of bases can be used, but 1,8-bis(dimethylamino)naphthalene (3, 22; 4, 35) is most satisfactory.

$$C_6H_5CH_2Cl + CO + ROH \xrightarrow[50-90\%]{\substack{Cat.\\ Base}} C_6H_5CH_2COOR$$

[1]J. K. Stille and P. K. Wong, *J. Org.*, **40**, 532 (1975).

**Bis(triphenylphosphine)palladium(II) halides,** $PdX_2[P(C_6H_5)_3]_2$, X = Cl, Br, and I.

The reagents can be prepared by heating potassium tetrachloropalladate $(K_2PdCl_4)$ with an excess of the potassium halide and an excess of triphenylphosphine in ethanol.[1] They can be crystallized from chloroform.

*Carboxyalkylation of halides.*[1a] Aryl, benzyl, and vinyl halides react with carbon monoxide and an alcohol at $100°$ (atmospheric pressure) in the presence of a tertiary amine and a dihalo(triphenylphosphine)palladium catalyst to form esters in good yields. In the case of vinylic halides, the reaction proceeds predominantly with retention of configuration.

$$RX + CO + R^1OH + R^2_3N \xrightarrow{cat.} RCOOR^1 + R^2_3NH^+X^-.$$

*Examples:*

$C_6H_5Br \longrightarrow C_6H_5COO\text{-}\underline{n}\text{-}C_4H_9 \quad (78\%)$

$C_6H_5CH_2Br \longrightarrow C_6H_5CH_2COO\text{-}\underline{n}\text{-}C_4H_9 \quad (45\%)$

69%    11%

*Amides.*[2] Amides can be prepared in a manner analogous to that described above for synthesis of esters; the alcohol is replaced by a primary or secondary amine. The reaction is limited to aryl, heterocyclic, and vinylic halides, and is

$$RX + CO + R^1NH_2 + R^2_3N \xrightarrow{cat.} RCONHR^1 + R^2_3NH^+X^-$$

highly stereoselective in the case of *cis-* and *trans*-vinylic halides.

*Aldehydes.*[3] Aryl, heterocyclic, and vinylic halides react at $80–150°$ under about 1000 psi of carbon monoxide and hydrogen (1:1) in the presence of these palladium complexes as catalysts and a tertiary amine (1 eq.) to give aldehydes, usually in good yields. The reaction is actually believed to involve a series of

$$RX + CO + H_2 + R^1_3N \xrightarrow{cat.} RCHO + R^1_3N^+HX^-$$

reactions in which the palladium complex is converted into Pd(0) and Pd(II) complexes. Actually, palladium(II) acetate with 2 eq. of triphenylphosphine produces identical catalysis.

*Examples:*

$C_6H_5Br \longrightarrow C_6H_5CHO \quad (94\%)$

$1\text{-}C_{10}H_7Br \longrightarrow 1\text{-}C_{10}H_7CHO \quad (82\%)$

3-Bromopyridine $\longrightarrow$ 3-Pyridinecarboxaldehyde (80%)

$$\underset{\underset{H}{\overset{C_6H_5}{\diagdown}}}{\overset{C_6H_5}{\diagup}}C=C\underset{\overset{Br}{\diagup}}{\overset{H}{\diagdown}} \longrightarrow \underset{\underset{H}{\overset{C_6H_5}{\diagdown}}}{\overset{C_6H_5}{\diagup}}C=C\underset{\overset{CHO}{\diagup}}{\overset{H}{\diagdown}} \quad (65\%)$$

$$\underset{\underset{I}{\overset{CH_3(CH_2)_3}{\diagdown}}}{\overset{CH_3(CH_2)_3}{\diagup}}C=CH_2 \longrightarrow \underset{\overset{OHC}{\diagup}}{\overset{CH_3(CH_2)_3}{\diagdown}}C=CH_2 + \underset{\underset{H}{\overset{CH_3(CH_2)_2}{\diagdown}}}{\overset{CH_3(CH_2)_2}{\diagup}}C=C\underset{\overset{CHO}{\diagup}}{\overset{CH_3}{\diagdown}}$$

*Substitution of aryl and vinylic halides.* Dieck and Heck[4] have reported a reaction of aryl and vinylic bromides and iodides with olefins catalyzed by palladium acetate and 2 eq. of triphenylphosphine that is related to the reaction mentioned above. New olefins are formed by replacement of the vinylic hydrogen of the original olefin by the Ar or R group of the halide.

*Examples:*

$$\underset{\underset{C_6H_5}{}}{\overset{Br}{\diagup}} + CH_2=CHCOOCH_3 + R_3N \xrightarrow[-R_3NH^+Br^-]{cat.} \underset{\underset{C_6H_5}{}}{\overset{CH=CHCOOCH_3}{}} + R_3NH^+Br^-$$

$$(36.5\%)$$

$$\underset{\underset{CH_3}{\overset{CH_3}{\diagdown}}}{\overset{CH_3}{\diagup}}C=C\underset{\overset{Br}{\diagup}}{\overset{H}{\diagdown}} + CH_2=CHCOOCH_3 + (C_2H_5)_3N \xrightarrow[75\%]{cat.} \underset{\underset{H_3C}{\overset{CH_3}{\diagdown}}}{\overset{CH_3}{\diagup}}C=C\underset{\overset{}{\diagdown}}{\overset{H}{\diagup}}\underset{\underset{COOCH_3}{}}{C=C}\underset{}{\overset{H}{\diagup}} + (C_2H_5)_3NH^+Br^-$$

*Acetylenes.*[5] This organometallic reagent, in combination with cuprous iodide, catalyzes the substitution of acetylenic hydrogen by iodoarenes and bromoolefins. It is particularly useful for synthesis of symmetrical acetylenes from acetylene.

*Examples:*

$$C_6H_5I + HC\equiv CH \xrightarrow[\substack{(C_2H_5)_2NH \\ 85\%}]{[(C_6H_5)_3P]_2PdCl_2, CuI} C_6H_5C\equiv CC_6H_5$$

$$C_6H_5CH=CHBr + HC\equiv CH \xrightarrow[95\%]{} C_6H_5CH=CHC\equiv CCH=CHC_6H_5$$

$$(C_6H_5)_2C=CHBr + C_6H_5C\equiv CH \xrightarrow[99\%]{} (C_6H_5)_2C=CHC\equiv CC_6H_5$$

[1](a) A. Schoenberg, I. Bartoletti, and R. F. Heck, *J. Org.*, **39**, 3318 (1974); (b) F. R. Hartley, *J. Organometal. Chem. Rev.*, **A6**, 119 (1970).
[2] A. Schoenberg and R. F. Heck, *ibid.*, **39**, 3327 (1974).
[3] *Idem*, *Am. Soc.*, **96**, 7761 (1974).
[4] H. A. Dieck and R. F. Heck, *Am. Soc.*, **96**, 1133 (1974).
[5] K. Sonogashira, Y. Tohda, and N. Hagihara, *Tetrahedron Letters*, 4467 (1975).

**9-Borabicyclo[3.3.1]nonane** (9-BBN), 2, 31; 3, 24–29; 4, 41; 5, 46–47.

Complete details for the preparation of 9-BBN by the reaction of diborane with 1,5-cyclooctadiene are now available. The reagent is obtained as a fine, white powder melting at 152–155° (sealed, evacuated capillary) in about 75% yield. It can be stored for over 2 years at room temperature under nitrogen. 9-BBN is dimeric both in the solid and vapor states and in solution.

Hydroboration of olefins with 9-BBN is normally complete after 1 hr. in refluxing THF or after 8 hr. in refluxing benzene or hexane. The regioselectivity in olefin hydroborations is higher with 9-BBN than with disiamylborane or dicyclohexylborane.[1]

*Examples:*

$$C_6H_5CH=CH_2 \longrightarrow CH_3CH(C_6H_5)OH + C_6H_5CH_2CH_2OH$$
$$1.5\% \qquad\qquad 98.5\%$$

*Reductions with ate complexes of 9-BBN.* Japanese chemists[2] have prepared the two ate complexes of 9-BBN (1) and (2) by reaction of B-*n*-butyl-9-BBN with *n*-butyllithium and methyllithium, respectively. The bridgehead

( 1,  $R^1 = R^2 = \underline{n}\text{-Bu}$ )
( 2,  $R^1 = CH_3$, $R^2 = \underline{n}\text{-Bu}$ )

hydrogens of (1) and (2) are exceptionally labile as hydride sources. Thus tertiary alkyl, benzyl, and allyl halides are reduced to hydrocarbons by reaction with (1) or (2) at 20° for 1-18 hr. followed by oxidation with alkaline hydrogen peroxide. Primary, secondary, and aryl halides are inert. These ate complexes therefore differ from metal hydrides in which the order of reactivity to alkyl halides is primary > secondary > tertiary.

*Examples:*

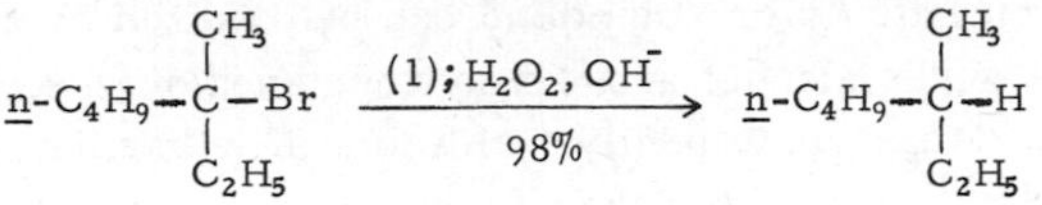

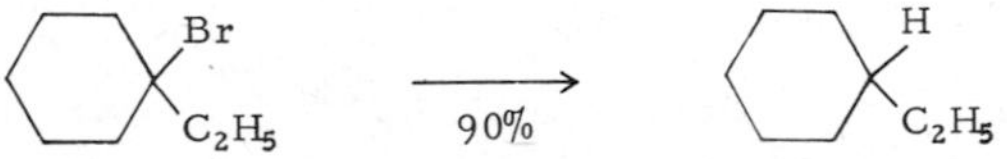

$$n\text{-}C_4H_9-\overset{\overset{\displaystyle CH_3}{|}}{\underset{\underset{\displaystyle C_2H_5}{|}}{C}}-Br \xrightarrow[98\%]{(1);\,H_2O_2,\,OH^-} n\text{-}C_4H_9-\overset{\overset{\displaystyle CH_3}{|}}{\underset{\underset{\displaystyle C_2H_5}{|}}{C}}-H$$

$$C_6H_5CH_2Cl \xrightarrow[100\%]{} C_6H_5CH_3$$

$$C_6H_5CH=CHCH_2Br \xrightarrow[90\%]{} C_6H_5CH=CHCH_3$$

*Oxidation of ate complexes of 9-BBN.* Kramer and Brown[3] have reported that the ate complex (1) undergoes an anomolous oxidation with alkaline hydrogen peroxide to give *cis*-bicyclo[3.3.0]octane-1-ol (3) as a major product together with cyclooctanone (4). Only traces of the expected product (5) are

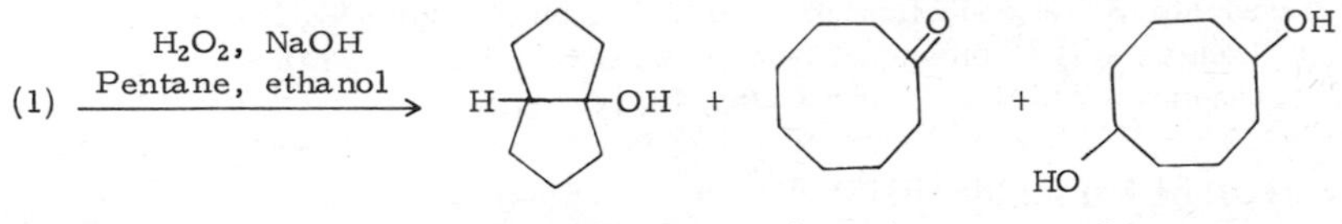

obtained. Kramer and Brown suggest that the formation of (3) involves attack by $H_2O_2$ on the bridgehead hydrogen of (1) with concurrent or stepwise migration of the boron–carbon bond to give (a), which then undergoes normal oxidation to (3).

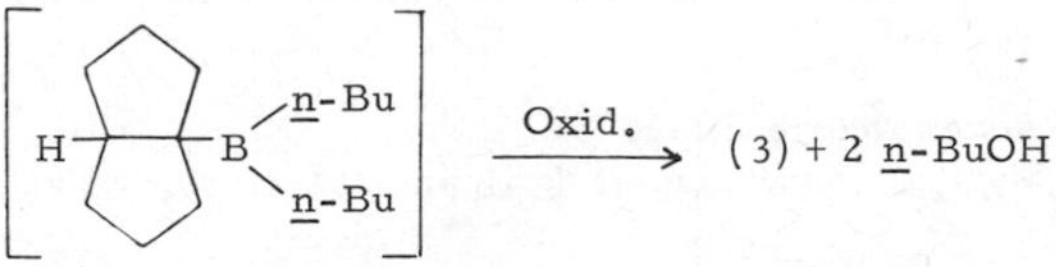

(a)

*Reduction of α,β-unsaturated carbonyl compounds.* Krishnamurthy and Brown[4] have reported that 9-BBN reduces carbonyl groups to hydroxyl groups faster than it hydroborates double bonds. For this reason it is valuable for reduction of $\alpha,\beta$-unsaturated aldehydes and ketones to allylic alcohols. The reaction is carried out in THF for 2–4 hr. at $0°$ and then for 1 hr. at $25°$. The reduction of cyclopentenone to cyclopentenol by this reagent is quantitative (glc). Diisobutylaluminum hydride (**3**, 101–102) has been used for such reductions. The intermediate boron compound can be oxidized by alkaline hydrogen peroxide to the allylic alcohol and 1,5-cyclooctanediol (the product from the 9-BBN moiety). More conveniently, addition of ethanolamine liberates the alcohol and precipitates 9-BBN as the adduct:

$$ \text{B—OR} + NH_2CH_2CH_2OH \xrightarrow{\text{Pentane}} ROH + \text{B} \underset{H_2}{\overset{O}{\diagdown N}} $$

One major advantage is that a wide variety of functional groups are not affected: nitro, halogen, sulfide, ester, tosylate. For instance, (1) is reduced to (2) in 86% yield (isolated).

$$ (1) \xrightarrow[86\%]{\text{9-BBN, THF}} (2) $$

[1] H. C. Brown, E. F. Knights, and C. G. Scouten, *Am. Soc.,* **96**, 7765 (1974).
[2] Y. Yamamoto, H. Toi, S.-I. Murahashi, and I. Moritani, *Am. Soc.,* **97**, 2558 (1975).
[3] G. W. Kramer and H. C. Brown, *J. Organometal. Chem.,* **90**, C1 (1975).
[4] S. Krishnamurthy and H. C. Brown, *J. Org.,* **40**, 1864 (1975).

**Borane–Dimethyl sulfide** (BMS); **4**, 124, 191; **5**, 47.

*Reductions.* Lane[1] has reviewed reductions with BMS. He notes that this reagent is somewhat less reactive than $BH_3 \cdot THF$ and usually requires a temperature of $20–25°$. He recommends that the reagent be added at this temperature. Reduction of aliphatic carboxylic acids proceeds readily, but reduction of benzoic acids is slow unless trimethyl borate is added.[2] The reagent reduces acids, esters, oximes, nitriles, and amides, but does not reduce halides or nitro groups.

[1] C. F. Lane, *Aldrichimica Acta,* **8**, 20 (1975).
[2] C. F. Lane, H. L. Myatt, J. Daniels, and H. B. Hopps, *J. Org.,* **39**, 3052 (1974).

**Boron tribromide, 1**, 66–67; **2**, 33–34; **3**, 30–31; **4**, 42; **5**, 49.

*Amides.* Esters are converted into carboxylic acids by boron tribromide (**4**, 42). Japanese chemists have modified this reaction to transesterification and to a synthesis of amides.[1] The ester in $CH_2Cl_2$ or $C_6H_6$ is treated with

1 eq. of $BBr_3$ at 20–40° for 1–5 hr. An amine or an alcohol (1 eq.) is then added and the mixture is stirred at 20–50° for 1–5 hr. Yields are satisfactory based on starting material that undergoes conversion. The reaction is useful in the case of unstable amines.

[1] H. Yazawa, K. Tanaka, and K. Kariyone, *Tetrahedron Letters*, 3995 (1974).

**Boron trichloride**, 1, 67–68; 2, 34–35; 3, 31–32; 4, 42–43; 5, 50–51.

*Displacement of acetoxy groups by halogen.* The acetoxy group of 3-acetoxymethyl-2- or 3-cephem-4-carboxylates (1) can be replaced by halogen

by treatment with boron trichloride or tribromide. Yields are lower if $AlCl_3$ or $TiCl_3$ is used. Three equivalents of the boron trihalides are required in the case of 3-cephem esters, whereas 1 eq. is sufficient in the reaction of 2-cephem esters.[1]

[1] H. Yazawa, H. Nakamura, K. Tanaka, and K. Kariyone, *Tetrahedron Letters,* 3991 (1974).

**Boron trifluoride etherate**, 1, 70–72; 2, 35–36; 3, 33; 4, 44–45; 5, 52–55.

*Reversal of regiospecificity in Diels–Alder reactions.*[1] 2,6-Dimethylbenzoquinone (1) undergoes "normal" addition to *trans*-1,3-pentadiene in refluxing benzene or toluene to give (3). However, if $BF_3 \cdot (C_2H_5)_2O$ is added, addition occurs at 0° and with complete reversed regiospecificity (but not stereospecificity) to give (2) in >80% yield. The reversal has also been noted with other substituted 1,3-dienes. The effect is not so marked with toluquinone; in this case both possible isomeric products are formed, but the Lewis acid changes the ratio of products significantly.

This reversal of orientation has been used to prepare (4) in a projected total synthesis of quassin (5), the bitter principle of *Quassia amara*, which contains

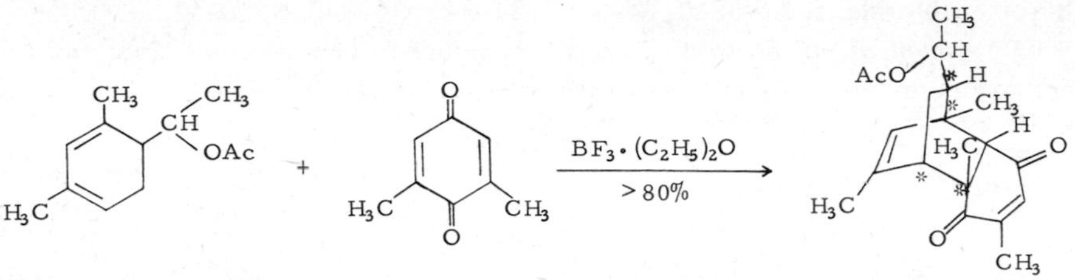

(4)

seven chiral centers, four of which are present in the desired orientation in (4).

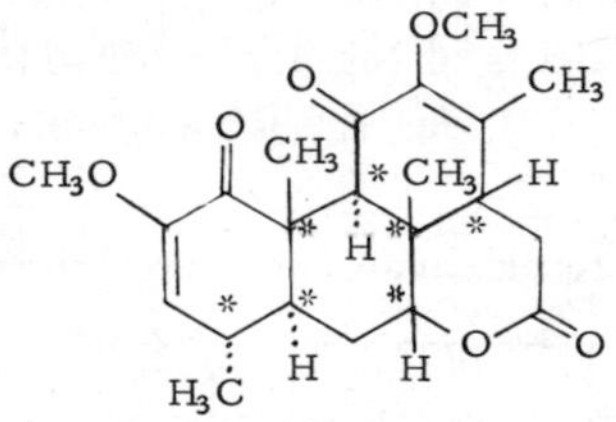

(5)

*Fluoroamine reagents.*[2] The reaction of $BF_3$ etherate with α-fluoroamines produces salts (1) that are similar to Vilsmeier reagents. These salts acylate electron-rich arenes.

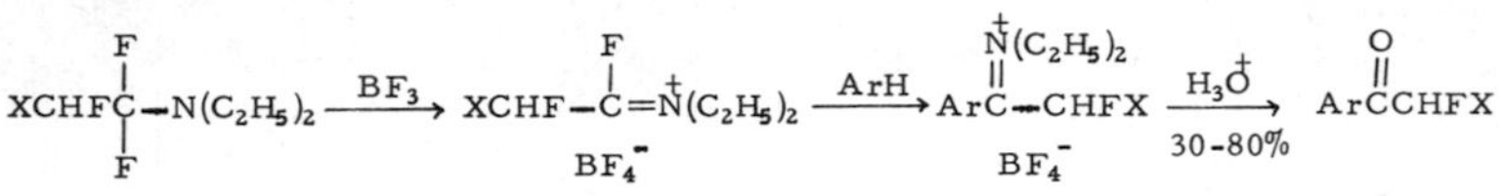

$$XCHF\overset{F}{\underset{F}{C}}-N(C_2H_5)_2 \xrightarrow{BF_3} XCHF-\overset{F}{C}=\overset{+}{N}(C_2H_5)_2 \xrightarrow{ArH} Ar\overset{\overset{+}{N}(C_2H_5)_2}{C}-CHFX \xrightarrow[30-80\%]{H_3\overset{+}{O}} Ar\overset{O}{C}CHFX$$

(X = Cl, F, CF₃)            BF₄⁻                    BF₄⁻

                             (1)                     (2)                     (3)

*Examples:*

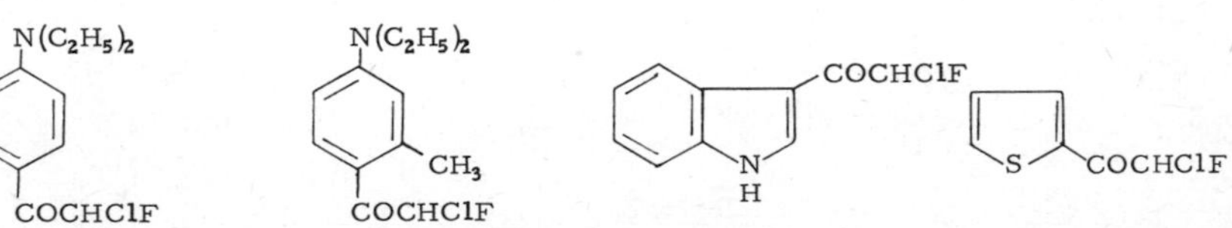

(37%)            (40%)            (78%)            (52%)

*Amidation of carboxylic acids.*[3]  Boron trifluoride etherate has been used as a reagent for reaction of carboxylic acids with primary or secondary amines to form amides. The reaction is accelerated by bases (triethylamine, DBU) and by azeotropic removal of the water formed. The reaction is conducted in refluxing benzene or toluene. Yields are generally in the range of 50–85%.

$$R^1COOH \ + \ HN\underset{R^3}{\overset{R^2}{\diagup}} \ \xrightarrow[\text{Base}]{BF_3 \cdot (C_2H_5)_2O} \ R^1\overset{O}{\overset{\|}{C}}-N\underset{R^3}{\overset{R^2}{\diagup}}$$

*Deoxygenation.*[4]  Quinoline or isoquinoline N-oxides are deoxygenated by irradiation in benzene in the presence of boron trifluoride etherate.

*Cyclopentenones.*  Smith and co-workers[5] have developed a new approach to the synthesis of cyclopentenones that involves treatment of $\beta,\gamma$-unsaturated

diazomethyl ketones (3), prepared as shown from ethyl 3,3-dimethylacrylate (1), with $BF_3$ etherate in nitromethane. The method was used in four cases, including a synthesis of *cis*-jasmone (5); RX = $BrCH_2C\equiv CCH_2CH_3$, followed by hydrogenation over Lindlar's catalyst. Mechanistic aspects of the cyclization are discussed briefly.

*Cyclopentenone annelation.*  Smith[6] has extended the acid-catalyzed decomposition of diazomethyl ketones (*see* **Trifluoracetic acid**, this volume) to the synthesis of bicyclopentenones. Thus treatment of the $\beta,\gamma$-unsaturated diazomethyl ketones (1) and (3) with $BF_3 \cdot (C_2H_5)_2O$ in nitromethane at 25° for 1.5 hr followed by treatment with refluxing aqueous hydrochloric acid gives

(2) and (4), respectively, in 50–70% yield (based on the acid chloride from which the diazomethyl ketones were prepared).

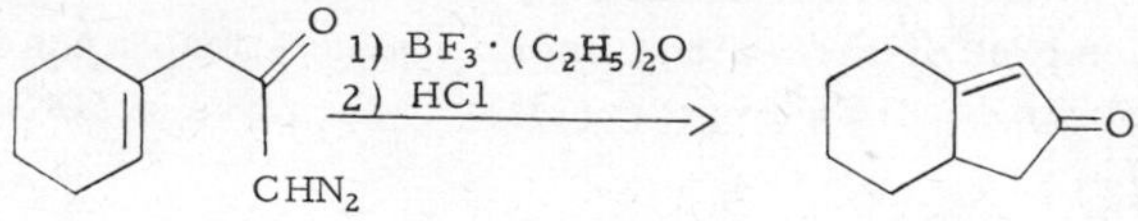

(1)    (2)

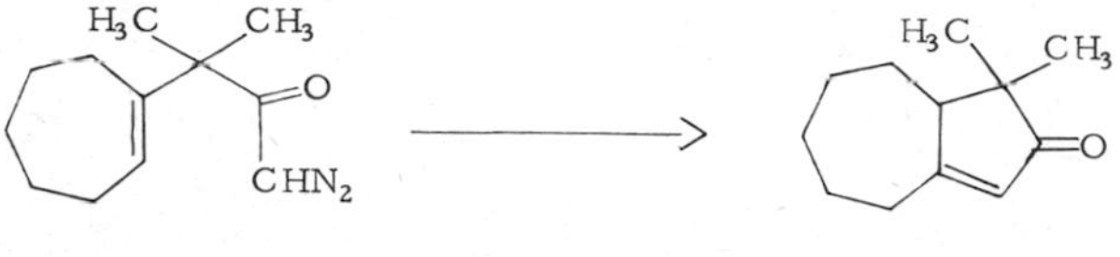

(3)    (4)

Note that the $\beta,\gamma$-unsaturated ketone (5) on treatment with 70% aqueous $HClO_4$ gives the cyclobutanone derivative (6).[8]

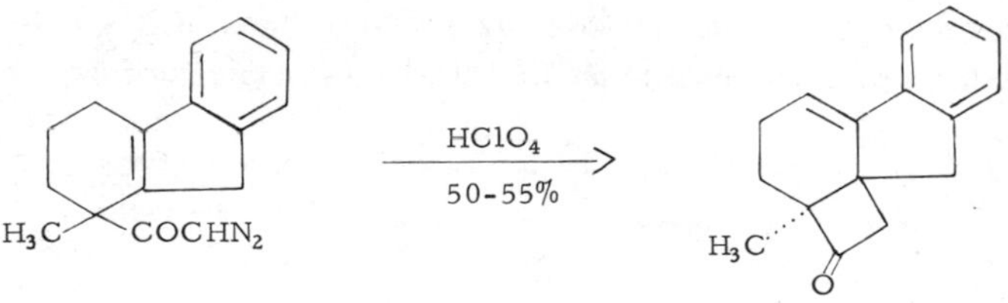

(5)    (6)

**trans-*Bishomobenzene*.** Taylor and Paquette[7] have reported a convenient synthesis of (4), *trans*-bishomobenzene (*trans*-tricyclo[5.1.0.0.$^{2,4}$]octene-5), from 3-norcarene (1). Addition of dibromocarbene leads to (2), which is cyclized to (3) by methyllithium. This hydrocarbon undergoes ring opening to

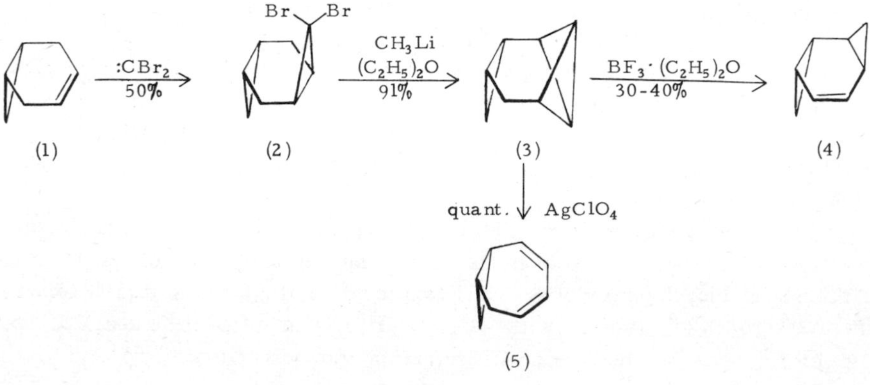

(1)    (2)    (3)    (4)

(5)

(4) when treated with $BF_3$ etherate. It is isomerized to 2,3-homotropilidene (5) by silver perchlorate.

*Cyclization of isocyanates.* In a total synthesis of the alkaloid (±)-lycoricidine (3), Ohta and Kimoto[9] effected cyclization of the isocyanate (1) to the lactam (2) with boron trifluoride etherate at room temperature. They state that this

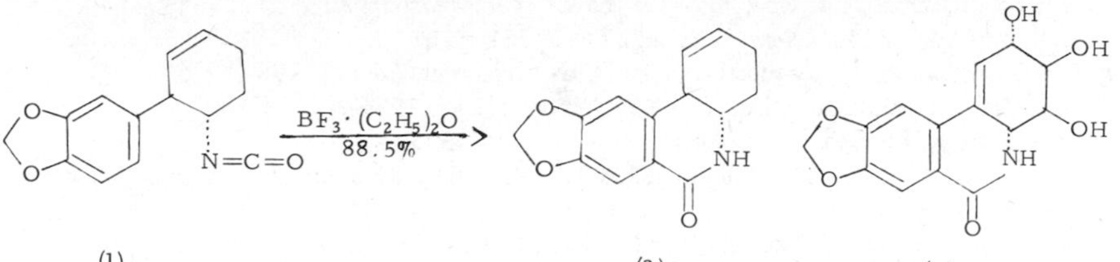

(1)                    (2)                    (3)

cyclization is applicable to other substituted phenethyl isocyanates.

*Bicyclic furanes.* French chemists[10] have described the synthesis of the furanes (3) from (1) by epoxidation and rearrangement of the resulting oxides (2) with boron trifluoride etherate. Acetylation of the hydroxyl group blocks rearrangement to the furane. The paper includes a possible mechanism.

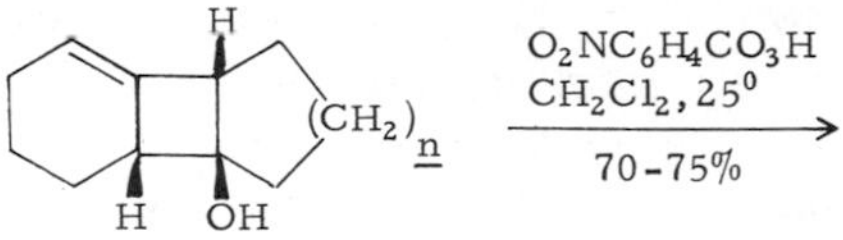

(1, $\underline{n} = 2, 3, 4$)

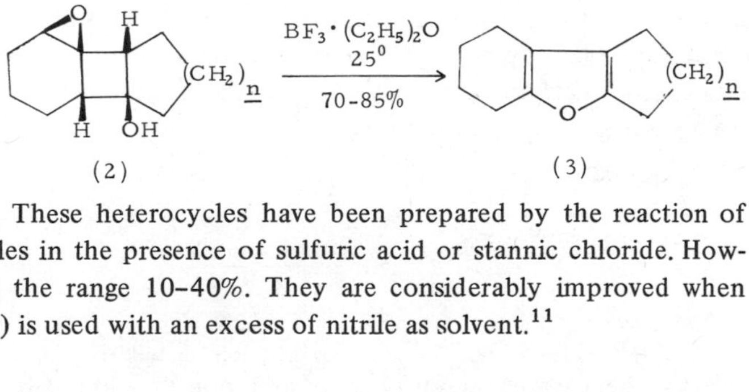

(2)                    (3)

$\Delta^2$-*Oxazolines.* These heterocycles have been prepared by the reaction of epoxides with nitriles in the presence of sulfuric acid or stannic chloride. However, yields are in the range 10–40%. They are considerably improved when $BF_3$ etherate (1 eq.) is used with an excess of nitrile as solvent.[11]

*Examples:*

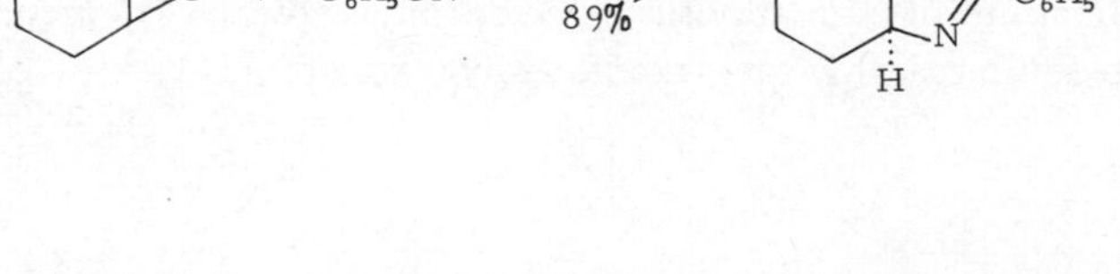

[1] Ž. Stojanac, R. A. Dickinson, N. Stojanac, R. J. Woznow, and Z. Valenta, *Canad. J. Chem.*, **53**, 616 (1975); N. Stojanac, A. Sood, Ž. Stojanac, and Z. Valenta, *ibid.*, **53**, 619 (1975).
[2] C. Wakselman and M. Tordeux, *J.C.S. Chem. Comm.*, 956 (1975).
[3] J. Tani, T. Oine, and I. Inoue, *Synthesis*, 714 (1975).
[4] N. Hata, I. Ono, and M. Kawasaki, *Chem. Letters*, 25 (1975).
[5] A. B. Smith, III, S. J. Branca, and B. H. Toder, *Tetrahedron Letters*, 4225 (1975).
[6] A. B. Smith, III, *J.C.S. Chem. Comm.*, 274 (1975).
[7] R. T. Taylor and L. A. Paquette, *Angew. Chem. internat. Ed.*, **14**, 496 (1975).
[8] U. R. Ghatak and B. Sanyal, *J.C.S. Chem. Comm.*, 876 (1974).
[9] S. Ohta and S. Kimoto, *Tetrahedron Letters*, 2279 (1975).
[10] B. Loubinoux, M. L. Viriot-Villaume, J. J. Chanot, and P. Caubere, *Tetrahedron Letters*, 843 (1975).
[11] J. R. L. Smith, R. O. C. Norman, and M. C. Stillings, *J.C.S. Perkin I*, 1200 (1975).

**Brij 35**, $C_{12}H_{25}(OCH_2CH_2)_{23}OH$. Mol. wt. 1199.57, m.p. 40–42°. Supplier: Aldrich.

*Surfactant catalysis.*[1] Surfactants disperse organic liquids in water and also form micelles.[2] Menger *et al.* have recently examined the application of surfactants to synthetic organic chemistry and have observed striking effects. No hydrolysis of $\alpha,\alpha,\alpha$-trichlorotoluene is observed in 20% NaOH at 80° after 1.5 hr. When 0.006 M Brij 35, a neutral surfactant, is added, hydrolysis to benzoic acid proceeds in 97% yield. Cetyltrimethylammonium bromide, $C_{16}H_{33}N^+(CH_3)_3Br^-$, a cationic surfactant, is somewhat more active in promoting this hydrolysis. Several facts are cited to show that these surfactants are not functioning as phase-transfer catalysts by solubilization of ionic reagents in organic phases. These two surfactants are shown to be active in a number of reactions, for example, the heterogeneous oxidation of piperonal to piperonylic acid (equation I). The paper suggests that the quaternary salt

(I) 
$$\xrightarrow[74\%]{KMnO_4, 55°}$$

is active because of emulsification and micellization, whereas Brij 35 serves as a dispersing agent only.

[1] F. M. Menger, J. U. Rhee, and H. K. Rhee, *J. Org.*, **40**, 3803 (1975).
[2] E. H. Cordes, Ed., *Reaction Kinetics in Micelles*, Plenum Press, New York, 1973.

**Bromine**, 3, 34; 4, 46–47; 5, 55–57.

$\beta$-*Bromo ketones.*[1] $\beta$-Bromo ketones (3) can be obtained in high yield by bromination of trimethylsilyl cyclopropyl ethers (2), prepared with the Simmons–Smith reagent from trimethylsilyl enol ethers (1).[2]

$$\underset{(1)}{\overset{\overset{\displaystyle (CH_3)_3SiO}{|}}{R^1-C}=CHR^2} \quad \xrightarrow{CH_2I_2-Zn} \quad \underset{(2)}{\overset{\overset{\displaystyle (CH_3)_3SiO}{|}}{R^1-C}\overset{\displaystyle CH_2}{\underset{\displaystyle CHR^2}{\diagdown}}} \quad \xrightarrow[CH_2Cl_2,\,-70^0]{Br_2} \quad \underset{\underset{\displaystyle R^2}{|}}{R^1\overset{\overset{\displaystyle O}{\|}}{C}CHCH_2Br} + (CH_3)_3SiBr$$

$$(3)$$

This reaction has been carried out on cyclopropanols[3]; these substances, however, are less available (2, 118–119) than trimethylsilyl cyclopropyl ethers and are readily decomposed by the hydrogen bromide formed on bromination. In the new synthesis, the by-product is trimethylbromosilane.

*Free-radical bromination of ketones.* Calò and Lopez[4] have reported dif-

$$\xrightarrow{Br_2,\,CCl_4,\,h\nu,} \quad C_6H_5CHBrCOCH_2CH_3$$

$$(1)$$

$$C_6H_5CH_2COCH_2CH_3 \xrightarrow{Br_2,\,CCl_4,\,h\nu} (1,\ 80\%) + C_6H_5CH_2COCHBrCH_3$$

$$(2,\ 20\%)$$

$$\xrightarrow{Br_2,\,CCl_4} (1,\ 60\%) + (2,\ 40\%)$$

ferences in the bromination of benzyl alkyl ketones depending on addition of cyclohexene epoxide as scavenger for the HBr formed. In this case irradiation is necessary for complete reaction and only benzylic bromination is observed. Bromination in the absence of light and scavenger leads to substitution at both possible positions. Apparently, in the presence of the epoxide, bromination does not involve the enolate, but involves a free-radical mechanism.

*Selective oxidation of cyclic tertiary amines.*[5] Bromine or NBS is recommended for the selective oxidation of certain cyclic tertiary amines (1) to the lactams (4). The reaction is believed to involve intermediate formation of the iminium salt (2), which can be isolated, and then of the α-hydroxy amine (3).

$$\xrightarrow[CH_2Cl_2]{Br_2,\,Na_2CO_3} \qquad \xrightarrow{OH^-}$$

$$(1) \qquad\qquad (2) \qquad\qquad (3)$$

$$\xrightarrow{Br_2,\,OH^-}$$

$$(4)$$

*Examples:*

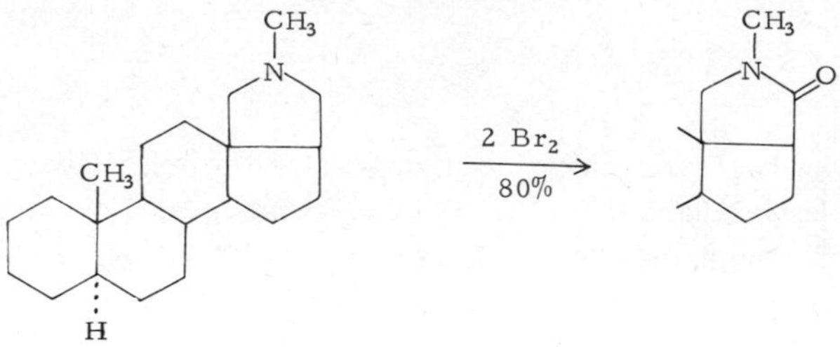

21-Norconanine

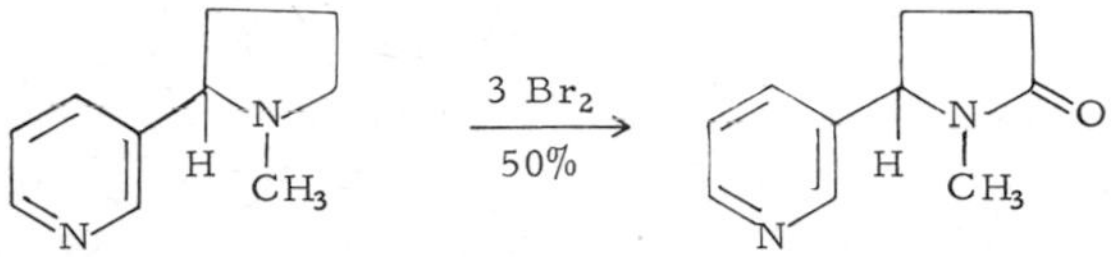

Nicotine

**α-Bromo acetals.**[6] Alkenylboronic acids are converted into *trans*-1-alkenyl iodides by treatment with iodine and base (5, 346–347):

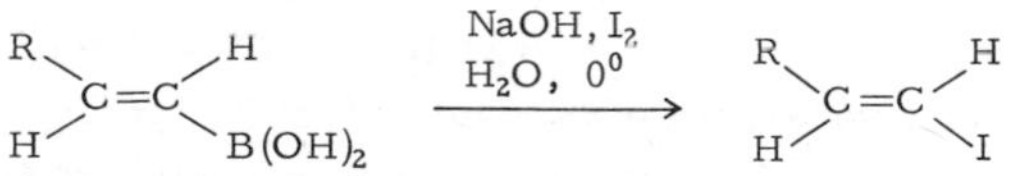

Replacement of iodine by bromine in this reaction leads to a mixture of *cis*- and *trans*-1-alkenyl bromides and the corresponding *n*-alkanal. After some experimentation, Hamaoka and Brown found that alkenylboronic acids can be converted into α-bromo dimethyl acetals by reaction with bromine and sodium methoxide in methanol at $-78°$:

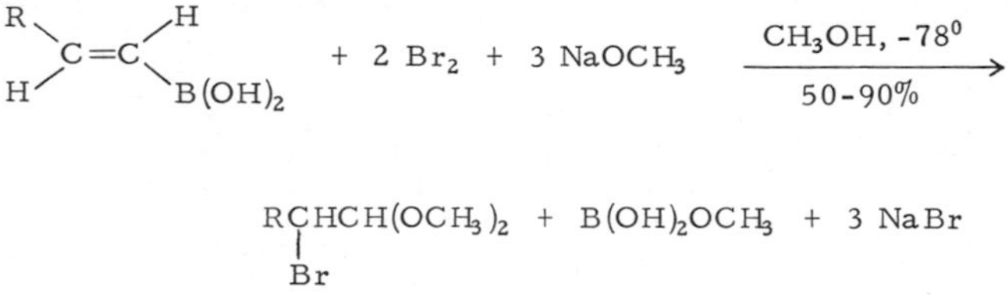

$$RCHCH(OCH_3)_2 \ + \ B(OH)_2OCH_3 \ + \ 3 \ NaBr$$
$$\underset{Br}{|}$$

In the absence of sodium methoxide only traces of the α-bromo acetal are obtained. Hamaoka and Brown suggest that the reaction involves methyl hypobromite.

**Side-chain bromination of alkylthiophenes.**[7] Alkylthiophenes are brominated in the side chain by NBS or by bromine under free-radical conditions ($h\nu$, azobisisobutyronitrile). One, two, or three bromines can be introduced before ring bromination competes.

*Examples:*

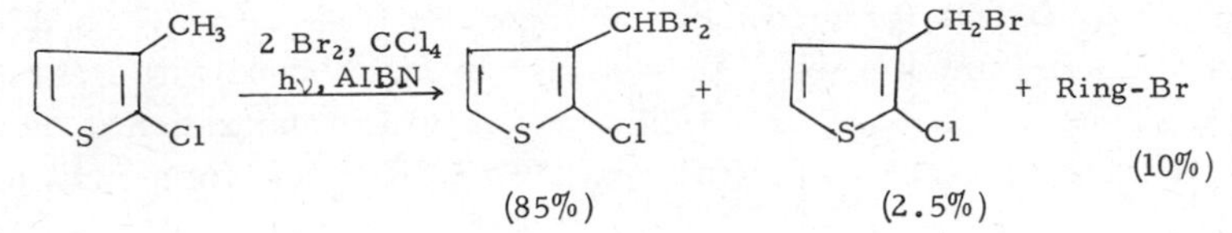

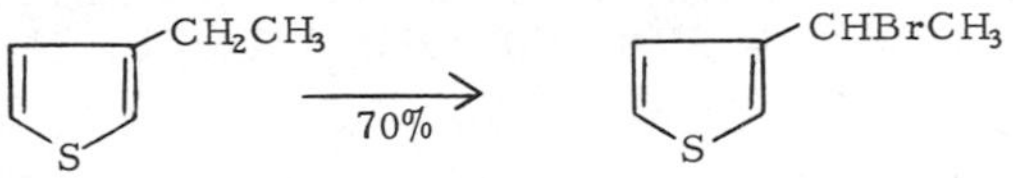

(85%)                    (2.5%)

+ Ring-Br

(10%)

$^1$S. Murai, Y. Seki, and N. Sonoda, *J.C.S. Chem. Comm.,* 1032 (1974).
$^2$S. Murai, T. Aya, and N. Sonoda, *J. Org.,* **38,** 4354 (1973); S. Murai, T. Aya, T. Renge,
  I. Ryu, and N. Sonoda, *ibid.,* **39,** 858 (1974).
$^3$C. H. DePuy, *Accts. Chem. Res.,* **1,** 33 (1968).
$^4$V. Calò and L. Lopez, *J.C.S. Chem. Comm.,* 212 (1975).
$^5$A. Picot and X. Lusinchi, *Synthesis,* 109 (1975).
$^6$T. Hamoaka and H. C. Brown, *J. Org.,* **40,** 1189 (1975).
$^7$J. A. Clarke and O. Meth-Cohn, *Tetrahedron Letters,* 4705 (1975).

**2-Bromo-2-cyano-N,N-dimethylacetamide,** $CNCHBrCON(CH_3)_2$ (1). Mol. wt.
191.05, m.p. 53–54°.

This reagent is prepared in 85% yield by bromination of N,N-dimethylcyano-
acetamide in HOAc–Ac$_2$O.$^1$

*Bromination.* The reagent is useful for selective α-bromination of ketones.$^2$
It is converted into N,N-dimethylcyanoacetamide, which can be recycled.

$$C_6H_5CH=CHCOCH_3 \xrightarrow[56\%]{\overset{(1)}{C_6H_6, \Delta}} C_6H_5CH=CHCOCH_2Br$$

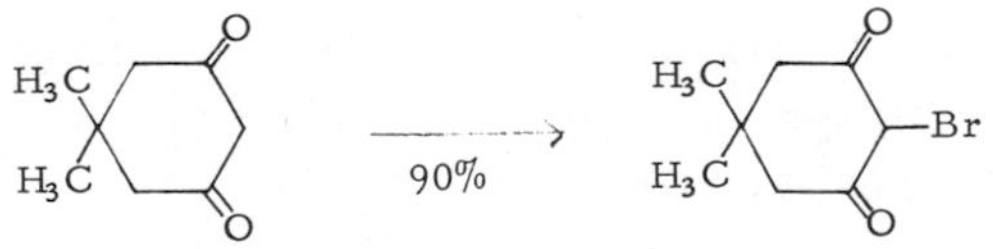

$$C_6H_5COCH_3 \xrightarrow[66\%]{} C_6H_5COCH_2Br$$

$$CH_3COCH_2CH_3 \xrightarrow[46\%]{} CH_3COCHBrCH_3$$

$^1$J. Suzuki, K. Suzuki, and M. Sekiya, *Chem. Pharm. Bull.,* **22,** 965 (1974).
$^2$M. Sekiya, K. Ito, and K. Suzuki, *Tetrahedron,* **31,** 231 (1975).

**Bromomethyllithium, 3,** 201.

*Epoxides.* Details for preparation of epoxides from carbonyl compounds with this reagent are available.[1] Diepoxides can be obtained with this reagent from unhindered 1,2-, 1,3-, or 1,4-diketones in which enolization is not possible. Hindered ketones [*e.g.*, $(CH_3)_3CCO—COC(CH_3)_3$] form only mono-epoxides.[2]

[1]G. Cainelli, A. Umani Ronchi, F. Bertini, P. Graselli, and A. Zubiani, *Tetrahedron,* **27,** 6109 (1971); G. Cainelli, N. Tangari, and A. Umani Ronchi, *ibid.,* **28,** 3009 (1972).
[2]M. Becker, H. Marschall, and P. Weyerstahl, *Ber.,* **108,** 2391 (1975).

**5-Bromopentanone-2 ethylene ketal,** Br (1). Mol. wt. 209.08, b.p. 98–101°/13 torr.

*Synthesis of sterol side chain.* Polish chemists[1] have prepared 25-hydroxy-cholesterol (6) from $\Delta^5$-3$\beta$-acetoxyandrostene-17-one (2) in 42% overall yield as formulated. The key step involves alkylation of the anion of the ester (3)

with (1) by method of Schlessinger *et al.* (**5,** 406). The product (4) has the desired 20R-configuration.

[1]J. Wicha and K. Bal, *J.C.S. Chem. Comm.,* 968 (1975).

**N-Bromosuccinimide, 1,** 78–80; **2,** 40–42; **3,** 34–36; **4,** 49–53; **5,** 65–66.

*Regeneration of carbonyl compounds from* **p-***toluenesulfonylhydrazones.*[1] Aldehydes and ketones can be regenerated from the *p*-toluenesulfonylhydrazones by treatment in acetone–methanol with excess NBS at 0–5°. Nitrogen is evolved rapidly; sodium hydrogen sulfite is then added to suppress bromination. The dimethyl ketal is an intermediate. Yields are generally in the range 75–90%;

however, in the case of tosylhydrazones of $\alpha,\beta$-unsaturated carbonyl groups, mixtures of products are generally obtained.

*Stereocontrolled addition of HOBr to* —CH=CH—. In a recent synthesis of (±)-perhydrohistrionicotoxin (5), a useful neurotoxin, Corey *et al.* found that the reaction of the oxime-olefin (1) with NBS (1.5 eq.) in 2:1 DME–$H_2O$ at $-20°$ proceeds with high positional specificity with respect to the introduced oxygen function.[2]

(1)

(2, 72%)

(3, ~10%)

(4, ~10%)          (5)

*α-Halo acyl chlorides* (3, 35).[3] α-Bromo acyl chlorides can be prepared in about 70–80% yield by reaction of acyl chlorides, prepared *in situ* by reaction of carboxylic acids with thionyl chloride, which can also be employed as solvent for the bromination reaction. NBS is considerably more efficient than bromine in this reaction. Use of NCS under similar conditions gives α-chloro acyl chlorides in comparable yields. Benzylic protons are not replaced by either reagent. α-Iodo acyl chlorides can be obtained in 70–80% yield by iodination of acyl chlorides with molecular iodine; in this case thionyl chloride must be used as solvent rather than $CCl_4$; and a higher temperature ($140°$) is required.

*Oxidation of hydrazines.*[4] 4-Arylurazoles (1) are oxidized to 4-aryl-1,2,4-triazoline-3,5-diones (2) by NBS in $CH_2Cl_2$ at room temperature in about 75% yield. The oxidation is also successful with 4-benzylurazole (84% yield). *t*-Butyl

(1)                (2)

hypochlorite has been used for this reaction (3, 223; 4, 381).

*Oxidative hydrolysis of an ethylene dithioketal.* Cain and Welling[5] report that the method of Corey and Erickson (4, 216) for oxidative hydrolysis of 2-acyl-1,3-dithianes is also useful for unmasking of ethylene dithioketals. Thus treatment of the bisethylene dithioketal (1) with 4 eq. of NBS in 10% aqueous acetone at 0° for 20 min. selectively removes the less hindered ethylene dithioketal group at $C_6$ to give (2) in 80% yield. Removal of the ethylene dithioketal group at $C_2$ with Raney nickel requires reflux in ethanol and is accompanied by reduction of the carbonyl group.

[1]G. Rosini, *J. Org.*, **39**, 3504 (1974).
[2]E. J. Corey, M. Petrzilka, and Y. Ueda, *Tetrahedron Letters*, 4343 (1975).
[3]D. N. Harpp, L. Q. Bao, C. J. Black, J. G. Gleason, and R. A. Smith, *J. Org.*, **40**, 3420 (1975).
[4]H. Wamhoff and K. Wald, *Org. Prep. Proc. Int.*, **7**, 251 (1975).
[5]E. N. Cain and L. L. Welling, *Tetrahedron Letters*, 1353 (1975).

**trans-1-Butadienyltriphenylphosphonium bromide** (1). Mol wt. 395.27.

*Preparation.*[1] The reagent (1) can be prepared in 92% overall yield from *trans*-1,4-dibromo-2-butene:

The reagent can also be prepared *in situ* by treatment of either the diphosphonium salt (2) or the bromophosphonium salt (3) in ether with potassium *t*-butoxide.[2]

$$Br^- \ (C_6H_5)_3\overset{+}{P}CH_2CH=CHCH_2\overset{+}{P}(C_6H_5)_3 \ Br^-$$

(2)

$$BrCH_2CH=CHCH_2\overset{+}{P}(C_6H_5)_3 \ Br^-$$

(3)

*1,3-Cyclohexadienes.* The reagent reacts with enolates of carbonyl compounds in THF (25°, 6–18 hr.) to form 1,3-cyclohexadienes in moderate yields (35–60%). It may react in the *s-cis*-form to give a *cis*-allyl ylide, which

closes to the cyclohexadiene, or an intermediate *trans*-allyl ylide may isomerize to the *cis*-isomer before ring closure.[1]

Annelation of dihydrocarvone (4) with (1) gives the cyclohexadiene (5) in 85% yield.[2] The work by Büchi and Pawlak complements a previous synthesis

(4)

$$(1), KOC(CH_3)_3$$
$$HOC(CH_3)_3, (C_2H_5)_2O$$
$$\xrightarrow{85\%}$$

(5)

(6)          +          (7)          →          (8)

of cyclohexadienes by the reaction of an $\alpha,\beta$-unsaturated ketone such as (7) with allyltriphenylphosphorane (6).[3]

[1] P. L. Fuchs, *Tetrahedron Letters*, 4055 (1974).
[2] G. Büchi and M. Pawlak, *J. Org.*, **40**, 100 (1975).
[3] G. Büchi and H. Wüest, *Helv.*, **54**, 1767 (1971).

**t-Butyl azidoformate (*t*-Butyloxycarbonyl azide), 1,** 84–85; **2,** 44–45; **3,** 36; **4,** 54.

*BOC-Amino acids.*[1] These derivatives can be prepared in high yield (70–98%) by the reaction of *t*-butyl azidoformate and amino acids in dioxane–water (1 : 1) in the presence of triethylamine (excess) for 2–40 hr.

*1-t-Butyloxycarbonyl-3-formylindole.*[2]  This compound (2) can be prepared

readily from 3-formylindole (1) as shown.

*Cyclic amidines.*[3]  Organic azides react with N-isopropylallenimine (1) to form cyclic amidines (2). In the case of *t*-butyl azidoformate, the product (2) can be converted into the parent N-isopropyl-$\beta$-lactamimide (3).

[1] Z. Grzonka and B. Lammek, *Synthesis,* 661 (1974).
[2] Y. Wolman, *Synthesis,* 732 (1975).
[3] J. K. Crandell and J. B. Komin, *J.C.S. Chem. Comm.,* 436 (1975).

*t*-**Butylcyanoketene, 4,** 55–56.

Details of the preparation and reactions of this ketene have been published.[1]

[1] W. Weyler, Jr., W. G. Duncan, and H. W. Moore, *Am. Soc.,* **97,** 6187 (1975).

*t*-**Butyldimethylchlorosilane, 4,** 57–58, 176–177; **5,** 74–75.

*Protection of hydroxyl groups.*[1]  The *t*-butyldimethylsilyl group is not suitable for protection of hydroxyl groups in the synthesis of 1,2- and 1,3-diacylglycerols because removal with tetra-*n*-butylammonium fluoride (4, 477–478) is accompanied by acyl migration. Thus desilylation of (1) results in formation of both (2) and (3). Acyl exchange has also been observed to accompany desilylation with HF and HF–pyridine. Acetic acid in THF or boric acid in trimethyl borate does not effect desilylation.

*Claisen rearrangement of allyl silyl ethers* (**4**, 307–308). Katzenellenbogen and Christy[2] have extended the rearrangement of silyl enol ether derivatives of allylic acetates to $\gamma,\delta$-unsaturated acids to systems in which a trisubstituted double bond is generated. Thus 3-acetoxy-2-methyl-1-nonene (1) was treated with lithium N-isopropylcyclohexylamide (LiICA) in THF at $-78°$ and then with *t*-butyldimethylchlorosilane to give the *t*-butyldimethylsiloxyvinyl ether (2). This was warmed at $70°$ for 2 hr. and hydrolyzed with acetic acid. The E-acid (3) was obtained in 80% yield and in high stereoselectivity ($>98\%$).

The trimethylsiloxyvinyl ether corresponding to (2) rearranges in lower yield ($\sim 53\%$).

This method was used in a highly stereoselective synthesis of a pheromone diol (6) of the queen butterfly. The key step involved rearrangement of the acetate (4), obtained in several steps from geraniol, to the acid (5). Lithium aluminum hydride reduction of (5) gave the diol (6).

[1]G. H. Dodd, B. T. Golding, and P. V. Ioannou, *J.C.S. Chem. Comm.*, 249 (1975).
[2]J. A. Katzenellenbogen and K. J. Christy, *J. Org.*, **39**, 3315 (1974).

**t-Butyldimethylsilyl cyanide,** $(CH_3)_3C-\underset{\underset{CH_3}{|}}{\overset{\overset{CH_3}{|}}{Si}}-CN.$  Mol. wt. 241.29.

*Preparation*[1]:

$$2\ (CH_3)_3C-\underset{\underset{CH_3}{|}}{\overset{\overset{CH_3}{|}}{Si}}-Cl\ +\ K_2Hg(CN)_4\ \xrightarrow{HMPT}\ 2\ (CH_3)_3C-\underset{\underset{CH_3}{|}}{\overset{\overset{CH_3}{|}}{Si}}-CN\ +\ 2\ KCl\ +\ Hg(CN)_2$$

*Cyanosilylation of ketones* (**4,** 542–543; **5,** 720). In a total synthesis of natural camptothecin (9), Corey *et al.*[2] used this *t*-butyldimethylsilyl derivative rather than trimethylsilyl cyanide (**5,** 720–722) to effect cyanosilylation of a ketone (1). Hydrolysis of the resulting cyano silyl ether to the required amide was not accompanied by desilylation with reversal of cyanohydrin formation. By use of carefully controlled conditions and with dicyclohexyl-18-crown-6–potassium cyanide as catalyst, they were able to convert (1) into the α-hydroxy

$$\underset{(1)}{R-\overset{\overset{CH_2CH_3}{|}}{C}=O}\ +\ (CH_3)_3C\underset{\underset{CH_3}{|}}{\overset{\overset{CH_3}{|}}{Si}}CN\ \xrightarrow[85\%]{Cat.}\ \underset{(2)}{R-\underset{\underset{CN}{|}}{\overset{\overset{CH_2CH_3}{|}}{C}}-OSi(CH_3)_2C(CH_3)_3}$$

$$\xrightarrow[CH_3OH,\ K_2CO_3]{H_2O_2}\ \underset{(3)}{R-\underset{\underset{CONH_2}{|}}{\overset{\overset{CH_2CH_3}{|}}{C}}-OSi(CH_3)_2C(CH_3)_3}\ +\ \underset{(4)}{R\underset{\underset{CONH_2}{|}}{\overset{\overset{CH_2CH_3}{|}}{C}}-OH}$$

73% overall $\Big\downarrow$ $H_2O-KOH-CH_3OH$

$$R=\ \overset{\text{furan with } CH_2OTHP}{\phantom{x}}\qquad\qquad R-\underset{\underset{COOH}{|}}{\overset{\overset{\overset{*}{CH_2CH_3}}{|}}{C}}-OH$$

(5)

carboxylic acid (5) in 73% overall yield. The product was resolved and converted into (6); combined coupling and cyclization with the diamine

(7)        +        (6)    $\xrightarrow{6.5\%}$    (8, R = COOCH₃)
(9, R = H)

(7) gave the methoxycarbonyl derivative (8) of camptothecin (9) in about 6.5% yield. One advantage of this synthetic approach is that the precursor (5) contains the asymmetric center of (9) and is resolvable by the natural base quinine.

[1]T. A. Bither, W. H. Knoth, R. V. Lindsey, Jr., and W. H. Sharkey, *Am. Soc.*, **80**, 4151 (1958).
[2]E. J. Corey, D. N. Crouse, and J. E. Anderson, *J. Org.*, **40**, 2140 (1975).

*t*-**Butyldiphenylsilyl chloride,** $(CH_3)_3C\underset{\underset{C_6H_5}{|}}{\overset{\overset{C_6H_5}{|}}{Si}}Cl$ . Mol. wt. 334.92, m.p. 92.5°.

The reagent is prepared by the reaction of dichlorodiphenylsilane and *t*-butyllithium in refluxing pentane (95% yield).

**t-***Butyldiphenylsilyl ethers.*[1] Primary and secondary alcohols are readily silylated by reaction with 1.1 eq. of the reagent in DMF containing 2.2 eq. of imidazole. The reaction proceeds at 25° with primary alcohols and at 25–60° for secondary alcohols (including axial). The *t*-BDPSi group offers some unique features: it is stable to 80% acetic acid (which cleaves *t*-butyldimethyl-silyl ethers) and to 50% aqueous trifluoracetic acid–dioxane (25°, 15 min.); it is stable to catalytic hydrogenolysis (20% palladium hydroxide-on-charcoal); it is cleaved in >90% yield by tetra-*n*-butylammonium fluoride in THF.

[1]S. Hanessian and P. Lavallee, *Canad. J. Chem.*, **53**, 2975 (1975).

*t*-**Butyl hydroperoxide, 1,** 88–89; **2,** 49–50; **3,** 37–38; **5,** 75–77.

*Decarboxylation* (**1,** 89; **2,** 49). Langhals and Rüchardt[1] have carried out

$$RCO_2-OC(CH_3)_3 \xrightarrow{\Delta} R\cdot + CO_2 + \cdot OC(CH_3)_3 \xrightarrow{\text{Solvent}} RH + HOC(CH_3)_3$$

the thermolysis or photolysis of peroxycarboxylates in ethyl phenylacetate, $C_6H_5CH_2COOC_2H_5$, b.p. 229°. The benzylic hydrogen is readily abstracted by the intermediate radicals. For preparative purposes the acid is converted into the acid chloride ($SOCl_2$) and then into the perester without purification of the intermediates. Yields of hydrocarbons are in the range of 23–55%. The solvent is eliminated from the hydrocarbon product by hydrolysis and extraction of the resulting acid into aqueous base.

*Examples:*

*Alkyl halides.* Jensen and Moder[2] have prepared alkyl *t*-butylperoxygly-oxalates (1)[3] as shown in the formulation and have found that when they are decomposed at 95° in the presence of $CCl_4$ or $BrCCl_3$, alkyl chlorides or

$$ROH \xrightarrow{ClCOCOCl} ROC\overset{O}{\overset{\|}{C}}-\overset{O}{\overset{\|}{C}}Cl \xrightarrow{(CH_3)_3COOH} ROC\overset{O}{\overset{\|}{C}}-\overset{O}{\overset{\|}{C}}OOC(CH_3)_3$$

(1)

bromides are obtained in fair yields (5–60%). This homolytic reaction can be useful when heterolytic reactions fail. Thus 2,2-dimethylcyclohexyl chloride was obtained in this way in 44% yield. Use of several reagents, including $(C_6H_5)_3PBr_2$ and $(C_6H_5)_3P-CCl_4$, failed to give any of the bromide or chloride. Thus this method may be useful in the case of sterically hindered alcohols or alcohols prone to rearrangement.

[1] H. Langhals and C. Rüchardt, *Ber.*, **108**, 2156 (1975).
[2] F. R. Jensen and T. I. Moder, *Am. Soc.*, **97**, 2281 (1975).
[3] *Caution:* The substances should not be isolated, but should be prepared and used only in dilute solutions.

*t*-Butyl hypochlorite, **1**, 90–94; **2**, 50; **3**, 38; **4**, 58–60; **5**, 77–78.

*Dehydrogenation.*[1] Tetraethynylethylenes (2) can be prepared from tetra-ethynylethanes (1) by conversion to the dilithio derivative followed by reaction with 2 eq. of *t*-butyl hypochlorite. Intermediate peroxides are decomposed by ferrous sulfate.

$$(1, \ R = C_6H_5, (CH_3)_3C, (CH_3)_3Si, CH_3)$$

[1] H. Hauptmann, *Angew. Chem. internat. Ed.*, **44**, 498 (1975).

*t*-Butyl γ-iodotiglate,

(1). Mol. wt. 282.12.

*Preparation.*[1] The phosphorane (2) is prepared as shown in equation I. Reaction of the phosphorane (2) with chloroacetaldehyde in methylene chloride

$$(III) \quad (3) + Na\,I \xrightarrow[>95\%]{\substack{acetone \\ 25^0}} (1)$$

gives almost exclusively the γ-chlorotiglate (3) with only traces of the angelate isomer (4) (equation II). The final step involves iodide exchange (equation III).

*Annelation of unsymmetrically substituted cyclohexanones.* Stotter and Hill[2] examined this reagent for a method of position-specific annelation of unsymmetrically substituted cyclohexanones, which can be converted specifically into either of the two possible lithium enolates (3, 310–311; 4, 299). They reasoned that an allylic iodide should react rapidly before equilibration of the enolate can occur and that the carboxylate substituent on the double bond should suppress 1,4-elimination of hydrogen iodide and also aid in the isolation of the alkylated ketone. These expectations were realized. Thus annelation of the lithium enolate (2) with (1) at 0° followed by hydrolysis gave the acid

(3) in high yield. This was converted to the N-vinyl carbamate (4) by the procedure of Weinstock.[3] The final step involved conversion to the intermediate δ-diketone and cyclization to (5) by treatment with 2% KOH in 4:1 CH₃OH–H₂O for 1 hr. at 25° and then for 2 hr. at 70°.

Application of the same sequence to the isomeric lithium enolate (6) gave

exclusively (7).

Enamines rather than lithium enolates can also be used.

Use of γ-chlorotiglates in this annelation was found to lead to complex mixtures of products.

[1]P. L. Stotter and K. A. Hill, *Tetrahedron Letters,* 1679, (1975).
[2]*Idem., Am. Soc.,* **96,** 6524 (1974).
[3]J. Weinstock, *J. Org.,* **26,** 3511 (1961).

*t*-Butyl lithioacetate, 5, 371.

**β,γ-Unsaturated esters.** Ruden and Gaffney[1] have prepared β,γ-unsaturated esters by a method based on the facile elimination of trimethylsilanol from β-hydroxysilanes.[2] Thus reaction of trimethylsilylacetone (1) with *t*-butyl lithioacetate (2) gives the hydroxy ester (3), which is converted into (4) by treatment with acid.

$$CH_3COCH_2Si(CH_3)_3 \;+\; LiCH_2COOC(CH_3)_3 \xrightarrow{\;75\%\;} \begin{array}{c} OH \\ | \\ CH_3CCH_2Si(CH_3)_3 \\ | \\ CH_2COOC(CH_3)_3 \end{array}$$

(1)                                (2)                                                    (3)

$$\xrightarrow[\;65\%\;]{\begin{array}{c} HClO_4,\, THF \\ 0^0 \end{array}} \begin{array}{c} CH_3C{=}CH_2 \\ | \\ CH_2COOC(CH_3)_3 \end{array} \;+\; (CH_3)_3SiOH$$

(4)

β,γ-Unsaturated amides and nitriles can be prepared in the same way by use of lithio-N,N-dimethylacetamide[3] and lithioacetonitrile.[4]

[1] R. A. Ruden and B. L. Gaffney, *Syn. Commun.*, 15 (1975).
[2] F. C. Whitmore, L. H. Sommer, J. Gold, and R. E. Van Strien, *Am. Soc.*, **69**, 1551 (1947).
[3] D. Seebach and D. N. Crouse, *Ber.*, **101**, 3113 (1968).
[4] E. M. Kaiser and C. R. Hauser, *J. Org.*, **33**, 3402 (1968).

$$Li^+$$

*t*-Butyl α-lithioisobutyrate, $(CH_3)_2\overset{-}{C}COOC(CH_3)_3$  (1).

The reagent is prepared from *t*-butyl isobutyrate and lithium diisopropylamide in hexane at −78°.

**β-Keto acids; ketones.**[1] This anion (1) reacts with benzoyl chlorides to give *t*-butyl α-benzoylisobutyrates (2) in fair to good yields. When (2) is treated with trifluoroacetic acid for a few minutes, the β-keto acids (3) are obtained in es-

$$ArCOCl + (1) \xrightarrow[\;\;]{\begin{array}{c}C_6H_6\\0^0{\to}25^0\end{array}} \begin{array}{c} O\;\; CH_3 \\ \| \;\;\;| \\ ArC-C-COOC(CH_3)_3 \\ | \\ CH_3 \end{array} \xrightarrow[\;quant.\;]{\begin{array}{c}CF_3COOH\\15\ min.,25^0\end{array}} \begin{array}{c} O\;\; CH_3 \\ \| \;\;\;| \\ ArC-CCOOH \\ | \\ CH_3 \end{array} + CF_3COOC(CH_3)_3$$

(2)                                                    (3)

$$\xrightarrow[quant.]{\begin{array}{c}CF_3COOH\\1\ hr.\ reflux\end{array}}$$

$$\begin{array}{c} O\;\; CH_3 \\ \| \;\;\;| \\ ArC-C-H \\ | \\ CH_3 \end{array} + CO_2$$

(4)

sentially quantitative yields. Prolonged reflux in trifluoroacetic acid results in decarboxylation to ketones (4).

[1] M. W. Logue, *J. Org.*, **39**, 3455 (1974).

**n-Butyllithium, 1**, 95–96; **2**, 51–53; **4**, 60–63; **5**, 78.

*3-Alkyl-1-alkynes.* Reaction of 1-bromo-1-alkynes (1) with at least 2 eq. of *n*-butyllithium in hexane gives 3-butyl-1-alkynes (2) in excellent yields. The favored mechanism involves metal–halogen exchange (a) and further reaction with *n*-butyllithium to give the dilithioalkyne (b), which is then alkylated at the propargylic carbon atom:

$$RCH_2C{\equiv}CBr \xrightarrow[\text{-\underline{n}-BuBr}]{\text{n-BuLi}} \left[ RCH_2C{\equiv}CLi \xrightarrow{\text{n-BuLi}} (RCHC{\equiv}C)Li_2 \right.$$

(1)                                    (a)                           (b)

$$\left. \xrightarrow{\underline{\text{n}}\text{-BuBr}} \begin{array}{c} RCHC{\equiv}CLi \\ | \\ \underline{\text{n}}\text{-Bu} \end{array} \right] \xrightarrow{H_3O^+} \begin{array}{c} RCHC{\equiv}CH \\ | \\ \underline{\text{n}}\text{-Bu} \end{array}$$

(c)                                    (2)

This reaction can be applied to a general synthesis of branched alkynes. A terminal alkyne is treated with 2 eq. of *n*-butyllithium in hexane to give the intermediate (b), which is then alkylated at the 3-position by an added alkyl bromide.[1]

*Wittig rearrangement of dithia[3.3]metacyclophanes.* Boekelheide *et al.*[2] have introduced some improvements in the original synthesis of cyclophanes (**4**, 114–115). Thus it was found that use of the Wittig rather than the Stevens rearrangement resulted in higher yields of the same product. For example, treatment of (1) with excess *n*-butyllithium in hexane or with lithium di-isopropylamide in hexane–THF at 0° followed by methylation of the inter-mediate thiolate gives (2) in 94% yield [mixture of 70 *anti*/30 *syn*]. The [2.2]-metacyclophane itself can be obtained from (2) by desulfurization; the cyclo-phane diene can be obtained by Hofmann elimination.

$$\xrightarrow[\text{94\%}]{\substack{\text{n-BuLi} \\ \text{CH}_3\text{I}}}$$

(1)                                    (2)

*Allylic alcohols.* Two laboratories[3,4] have reported $C_6H_5$Se–Li exchange of selenoketals on treatment with *n*- or *t*-butyllithium in THF at –80°. Thus (1), prepared from acetone and $C_6H_5$SeH, is converted by *n*-butyllithium into

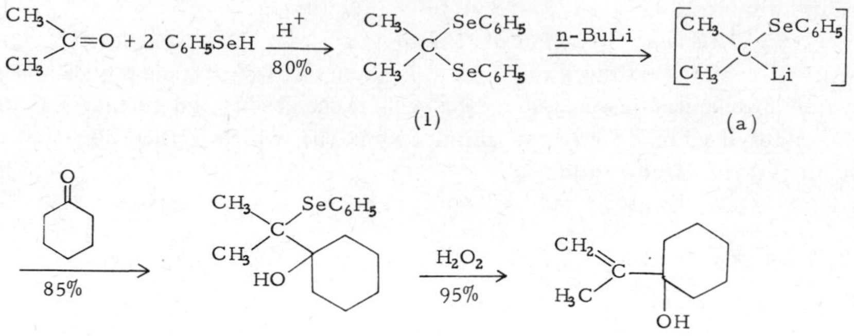

(1)                                                        (a)

(2)

(3)

(a), which reacts with cyclohexanone to give the β-hydroxy selenide (2). Selen-oxide fragmentation of (2) (**5**, 272–276) then gives the allylic alcohol (3) in 65% overall yield. The noteworthy feature is that the synthesis involves coupling of two aldehydes or ketones.

The reaction of selenides (4) with *n*-butyllithium and then with a carbonyl compound affords alcohols (5) resulting mainly from migration of a substituent originally attached to selenium.

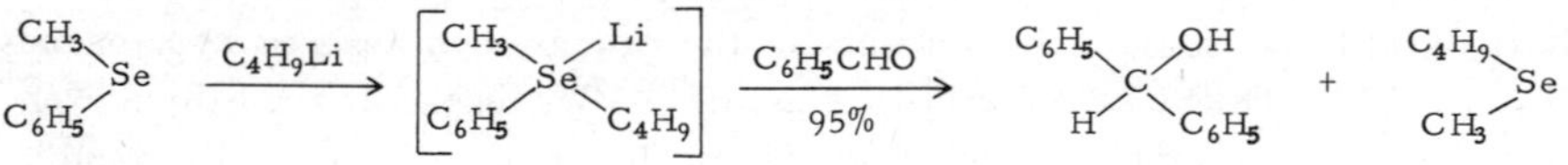

(4)                (b)                (5)

*Lithiation reactions.* Narasimhan and Mali have reported useful syntheses of naphtho[1,8-*bc*]pyrane[5] (equation I) and isocoumarin[6] (equation II).

(I)

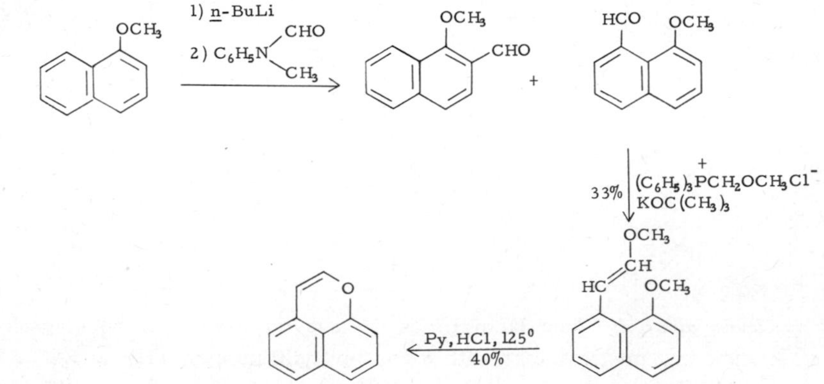

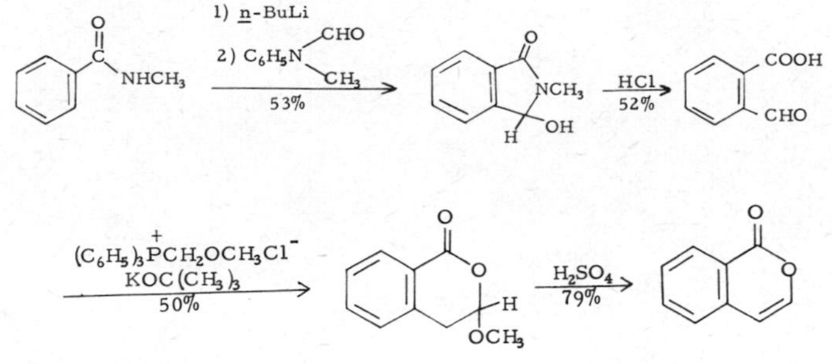

*Intramolecular cyclization to macrocyclic terpenoids.* Treatment of the allylic thioether (1), prepared from *trans,trans*-geranyllinalool, with *n*-butyl-lithium in THF at $-78°$ in the presence of DABCO, leads to a carbanion that

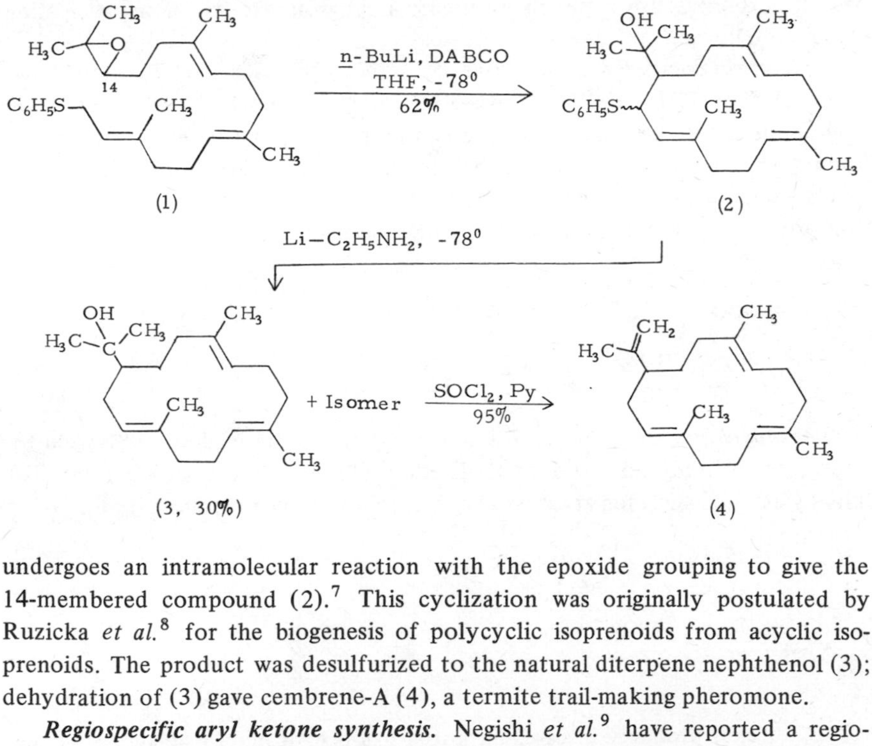

(1)

(2)

(3, 30%)

(4)

undergoes an intramolecular reaction with the epoxide grouping to give the 14-membered compound (2).[7] This cyclization was originally postulated by Ruzicka *et al.*[8] for the biogenesis of polycyclic isoprenoids from acyclic iso-prenoids. The product was desulfurized to the natural diterpene nephthenol (3); dehydration of (3) gave cembrene-A (4), a termite trail-making pheromone.

*Regiospecific aryl ketone synthesis.* Negishi *et al.*[9] have reported a regio-specific synthesis of aryl ketones from aryl bromides in which the new C—C bond is specifically at the carbon atom that originally was substituted by bromine. Thus the aryllithium compound (2) is prepared by halogen–metal exchange; this reacts with a trialkylborane to form a lithium aryltrialkylborate (3). The reac-

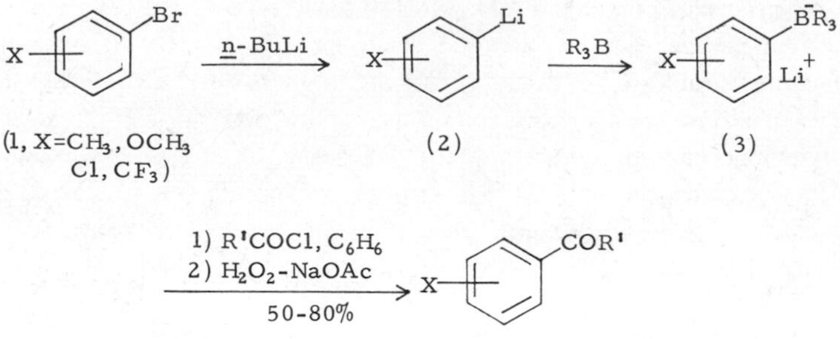


(1, X=$CH_3$, $OCH_3$, Cl, $CF_3$) (2) (3) (4)

tion of (3) with an acyl chloride followed by oxidation gives the ketone (4), usually in good yield. Both aromatic and aliphatic acyl halides can be used; (3) can contain a deactivating group such as *m*-Cl. Secondary alkyl groups (cyclopentyl, norbornyl) are better than primary alkyl groups for the nonparticipating R group on boron.

***Dialkylmalonic acids.***[10] Dianions of carboxylic acids have been prepared using LDA in THF–HMPT (**4**, 301–302). Krapcho reports that trianions (2) of monosubstituted malonic acids (1) can be prepared with 3 equiv. of *n*-butyllithium in THF at 0°. These can be alkylated by primary halides to give dialkylmalonic acids (3). Alkylation cannot be achieved with secondary halides. This procedure is not applicable to malonic acid itself.

$$RCH(COOH)_2 \xrightarrow[\text{THF, }-50^0]{3\ \text{n-BuLi}} RC(Li)(CO\bar{O})_2 2\ Li^+ \xrightarrow[75-85\%]{R'X,\ 0^0} RR'C(COOH)_2$$

(1, R = $C_6H_5$, $CH_3$) (2) (3)

***Benzocyclopropene*** (2).[11] This hydrocarbon can be made in 30% yield by treatment of *o*-bromobenzyl methyl ether (1) with *n*-butyllithium in THF initially at −40° and then at reflux for 1 hr. followed by aqueous work-up.

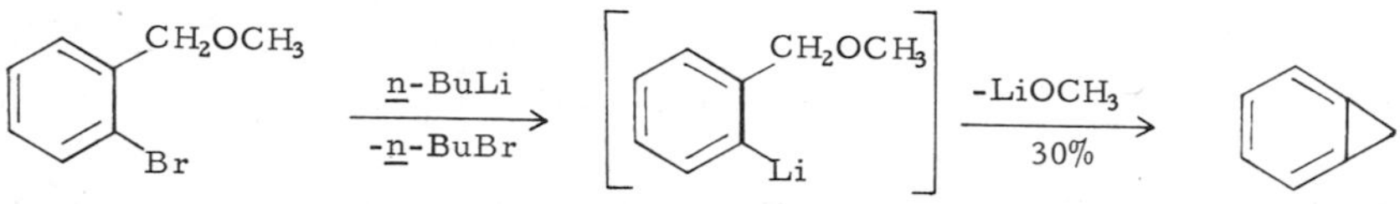


(1) (a) (2)

The reaction probably involves an initial halogen–lithium exchange followed by 1,3-elimination of $LiOCH_3$.

Birch reduction of (2) leads only to products formed by cleavage of the cyclopropene ring: toluene, 2,5-dihydrotoluene, and 1,2-diphenylethane (62 : 28 : 10).

*Site-specific tritium-labeled arenes.* Taylor[12] reports that tritium-labeled aromatics can be obtained by wetting dried ether with tritiated water, adding an aryl halide, cooling to $-70°$, and then adding *n*-butyllithium. Trimethylsilyl derivatives of aromatics can be prepared using trimethylchlorosilane in the same way. When the organometallic intermediate is formed first and then treated with tritiated water, only the unlabeled hydrocarbon is obtained in some cases. Apparently, the cross-metalation reaction is faster than the reaction of *n*-butyllithium with either water or trimethylchlorosilane.

*1-Lithiocyclopropyl bromides.* The reaction of 7,7-dibromonorcarane (1) with *n*-butyllithium in THF at low temperatures under carefully controlled conditions results in exclusive formation of *anti*-7-bromo-*syn*-7-lithionorcarane (2).[13] The reaction of (2) with a number of reagents [*e.g.*, $CO_2$, $(CH_3)_3 SiCl$] is reported in the same paper.

(1)                    (2)

Japanese chemists[14] have now reported the alkylation of 1-lithiocyclopropyl bromides, prepared by treatment of *gem*-dibromocyclopropanes with *n*-butyllithium, by reaction with an alkyl halide. They were able to prepare *syn*-7-methyl-*anti*-7-bromonorcarane (3) as the exclusive product by addition of methyl iodide to (2), prepared essentially by Seyferth's procedure.[13] As applied to other *gem*-dibromocyclopropanes, the reaction is usually stereoselective rather

(3)

than stereospecific. Several factors are involved, such as the solvent and the time at which the alkyl halide is added to the reaction. For example, the reaction as

(4)                    (5)        +        (6)

applied to (4) can lead to either (5) or (6) as the predominant product. Under conditions of kinetic control (addition of HMPT), (6) is the predominant product. If the alkyl halide is added after a period of aging of the intermediate carbanion, (5) is the exclusive product.

The same paper reports some interesting reactions of (5) as shown in the formulation. In addition, the bromine atom of the products can be replaced by

hydrogen with retention of configuration by reduction with sodium in liquid ammonia.[15]

*Reaction with cyclic vinyl halides.* Reaction of 2-chlorobicyclo[2.2.1]heptene-1 (1) with *n*-butyllithium in THF at 25° gives (2) and (3) as the major products. Experiments with optically inactive and active (1) labeled with deuterium indicate that a symmetrical species is an intermediate in the formation of

(2) and that the first step involves loss of the vinylic proton. This evidence strongly suggests that norbornyne (a) is the major intermediate.[16]

The reaction of (1) with methyllithium gives (4) in 73% yield. In this case a cycloalkyne is not involved, because optically active (1) gives optically active (2) with retention of stereochemistry.[17]

[1] A. J. Quillinan, E. A. Khan, and F. Scheinmann, *J.C.S. Chem. Comm.*, 1030 (1974).
[2] R. H. Mitchell, R. Otsubo, and V. Boekelheide, *Tetrahedron Letters*, 219 (1975).
[3] W. Dumont, P. Bayet, and A. Krief, *Angew. Chem. internat. Ed.*, 13, 804 (1974).
[4] D. Seebach and A. K. Beck, *ibid.*, 13, 806 (1974).
[5] N. S. Narasimhan, and R. S. Mali, *Synthesis*, 796 (1975).
[6] *Idem, ibid.*, 797 (1975).
[7] M. Kodama, Y. Matsuki, and S. Ito, *Tetrahedron Letters*, 3065 (1975).

[8] L. Ruzicka, with A. Eschenmoser, and H. Heusser, *Experientia*, 9, 357 (1953).

[9] E. Negishi, A. Abramovitch, and R. E. Merrill, *J.C.S. Chem. Comm.*, 138 (1975).

[10] A. P. Krapcho and D. S. Kashdan, *Tetrahedron Letters*, 707 (1975).

[11] P. Radlick and H. T. Crawford, *J.C.S. Chem. Comm.*, 127 (1974).

[12] R. Taylor, *Tetrahedron Letters*, 435 (1975).

[13] D. Seyferth and R. L. Lambert, Jr., *J. Organometal. Chem.*, C53 (1973).

[14] K. Kitatani, T. Hiyama, and H. Nozaki, *Am. Soc.*, 97, 949 (1975).

[15] H. M. Walborsky, F. P. Johnson, and J. B. Pierce, *ibid.*, 90, 5222 (1968).

[16] P. G. Gassman and J. J. Valcho, *Am. Soc.*, 97, 4769 (1975).

[17] P. G. Gassman and T. J. Adkins, *Tetrahedron Letters*, 3035 (1975).

### *t*-Butyloxycarbonyl fluoride (BOC—F); 4, 65.

*Preparation.* Wackerle and Ugi[1] have reported an improved procedure in which carbonyl chloride fluoride is generated from trichlorofluoromethane and oleum:

$$Cl_3CF \xrightarrow{H_2SO_4/SO_3} ClCF(=O) \xrightarrow[90-93\%]{\substack{(CH_3)_3COH \\ (C_2H_5)_3N, \ 20^0}} (CH_3)_3COCF(=O)$$

*t*-**Butyloxylcarbonyl-3-formylindole.**[1] This reagent was prepared by reaction of 3-formylindole with BOC—F and triethylamine in methylene chloride in 93% yield. The yield was only 70% using BOC—Cl.

$$O = C - OC(CH_3)_3$$

[1] L. Wackerle and I. Ugi, *Synthesis*, 598 (1975).

2-*t*-**Butyloxycarbonyloxyimino-2-phenylacetonitrile,** $(CH_3)_3COCO(=O)N=C(CN)(C_6H_5)$    (1).

Mol. wt. 246.27, m.p. 84–86°, stable.

*Preparation:*

$$HON=C(CN)(C_6H_5) + COCl_2 \xrightarrow{DMA} ClCOON=C(CN)(C_6H_5) \xrightarrow[Py]{(CH_3)_3COH} (1)$$

*t*-*Butyloxycarbonylation.*[1] The reagent reacts with amino acids in the presence of triethylamine to give BOC derivatives in high yield (4–5 hr. at 20° or 1 hr. at 45°).

$$(1) + H_2NCHCOOH \cdot N(C_2H_5)_3 \overset{R}{\underset{}{}} \longrightarrow (CH_3)_3COCNHCHCOOH \cdot N(C_2H_5)_3 \overset{O \ \ R}{\underset{}{}} + HON=C(CN)(C_6H_5)$$

[1] M. Itoh, D. Hagiwara, and T. Kamiya, *Tetrahedron Letters*, 4393 (1975).

*n*-**Butylphenyltin dihydride, polymeric,** (P)$-C_6H_4\overset{\underset{\displaystyle H}{|}}{\overset{\displaystyle H}{\overset{|}{S}}}n-C_4H_9$-$\underline{n}$  (1). Crosby *et al.*[1]

have reported a preparation of a polymeric form of an organotin dihydride that contains an average of 2.0 moles of hydride per gram of reagent. The polymeric reagent is more stable than a simple organotin dihydride, but is best stored as the precursor, (P)$-C_6H_4-Sn(Cl)_2C_4H_9$-$\underline{n}$, from which it is formed by reduction with $LiAlH_4$. The polymeric reagent reduces aldehydes and ketones, but does so slowly at the reflux temperature of toluene (111°); satisfactory yields of alcohols can be obtained if the reagent is added in portions. The reagent reduces alkyl and aryl halides at temperatures of 20–111°. Unfortunately, the used resin from the reduction can be regenerated only to the extent of about 60% of the original active hydride content.

[1] N. M. Weinshenker, G. A. Crosby, and J. Y. Wong, *J. Org.*, **40**, 1966 (1975).

# C

**$\pi$-(2-Carboethoxyallyl)nickel bromide,**

$$\underset{(1)}{C_2H_5O\overset{O}{\overset{\|}{C}}-\overset{CH_2}{\underset{CH_2}{C}}\overset{Br}{\underset{Br}{Ni}}\overset{CH_2}{\underset{CH_2}{C}}-\overset{O}{\overset{\|}{C}}OC_2H_5} \text{. Mol. wt. 414.11.}$$

The complex is prepared from 2-carboethoxyallyl bromide and nickel carbonyl in benzene at $40°$.

*α-Methylene-γ-butyrolactones.*[1] The complex reacts with aldehydes, alicyclic ketones, and α-diketones to form α-methylene-γ-butyrolactones, generally in high yield. Aliphatic ketones and α,β-unsaturated ketones are essentially unreactive to (1).

$$\underset{R^1}{\overset{R}{}}C=O + (1) \xrightarrow[25-55°]{DMF} \left[\underset{R^1}{\overset{R}{}}\underset{OH}{\overset{CH_2C=CH_2}{\underset{OC_2H_5}{\underset{\|}{C=O}}}}\right] \xrightarrow[70-90\%]{-C_2H_5OH} \underset{R^1}{\overset{R}{}}\overset{}{\underset{O}{}}=CH_2$$

[1] L. S. Hegedus, S. D. Wagner, E. L. Waterman, and K. Siirala-Hansen, *J. Org.*, **40**, 593 (1975).

**Carboethoxycyclopropyltriphenylphosphonium tetrafluoroborate** (1), **5**, 90–91.

*Spiroannelation.*[1] Successive treatment of 2-formylcyclohexanone (2) with sodium hydride and (1) results in formation of (3) as the major product. No product resulting from reaction at the keto group has been detected.

$$\underset{(2)}{} + \underset{(1)}{} \xrightarrow[40\%]{\substack{NaH \\ HMPT}} \underset{(3)}{}$$

This reaction has been developed into a synthesis of a number of spirovetivones. The first step involves the reaction of the sodium enolate of (4) with (1) to give the spiro ester (5) in 38% yield. The product was converted by known reactions into several known spirovetivones, for example, *dl*-β-vetivone (6) and *dl*-α-vetispirene (7).

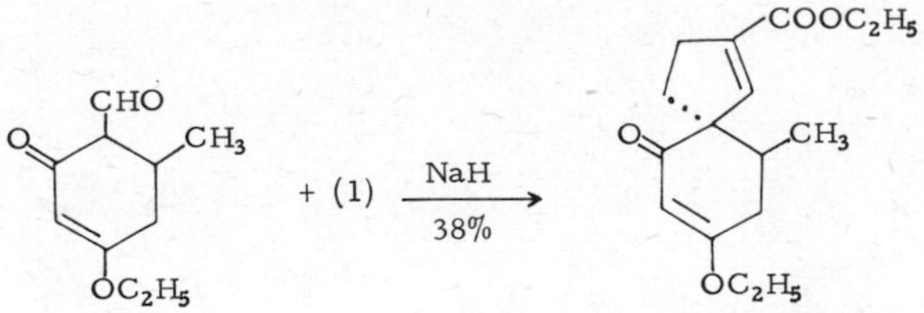

(4)    + (1)  $\xrightarrow[\text{38\%}]{\text{NaH}}$  (5)

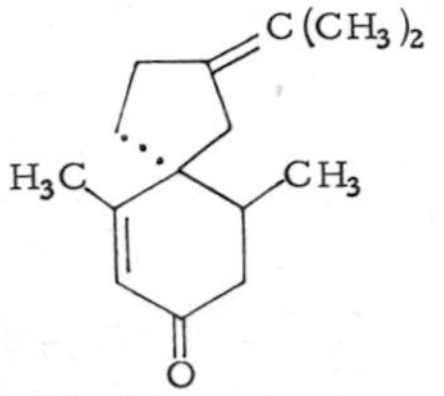

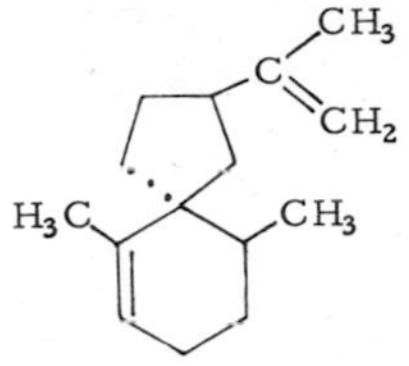

(6)

(7)

*2,3-Dihydrofuranes.*[2] Sodium salts of carboxylic acids (1) react with the reagent (2) in HMPT at 25° or in $CHCl_3$ at reflux to form 5-substituted-4-carboethoxy-2,3-dihydrofuranes (3). The reaction involves an intramolecular Wittig reaction.

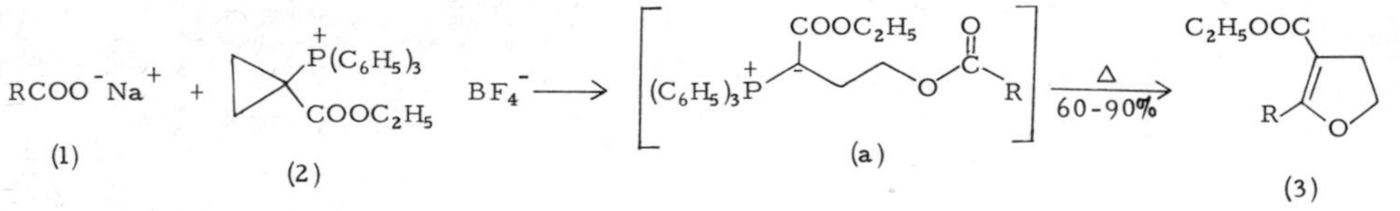

**Other examples:**

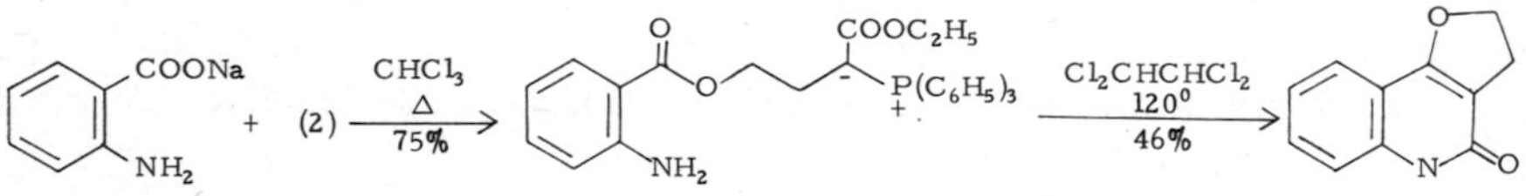

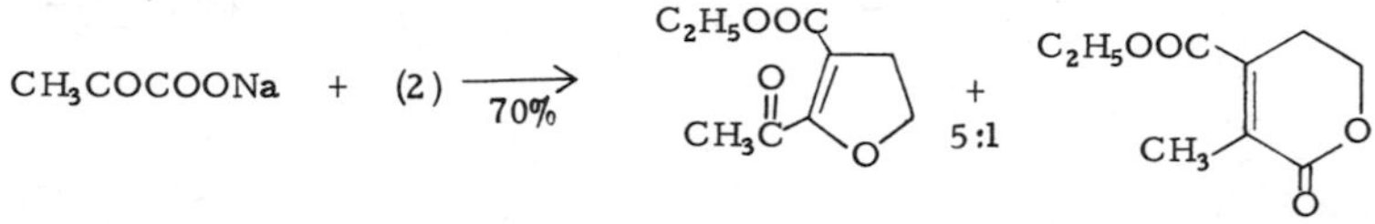

[1] W. G. Dauben and D. J. Hart, *Am. Soc.*, **97**, 1622 (1975).
[2] *Idem, Tetrahedron Letters*, 4353 (1975).

## Carbon dioxide, 3, 40–41; 5, 93–94.

*Carboxylation of active methylene compounds.*[1] Relatively acidic active methylene compounds can be carboxylated by the combination of $CO_2$ and DBU in DMSO followed by hydrolysis with aqueous HCl.

Yields are lower with less acidic substrates such as acetophenone. Carboxylation of ketones generally is carried out in higher yields with $CO_2$ and lithium 4-methyl-2,6-di-$t$-butylphenoxide as base.[2]

Ferric ethoxide has also been used for carboxylation of ketones, but yields are not impressive.[3]

$$C_6H_5COCH_3 + CO_2 \xrightarrow[\text{7\%}]{\underset{\text{DMF, }80^0}{Fe(OC_2H_5)_3}} C_6H_5COCH_2COOH$$

*Carboxylic acids; allenes.*[4] Alkylidenetriphenylphosphoranes (1) react with $CO_2$ to give ylides (2). These are crystalline, but melt with decomposition to give allenes (3) in rather low yield. They are converted into carboxylic acids in high yield on alkaline hydrolysis.

[1]E. Haruki, M. Arakawa, N. Matsumura, Y. Otsuji, and E. Imoto, *Chem. Letters*, 427 (1974).
[2]E. J. Corey and R. H. K. Chen, *J. Org.*, 38, 4086 (1973).
[3]T. Ito and Y. Takami, *Chem. Letters*, 1035 (1974).
[4]H. J. Bestmann, T. Denzel, and H. Salbaum, *Tetrahedron Letters*, 1275 (1974).

**Carbon disulfide, 1,** 114.

*Isothiocyanates.* Alkyl and aryl isothiocyanates can be prepared in 55–99% yield by treatment of an amine with $n$-butyllithium in THF; after 30 min., carbon disulfide is added. The resultant lithium dithiocarbamate is then allowed to react again with *n*-butyllithium and carbon disulfide; the product loses $Li_2CS_3$ to form an isothiocyanate. Successive additions of *n*-butyllithium and carbon disulfide are essential for satisfactory yields. The method fails in the case of amines that contain groups that react with *n*-butyllithium.[1]

*2-Alkoxy-1,3-benzodithioles* (1). Carbon disulfide undergoes 1,3-dipolar cycloaddition to benzyne [generated from benzenediazonium-2-carboxylate (1, 46)] to give 1,3-benzodithiole-2-carbene (a). This carbene reacts with alcohols to form (1) in 35–50% yields. The reaction can be carried out in one-step from anthranilic acid, since carbon disulfide is more reactive than alcohols toward benzyne.[2]

(a)    (1)

[1] S. Sakai, T. Aizawa, and T. Fujinami, *J. Org.*, **39**, 1970 (1974).
[2] J. Nakayama, *Synthesis*, 38 (1975).

## Carbon suboxide, $O=C=C=C=O$.

The preparation and synthetic uses of carbon suboxide have been reviewed by Kappe and Ziegler.[1] This reactive "bisketene" is particularly useful for synthesis of heterocycles.

[1] T. Kappe and E. Ziegler, *Angew. Chem. internat. Ed.*, **13**, 491 (1974).

## Carbonylcyclopropane (Cyclopropylidenemethenone) (1).

*Generation.*[1] This ketene can be generated by pyrolysis of (2)[2] at 500°/0.05 torr through a silica tube. It is converted into the dimer (3) when warmed to 20°.

(2)    (1)    (3)

[1] G. J. Baxter, R. F. C. Brown, F. W. Eastwood, and K. J. Harrington, *Tetrahedron Letters*, 4283 (1975).
[2] **2,2-Dimethyl-1,3-dioxane-4,6-dione-5-spiropropane**, this volume.

**N,N'-Carbonyldiimidazole, 1**, 114–116; **2**, 61; **5**, 97–98.

*Cyclic carbonates.*[1] N,N'-Carbonyldiimidazole is an excellent reagent for conversion of *cis*-1,2-diols into cyclic carbonates:

(1)                                                                                    (a)

(2, 94%)

The by-product is the water-soluble imidazole. The reaction proceeds in high yield in the case of *cis*-1,2-diols. Cyclic carbonates of *trans*-1,2-diols are formed in much lower yields.

[1] J. P. Kutney and A. H. Ratcliffe, *Syn. Commun.*, 47 (1975).

**2-Carboxyethyltriphenylphosphonium perbromide,** $(C_6H_5)_3\overset{+}{P}CH_2CH_2COOH\ Br_3^-$ (1). Mol. wt. 335.35, m.p. 139°.

*Preparation:*

$$CH_2=CHCOOH + (C_6H_5)_3P + HBr \xrightarrow[91\%]{100°} (C_6H_5)_3\overset{+}{P}CH_2CH_2COOH\ Br^- \xrightarrow[87\%]{\underset{HOAc}{Br_2}} \quad (1)$$

m. p. 196-198°

*Bromination.* This perbromide is useful for selective bromination of a ketone in the presence of a double bond (*cf.* pyrrolidone-2-hydrotribromide, **3**, 240–241). In three cases reported yields are in the range 60–80%. Of course, this reagent can also be used for α-bromination of saturated ketones. No reaction was observed with camphor, however, or with esters. The intermediate monobromide formulated above in the preparation of (1) is formed on brominations with (1). In effect, the perbromide can be considered as a bromine-transfer agent.[1]

[1] V. W. Armstrong, N. H. Chishti, and R. Ramage, *Tetrahedron Letters*, 373 (1975).

**Caro's acid (Sulfomonoperacid), 1**, 118–119; **3**, 43.

*Nitroso compounds.*[1] The oxidation of arylamines to nitroso compounds requires carefully neutralized reagent.[2] Cosolvents (glyme and dioxane) cannot be used.

*Examples:*

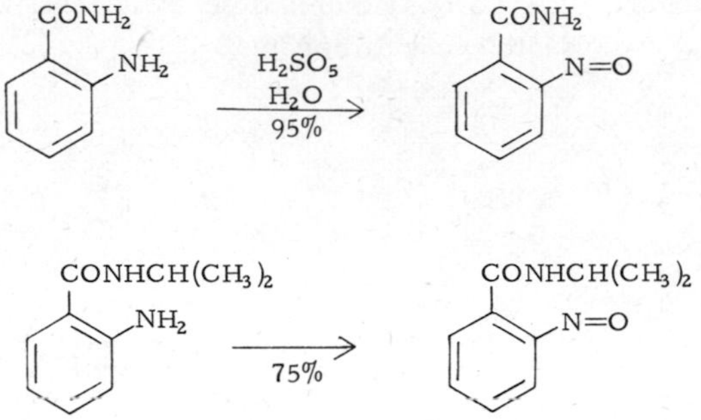

[1] G. P. Nilles, personal communication; *see also, J. Agricultural and Food Chemistry,* **23,** 410 (1975).

[2] Potassium persulfate (10 g.) is stirred in ice-cold concd. sulfuric acid (7 ml.) until a thick paste is formed and it is then stirred with 100 g. of ice. The pH of the solution is adjusted to pH 2 with solid $K_2CO_3$. Insoluble salts are removed, and the filtrate is adjusted to 75 ml. This solution is used for oxidation of 3.00 mmole of the amine.

## Catecholborane (1,3,2-Benzodioxaborole), 4, 25, 69–70. Supplier: Aldrich.

*Reduction of* >C=O *to* >CH$_2$. Tosylhydrazones are reduced to the corresponding methylene compounds by reaction with catecholborane in CHCl$_3$ or CH$_2$Cl$_2$ at $-10°$ for 20 min. followed by addition of sodium acetate (yields are lower in the absence of the acetate). The reaction mixture is then refluxed for 1 hr. before work-up. Yields are fair to good.[1]

*Examples:*

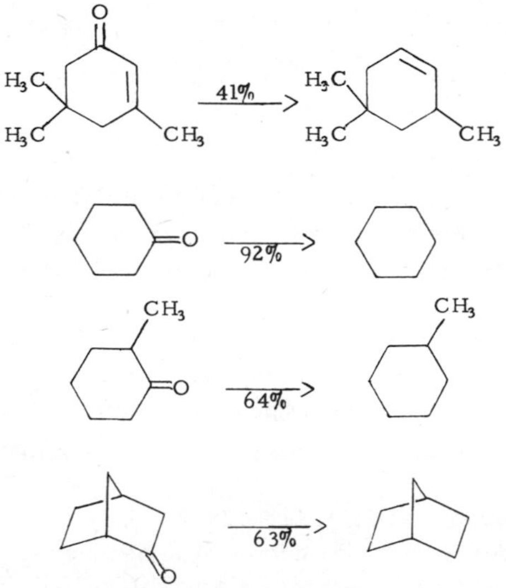

[1] G. W. Kabalka and J. D. Baker, Jr., *J. Org.,* **40,** 1834 (1975).

**Ceric ammonium nitrate (CAN), 1**, 120–121; **2**, 63–65; **3**, 44–45; **4**, 71–74; **5**, 101–102.

*Oxidation of enamino ketones.*[1] The major product (2, 30% yield) of the oxidation of (1), ethyl 1,4,5,6-tetrahydronicotinate, with CAN contains the furodipiperidine ring system.

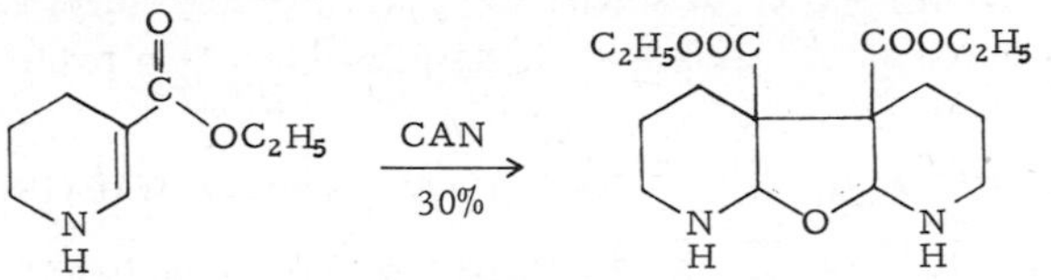

(1)                                    (2)

[1] B. W. Herten and G. A. Poulton, *J.C.S. Chem. Comm.*, 456 (1975).

**Cerium(IV) oxide–Hydrogen peroxide, $CeO_2$–$H_2O_2$.** Cerium(IV) oxide is supplied by Alfa.

*Oxidation of phenols.* Barton *et al.*[1] carried out some studies on the oxidation of phenols with hydrogen peroxide and samples of old cerium(IV) oxide. It was later found that the oxidations reported require activation if freshly prepared pure dioxide is used. It is dissolved in hot $H_2SO_4$, precipitated at pH 12 with sodium hydroxide, and then heated at *ca.* 900° for 24 hr. This material in combination with 30% $H_2O_2$ oxidizes phenols such as (1) to hydroperoxy-cyclohexadienones (2) in good yield. Other reported reactions are the oxidation of (4) to the oxide (5) and of (6) to juglone (7). This oxidation system probably

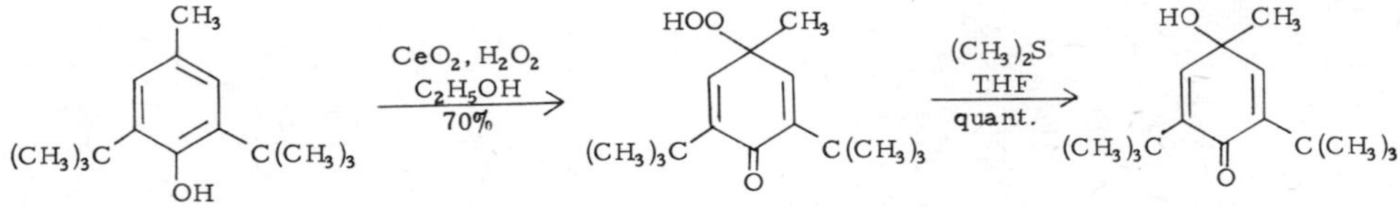

(1)                    (2)                    (3)

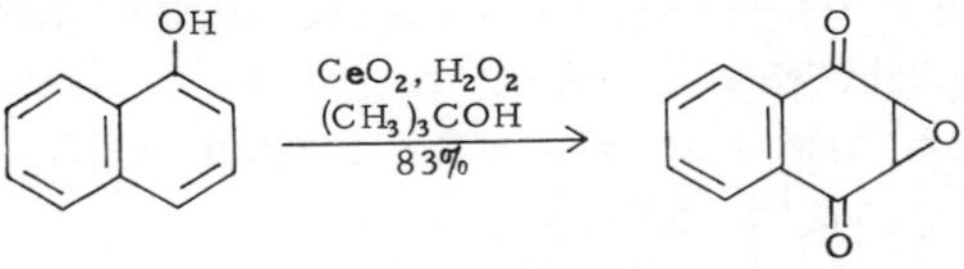

(4)                    (5)

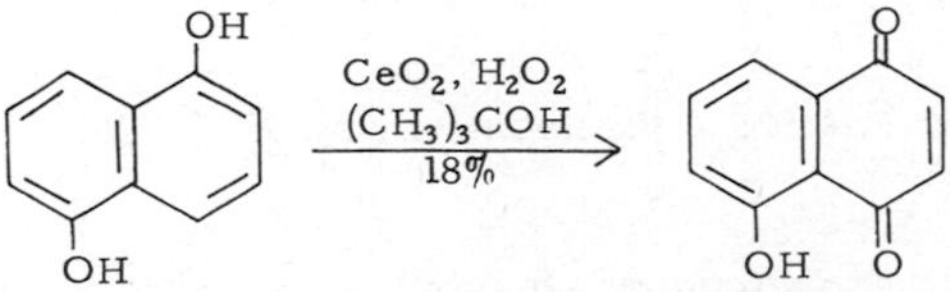

(6)                    (7)

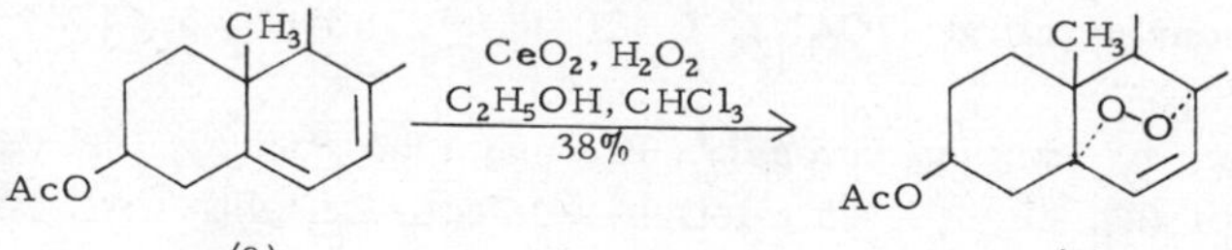

(8)                                    (9)

functions as an *in situ* source of singlet oxygen, since ergosteryl acetate is converted by this combination into the 5α,8α-peroxide (9), a reaction known to involve singlet oxygen.

[1] D. H. R. Barton, P. D. Magnus, and J. C. Quinney, *J.C.S. Perkin I*, 1610 (1975).

**Cesium fluoride**, CsF. Mol. wt. 168.36. Suppliers: Alfa, ROC/RIC.

*Cyclopropenes.* Chan and Massuda[1] have prepared cyclopropenes by fluoride ion-promoted elimination of trimethylhalosilane from dihalocarbene adducts of vinylsilanes. Thus 1-chlorocyclopropene (3) was prepared from trimethylvinylsilane (1) by addition of dichlorocarbene (Seyferth's method, **1**, 852). Elimination of trimethylchlorosilane was effected with cesium fluoride

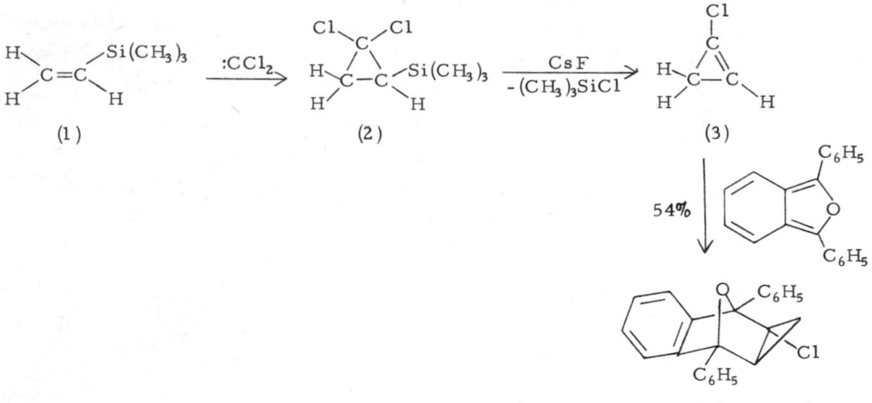

(1)                      (2)                            (3)

54%

(4)

in diglyme at 80°. The product (3) was identified by NMR and by cycloaddition to 1,3-diphenylisobenzofurane to give the adduct (4). 7-Chlorobicyclo-[4.1.0]heptene-6 (5) was also obtained in the same way.

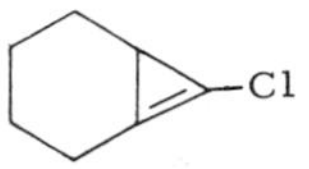

(5)

[1] T. H. Chan and D. Massuda, *Tetrahedron Letters*, 3383 (1975).

**Chloral, 1**, 122; **3**, 45.

*4-Hydroxycyclopentenones.* In connection with a synthesis of prostaglandins, Stork *et al.*[1] have developed a general synthesis of 2-alkyl-4-hydroxycyclopentenones (4). Base-catalyzed rearrangement of epoxy ketones of type (1)

(1)                    (2)                    (3)                    (4)

leads to hydroxycyclopentenones (2). These can be rearranged to the more useful isomers (4) by dilute base (1% NaOH), but the yields are generally only fair unless the hydroxyl group and the alkyl group in (2) are *cis* to each other. The problem can be solved by conversion of (2) into the chloral acetal (3), which is rearranged rapidly to (4) by base. Actually, it is not necessary to isolate (2); the oxide (1) can be converted into (4) by treatment with triethylamine (2.5 eq.) and then with chloral (1.1 eq.). When R was $-CH_2C\equiv C(CH_2)_3CH_3$, this transformation was accomplished in 69% yield. The intermediate (5) for the synthesis of prostaglandin $E_2$ (6) was prepared in the same manner.

(5)                                        (6)

[1]G. Stork, C. Kowalski, and G. Garcia, *Am. Soc.*, **97**, 3258 (1975).

**Chlorine, 5,** 105–106.

*Reaction with tyrosine.*[1] Dilute chlorine water converts tyrosine (1) into 3-chloro-4-hydroxybenzyl cyanide (2). Increased amounts of the reagent result in conversion mainly into (3) and (4). On further reaction with chlorine water

(4) is converted into 2,6-dichloro-*p*-benzoquinone (5). Similar results were obtained on reaction of chlorine water with *p*-hydroxycinnamic acid.

These results are significant in view of the widespread use of chlorine to purify water.[2]

[1] Y. Shimizu and R. Y. Hsu, *Chem. Pharm. Bull. Japan,* **23,** 2179 (1975).
[2] J. L. Marx, *Science,* **186,** 809 (1974); *Chem. Eng. News,* (45), 5 (1974).

## Chlorine bromide, BrCl.

*Bromination.* The combination of bromine and chlorine effects, under irradiation, bromination at the site where chlorine normally appears in chlorinations. The reaction is catalyzed by the hydrogen chloride formed. Straight

bromination, in which hydrogen bromide is formed, is slower.[1]

[1] T. R. Nelsen, J. E. Babiarz, J. T. Bartholomew, and D. C. Dittmer, *Org. Syn.,* submitted (1975).

## Chloroacetonitrile, 1, 129–130.

*Carboxylic acids from ketones.* White and Wu[1] have described a new conversion of ketones into carboxylic acids with addition of one carbon atom. The first step is a Darzens condensation of the ketone with chloroacetonitrile,[2] with sodium 2-methyl-2-butoxide (**1**, 1096; **2**, 386) as base. Glycidonitriles (1) are obtained in about 95% yield. These can be rearranged to α-ketonitriles (2) by a catalytic amount of potassium hydrogen sulfide, lithium trifluoroacetate, or lithium perchlorate in refluxing toluene or xylene (*caution:* HCN is evolved). Hydrolysis of (2) with aqueous base gives the carboxylic acid.

This route is preferred for aromatic ketones ($R^1$ = Ar). A less direct route involving treatment with dry hydrogen chloride, acetylation, and dehydration to an α-acetoxyacrylonitrile (4) is preferred for aliphatic ketones.[3] The product can be converted into the carboxylic acid (3) or into an α-keto acid (5). Overall yields of the carboxylic acids by this route are in the range 50–75%.

[1]D. R. White and D. K. Wu, *J.C.S. Chem. Comm.*, 988 (1974).
[2]G. Stork, W. S. Worrall, and J. J. Pappas, *Am. Soc.*, **82**, 4315 (1960).
[3]B. Rickborn and R. M. Gerkin, *ibid.*, **93**, 1693 (1971); B. C. Hartman and B. Rickborn, *J. Org.*, **37**, 943 (1972).

**Chloroacetylium hexafluoroantimonate,** $ClCH_2\overset{+}{C}{=}O\ SbF_6^-$. Mol. wt. 313.25, m.p. 15°.  **Bromoacetylium hexafluoroantimonate,** $BrCH_2\overset{+}{C}{=}O\ SbF_6^-$. Mol. wt. 357.71, m.p. 82°.

These stable, crystalline salts are prepared by the reaction of chloroacetyl fluoride and bromoacetyl fluoride with antimony pentafluoride in liquid $SO_2$ or $FCl_2C{-}CClF_2$ (Freon 113).

$$XCH_2COF\ +\ SbF_5\ \longrightarrow\ XCH_2\overset{+}{C}{=}O\ SbF_6^-$$
$$X = Cl,\ Br$$

*Haloacetylation of arenes.* The salts react with arenes to give high yields of halomethyl aryl ketones:

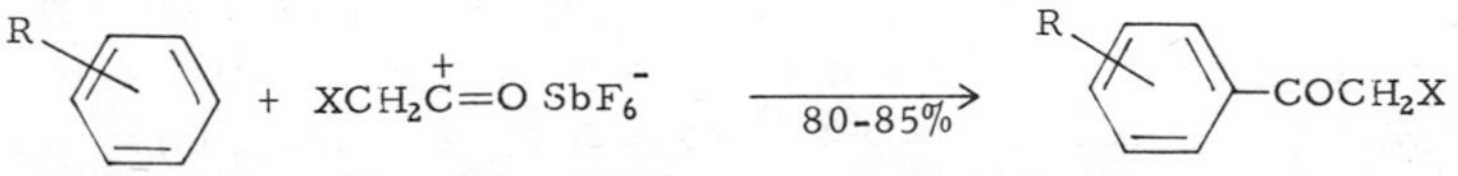

The expected isomer distribution is observed; thus in the case of toluene the isomer distribution of the products is: *o/m/p* = 4–8: 1–3: 90–95%.[1]

[1]G. A. Olah, H. C. Lin, and A. Germain, *Synthesis*, 895 (1974).

**π-(Chlorobenzene)chromium tricarbonyl,** 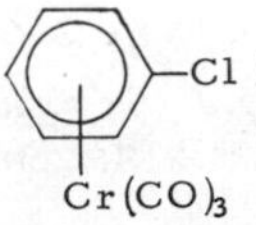 Cl (1). Mol. wt. 248.60, m.p. 102–103°.

This organometallic substance is prepared by the reaction of chlorobenzene with chromium hexacarbonyl in a special apparatus that allows recycling.[1]

*Phenylation of carbanions.* Halobenzenes do not undergo nucleophilic displacement reactions unless substituted by electron-withdrawing substituents in the *ortho-* or *para*-positions. However, Nicholls and Whiting[2] observed that the chlorine atom of (1) is readily replaced by a methoxyl group in 90% yield on reaction with sodium methoxide in refluxing methanol (24 hr.).

Semmelhack and Hall[3] now report that (1) phenylates some carbanions, particularly tertiary carbanions. Thus (1) reacts at 25° with the anion of isobutyronitrile, generated by LDA in THF below 0°. Treatment of the intermediate complex (2) with iodine liberates the free organic ligand (3).

(2)

*Other examples:*

$$(1) + {}^-C(CH_3)_2COOC_2H_5 \xrightarrow[71\%]{THF} C_6H_5C(CH_3)_2COOC_2H_5$$

$$(1) + {}^-CH(COOC_2H_5)_2 \xrightarrow[51\%]{HMPT} C_6H_5CH(COOC_2H_5)_2$$

Note that the anion of diethyl malonate requires a more polar solvent than THF. $\pi$-(Fluorobenzene)chromium tricarbonyl is more reactive than (1); the corresponding iodo complex is less reactive than (1). The reaction appears to be fairly limited. Thus the anions of 1,3-dithiane, $t$-butyl acetate, acetonitrile, and acetophenone fail to react with (1).

The mechanism of this phenylation reaction appears to be more complicated than that of classical nucleophilic aromatic substitution.

[1] W. Strohmeier, *Ber.,* **94,** 2490 (1961).
[2] B. Nicholls and M. C. Whiting, *J. Chem. Soc.,* 551 (1959).
[3] M. F. Semmelhack and H. T. Hall, *Am. Soc.,* **96,** 7091, 7092 (1975).

**1-Chloro-4-bromomethoxybutane,** $ClCH_2(CH_2)_2CH_2OCH_2Br$. Mol. wt. 201.51, b.p. 94°/5 torr; **1-Chloro-4-chloromethoxybutane,** $ClCH_2(CH_2)_2CH_2OCH_2Cl$. Mol. wt. 156.04, b.p. 70°/5 torr.

These reagents are prepared by the reaction of 4-chloro-1-butanol with paraformaldehyde and hydrogen bromide or hydrogen chloride in about 50% yield.

*Halomethylation of arenes.*[1]  Arenes, under catalysis with stannic chloride or zinc bromide, undergo a Friedel–Crafts reaction with these reagents to give halomethylarenes.

1,4-Bis(halomethoxy)butanes, prepared from 1,4-butanediol, paraformaldehyde, and a hydrogen halide, can be used in the same way.

[1]G. A. Olah, D. A. Beal, S. H. Yu, and J. A. Olah, *Synthesis*, 560 (1974).

**Chloro(carbonyl)bis(triphenylphosphine)rhodium(I), [Bis(triphenylphosphine)-rhodium carbonyl chloride],** $[(C_6H_5)_3P]_2Rh(CO)Cl$. Mol. wt. 690.96, air stable. Suppliers: Alfa, ROC/RIC.

*Preparation.*[1]

*Alkylation of acid chlorides to form ketones.* Hegedus and co-workers[2] have developed a method for synthesis of ketones from acid chlorides with this complex (1). Treatment of (1) with an organolithium (or Grignard reagent)

gives a Rh(I) complex (2). An acid chloride adds oxidatively to (2) to form a Rh(III) complex (3), which eliminates a ketone with regeneration of the original rhodium(I) complex.

This method has two limitations. Yields of ketones are low unless R is a primary alkyl, aryl, or allyl group. The rhodium reagent is expensive, even if prepared rather than purchased. However, it is attractive for synthesis of sensitive ketones, since conditions are mild. It was used, for example, for synthesis of (S)-(+)-4-methyl-3-heptanone, the alarm pheromone of a fungus-growing ant of the genus *Atta*, with complete retention of stereochemistry (equation I).

[1]D. Evans, J. A. Osborn, and G. Wilkinson, *Inorg. Syn.*, **11**, 99 (1968).
[2]L. S. Hegedus, P. M. Kendall, S. M. Lo, and J. R. Sheats, *Am. Soc.*, **97**, 5448 (1975).

**6-Chloro-1-*p*-chlorobenzenesulfonyloxybenzotriazole**   (1). Mol.   wt.   344.17, m.p. 125–127°

(1)

The sulfonate (1) is prepared by the reaction of 6-chloro-N-hydroxybenzo-triazole with *p*-chlorobenzenesulfonyl chloride under Schotten–Baumann conditions.[1]

*Amides.*[1] Sulfonates of this type have been used as coupling reagents for formation of peptides. The by-products, being acidic, are readily removed by

bicarbonate. N-Methylmorpholine is preferred as the base. Yields of dipeptides are generally greater than 70%. A preliminary study indicates that racemization is less than that observed with DCC as the coupling agent.

*Esterification.*[2] Carboxylic acids react with (1) in $CHCl_3$ or acetonitrile in the presence of 1 eq. of triethylamine to form an active ester (2), which can be isolated if desired. The esters (2) react with an alcohol, again in the presence of 1 eq. of base, to form an ester of the carboxylic acid. The reaction can be carried out in one step by mixing the acid, the alcohol, the coupling reagent, and 2 eq. of base in ether. In either case, the reaction takes place at 20°. Yields are generally in the range of 65–85%.

(2)

1-Methanesulfonyloxybenzotriazole can also serve as the coupling agent.

[1]M. Itoh, H. Nojima, J. Notani, D. Hagiwara, and K. Takai, *Tetrahedron Letters,* 3089 (1974).
[2]M. Itoh, D. Hagiwara, and J. Notani, *Synthesis,* 456 (1975).

**α-Chloro-N-cyclohexylpropanaldonitrone, 4,** 80–82; **5,** 110–113.

*Substitution reaction with 2-acetonylfurane.* A synthesis of nonactic acid ester (4), a building block of the macrotetrolide nonactin, involves the reaction of 2-acetonylfurane (1) with Eschenmoser's reagent (2). The resulting unstable aldehyde (b) is immediately oxidized and then esterified to give (3). This is converted into the methyl ester of nonactic acid (4) in two steps.[1]

$$CH_3COCH_2 - [furan] \quad + \quad Cl \overset{H}{\underset{|}{C}} \overset{H}{\underset{|}{C}} = \overset{-O}{\overset{+}{N}} - [cyclohexyl] \quad + AgBF_4 \quad \xrightarrow[-AgCl, -HBF_4]{CH_2Cl_2, -20^0}$$

(1)                                    (2)

$$\left[ CH_3COCH_2 - [furan] \overset{H}{\underset{CH_3}{\overset{|}{C}}} \overset{H}{\underset{|}{C}} = \overset{+}{N} - [cyclohexyl] \xrightarrow[-C_6H_{11}NHOH]{H_2O, HCl} CH_3COCH_2 - [furan] \overset{H}{\underset{CH_3}{\overset{|}{C}}} CHO \right]$$

(a)                                                              (b)

$$\xrightarrow[27\%]{\begin{array}{l}1) CrO_3 \\ 2) CH_2N_2\end{array}} CH_3COCH_2 - [furan] \overset{H}{\underset{CH_3}{\overset{|}{C}}} COOCH_3 \xrightarrow{\begin{array}{l}1) Rh/H_2 \\ 2) NaBH_4\end{array}} CH_3\overset{OH}{\underset{H}{\overset{|}{C}}}CH_2 - [tetrahydrofuran] \overset{H}{\underset{CH_3}{\overset{|}{C}}} COOCH_3$$

(3)                                              (4)

*Propellane lactones.* 12-Oxo-11-oxa[4.4.3] propella-3,8-diene (5) has been synthesized by the reaction of isotetralin (1) with α-chloro-N-cyclohexylacetaldonitrone (2).[2]

(1)          (2)                                              (3)

$$\xrightarrow[20^0]{KOC(CH_3)_3, HOC(CH_3)_3} \quad \xrightarrow[74\% \text{ from (3)}]{HCl, H_2O, CH_3OCH_2CH_2OCH_3 \atop 90^0}$$

(4)                                              (5)

[1] H. Gerlach and H. Wetter, *Helv.*, **57**, 2306 (1974).
[2] A. Rüttimann and D. Ginsburg, *Helv.*, **58**, 2237 (1975).

**Chlorodicarbonylrhodium(I) dimer,** $[Rh(CO)_2Cl]_2$. Mol. wt. 388.77, red, m.p. 121°. Suppliers: ROC/RIC, Strem.

*Carbonylation.* Israeli chemists[1] have reported a convenient synthesis of

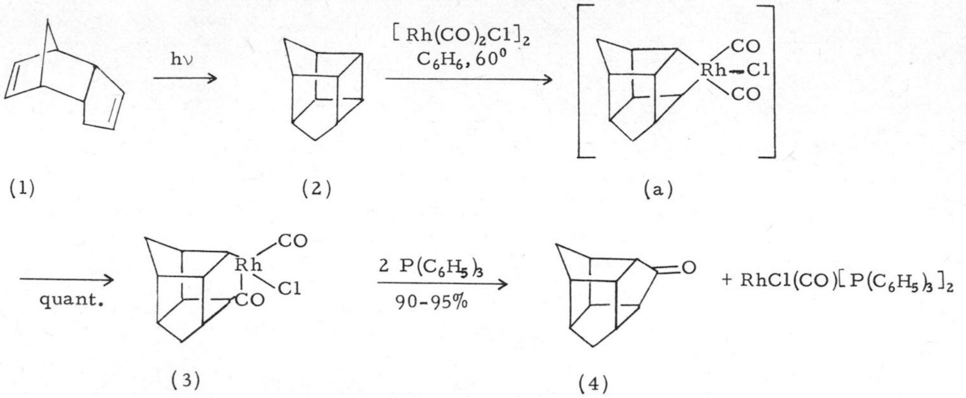

(1)                    (2)                         (a)

(3)                            (4)

trishomocubanone (4) from dicyclopentadiene (1). Irradiation of (1) gives 1,3-bishomocubane (2). This is then heated under nitrogen with 0.5 eq. of $[Rh(CO)_2Cl]_2$; the complex (3) is formed quantitatively. This is stable in refluxing xylene, but on pyrolysis at 250° trishomocubanone (4) is obtained in 40–50% yield. Higher yields of (4) are obtained by treatment of (3) with 2 moles of triphenylphosphine at 60° for 15 min.

[1] J. Blum, C. Zlotogorski, and A. Zoran, *Tetrahedron Letters,* 1117 (1975).

**Chlorodiphenylmethylium hexachloroantimonate,** $(C_6H_5)_2\overset{+}{C}Cl\ \overset{-}{S}bCl_6$ (1). Mol. wt. 536.16, m.p. 164–165° dec. Very sensitive to moisture.

The reagent is prepared[1] from dichlorodiphenylmethane and antimony pentachloride.

*Conversion of alcohols into amides.*[2] The reagent reacts with alcohols in nitrile solvents to give nitrilium ions derived from the alcohol; on quenching with water, the corresponding amides are formed.

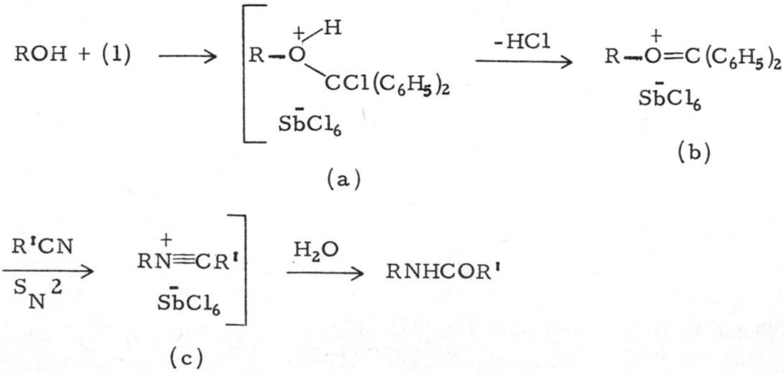

(a)

(b)

(c)

Cholesterol reacts with retention of configuration (through the $i$-cholesteryl ion), but other asymmetric alcohols give the racemic amide. Benzophenone is the only by-product.

[1]G. A. Olah and J. J. Svoboda, *Synthesis,* 307 (1972).
[2]D. H. R. Barton, P. D. Magnus, J. A. Garbarino, and R. N. Young, *J.C.S. Perkin I,* 2101 (1974).

**Chloromethyl methyl ether, 1,** 132–135; **5,** 120.

*Methoxymethyl aryl ethers,* $ArOCH_2OCH_3$. These ethers can be prepared by treatment of phenols with NaH in ether–DMF followed by chloromethyl methyl ether. Ronald[1] reports that these ethers show enhanced susceptibility at the *ortho*-position to halogen–metal exchange, particularly with $t$-butyllithium. Thus (1) can be converted into (2) in high yield by metalation, reaction with $CO_2$, and then reaction with diazomethane. In contrast, the methyl ether of $m$-cresol under similar conditions gives less than 1% yield of carboxylic acids.

$$\begin{array}{ccc} (1) & \xrightarrow[\substack{2)\ CO_2\ ;\ CH_2N_2 \\ 90-95\%}]{1)\ (CH_3)_3CLi,\ 0^0} & (2) \end{array}$$

[1]R. C. Ronald, *Tetrahedron Letters,* 3973 (1975).

**Chloromethyl methyl sulfide,** $ClCH_2SCH_3$. Mol. wt. 96.58, b.p. 105°. Stench. Suppliers: Aldrich, Fluka.

*Methylation of amides.*[1] Treatment of an amide, for example, benzamide, with chloromethyl methyl sulfide in TFA or methanesulfonic acid gives the substituted amide in good yield. This product is converted into the N-methyl-amide when refluxed in 90% ethanol in the presence of a large excess of Raney nickel.

$$C_6H_5CONH_2 \ + \ ClCH_2SCH_3 \ \xrightarrow[83\%]{\substack{CH_3SO_3H,\ 0^0 \\ 76\ hours}} \ C_6H_5CONHCH_2SCH_3$$

$$\xrightarrow[80\%]{\substack{Raney\ Ni \\ C_2H_5OH}} \ C_6H_5CONHCH_3$$

This method was used to methylate an amide group of a tetracycline; attempted hydrolysis of amide groups of tetracyclines leads to complete decomposition.

*Protection of hydroxyl groups.*[2] Hydroxyl groups can be protected as the methylthiomethyl ether, prepared by reaction of a sodium alkoxide with chloromethyl methyl sulfide and sodium iodide (1 eq. of each) at 0° for 1 hr. and

$$ROH \xrightarrow[\text{DME}]{\text{NaH}} RONa \xrightarrow[\text{NaI}]{\text{ClCH}_2\text{SCH}_3} ROCH_2SCH_3$$

$25°$ for 1.5 hr. The protecting group is cleaved by mercuric chloride (2 eq.), calcium carbonate (3 eq.) in $CH_3CN-H_2O$ at $25°$ (1-2 hr.). Under these neutral conditions silyl ethers and 1,3-dithianes are not affected.

[1] L. Bernardi, R. de Castiglione, and U. Scarponi, *J.C.S. Chem. Comm.*, 320 (1975).
[2] E. J. Corey and M. G. Bock, *Tetrahedron Letters*, 2643 (1975).

**1-Chloro-3-pentanone, 3, 49.**

*Annelation.*[1] 1,10-Dimethyl-1(9)-octalone-2 (2) has been prepared by the reaction of 2-methylcyclohexanone with the chloro ketone and an acid catalyst, *p*-toluenesulfonic acid or sulfuric acid. The acid presumably generates ethyl

vinyl ketone and also serves as a catalyst in the Michael and aldolization steps.

[1] P. A. Zoretic, B. Branchaud, and T. Maestrone, *Tetrahedron Letters*, 527 (1975).

***m*-Chloroperbenzoic acid. 1**, 135-139; **2**, 68-69; **3**, 45-50.

*Oxidation of hydroxyl groups.*[1] Alcohols are not ordinarily oxidized by

peracids; hence the observation that (1) is oxidized by *m*-chloroperbenzoic acid to (2)[2] suggests that this oxidation is connected with the fact that (1) is a nitroxyl radical. Indeed, a variety of alcohols have been found to be oxidized by *m*-chloroperbenzoic acid in the presence of a catalytic amount of 2,2,6,6-tetra-methylpiperidine-1-oxyl (3).[3] Actually, the nitroxyl catalyst can be generated *in situ* by the addition of 2,2,6,6-tetramethylpiperidine (TMP) or the hydrochloride to the oxidation reaction. A variety of other stable radicals are ineffective. Baeyer-Villiger oxidation of the ketonic product can be avoided by use of only 1 eq. of oxidant and short reaction periods. On the whole, yields of ketones are good to excellent; yields of aldehydes appear to be somewhat lower (40 and 76%, two examples).

*Examples:*

$$\text{cyclopentyl-OH} \xrightarrow[\substack{77\%}]{\substack{\text{Cl-C}_6\text{H}_4\text{COOOH} \\ \text{TMP}\cdot\text{HCl, CH}_2\text{Cl}_2}} \text{cyclopentanone}=O$$

$$\underset{\underset{\text{OH}}{|}}{\text{C}_6\text{H}_5\text{CH}_2\text{CHCH}_3} \xrightarrow[87\%]{} \underset{\underset{\text{O}}{||}}{\text{C}_6\text{H}_5\text{CH}_2\text{CCH}_3}$$

$$\xrightarrow{94\%}$$

$$\text{C}_6\text{H}_5\text{CH}_2\text{CH}_2\text{OH} \xrightarrow[76\%]{} \text{C}_6\text{H}_5\text{CH}_2\text{CHO}$$

$$\text{CH}_3(\text{CH}_2)_3\text{CH}_2\text{OH} \xrightarrow[90\%]{} \text{CH}_3(\text{CH}_2)_3\text{COOH}$$

This oxidation is obviously limited to alcohols that do not bear functional groups reactive to peracids; on the other hand, it can be useful for multistage oxidations. Thus (5) can be prepared from (4) by a one-pot reaction.

$$\xrightarrow[86\%]{\substack{1)\ \text{ClC}_6\text{H}_4\text{COOOH} \\ 2)\ \text{ClC}_6\text{H}_4\text{COOOH, TMP}\cdot\text{HCl}}}$$

(4)                                                   (5)

Ganem[4] has also reported oxidation of secondary alcohols to ketones in about 50–70% yield with 2,2,6,6-tetramethylpiperidine and 2 molar eq. of perbenzoic acid in $CH_2Cl_2$. Yields are improved by 5–10% by use of preformed 2,2,6,6-tetramethylpiperidine-1-oxyl and 1 eq. of the peracid. Ganem also observed that in some cases the ketone is not isolable directly, but only after the reaction mixture is refluxed for several hours.

In a more recent paper Cella and McGrath[5] report that alcohols can be oxidized in MCPBA in purified THF, with hydrochloric acid (10 mole %) as the only catalyst. Yields of ketones are high (75–95%) for unhindered secondary alcohols, but significantly lower for even slightly hindered alcohols (borneol → camphor, 23% yield). A further limitation is that the method is not suitable for acid-sensitive substrates.

**α-Hydroxy aldehydes and ketones.** α-Acetoxylation of aldehydes can be effected by conversion to the silyl enol ether (1) followed by reaction with

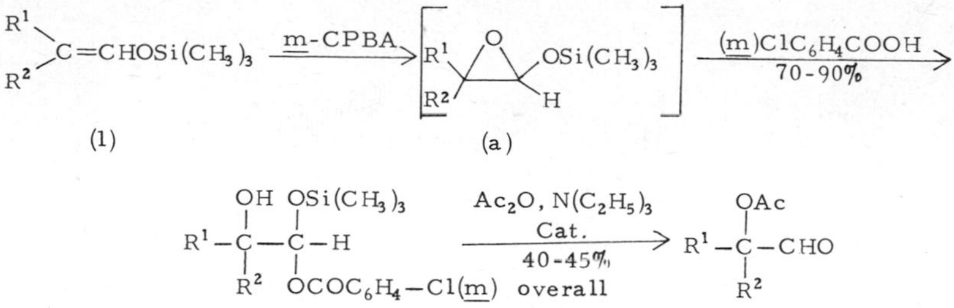

(1)                                        (a)

(2)                                        (3)

*m*-chloroperbenzoic acid. The intermediate epoxide (a) is opened by reaction with *m*-chlorobenzoic acid to give (2). A variety of reactions failed to convert (2) to protected α-hydroxy aldehydes, but the desired reaction was eventually effected by reaction of (2) with acetic anhydride and triethylamine in the presence of 4-pyrrolidinopyridine as catalyst. Overall yields of α-acetoxy aldehydes were on the order of 40–45%.[6]

α-Siloxy ketones (5)[6,7] can be prepared directly by epoxidation of silyl enol ethers of ketones (4), with migration of the silicon group.

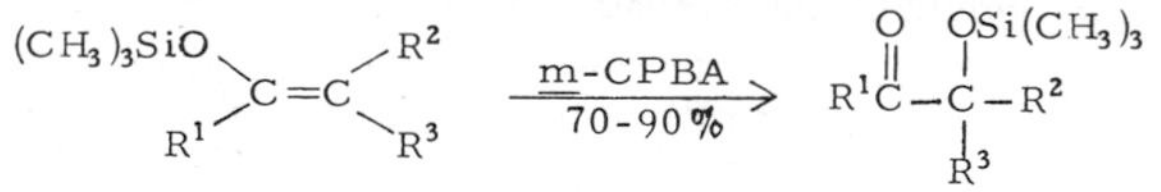

(4)                                        (5)

*Examples:*

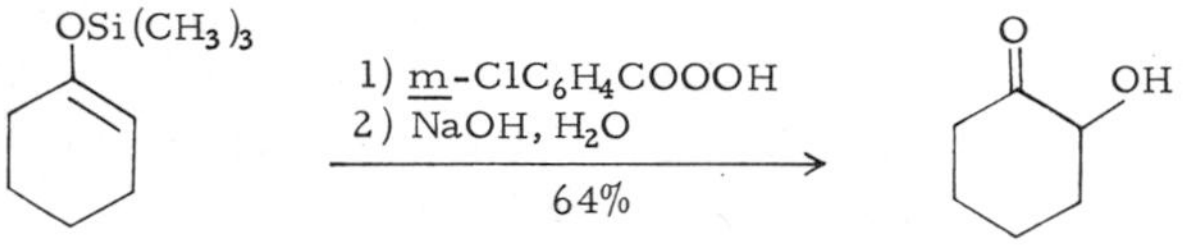

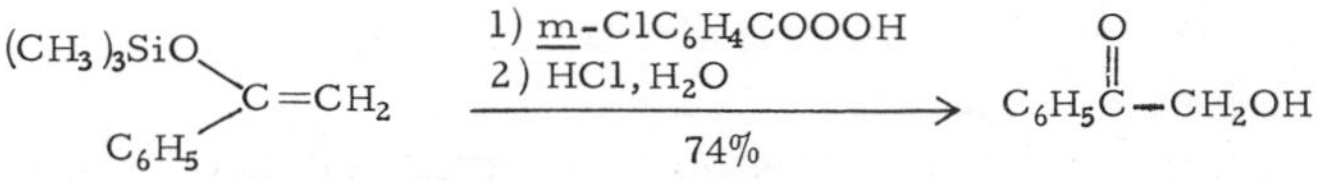

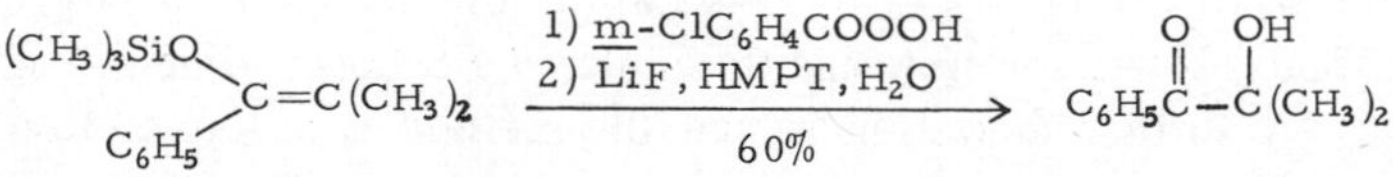

1) $\underline{m}$-ClC$_6$H$_4$COOOH
2) NaOH, H$_2$O

64%

1) $\underline{m}$-ClC$_6$H$_4$COOOH
2) HCl, H$_2$O

74%

C$_6$H$_5$C—CH$_2$OH

1) $\underline{m}$-ClC$_6$H$_4$COOOH
2) LiF, HMPT, H$_2$O

60%

C$_6$H$_5$C—C(CH$_3$)$_2$

*α-Hydroxy carboxylic acids.*[8] Oxidation of ketene bis(trimethylsilyl)acetals (1)[9] with *m*-chloroperbenzoic acid followed by treatment with 1.5 *N* HCl gives α-hydroxy carboxylic acids (2).

*α-Diketones.* Curci *et al.*[10] found that diazodiphenylmethanes are oxidized by perbenzoic acid to benzophenones in high yield. This reaction has now been extended to oxidation of cyclic α-diazo ketones to α-diketones with *m*-chloroperbenzoic acid.[11]

$$(Ar)_2C=N_2 \xrightarrow[\text{90-99\%}]{C_6H_5COOOH} (Ar)_2C=O$$

Ring size has little influence on the oxidation. In some cases the α-diketones are readily oxidized to the corresponding anhydrides (*n* = 6 and 7). Yields of diketones are around 90% when *n* = 4, 8, and 10.

*Deamination of amines to ketones.* Dinizo and Watt[12] have reported oxidative deamination by conversion to an imine (1), oxidation to an oxaziridine

(2), and, finally, base-catalyzed ring opening to a ketone (3). The procedure was suggested by the enzymatic oxidative deamination of α-amino acids mediated by pyridoxal pyrophosphate.

*3,3-Dimethyloxaziridines.* Primary and secondary amines can be converted into carbonyl groups by the sequence shown in equation I.[13]

$$
\text{(I)} \quad
\begin{array}{c} R^1 \\ \diagup \\ R^2 \end{array}\!\!\!CH\!-\!NH_2 \ + \ (CH_3)_2C\!=\!O \xrightarrow[\sim 80\%]{} \begin{array}{c} R^1 \\ \diagup \\ R^2 \end{array}\!\!\!CH\!-\!N\!=\!C(CH_3)_2 \xrightarrow[65-75\%]{ClC_6H_4CO_3H}
$$

$$
\begin{array}{c} R^1 \\ \diagup \\ R^2 \end{array}\!\!\!CH\!-\!N\!-\!C\!\!\begin{array}{c} CH_3 \\ \diagdown \\ CH_3 \end{array} \xrightarrow[\sim 100\%]{\begin{array}{c}HCl, H_2O\\20^0\end{array}} \begin{array}{c} R^1 \\ \diagup \\ R^2 \end{array}\!\!\!C\!=\!O \ + \ NH_3 \ + \ (CH_3)_2C\!=\!O
$$

[1] J. A. Cella, J. A. Kelley, and E. F. Kenehan, *J. Org.,* **40,** 1860 (1975).
[2] *Idem, J.C.S. Chem. Comm.,* 943 (1974).
[3] E. G. Rozantzev and M. B. Neiman, *Tetrahedron,* **20,** 131 (1964).
[4] B. Ganem, *J. Org.,* **40,** 1998 (1975).
[5] J. A. Cella, and J. P. McGrath, *Tetrahedron Letters,* 4115 (1975).
[6] A. Hassner, R. H. Reuss, and H. W. Pinnick, *J. Org.,* **40,** 3427 (1975).
[7] G. M. Rubottom, M. A. Vazquez, and D. R. Pelegrina, *Tetrahedron Letters,* 4319 (1974).
[8] G. M. Rubottom and R. Marrero, *J. Org.,* **40,** 3783 (1975).
[9] C. Ainsworth and Y.-N. Kuo, *J. Organometal. Chem.,* **46,** 73 (1972).
[10] R. Curci, F. Difuria, and F. Marcuzzi, *J. Org.,* **36,** 3774 (1971).
[11] R. Curci, F. DiFuria, J. Ciabattoni, and P. W. Concannon, *ibid.,* **39,** 3295 (1974).
[12] S. E. Dinizo and D. S. Watt, *Am. Soc.,* **97,** 6900 (1975).
[13] D. St. C. Black, and N. A. Blackman, *Australian J. Chem.,* **28,** 2547 (1975).

*o*-**Chlorophenyl phosphorodichloridite,** —OPCl₂ (1). Mol. wt. 229.47, b.p. 111–112°/10 mm.

*Preparation.*[1] The phosphorodichloridite is prepared in 98% yield by heating *o*-chlorophenol with excess phosphorus trichloride for 10 hr.; anhydrous magnesium chloride is added after about one-third of the hydrogen chloride has been formed.

*Polynucleotides.*[2] Aryl phosphorodichloridates (**1,** 847) are useful for synthesis of short oligonucleotides by the phosphotriester procedure; however, the reactions are slow and yields are low as the chain length is increased. However, aryl phosphorodichloridites react with hydroxyl groups in THF at $-78°$

$$
ROH + (1) \longrightarrow \underset{Cl}{\overset{OR}{P}}\!-\!OC_6H_4Cl \xrightarrow{R'OH} \underset{OR'}{\overset{OR}{P}}\!-\!OC_6H_4Cl \xrightarrow[H_2O]{I_2} O\!=\!\underset{OR'}{\overset{OR}{P}}\!-\!OC_6H_4Cl
$$

$$
\xrightarrow[3 \text{ hr.}, 25^0]{NaOH, H_2O} O\!=\!\underset{OR'}{\overset{OR}{P}}\!-\!OH
$$

in the presence of pyridine. The resulting phosphites are oxidized to phosphates by iodine in aqueous THF at $-10°$ in minutes. The $o$-chlorophenyl group can be removed by 0.1 $M$ NaOH at $25°$ (3 hr.). Hydroxyl groups can be blocked as the phenoxyacetyl, methoxytrityl, benzoyl, and $\beta$-benzoylpropionyl derivatives. Letsinger et al. have reported the synthesis of triesters with a $5'$-$5'$, a $3'$-$3'$, and a $3'$-$5'$ internucleotide link with this reagent in 76, 66, and 65% yields, respectively.

[1]H. Tolkmith, J. Org., 23, 1682 (1958).
[2]R. L Letsinger, J. L. Finnan, G. A. Heavner, and W. B. Lunsford, Am. Soc., 97, 3278 (1975).

**N-Chloro-N-sodiourethane,** $C_2H_5O\overset{\overset{\displaystyle O}{\|}}{C}\underset{\underset{\displaystyle Na}{|}}{N}Cl$ (1). Mol. wt. 145.52, hygroscopic.

*Preparation*[1] :

$$C_2H_5O\overset{\overset{\displaystyle O}{\|}}{C}NH_2 + Cl_2 \xrightarrow[\sim\ 50\%]{-HCl} C_2H_5O\overset{\overset{\displaystyle O}{\|}}{C}NHCl \xrightarrow[98.7\%]{\overset{\displaystyle NaOH}{CH_3OH,\ 0°}} \quad (1)$$

*6,6-Disubstituted penams; 7,7-disubstituted cephams.*[2] Reaction of the penicillanate ester (2) with (1) results in incorporation of the urethane group at $C_6$ to give (3) in high yield. The penam (3) can be converted into the cepham (5) by way of the sulfoxide (4).

(2)    $\xrightarrow[80-90\%]{(1)}$    (3)

$\xrightarrow{ClC_6H_4CO_3H}$    (4)    $\xrightarrow[50\%]{\overset{\displaystyle DMF,\ Ac_2O}{30°}}$    (5)

[1]D. Saika and D. Swern, J. Org., 33, 4548 (1968).
[2]M. M. Campbell and G. Johnson, J.C.S. Chem. Comm., 479 (1975).

**N-Chlorosuccinimide, 1,** 139; **2,** 69–70; **5,** 127–129.

*Oxidation of alcohols.*[1] In a new method for oxidation of primary and secondary alcohols to aldehydes and ketones, respectively, the alcohol is converted into an alkoxymagnesium bromide by reaction with ethylmagnesium bromide in THF at $20°$; a solution of lithium $t$-butoxide[2] (2.4 eq.) is added and then, after 1 hr., NCS (2.4 eq.) is added. The suspension is stirred for 1 hr.

before usual work-up. NBA and NBS can be used in place of NCS, but yields are lower. Yields are also lower in the absence of a base; optimum yields are obtained with lithium $t$-butoxide.

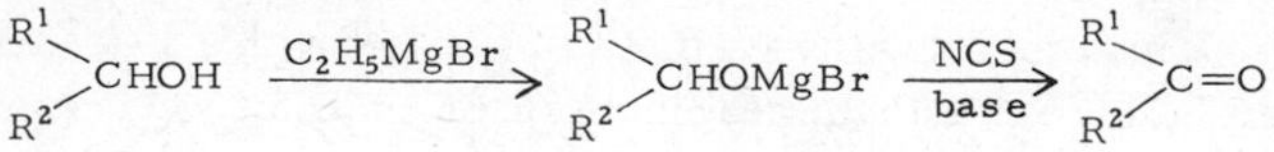

$$\underset{R^2}{\overset{R^1}{>}}CHOH \xrightarrow{C_2H_5MgBr} \underset{R^2}{\overset{R^1}{>}}CHOMgBr \xrightarrow[base]{NCS} \underset{R^2}{\overset{R^1}{>}}C{=}O$$

*Examples:*

    Cinnamyl alcohol $\longrightarrow$ cinnamyl aldehyde (98%)

    Octanol-1 $\longrightarrow$ octanal (92%)

    Octanol-2 $\longrightarrow$ octanone-2 (92%)

    Cyclohéxanol $\longrightarrow$ cyclohexanone (80%)

*Oxidative decarboxylation.* Trost and Tamura[3] have described a two-step procedure for conversion of a carboxylic acid to a ketone with loss of one carbon atom. The acid is converted into the dianion with lithium diisopropyl-amide (THF or THF–HMPT) followed by sulfenylation with dimethyl disulfide. In the second step, solid NCS is added to a mixture of the sulfenylated acid (1) and sodium bicarbonate in an alcohol (usually ethanol); $CO_2$ is evolved immediately. Aqueous work-up after 1–3 hr. gives the ketal (2), which is hydrolyzed to the ketone (3).

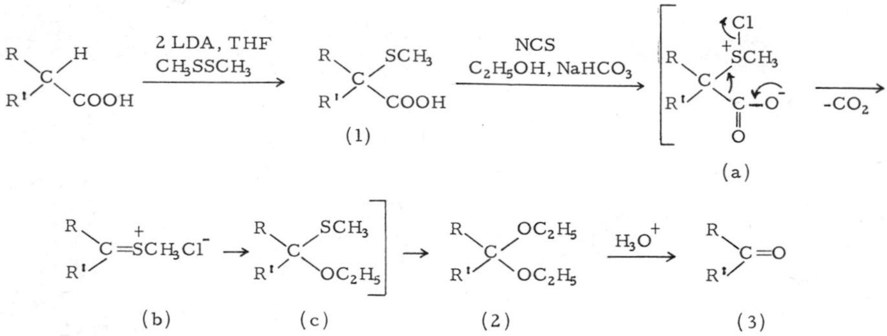

*Examples:*

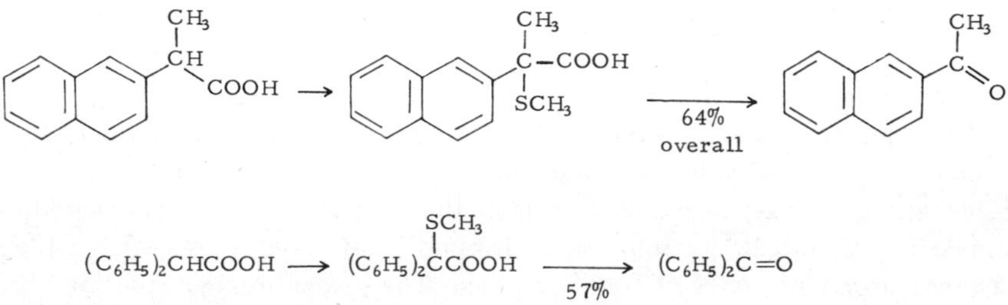

The new process was used to prepare the Corey prostaglandin intermediate (4):

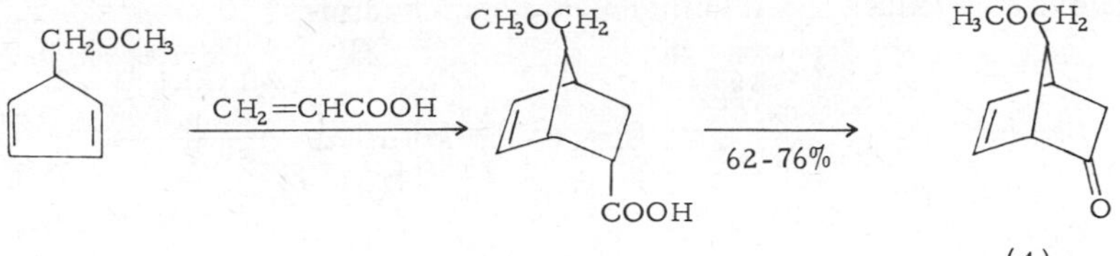

(4)

*α-Chlorination of acid chlorides.*[4]   The α-bromination of acid chlorides with NBS (3, 35) has been extended to α-chlorination with NCS. The acid chlorides can be formed *in situ*.

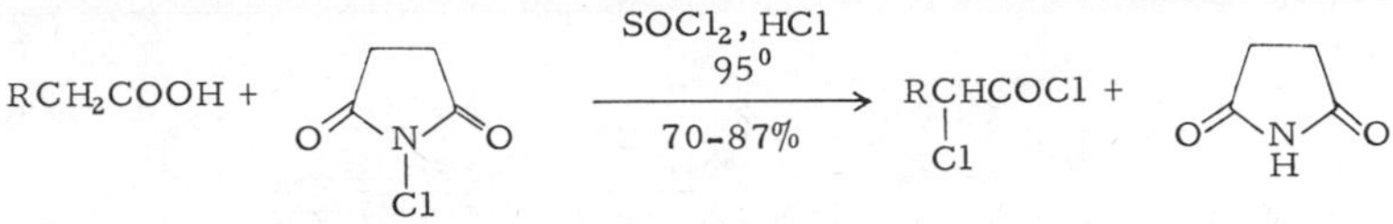

α-Iodoacyl chlorides can be prepared by reaction of acid chlorides with iodine and a trace of HI.

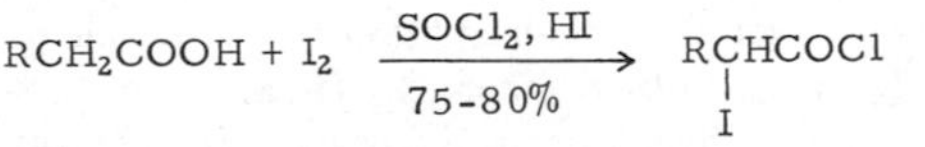

*α,α-Dihalogenated aldehydes.*   Halogenation of aliphatic aldehydes usually results in formation of acyl halides. Belgian chemists[5] have been able to prepare α,α-dihalo aldehydes by conversion to the Schiff base with *t*-butylamine in $CCl_4$ followed by treatment with 2 eq. of NCS or NBS at room temperature. The dihalogenated aldehydes are obtained on acid hydrolysis. The intermediates need not be purified.

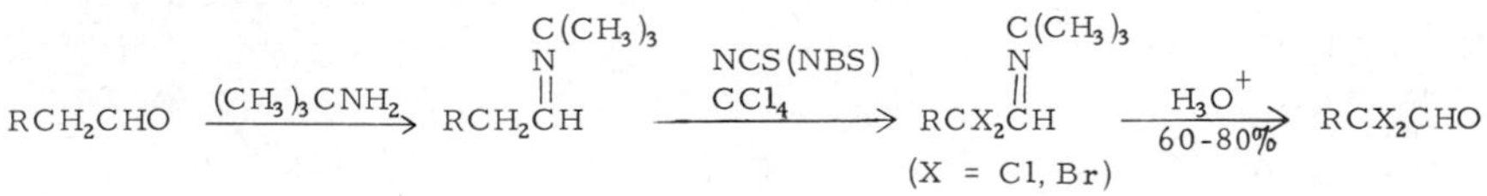

*β-Sultines* (5, 127–128).   Durst and Gimbarzevsky[6] have been able to isolate a crystalline β-sultine (1) that is stable at room temperature for several days; the sultine decomposes at 30° in $CH_2Cl_2$ to an olefin and $SO_2$.

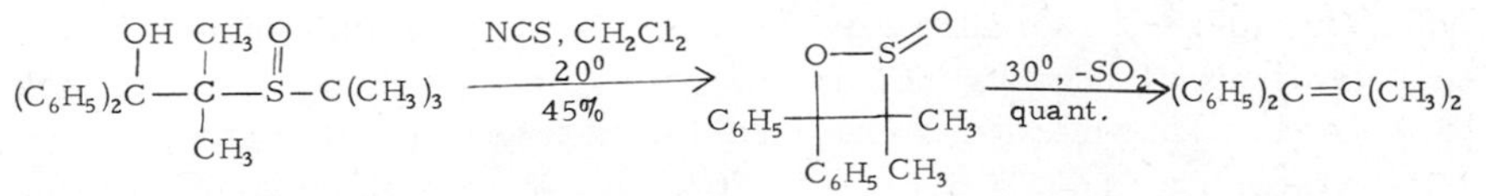

(1)

*2-Alkyl-1-t-butylaziridines.* These 1,2-disubstituted aziridines (3) can be obtained by chlorination of imines of aldehydes (1) with 2 eq. of NCS to form (2) followed by reduction with lithium aluminum hydride.[7]

$$RCH_2CH{=}N{-}C_4H_9{-}\underline{t} \xrightarrow[\substack{0-20^0 \\ 80-95\%}]{2\,NCS,\,CCl_4} R{-}\overset{\overset{\displaystyle Cl}{|}}{\underset{\underset{\displaystyle Cl}{|}}{C}}{-}CH{=}N{-}C_4H_9{-}\underline{t} \xrightarrow[\substack{ether,\,0^0 \\ {\sim}100\%}]{2\,LiAlH_4}$$

(1)                    (2)

(3)

*Solvolysis of ketene thioacetals.* Reaction of ketene thioacetals (2), prepared as shown, with NCS, an alcohol, and acetonitrile provides a useful route to α-chlorocarboxylic acid esters (3). NCS can be replaced by NBS with equally satisfactory results.[8]

(1)                    (2)                    (3)

[1] T. Mukaiyama, M. Tsunoda, and K. Saigo, *Chem. Letters,* 691 (1975).
[2] Prepared by addition of a solution of *n*-butyllithium in *n*-hexane to *t*-butanol in THF.
[3] B. M. Trost and Y. Tamura, *Am. Soc.,* 97, 3528 (1975).
[4] D. N. Harpp, L. Q. Bao, C. J. Black, and J. G. Gleason, *Tetrahedron Letters,* 3235 (1974).
[5] R. Verhé, N. De Kimpe, L. De Buyck, and N. Schamp, *Synthesis,* 455 (1975).
[6] T. Durst and B. P. Gimbarzevsky, *J.C.S. Chem Comm.,* 724 (1975).
[7] N. De Kimpe, R. Verhé, L. De Buyck, and N. Schamp, *Syn. Commun.,* 5, 269 (1975).
[8] B.-T. Gröbel, R. Bürstinghaus, and D. Seebach, *Synthesis,* 121 (1976).

## N-Chlorosuccinimide–Dialkyl sulfides, 4, 87–90.

*N-Aryl sulfimides.* N-Aryl sulfimides can be prepared in satisfactory yields by reaction of anilines with sulfides and N-chlorosuccinimide, *t*-butyl hypochlorite, or sulfuryl chloride.[1]

$$ArNH_2\ +\ S\overset{\diagup R^1}{\diagdown R^2}\ +\ NCS \xrightarrow[\substack{60-90\%}]{\substack{1)\ CH_2Cl_2,\,-20^0 \\ 2)\ NaOH}} ArN{\leftarrow}S\overset{\diagup R^1}{\diagdown R^2}$$

[1] P. K. Claus, W. Rieder, P. Hofbauer, and E. Vilsmaier, *Tetrahedron,* 31, 505 (1975).

## N-Chlorosuccinimide–Dimethyl sulfide.

*Benzofuranes.* Gassman and Amick[1] have extended their synthesis of indoles (4, 190–191) by *o*-alkylation of anilines with chlorosulfonium chlorides to a synthesis of benzofuranes by the corresponding reaction with phenols (5, 131–132). Thus reaction of the phenol salt derived from NCS and 1-methylthio-2-propanone (1) in $CH_2Cl_2$ at $-70^0$ gives the azasulfonium salt (2, not isolated). Addition of triethylamine gives (3) in low yield through a series of intermediates discussed previously. The final step involves desulfurization to 2-methylbenzofurane (4). Although the overall yield is low, this procedure is probably general.

(1)                                                     (2, not isolated)

(3)                                    (4)

[1]P. G. Gassman and D. R. Amick, *Syn. Commun.*, 325 (1975).

**N-Chlorosuccinimide–1,3-Dithiane complex,** (1). Mol. wt. 253.77.

The reagent is prepared *in situ* by the reaction of NCS and 1,3-dithiane at −70° in methylene chloride.

*o-Formylation of phenols.* Gassman and Amick[1] have used this complex (1) for *ortho*-formylation of phenols. The method is related to a selective *o*-formylation of anilines (5, 131). The probable mechanism is formulated. Overall yields range from 20 to 35%, being lowest with phenol itself.

(a)                                    (b)

(c)                     (2)                     (3)

An alternative approach is also feasible. The phenol is treated with the complex of NCS and dimethyl sulfide (4, 87–90) to give the azasulfonium salt (d), which is converted as shown to the *o*-methylthiomethyl phenol (4).[2]

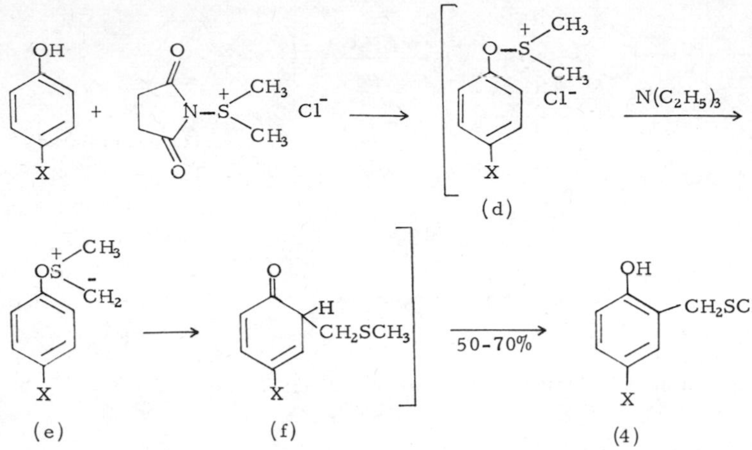

(d)

(e)    (f)    50–70%    (4)

These products (4) can be converted into *o*-formylphenols by protection of the hydroxyl group, oxidation with NCS, and hydrolysis of the oxidation product (without isolation) with aqueous THF at 100°.

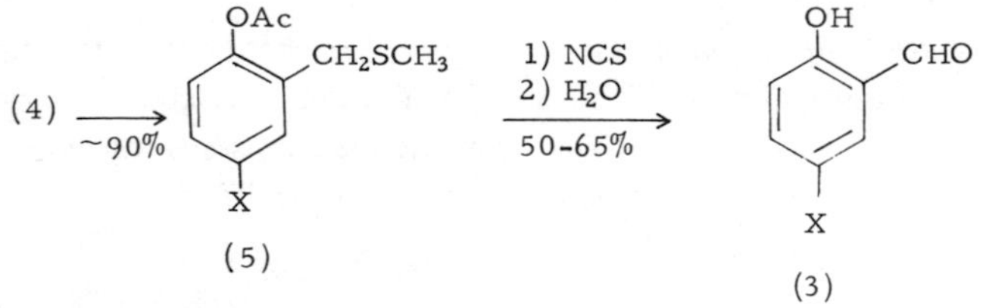

(4) →~90%→ (5) →1) NCS 2) H₂O / 50–65%→ (3)

Overall yields obtained by the second method are generally somewhat higher than those obtained by the first method. In addition, the second route is less sensitive to the nature of the *para*-substituent.

[1] P. G. Gassman and D. R. Amick, *Tetrahedron Letters*, 3463 (1974).
[2] *Idem, ibid.*, 889 (1974).

## N-Chlorosuccinimide–Triethylamine.

*Oxidation.* Durst and co-workers[1] find that this combination rapidly oxidizes catechols to *ortho*-quinones and hydroquinones to *para*-quinones at −25 to 0°. Based on NMR and IR, yields are quantitative. In addition, anthrone is converted into bianthrone (75% isolated yield) and benzophenone hydrazone is oxidized to diphenyldiazomethane.

The combination of NCS, dimethyl sulfide, and triethylamine has been used for oxidation of catechols and hydroquinones to quinones.[2] The present report suggests that dimethyl sulfide is probably not necessary.

Aromatic hydroxyl groups, unlike aliphatic hydroxyl groups (4, 88–89), do not react with NCS and dimethyl sulfide in the absence of a base.

[1]H. D. Durst, M. P. Mack, and F. Wudl, *J. Org.*, **40**, 268 (1975).
[2]J. P. Marino and A. Schwartz, *J.C.S. Chem. Comm.*, 812 (1974).

**Chlorosulfuric acid, 1,** 140–141; **2,** 70.

*4,9-Disubstituted diamantanes.*[1] The tetrahydro derivative (1) of Bisnor-S (**2,** 123–124; **3,** 7; **4,** 140) on treatment with neat chlorosulfuric acid (mole ratio 1:5.7) at −15° is converted into 4,9-dichlorodiamantane (2) in 82% isolated yield. The mechanism for formation of (2) is not known, but it does not

(2)

involve formation of diamantane, which is formed as a minor product. The dichloro derivative (2) was converted into the corresponding diol, dicarboxylic acid (m.p. 456°), and diamine.

[1]F. Blaney, D. E. Johnston, M. A. McKervey, and J. J. Rooney, *Tetrahedron Letters*, 99 (1975).

**Chlorosulfonyl isocyanate (CSI)**, **1**, 117–118; **2**, 70; **3**, 51–53; **4**, 90–94; **5**, 132–136.

*Addition to aromatic ketones.*[1] An extensive study of the reaction of CSI with aromatic ketones (1) presents evidence that the first step is formation of a β-keto carboxamide (2), which can react with another molecule of CSI to form a malonamide (3). Generally, however, (2) reacts with CSI to form, after reductive hydrolysis with sodium sulfite, either an oxathiazine 2,2-dioxide (4) and/or an oxazinedione (5). The ratio of (4) to (5) depends on the keto-enol equilibrium of the intermediate (2) and on the reaction conditions.

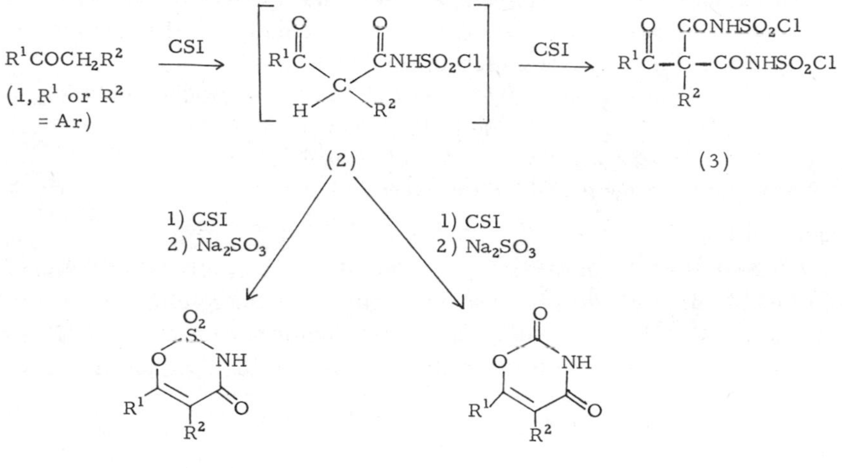

(2)

(3)

(4)

(5)

*Addition to vinyl esters.*[2] The reagent reacts with vinyl esters (1) to give, after reduction, β-lactams (2) in 35–65% yield.

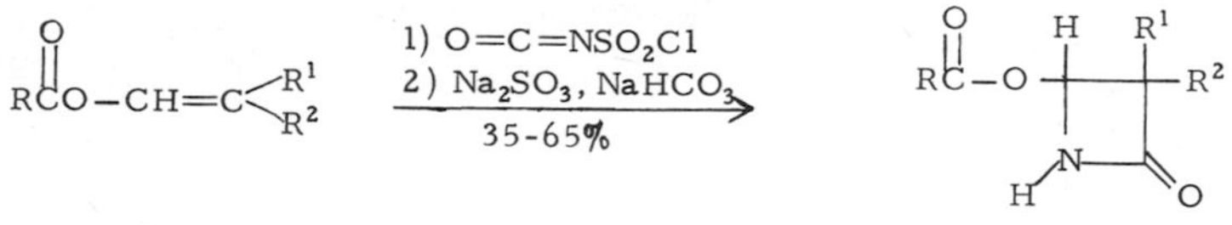

(1)

(2)

[1] A. Hassner and J. K. Rasmussen, *Am. Soc.*, **97**, 1451 (1975).
[2] K. Clauss, D. Grimm, and G. Prossel, *Ann.*, 539 (1974).

**1-Chloro-N,N,2-trimethylpropenylamine**    (**N,N-Dimethyl-1-chloro-2-methyl-1-propenylamine**), **4**, 94–95; **5**, 136–138.

*Preparation* (**4**, 94).[1] Details for the preparation have been submitted to *Organic Synthesis*. The phosgenation of amides is general; *caution*: the reaction can be exothermic. With less reactive amides DMF can be used as catalyst. All α-haloenamines are highly hygroscopic.

*Cyclobutenones.*[2] The salts (1), obtained by reaction of the amine with AgBF$_4$ or ZnCl$_2$, form [2 + 2] cycloadducts (2) at 20° with acetylenes in high yield. These adducts are hydrolyzed in >90% yield by dilute NaOH to cyclobutenones (3). The salts (1) also undergo cycloaddition with acetylene itself, but

$$(CH_3)_2 C=C=\overset{+}{N}(CH_3)_2 \ X^-$$

$$(1, \ X=BF_4^-, \ ZnCl_3^-)$$

77-80% | HC≡CH

$$\xrightarrow[>80\%]{R-C\equiv C-R}$$

(2)

$$\xrightarrow[90-100\%]{NaOH, \ H_2O}$$

(3)

(4)                    84%                    (5)

the adduct undergoes ring opening on alkaline hydrolysis. Two regioisomers are formed in the addition of (1) to 1-alkynes, RC≡CH.

The cyclobutenylideneammonium salts (2) and (4) are reactive dienophiles. Thus (4, X = BF$_4^-$) reacts with cyclopentadiene at 20° to give the adduct (5) in 84% yield. The adduct can be hydrolyzed to the corresponding cyclobutanone.

[1] B. Haveaux, M. Rens, A. R. Sidani, J. Toye, and L. Ghosez, *Org. Syn.*, submitted (1976).
[2] C. Hoornaert, A. M. Hesbain-Frisque, and L. Ghosez, *Angew. Chem. internat. Ed.*, **14**, 569 (1975).

**Chromic acid**, 1, 142–144; 2, 70–72; 3, 54; 4, 95–96; 5, 138–140.

*Jones reagent* (1, 142–143). Primary allylic or benzylic alcohols are oxidized in high yield with chromic acid in acetone.[1] Thus cinnamyl alcohol is oxidized to cinnamaldehyde in 84% yield, and benzaldehyde is obtained from benzyl alcohol in 76% yield. Manganese dioxide has usually been used in such oxidations.

Terminal olefins are oxidized by Jones reagent in the presence of catalytic amounts of mercuric acetate to methyl ketones in 80–90% yield. The mercury salt is essential (probable mechanism shown in I); salts of other metals are less efficient. Mercuric propionate can also be used.

$$(I) \qquad RCH=CH_2 \ + \ HgOAc^+ \ \underset{}{\overset{H_2O}{\rightleftharpoons}} \ RCHCH_2HgOAc \ \xrightarrow{[O]} \ RCCH_2HgOAc$$

$$\xrightarrow[80-90\%]{H^+} \ RCCH_3 \ + \ HgOAc^+$$

1,2-Disubstituted olefins are oxidized to ketones in variable yields. In this case yields can be improved by use of sodium dichromate as oxidant and trifluoro-acetic acid as acid.[2]

$$\underline{cis}\text{-2-Octene} \longrightarrow \underset{(64\%)}{\text{2-Octanone}} \quad + \quad \underset{(36\%)}{\text{3-Octanone}}$$

$$\text{Cyclohexene} \xrightarrow[41\%]{} \text{Cyclohexanone}$$

This procedure is a useful alternative to Wacker oxidation[3] of olefins to ketones.

Steroidal 18(20)-nitrones are cleaved by Jones reagent in 1:1 (v/v) aqueous acetone; for example, (1) → (2).[4]

(1, 57 mg.)                    (2, 46 mg.)

*Deoximation.* Jones reagent and chromic anhydride–pyridine (**3**, 55–56; **4**, 96) have been added to the known reagents for oxidative deoximation. The former reagent was used in acetic acid rather than acetone for solubility reasons. Periodic acid also can be used, but it sometimes gives rise to iodinated products.[5]

*Butyrolactones.*[6] Cyclobutanols and cyclobutanones are oxidized by potassium dichromate in aqueous $H_2SO_4$ to butyrolactones in yields of 50–90%. Acetic acid is added if necessary to effect solution.

[1] K. E. Harding, L. M. May, and K. F. Dick, *J. Org.*, **40**, 1664 (1975).
[2] H. R. Rogers, J. X. McDermott, and G. M. Whitesides, *J. Org.*, **40**, 3577 (1975).
[3] E. W. Stern, *Transition Metals in Homogeneous Catalysis*, G. Schranzer, Ed., Dekker, New York, 1972, Chapt. 4.
[4] D. H. R. Barton, N. K. Basu, M. J. Day, R. H. Hesse, M. M. Pechet, and A. N. Starratt, *J.C.S. Perkin I*, 2243 (1975).
[5] H. C. Araújo, G. A. L. Ferreira, and J. R. Mahajan, *J. C. S. Perkin I*, 2257 (1974).
[6] R. Jeanne-Carlier and F. Bourelle-Wargnier, *Tetrahedron Letters*, 1841 (1975).

**Chromium hexacarbonyl**, $Cr(CO)_6$, 5, 142–143.

*Cr(CO)$_3$(ArH) complexes.* Many aromatic compounds react with chromium hexacarbonyl, preferably in diethylene glycol dimethyl ether as solvent, to form tricarbonylchromium complexes; yields are generally in the range of 40–90%. Arenes substituted with the groups COOH, CHO, CN, and $NO_2$ fail to form complexes. The complexes from phenol and hydroquinone are sensitive to oxidation. However, generally the complexes are stable, yellow or orange, and crystalline. The rate of formation of the complexes is retarded by electron-attracting groups and by bulky substituents (*t*-butylbenzene reacts very slowly).[1]

*Alkylation of arenes.*[2] Arenes can be activated to alkylation by complexation with the $Cr(CO)_3$ group, which is strongly electron withdrawing. For example, treatment of acetophenone itself with methyl iodide and sodium hydride in DMF gives $C_6H_5COCH_2CH_3$, $C_6H_5COCH(CH_3)_2$, and starting material.

$$(CO)_3CrC_6H_5\overset{O}{\overset{\|}{C}}CH_3 \xrightarrow[90\%]{\substack{NaH,\ CH_3I\\DMF,\ 25°}} (CO)_3CrC_6H_5\overset{O}{\overset{\|}{C}}CH(CH_3)_2 \longleftarrow (CO)_3CrC_6H_5\overset{O}{\overset{\|}{C}}CH_2CH_3$$

(1)                    (2)                    (3)

$$\xrightarrow[DMSO]{KOC(CH_3)_3,\ CH_3I} (CO)_3CrC_6H_5\overset{O}{\overset{\|}{C}}C(CH_3)_3 \longleftarrow$$

(4)

When complexed with $Cr(CO)_3$ as in (1), the product of methylation is (2), obtained in 90% yield. This is also the product of similar methylation of (3). Complexes (1), (2), and (3) are all converted into (4) by methylation in DMSO with potassium *t*-butoxide as base.

An even more striking example is furnished by methyl phenylacetate. This ester is inert to $CH_3I$ and NaH in DMF at 25°, but complex (5) is converted into (6) under these conditions in 97% yield. Complex (5) has also been converted into the cyclobutane derivative (7).

$$(CO)_3CrC_6H_5CH_2COOCH_3 \xrightarrow[97\%]{\substack{CH_3I,\ NaH\\DMF}} (CO)_3CrC_6H_5\overset{CH_3}{\underset{CH_3}{\overset{|}{\underset{|}{C}}}}COOCH_3$$

(5)                                    (6)

$$\xrightarrow[87\%]{\substack{Br(CH_2)_3Br\\NaH,\ DMF}}$$

(7)

$$\text{(CO)}_3\text{CrC}_6\text{H}_5\text{CH}_2\text{CH}_2\text{COOCH}_3 \xrightarrow[\text{DMF},25^\circ]{\text{CH}_3\text{I, NaH}} \left\{ \begin{array}{c} \text{(CO)}_3\text{CrC}_6\text{H}_5\text{CH}_2\overset{\overset{\text{CH}_3}{|}}{\text{C}}\text{HCOOCH}_3 \\ (9,\ 65\%) \\ + \\ \text{(CO)}_3\text{CrC}_6\text{H}_5\text{CH}_2\overset{\overset{\text{CH}_3}{|}}{\underset{\underset{\text{CH}_3}{|}}{\text{C}}}\text{COOCH}_3 \\ (10,\ 10\%) \end{array} \right.$$

$$\text{(8)}$$

Alkylation of (8) shows that the carbon $\beta$ to the phenyl group rather than $\alpha$ is specifically activated to alkylation.

These products can be decomplexed by exposure in ether solution to sunlight in air.[3]

[1] B. Nicholls and M. C. Whiting, *J. Chem. Soc.*, 551 (1959).
[2] G. Jaouen, A. Meyer, and G. Simonneaux, *J.C.S. Chem. Comm.*, 813 (1975).
[3] G. Jaouen and R. Dabard, *Tetrahedron Letters*, 1015 (1971).

**Chromyl chloride, 1, 151; 2, 79; 3, 62; 4, 98–99; 5, 144–145.**

*Oxidation of alcohols.* Sharpless and Akashi[1] have prepared an effective oxidation reagent by addition of chromyl chloride[2] to 3 eq. of pyridine in methylene chloride at $-78^\circ$. This material oxidizes primary alcohols to aldehydes even at $-70^\circ$ and only 1.1 eq. of reagent is necessary (Collins oxidation requires about 6 eq. of oxidant, **4,** 216). Yields of simple, saturated aldehydes are high, 90% or better. Unfortunately, this method is so vigorous that it is only suitable for small-scale (10 mmole) oxidations. When the chromyl chloride–pyridine reagent is modified by addition of 2 eq. of *t*-butanol, a new reagent (1) is formed that is suitable for small- and large-scale oxidations. For example, 1-decanol (110 mmole) has been oxidized with this reagent to 1-decanal

$$\text{(1)}$$

in 79% yield. Reagent (1) is presumably the pyridine complex of Oppenauer's *t*-butyl chromate (**1,** 86–87). However, use of material prepared according to Oppenauer's procedure and in the presence of pyridine gives only low yields of aldehydes.

This chromyl chloride based reagent is inferior to Collins reagent for oxidation of allylic alcohols since *cis–trans* isomerization occurs. It has also been found to be unsatisfactory for oxidation of *m*-hydroxybenzyl alcohol (38%

yield of aldehyde). However, it is useful for oxidation of simple, saturated, primary alcohols on a large scale.

[1] K. B. Sharpless and K. Akashi, *Am. Soc.*, **97**, 5927 (1975).
[2] Commercial material (Alfa, ROC/RIC) can be used without purification. The reagent can also be prepared by the method described in Brauer's *Handbook of Preparative Inorganic Chemistry*, Vol. 2, Academic Press, New York, 1965, p. 1384.

### Claisen's alkali, 1, 153–154.

*Claisen rearrangements.*[1] A recent review stresses the value of Claisen's alkali for separation of neutral and phenolic products of Claisen rearrangements. Extraction with aqueous sodium hydroxide is effective for fairly acidic phenols, but incomplete for weakly acidic phenols.

[1] S. J. Rhoads and N. R. Raulins, *Org. React.*, **22**, 1 (1975).

### Cobalt(III) acetate, $Co(OOCCH_3)_3$. Mol. wt. 236.07.

*Oxidative decarboxylation.*[1] Arylacetic acids are converted into benzyl acetates by cobalt(III) acetate in refluxing acetic acid containing added copper(II) acetate. The reaction was shown to involve benzylic radical cation

$$R \cdot C_6H_4CH_2COOH \xrightarrow[\substack{HOAc, \Delta \\ 70-100\%}]{Co(OAc)_3, Cu(OAc)_2} R \cdot C_6H_4CH_2OCOCH_3$$

$$(R = H, OCH_3, CH_3, Cl)$$

intermediates.

Under the same conditions $\gamma$-phenylbutyric acid (1) is converted into $\gamma$-phenylbutyrolactone (2).

$$C_6H_5CH_2CH_2CH_2COOH \xrightarrow{50\%}$$

(1)

(2)

[1] R. M. Dessau and E. I. Heiba, *J. Org.*, **40**, 3647 (1975).

### Cobalt(II) chloride, $CoCl_2$. Mol. wt. 129.84. Suppliers: Alfa, Fisher, ROC/RIC.

*Isolation of a 1-hydroxypyrazole 2-oxide by chelation.* Nitrosation of benzalacetone oxime with *n*-butyl nitrite in aqueous ethanol containing pyridine and $CoCl_2$ (0.5 eq.) leads to the chelate (1), a violet solid, which on treatment with acid gives the oxide (2).[1] The oxide had not been isolated previously[2] because under nitrosating conditions it is converted into (3) and (4).

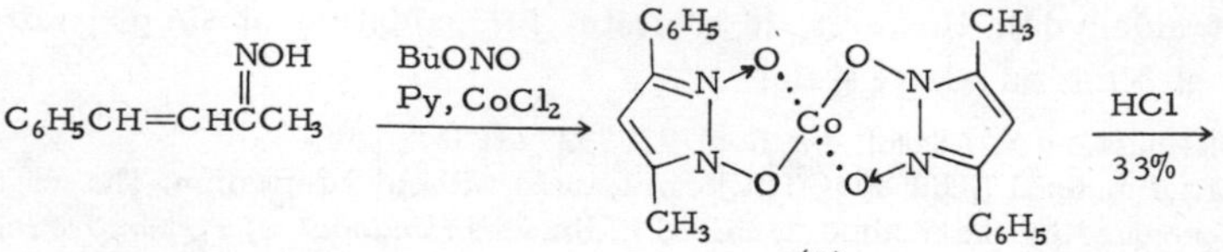

$$C_6H_5CH=CHCCH_3 \quad \overset{NOH}{\underset{}{\|}} \quad \xrightarrow[\text{Py, CoCl}_2]{\text{BuONO}}$$

(1)

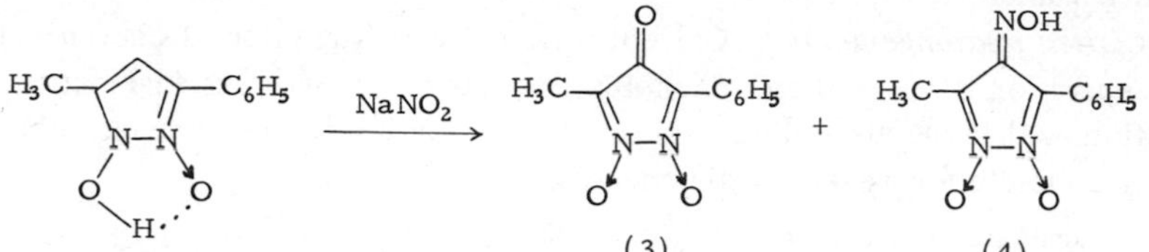

$$\xrightarrow[33\%]{\text{HCl}}$$

(2)                    (3)                    (4)

$\xrightarrow{\text{NaNO}_2}$

+

[1] J. F. Hansen and D. E. Vietti, *J. Org.*, **40**, 816 (1975).
[2] J. P. Freeman and J. J. Gannon, *ibid.*, **34**, 194 (1969).

**Cobaltocene (Di-$\pi$-cyclopentadienylcobalt),** 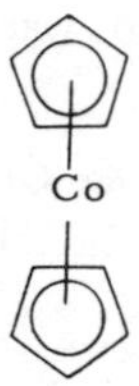 (1). Mol. wt. 189.12. Sup-

pliers: Aldrich, Orgmet, ROC/RIC.

*Pyridine synthesis.*[1] This organometallic reagent catalyzes the reaction of acetylene and monosubstituted acetylenes with nitriles to form pyridines (equations I and II). Disubstituted acetylenes do not undergo this reaction.

(I) 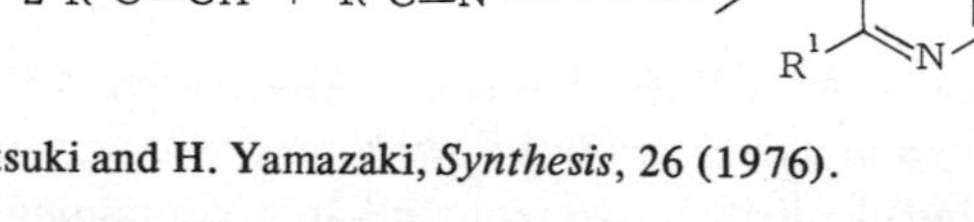    $2\ HC{\equiv}CH + RC{\equiv}N \xrightarrow[\text{30--70\%}]{\overset{(C_5H_5)_2Co}{150^0}}$

(II) 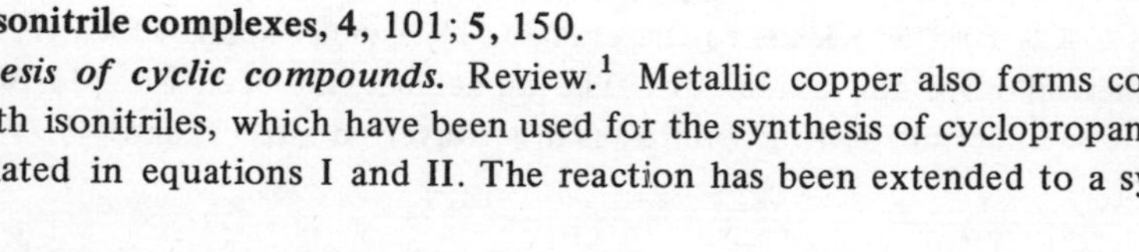    $2\ R^1C{\equiv}CH + R^2C{\equiv}N \xrightarrow{\overset{(C_5H_5)_2Co}{150^0}}$ +

[1] Y. Wakatsuki and H. Yamazaki, *Synthesis*, 26 (1976).

**Copper–Isonitrile complexes, 4, 101; 5, 150.**

*Synthesis of cyclic compounds.* Review.[1] Metallic copper also forms complexes with isonitriles, which have been used for the synthesis of cyclopropanes, as formulated in equations I and II. The reaction has been extended to a syn-

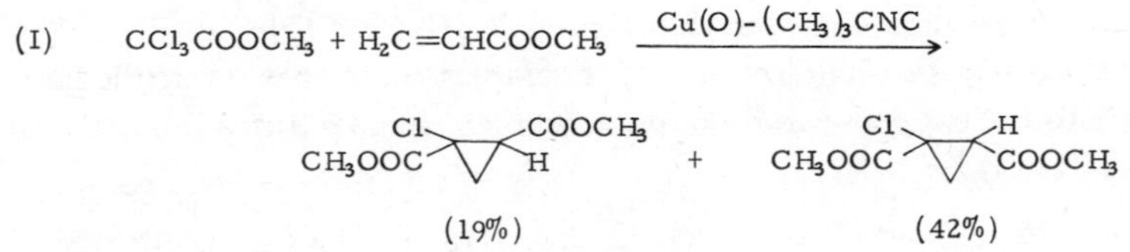

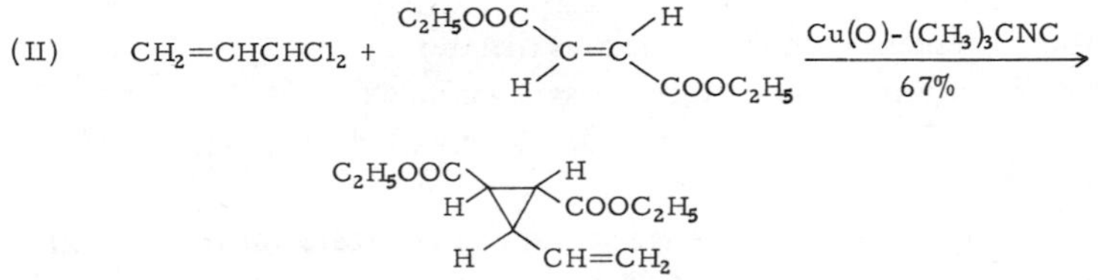

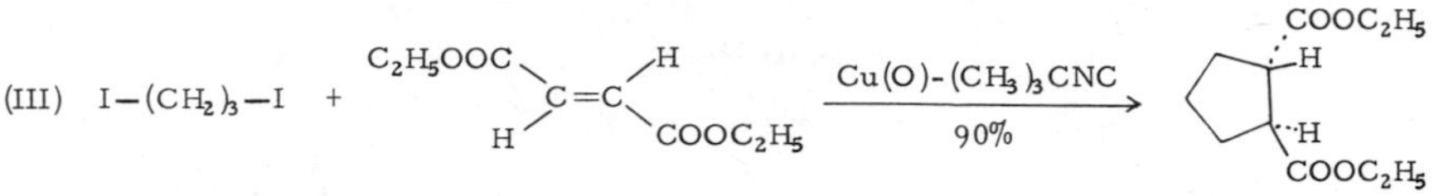

(I)    $CCl_3COOCH_3 + H_2C=CHCOOCH_3 \xrightarrow{Cu(O)-(CH_3)_3CNC}$

(19%)                (42%)

(II)    $CH_2=CHCHCl_2 +$

$\xrightarrow[67\%]{Cu(O)-(CH_3)_3CNC}$

(III)    $I-(CH_2)_3-I +$

$\xrightarrow[90\%]{Cu(O)-(CH_3)_3CNC}$

thesis of cyclopentanedicarboxylates (equation III). These reactions are considered to involve copper carbenoid intermediates.

*Cyclopentanecarboxylates.* These substances can be prepared by the reaction of 1,3-diiodopropane with an $\alpha,\beta$-unsaturated ester in the presence of copper and an isonitrile ($t$-butyl isonitrile or cyclohexyl isonitrile).[2]

*Example:*

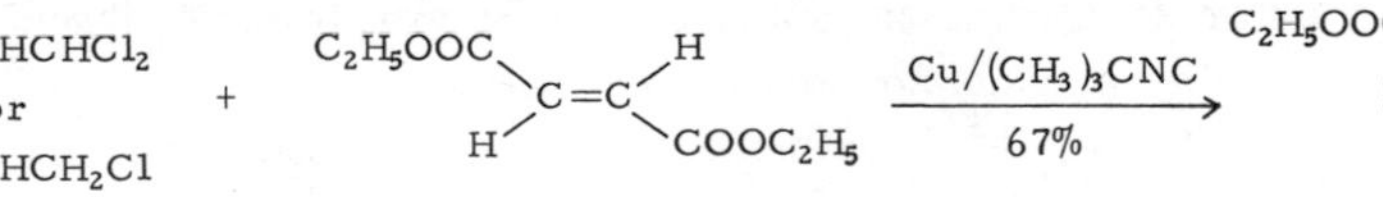

$I(CH_2)_3I + CH_2=CHCOOCH_3 \xrightarrow[58\%]{Cu, (CH_3)_3CNC \\ 80^\circ, 12\ hr.}$

*Vinylcyclopropanes.*[3] Vinylcyclopropanes can be obtained in fair to moderate yields by the reaction of 1,3 or 3,3-dichloropropenes with $\alpha,\beta$-unsaturated esters in the presence of this complex in benzene at $80^\circ$.

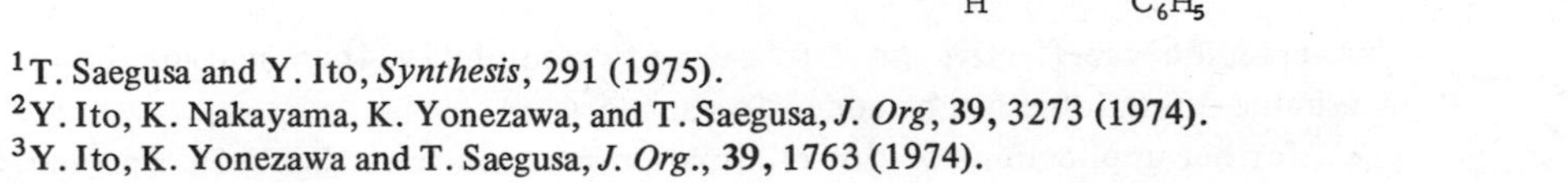

$CH_2=CHCHCl_2$
or                +
$ClCH=CHCH_2Cl$

$\xrightarrow[67\%]{Cu/(CH_3)_3CNC}$

$+ CH_2=CHCOOCH_3 \xrightarrow{66\%}$

[1] T. Saegusa and Y. Ito, *Synthesis*, 291 (1975).
[2] Y. Ito, K. Nakayama, K. Yonezawa, and T. Saegusa, *J. Org*, **39**, 3273 (1974).
[3] Y. Ito, K. Yonezawa and T. Saegusa, *J. Org.*, **39**, 1763 (1974).

**Copper(I) phenylacetylide–Tri-*n*-butylphosphine,** $C_6H_5C\equiv CCu \cdot P(n\text{-}Bu)_3$.

*Reversible carbon dioxide fixation.* This complex undergoes reversible insertion of $CO_2$ into the copper—carbon bond at ambient temperatures and ordinary pressure (equation I).[1]

$$(I) \quad C_6H_5C\equiv CCu \cdot P(n\text{-}Bu)_3 + CO_2 \rightleftharpoons C_6H_5C\equiv CCO_2Cu \cdot P(n\text{-}Bu)_3$$

[1] T. Tsuda, Y. Chujo, and T. Saegusa, *J. C. S. Chem. Comm.* 963 (1975).

**Copper(I) trifluoromethanesulfonate (Cuprous triflate), 5,** 151–152.

The salt can be prepared,[1] undoubtedly as a complex with acetonitrile, by heating cupric triflate[2] with copper powder at reflux in acetone containing some acetonitrile.

*Ullmann reaction.*[1] When *o*-iodonitrobenzene (1) is treated at room temperature with cuprous triflate in acetone solution containing 5% aqueous ammonia,

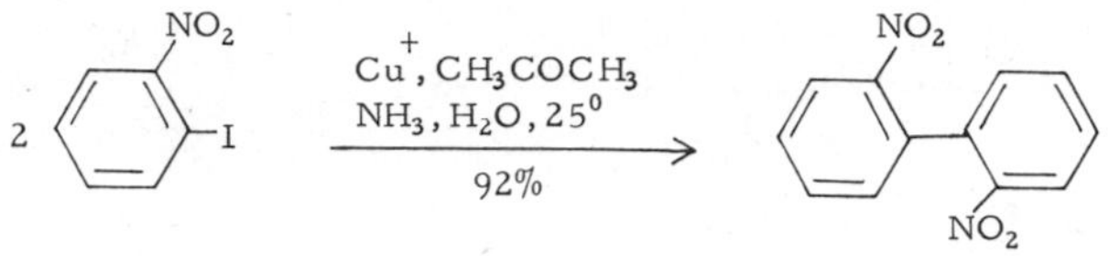

$$2 \quad (1) \quad \xrightarrow[\displaystyle \underset{92\%}{\text{NH}_3,\text{H}_2\text{O},25^0}]{\overset{+}{\text{Cu}},\text{CH}_3\text{COCH}_3} \quad (2)$$

$2,2'$-dinitrobiphenyl (2) is formed in less than 5 min. in 92% yield. No reaction occurs in refluxing solvent after 24 hr. in the absence of ammonia. No reaction occurs at room temperature when cuprous triflate is replaced by copper powder or cuprous oxide. However, tetrakisacetonitrilecopper(I) perchlorate[3] is as effective as cuprous triflate.

*o*-Bromonitrobenzene reacts more slowly than (1) in this Ullmann-like reaction, and only 15% of (2) is formed. The other major product is *o*-nitroaniline. *p*-Iodonitrobenzene, *o*-iodofluorobenzene, *o*-iodoanisole, and iodobenzene fail to react at 25°. Methyl *o*-iodobenzoate reacts, but is converted into methyl anthranilate (80% yield).

The reactivity order is identical to that observed in the classical Ullmann reaction.[4] However, the present conditions are unusually rapid. The ready displacement of iodine by ammonia is also noteworthy.

*Vinyl phenyl sulfides.*[5] Treatment of diphenyl thioacetals and thioketals with this soluble Cu(I) salt induces elimination of thiophenol at 25°:

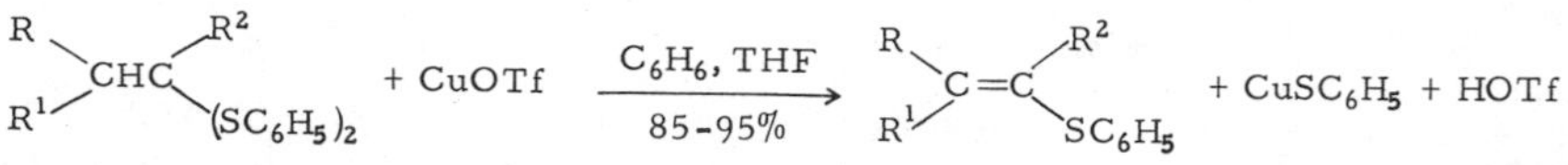

Cuprous iodide is ineffective, possibly owing to insolubility. If an acid medium is undesirable 2,6-lutidine can be added to the reaction.

Two further applications of this elimination reaction have also been reported.

One is a high-yield synthesis of *trans*-stilbene (equation I); the other is a synthesis of a furane (equation II).

(I)    $C_6H_5CH_2Br + C_6H_5\overset{\underset{|}{Li}}{C}HSC_6H_5 \longrightarrow C_6H_5\overset{\underset{|}{C}H_2C_6H_5}{C}HSC_6H_5 \xrightarrow[>95\%]{CuOTf} \overset{C_6H_5}{\underset{H}{}}C=C\overset{H}{\underset{C_6H_5}{}}$

(II)    $[\,C_6H_5C(SC_6H_5)_2\,]_2CuLi + CH_3COCH=CH_2 \longrightarrow C_6H_5C(SC_6H_5)_2CH_2CH_2COCH_3$

$$\xrightarrow[80\%]{CuOTf}$$

This elimination reaction has been applied[6] to the adducts (2) obtained by reaction of lithio dithiophenoxymethane (1)[7] with aldehydes or ketones. Treatment of (2) with cuprous triflate and a base gives the homologous α-thiophenoxy ketone (3) with preferential migration either of hydrogen or of the more highly substituted alkyl group.

$(C_6H_5S)_2CHLi \quad + \quad \overset{R^1}{\underset{R^2}{}}C=O \xrightarrow[80-90\%]{} \overset{HO}{\underset{R^1}{}}\underset{R^2}{C}\overset{CH(SC_6H_5)_2}{} \xrightarrow[-CuSC_6H_5]{\overset{CuOTf,\ 78^0}{C_2H_5N(i-Pr)_2}}$

(1)                                                        (2)

$$\left[\ \overset{HO}{\underset{R^1}{}}\underset{R^2}{C}\overset{\overset{+}{C}HSC_6H_5}{\phantom{x}}\ {}^-OTf\ \right] \xrightarrow[65-85\%]{-HOTf} \overset{O}{R^1C}\overset{\parallel}{C}H\overset{R^2}{\underset{SC_6H_5}{}}$$

(a)                                        (3)

Application of this sequence to alicyclic four-, five-, and six-membered ketones results in ring expansion, as illustrated for the case of cyclopentanone (4). An intermediate in the conversion of (5) to (6) was detected by thin layer chromatography and was eventually isolated in the case of the adduct (7) from (1)

(4)                                 (5)                                 (6)

and cycloheptanone. The intermediate (8) is an α-epoxythioether and can also be prepared by reaction of cycloheptanone with the ylide (11). If the reaction of (7) with cuprous triflate is conducted at 62°, 1-thiophenoxycycloheptanecarboxaldehyde (9) is obtained as the major product.

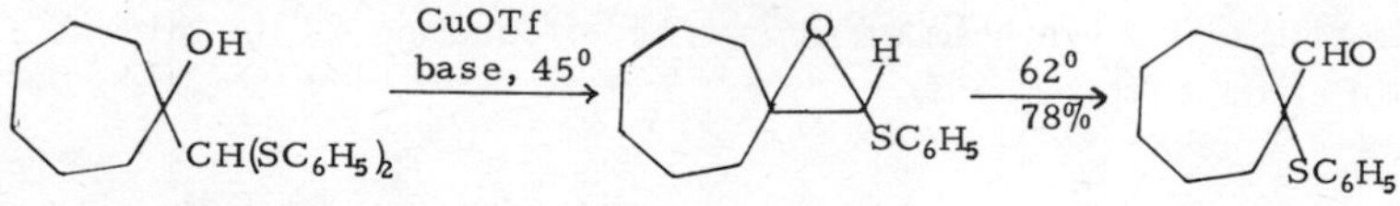

(7)                          (8)                            (9)

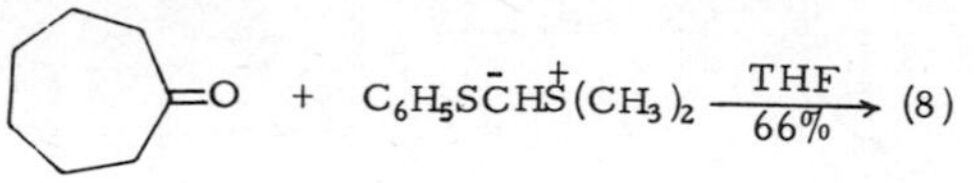

(10)                (11)

*Reductive coupling, hydrolysis, and dehalogenation.*[8] Homogeneous Ull-mann coupling can be conducted with copper(I) triflate in acetone–acetonitrile containing aqueous ammonia (equation I).[9]

(I)   2

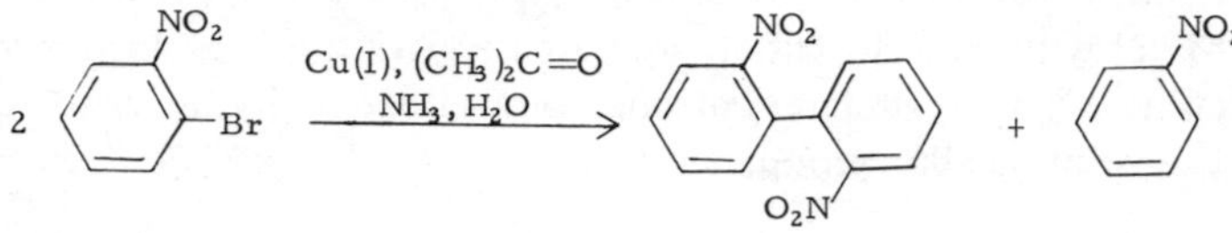

(90%)              (8%)

(II)

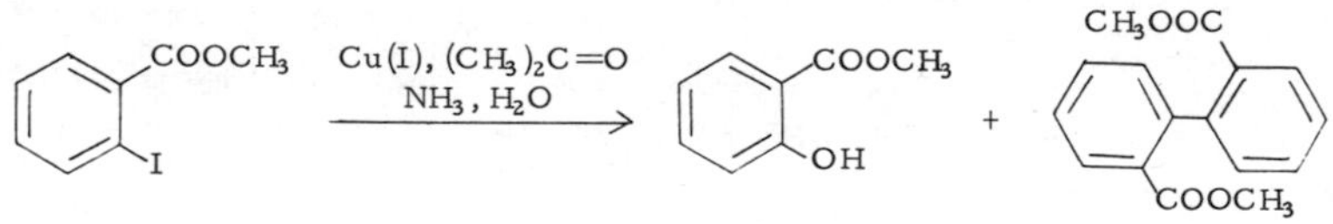

(main)              (minor)

(III)

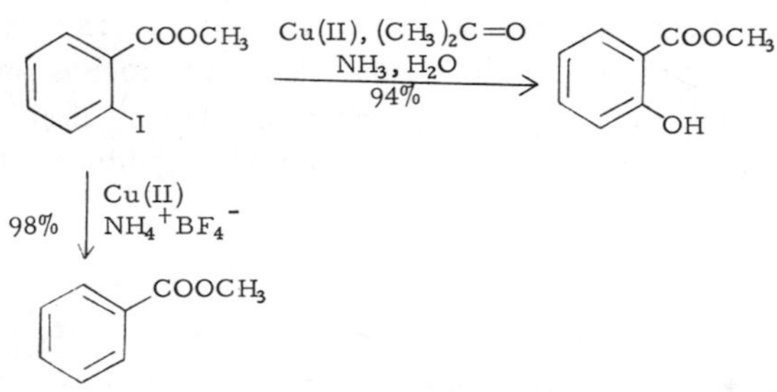

Methyl *o*-iodobenzoate is converted under these conditions (equation II) into methyl salicylate (main product) and dimethyl diphenate (minor product). If the cuprous salt is replaced by the cupric salt, hydrolysis is essentially the only re-action (equation III). If ammonium tetrafluoroborate is added, methyl benzoate is the only product.

Cuprous triflate can also be used for coupling of vinyl halides (*cf.* **Use of copper powder, 4, 102–103**). In this case acetone saturated with ammonia gas is superior to aqueous ammonia.

$$
2\ \underset{H}{\overset{C_2H_5OOC}{>}}C=C\underset{COOCH_3}{\overset{I}{<}}\ \xrightarrow[\substack{NH_3 \\ 93\%}]{Cu(I),\ (CH_3)_2C=O}\ \underset{C_2H_5OOC}{\overset{CH_3OOC}{>}}C=C\ C=C\underset{COOCH_3}{\overset{H}{<}}\underset{COOC_2H_5}{}
$$

$$
2\ \underset{H}{\overset{C_2H_5OOC}{>}}C=C\underset{I}{\overset{COOCH_3}{<}}\ \xrightarrow[\substack{NH_3,\ 40^\circ \\ 86\%}]{Cu(I),\ (CH_3)_2C=O}\ \underset{H}{\overset{CH_3OOC}{>}}C=C\ C=C\underset{COOCH_3}{\overset{COOC_2H_5}{<}}\underset{}{C_2H_5OOC}
$$

[1] T. Cohen and J. G. Tirpak, *Tetrahedron Letters*, 143 (1975).
[2] C. L. Jenkins and J. K. Kochi, *Am. Soc.*, **94**, 843 (1972).
[3] B. J. Hathaway, D. G. Holah, and J. D. Postlethwaite, *J. Chem. Soc.*, 3215 (1961).
[4] P. E. Fanta, *Synthesis*, 9 (1974).
[5] T. Cohen, G. Herman, J. R. Falck, and A. J. Mura, Jr., *J. Org.*, **40**, 812 (1975).
[6] T. Cohen, D. Kuhn, and J. R. Falck, *Am. Soc.*, **97**, 4749 (1975).
[7] E. J. Corey and D. Seebach, *J. Org.*, **31**, 4097 (1966).
[8] T. Cohen, and I. Cristea, *J. Org.*, **40**, 3649 (1975).
[9] Solutions of this salt can be obtained by the reaction of cupric triflate hydrate, $Cu(CF_3SO_3)_2 \cdot 5.5\ H_2O$ (2.70 mmole) with copper powder (2.50 mmole) in acetone (25 ml.) and acetonitrile (1.25 ml.). The reaction does not proceed in the absence of acetonitrile.

## Crown ethers, 4, 142–143; 5, 152–155.

*Caution:* Leong[1] reports that the cyclic tetramer of ethylene oxide is acutely toxic to rats and rabbits. He suggests that all macrocyclic ethers should be handled in a manner that precludes exposure until their toxicological properties have been evaluated.

*New crown ethers.* Some new crown ether-esters have been prepared, for example, 2,4-diketo(16-crown-5), (1), and 2,4-diketo(19-crown-6), (2), by reaction of malonyl dichloride with tetraethylene glycol and pentaethylene

(1)

(2)

glycol, respectively. Both (1) and (2) form complexes with ions of alkaline earth metals (magnesium, calcium, and strontium).[2]

Högberg and Cram[3] have reported the preparation of 16-benzocrown amine ethers of type (1).

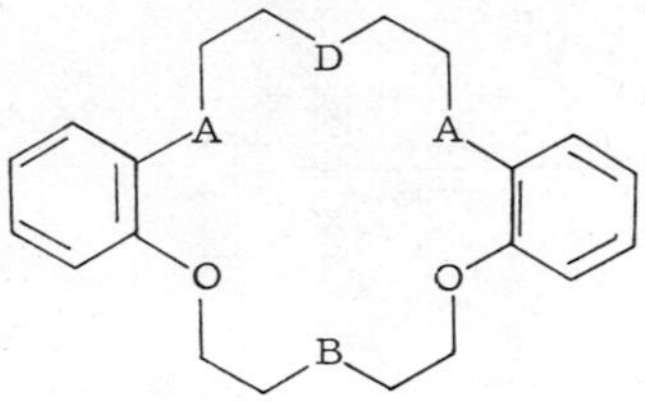

(1)

A, B, and D =

O, NH, NTs

*12-Crown-4 (1) and 15-crown-5 (2).* Liotta *et al.*[4] have published preparations of these crown ethers by modified Williamson syntheses from readily available starting materials. 12-Crown-4 is useful for complexing lithium ions; 15-crown-5 resembles 18-crown-6 in ability to complex potassium and sodium ions.

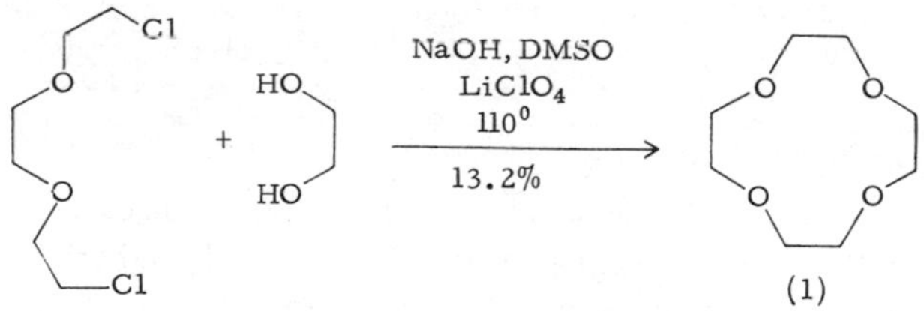

(1)

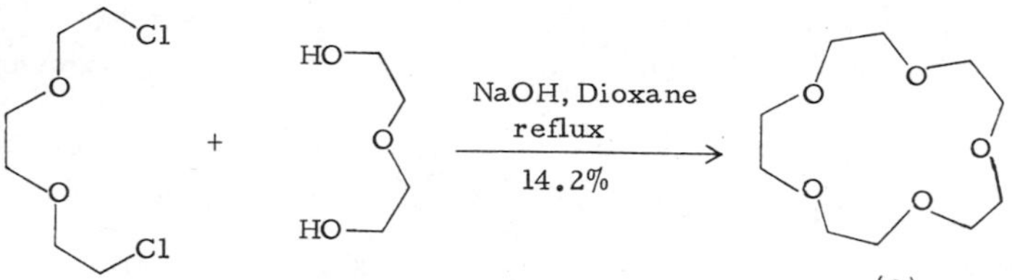

(2)

*Chiral 18-crown-6-polyethers.* British chemists[5] have prepared optically pure chiral crown ethers, such as LL-(1) and DD-(2) from L-tartaric acid and D-mannitol, respectively, and have shown[6] by NMR spectra that they exhibit differentiation in complex formation between R- and S-α-phenylethylammonium hexafluorophosphate. Thus DD-(2) complexes preferentially with the cation of the R-salt.

LL-(1), $a_D$ -3.9°          DD-(2), $a_D$ + 7.6°

This work represents an extension of simpler chiral cyclic polyethers of Cram[7] (**5**, 103).

Deber and Blout[8] have reported enantiomeric differentiation between D- and L-amino acid salts in complexes with cyclo (L-Pro-Gly)$_n$ peptides ($n$ = 3, 4).

*Chiral crown ether derived from L-(+)-tartaric acid.* French chemists[9] have prepared the chiral crown ether (1) from L-(+)-tartaric acid using the new etherification procedure of Crass *et al.* (*see* **Thallium ethoxide**, this volume).

(1, $a_D$ + 108°)

*Nitriles; nitro compounds.* Benzylic chlorides or bromides react in aceto-nitrile or methylene chloride with dry KCN in the presence of 18-crown-6 to give benzyl nitriles in 85–95% yield. This procedure has been used to convert tri-methylchlorosilane into trimethylsilyl cyanide (**4**, 542–543) in about 45% yield.

In the same way alkyl halides react with potassium nitrite in acetonitrile con-taining catalytic amounts of the crown ether to give nitro compounds. For example, 1-nitrooctane is obtained from 1-bromooctane in 65–70% yield. Generally, yields are lower than this, owing to formation of nitrite esters.

*Wittig reactions.* In the original form, the Wittig reaction shows slight stereo-selectivity, with *cis*-alkenylation somewhat more dominant. Boden[11] has re-ported a remarkable effect of complexation of potassium *t*-butoxide or potas-sium carbonate with 18-crown-6. For nonstabilized ylides, *cis*-alkenylation is favored in THF, whereas *trans*-alkenylation is favored in CH$_2$Cl$_2$. For stabilized ylides *trans*-alkenylation is favored in both solvents (about 75 : 25).

*Generation of carbenes* (**5**, 155). Moss and Mallon[12] have published details for differentiation of carbenoids and free carbenes by olefin competition exper-iments in which the binary cyclopropane mixtures from pairs of competing olefins are analyzed by gas chromatography. The relative reactivities of the olefins can then be calculated. This analysis coupled with crown ether complexation permits a more accurate differentiation between free carbenes and carbenoids.[12] Thus chloromethylthiocarbene, $CH_3 \ddot{S}CCl$, has the same relative reactivity toward five different olefins when it is generated from $CH_3SCHCl_2$ with KO-$t$-Bu regardless of the presence of an equivalent of 18-crown-6. On the other hand, phenylfluorocarbene, $C_6H_5\ddot{C}F$, generated by base from $C_6H_5CHBrF$, shows markedly different reactivity to olefins, depending on whether it is generated in the presence or absence of 18-crown-6.

Moss and co-workers[14] have also examined the effect of 18-crown-6 on addition of dibromocarbene to alkenes. Curiously enough this crown ether usually has no effect on the reaction. They also report that, contrary to original statements of Doering and Henderson,[15] dibromocarbene resembles dichlorocarbene in relative reactivity to alkenes.

*"Naked" cyanide ion.*[16] Alkyl halides are converted into nitriles by potassium cyanide in the presence of a catalytic amount of 18-crown-6. Acetonitrile (or benzene) is used as solvent, and the two-phase system is stirred vigorously at 25–83°. Little or no reaction occurs in the absence of the crown ether, an indication that the ether is functioning also as a phase-transfer catalyst. Primary halides are also converted quantitatively into nitriles; chlorides react much faster than bromides. A few percent of elimination products are formed in the reaction of secondary halides. Cyclohexyl halides give only cyclohexene by elimination. $o$-Dichlorobenzene fails to react. Methacrylonitrile undergoes hydrocyanation to 1,2-dicyanopropane (92% yield).

*Examples:*

$$ClCH_2CH_2Cl \longrightarrow NCCH_2CH_2CN \quad (97\%)$$

$$CH_3(CH_2)_4CH_2Br \longrightarrow CH_3(CH_2)_4CH_2CN \quad (100\%)$$

$$C_6H_5CH_2Cl \longrightarrow C_6H_5CH_2CN \quad (94\%)$$

(32%)

*Nucleophilicities of "naked" anions.* (5, 153–154). Liotta and Grisdale[17] have determined the relative reactivities of various nucleophiles in acetonitrile in the presence of 18-crown-6. In this study the halides appear to have virtually identical reactivities. (In water the reactivities vary by a factor of 200.) Surprisingly, in this system·potassium acetate is the most reactive nucleophile; it is about 30 times as reactive as KSCN, which is a potent nucleophile in aqueous solutions. The differences are attributed to the weaker solvation of anions in $CH_3CN$ than in protic solvents.

*α,β-Unsaturated γ-butyrolactones.* Substituted lactones [2 (5H)-furanones] of this type (2) can be prepared in generally high yields by reaction of potassium phenylacetate with α-bromo aldehydes or ketones in acetonitrile or benzene in the presence of catalytic amounts of 18-crown-6. If the reaction is conducted at 20° for a short period, the intermediate keto esters can be isolated in high yield. These can be cyclized to (2) by sodium hydride in DMSO, but it is usually more convenient to reflux the reaction for 1–3 days, in which case (2) is obtained in one operation.[18]

(1)                                              (2)

[1] B. K. J. Leong, *Chem. Eng. News,* Jan. 27, 5 (1975).
[2] J. S. Bradshaw, L. D. Hansen, S. F. Nielsen, M. D. Thompson, R. A. Reeder, R. M. Izatt, and J. J. Christensen, *J. C. S. Chem. Comm.*, 874 (1975).
[3] S. A. G. Högberg and D. J. Cram, *J. Org.*, **40**, 151 (1975).
[4] F. L. Cook, T. C. Caruso, M. P. Byrne, C. W. Bowers, D. H. Speck, and C. L. Liotta, *Tetrahedron Letters*, 4029 (1974).
[5] W. D. Curtis, D. A. Laidler, J. F. Stoddart, and G. H. Jones, *J. C. S. Chem. Comm.*, 833 (1975).
[6] *Idem, ibid.*, 835 (1975).
[7] D. J. Cram and J. M. Cram, *Science*, **183**, 803 (1974).
[8] C. M. Deber and E. R. Blout, *Am. Soc.*, **96**, 7566 (1974).
[9] J.-M. Girodeau, J.-M. Lehn, and J.-P. Sauvage, *Angew. Chem. internat. Ed.*, **14**, 764 (1975).
[10] J. W. Zubrick, B. I. Dunbar, and H. D. Durst, *Tetrahedron Letters,* 71 (1975).
[11] R. M. Boden, *Synthesis*, 784 (1975).
[12] R. A. Moss and C. B. Mallon, *Am. Soc.*, **97**, 344 (1975).
[13] R. A. Moss, M. A. Joyce, and F. G. Pilkiewicz, *Tetrahedron Letters*, 2425 (1975).
[14] R. A. Moss, M. A. Joyce, and J. K. Huselton, *ibid.*, 4621 (1975).
[15] W. v. E. Doering and W. A. Henderson, Jr., *Am. Soc.*, **80**, 5276 (1958).
[16] F. L. Cook, C. W. Bowers, and C. L. Liotta, *J. Org.*, **39**, 3416 (1974).
[17] C. L. Liotta and E. E. Grisdale, *Tetrahedron Letters*, 4205 (1975).
[18] A. Padwa and D. Dehm, *J. Org.*, **40**, 3139 (1975).

**Cryptates, 5,** 156.
Graf and Lehn[1] have synthesized the particularly interesting cryptate (1). This macrotricyclic tetramine hexaether is soluble in all solvents from petroleum ether to water. The cavity radius is about 1.8Å and is fairly rigid. It forms very

stable complexes with the cations $K^+$, $Rb^+$, and $Cs^+$; the complex with $Cs^+$ is probably the most stable known complex of this cation.

(1)

*Esters.* Esters can be prepared in high yield by the reaction of potassium alkanoates with alkyl halides in the presence of cryptate 222.[2]

$$RCOOK \ + \ R'Br \ \xrightarrow[90-99\%]{C_6H_6} \ RCOOR'$$

[1] E. Graf, and J.-M. Lehn, *Am. Soc.*, 97, 5022 (1975).
[2] S. Akabori and M. Ohtomi, *Bull. Chem. Soc. Japan*, 48, 2991 (1975).

**Cupric acetate, 1,** 159–160; **2,** 84; **3,** 65; **4,** 105; **5,** 156–157.

*Intramolecular oxidative coupling of acetylenes* (1, 159; 4, 105). Two recent syntheses[1,2] of the first known bicyclic annulenes (1) analogous to naphthalene involved this reaction in two steps. These substances are relatively

[1, R = CH$_3$, C(CH$_3$)$_3$]

stable, both as solids and in ether solution. As expected, they are diatropic ("aromatic").

[1] T. M. Cresp and F. Sondheimer, *Am. Soc.*, 97, 4412 (1975).
[2] T. Kashitani, S. Akiyama, M. Iyoda, and M. Nakagawa, *ibid.*, 97, 4424 (1975).

**Cupric bromide, 1,** 161–162; **2,** 84; **5,** 158.

*Xanthones.*[1] Xanthone is formed in 50% yield when *o*-phenoxybenzaldehyde is heated in nitrobenzene with either cupric bromide or cupric chloride. The radical (a) is a probable intermediate. Xanthones are also formed, but in low

(a)

yield (1–8%), by the reaction of salicylaldehyde–copper complexes with aromatic halides.

[1] J. I. Okogun and K. S. Okwute, *J.C.S. Chem. Comm.*, 8 (1975).

**Cupric chloride,** **1**, 163; **2**, 84–85; **3**, 66; **4**, 105–107; **5**, 158–160.

*1,4-Diketones.* Saegusa *et al.*[1] have reported the synthesis of 1,4-diketones by treatment of lithium enolates of methyl ketones with cupric chloride (1 eq.) in DMF at $-78°$. Yields of coupled products are high in the case of ketones with only one enolizable hydrogen (pinacolone and acetophenone):

$$2\ (CH_3)_3CCOCH_3 \xrightarrow[-78°]{LDA, THF} (CH_3)_3C\overset{\overset{\displaystyle O^-Li^+}{|}}{C}=CH_2 \xrightarrow[95\%]{CuCl_2, DMF \\ -78°} (CH_3)_3CCOCH_2CH_2COC(CH_3)_3$$

$$2\ C_6H_5COCH_3 \xrightarrow[83\%]{1)\ LDA, THF, -78° \\ 2)\ CuCl_2, DMF, -78°} C_6H_5COCH_2CH_2COC_6H_5$$

Coupling of ketones with two different enolizable hydrogen atoms gives a mixture of 1,4-diketones in which the less crowded product predominates:

$$CH_3COC_2H_5 \longrightarrow \underset{(58\%)}{C_2H_5COCH_2CH_2COC_2H_5} + \underset{(12\%)}{C_2H_5COCH_2\overset{\overset{\displaystyle CH_3}{|}}{C}HCOCH_3}$$

In addition, unsymmetrical 1,4-diketones and $\gamma$-keto esters can be obtained by cross-coupling of enolates of two different methyl ketones and of the enolate of a methyl ketone with that of an alkyl acetate, respectively. Only traces of condensation products are formed.

$$\underset{3:1}{(CH_3)_3CCOCH_3 + CH_3COC_6H_5} \xrightarrow{60\%} (CH_3)_3CCOCH_2CH_2COC_6H_5$$

$$\underset{1:3.5}{(CH_3)_3CCOCH_3 + CH_3COOCH_3} \xrightarrow{70\%} (CH_3)_3CCOCH_2CH_2COOCH_3$$

*Cleavage of methylvitamin* $B_{12}$.[2] Cupric chloride in conjunction with lithium chloride cleaves the $Co-CH_3$ bond of methylvitamin $B_{12}$ (methylcobalamin) to give aquovitamin $B_{12}$. A complex of the type $Cu^{2+}(H_2O)_3(Cl^-)_2$ is inferred as the active species. The cleavage is regarded as a methyl transfer (equation I).

$$(I) \quad Me-Co\diagdown + CuCl_2 \longrightarrow CH_3-CuCl + Cl-Co\diagdown$$
$$\downarrow$$
$$CH_3Cl + Cu$$

*Cyclopropanation.*[3] The reaction of ethyl cyanoacetate (1) with cupric chloride and cupric acetate in the presence of an olefin such as cyclohexene at $100-110°$ in DMF leads to a mixture of two cyclopropanes and some unsaturated products of addition. The latter are eliminated by oxidation with $KMnO_4$. The cyclopropanes were identified as *exo-* and *endo-*isomers (2) and (3). Cyclopropanes were also obtained under similar conditions from styrene and from decene-1.

Ethyl cyanoacetate can be replaced by dimethyl malonate, but yields of cyclopropane derivatives are lower.

This synthesis of cyclopropanes may involve chlorination $\alpha$ to the carbonyl group as an initial step, although ethyl cyanoacetate is not chlorinated in the absence of an olefin under these conditions.

*α-Hydroxylation of a lactone.* The final step in a total synthesis of *dl*-camptothecin (2) in two laboratories[4,5] involved hydroxylation of the lactone ring.

This reaction was accomplished by passing oxygen through a solution of (1) in DMF containing a trace of 25% aqueous dimethylamine and a catalytic amount of freshly prepared cupric chloride (5 hr.).

*Dichloroalkenes.*[6] The reaction of alkylphenylacetylenes with $CuCl_2$ in acetonitrile containing LiCl gives a mixture of the E- and Z-dichloroalkenes in good yield (equation I). The E-isomer is formed predominately except when $R = C(CH_3)_3$. Chloroiodination with $CuCl_2$ and $I_2$ proceeds even more smoothly and is more stereospecific.

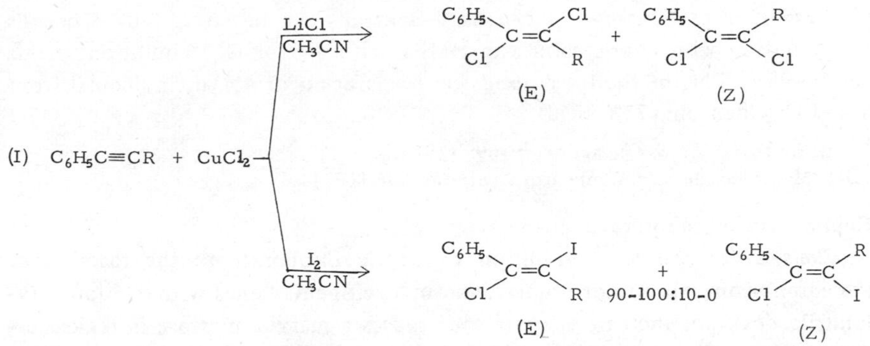

(I)  $C_6H_5C\equiv CR + CuCl_2$

[reaction with LiCl / CH$_3$CN giving]

(E)  ... (Z)

[reaction with I$_2$ / CH$_3$CN giving]

(E)  90-100:10-0  (Z)

[1] Y. Ito, T. Konoike, and T. Saegusa, *Am. Soc.*, **97**, 2912 (1975).
[2] H. Yamamoto, T. Yokoyama, and T. Kwan, *Chem. Pharm. Bull. Japan*, **23**, 2186 (1975).
[3] M. Barreau, M. Bost, M. Julia, and J.-Y. Lallemand, *Tetrahedron Letters*, 3465 (1975).
[4] M. Bock, T. Korth, J. M. Nelke, D. Pike, H. Radunz, and E. Winterfeldt, *Ber.*, **105**, 2126 (1972).
[5] C. S. F. Tang, C. J. Morrow, and H. Rapoport, *Am. Soc.*, **97**, 159 (1975).
[6] S. Uemura, A. Onoe, and M. Okano, *J.C.S. Chem. Comm.*, 925 (1975).

**Cupric sulfate,** **1**, 164–165; **2**, 89; **5**, 162–163.

*Rearrangement of α-diazoketones.* Synthesis of cyclopropanes by copper-catalyzed decomposition of α-diazoketones has been a useful reaction, particularly as a route to natural products (**2**, 82–84; **3**, 63–65, 69–71; **4**, 103).

Decomposition of β,γ-unsaturated diazoketones of type (1) with copper sulfate in methanol proceeds by a skeletal rearrangement to γ,δ-unsaturated esters (2):

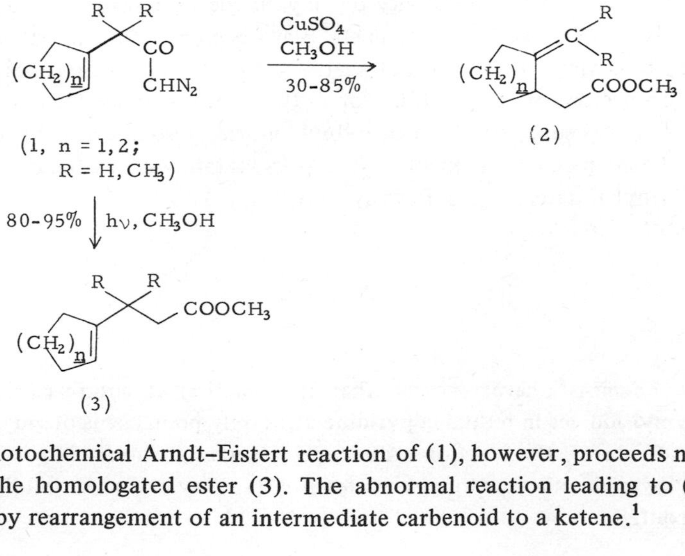

(1, n = 1, 2;
R = H, CH$_3$)

(2)

80-95% | hν, CH$_3$OH

(3)

The photochemical Arndt–Eistert reaction of (1), however, proceeds normally to give the homologated ester (3). The abnormal reaction leading to (2) may proceed by rearrangement of an intermediate carbenoid to a ketene.[1]

*Oxidation of benzoins to benzils.*[2] Benzoins can be converted to benzils by oxidation with cupric sulfate in HMPT with air (or $O_2$) ebullition; yields are 60–95%. This method was used for preparation of 4,4'-divinylbenzil from 4,4'-divinylbenzoin (73% yield).

[1] A. B. Smith, III, *J.C.S. Chem. Comm.*, 695 (1974).
[2] D. P. Macaione and S. E. Wentworth, *Synthesis*, 716 (1974).

**Cupric tetrafluoroborate, 3, 66–67.**

*Diels–Alder reaction.*[1] Addition of cupric fluoborate to the reaction of the equilibrium mixture of isomeric methylcyclopentadienes with α-chloroacrylonitrile does not increase the rate, but causes a marked increase in regioselectivity. Thus the catalyzed reaction gives, after treatment with base, the four products shown in the yields given on the first line below each structure. The yields obtained without catalysis are indicated in brackets on the second line.

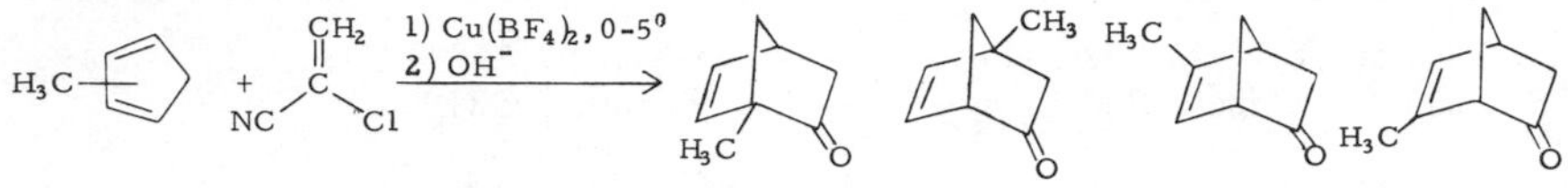

57%        9%        32%        2%

[13.5%]    [16%]    [38%]    [32.5%]

[1] H. L. Goering and C.-S. Chang, *J. Org.*, **40**, 2565 (1975).

**Cuprous acetate,** $CuOCOCH_3$. Mol. wt. 122.59, m.p. 115°. Sensitive to moisture. Supplier: Alfa.

*Vinyl acetates.* The unreactivity of vinyl halides to bimolecular substitution reactions is well-known. Dutch chemists[1] have been able to convert cyclic vinyl bromides into vinyl acetates (enol acetates) by reaction with excess cuprous acetate[2] in acetonitrile at 80–110° for 2–10 hr. Yields range from 15 to 70%. The reaction was successful with one vinyl chloride tested. The cuprous acetate can also be prepared by reduction of cupric acetate with hydrazine, but then yields of vinyl acetates are significantly lower.

*Example:*

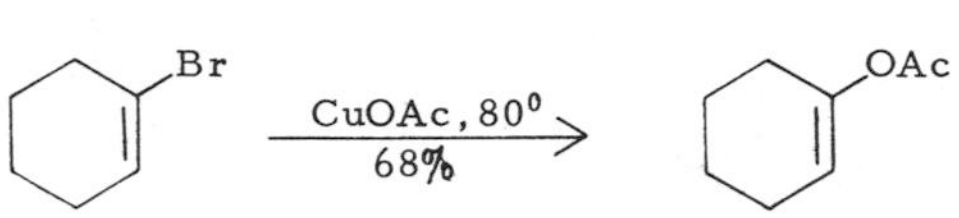

French chemists[3] have reported that the reaction of cuprous acetate with acyclic vinyl iodides in refluxing pyridine gives only poor yields of vinyl acetates (~25%). However, the reaction with cupric acetate (commercial) in refluxing N-methylpyrrolidone gives vinyl acetates in 60–70% yield with complete retention of configuration:

$$\underset{C_2H_5}{\overset{n\text{-}Bu}{>}}C=C\underset{I}{\overset{H}{<}} \quad \xrightarrow[70\%]{Cu(OAc)_2} \quad \underset{C_2H_5}{\overset{n\text{-}Bu}{>}}C=C\underset{OAc}{\overset{H}{<}}$$

Yields are lower and no selectivity is observed in refluxing pyridine. Commercon *et al.* effected similar substitution reactions of vinyl iodides with a variety of copper(I) derivatives: CuCl, CuCN, $CuC_6H_5$, $CuC\equiv CC_4H_9$-$n$. They do not believe that an intermediate vinylcopper (I) species is involved.

[1]G. W. Klumpp, H. Bos, R. F. Schmitz, and J. J. Vrielink, *Tetrahedron Letters*, 3429 (1975).

[2]Prepared according to D. A. Edwards, and R. Richards, *J.C.S. Dalton*, 2463 (1973). Anhydrous acetonitrile is added to anhydrous cupric acetate and copper foil. Acetic acid-acetic anhydride (4:1) is added and the reactants are stirred under $N_2$ for 48 hr. Excess copper is removed, and the cuprous acetate is precipitated with ether. Yield, 85%.

[3]A. Commercon, J. Normant, and J. Villieras, *J. Organometal. Chem.*, 93, 415 (1975).

**Cuprous bromide, 1**, 165–166; **2**, 90–91; **3**, 67; **4**, 108; **5**, 163–164.

*Allenediynes.*[1] These compounds can be synthesized by coupling of allenic halides with butadiynyl(trimethyl)silane, with cuprous bromide as catalyst (equation I). This method was used for synthesis of a natural allenediynol

(I) $\quad \underset{C_2H_5}{\overset{CH_3}{>}}C=C=CHBr \quad + \quad H(C\equiv C)_2Si(CH_3)_3 \quad \xrightarrow[\substack{(n\text{-}Bu)_3N, DMF \\ 46\%}]{Cu_2Br_2} $

$\underset{C_2H_5}{\overset{CH_3}{>}}C=C=CH(C\equiv C)_2Si(CH_3)_3 \quad \xrightarrow[\substack{CH_3OH \\ 84\%}]{NaOH} \quad \underset{C_2H_5}{\overset{CH_3}{>}}C=C=CH(C\equiv C)_2H$

in racemic form (equation II).

(II) $\quad (CH_3)_3SiOCH_2CH=C=CHBr \quad + \quad H(C\equiv C)_2Si(CH_3)_3 \quad \xrightarrow{18\%}$

$HOCH_2CH=C=CH(C\equiv C)_2Si(CH_3)_3 \quad \xrightarrow[\substack{CH_3OH \\ 30\%}]{NaOH} HOCH_2CH=C=CH(C\equiv C)_2H$

*Allenes.*[2] A wide variety of allenes can be prepared by reaction in THF of acetylenic tosylates with an organocopper reagent "RCu" prepared from a Grignard reagent and an equivalent amount of copper(I) bromide. When

$$R^1C\equiv C-CHR^2-OTs \quad \xrightarrow[\substack{THF \\ 80-90\%}]{RMgX + CuBr} \quad RR^1C=C=CHR^2$$

$R = C(CH_3)_3$, however, only a catalytic amount of CuBr is required. The allenes are contaminated with less than 5% of acetylenes, $R^1C\equiv CCHRR^2$, when THF is used as solvent. In the absence of CuBr, only resinous products are formed.

*Arylation of β-dicarbonyl compounds* (5, 163–164). Experimental details are available for the preparation of homophthalic acids by arylation of β-keto esters with 2-bromobenzoic acids with copper catalysts.[3]

[1] P. D. Landor, S. R. Landor, and P. Leighton, *J.C.S. Perkin I*, 1628 (1975).
[2] P. Vermeer, J. Meijer, and L. Brandsma, *Rec. trav.*, **94**, 112 (1975).
[3] A. Bruggink and A. McKillop, *Tetrahedron*, **31**, 2607 (1975).

## Cuprous bromide–Sodium bis(2-methoxyethoxy)aluminum hydride (cf. 5, 169).

*1,4-Reduction of α,β-unsaturated ketones and esters.*[1] This copper hydride reagent in the presence of 2-butanol (proton donor) reduces α,β-unsaturated ketones and esters at $-20°$ (THF, 1 hr.).

*Examples:*

$$CH_3CH=CHCOOCH_3 \xrightarrow[84\%]{} CH_3CH_2CH_2COOCH_3$$

$$C_6H_5CH=CHCOOCH_3 \xrightarrow[82\%]{} C_6H_5CH_2CH_2COOCH_3$$

[1] M. F. Semmelhack and R. D. Stauffer, *J. Org.*, **40**, 3619 (1975).

## Cuprous *t*-butoxide, 4, 109.

*Cyclopropanes.*[1] In the presence of cuprous *t*-butoxide complexed with tri-*n*-butylphosphine, methyl α-chloropropionate and methacrylonitrile react to form the isomeric cyclopropanes (1) and (2) via organo-copper intermediates. No reaction occurs in the absence of the ligand. This reaction was carried out

in 1967 with sodium hydride.[2] The two systems exhibit a striking difference. In the case of sodium hydride, the steric course depends on the polarity of the solvent. In DMF the *trans*-isomer (1) is formed preferentially; in benzene the *cis*-isomer is the predominant product. In the copper system, the polarity of the solvents has only a slight effect on the steric course, and the more stable *trans*-isomer predominates in both DMF and benzene. One explanation is that chelation is involved in the sodium system but not in the copper system.

[1]T. Tsuda, F. Ohoi, S. Ito, and T. Saegusa, *J.C.S. Chem. Comm.*, 327 (1975).
[2]Y. Inouye, M. Horiike, M. Ohno, and H. M. Walborsky, *Tetrahedron,* **24**, 2907 (1967).

**Cuprous chloride, 1,** 166–169; **2,** 91–92; **3,** 67–69; **4,** 109–110; **5,** 164–165.

*Dealkylation of N,N-dichloro-t-alkyl primary amines.* The reaction of N,N-dichloro-1-methylcyclohexylamine (1) with $Cu_2Cl_2$ in DMSO at $20°$ leads to (2) and (3) in a total yield of 96%.[1] The reaction appears to be limited to amines

(1)                    (2, 64%)           (3, 32%)

(4)                    (5, 68-79%)

(6)                    (7, 65-71%)

of this type. For example N,N-dichlorohexylamine is converted under these conditions mainly into hexanenitrile (60% yield).

A similar reaction was reported in 1968.[2]

*Dimerization of β-dicarbonyl compounds.* The dicarbanions of β-dicarbonyl compounds are dimerized by treatment with iodine (or bromine) when catalyzed by cuprous chloride or cobaltous chloride.[3]

*1,2,4,5-Tetraenes.* The conjugated diallene (2) can be prepared by coupling of the bromoallene (1) with cuprous chloride in DMF at $25°$. The diallene (2)

can also be obtained under the same conditions from the prop-2-ynyl acetate (3). Both routes are assumed to involve coupling of an allene radical, $(C_6H_5)_2C=C=\overset{\cdot}{C}C_6H_5$.[4]

*1-Alkynes.* 1-Alkynes are formed in 75–95% yield when methoxyallene is treated with a Grignard reagent in ether in the presence of a catalytic amount of a cuprous halide (CuCl, CuBr, CuI). The active species is probably [RCuX] MgX.[5]

$$CH_2=C=CHOCH_3 + RMgX \xrightarrow[75-95\%]{CuX} RCH_2C\equiv CH$$

*Azoxy compounds.*[6] Azoxy compounds can be prepared by condensation of nitroso compounds with N,N-dichloro substrates in the presence of cuprous chloride.

[1] J. A. Tonnis, P. Donndelinger, C. K. Daniels, and P. Kovacic, *J.C.S. Chem. Comm.*, 560 (1975).
[2] C. M. Sharts, *J. Org.*, **33**, 1008 (1968).
[3] K. G. Hampton and J. J. Christie, *J. Org.*, **40**, 3887 (1975).
[4] F. Toda and Y. Takehira, *J.C.S. Chem. Comm.*, 174 (1975).
[5] J. Meijer and P. Vermeer, *Rec. trav.*, **93**, 183 (1974).
[6] V. Nelson and P. Kovacic, *J.C.S. Chem. Comm.*, 312 (1975).

**Cuprous cyanide, 5,** 166–167.

*Reaction with 1-bromopropyne-3-ols* (1). Cuprous cyanide and 1-bromo-propyne-3-ols in dry DMF form rather unstable complexes of the approximate composition $(CuCN)_4 \cdot (DMF)_2 \cdot RR^1C(OH)C\equiv CBr$. When the complex is heated at 45–50° in the presence of water and excess DMF, 4-hydroxybutyne-nitriles (2) are obtained in about 60% yield. If the complex is isolated and treated with water alone, the diynediol (3) is obtained as the main product.[1]

$$\underset{(1)}{\overset{R}{\underset{R^1}{\diagdown}}\hspace{-0.3em}\underset{OH}{\overset{|}{C}}-C\equiv C-Br} \quad \xrightarrow[\sim 60\%]{\substack{CuCN \\ DMF,\,H_2O}} \quad \underset{(2)}{\overset{R}{\underset{R^1}{\diagdown}}\hspace{-0.3em}\underset{OH}{\overset{|}{C}}-C\equiv C-CN}$$

$$\downarrow$$

$$Complex \quad \xrightarrow[30-50\%]{H_2O} \quad \underset{(3)}{\overset{R}{\underset{R^1}{\diagdown}}\hspace{-0.3em}\underset{OH}{\overset{|}{C}}-C\equiv C-C\equiv C-\underset{OH}{\overset{|}{C}}\hspace{-0.3em}\overset{\diagup R}{\diagdown_{R^1}}}$$

[1] S. R. Landor, B. Demetriou, R. Grzeskowiak, and D. F. Pavey, *J. Organometal. Chem.*, **93**, 129 (1975).

**Cuprous iodide, 1**, 169; **2**, 92; **3**, 69–71; **5**, 167–168.

*Phloroglucinol.*[1] This phenol (3) can be prepared from 1,3,5-tribromo-benzene by treatment with sodium methoxide in methanol and DMF in the presence of cuprous iodide as catalyst. The resulting ether (2) is hydrolyzed to phloroglucinol (3) by concentrated hydrochloric acid. The overall yield is 85–95%.

[1] A. McKillop, B. D. Howarth, and R. J. Kobylecki, *Syn. Commun.*, **4**, 35 (1974).

**Cuprous iodide–Grignard reagents.**

*Allenic alcohols.* Claesson *et al.*[1] have prepared organic cuprate reagents by the reaction of Grignard reagents with CuI in the ratio of 4–5 : 1 eq. in ether at –30° under $N_2$. These cuprates displace methyl and tetrahydropyranyl ether groups of acetylenic alcohols with shift of the unsaturated linkage to give α- and β-allenic alcohols as shown in the examples.

*Examples:*

$$(CH_3)_2\overset{\overset{\displaystyle OTHP}{|}}{C}-C\equiv CCH_2CH_2OH \quad \xrightarrow[56\%]{\substack{CH_3MgBr,\,CuI \\ (C_2H_5)_2O,\,20^0}} \quad (CH_3)_2C=C=C\underset{\underset{\displaystyle CH_3}{|}}{C}H_2CH_2OH$$

$$\underset{\substack{|\\ \text{OCH}_3}}{\text{C}_3\text{H}_7\text{CH}}-\text{C}\equiv\text{CCH}_2\text{OH} \xrightarrow[\substack{(\text{C}_2\text{H}_5)_2\text{O},\,20^0}]{\text{C}_2\text{H}_5\text{MgBr},\,\text{CuI}} \text{C}_3\text{H}_7\text{CH}=\text{C}=\underset{\substack{|\\ \text{C}_2\text{H}_5}}{\text{CCH}_2\text{OH}} + \text{C}_3\text{H}_7\text{CH}=\text{C}=\text{CHCH}_2\text{OH}$$

$$(71\%) \qquad\qquad (5\%)$$

$$\underset{\substack{|\quad|\\ \text{H}\ \ \text{H}}}{\underset{\substack{|\\ \text{OCH}_3}}{\text{C}_3\text{H}_7\text{CH}}-\text{C}=\text{C}-\text{CH}_2\text{OH}} \xrightarrow[59\%]{\substack{\text{CH}_3\text{MgBr},\,\text{CuI}\\ (\text{C}_2\text{H}_5)_2\text{O},\,\text{THF},\,20^0}} \underset{\substack{\ \ \text{H}\ \text{H}}}{\text{C}_3\text{H}_7\text{C}=\text{C}-\text{CH}(\text{CH}_3)\text{CH}_2\text{OH}}$$

As shown in the last example, the method is applicable to olefinic methyl ethers. The free alcohol group is probably not essential in these reactions.

[1] A. Claesson, I. Tàmnefors, and L.-I. Olsson, *Tetrahedron Letters*, 1509 (1975).

**Cuprous trimethylsilylacetylide,** $\text{CuC}\equiv\text{CSi}(\text{CH}_3)_3$. Mol. wt. 160.75, unstable even at $-20^\circ$.

The reagent can be prepared *in situ* from $\text{CuOC}(\text{CH}_3)_3$ and $\text{HC}\equiv\text{CSi}(\text{CH}_3)_3$ in THF at $0^\circ$.

*Trimethylsilylethynyl ketones.* The reagent reacts with acid chlorides to give trimethylsilylethynyl ketones, $\text{RCOC}\equiv\text{CSi}(\text{CH}_3)_3$, in 30–65% yield.[1]

[1] M. W. Logue and G. L. Moore, *J. Org.*, **40**, 131 (1975).

**Cyanogen bromide, 1,** 174–176; **2,** 93; **4,** 110; **5,** 169–170.

*Alkene synthesis.*[1] A new synthesis of alkenes involves treatment of dilithium ethynylbis(trialkylborates) (1) with cyanogen bromide and a base (sodium methoxide). If cyanogen bromide and sodium methoxide are used in equimolecular amounts, *trans*-alkenes (2) are formed mainly. When 3 eq. of cyanogen bromide and 1 eq. of base are used, tri- and tetrasubstituted alkenes, (3) and (4), are formed.

$$\text{LiC}\equiv\text{CLi} + 2\,\text{B}(\text{C}_3\text{H}_7\text{-}\underline{\text{n}})_3 \xrightarrow[\text{THF},\,0^0]{\text{Ether}} \text{Li}^+(\text{C}_3\text{H}_7\text{-}\underline{\text{n}})_3\bar{\text{B}}\text{C}\equiv\text{C}\bar{\text{B}}(\text{C}_3\text{H}_7\text{-}\underline{\text{n}})_3\text{Li}^+$$

$$(1)$$

$$(1) + \text{BrCN} \xrightarrow{\text{CH}_3\text{ONa}} \underset{(2,\,88\%)}{\underset{\text{H}}{\overset{\text{n-C}_3\text{H}_7}{>}}\text{C}=\text{C}\underset{\text{C}_3\text{H}_7\text{-}\underline{\text{n}}}{\overset{\text{H}}{<}}} + \underset{(3,\,3\%)}{\underset{\text{H}}{\overset{\text{n-C}_3\text{H}_7}{>}}\text{C}=\text{C}\underset{\text{C}_3\text{H}_7\text{-}\underline{\text{n}}}{\overset{\text{C}_3\text{H}_7\text{-}\underline{\text{n}}}{<}}}$$

$$(1) + 3\,\text{BrCN} \xrightarrow{\text{CH}_3\text{ONa}} (2,\,8\%) + (3,\,43\%) + \underset{(4,\,30\%)}{\underset{\text{n-C}_3\text{H}_7}{\overset{\text{n-C}_3\text{H}_7}{>}}\text{C}=\text{C}\underset{\text{C}_3\text{H}_7\text{-}\underline{\text{n}}}{\overset{\text{C}_3\text{H}_7\text{-}\underline{\text{n}}}{<}}}$$

*Solid-phase peptide synthesis.* In the Merrifield solid-phase method for peptide synthesis (**1,** 516–518), vigorous conditions are usually necessary for liberation of the free peptide from the support. Hancock and Marshall[2] have reported that cleavage can be effected under mild conditions by the specific cleavage of methionine peptides with cyanogen bromide (1, 175). At the end of the synthesis, the peptide resin is treated with cyanogen bromide, and this reaction liberates the peptide with homoserine at the C-terminus. This peptide is then cleaved with carboxypeptidase A, which selectively cleaves the homoserine unit. The cyanogen bromide cleavage occurs most readily if the methionine residue is at least two amino acids removed from the point of attachment to the resin. The cleavage is promoted by use of propionic acid, which is superior to acetic acid and formic acid. Yields in the cyanogen bromide reaction vary from 55 to 98%; the enzymatic reaction proceeds in 80–98% yield. BOC groups are also cleaved by cyanogen bromide and the acid, but CbO, nitro, and benzyl groups are stable.

[1] N. Miyaura, S. Abiko, M. Itoh, and A. Suzuki, *Synthesis*, 669 (1975).
[2] W. S. Hancock and G. R. Marshall, *Am. Soc.*, 97, 7488 (1975).

**Cyanuric chloride, 3, 72; 5, 522–523.**

*Conversion of phenols to arenes.*[1] Phenolic hydroxyl groups can be replaced by hydrogen by a two-step procedure:

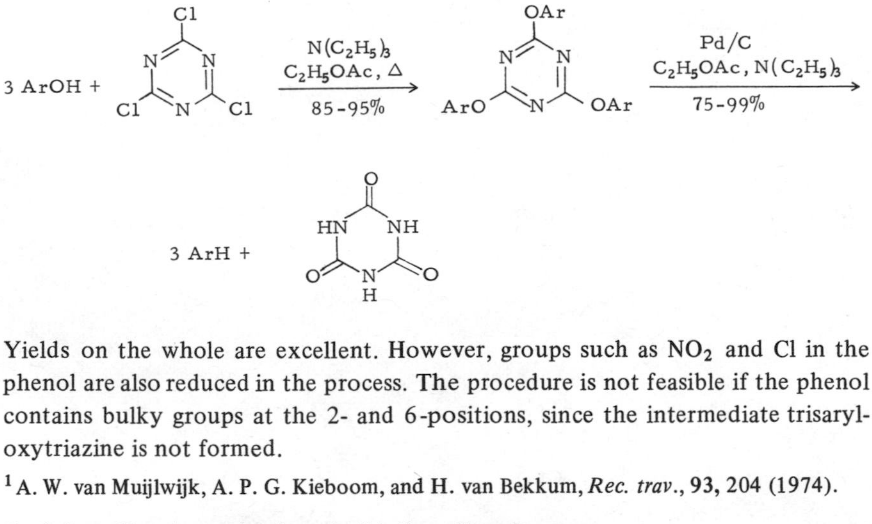

Yields on the whole are excellent. However, groups such as $NO_2$ and Cl in the phenol are also reduced in the process. The procedure is not feasible if the phenol contains bulky groups at the 2- and 6-positions, since the intermediate trisaryloxytriazine is not formed.

[1] A. W. van Muijlwijk, A. P. G. Kieboom, and H. van Bekkum, *Rec. trav.*, 93, 204 (1974).

**Cyclobutadieneiron tricarbonyl, (1). Supplier: Strem.**

*Preparation, 2, 140; 3, 101.*

*Diels–Alder reactions* (2, 95; 4, 72). Preparation of derivatives of Dewar benzene from cyclobutadiene and various acetylenes was first reported by Pettit *et al.*[1] Addition of ceric ammonium nitrate (CAN) to an ice-cold acetone solution of equimolar amounts of cyclobutadieneiron tricarbonyl (1) and dibenzoylacetylene gives as the major product 2,3-dibenzoylbicyclo[2.2.0]hexa-2,5-diene (2). When 2 eq. of cyclobutadieneiron tricarbonyl are used, a 2:1

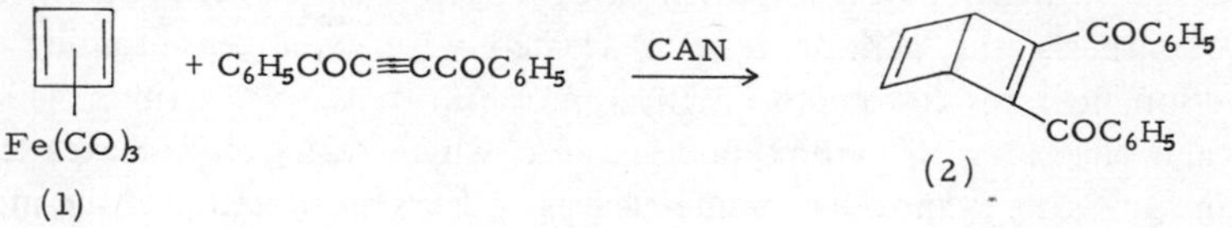

(1)

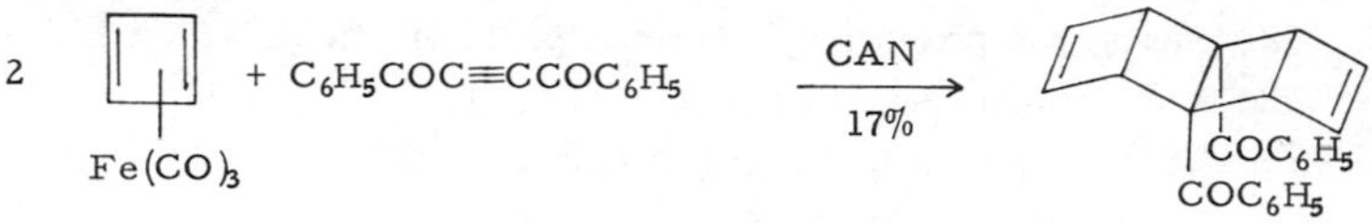

adduct is obtained, identified as 1,6-dibenzoyltetracyclo[$4.4.0.0^{2,5}0^{7,10}$]deca-3,8-diene (3). The reaction has also been carried out with dimethyl acetylene-dicarboxylate as the dienophile.[2]

*Tricyclo[$3.1.1.0^{3,6}$]heptanes.* Meinwald and Mioduski[3] have reported the first synthesis of this ring system by the following scheme:

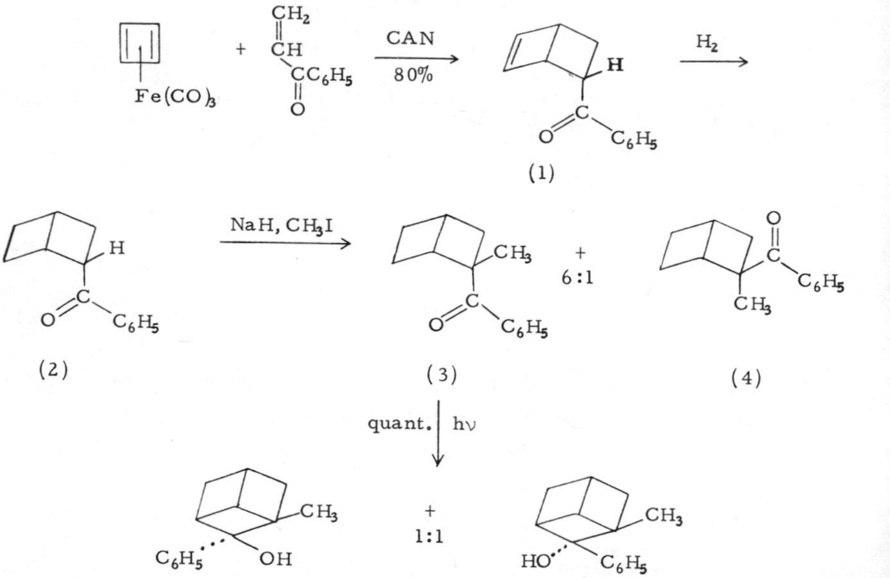

Irradiation of (2) itself results only in fragmentation to a cyclobutene derivative.

[1] L. Watts, J. D. Fitzpatrick, and R. Pettit, *Am. Soc.*, 87, 3253 (1965).
[2] J. Meinwald and J. Mioduski, *Tetrahedron Letters*, 3839 (1974).
[3] *Idem, ibid.*, 4137 (1974).

1-Cyclobutenylmethyllithium,     $Li^+$. Mol. wt. 74.05.

(1)

This ambident anion is prepared by metalation of methylenecyclobutane with the $n$-butyllithium–TMEDA complex (**2**, 403; **3**, 284; **4**, 485–489; **5**, 80–86).

*Isoprenoid synthesis.* This reagent has been used[1] as an isoprene synthon, since cyclobutenes are converted into 1,3-dienes on pyrolysis. For example, the anion (1) reacts with prenyl bromide (**5**, 64) to give a mixture of (2) and (3). When heated for a short time, (2) is converted into myrcene (4).

The anion was also used in a synthesis of one component (6) of the pheromone of the male bark beetle *Ips confusus*.

[1] S. R. Wilson and L. R. Phillips, *Tetrahedron Letters*, 3047 (1975).

**β-Cyclodextrin (Cycloheptaamylose, β-Schardinger dextrin).** Mol. wt. 1135.01, m.p. 298–300° dec. Supplier: Aldrich.

This dextrin is a cyclic oligosaccharide composed of seven D-glucose units; it forms inclusion complexes.[1]

*Photochemical Fries rearrangements.*[2] Photochemical rearrangement of phenyl acetate (1) gives the *o*-and *p*-hydroxy ketones (2) and (3) in equal amounts and phenol. When β-cyclodextrin is present, the rearrangement becomes

|  | (25.7%) | (25.7%) | (48.6%) |
| added β-cyclodextrin | (11.2%) | (69%) | (19.8%) |

stereoselective, with *p*-hydroxyacetophenone being formed as the major product.

[1] F. Cramer, *Einschluss Verbindungen*, Springer-Verlag, Berlin, 1954; C. A. Hampel and G. G. Hawley, *The Encyclopedia of Chemistry*, 271–272, Van Nostrand Reinhold, New York, 1973.

[2] M. Ohara and K. Watanabe, *Angew. Chem. internat. Ed.*, **14**, 820 (1975).

**Cyclohexyl isocyanide,** (1). Mol. wt. 109.17, $n_D$ 1.4502. Supplier: Aldrich. *Caution*: Severe poison.

*Phosphorylation.*[1] Treatment of phenyl phosphate with absolute methanol and excess cyclohexyl isocyanide in pyridine at 40° for 1 day affords methyl phenyl phosphate in 87% yield.

The procedure was used for preparation of 5′-phosphates of nucleotides. Nucleotide 2′(or 3′)-phosphates are converted by (1) in pyridine into 2′,3′-cyclic phosphates:

[1] Y. Mizuno and J. Kobayashi, *J.C.S. Chem. Comm.*, 997 (1974).

**π-Cyclopentadienylcobalt dicarbonyl,** –Co(CO)$_2$ . **5,** 172–173. Mol. wt. 180.05, b.p. 37–38.5°/2 mm., dark red, sensitive to air and heat. Suppliers: Alfa, ROC/RIC, Strem.

*Benzocyclobutenes.* Slow addition of 1,5-hexadiyne (1) in *n*-octane to a

refluxing solution of a large excess of bis(trimethylsilyl)acetylene (2) and a catalytic amount of this cobalt complex leads to cooligomerization to (3) and (4). Benzocyclobutene (6) can be obtained by treatment of (3) with a protic

acid by way of the monosilyl derivative (5). Reaction of (3) with other electrophiles (Br$_2$, AcCl) leads to functionalized derivatives of (6).[1]

*Indanes and tetralins.* A general synthesis of indanes (3a) and tetralins (3b) involves the cooligomerization of 1,6-heptadiyne (1a) or 1,7-octadiyne (1b) with substituted acetylenes (2) catalyzed by cyclopentadienylcobalt dicarbonyl in *n*-octane. In general, yields of (3) are only modest (15–25%) because of

competing cyclotrimerization; however, with bis(trimethylsilyl)acetylene, R$^1$ = R$^2$ = Si(CH$_3$)$_3$, yields of (3a) and (3b) are about 50%.[2]

*1,2-Cyclopropa-4,5-cyclobutabenzene (5).* A crucial step in a synthesis of this strained hydrocarbon involves cooligomerization of (1) and (2) promoted by

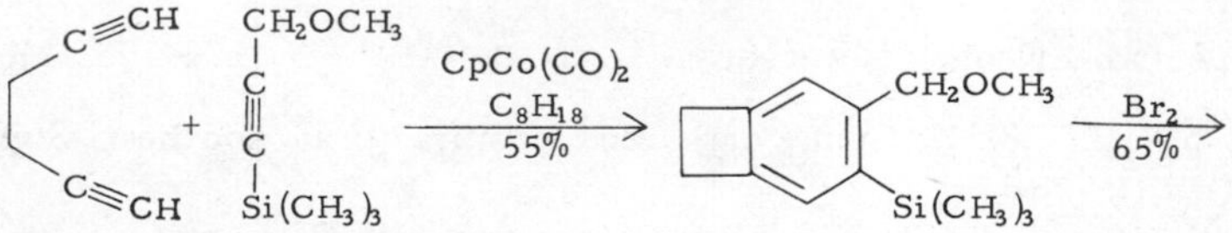

(1)    (2)    (3)

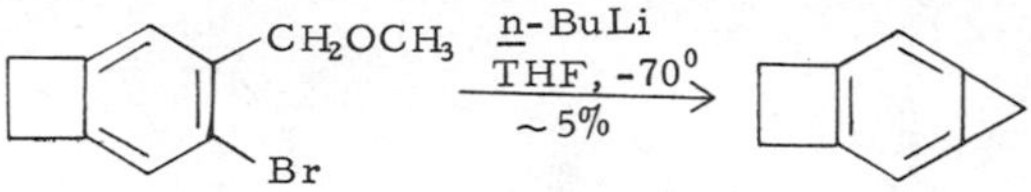

(4)                                    (5)

π-cyclopentadienylcobalt dicarbonyl using high dilution.[3]

[1] W. G. L. Aalbersberg, A. J. Barkovich, R. L. Funk, R. L. Hillard, III, and K. P. C. Vollhardt, *Am. Soc.*, **97**, 5600 (1975).
[2] R. L. Hillard, III, and K. P. C. Vollhardt, *Angew. Chem. internat. Ed.*, **14**, 712 (1975).
[3] C. J. Saward and K. P. C. Vollhardt, *Tetrahedron Letters*, 4539 (1975).

**π-(Cyclopentadienyl)(isobutenyl)iron dicarbonyl tetrafluoroborate**, [Fp(olefin)]$^+$BF$_4^-$ (1). Mol. wt. 319.87, yellow-orange crystals, stable for some time at 20°.

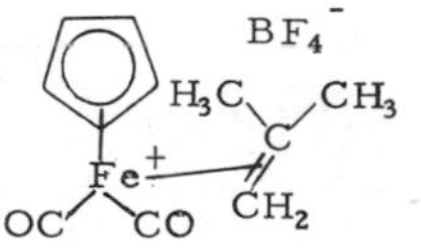

(1)

*Preparation.*[1]

*Protection of C–C double bonds.*[2] The olefinic ligand of this cationic complex can be exchanged when heated above 20° with an excess of certain other olefins. Nicholas has used this property for protection of C–C double bonds, since the olefins can be regenerated by treatment of the complex with sodium iodide in acetone. The method is particularly useful for selective protection of dienes and polyenes. For example, the complex of norbornadiene undergoes electrophilic additions without isomerization to nortricyclane derivatives (equation I). The complex coordinates selectively with the less substituted and/or

(I)

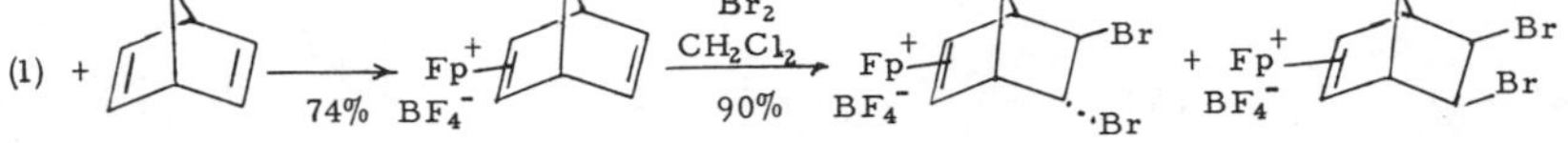

(II)

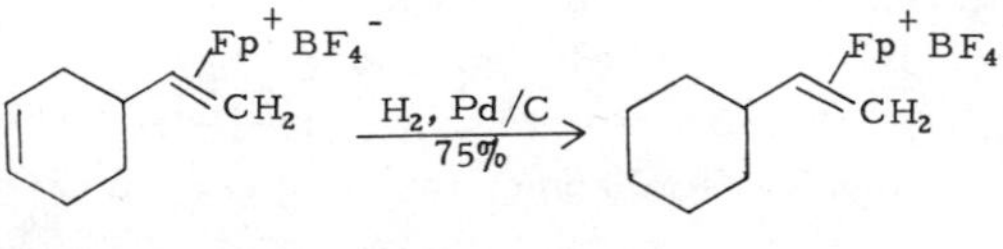

(III)    $H_2C \overset{Fp^+ BF_4^-}{=CHCH_2C \equiv C-C_3H_7}$    $\xrightarrow{H_2, Pd/C}$    $H_2C \overset{Fp^+ BF_4^-}{=CHCH_2CH_2CH_2C_3H_7}$

(IV)

strained double bond. Thus selective hydrogenation of 4-vinylcyclohexene (equation II) and of 1-octene-4-yne (equation III) can be effected. This method of protection has also been used to effect selective bromination of the aromatic ring of eugenol (equation IV).

[1] W. P. Giering, and M. Rosenblum, *J. Organometal. Chem.*, **25**, C71 (1970); *J.C.S. Chem. Comm.*, 441 (1971); M. Rosenblum, *Accts. Chem. Res.*, **7**, 123 (1974).

[2] K. M. Nicholas, *Am. Soc.*, **97**, 3254 (1975).

## Cyclopropyldiphenylsulfonium tetrafluoroborate, 4, 213.
### *Preparation.*[1]

[1] M. J. Bogdanowicz and B. M. Trost, *Org. Syn.*, **54**, 27 (1974).

# D

**Diacetatobis(triphenylphosphine)palladium(II), 5, 497–498.**

*Arylation and vinylation of olefins.*[1]  Aryl and vinylic bromides and iodides react with mono- and disubstituted olefins in the presence of a tertiary amine and a catalyst composed of palladium acetate and 2 eq. of triphenylphospine to form new olefins in which the vinylic hydrogen of the original olefin has been replaced by the aryl or vinyl group of the halide.

*Examples:*

$$C_6H_5Br \ + \ CH_2{=}CHCOOCH_3 \ + \ R_3N \ \xrightarrow[85\%]{\substack{Pd(OAc)_2 \\ 2\ P(C_6H_5)_3}} \ \underset{H}{\overset{C_6H_5}{>}}C{=}C\underset{COOCH_3}{\overset{H}{<}} \ + \ R_3NH^+ \ Br^-$$

$$(CH_3)_2C{=}C\underset{Br}{\overset{H}{<}} \ + \ CH_2{=}CHCOOCH_3 \ \xrightarrow{75\%} \ (CH_3)_2C{=}C\overset{H}{\underset{C{=}C\underset{COOCH_3}{\overset{H}{<}}}{}}$$

$$C_6H_5I \ + \ \underset{C_6H_5}{\overset{CH_3}{>}}C{=}CH_2 \ \longrightarrow \ \underset{C_6H_5}{\overset{CH_3}{>}}C{=}C\underset{C_6H_5}{\overset{H}{<}} \ + \ \underset{C_6H_5}{\overset{CH_3}{>}}C{=}C\underset{H}{\overset{C_6H_5}{<}}$$
$$\quad\quad\quad\quad\quad\quad\quad\quad\quad\quad (5.3\%) \quad\quad\quad\quad\quad\quad (88\%)$$

$$C_6H_5I \ + \ \underset{CH_3}{\overset{H}{>}}C{=}C\underset{C_6H_5}{\overset{H}{<}} \ \longrightarrow \ \underset{C_6H_5}{\overset{CH_3}{>}}C{=}C\underset{C_6H_5}{\overset{H}{<}} \ + \ \underset{CH_3}{\overset{C_6H_5}{>}}C{=}C\underset{C_6H_5}{\overset{H}{<}}$$
$$\quad\quad\quad\quad\quad\quad\quad\quad\quad\quad (79.6\%) \quad\quad\quad\quad\quad\quad (14.6\%)$$

$$C_6H_5I \ + \ \underset{H}{\overset{CH_3}{>}}C{=}C\underset{C_6H_5}{\overset{H}{<}} \ \longrightarrow \ \underset{CH_3}{\overset{C_6H_5}{>}}C{=}C\underset{C_6H_5}{\overset{H}{<}} \quad \text{no \underline{cis}-isomer formed}$$
$$\quad\quad\quad\quad\quad\quad\quad\quad\quad\quad (71\%)$$

The last two examples show that the reaction is stereoselective.  Less stereoselectivity is observed when triphenylphosphine is replaced by other ligands.

*Disubstituted acetylenes and enynes.*[2]  Monosubstituted acetylenes are converted into disubstituted acetylenes by reaction with aryl, heterocylic, or vinylic bromides or iodides at $100°$ in the presence of a basic amine (triethylamine, piperidine) and this palladium complex.  The major limitation is that yields are low with halides having strongly electron-donating substituents.

*Examples:*

$$C_6H_5I \ + \ C_4H_9C{\equiv}CH \xrightarrow[60\%]{\underset{R_3N}{\text{Cat.}}} C_4H_9C{\equiv}CC_6H_5 \ + \ R_3NH^+ \ I^-$$

$$CH_3\underset{\underset{Br}{|}}{C}{=}CH_2 \ + \ C_6H_5C{\equiv}CH \xrightarrow[88\%]{} C_6H_5C{\equiv}C-\underset{\underset{CH_3}{|}}{C}{=}CH_2$$

$$O_2N-\!\!\left\langle\underset{\phantom{.}}{\phantom{xx}}\right\rangle\!\!-Br \ + \ (CH_3)_3CC{\equiv}CH \xrightarrow[88.4\%]{} (CH_3)_3CC{\equiv}C-\!\!\left\langle\underset{\phantom{.}}{\phantom{xx}}\right\rangle\!\!-NO_2$$

[1] H. A. Dieck and R. F. Heck, *Am. Soc.*, **96**, 1133 (1975).
[2] *Idem*, *J. Organometal. Chem.*, **93**, 259 (1975).

## 1,5-Diazabicyclo[4.3.0]nonene-5 (DBN), 1, 189–190; 2, 98–99; 4, 116–119; 5, 176.

*α-Methylene-γ-lactones.*[1] A new synthesis of these lactones involves cyclo-addition of methylbromoketene to various cyclic olefins to form bromocyclobu-tanones (1). These are then oxidized to lactones (2) by *m*-chloroperbenzoic acid (high yield). The last step involves dehydrohalogenation with DBN in boiling toluene. The route is applicable to synthesis of α-methylene-γ-lactones posses-

(1)                    (2)                    (3)

sing *cis*-fused five- and six-membered, saturated and unsaturated rings.

[1] S. M. Ali and S. M. Roberts, *J.C.S. Chem. Comm.*, 887 (1975).

## Diazabicyclo[2.2.2]octane (DABCO) 2, 99–101; 4, 119; 5, 176–177.

*1-Azirines.* Azirines (2) are formed from vinyl azides (1) when the azides are refluxed in toluene in the presence of a tertiary amine.[1] DABCO is particularly effective; N,N-diethylaniline is the least effective amine. Yields are higher than those obtained by thermolysis in aprotic solvents.

(1)                    (2)

*Trinitrobenzene-arylamine* ρ-*complexes.* 1,3,5-Trinitrobenzene readily forms Meisenheimer-type complexes with primary and secondary aliphatic amines, but not with aryl amines. Buncel and Leung[2] have found that aniline does form a red Meisenheimer-type complex in the presence of DABCO or triethylamine in DMSO at 25°.

$$+ C_6H_5NH_2 \xrightarrow{NR_3, DMSO}$$

[1]M. Komatsu, S. Ichijima, Y. Ohshiro, and T. Agawa, *J. Org.*, **38**, 4341 (1973).
[2]E. Buncel and H. W. Leung, *J.C.S. Chem. Comm.*, 19 (1975).

## 1,5-Diazabicyclo[5.4.0]undecene-5 (DBU), 2, 101; 4, 16–18; 5, 177–178.

*Knoevenagel condensation.* DBU was used as the base for the condensation of dimethyl 3-hydroxyhomophthalate (1) with *p*-benzyloxybenzaldehyde (2). Note that this reaction is not successful with *p*-hydroxybenzaldehyde. The condensation was the first step in a synthesis of hydrangenol (4), a component of a variety of *Hydrangea Hortensia D.C.*[1]

*Pyrrole-2,4-dicarboxylates.*[2]   In the presence of this base, aldehydes condense with alkyl isocyanoacetates to form alkyl pyrrole-2,4-dicarboxylates in 50–70% yield.

[1]Y. Naoi, S. Higuchi, T. Nakano, K. Sakai, A. Nishi, and S. Sano, *Syn. Commun.*, 387 (1975).
[2]M. Suzuki, M. Miyoshi, and K. Matsumoto, *J. Org.*, **39**, 1980 (1974).

**Diazomethane, 1,** 191–195; **2,** 102–104; **3,** 74; **4,** 120–122; **5,** 179–182.

*Generation* **in situ (5,** 179–180). *Caution:* N-Nitroso-N-methylurea is potentially carcinogenic. To minimize decomposition of diazomethane, do not use ground-glass equipment, sharp surfaces, or strong light.

*Examples*[1] :

Mechanistic studies of this reaction have been published.[2]

[1]J. W. Kozarich, *Org. Syn.*, submitted (1975).

[2]S. M. Hecht and J. W. Kozarich, *Tetrahedron Letters*, 5147 (1973); *idem, J. Org.*, **38,** 1821 (1973).

**Dibenzo-18-crown-6, 4,** 142.

*Catalysts in reactions of carbanions and halocarbenes.* Mąkosza and M. Ludwikow[1] report that crown ethers, particularly dibenzo-18-crown-6 (1), can function as phase-transfer catalysts in two-phase reactions of carbanions and halocarbenes. An example is the reaction of diphenylacetonitrile (2), which contains an acidic hydrogen, with an alkyl halide in 50% aqueous NaOH:

High yields are obtained in the reaction of dichlorocarbene with alkenes in the presence of this catalyst:

This procedure can also be used in Darzens condensations:

(2 isomers)

*Decarboxylation.* Hunter *et al.*[2] report that addition of 1 eq. of dibenzo-18-crown-6 to a solution of sodium 3-(fluoren-9-ylidene)-2-phenylacrylate (1) in THF greatly accelerates ($>10^5$) the decarboxylation to give the orange-red carbanion (2). A mixture of (3) and (4) is obtained on work-up. The catalytic effect of the crown ether is believed to involve formation of a complex with the sodium ion with production of a highly reactive separated ion pair.

The rate of decarboxylation of (1) is also sensitive to the solvent, being particularly rapid in acetonitrile and DMSO. Even the parent acid of (1) is decarboxylated instantly in DMSO at 25° to give a mixture of (3) and (4) in the ratio 3 : 7. This ratio is comparable to that obtained by decarboxylation of the sodium salt (1) at 80°.

[1] M. Mąkosza and M. Ludwikow, *Angew. Chem. internat. Ed.*, **13**, 665 (1974).
[2] D. H. Hunter, W. Lee, and S. K. Sim, *J.C.S. Chem. Comm.*, 1018 (1974).

**Dibenzoyl peroxide, 1,** 196–198; **4,** 122–123; **5,** 182–183.

*Benzoyloxylation of pyrimidines.*[1] Dibenzoyl peroxide in acetic acid oxidizes 2,4,6-triaminopyrimidines at $C_5$. Yields are lower in solvents such as $CH_2Cl_2$, $CHCl_3$, and $CH_3CN$. The conversion of (1) to (2) is typical. The benzoyl group of (2) can be removed by dimethylamine in methanol.

*Benzoyloxylation of 2-phenylindole.*  2-Phenylindole (1) can be converted into 2-phenylindoxyl O-benzoate by reaction with potassium or sodium metal in HMPT or in THF followed by addition of dibenzoyl peroxide.[2]

1) K, HMPT
2) $(C_6H_5COO)_2$
31.2%

(1)                                                            (2)

[1] J. M. McCall and R. E. TenBrink, *Synthesis*, 443 (1975).
[2] T. Nishio, M. Yuyama, and Y. Omote, *Chem. Ind.*, 480 (1975).

**Diborane**, **1**, 199–207; **2**, 106–108; **3**, 76–77; **4**, 124–126; **5**, 184–186.

*Conversion of—COOH to —$CH_3$.*[1]  Acyl hydrazides (1, prepared from carboxylic acids and *p*-tosylhydrazine) are reduced by diborane in THF to tosylhydrazines (2); these are converted to hydrocarbons when heated in methanolic KOH.

$$R-CO-NHNHTs \xrightarrow[75-80\%]{B_2H_6, 25^0} RCH_2NHNHTs \xrightarrow[65-70\%]{\substack{KOH \\ CH_3OH, \Delta}} RCH_3 + N_2 + TsH$$

(1)                                    (2)                                    (3)

Aromatic tosylhydrazides are reduced by diborane; but they require higher temperatures and a longer reaction period and the reaction is more complex. However, if $AgNO_3$ or $KIO_4$ is added to the reaction mixture, hydrocarbons can be obtained, although in somewhat low yield:

$$ArCONHNHTs \xrightarrow[30-45\%]{\substack{1) B_2H_6 \\ 2) KIO_4}} ArCH_3$$

*Reduction of chiral amines.*[2]  Reduction of the 20-iminopregnane derivative (1) with either diborane or ($\pm$)-di-3-pinanylborane gives, after debenzylation, about equal amounts of the 20$\alpha$- and 20$\beta$-amine (2a and 2b). However, reduction

1) $B_2H_6$
2) $H_2$, Pd/C
quant.

(1)                                                (2a, 20$\alpha$)
                                                    (2b, 20$\beta$)

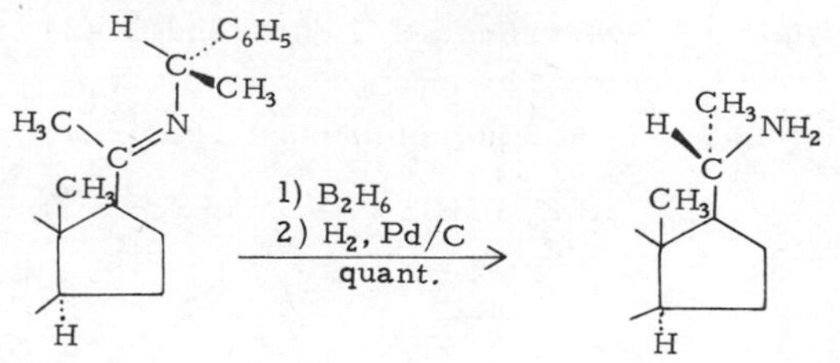

(3)

(2a, 20a)

of the chiral amine (3), prepared by reaction of the 20-ketopregnane with (–)-S-α-phenylethylamine, followed by debenzylation, gives as the unique product the 20α-amine (2a, funtaphyllamine). Similar reduction of the imine prepared from (+)-R-α-phenylethylamine gives essentially the epimeric 20β-amine (2b).

*Hydroboration of cyclic trimethylsilyl enol ethers; synthesis of cycloalkenes.*[3] Cycloalkanones can be converted into the corresponding cycloalkenes by conversion to the trimethylsilyl enol ether, hydroboration, and treatment with acid, usually aqueous HCl. The sequence is illustrated for cyclohexanone (equation I).

(I)

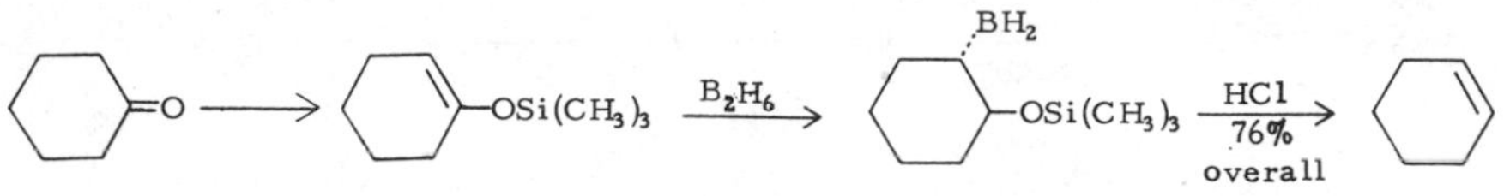

The method is useful for synthesis of unsymmetrical cycloalkenes from ketones in which the isomeric trimethylsilyl ethers can be obtained (*e.g.*, 2-methylcyclohexanone).

[1] O. Attanasi, L. Caglioti, F. Gasparrini, and D. Misiti, *Tetrahedron,* **31**, 341 (1975).
[2] G. Demailly and G. Solladié, *Tetrahedron Letters,* 2471 (1975).
[3] G. L. Larson, E. Hernández, C. Alonso, and I. Nieves, *ibid.,* 4005 (1975).

**Dibromomethyllithium,** $Br_2CHLi$. Mol. wt. 179.79.
   The reagent is formed *in situ* by the reaction of methylene bromide and lithium dicyclohexylamide in THF at $-78°$ (**5**, 403).
   *Ring expansion of ketones.* The reagent forms adducts of type (1) with cyclic ketones. Japanese chemists find that treatment of the adduct (1) of cyclododecanone with *n*-butyllithium in hexane at $-78°$ for 30 min. and then at $0°$

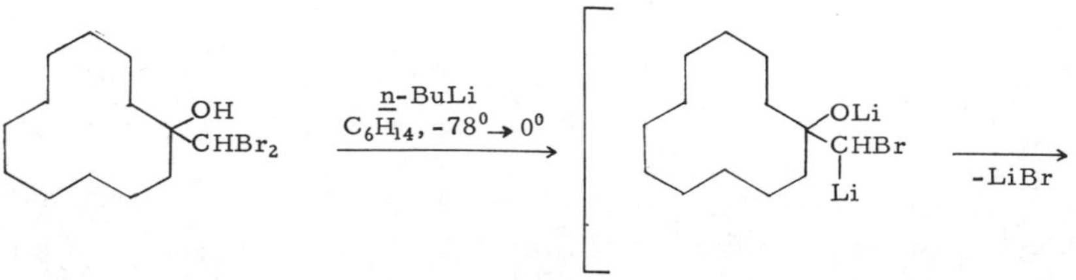

(1)

(a)

$$\text{(b)} \qquad \text{(c)} \qquad \xrightarrow[89\%]{H^+} \qquad \text{(2)}$$

for 5 min. gives, after acid treatment, cyclotridecanone (2) in 89% yield. The ring enlargement involves halogen–lithium exchange (a) and decomposition to the carbene (b), which forms the lithium enolate (c) by insertion.[1]

[1]H. Taguchi, H. Yamamoto, and H. Nozaki, *Am. Soc.*, **96**, 6510 (1974).

**Di-*n*-butylcopperlithium, 2, 151; 3, 79; 4, 127–128; 5, 187–188.**

*1,4-Addition to α,β-unsaturated acetylenic esters* (3, 108). Henrick *et al.*[1] have published details of an extended study of the conjugate addition of various organocopper (I) reagents to α,β-unsaturated acetylenic esters, particularly of various *n*-butylcopper(I) reagents to methyl 2-butynoate (1). In all cases the product of *cis*-addition to the triple bond (2) predominates, but the stereoselectivity depends on the reagent, solvent, ligands, temperature, and time. In the

$$CH_3C{\equiv}CCOOCH_3 \; + \; \text{"}\underline{n}\text{-BuCu"} \longrightarrow \underset{\underline{n}\text{-Bu}}{\overset{CH_3}{>}}C{=}C\underset{H}{\overset{COOCH_3}{<}} \; + \; \underset{\underline{n}\text{-Bu}}{\overset{CH_3}{>}}C{=}C\underset{COOCH_3}{\overset{H}{<}}$$

$$\text{(1)} \qquad\qquad\qquad\qquad (2,E) \qquad\qquad\qquad (3,Z)$$

case of di-*n*-butylcopperlithium, both alkyl groups are utilized. When ether is used as solvent, (2) and (3) are obtained in a ratio of 74:26. If tetramethylethylenediamine (TMEDA) is added as a ligand, the ratio of (2) to (3) changes to 97:3. The E-isomer is obtained almost exclusively when the solvent is THF. The highest stereoselectivity, however, is observed with polymeric *n*-butylcopper reagents, particularly a reagent prepared from *n*-butyllithium and cuprous iodide in THF at $-40°$. This reagent, *n*-BuCu · LiI, reacts with (1) to give (2) in 96% yield in > 99% purity. The polymeric *n*-butylcopper reagent *n*-BuCu · MgBrI, prepared from *n*-BuMgBr and CuI, is less selective unless TMEDA is added as ligand.

This study of the stereoselective synthesis of trisubstituted double bonds was undertaken in connection with synthesis of substances of the $C_{18}$-*Cecropia* juvenile hormone type, such as (5), the $C_3$-homologue of the natural hormone. The first step involved stereospecific synthesis of the E-acetal ester (4, 99.5% E-isomer) by addition of an organocopper reagent to an acetylenic ester. The product was converted by several steps into (5).

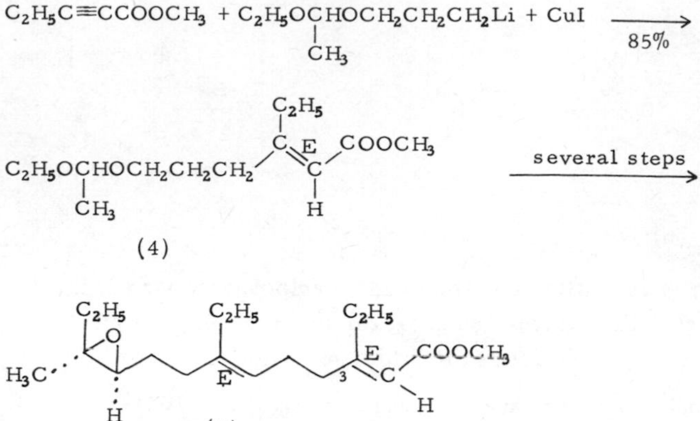

$$C_2H_5C \equiv CCOOCH_3 + C_2H_5OCHOCH_2CH_2CH_2Li + CuI \xrightarrow[\;85\%\;]{}$$
$$\qquad\qquad\qquad\qquad\qquad |$$
$$\qquad\qquad\qquad\qquad\quad CH_3$$

(4)

several steps →

(5)

***β-Diketones.***[2]  Copper enolates generated by conjugate addition of organo-copperlithium reagents to α,β-unsaturated ketones react with an acyl chloride to give α-acylated-β-alkylated ketones. Only a trace of products of O-acylation is formed if HMPT is present as a ligand.

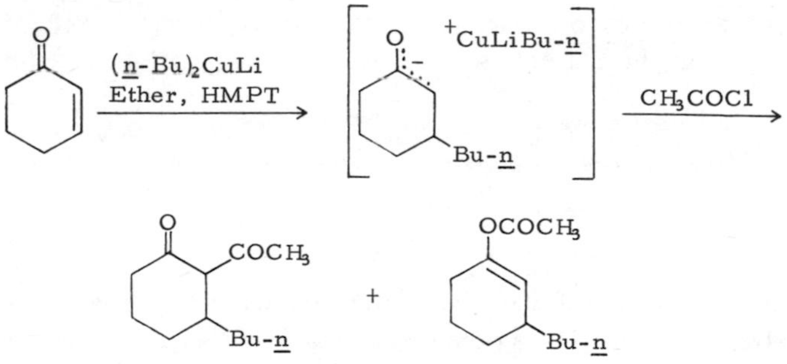

This reaction was used in a synthesis of two 7-oxoprostaglandins.

***Reaction with α-bromoketones.***[3]  This cuprate reacts with some α-bromoke-tones, such as (1), to give the product (2) of β-alkylation together with the product of reduction (3) (equation I). The addition of methyl iodide before work-up results in formation of (4) and (5) (equation II). The products (2) and (4) are evidently formed from an intermediate α,β-unsaturated ketone. The reaction furnishes a route to rather inaccessible ketones.

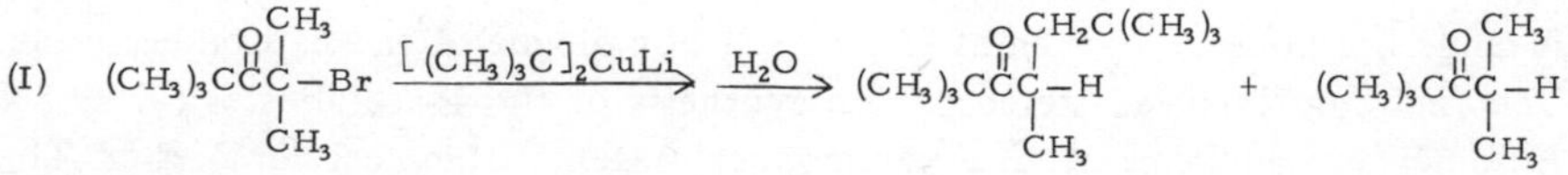

(II)    (1) $\xrightarrow{[(CH_3)_3C]_2CuLi,\ CH_3I}$ $(CH_3)_3\overset{\overset{\displaystyle CH_2C(CH_3)_3}{|}}{\underset{\underset{\displaystyle CH_3}{|}}{C}}\overset{\overset{\displaystyle O}{\|}}{C}-CH_3$    +    $(CH_3)_3C\overset{\overset{\displaystyle O}{\|}}{C}C(CH_3)_3$

(4, 21%)                          (5, 51%)

*Cyclization of* $\omega$-*halo-1-phenyl-1-alkynes.*[4] The bromide (1) reacts with this dialkyl cuprate reagent in pentane–ether (10:1) at $-30°$ to reflux to give the

(I)  $C_6H_5C\equiv C(CH_2)_4Br$ $\xrightarrow[\text{pentane, ether}]{Bu_2CuLi}$

(1)

$C_6H_5\diagdown\underset{\|}{C}\diagup^H$
$(CH_2)_4$
(2, 79%)

+

$C_6H_5\diagdown\underset{\|}{C}\diagup^{Bu}$
$(CH_2)_4$
(3, 13%)

+ $C_6H_5C\equiv C(CH_2)_4Bu$
(4, 1%)

+  $C_6H_5C\equiv C(CH_2)_4H$  +  $C_6H_5C\equiv C(CH_2)_2CH=CH_2$

(5, 5%)                        (6, 3%)

cyclized compound (2) together with the minor products (3)–(6). The product distribution is strongly influenced by the solvent. Thus pentane–ether (1:1) gives (4) as the major product. The iodide corresponding to (1) is converted only into (2) (91% yield) and (3) (8% yield).

The choice of the organo-copper reagent is important. Thus reaction of (1) with dimethylcopperlithium yields only the linear product corresponding to (4), Bu = $CH_3$.

The cyclization depicted in equation I probably proceeds through an acyclic cuprate intermediate such as (a), which cyclizes to (b). This can react with methyl iodide to form (7).

$C_6H_5C\equiv C(CH_2)_4CuBuLi \longrightarrow$

(a)

$C_6H_5\diagdown\underset{\|}{C}\diagup^{CuBuLi}$
$(CH_2)_4$
(b)

$\longrightarrow$    (2)

$53\%\ \Big\downarrow CH_3I$

$C_6H_5\diagdown\underset{\|}{C}\diagup^{CH_3}$
$(CH_2)_4$
(7)

[1] R. J. Anderson, V. L. Corbin, G. Cotterrell, G. R. Cox, C. A. Henrick, F. Schaub, and J. B. Siddall, *Am. Soc.,* 97, 1197 (1975).
[2] T. Tanaka, S. Kurozumi, T. Toru, M. Kobayashi, S. Miura, and S. Ishimoto, *Tetrahedron Letters,* 1535 (1975).
[3] J.-E. Dubois, P. Fournier, and C. Lion, *ibid.,* 4263 (1975).
[4] J. K. Crandall, P. Battioni, J. T. Wehlacz, and R. Bindra, *Am. Soc.,* 97, 7171 (1975).

**Di-*t*-butyl diperoxycarbonate,** $(CH_3)_3COO\overset{\overset{O}{\|}}{C}OOC(CH_3)_3$ (1). Mol. wt. 206.24, b.p. 48°/0.3 mm.

The perester is prepared by reaction of *t*-butyl hydroperoxide with phosgene at low temperatures (44% yield). It decomposes above 90° with liberation of $CO_2$ and formation of two radicals (equation I).[1]

$$(I) \qquad (1) \xrightarrow{\Delta} (CH_3)_3CO\cdot \;+\; (CH_3)_3COO\overset{\overset{O}{\|}}{C}O\cdot \longrightarrow CO_2 \;+\; (CH_3)_3COO\cdot$$

*Hydroxylation of desoxycholic acid.*[2] This perester (1) and desoxycholic acid (2) form a 1:4 molecular complex, which when heated (90°) or irradiated with $\lambda > 300$ nm (25°) decomposes with formation of 3α,5β,12α-trihydroxycholanic acid (3) and 3-keto-12α-hydroxycholanic acid (4) as the major products. A trace of 3α,12α-dihydroxy-5β-androstane is formed.

(2)                              (3, 15%)              (4, 15%)

An X-ray crystal structure analysis of the complex shows that the tertiary 3-, 5-, and 20-hydrogens of the steroid in particular are exposed to the guest (1) in the inclusion channel.

[1] M. M. Martin, *Am. Soc.*, **83**, 2869 (1961).
[2] N. Friedman, M. Lahav, L. Leiserowitz, R. Popovitz-Biro, C.-P. Tang, and Z. Zaretzkii, *J.C.S. Chem. Comm.*, 864 (1975).

**Di-*t*-butylperoxyoxalate,** $[(CH_3)_3CO \cdot O \cdot OC]_2$ (1).

*Cyclization of unsaturated hydroperoxides.* Porter *et al.*[1] have effected cyclization of unsaturated hydroperoxides with *t*-butoxy radicals. The success of this procedure depends on the fact that the hydrogen of hydroperoxides (ROO*H*) is abstracted by *t*-butoxy radicals much more readily than an allylic hydrogen. For example, treatment of the unsaturated hydroperoxide (2) with di-*t*-butylperoxyoxalate (1) in oxygen-saturated benzene at 23° for 2 days gives a mixture of two cyclic peroxides, (3) and (4). Reduction of (3) with triphenylphosphine permits isolation of (4) in 30% overall yield. The original paper should be consulted for a probable mechanism.

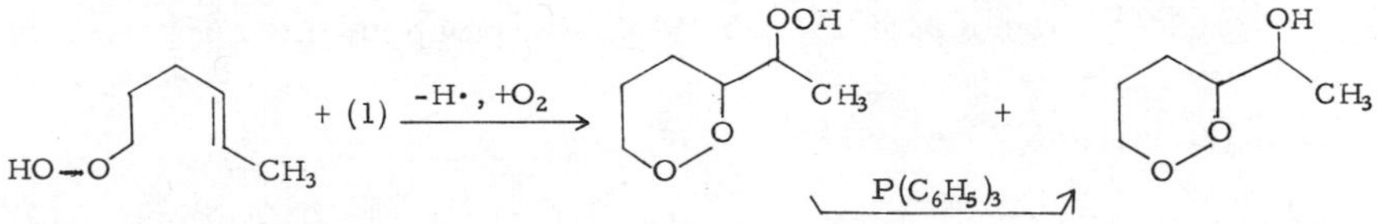

(2)      (3)      (4)

Under similar conditions (5) was cyclized to (6).

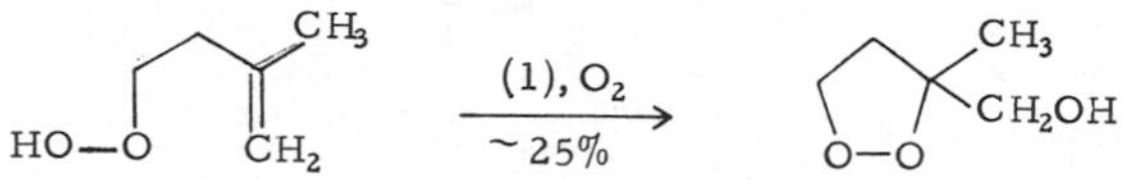

(5)      (6)

The unsaturated lipid hydroperoxide (7) is converted into the prostaglandin-type compound (9) by incubation with di-*t*-butylperoxyoxalate in oxygen-saturated benzene at 25° (16 hr.) followed by reduction with $NaBH_4$.[2]

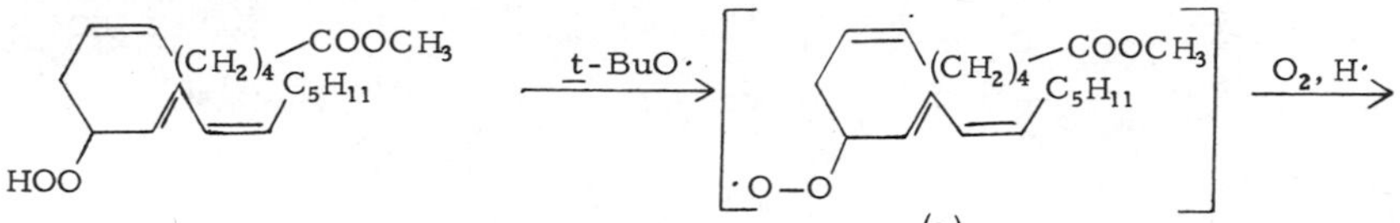

(7)      (a)

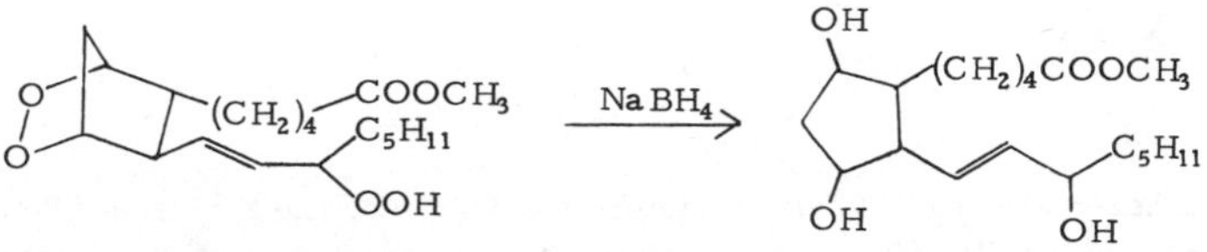

(8)      (9)

[1]M. O. Funk, R. Isaac, and N. A. Porter, *Am. Soc.*, **97**, 1281 (1975).
[2]N. A. Porter and M. O. Funk, *J. Org.*, **40**, 3614 (1975).

**Di(2-*t*-butylphenyl) phosphorochloridate** (1). Mol. wt. 380.84; hygroscopic.
*Preparation:*

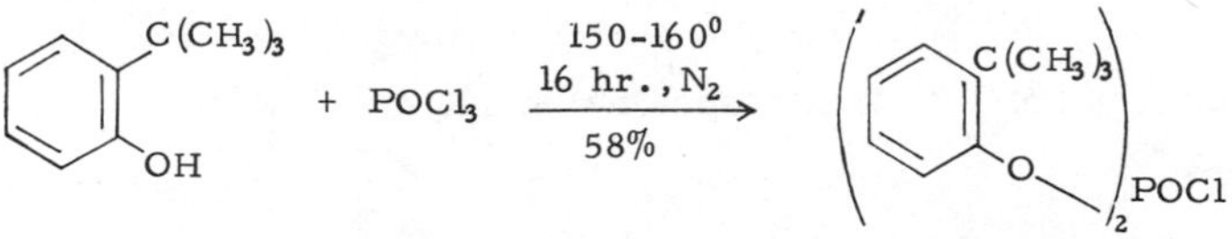

(1)

*Selective phosphorylation.* The reagent selectively reacts with the 5′-hydroxyl groups of unprotected nucleotides. 5′-Nucleotides are obtained in

54-63% yield by removal of the *t*-butylphenyl group by hydrogenolysis in glacial acetic acid.[1]

[1] J. Hes and M. P. Mertes, *J. Org.*, **39**, 3767 (1974).

### 2,3-Dichloro-5,6-dicyano-1,4-benzoquinone (DDQ), 1, 215–219; 2, 112–117; 3, 83–86; 4, 130–134; 5, 193–194.

*Cycloaddition reactions.*[1,2] Cycloaddition reactions of DDQ have been reported recently (equations I and II).

*Oxidation of hydroxytropolones.* Cyclohepta-3,6-diene-1,2,5-trione ("*p*-tropoquinone") (2) can be obtained in quantitative yield by oxidation of 5-hydroxytropolone (1) with DDQ or *p*-chloranil.[3]

Similar oxidation of 3-hydroxytropolone in methanol affords the methyl hemiketal of cyclohepta-4,6-diene-1,2,3-trione (4). In the presence of water the 2-hydrate (5) is obtained. All efforts to convert (5) into the 1,2,3-trione have failed.[4]

**Benzylic hydroxylation.**[5] The 11-oxoestrone derivative (1) is oxidized by DDQ in aqueous dioxane to the ketol (2) in 50% yield. The hydroxyl group has been shown to originate from water. The oxidation is believed to involve an intermediate quinone methide (a).

(1)            (a)            (2)

The oxidation of guaiacylacetone (3) in methanolic dioxane is similar.

(3)                          (4)

**Selective dehydrogenation.**[6] 2-Acetyl-3,4-dihydro-6-methoxynaphthalene (4) is prepared conveniently by Birch reduction of (1) to a mixture of (2) and (3). Reaction of the mixture with DDQ gives (4); other reagents, for example, NBS, effect complete aromatization.

(1)            (2, 50%)        (3, minor product)

58%    DDQ, C₆H₆
15-20°, 20 hrs.

(4)

*A 1,5-naphthoquinone.* Schmand and Boldt[7] have prepared the first known 1,5-naphthoquinone (2) by dehydrogenation of 1,5-dihydroxy-3,7-di-*t*-butyl-naphthalene (1) with DDQ in $CH_2Cl_2(N_2)$. The yield is nearly quantitative.

(1)                                              (2)

Only brown amorphous substances are formed from 1,5-dihydroxynaphthalene itself.

[1] R. Noyori, N. Hayashi, and M. Katô, *Tetrahedron Letters*, 2983 (1973).
[2] S. Kuroda, M. Funamizu, and Y. Kitahara, *ibid.*, 1973 (1975).
[3] S. Ito, Y. Shoji, H. Takeshita, M. Hirama, and K. Takahashi, *ibid.*, 1075 (1975)
[4] M. Hirama and S. Ito, *ibid.*, 1071 (1975).
[5] G. M. Buchan, J. W. A. Findlay, and A. B. Turner, *J.C.S. Chem. Comm.*, 126 (1975).
[6] V. M. Kapoor and A. M. Mehta, *Synthesis*, 471 (1975).
[7] H. L. K. Schmand and P. Boldt, *Am. Soc.*, 97, 447 (1975).

## Dichloromethylenedimethylammonium chloride, 4, 135–138; 5, 195–198.

*Nitriles.*[1] Aldoximes are converted into nitriles in high yield when refluxed with the reagent (1) in $CHCl_3$ or $CH_2Cl_2$ for 2–4 hr.:

$$R-CH=NOH + (CH_3)_2\overset{+}{N}=CCl_2 \quad Cl^- \xrightarrow{82-98\%} RC\equiv N + (CH_3)_2\overset{O}{\overset{||}{N}}CCl + 2\ HCl$$

(1)

Under these conditions ketoximes undergo Beckmann rearrangement:

$$C_6H_5\overset{NOH}{\overset{||}{C}}CH_3 \xrightarrow[86\%]{\substack{1)\ (1) \\ 2)\ Na_2CO_3,\ H_2O}} C_6H_5NH\overset{O}{\overset{||}{C}}CH_3$$

[1] V. P. Kukhar and V. I. Pasternak, *Synthesis*, 563 (1974).

## Dichloromethyllithium, $Cl_2CHLi$, 1, 223–224; 2, 119; 3, 29; 4, 138–139; 5, 199–200.

*Reaction with fulvenes.* Amaro and Grohmann[1] have studied the reaction of fulvenes with dichloromethyllithium under two conditions: preformed reagent or reagent generated in the presence of the fulvene with lithium diisopropylamide. In both cases the reaction is carried out first at $-75$ to $-95°$ and then eventually at $0°$. In the case of fulvene itself or of 6,6-dimethylfulvene (1, R = $CH_3$), yields of chlorospiro[2.4]heptadienes (2) are 55–80%. In the case of 6-mono-alkylfulvenes, two isomeric products can be formed; usually the thermodynamically less stable *cis*-isomer is obtained preferentially.

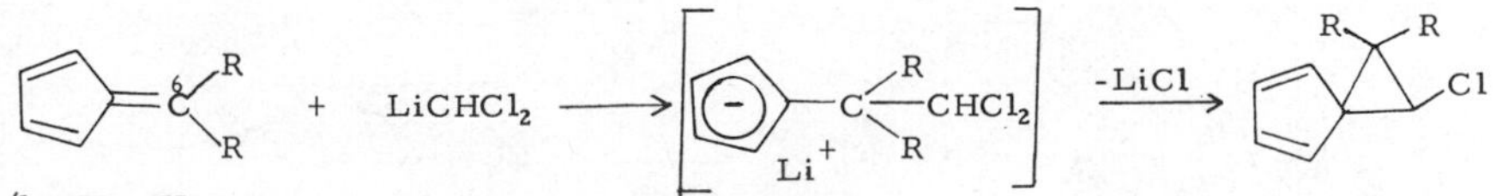

(1, R = H or alkyl)

(2)

These spiroheptadienes are not formed from the reaction of fulvene with chlorocarbene generated from dichloromethane and methyllithium at $-20°$ in ether. Thus 6,6-dimethylfulvene reacts with chlorocarbene to give α-methyl-styrene (24%) as the only identified product.

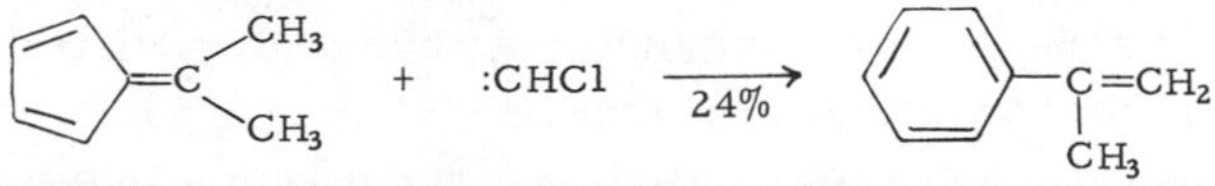

*Reaction with benzoyl chloride.* Benzoyl chloride reacts with dichloro-methyllithium to give the oxirane (1) or the carbinol (2), depending on the reaction conditions.[2]

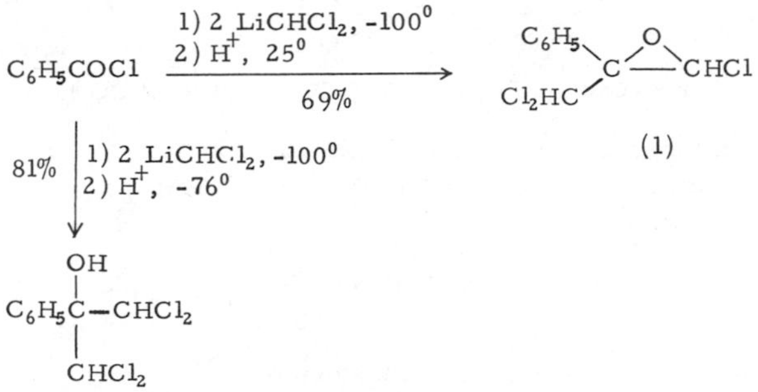

[1] A. Amaro and K. Grohmann, *Am. Soc.*, **97**, 3830 (1975).
[2] J. Grosser and G. Köbrich, *Ber.*, **108**, 328 (1975).

## α,α-Dichloromethyl methyl ether, 2, 120; 5, 200–203.

*Olefins.* The reaction of tertiary α-chloroboronic esters (preparation, 5, 201) with silver nitrate in aqueous ethanol at 25° results in formation of olefins in good yield. For example, (1) is converted into cyclohexylidenecyclo-hexane (2) in 83% yield, and (3) is converted into (4).

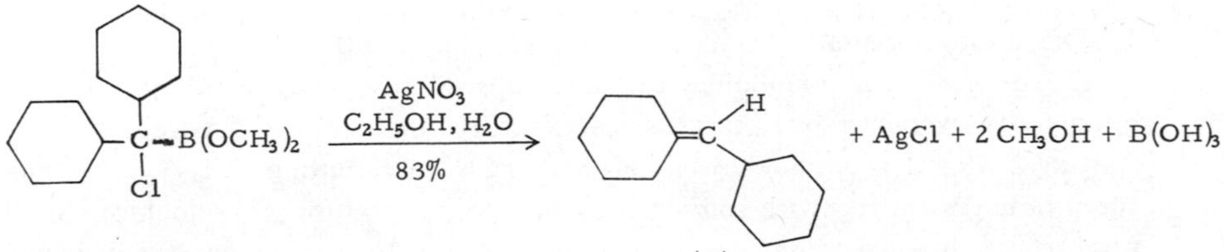

$$
\underset{(3)}{\overset{\begin{array}{c}\text{CH}_3\ \text{CH}_3\end{array}}{\underset{\begin{array}{c}\text{CH}_3\ \text{CH}_3\end{array}}{\text{HC}-\text{C}-\text{C}-\text{B(OCH}_3)_2}}}\ \underset{82\%}{\xrightarrow{\hspace{1cm}}}\ \underset{(4)}{\text{HC}-\text{C}=\text{C}}
$$

Brown[1] considers that these reactions involve ionization of the chlorine substituent to give a carbonium ion, which rearranges and then undergoes β-elimination. If this elimination is undesirable (as in the synthesis of ketones, 5, 202), it can be avoided by substitution of the chlorine by methoxyl by reaction with 1 eq. of sodium methoxide in methanol.

*α-Keto acid chlorides.*[2] The preparation of chlorides of α-keto acids proceeds in low yields with the usual reagents (phosphorus trichloride, phosgene). Dutch chemists have used dichloromethyl methyl ether successfully for this purpose in four cases. Yields ranged from 10 to 78%.

$$
\overset{O}{\overset{\|}{\text{RCCOOH}}} + \text{Cl}_2\text{CHOCH}_3 \longrightarrow \overset{O}{\overset{\|}{\text{RCCOCl}}} + \text{HCOOCH}_3 + \text{HCl}
$$

[1] H. C. Brown, J.-J. Katz, and B. A. Carlson, *J. Org.*, **40**, 814 (1975).
[2] H. C. J. Ottenheijm and J. H. M. De Man, *Synthesis*, 163 (1975).

**Dichlorovinylene carbonate, 2, 122–123.**
Preparation and synthetic uses of halovinylene carbonates have been reviewed by Scharf.[1]

[1] H.-D. Scharf, *Angew. Chem. internat. Ed.*, **13**, 520 (1974).

**Dicobalt octacarbonyl, 1, 224–225; 3, 89; 4, 139; 5, 204–205.**
*Catalytic hydrogenation of α,β-unsaturated carbonyl compounds.* α,β-Unsaturated aldehydes and ketones are reduced to the corresponding saturated carbonyl compounds with high selectivity under conditions of the oxo reaction [$H_2$, CO, $Co_2(CO)_8$, 1, 225–226].[1]
*Reduction of arenes* (1, 228). A mechanism involving free-radical intermediates has been proposed.[2]

[1] E. Ucciani, R. Laï, and L. Taoguy, *Compt. Rend.*, **281 (C)**, 877 (1975).
[2] H. M. Feder and J. Halpern, *Am. Soc.*, **97**, 7186 (1975).

**1,2-Dicyanocyclobutene** (1). Mol. wt. 104.11, b.p. 55–60°/0.06 torr.
*Preparation.* This dienophile can be prepared from *cis,trans*-dicyanocyclo-butane by several routes; the most satisfactory procedure is illustrated.[1] The starting material is monochlorinated with $PCl_5$ in refluxing $CCl_4$–$CHCl_3$ and the product is treated with sodium bicarbonate. A mixture of monochlorinated products is obtained, which is dehydrochlorinated with triethylamine in benzene

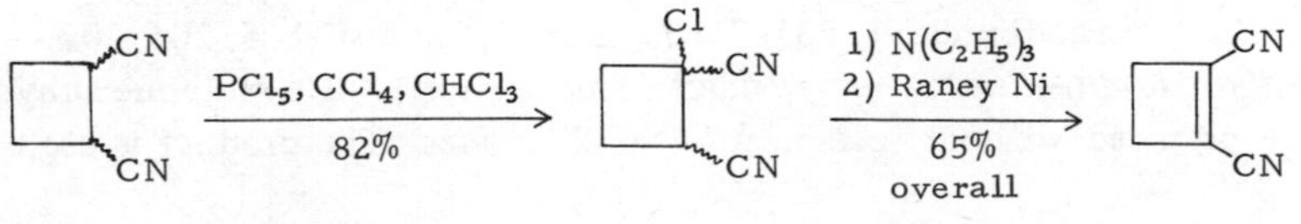

(1)

at 80° to give slightly crude (1). Pure (1) is obtained by treatment with Raney nickel or Raney cobalt at 70°.

*Diels–Alder reactions.* 1,2-Dicyanocyclobutene (1) is a reactive dienophile; the reaction can be carried out in ether at 25–30° or at 110° with no solvent.

*Examples:*

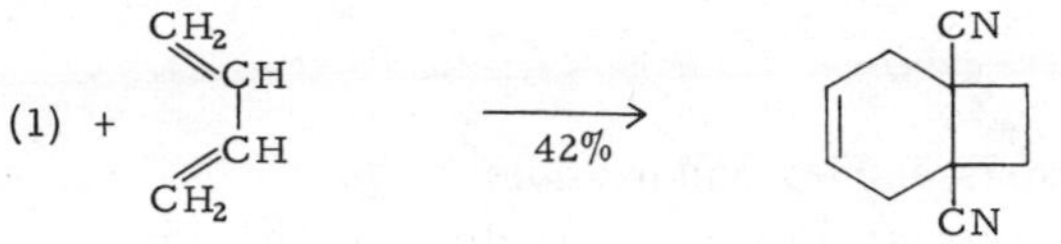

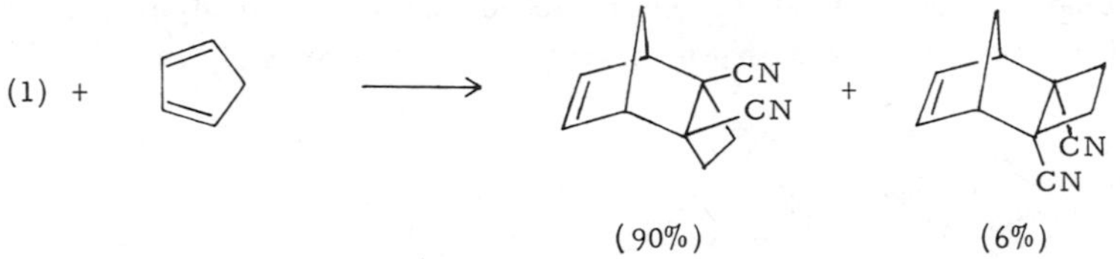

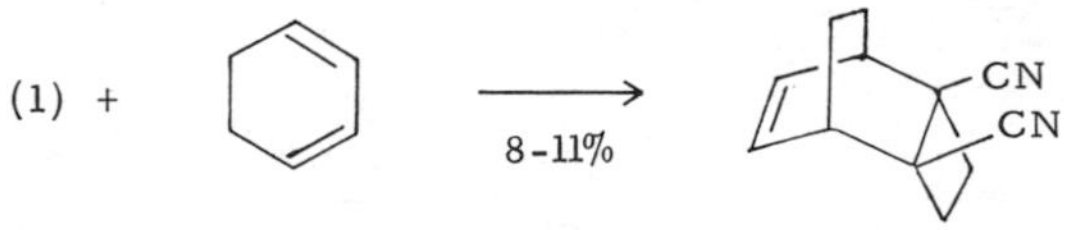

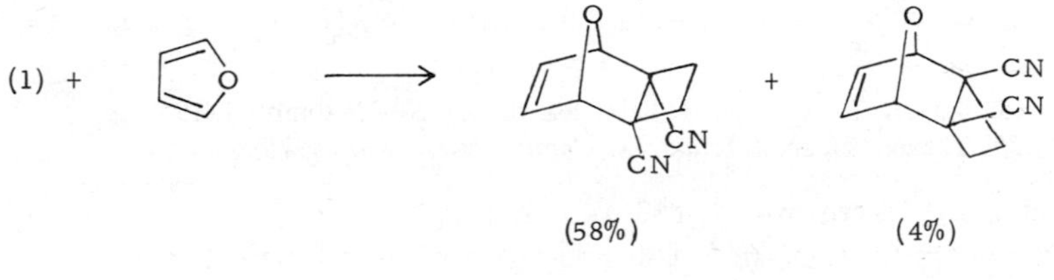

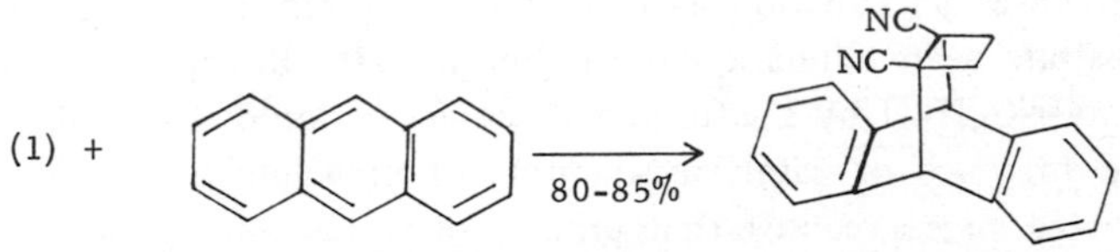

(1) +    $\xrightarrow{42\%}$

(1) +    $\longrightarrow$    (90%)    +    (6%)

(1) +    $\xrightarrow{8-11\%}$

(1) +    $\longrightarrow$    (58%)    +    (4%)

(1) +    $\xrightarrow{80-85\%}$

Another useful reaction of (1) is conversion to 2,3-dicyano-1,3-butadiene by pyrolysis at 380–440° and $10^{-1}$ to $10^{-2}$ torr (82% yield).

[1] D. Belluš, K. von Bredow, H. Sauter, and C. D. Weis, *Helv.*, **56**, 3004 (1973).

**Dicyclohexylcarbodiimide, 1,** 231–236, **2,** 126; **3,** 91; **4,** 141; **5,** 206–207.

*N-Phthaloyl-L-glutamic anhydride.*[1] This derivative of L-glutamic anhydride can be prepared without racemization as illustrated. The product is useful for

preparation of γ-glutamyl esters, amides, and peptides.

*Polymeric carbodiimide reagent for peptide synthesis.* Japanese chemists[2] have prepared a carbodiimide reagent attached to a styrene-divinylbenzene polymer as formulated and have used it successfully for synthesis of dipeptides.

[1] J. A. Elberling, R. T. Zera, and H. T. Nagasawa, *Org. Syn.*, submitted (1975).
[2] H. Ito, N. Takamatsu, and I. Ichikizaki, *Chem. Letters,* 577 (1975).

**Dicyclohexyl-18-crown-6, 4,** 142–143; **5,** 207.

*Phase-transfer catalysis.*[1] Dibenzo-18-crown-6, dibenzo-15-crown-5, and dicyclohexyl-18-crown-6 can function as phase-transfer catalysts in reactions of anions if the substrate bears aliphatic chains. Thus $n\text{-}C_8H_{17}Br$ can be converted quantitatively into $n\text{-}C_8H_{17}I$ by a saturated aqueous solution of KI (Finkelstein reaction) in the presence of catalytic amounts of dicyclohexyl-18-crown-6. Secondary substrates are less reactive than primary ones. The paper reports other phase-transfer-catalyzed reactions as illustrated:

$$\underline{n}\text{-C}_6\text{H}_{13}\text{CH}=\text{CH}_2 + \text{KMnO}_4 \xrightarrow[\text{50\%}]{\overset{\text{cat.}}{\text{C}_6\text{H}_6\text{-H}_2\text{O}}} \underline{n}\text{-C}_6\text{H}_{13}\text{COOH}$$

$$\underline{n}\text{-C}_6\text{H}_{13}\text{COCH}_3 + \text{NaBH}_4 \xrightarrow[\text{92\%}]{\overset{\text{cat.}}{\text{C}_6\text{H}_6\text{-aq. NaOH}}} \underline{n}\text{-C}_6\text{H}_{13}\text{CHOHCH}_3$$

*Chlorofluorocarbene.* Schlosser *et al.*[2] have found that this crown ether is significantly superior to the commonly used ammonium salts for generation of chlorofluorocarbene from dichlorofluoromethane by the two-phase technique. They used potassium hydroxide as base rather than sodium hydroxide, since crown ethers bind potassium ions more selectively than sodium ions. They used the carbene for synthesis of fluorodienes. The method is illustrated for the conversion of methallyl chloride (1) into 2-fluoro-3-methyl-1,3-butadiene (3-fluoro-isoprene), (4). The conversion of (3) into (4) involves a 1,4-elimination of "ICl"

accompanied by ring opening.

[1] D. Landini, F. Montanari, and F. M. Pirisi, *J.C.S. Chem. Comm.*, 879 (1974).
[2] M. Schlosser, B. Spahic, C. Tarchini, and Le Van Chau, *Angew. Chem. Internat. Ed.*, **14**, 365 (1975).

**Di($\eta^5$-cyclopentadienyl) (chloro)hydridozirconium(IV), [(C$_5$H$_5$)$_2$HZrCl] .**
Mol. wt. 257.87; nonvolatile, probably polymeric, sensitive to moisture and to light. Supplier: Alfa (Schwartz Reagent).

(1)

*Preparation.* This hydride is prepared by reduction of dicyclopentadienyl-zirconium dichloride (zirconocene dichloride) with either LiAlH(O-*t*-Bu)$_3$[1] or NaAlH$_2$(OC$_2$H$_4$OCH$_3$)$_2$.[2]

*Hydrozirconation of alkenes.*[2] This organic metallic reagent reacts with olefins at 25–40° to form alkylzirconium complexes of the type (C$_5$H$_5$)$_2$Zn(R)Cl, in which the transition metal moiety is attached to the least hindered position of the olefin. For example, when a suspension of (1) in benzene is shaken with 1-octene or either *cis*- or *trans*-4-octene, (2) is formed in quantitative yield. Thus internally metalated zirconium compounds undergo rearrangement much more rapidly than organoboron or aluminum analogs. Other examples of hydrozirconation are formulated. Very hindered olefins (*e.g.*, tetramethylethylene) fail to react at 40°.

The alkylzirconium compounds are hydrolyzed to alkanes by dilute aqueous acid. They are converted by halogens into alkyl halides and are acylated in high yield by acetyl chloride with liberation of a (C$_5$H$_5$)$_2$Zr(Cl)X species, which can be recycled.

$$(2) \xrightarrow[100\%]{\substack{HCl, H_2O \\ 0^0}} \underline{n}\text{-Octane}$$

$$(2) \xrightarrow[96\%]{Br_2, 0^0} \text{1-Bromooctane}$$

$$(2) \xrightarrow[80\%]{\substack{CH_3COCl, C_6H_6 \\ 25^0}} C_8H_{17}COCH_3$$

An even more striking reaction of the alkylzirconium compounds formed by hydrozirconation is insertion of carbon monoxide into the C—Zr bond to form zirconium-acyl complexes.[3] For example, 1-hexene or an internal hexene is treated in benzene with (1) to form the alkylzirconium complex (6), which is then allowed to react under 20 psi of carbon monoxide at room temperature. The colorless acyl complex (7) is formed within several hours. An intermediate zirconium carbonyl species is not detected. The acyl complex (7) is converted into an aldehyde (8) on protonolysis, into a carboxylic acid (9) by treatment first with aqueous NaOH and then 30% $H_2O_2$, or into an ester (10) by treatment with NBS and then with an alcohol. This procedure differs significantly from

classical hydroformylation in that only the terminal aldehyde is formed, even with internal olefins as starting materials.

*Hydrozirconation of alkynes.*[4] The reagent (1) undergoes facile *cis*-addition to acetylenes. Thus (1) reacts with 1-butyne to give the *trans*-vinylzirconium compound:

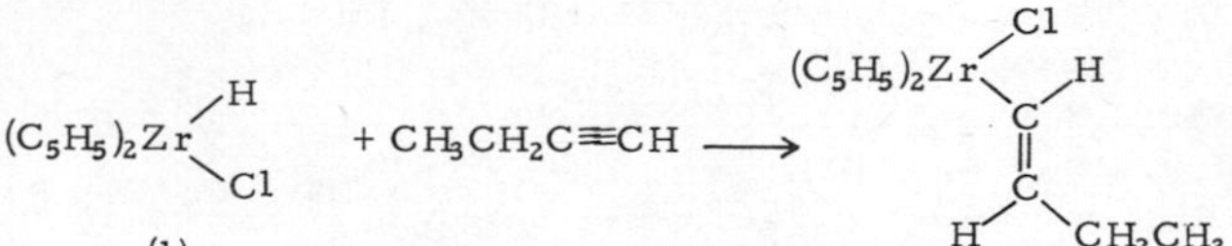

$$(C_5H_5)_2Zr\begin{matrix}H\\Cl\end{matrix} + CH_3CH_2C\equiv CH \longrightarrow$$

(1)

In the case of unsymmetrically disubstituted alkynes, mixtures of two *cis*-adducts are obtained in a ratio that depends on steric effects. For example, hydrozirconation of 5-methyl-2-hexyne with (1) for 2 hr. gives a mixture of (2) and (3) in the ratio 55:45. When a few milligrams of (1) is added to the original

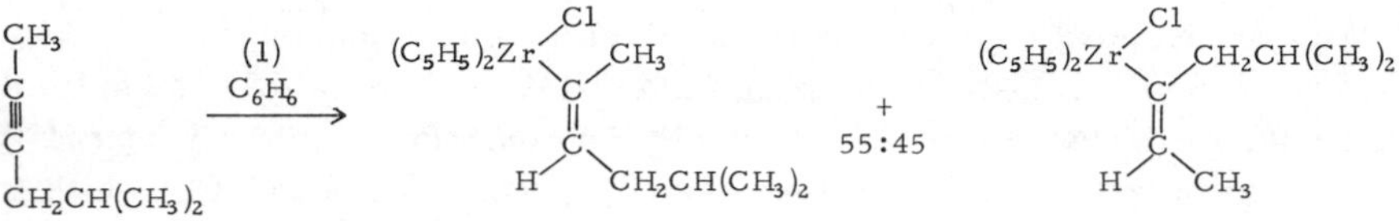

$$\begin{array}{c}CH_3\\|\\C\\|||\\C\\|\\CH_2CH(CH_3)_2\end{array} \xrightarrow[C_6H_6]{(1)}$$

(2)   +   (3)
55:45

(2)   +   (3)   $\xrightarrow{(1)}$   (2)   +   (3)
55:45                              >95%      <5%

mixture in benzene after a few hours, the ratio of (2) to (3) is approximately 95:5.

Treatment of the mixture of vinylic complexes obtained in this way with NBS, NCS, or iodine gives the corresponding vinylic halides in 70–100% yield with retention of stereochemistry and with the same composition of positional isomers:

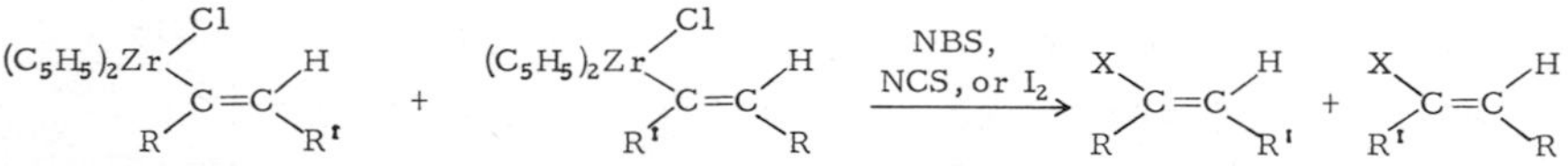

This reaction followed by reaction with lithium dialkyl cuprates is an attractive route to trisubstituted olefins from disubstituted acetylenes.

*Terminal alcohols.*[5] The alkylzirconium (IV) complexes $(C_5H_5)_2Zr(R)Cl$ are oxidized by peroxides or peracids to alcohols in about 40–70% yield. The only by-products are alkanes arising from protonolysis of the C–Zr bond. Actually, in many cases oxygen (dry) is the best reagent (60–80% yield).

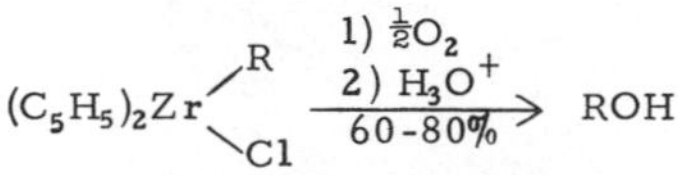

$$(C_5H_5)_2Zr\begin{matrix}R\\Cl\end{matrix} \xrightarrow[60-80\%]{\begin{matrix}1)\ \frac{1}{2}O_2\\2)\ H_3O^+\end{matrix}} ROH$$

This oxygenation was shown to proceed through the alkoxide $(C_5H_5)_2Zr(OR)Cl$. If R is chiral, oxidation proceeds with 50% retention and 50% racemization of configuration. But oxidation with $t$-butyl hydroperoxide results in 100% reten-tion of configuration. The paper includes a possible mechanism for the oxidation.

[1] B. Kautzner, P. C. Wailes, and H. Weigold, *Chem. Comm.*, 1105 (1969).
[2] D. W. Hart and J. Schwartz, *Am. Soc.*, **96**, 8115 (1974).
[3] C. A. Bertelo and J. Schwartz, *ibid.*, **97**, 228 (1975).
[4] D. W. Hart, T. F. Blackburn, and J. Schwartz, *ibid.*, **97**, 679 (1975).
[5] T. F. Blackburn, J. A. Labinger, and J. Schwartz, *Tetrahedron Letters*, 3041 (1975).

$\eta^5$-**Dicyclopentadienyltitanium dichloride (Titanocene dichloride)**, $Cp_2 TiCl_2$. Mol. wt. 249.00. Suppliers: Alfa, ROC/RIC.

*Reduction of bromides.*[1]  In the presence of catalytic amounts of this or-ganometallic reagent, isopropylmagnesium bromide reduces vinyl, aryl, and alkyl bromides in ether at $20°$. Chlorides are not reduced by this system; thus $p$-bromochlorobenzene can be reduced to chlorobenzene in almost quantitative yield. The titanium hydride $Cp_2 Ti—H$ is considered to be involved.

*Examples:*

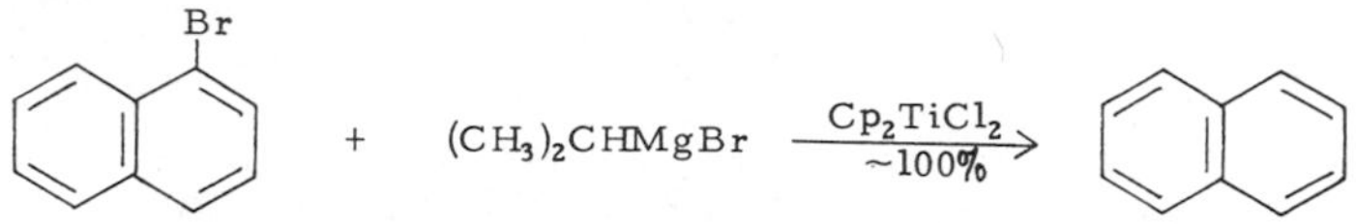

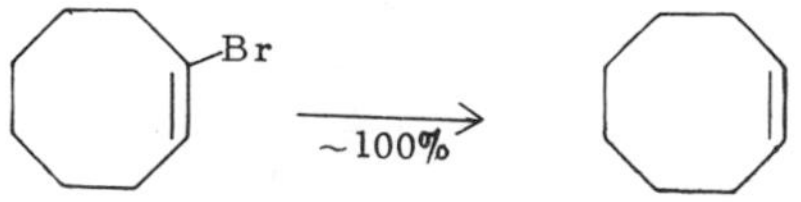

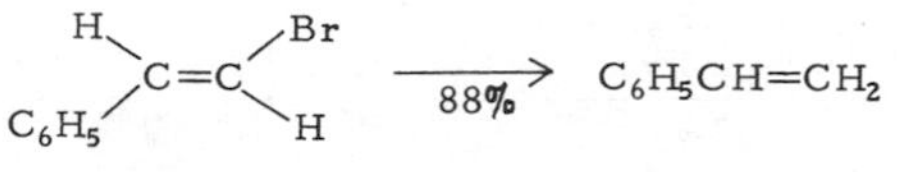

$$\underline{n}\text{-}C_{12}H_{25}Br \xrightarrow[74\%]{} \underline{n}\text{-}C_{12}H_{26}$$

*Reduction of alkoxysilanes.*[2]  The same system readily reduces alkoxysi-lanes, even the highly hindered silane (1).

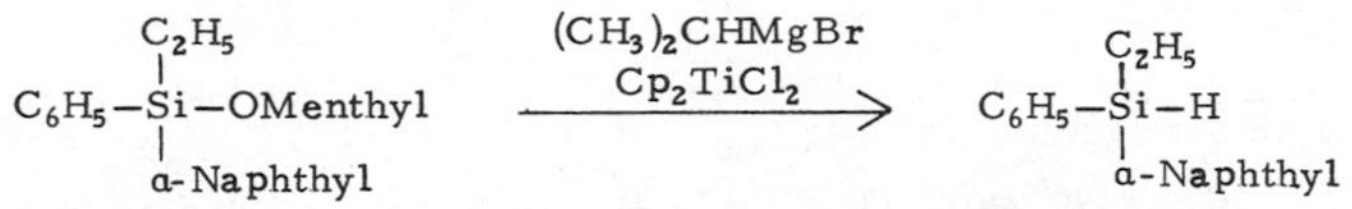

(1)

[1] E. Colomer and R. Corriu, *J. Organometal. Chem.*, **82**, 367 (1974).
[2] R. J. P. Corriu and B. Meunier, *J. Organometal. Chem.*, **65**, 187 (1974).

**Diethoxyaluminum chloride,** $Al(OC_2H_5)_2Cl$. Mol. wt. 152.56.

The reagent is prepared *in situ* by mixing one part of $AlCl_3$ with two parts of aluminum ethoxide (ROC/RIC).

*Dehydration of β-hydroxy esters.*[1]  β-Hydroxy esters are dehydrated to α,β-unsaturated esters in 44–56% yield by treatment with aluminum ethoxide or diethoxyaluminum chloride and then with lithium diisopropylamide. Although the yields are moderate, no β,γ-isomers are formed and, in suitable cases, only the E-isomer is obtained. The reaction is considered to proceed through a β-alanoxy intermediate, which can form a six-membered ring transition state on enolization.

*Examples:*

[1] J. A. Katzenellenbogen and T. Utawanit, *Am. Soc.*, **96**, 6153 (1974).

**Diethylaluminum cyanide,** 1, 244; 2, 127–128; 4, 146–147.

*Stereospecific conjugate cyanation.* By adjusting the reaction conditions, Kelly *et al.*[1] were able to convert the octalone (1) into either the *cis*-cyano-decalone (2) or the *trans*-isomer (3).

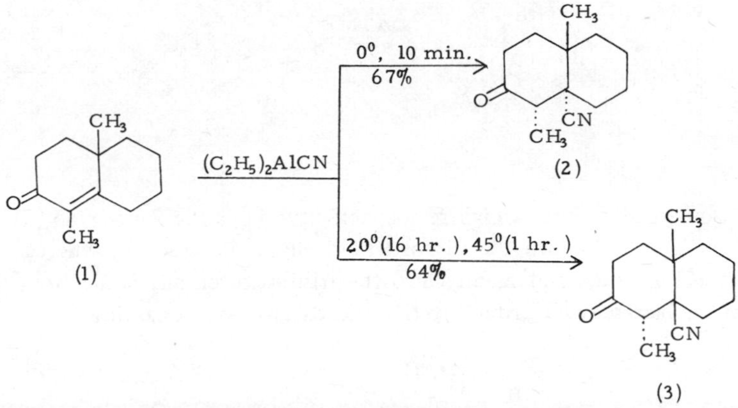

(1)   (2)   (3)

$^1$R. B. Kelly, S. J. Alward, and J. Zamecnik, *Canad. J. Chem.*, **53**, 244 (1975).

**Diethylaluminum 2,2,6,6-tetramethylpiperidide (DATMP),**

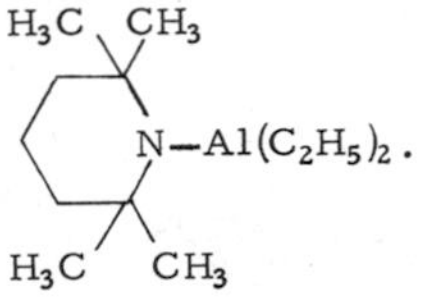

(1)

Mol. wt. 225.35.

This diethylaluminum dialkylamide is prepared *in situ* from diethylaluminum chloride and lithium 2,2,6,6-tetramethylpiperidide (LiTMP, **4**, 310–311) in benzene at $0°$.

*Isomerization of epoxides to allylic alcohols.* This reaction is usually carried out with base (lithium diethylamide, **1**, 610–611; **2**, 247–248; **4**, 298). Japanese chemists$^1$ have used instead this new reagent of the type $R^1R^2NAl(C_2H_5)_2$, which is effective because of the affinity of aluminum for oxygen. Of several species, DATMP proved most satisfactory. It effects the isomerization of (1) and of (3) in high yield under very mild conditions. Note, however, that the oxides of cyclopentene, cyclohexene, and cycloheptene are stable under the same conditions.

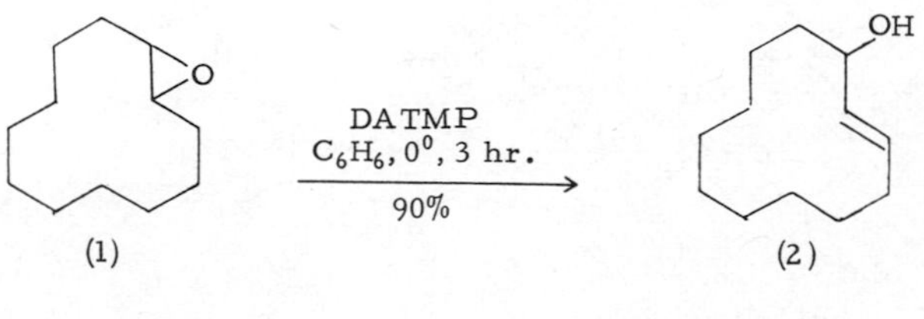

(1)   DATMP $C_6H_6, 0°, 3$ hr. 90%   (2)

(3)   DATMP 0.5 hr., 0° 88%   (4)

The isomerization shows high regiospecificity. Thus the Z-epoxide (5) is converted in high yield into the disubstituted allylic alcohol (6), whereas the E-epoxide (7) is converted mainly into the trisubstituted allylic alcohol (8). The bulk of the base may contribute to the selectivity of isomerization.

(5)   DATMP 0°, 2 hr. 90%   (6)

(7)   DATMP 0°, 2 hr.   (8, 78%) + 6 (12%)

*1,3-Dienes.* Yamamoto et al.[2] have recorded a sequence of reactions that effects conversion of an allylic alcohol into a 3-ene-1,2-diol and then into a 1,3-diene. The method is illustrated for the case of nerol (1) and geraniol (5).

(1)   1) $(CH_3)_3C-OOH$, V(acac)$_2$, 25° 2) $(CH_3)_3SiCl$, $HN[Si(CH_3)_3]_2$, Py   (2)   1) 2) KF, $CH_3OH$, $H_2O$ 79% from (1)

(3)   1) $PBr_3$, $Cu_2Br_2$ -78° → 0° 2) Zn 58%   (4)

$$(5) \xrightarrow{65\%} (6) \xrightarrow{66\%} (7)$$

The starting alcohol is epoxidized (**t-butyl hydroperoxide, vanadium acetyl-acetonate,** this volume) and then the hydroxyl group is converted into the silyl ether. The resulting epoxy silyl ether reacts with the organometallic reagent to give, after desilylation, almost exclusively the enediol (3). Even higher specificity can be achieved by use of $t$-butyldimethylsilyl as the protecting group. Application of the sequence to geraniol leads to (6) in 65% yield with less than 6% of other isomers.

The diols (3) and (6) are converted into 1,3-dienes by treatment with phosphorus tribromide and cuprous bromide (essential) at $-78°$ for 5 min. and then at $0°$ for 1 hr. followed by addition of excess zinc powder at $0°$ for 2 hr. In this way (3) is converted into myrcene (4), and (6) is converted into *trans-β*-ocimene (7). The paper reports the synthesis of four other conjugated dienes.

[1] A. Yasuda, S. Tanaka, K. Oshima, H. Yamamoto, and H. Nozaki, *Am. Soc.,* **96**, 6513 (1974).

[2] S. Tanaka, A. Yasuda, H. Yamamoto, and H. Nozaki, *Am. Soc.,* **97**, 3252 (1975).

**Diethylaminosulfur trifluoride** $(C_2H_5)_2NSF_3$. Mol. wt. 161.19, b.p. $46-47°/10$ torr. Stable for several months in an inert plastic bottle. *Caution: Handle with gloves; this material can cause burns.*

The reagent is prepared from N,N-diethylaminotrimethylsilane (PCR, Inc.):

$$(C_2H_5)_2NSi(CH_3)_3 + SF_4 \longrightarrow (C_2H_5)_2NSF_3 + FSi(CH_3)_3$$

$$(1, 80-90\%)$$

*Fluorination of alcohols.*[1] Primary, secondary, and tertiary alcohols are converted by (1) into the corresponding fluorides; the reaction occurs readily at $10°$ or below. One advantage of this reagent is that products of dehydration and carbonium ion rearrangement are formed in smaller amounts than with $SF_4$, $SeF_4 \cdot$ pyridine (**5**, 576–577), and $(C_2H_5)_2NCF_2CHClF$ (**5**, 214–216). It is also useful for fluorination of aldehydes and ketones that are sensitive to acid.

*Examples:*

$$\underline{p}\text{-}NO_2C_6H_4CH_2OH + (1) \xrightarrow[65-70\%]{CH_2Cl_2, 10°} \underline{p}\text{-}NO_2C_6H_4CH_2F + (C_2H_5)_2NSOF + HF$$

$$\text{Cyclooctanol} \xrightarrow[\quad]{(1) \atop CCl_3F} \underset{70\%}{\text{Cyclooctyl fluoride}} + \underset{30\%}{\text{Cyclooctene}}$$

$$(CH_3)_2CHCH_2OH \xrightarrow[\text{}]{\overset{(1)}{CH_3O(CH_2CH_2O)_2CH_3}} (CH_3)_2CHCH_2F + (CH_3)_3CF$$
$$49\% \qquad\qquad 21\%$$

$$C_6H_5CHO \xrightarrow[75\%]{\overset{(1)}{CH_2Cl_2}} C_6H_5CHF_2 + (C_2H_5)_2NSOF$$

$$C_6H_5COCH_3 \xrightarrow[66\%]{\overset{(1)}{Glyme}} C_6H_5CF_2CH_3$$

*Acid fluorides.*[2]  Dialkylaminosulfur trifluoride converts acid chlorides into acid fluorides in 70–80% yield.

$$R'COCl + R_2NSF_3 \xrightarrow[70-80\%]{20 \to 60^0} R'COF + R_2NSClF_2$$

*Related reactions:*

$$R'SOCl \xrightarrow[85-90\%]{} R'SOF$$

$$R'SO_2Cl \xrightarrow[70-80\%]{} R'SO_2F$$

$$(R'O)_2POCl \xrightarrow[70\%]{} (R'O)_2POF$$

[1] W. J. Middleton, *J. Org.*, **40**, 574 (1975).
[2] L. N. Markovski and V. E. Pashinnik, *Synthesis*, 801 (1975).

## Diethylamino(trimethyl)silane, $(CH_3)_3SiN(C_2H_5)_2$.

*Ullmann reaction.* The presence of free amino, hydroxyl, or carboxyl groups prevents formation of biaryls in the Ullmann reaction. The reaction proceeds normally (35–70% yield) when these groups are protected by silylation.[1]

*Examples:*

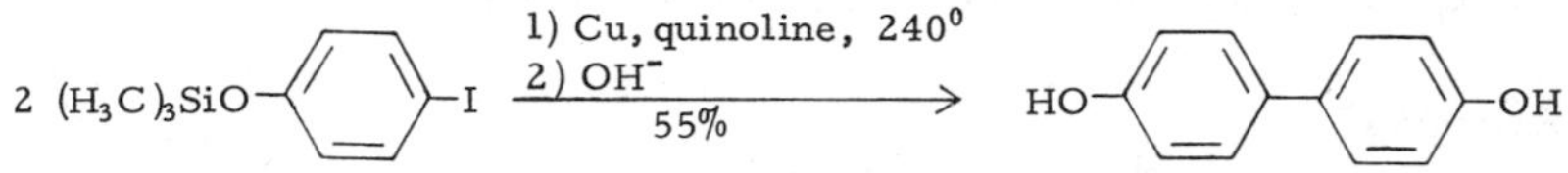

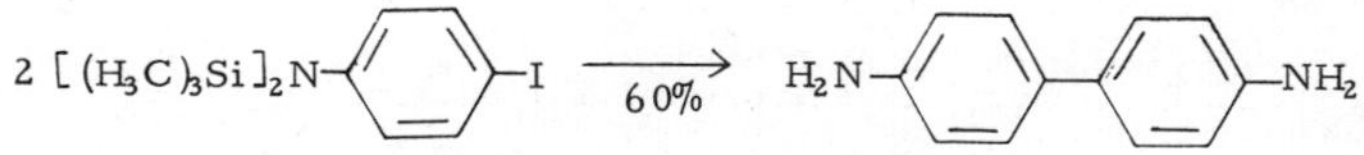

[1] F. D. King and D. R. M. Walton, *Synthesis,* 40 (1976).

**Diethyl azodicarboxylate (DAD), 1,** 245–247; **2,** 128–129; **4,** 148–149; **5,** 212–213.

*Purines.*[1] Fusion of the 6-benzylaminouracil (1) with 3 eq. of diethyl azodicarboxylate (DAD) (2) at 170–180° for 2 hr. gives the purine 8-phenyltheophylline (4) in 71% yield. Use of 1 eq. of DAD gives a Michael-type adduct (3), which is not converted by thermolysis into (4). However, when it is heated with DAD, (4) is obtained, but in lower yield. It is suggested that DAD acts as a dehydrogenation reagent to form (a), which cyclizes immediately to the purine. The reagent also supplies the new nitrogen atom in (4). Yields of purines

are lower when 6-alkylaminouracils are used as precursors.

[1] F. Yoneda, S. Matsumoto, and M. Higuchi, *J.C.S. Chem. Comm.,* 146 (1975).

**Diethyl(2-chloro-1,1,2-trifluoroethyl)amine   (CTT), 1, 249; 2, 130; 3, 95–96; 5, 214–216.**

*Fluorination of an allylic alcohol.* Reaction of methyl gibberellate (1) with this fluoroamine in $CH_2Cl_2$ at $0°$ gives (2) and (3) as the major products.[1]

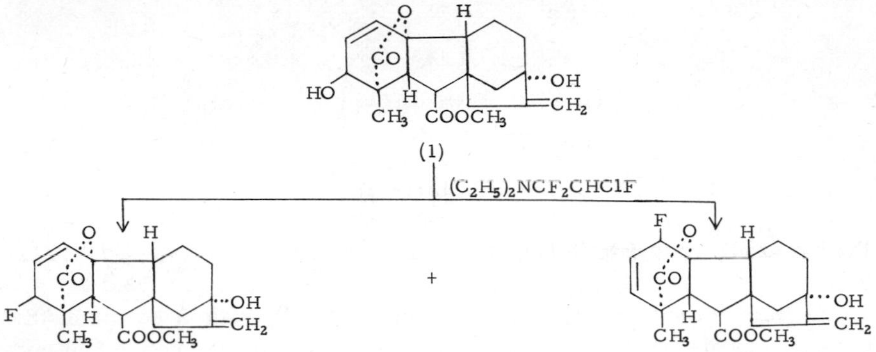

(1)

$(C_2H_5)_2NCF_2CHClF$

(2, 44%)          +          (3, 31%)

*Fluorinated cephalosporins.*[2] A typical reaction of CTT with a cephalosporin is formulated.

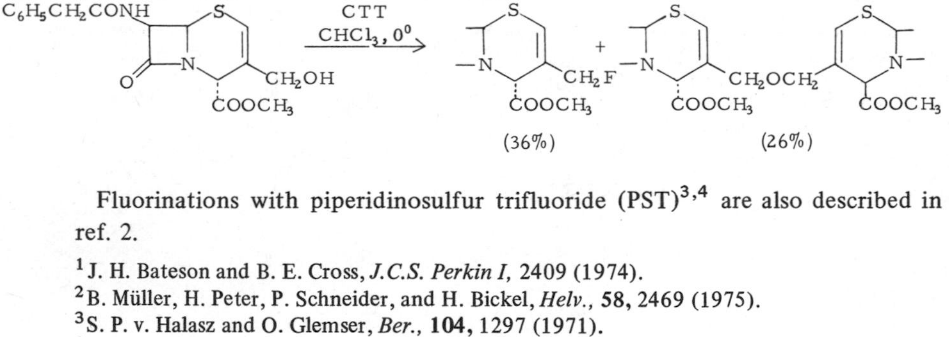

(36%)          (26%)

Fluorinations with piperidinosulfur trifluoride (PST)[3,4] are also described in ref. 2.

[1] J. H. Bateson and B. E. Cross, *J.C.S. Perkin I*, 2409 (1974).
[2] B. Müller, H. Peter, P. Schneider, and H. Bickel, *Helv.*, **58**, 2469 (1975).
[3] S. P. v. Halasz and O. Glemser, *Ber.*, **104**, 1297 (1971).
[4] L. N. Markovskij, V. E. Pashinnik, and A. V. Kirsanow, *Synthesis*, 787 (1973).


**Diethylene orthocarbonate,** 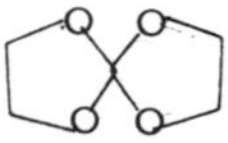 (1). Mol. wt. 132.15, m.p. 143.5°.

*Preparation.* The reagent can be prepared (33% yield) by the reaction of sodium glycolate with $CCl_3NO_2$.[1] It has also been prepared by the reaction of carbon disulfide with bis(tributyltin)ethylene glycolate (26% yield)[2] and with dithallous ethylene glycolate (60% yield).[3]

*Transacetalization.*[1] The reagent is effective for transacetalization. The reaction is carried out at $20°$ in chloroform with catalysis by *p*-toluenesulfonic acid or by slightly wet $BF_3$ etherate. Anhydrous $BF_3$ etherate is ineffective.

Acetophenone is converted in this way into the diethylene ketal in 74% yield. Under the same conditions benzophenone is unreactive. The reagent appears particularly useful for preparation of acetals of aromatic *ortho*-hydroxy aldehydes. Ketalization of $\Delta^4$-3-keto steroids is accompanied by migration of the double bond [(1) → (2)]. The tetrathio orthocarbonate (3)[4] can be used in the

(1)                                                  (2)

same way to prepare ethylene dithioacetals.

(3)

[1] D. H. R. Barton, C. C. Dawes, and P. D. Magnus, *J.C.S. Chem. Comm.*, 432 (1975).
[2] S. Sakai, Y. Kiyohara, K. Itohi, and Y. Ishii, *J. Org.*, **35**, 2347 (1970).
[3] S. Sakai, Y. Kuroda, and Y. Ishii, *ibid.*, **37**, 4198 (1972).
[4] J. J. D'Amico and R. H. Campbell, *ibid.*, **32**, 2567 (1967).

**N,N-Diethylhydroxylamine**, $HON(C_2H_5)_2$. Mol. wt. 89.14. Supplier: Aldrich.
    The preparation has been reported only in a patent.[1]
    *Reduction of quinones to hydroquinones.*[2] 1,4-Benzoquinones and 1,4-naphthoquinones are reduced to the hydroquinones by this reagent, which is oxidized to the nitrone $CH_3CH=N^+(O^-)C_2H_5$; this product forms a complex with the hydroquinone, which is destroyed by treatment with dilute hydrochloric acid. Yields are in the range of 65–95%. One advantage is that halo, acetoxy, and azo groups are not affected.

[1] W. Ruppert, *Ger. Pat.*, 1,004,191 (1957) [*C.A.*, **54**, 584c (1960)].
[2] S. Fujita and K. Sano, *Tetrahedron Letters*, 1695 (1975).

**Diethyl isocyanomethylphosphonate, 4**, 271.
    Experimental details for the preparation and reactions of this phosphonate are available.[1]
    The reagent reacts with aldehydes (but not ketones) in the presence of sodium cyanide as catalyst to give oxazolines (1). These products are converted into vinyl formamides (2) by potassium *t*-butoxide in THF.

(1)                                                  (2)

[1] U. Schöllkopf, R. Schröder, and D. Stafforst, *Ann.*, 44 (1974).

**Diethyl ketomalonate,** $O{=}C(COOC_2H_5)_2$. Mol. wt. 174.13, b.p. 208–210°. Supplier: Aldrich.

*β,γ-Unsaturated-δ-valerolactones.*[1] This reagent undergoes Diels–Alder reactions with 1,3-dienes (1) at 130° to give, after hydrolysis, the adducts (2) in 63–70% yield. These are converted into lactones (3) by sodium azide degradation without isolation of intermediates. Diethyl ketomalonate therefore functions as an equivalent to carbon dioxide, which does not undergo Diels–Alder reactions.

(1, $R^1$, $R^2 = H$, $CH_3$)

The lactone (3, $R^1$, $R^2 = CH_3$) is converted into (4), (5), and (6) as formulated.

[1] R. A. Ruden and R. Bonjouklian, *Am. Soc.*, **97**, 6892 (1975).

**Diethyl lithiodichloromethylphosphonate,** (1). Mol. wt. 226.96.

*Preparation.*[1,2] This reagent is prepared from chloromethylphosphonyl dichloride (Aldrich):

This halomethyllithium reagent can also be obtained by halogen–lithium exchange in THF at $-80°$ between $n$-butyllithium and diethyl trichloromethylphosphonate,[2] $CCl_3P(O)(OC_2H_5)_2$.

*1,1-Dichloro-1-alkenes.*[2,3] The reagent has been shown to react with aldehydes or ketones at $-80°$ in THF to give moderate yields of 1,1-dichloro-1-alkenes:

$$\underset{R^1}{\overset{R}{>}}C=O + (1) \xrightarrow{THF, -80°} R-\underset{\underset{R^1}{|}}{\overset{\overset{OLi}{|}}{C}}-CCl_2\overset{\overset{O}{||}}{P}(OC_2H_5)_2 \xrightarrow[47-69\%]{THF\ reflux}$$

$$\underset{R^1}{\overset{R}{>}}C=CCl_2 + Li^+(C_2H_5O)_2\overset{\overset{O}{||}}{P}O^-$$

French chemists[4] have been able to increase the yields to 85–90% by use of a 40:60 mixture of THF–ether as solvent instead of pure THF. Note that Corey and Fuchs (4, 550) have prepared 1,1-dibromo-1-alkenes by the reaction of an aldehyde with triphenylphosphine, carbon tetrabromide, and zinc in methylene chloride.

*Preparation of alkynes.*[4] The 1,1-dichloro-1-alkenes have been converted into 1-alkynes or internal alkynes by treatment with $n$-butyllithium at $-10°$ to $-70°$. The resulting lithium alkynides can be hydrolyzed or alkylated (equation I).

(I)  $RCH=CCl_2 \xrightarrow[THF]{2\ n\text{-}BuLi} RC\equiv CLi$

$\xrightarrow[75-95\%]{H_3O^+} RC\equiv CH$

$\xrightarrow[75-90\%]{ClCH_2OCH_3} RC\equiv CCH_2OCH_3$

1,1-Dichloro-1-alkenes can also be converted into 1-chloro-1-alkynes by reaction with lithium diethylamide in ether–THF (equation II):

(II)  $RCH=CCl_2 \xrightarrow[65-90\%]{\substack{1)\ Li[N(C_2H_5)_2]\\2)\ H_3O^+}} RC\equiv CCl$

[1] P. Savignac, M. Dreux, and P. Coutrot, *Tetrahedron Letters,* 609 (1975).
[2] P. Savignac, J. Petrova, M. Dreux, and P. Coutrot, *Synthesis,* 535 (1975).
[3] D. Seyferth and R. S. Marmor, *J. Organometal. Chem.,* **59**, 237 (1973).
[4] J. Villieras, P. Perriot, and J. F. Normant, *Synthesis,* 458 (1975).

**Diethyl mercurybisdiazoacetate,** $Hg(N_2CCOOC_2H_5)_2$  (1).   Mol.  wt.  426.80, m.p. 103–104°, sulfur-yellow.

*Preparation.* This substance is prepared by the reaction of yellow mercuric oxide and ethyl diazoacetate (85% yield).[1]

*Carboethoxymethyne.* Strauz *et al.*[2] reasoned that photolysis of (1) should

$$Hg(N_2CCOOC_2H_5)_2 \xrightarrow{\text{h}\nu} 2\ N_2\ +\ Hg\ +\ 2\ :\dot{C}COOC_2H_5$$

(1)                                                              (2)

generate a carbyne (2), and indeed photolysis at a wavelength shorter than 290 nm in the presence of an olefin yields an array of products, some of which evidently are formed from (2). Two general reactions take place: addition to form a cyclopropyl radical and insertion into the allylic C—H bonds to form an alkyl radical.

Patrick and Kovitch[3] have shown that the carbyne (2) reacts with chloro-alkanes to give mainly products formed by insertion into a C—Cl bond. In contrast, carboethoxycarbene, :CHCOOC$_2$H$_5$, under the same conditions does not give products of insertion into C—Cl or C—H bonds. It follows that (2) does not tend to abstract a hydrogen atom to form the carbene.

[1] E. Buchner, *Ber.,* **28,** 215 (1895).
[2] O. P. Strauz, G. J. A. Kennepohl, F. X. Garneau, T. DoMinh, B. Kim, S. Valenty, and P. S. Skell, *Am. Soc.,* **96,** 5723 (1974).
[3] T. B. Patrick and G. H. Kovitch, *J. Org.,* **40,** 1527 (1975).

**Diethylmethyl(methylsulfonyl)ammonium    fluorosulfonate**    ("Easy   Mesyl"), $CH_3SO_2\overset{+}{N}CH_3(C_2H_5)_2\ FSO_3^-$  (1). Mol. wt. 251.32, m.p. ~115° dec. Must be protected from moisture.

*Preparation:*

$$CH_3SO_2N(C_2H_5)_2\ +\ CH_3OSO_2F\ \xrightarrow[64\%]{50°,\ 3\ days}\ (1)$$

*Mesylation.* King and du Manoir[1] have prepared four quaternary methyl-sulfonylammonium salts of type (1) and have found that in the presence of catalytic amounts of base (pyridine or N,N-dimethylaminoacetonitrile) they react with alcohols at 0° within 10 min. to form mesylates in high yields (usually 75–100%). Optimum yields were obtained with (1). Use of strong bases such as triethylamine results in lower yields of mesylates. Amines are also converted by (1) into methanesulfonamides in high yield. The paper by King and du Manoir also presents evidence that mesylations with (1) involve the sulfene $CH_2=SO_2$.

[1] J. F. King and J. R. du Manoir, *Am. Soc.,* **97,** 2566 (1975).

**Diethyl phenylsulfinylmethylphosphonate,** $(C_2H_5O)_2P(O)CH_2S(O)C_6H_5$ (1). Mol. wt. 276.29.

*Phase-transfer catalysis.* A Polish group[1] reported that the Wittig–Horner reaction with α-phosphoryl sulfoxides, sulfones, and sulfides could be conducted in a two-phase system (aqueous NaOH—methylene chloride) with benzyltriethyl-ammonium chloride as catalyst. Later work showed that a catalyst was not necessary because these sulfur compounds themselves can function as catalysts for phase-transfer reactions. Thus (1) is an effective catalyst for alkylation of ketones by alkyl halides in the presence of 50% aqueous NaOH. Related, but somewhat less active, catalysts are sulfones such as (2), α-disulfoxides (3), and bisphosphonates (4).

$$
\begin{array}{ccc}
\underset{\text{(2)}}{(C_2H_5O)_2\overset{O}{\overset{\|}{P}}CHR\underset{\underset{O}{\|}}{\overset{O}{\overset{\|}{S}}}C_6H_5} & \underset{\text{(3)}}{C_6H_5\overset{O}{\overset{\|}{S}}CH_2\overset{O}{\overset{\|}{S}}C_6H_5} & \underset{\text{(4)}}{(C_2H_5O)_2\overset{O}{\overset{\|}{P}}CH_2\overset{O}{\overset{\|}{P}}(OC_2H_5)_2}
\end{array}
$$

Catalysts of this new type do not catalyze exchange of alkyl bromides with KI or with cyanide ion. Presumably they function in a way different from that of onium salts.[2]

[1] M. Mikolajczyk, S. Grzejszczak, W. Midura, and A. Zatorski, *Synthesis*, 278 (1975).
[2] M. Mikolajczyk, S. Grzejszczak, A. Zatorski, F. Montanari, and M. Cinquini, *Tetrahedron Letters*, 3757 (1975).

**Diethyl phenylthiomethanephosphonate,** (1). Mol. wt. 260.30, b.p. 145–150°/ 0.5 torr.

*Preparation:*

*Wittig–Horner reaction.*[1] This phosphonate has been used for carbonyl olefination with α-chlorination, as formulated. It is advantageous to add a stoichiometric amount of lithium bromide to the reaction.

[1] P. Coutrot, C. Laurenco, J. Petrova, and P. Savignac, *Synthesis*, 107 (1976).

## Diethyl phosphonate, 1, 251–253; 2, 132–133.

*Arylphosphonic acids.* Diethyl arylphosphonates can be prepared by re-action of aryl iodides (bromides react more slowly) with this reagent in the presence of sodium in liquid ammonia with irradiation of about 300–350 nm. This arylation is believed to involve radical and radical anion intermediates ($S_{RN}1$ mechanism).[1]

$$C_6H_5I \ + \ (C_2H_5O)_2PO^-Na^+ \xrightarrow[83\%]{h\nu,\ NH_3} C_6H_5\overset{\overset{O}{\|}}{P}(OC_2H_5)_2 \ + \ NaI$$

[1] J. F. Bunnett, and X. Creary, *J. Org.*, **39**, 3612 (1974); J. F. Bunnett, and R. H. Weiss, *Org. Syn.*, submitted (1975).

## Diethyl phosphorochloridate, 1, 248; 3, 98; 5, 217.

*Thiol esters.*[1] Thiol esters can be prepared by reaction of a carboxylic acid (aliphatic or aromatic) with this chloridate and triethylamine in THF to form a rather unstable intermediate (a), which reacts with the Tl(I) salts of alkane-thiols[2] to form thiol esters (equation I). Note that hydroxy acids react selec-tively to form thiol esters. Selected examples of thiol esters prepared in this way are formulated with the yields shown in parentheses.

$$\text{(I)} \quad RCOOH \xrightarrow[\text{N(C}_2\text{H}_5)_3 \text{ in THF}]{(C_2H_5O)_2P(O)Cl} \left[ RCOO\overset{\overset{O}{\uparrow}}{P}(OC_2H_5)_2 \right] \xrightarrow[85-95\%]{Tl(I)SR^1} RCOSR^1$$
$$\text{(a)}$$

$$CH_3(CH_2)_5CH(OH)(CH_2)_{10}COSC(CH_3)_3$$
$$(84\%)$$

$$(91\%)$$

$$(90\%)$$

$$(89\%)$$

[1] S. Masamune, S. Kamata, J. Diakur, Y. Sugihara, and G. S. Bates, *Canad. J. Chem.*, **53**, 3693 (1975).
[2] Prepared by reaction of Tl(I) ethoxide with thiols.

## Diethyl phosphorocyanidate (Diethylphosphoryl cyanide), 5, 217.

*Thiol esters.* Thiol esters can be prepared directly from carboxylic acids and thiols with diethyl phosphorocyanidate (2 eq.) and triethylamine (2 eq.)

$$\overset{O}{\overset{\|}{}}$$

in DMF at $25°$ (3 hr.). Diphenyl phosphoroazidate, $N_3P(OC_6H_5)_2$ (4, 210–211; 5, 280), can also be used, but it is less satisfactory except for preparation of

$$RCOOH + R'SH \longrightarrow RCOSR'$$

thiol esters of $\alpha$-amino acids. These are formed with little, if any, racemization.[1]

*Synthesis of amides.* Amides can be obtained in about 50–95% yield by mixing carboxylic acids and amines with this reagent and triethylamine in DMF. Yields are lower in the absence of the base. The reagent can also be used for synthesis of peptides with virtually no racemization. Thus benzoyl-L-leucylglycine ethyl ester was obtained in 86% yield from benzoyl-L-leucine and glycine ethyl ester hydrochloride; the optical purity was 96%.[2]

In connection with synthesis of quinomycin model systems, Chen and Olsen[3] found that diethylphosphoryl cyanide is the reagent of choice for coupling amino acid derivatives to *cis*- or *trans*-4-aminocyclohexanecarboxylic acid. Other procedures, including the carbodiimide, *p*-nitrophenyl active ester, and symmetrical anhydride methods, were less satisfactory.

*Solid-phase peptide synthesis.*[4] Both diethyl phosphorocyanidate and the related diphenylphosphoryl azide have been shown to be useful in solid-phase peptide synthesis. They have been used successfully for synthesis of porcine motilin (composed of 22 amino acids).

[1] S. Yamada, Y. Yokoyama, T. Shioiri, *J. Org.*, **39**, 3302 (1974).
[2] S.-I. Yamada, Y. Kasai, and T. Shioiri, *Tetrahedron Letters*, 1595 (1973).
[3] W.-Y. Chen and R. K. Olsen, *J. Org.*, **40**, 350 (1975).
[4] S. Yamada, N. Ikota, and T. Shioiri, *Am. Soc.*, **97**, 7174 (1975).

**Diethyl phosphorylmethyl methyl sulfoxide**, $(C_2H_5O)_2\overset{O}{\overset{\|}{P}}CH_2\overset{O}{\overset{\|}{S}}CH_3$ (1). Mol. wt. 214.22.

*Preparation*[1] :

$$(C_2H_5O)_3P + CH_3SCH_2Cl \xrightarrow[49\%]{} (C_2H_5O)_2\overset{O}{\overset{\|}{P}}CH_2SCH_3 \xrightarrow[96.5\%]{NaIO_4} (C_2H_5O)_2\overset{O}{\overset{\|}{P}}CH_2\overset{O}{\overset{\|}{S}}CH_3$$

$\alpha,\beta$-*Unsaturated sulfoxides.*[2] $\alpha,\beta$-Unsaturated sulfoxides (2) can be prepared in 50–80% yield by the Horner–Wittig reaction of an aldehyde or ketone with the anion of (1) at $-78°$. The reaction is not stereospecific; E- and Z-isomers

$$(1) \xrightarrow[\text{THF, } -78°]{\text{n-BuLi}} (C_2H_5O)_2\overset{O}{\overset{\|}{P}}\underset{\underset{Li}{|}}{C}H\overset{O}{\overset{\|}{S}}CH_3 \xrightarrow[50-80\%]{\overset{R^1}{\underset{R^2}{}}C=O} CH_3\overset{O}{\overset{\|}{S}}CH=C\overset{R^1}{\underset{R^2}{}} + (C_2H_5O)_2\overset{O}{\overset{\|}{P}}OLi$$
$$(2)$$

of (2) are obtained from aldehydes and unsymmetrical ketones.

[1] M. Mikolajczyk and A. Zatorski, *Synthesis*, 669 (1973).
[2] M. Mikolajczyk, S. Grzejszczak, and A. Zatorski, *J. Org.*, **40**, 1979 (1975).

**Diethylzinc-Bromoform-Oxygen.**

*Monobromocyclopropanes*[1] (*cf.* **1**, 253; **2**, 134; **4**, 153). The carbenoid reagent derived from diethylzinc and bromoform in the presence of oxygen converts olefins into monobromocyclopropanes in 70–85% yield. Little reaction occurs in the absence of $O_2$. *syn*-Isomers are formed selectively. In the reaction with *cis*-butene-2 only traces of the *anti*-isomer were detected by glc.

This system is also useful for ring expansion of arenes to 7-ethylcyclohepta-1,3,5-trienes (**4**, 153).

[1] S. Miyano, Y. Matsumoto, and H. Hashimoto, *J.C.S. Chem. Comm.*, 364 (1975).

**Diethylzinc-Iodoform, 4**, 153.

*Purification of diethylzinc.*[1]

*Iodocyclopropanation.*[2] Cyclohexene, 1-hexene, *cis*-2-butene, isobutene, and styrene are converted into the corresponding iodocyclopropanes by this system in yields of 70, 34, 63, 55, and 44%, respectively.

[1] S. Miyano and H. Hashimoto, *Bull. Chem. Soc. Japan*, **46**, 3257 (1973).
[2] *Idem, ibid.*, **47**, 1500 (1974).

**$p$-Dihydroxyborylbenzyloxycarbonyl chloride,** $(HO)_2B$—⟨benzene ring⟩—$CH_2OCOCl$ (1).

Mol. wt. 214.41.

*Preparation.* This substance is prepared by reaction of $p$-hydroxymethyl-phenylboronic acid, m.p. 194–195°, with excess phosgene in dry dioxane. It can be protected as the catechol complex, m.p. 135°.

*Protection of amino groups.* Kemp and Roberts[1] have used this boron-derived reagent for protection of amino groups during peptide synthesis. It is compatible with $p$-nitrophenyl ester, acyl azide, and DCC coupling procedures. The Dobz group is cleaved by standard conditions for the carbobenzoxy group. More interestingly, deprotection can be achieved by two unique reactions: hydrolysis catalyzed by a metal ion (equation I) and peroxide oxidation to a phenol (equation II).

(I)

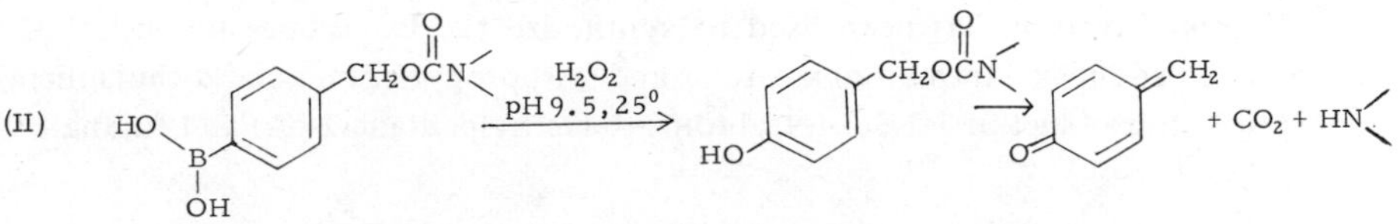

(II)

Another interesting property is that Dobz amino acids can be solubilized in $CH_3CN$, $CCl_4$, and $CH_2Cl_2$ by addition of appropriate complexing agents.

[1] D. S. Kemp and D. C. Roberts, *Tetrahedron Letters*, 4629 (1975).

**Diimide, 1**, 257–258; **2**, 139; **3**, 99–101; **4**, 154–155; **5**, 220.

*Reduction of alkenes.* Siegel et al.[1] have discussed the stereochemistry of diimide reduction of alkenes as compared with catalytic hydrogenation. The same laboratory[2] investigated the reduction of conjugated dienes to enes; no evidence for 1,4-reduction was obtained.

[1] S. Siegel, G. M. Foreman, and D. Johnson, *J. Org.*, **40**, 3589 (1975).
[2] S. Siegel, M. Foreman, R. P. Fisher, and S. E. Johnson, *ibid.*, **40**, 3599 (1975).

**Diiron nonacarbonyl, 1**, 259–260; **2**, 139–140; **3**, 101; **4**, 157–158; **5**, 221–224.

*Troponoids* (**5**, 222–223). The reaction of $\alpha,\alpha'$-dibromo ketones with diiron nonacarbonyl to generate an oxyallyl-Fe(II) species originally suffered one limitation: only secondary and tertiary dibromo ketones reacted satisfactorily. For example, reaction of $\alpha,\alpha'$-dibromoacetone, $BrCH_2COCH_2Br$, fails. Noyori et al.[1] have presented a solution to this limitation. The reaction is carried out with a polybromo ketone and the bromine atoms in the adduct are removed with Zn–Cu couple. The synthesis of 8-oxabicyclo[3.2.1]oct-6-ene-3-one (1) is

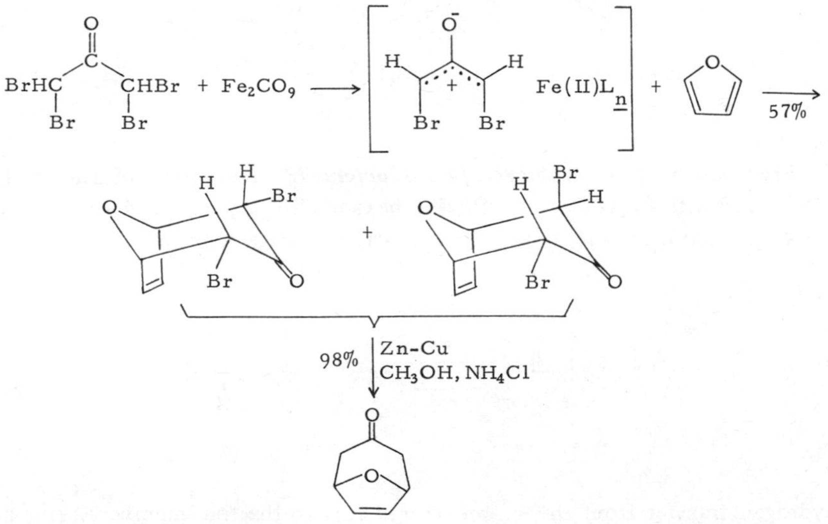

(1)

typical. Tribromides of methyl ketones are equally satisfactory.

This modification has been used to synthesize the tropolones hinokitiol (2, from the adduct of tetrabromoacetone and 2-isopropylfurane) and α-thujaplicin (3, from the adduct of 1,1,3,3-tetrabromo-4-methylpentane-2-one and furane).

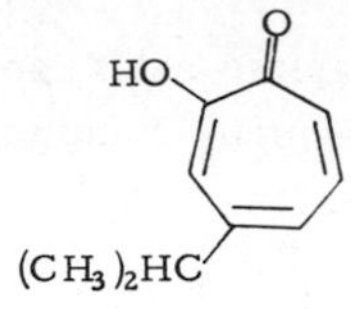

(2) 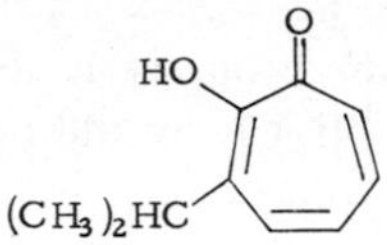          (3)

The reaction of α,α,α′,α′-tetrabromoacetone, 3-isopropylfurane, and diiron nonacarbonyl was used as the first step in a synthesis of the tropone nezukone (4).

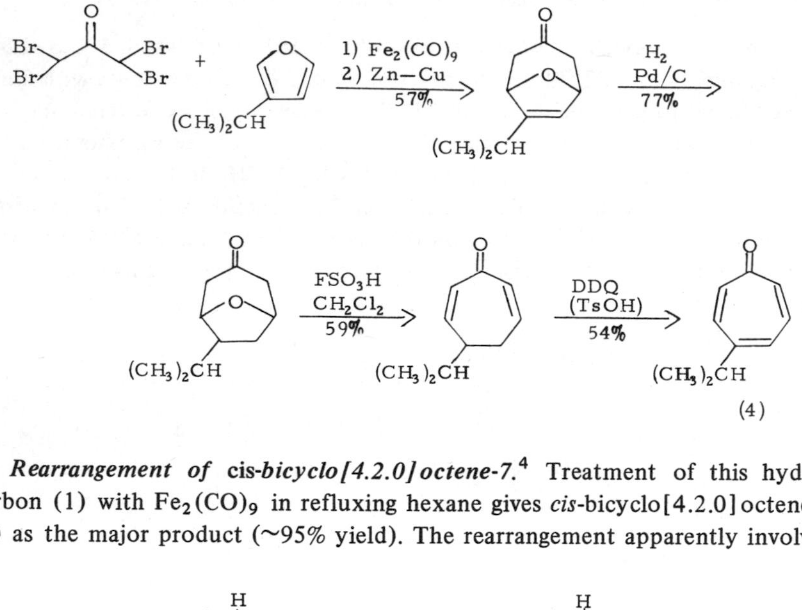

(4)

*Rearrangement of* cis-*bicyclo[4.2.0]octene-7.*[4] Treatment of this hydrocarbon (1) with $Fe_2(CO)_9$ in refluxing hexane gives *cis*-bicyclo[4.2.0]octene-2 (2) as the major product (~95% yield). The rearrangement apparently involves

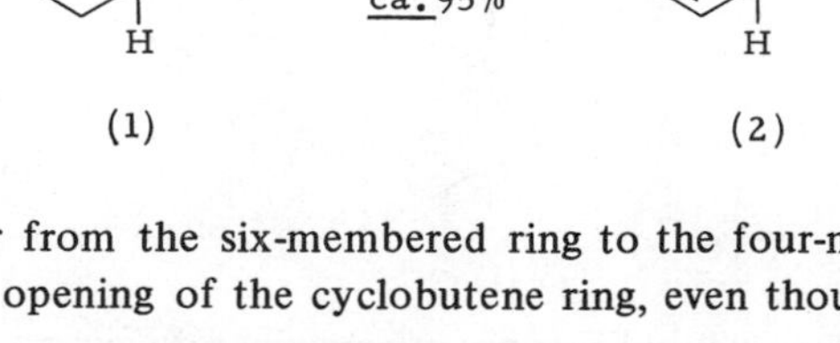

(1)          (2)

hydrogen transfer from the six-membered ring to the four-membered ring and does not involve opening of the cyclobutene ring, even though this reaction is theoretically possible. *cis*-Bicyclo[4.2.0]octene-3 is also isomerized to (2) under the same reaction conditions.

*Reaction with thiobenzophenones.* Treatment of thiobenzophenones (1) with diiron nonacarbonyl in benzene at 25° gives *ortho*-metalated complexes (2) in reasonable to high yield. The purple-red complexes are stable to air. One

(1, R=H, CH$_3$, OCH$_3$)          (2)                    (3)

interesting reaction of the complexes is oxidation with ceric ammonium nitrate (CAN) to thiolactones (3), which are dihydro derivatives of the rather inaccessible isobenzothiophene system. Thiolactones (3) are also formed by irradiation of (2) in ethanol.[5]

Oxidative cleavage of (2) with excess 30% hydrogen peroxide in acetic anhydride or with *m*-chloroperbenzoic acid in benzene gives lactones (4) in 45–75% yield.[6]

(4)

The complexes (2) are also cleaved by mercuric acetate in an alcohol, which effects *ortho*-mercuration and conversion to an ether, for example, (5).[7]

$$(2) \quad \xrightarrow[\;60-70\%\;]{\begin{array}{c}\text{Hg}(\text{OAc})_2\\ \text{R}'\text{OH}\end{array}}$$

(5)

[1] R. Noyori, S. Makino, T. Okita, and Y. Hayakawa, *J. Org.*, **40**, 806 (1975).
[2] E. LeGoff, *J. Org.*, **29**, 2048 (1964).
[3] Y. Hayakawa, M. Sakai, and R. Noyori, *Chem. Letters*, 509 (1975).
[4] K. Cann and J. C. Barborak, *J.C.S. Chem. Comm.*, 190 (1975).
[5] H. Alper and A. S. K. Chan, *Am. Soc.*, **95**, 4905 (1973).
[6] H. Alper and W. G. Root, *J.C.S. Chem. Comm.*, 956 (1974).
[7] *Idem*, *Tetrahedron Letters*, 1611 (1974).

**Diisobutylaluminum hydride, 1**, 260–262; **2**, 140–142; **3**, 101–102; **4**, 158–161; **5**, 224–225.

*Review.*[1] Winterfeldt has reviewed reductions with DIBAH and triisobutyl-aluminum (**1**, 260; **4**, 535). He notes that, although the neat reagents are highly reactive and must be handled with care, the 20 and 25% solutions available commercially are more stable and require only usual precautions (protection from water and oxygen). He notes that DIBAH is still the reagent of choice for reduction of esters and nitriles to aldehydes. He also considers DIBAH the most useful reagent for reduction of conjugated enones to unsaturated alcohols and for reduction of lactones to lactols and of unsaturated lactones to furanes. In the alkaloid field DIBAH has been very useful for reduction of amides and lactams. Winterfeldt cites a Schering patent that reports cleavage at 70–80° of methyl ethers in the estrone series to the corresponding phenols with DIBAH in yields of 80–90% without rearrangements encountered in acid-catalyzed cleavage.

*trans-1,4-Dienes.* Lynd and Zweifel[2] have extended the synthesis of *trans,trans*-1,3-dienes from vinylalanes (**4**, 158–159) to a synthesis of pure *trans*-1,4-dienes. Thus sequential addition of allyl bromide (1 eq.) and dry cuprous chloride (1 eq.)[3] to the vinylalane (1) in *n*-hexane gives *trans*-1,4-undecadiene (2) in about 70% yield. If the reaction is carried out in THF, some of the 1,3-diene is formed by dimerization. Allyl bromide does not react with the alane in the absence of the copper salt. In the case of more reactive vinylalanes (3), the

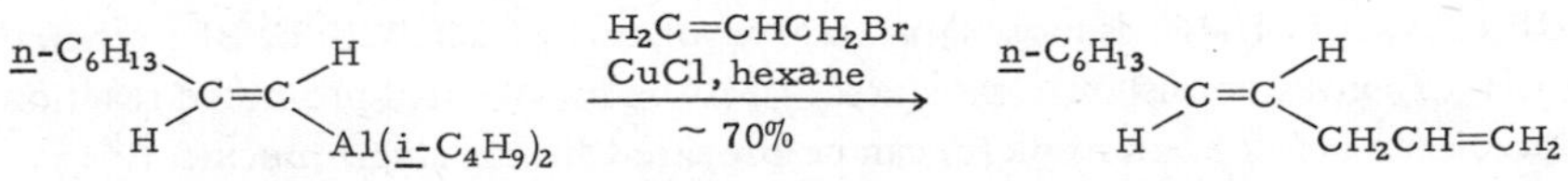

$$n\text{-}C_6H_{13}, \quad H \quad \xrightarrow[\sim\ 70\%]{\begin{array}{c} H_2C=CHCH_2Br \\ CuCl,\ hexane \end{array}} \quad n\text{-}C_6H_{13}, \quad H$$

(1)          (2)

reaction can be moderated by use of cuprous iodide rather than cuprous chloride as in the second example formulated.

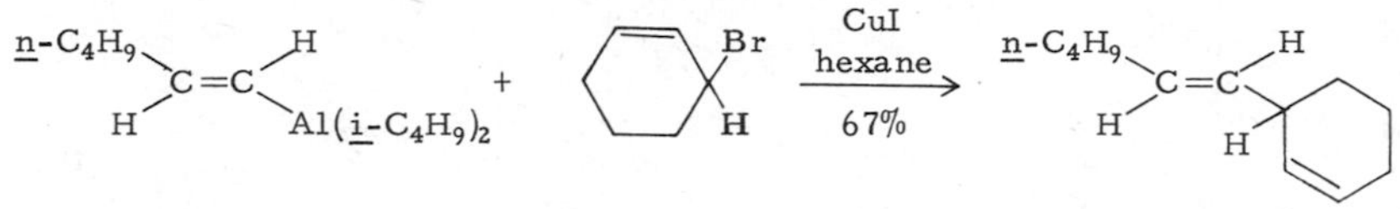

(3)          (4)

*trans-3-Alkenols.*[4] These unsaturated alcohols (2) can be prepared by reaction of ethylene oxide with the vinylalane ate complex (1). *n*-Butylamine is added to retard polymerization of the oxide.

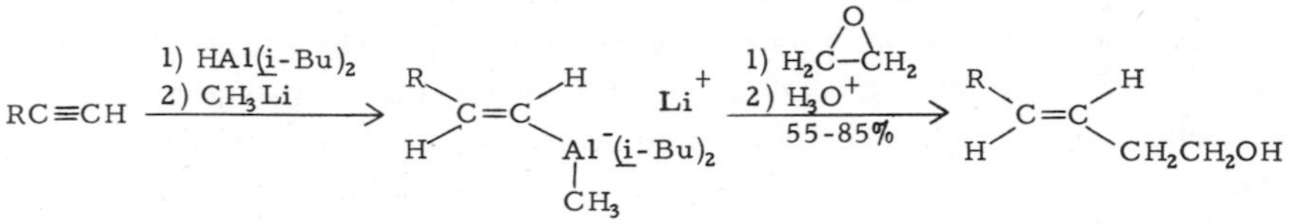

(1)          (2)

*β-Hydroxyalkylsilanes.* This reducing agent reduces the β-ketosilane (1) in pentane at $-120°$ to only one diastereoisomer of the β-hydroxyalkylsilane (2), presumably the *threo*-form. Lithium aluminum hydride or aluminum hydride gives both the *threo*- and *erythro*-isomers; loss of the trimethylsilyl group accompanies reduction with $NaBH_4$ in HMPT or with $NaBH_2S_3$. Elimination of

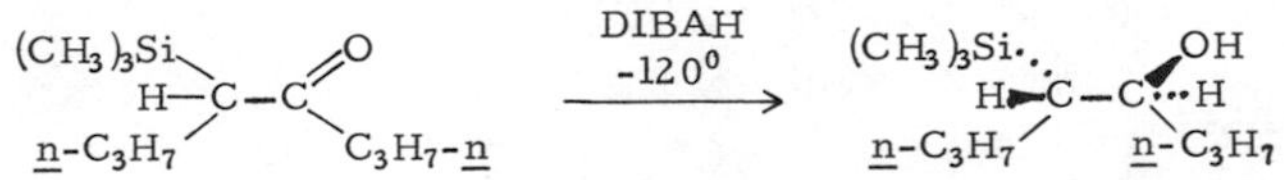

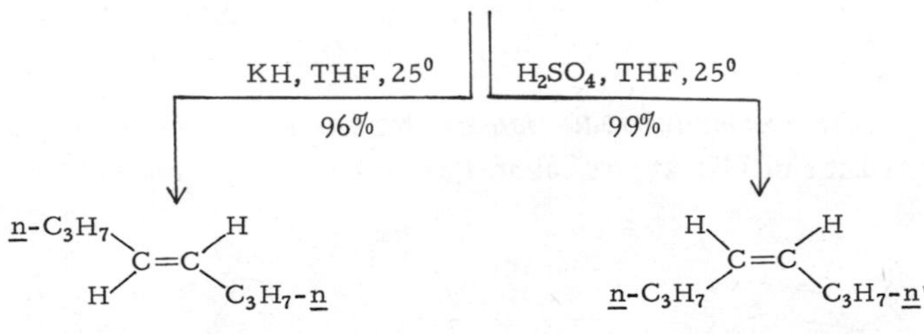

(3, 95% trans)          (4, 92% cis)

$(CH_3)_3SiOH$ with base (*syn*) gives *trans*-4-octene in high yield; elimination under acidic conditions (*anti*) results in *cis*-4-octene. Of various bases KH in

THF or NaH in HMPT is most satisfactory; sulfuric acid in THF or $BF_3$ etherate in $CH_2Cl_2$ gives consistently satisfactory results for the acid-promoted reaction. Thus either a *cis*- or a *trans*-olefin can be prepared from a single precursor.[5]

*Spirobenzylisoquinolines.* Treatment of dehydrocordrastine (1) with DIBAL at $-78°$ leads to (2) and (3), presumably by reductive opening of the lactone ring followed by reclosure.[6] The products are analogs of known alkaloids, which

(1)                    (2)                    (3)

have not been synthesized to date.

*Reduction of α,β-unsaturated γ-lactones to furanes* (1, 262). Pelletier et al.[7] have developed a general method for conversion of 2(5H)-furanones (1) into substituted furanes (4). The method involves cycloaddition of diazoalkanes, diazo esters, and diazoketones, followed by decomposition to alkylated 2(5H)-furanones (3). The final step involves reduction with diisobutylaluminum hydride.

(1, R=H, CH₃)              (2)                    (3)

(4)

*Stereoselective reduction of tropinone.* Tropinone (1) is reduced with this hydride in THF at $-78°$ almost exclusively to tropine (2).[8]

(1)                    (2)                    (3)

*3,5-Dioxocyclopentenes* (2).[9] These substances, which have been used in a prostaglandin synthesis,[10] can be prepared by reduction of (1) with diisobutyl-aluminum hydride at $-60°$. Reduction of (1) with lithium aluminum hydride affords (3) as the major product.

*α-Amino aldehydes.* Benzyloxycarbonyl amino acid esters are reduced to α-amino aldehydes by diisobutylaluminum hydride. Direct chromatography of the products on silica gel results in considerable racemization; however, the semicarbazones can be purified safely.[11] The α-amino aldehydes have been used to synthesize peptide aldehydes, some of which have interesting physiological properties.[12]

[1] E. Winterfeldt, *Synthesis*, 617 (1975).

[2] R. A. Lynd and G. Zweifel, *Synthesis*, 658 (1974).

[3] Commercial cuprous chloride is stirred with sodium sulfite and then precipitated with concd. HCl. The precipitate is washed successively with aqueous sodium sulfite, concd. HCl, HOAc, $C_2H_5OH$, and ether, and is dried over $P_2O_5$ *in vacuo* at $110°$.

[4] S. Warwel, G. Schmitt, and B. Ahlfaenger, *Synthesis*, 632 (1975).

[5] P. F. Hudrlik and D. Peterson, *Am. Soc.*, **97**, 1464 (1975).

[6] H. L. Holland, D. B. MacLean, R. G. A. Rodrigo, and R. F. H. Manske, *Tetrahedron Letters*, 4323 (1975).

[7] S. W. Pelletier, Z. Djarmati, S. P. Lajšić, I. V. Mićović, and D. T. C. Yang, *Tetrahedron*, **31**, 1659 (1975).

[8] Y. Hayakawa and R. Noyori, *Bull. Chem. Soc. Japan*, **47**, 2617 (1974).

[9] W. Van Brussel and M. Vandewalle, *Synthesis*, 39 (1976).

[10] C. J. Sih, R. G. Saloman, P. Price, R. Sood, and G. Peruzzotti, *Am. Soc.*, **97**, 857 (1975).

[11] A. Ito, R. Takahashi, and Y. Baba, *Chem. Pharm. Bull*, **23**, 3081 (1975).

[12] A. Ito, R. Takahashi, C. Miura, and Y. Baba, *ibid.*, **23**, 3106 (1975).

**Diisopinocampheylborane (Tetra-3-pinanyldiborane), 1**, 262–263; **4**, 161–162.
*(25S)-26-Hydroxycholesterol.* The reaction of the tetrahydropyranyl ether

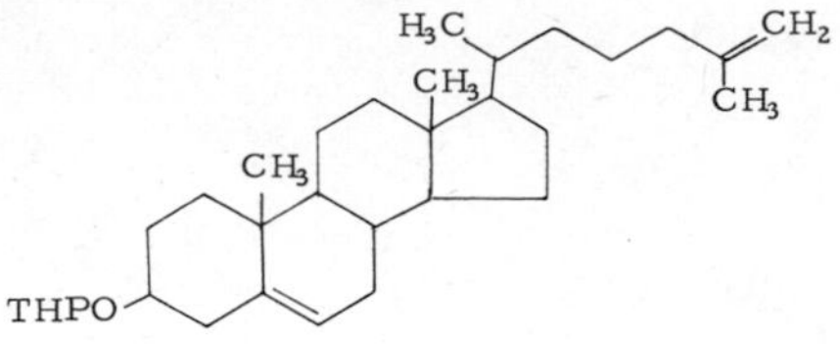

(1)

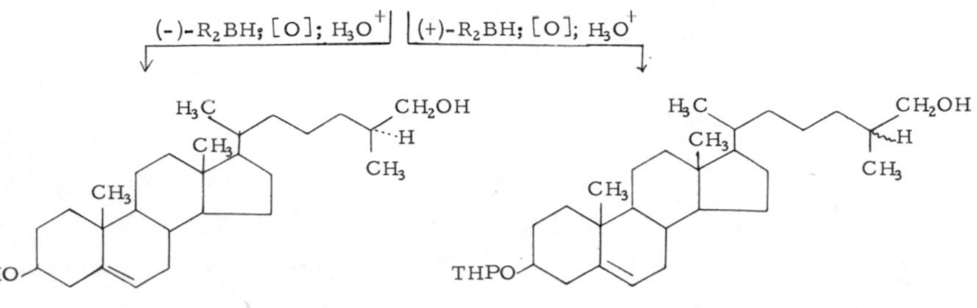

(-)-R₂BH; [O]; H₃O⁺    (+)-R₂BH; [O]; H₃O⁺

$(2, 25S; \alpha_D-38°, 83\%$
optical purity)

$(3, 25RS)$

of $\Delta^{5,25}$-cholestadiene-3β-ol (1) shows considerable chiral recognition to (-)-diisopinocampheylborane, but, rather unexpectedly, the reaction with (+)-diisopinocampheylborane gives the racemic product (3). The results are rationalized on the grounds that the steroid nucleus has a "25S-inducing effect".[1]

[1] R. K. Varma, M. Koreeda, B. Yagen, K. Nakanishi, and E. Caspi, *J. Org.*, **40**, 3680 (1975).

**Diketene, 1**, 264–266; **5**, 225–226.

*Esters of acetoacetic acid.* Diketene reacts exothermally with alcohols in the presence of triethylamine as catalyst to form esters of acetoacetic acid.[1]

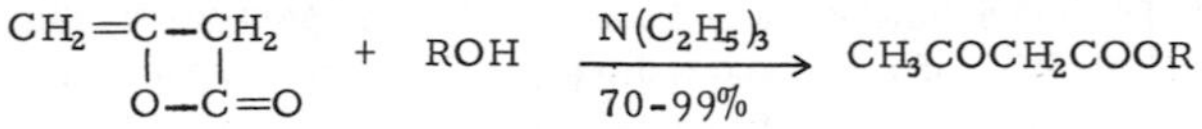

[1] O. Mauz, *Ann.*, 345 (1974).

**1,3-Dilithiopropyne, LiCH₂C≡CLi.** Mol. wt. 51.93.

*Acetylenes.* Propyne can be converted into the 1,3-dilithio derivative by reaction with 2 eq. of *n*-butyllithium in the presence of 1 eq. of TMEDA or DABCO (-60°, hexane–ether). This derivative reacts with electrophiles initially at the propargylic carbon, then at the terminal acetylide site. The reactions can be carried out in sequence in one pot.[1]

*Example:*

$$LiCH_2C \equiv CLi \xrightarrow[\text{HMPT}]{\text{n-BuBr}} \left[\underline{n}\text{-BuCH}_2C \equiv CLi\right] \xrightarrow[80\%]{\text{HCHO}} \underline{n}\text{-BuCH}_2C \equiv CCH_2OH$$

$$35\% \downarrow H_3O^+$$

$$\underline{n}\text{-BuCH}_2C \equiv CH$$

[1] S. Bhanu and F. Scheinmann, *J.C.S. Chem. Comm.*, 817 (1975).

## Dilithium tetrachlorocuprate, 4, 163–165.

*Halopolycarbon homologation of Grignard reagents.*[1]  This reagent catalyses the coupling of alkyl and aryl Grignard reagents with $\alpha,\omega$-dibromoalkanes:

$$RMgX + Br(CH_2)_{\underline{n}}Br \xrightarrow{\text{Cat.}} R(CH_2)_{\underline{n}}Br \quad [+R(CH_2)_{\underline{n}}R]$$

$$\underline{n}=3,4,5,6,10 \qquad\qquad 35\text{-}95\%$$

The reaction is carried out at 5–10° in THF or $C_6H_6$–THF. The choice of reactant partners is dictated primarily by cost and availability. One limitation is that ethylene dibromide cannot be used because it is converted into ethylene with formation of R–R.

An alternative synthesis involves the reaction of aryl- or alkyllithium reagents with $\alpha,\omega$-dibromoalkanes in the presence of TMEDA, which is required to prevent extensive halogen–metal exchange.

[1] L. Friedman and A. Shani, *Am. Soc.*, **96**, 7101 (1974).

## Dilithium tris(pent-1-ynyl)cuprate, $(CH_3CH_2CH_2C \equiv C)_3CuLi_2$ (1). Mol. wt. 278.77.

The cuprate is generated by stepwise addition of 1-pentynyllithium to purified cuprous iodide.[1]

*Reaction with cyclic enones.*[2]  In the presence of HMPT, this cuprate undergoes 1,2-addition to cyclic enones in high yield (equation I). The reaction may

$$(I) \qquad + \quad (1) \xrightarrow[95\%]{\substack{\text{THF, HMPT} \\ -78° \rightarrow 20°}}$$

be promising for synthesis of modified prostaglandins (equation II).

$$(II) \qquad + \quad (1) \longrightarrow \qquad \xrightarrow[\text{high}]{\substack{\text{Jones oxid.} \\ -10°}}$$

[1] H. O. House and W. F. Fischer, Jr., *J. Org.*, **34**, 3615 (1969).
[2] G. Palmisano and R. Pellegata, *J.C.S. Chem. Comm.*, 892 (1975).

*trans*-**2,4-Dimethoxymethyl-5-phenyloxazoline (1)**. Mol. wt. 235. 28.
**Preparation:**

$$(-)-(1)$$

*Asymmetric synthesis of 2-methoxyalkanoic acids.* Meyers *et al.*[1] have
prepared optically active 2-methoxyalkanoic acids by treatment of (1) with
LDA followed by alkylation with an alkyl iodide, sulfate, or tosylate, (2).
Acidification of (2) with 3 *N* HCl (95°, 3–4 hr.) gives the alkanoic acid (3).
The yields of (3) are high, but optical yields are only 45–65%.

2-Chloroalkanoic acids were prepared similarly from (−)-2-chloromethyl-4-
methoxymethyl-5-phenyloxazoline. In this case HMPT was added to facilitate
the alkylation step.

[1] A. I. Meyers, G. Knaus, and P. M. Kendall, *Tetrahedron Letters*, 3495 (1974).

**Di(α-methoxyvinyl)copperlithium [Lithium di(α-methoxyvinyl)cuprate]**,

CuLi (1). Mol. wt. 185.65, yellow.

The reagent is prepared[1] from the reaction of α-methoxyvinyllithium (this
volume) in THF with purified cuprous iodide[2] and dimethyl sulfide (complexing
agent, this volume) at −40° for 30 min.

*1,4-Diketones; γ-keto esters.*[1] This cuprate undergoes normal addition to
α,β-unsaturated ketones, for example (2), to form adducts (3). These can be
converted by acid hydrolysis into 1,4-diketones (4) or into γ-keto esters (5) by
ozonolysis.

One limitation is that addition of the cuprate to β,β-disubstituted enones is markedly inhibited.

*Reaction with halides.*[1] The reagent, unlike α-methoxyvinyllithium, reacts with benzyl bromide as shown:

The cuprate does not react with 2-bromooctane (0° for 4 hr. then 20° for 15 hr.).

*Di(α-ethoxyvinyl)copperlithium.*[3] This substance has been prepared in the same way from α-ethoxyvinyllithium and has been used to effect conjugate addition to 3-cyclohexenones. Halide couplings have also been reported for this reagent. Thus 3-bromocyclohexene reacts to give (1), which on hydrolysis gives the β,γ-enone (2):

This cuprate does not react with epoxycyclohexane.

[1] C. G. Chavdarian and C. H. Heathcock, *Am. Soc.*, **97**, 3822 (1975).
[2] Crystallized from aqueous potassium iodide and then refluxed.
[3] R. K. Boeckman, Jr., K. J. Bruza, J. E. Baldwin, and O. W. Lever, Jr., *J.C.S. Chem. Comm.*, 519 (1975).

**Dimethyl acetylenedicarboxylate** (DMAD), **1**, 272–273; **2**, 145–146; **3**, 103–104; **4**, 168–172; **5**, 227–230.

*Reaction with benzothiazole.* McKillop and Sayer[1] have clarified the reaction of benzothiazole (1) with DMAD for which conflicting results had been reported. Under strictly anhydrous conditions the 1:2 adduct (2) is formed, albeit in low yield. If water is present, (3) is formed in virtually quantitative yield. The chemists suggest that an initial dipolar adduct (a) is formed, which can react with DMAD to form (2) or with water to form (b), which rearranges to (3).

[1] A. McKillop and T. S. B. Sayer, *Tetrahedron Letters*, 3081 (1975).

**(+)-(2S,3R)-4-Dimethylamino-3-methyl-1,2-diphenyl-2-butanol (R\*OH), 5, 231.**

*Asymmetric reductions of α-acetoxy ketones.*[1] The asymmetric reduction of ketones has been extended to asymmetric reduction of α-acetoxy ketones to *cis*- and *trans*-1,2-diols. Thus (1) is reduced to a mixture of (2) and (3). On the

basis of oxidation to known compounds, (2) is formed with an enantiomeric excess of 64% and (3) with an excess of 76%. This asymmetric reduction is of interest because chiral 1,2-dihydrodiols are known metabolites of polycyclic carcinogenic hydrocarbons.

[1] K. Kabuto and H. Ziffer, *J. Org.*, **40**, 3467 (1975).

**2-(N,N-Dimethylamino)-4-nitrophenyl phosphate (1).** Mol. wt. 260.14.
   *Preparation*[1]:

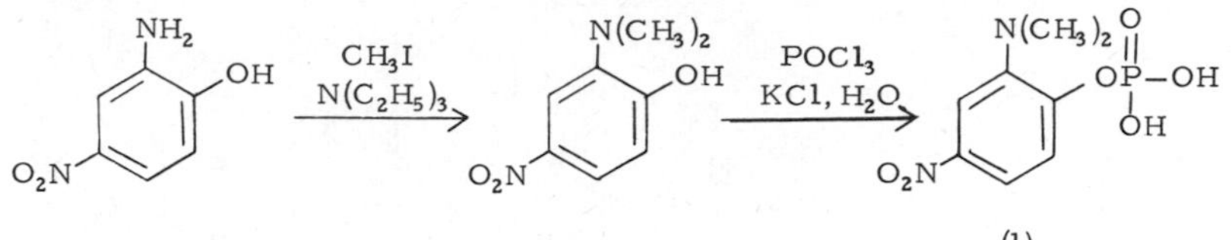

(1)

*Phosphorylation.*[1] The monotriethylammonium salt of (1) reacts with an alcohol (ROH) in the presence of an acid catalyst (HOAc, TFA) and triethylamine in pyridine to give monoesters, $ROP(O)(OH)_2$, in about 75–85% yield. This reaction is particularly useful for selective phosphorylation of the 5′-hydroxyl group of unprotected nucleosides. Thus adenosine can be converted into adenosine 5′-phosphate in 77% yield.

[1] Y. Taguchi and Y. Mushika, *Tetrahedron Letters*, 1913 (1975).

**2,3-Dimethyl-2-butylborane (Thexylborane), 1,** 276; **2,** 148–149; **4,** 175–176; **5,** 232–233.
   *Thexylmonoalkylboranes.* Thexylmonoalkylboranes can be prepared from disubstituted and trisubstituted olefins in $\geqslant 90\%$ yield by hydroboration at $-25°$. At higher temperatures dehydroboration to tetramethylethylene and a monoalkylborane becomes a competing reaction:

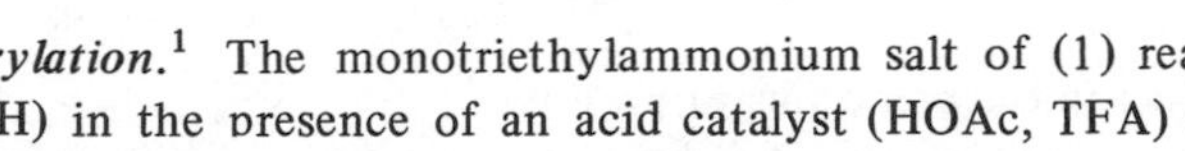

In fact, monoalkylboranes can be prepared in high yield by treatment of thexylmonoalkylboranes with excess triethylamine.[1]

   Thexylmonoalkylboranes react with unhindered olefins to form thexyldialkylboranes ($\vdash\hspace{-3pt}+\hspace{-3pt}B\!\!<^{R}_{R'}$) in $\geqslant 85\%$ yield. With hindered olefins or trisubstituted olefins, dehydroboration to tetramethylethylene predominates to give "mixed" dialkylboranes, RR′BH. These can be converted into "mixed" trialkylboranes, $RR^1R^2B$, by reaction with less hindered olefins. Trialkylboranes of this type have previously been difficult to prepare.[2]

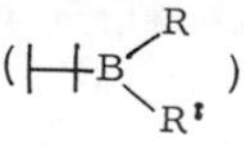

   *Trisubstituted olefins.* The ate complexes (1), obtained from thexyldialkylboranes and an alkynyllithium, undergo alkylation with migration from boron to carbon to give vinylboranes (2) and (3) in which the former isomer predominates. After acid hydrolysis the trisubstituted olefins (4) and (5) are obtained in a ratio of about 9:1. That is, the alkylating group and the migrating group are

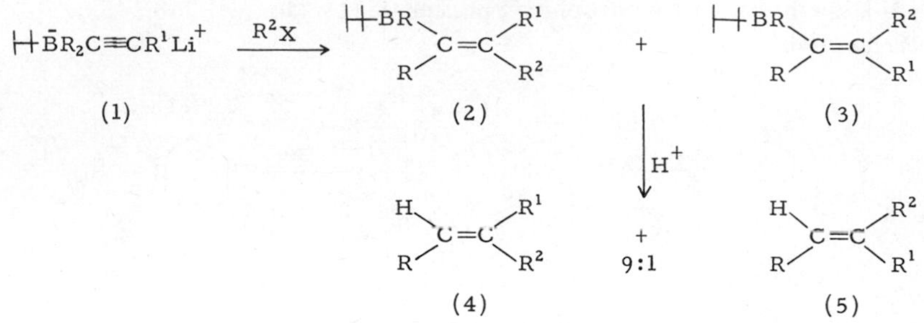

(1)     (2)     +     (3)

H⁺  9:1

(4)     (5)

*cis* to each other in the predominant product.[3]

Note that application of this sequence to trialkylalkynylborates, $R_3\bar{B}C\equiv CR^1Li^+$, also gives olefins (4) and (5), but this synthesis is not stereoselective.[4]

[1] H. C. Brown, E. Negishi, and J.-J. Katz, *Am. Soc.*, **97**, 2791 (1975).
[2] H. C. Brown, J.-J. Katz, C. F. Lane, and E. Negishi, *ibid.*, **97**, 2799 (1975).
[3] A. Pelter, C. Subrahmanyam, R. J. Laub, K. J. Gould, and C. R. Harrison, *Tetrahedron Letters*, 1633 (1975).
[4] A. Pelter, C. R. Harrison, and D. Kirkpatrick, *J.C.S. Chem. Comm.*, 544 (1973).

## 2,2-Dimethyl-6-(*p*-chlorophenylthiomethylene)cyclohexanone (1). Mol. wt. 208.75, m.p. 104–105°.

*Preparation:*

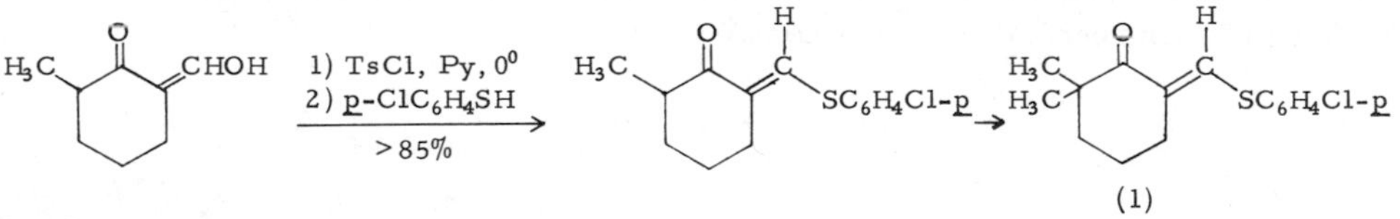

1) TsCl, Py, 0°
2) p-ClC₆H₄SH
>85%

(1)

Compare *n*-**Butyl mercaptan, 2,** 53–54; **4,** 64–65.

*Pentannelation.*[1] This term is used for annelation of a five-membered ring to another ring. This vinyl sulfide has been used for an efficient synthesis of hydrindanones. Treatment of (1) with base gives the allyl anion (2), which undergoes

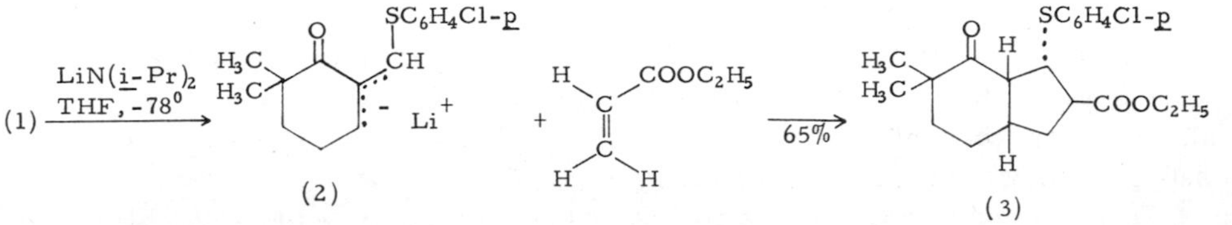

(1)  LiN(i-Pr)₂
THF, -78°

(2)     +     65%     (3)

[2 + 4] cycloaddition with an α,β-unsaturated ester, for example ethyl acrylate, to give the adduct (3) in 95% yield based on recovered (1). The adduct is considered to have the more stable *cis* ring fusion. The *trans* arrangement of the —SAr and —COOC₂H₅ groups is consistent with the NMR spectrum.

The product (3) is converted into the dione (6) by oxidation to a mixture of

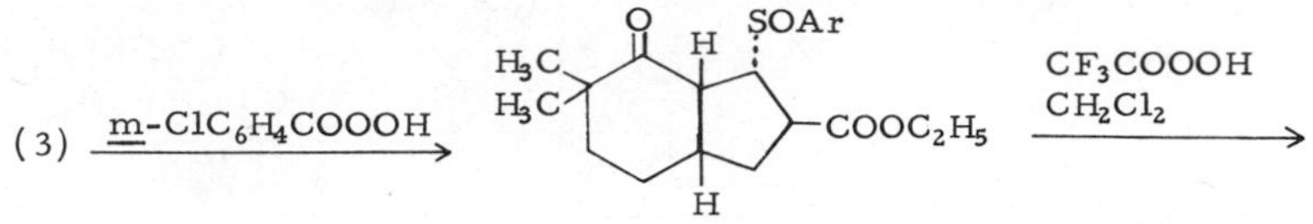

(4, 2 isomers)

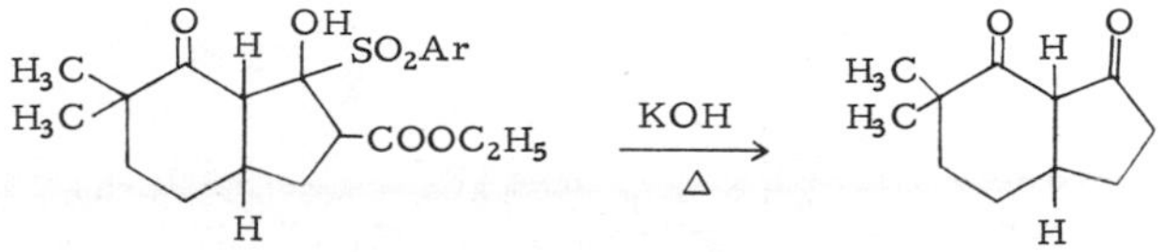

(5)                 (6)

two sulfoxides (4); these are oxidized to the same hydroxysulfone (5), possibly by an initial Pummerer rearrangement. Treatment of (5) with KOH in refluxing chloroform generates the carbonyl group and effects decarboxylation to give (6).

[1] J. P. Marino and W. B. Mesbergen, *Am. Soc.*, **96**, 4050 (1974).

**Dimethylcopperlithium (Lithium dimethyl cuprate)**, 2, 151–153; 3, 106–113; 4, 177–183; 5, 234–244.

*Preparation of hindered ketones.* Reaction of dialkylcopperlithium reagents with α,α'-dibromoketones can be used for monoalkylation of ketones; alkylation of the reaction intermediate leads to dialkylation:

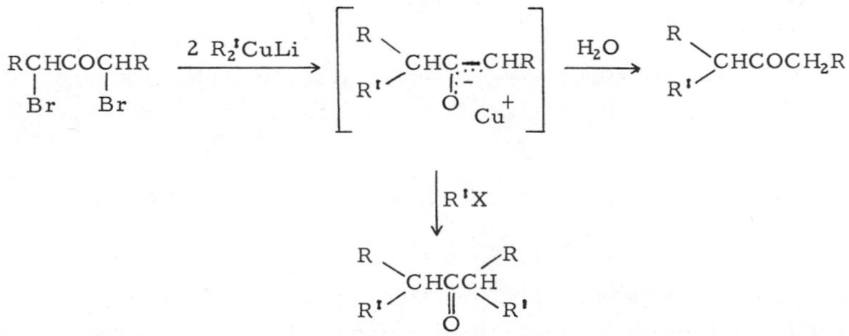

The method is particularly useful for preparation of hindered ketones; however, it is limited by the difficulty of preparing α,α'-dibromoketones from hindered ketones; it is also impossible to introduce two *t*-butyl groups into a dibromoketone.[1]

*Examples:*

$$(CH_3)_2CHCHCOCHCH(CH_3)_2 \xrightarrow[2) CH_3I]{1)(CH_3)_2CuLi}$$

with Br, Br substituents

products:

$$\begin{array}{c} (CH_3)_2CH \\ \diagdown \\ CH_3 \end{array} CHCOCH_2CH(CH_3)_2$$

(76%)

$$\begin{array}{c} (CH_3)_2CH \\ \diagdown \\ CH_3 \end{array} CHCOCH \begin{array}{c} CH(CH_3)_2 \\ \diagup \\ CH_3 \end{array}$$

(16%)

$$(CH_3)_3CCHCOCHC(CH_3)_3 \xrightarrow[80\%]{\substack{1)(CH_3)_2CuLi \\ 2)\ CH_3I}} \begin{array}{c} (CH_3)_3C \\ \diagdown \\ CH_3 \end{array} CHCOCH \begin{array}{c} C(CH_3)_3 \\ \diagup \\ CH_3 \end{array}$$

with Br, Br substituents

Hindered ketones can also be obtained by reaction of a cuprate reagent with an α-bromo ketone. This method permits regiospecific introduction of a primary, secondary, or even tertiary alkyl group at the site of the bromine atom.[2]

$$\begin{array}{c} R \\ \diagdown \\ R^1 \diagup \end{array} \underset{\underset{Br}{|}}{C} {-}COR^2 \xrightarrow{R^3_2CuLi} R^1{-}\underset{\underset{R^3}{|}}{\overset{\overset{R}{|}}{C}}{-}COR^2$$

*Examples:*

$$(CH_3)_2\underset{\underset{Br}{|}}{C}COCH(CH_3)_2 \xrightarrow[90\%]{(CH_3)_2CuLi} (CH_3)_3CCOCH(CH_3)_2$$

$$(CH_3)_3C\underset{\underset{Br}{|}}{C}OCHCH(CH_3)_2 \xrightarrow[31\%]{(CH_3)_2CuLi} (CH_3)_3CCOCH \begin{array}{c} CH(CH_3)_2 \\ \diagup \\ CH_3 \end{array}$$

$$(CH_3)_3C\underset{\underset{Br}{|}}{C}OCHC(CH_3)_3 \xrightarrow[8\%]{[(CH_3)_3C]_2CuLi} (CH_3)_3CCOCH[C(CH_3)_3]_2$$

*Alkylation of ketones.*[3] The reaction of dimethylcopperlithium (large excess) with α,α-dibromoketones effects monoalkylation in good yield as shown in equation I. Addition of an alkyl halide to the reaction results in dialkylation

$$(I) \quad (CH_3)_3C\overset{\overset{\displaystyle O}{\|}}{C}CHBr_2 \xrightarrow[92\%]{\substack{1)\ (CH_3)_2CuLi \\ 2)\ H_2O}} (CH_3)_3C\overset{\overset{\displaystyle O}{\|}}{C}CH_2CH_3$$

(II)   $(CH_3)_3CCCHBr_2$   $\xrightarrow[\text{91\%}]{\begin{array}{l}\text{1) }(CH_3)_2CuLi\\\text{2) }CH_3I\\\text{3) }H_2O\end{array}}$   $(CH_3)_3CCCH\begin{array}{l}CH_3\\CH_3\end{array}$

(equation II).

*Isopropylidenation of ketones,* $>C=O \rightarrow >C=C(CH_3)_2$. Posner *et al.*[4] have developed an efficient two-step method for effecting isopropylidenation of ketones. The ketone is treated with dibromomethylenetriphenylphosphorane, generated *in situ* from triphenylphosphine and carbon tetrabromide (**4,** 550–551). This reaction proceeds in good yield (80–85%) in the case of unhindered ketones; hindered ketones such as 2-methylcyclohexanone and 5-nonanone do not undergo dibromomethylenation. The second step involves reaction

$$>C=O \xrightarrow[C_6H_6, \Delta]{(C_6H_5)_3P, CBr_4} \quad >C=C \begin{array}{l}Br\\Br\end{array} \xrightarrow{(CH_3)_2CuLi} \quad >C=C(CH_3)_2$$

of the 1,1-dibromo-1-alkene with excess dimethylcopperlithium. Both bromine atoms are replaced by methyl groups in high yield (80–95%). Posner has suggested two possible paths for this reaction.

The process was used to synthesize a selinadiene sesquiterpene (**3**).

(1)   $\xrightarrow[\text{86\%}]{(C_6H_5)_3P, CBr_4}$   (2)   $\xrightarrow[\text{93\%}]{(CH_3)_2CuLi}$

(3)

*γ,δ-Unsaturated ketones.*[5] Dimethylcopperlithium undergoes conjugate addition to 1-acyl-2-vinylcyclopropanes to give, after hydrolysis, γ,δ-unsaturated ketones. For example, (E)-3-hexenyl phenyl ketone (1) is formed in high

$$CH_2=CH-\!\!\!\!\triangle\!\!\!\!-COC_6H_5 \xrightarrow[\text{98\%}]{\begin{array}{l}\text{1)}(CH_3)_2CuLi, (C_2H_5)_2O\\\text{2) }NH_4Cl, H_2O\end{array}} \begin{array}{cc}CH_3CH_2 & H\\ & C=C\\ H & CH_2CH_2COC_6H_5\end{array}$$

(1)

yield from the reaction of dimethylcopperlithium with 1-benzoyl-2-vinylcyclo-propane. Vinyl and aryl groups can also be introduced:

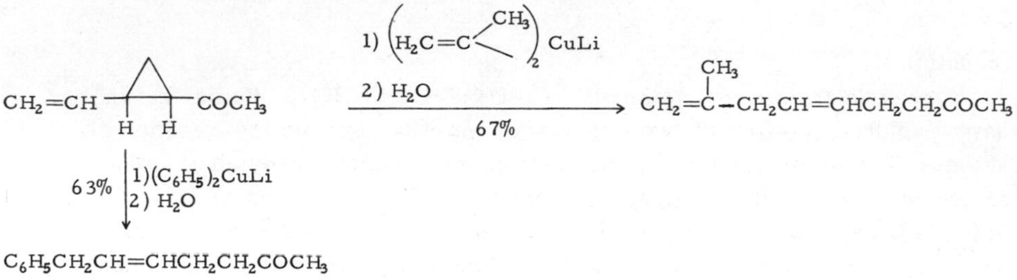

$CH_2$=CH—△—$COCH_3$
       H   H

1) $\left(H_2C=C\diagup^{CH_3}\right)_2$ CuLi

2) $H_2O$

67%

$CH_2$=$\overset{\overset{\displaystyle CH_3}{|}}{C}$—$CH_2CH$=$CHCH_2CH_2COCH_3$

63% | 1)$(C_6H_5)_2$CuLi
     | 2) $H_2O$
     ↓

$C_6H_5CH_2CH$=$CHCH_2CH_2COCH_3$

Trialkylboranes undergo this 1, 6-addition, but this reaction requires oxygen as catalyst and the yields are low (5–30%).

*Conjugate addition to cyclopropyl lactones and β-keto esters.* (4, 219–220).[6] The reaction of the cyclopropyl lactone (1) with 1.5 eq. of dimethylcopper-lithium at −30 to 0° gives the isomeric spirolactones (2) in high yield. The related cyclopropyl β-keto ester (3) reacts similarly to give (4).

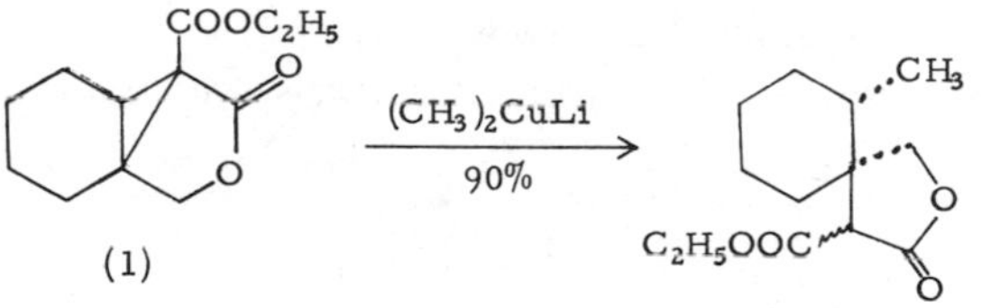

(1)     $\xrightarrow[90\%]{(CH_3)_2CuLi}$     (2)

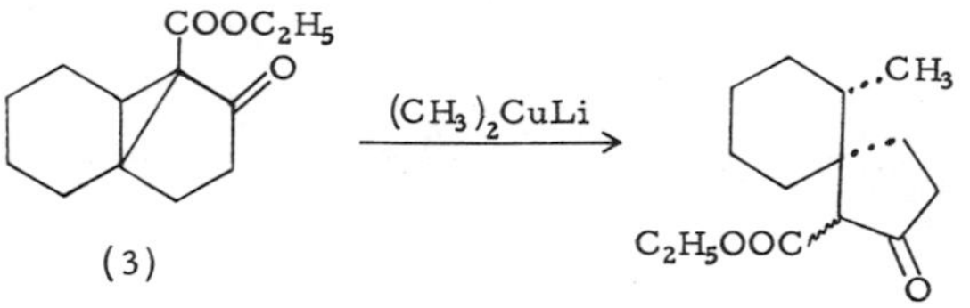

(3)     $\xrightarrow{(CH_3)_2CuLi}$     (4)

Cyclopropyl ketones apparently do not undergo this reaction; (5) and (6) are unreactive to $(CH_3)_2$CuLi.

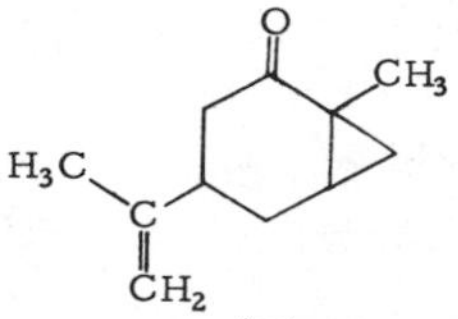

(5)

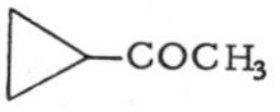

(6)

The direction of reductive cleavage of the cyclopropyl ring in (1) and (3) is apparently controlled by electronic factors (cf. **4**, 288–290).

**β-Vetivone.** Swiss chemists [7] have developed a stereoselective synthesis of β-vetivone by addition of dimethylcopperlithium to the 2-formylcyclohexadienone (1), which gives, after deformylation ($CH_3OH–H_2O$, 195°), racemic β-vetivone (2) and 10-epi-β-vetivone (3) in a 2:1 ratio. In the absence of the

formyl groups (2) and (3) are obtained in approximately equal amounts. Several related reagents give an even more favorable mixture of (2) and (3); a ratio of 5:1 of (2) and (3) was obtained with a lithium methylbromocuprate-diisobutylamine complex, $[CH_3CuBr]^-Li^+[(i\text{-}Bu)_2NH]_2$.[8]

*Conjugate addition followed by aldol condensation.* The anionic copper intermediate of conjugate addition of dimethylcopperlithium undergoes aldol condensation with aliphatic aldehydes in the presence of zinc chloride (**5**, 763).[9]

*Examples:*

*Intramolecular aldol condensations.* Swiss chemists[10] have reported two examples of preparation of cyclic aldols by reaction of ζ-oxo-α,β-enones by conjugate addition of dimethylcopperlithium followed by a directed intramolecular aldol condensation. Both aldols are products of kinetic control; they are not formed if the reaction is equilibrated at $22°$ (24 hr.).

*Reaction with aldehydes.* Crabbé *et al.*[11] report that dimethylcopperlithium (2 eq.) usually reacts with aldehydes to give products of 1,2-addition; sometimes the primary alcohol is also formed by a reductive process.

*Reaction with $\alpha,\beta$-epoxy ketoximes.*[12] The oxime of 2,3-epoxycyclohexa-none (1) reacts with dimethylcopperlithium (5 eq.) in ether at $-25°$ to give (2),

which was not purified, but was hydrolyzed by the method of McMurry and Melton (5, 670–671) to *trans*-3-hydroxy-2-methylcyclohexanone (3).

This reaction bears an analogy to the reaction of $\alpha$-halo oximes with a nucleo-phile, for example, the conversion of (A) to (B).[13]

This new reaction was shown to be applicable to various $\alpha,\beta$-epoxy ketoximes and to other cuprates. The products are easily dehydrated; hence this method can be used to effect $\alpha$-alkylation of $\alpha,\beta$-enones.

This process was developed in connection with a projected total synthesis of the macrolide erythronolide B.

[1]C. Lion and J.-E. Dubois, *Tetrahedron*, **31**, 1223 (1975).
[2]J.-E. Dubois and C. Lion, *ibid.*, **31**, 1227 (1975).
[3]*Idem, Compt. rend.*, **280** (C), 217 (1975).
[4]G. H. Posner, G. L. Loomis, and H. S. Sawaya, *Tetrahedron Letters*, 1373 (1975).
[5]N. Miyaura, M. Itoh, N. Sasaki, and A. Suzuki, *Synthesis*, 317 (1975).
[6]R. D. Clark and C. H. Heathcock, *Tetrahedron Letters*, 529 (1975).
[7]G. Bozzato, J.-P. Bachmann, and M. Pesaro, *J.C.S. Chem. Comm.*, 1005 (1974).
[8]Prepared by addition of 1 eq. of $CH_3Li$ to an ether solution of 1.1 eq. of $[(i\text{-}Bu)_2NH]_2 \cdot$ CuBr; *see* H. O. House and W. F. Fischer, Jr., *J. Org.*, **33**, 949 (1968).
[9]K. K. Heng and R. A. J. Smith, *Tetrahedron Letters*, 589 (1975).
[10]F. Näf, R. Decorzant, and W. Thommen, *Helv.*, **58**, 1808 (1975).
[11]E. Barreiro, J. L. Luche, J. Zweig, and P. Crabbé, *Tetrahedron Letters*, 2353 (1975).
[12]E. J. Corey, L. S. Melvin, Jr., and M. F. Haslanger, *ibid.*, 3117 (1975).
[13]M. Ohno, N. Naruse, and I. Terasawa, *Org. Syn. Coll. Vol.*, **5**, 266 (1973).

**N,N-Dimethyl dichlorophosphoramide**, $Cl_2PN(CH_3)_2$ . Mol. wt. 161.96, b.p. $45°/$ 1 mm.

*Preparation*[1] *:*

$$(CH_3)_2NH + POCl_3 \xrightarrow[88-95\%]{} (CH_3)_2NPOCl_2 + HCl$$

*Olefins from vic-glycols.*[2] Conversion of *vic*-glycols into the cyclic phosphoric amide followed by treatment with sodium in refluxing xylene leads to olefins via preferential *syn*-elimination.

*Examples:*

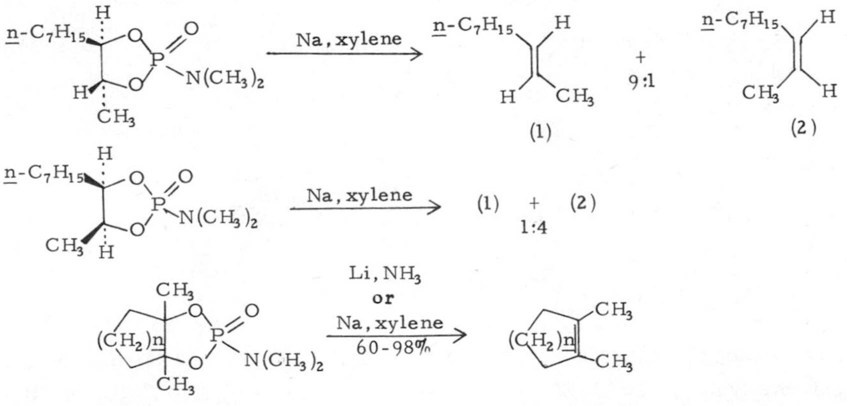

[1]E. N. Walsh and A. D. F. Toy, *Inorg. Syn.*, **7**, 69 (1963).
[2]J. A. Marshall and M. E. Lewellyn, *Syn. Commun.*, **5**, 293 (1975).

**7,8-Dimethyl-1,5-dihydro-2,4-benzodithiepin** (1). Mol. wt. 210.37, m.p. 185–187°.

*Preparation*[1,2]:

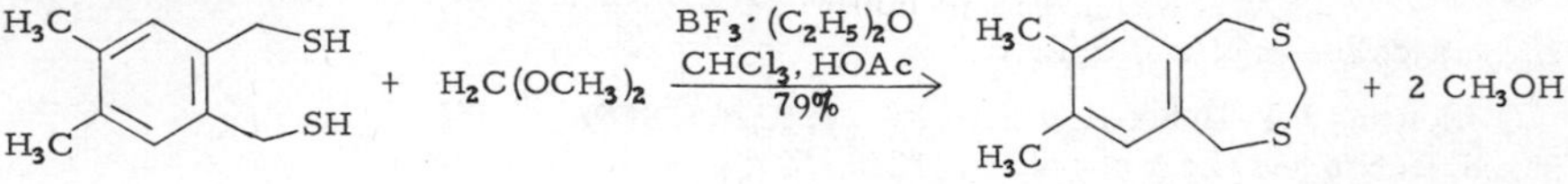

$$+ \ 2 \ CH_3OH$$

(1)

*Aldehyde synthesis.*[2] The anion of (1) reacts with alkyl bromides to give products (2), which are cleaved to aldehydes by $CuO\text{–}CuCl_2$:

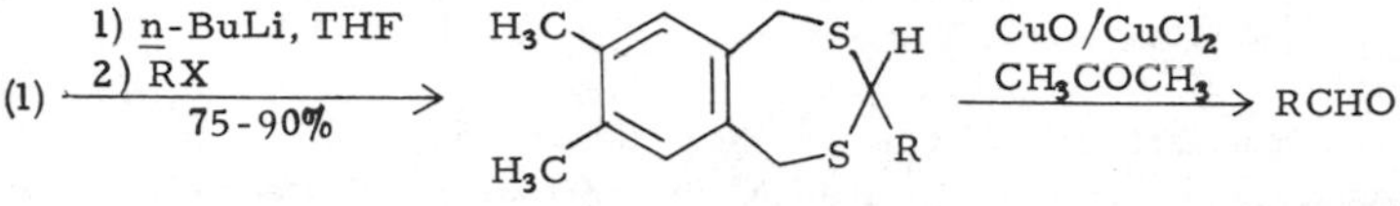

(2)

Methyl ketones can be synthesized by use of the related dithiane (3).

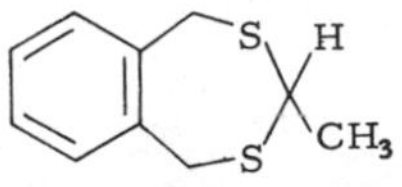

(3)

The main advantage of (1) over 1,3-dithiane is that it does not have a disagreeable odor.

[1] I. Shahak and E. D. Bergmann, *J. Chem. Soc.*, 1005 (1966).
[2] K. Mori, H. Hashimoto, Y. Takenaka, and T. Takigawa, *Synthesis*, 720 (1975).

**6,6-Dimethyl-5,7-dioxaspiro[2.5]octane-4,8-dione** (1). Mol. wt. 170.16, m.p. 63.5–64.5°.

*Preparation*[1]:

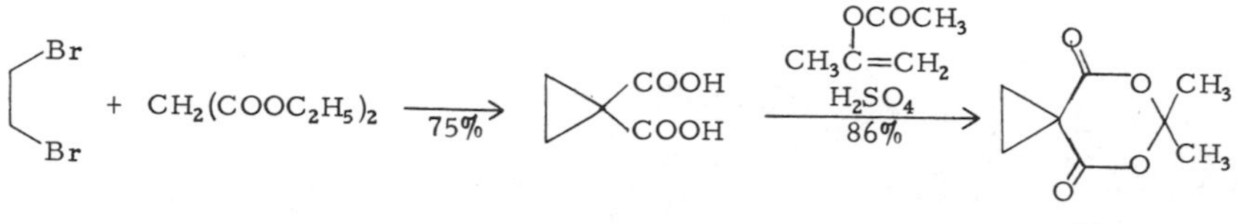

(1)

*Homoconjugate reactions.*[2] In comparison to cyclopropane-1,1-dicarboxylic acid, the cyclic acylal (1) shows pronounced double-bond character in reactions with nucleophiles. Thus it reacts with piperidine in benzene at 20° to give the zwitterion (2) in quantitative yield. The second reaction (II) is an example of the

(I)    (1) + HN⟨⟩ $\xrightarrow{\text{quant.}}$ [structure: $\overset{+}{N}HCH_2CH_2$— with dioxolane CH₃ CH₃ ring]  (2)

(II)    (1) + $C_6H_5NH_2$ $\xrightarrow{20^0, 11\ hr}$ [$C_6H_5\overset{..}{\underset{H}{N}}$ → structure with CH₃ CH₃] $\xrightarrow{\text{quant.}}$ $C_6H_5N$—[ring with COOH, O]  (3)

(III)    (1) + [structure: ONa, COOCH₃ cyclohexene] $\xrightarrow[88\%]{\substack{DME \\ 20^0}}$

[structure: O, COOCH₃, (CH₂)₂— dioxolane CH₃ CH₃ O] $\xrightarrow[92\%]{\substack{H_2SO_4, H_2O \\ \Delta}}$ [structure: O, (CH₂)₃COOH cyclohexanone]  (4)

use of (1) for synthesis of a heterocycle. The last example (equation III) formulates the reaction of (1) with the sodium enolate of 2-carbomethoxycyclohexanone to give, after hydrolysis and decarboxylation, 4-(2-cyclohexanone)butyric acid (4). Compound (1) is thus the synthetic equivalent of $\overset{+}{C}H_2CH_2CH(COOH)_2$.

[1] R. K. Singh and S. Danishefsky, *J. Org.*, **40**, 2969 (1975).
[2] S. Danishefsky and R. K. Singh, *Am. Soc.*, **97**, 3239 (1975).

## Dimethyl disulfide, 5, 246–247.

*α,β-Unsaturated esters.* Trost and Salzmann[1] report that sulfenylation of ester enolates with dimethyl disulfide can be accomplished in high yield if a mixture of THF and HMPT is used as solvent. The reaction is used in a synthesis of queen's substance (4) from (1), available from azelaic acid monomethyl

[structure: $H_3C$, $(CH_2)_6CH_2COOCH_3$ dioxolane ring]  (1) $\xrightarrow{\substack{LiICA, THF, -78^0 \\ CH_3SSCH_3, HMPT, 25^0}}$ [structure: $H_3C$, $(CH_2)_6\overset{|}{\underset{SCH_3}{C}}HCOOCH_3$ dioxolane ring]  (2)

$\xrightarrow[\substack{69\% \\ overall}]{\substack{HOOCCOOH \\ H_2O}}$ [structure: $CH_3\overset{O}{\overset{||}{C}}(CH_2)_5CH_2\overset{|}{\underset{SCH_3}{C}}HCOOCH_3$]  (3) $\xrightarrow[86\%]{\substack{1)\ NaIO_4 \\ 2)\ 110^0}}$ [structure: $H_3C\overset{O}{\overset{||}{C}}(CH_2)_5$—$\overset{H}{\underset{H}{C}}=CCOOCH_3$]  (4)

ester. Sulfenylation proceeds in about 30% yield on the free keto acid, but in satisfactory yield with the corresponding ethylene ketal (1). On desulfenylation (4) is obtained in 47% overall yield from (1).

[1] B. M. Trost and T. N. Salzmann, *J. Org.*, **40**, 148 (1975).

### N-(1,2-Dimethylethenylenedioxyphosphoryl)imidazole, (1). Mol. wt. 200.13, m.p. 62–64°.

*Preparation*[1]:

(1)

*Unsymmetrical phosphodiesters.* Ramirez et al.[2] have developed a high-yield, one-flask synthesis of unsymmetrical phosphodiesters (4). The reagent

(1) is allowed to react with an alcohol ($R^1OH$) at room temperature for 45 min. and then $R^2OH$ is introduced; (3) is formed after a time and is then hydro-lyzed to (4). If possible, $R^1OH$ should be the bulkier alcohol. This synthesis is successful because imidazole is formed and serves as catalyst for the reactions of both $R^1OH$ and $R^2OH$.

A one-flask synthesis is also possible starting with di(1,2-dimethylethenylene) pyrophosphate (5). In this case triethylamine is used as catalyst. Interestingly enough this base catalyzes the reaction of primary alcohols ($R^2OH$) with (6), but not of secondary alcohols. Therefore triethylamine is less efficient as catalyst than imidazole.

[1] F. Ramirez, H. Okazaki, and J. F. Marecek, *Synthesis,* 637 (1975).
[2] F. Ramirez, J. F. Marecek, and H. Okazaki, *Am. Soc.,* 97, 7181 (1975).

(–)-N,N-Dimethylephedrinium bromide,

$$
\begin{array}{c}
CH_3 \quad Br^- \\
H\!-\!\overset{+}{C}\!-\!N(CH_3)_3 \\
| \\
H\!-\!C\!-\!OH \\
| \\
C_6H_5
\end{array}
$$

(1). Mol. wt. 274.22,

m.p. 215°, $\alpha_D$ –22.5°.

The salt is prepared by quaternization of (–)-ephedrine with methyl bromide.

*Asymmetric oxirane synthesis.* Japanese chemists[1] have reported asymmetric synthesis of 2-phenyloxirane from benzaldehyde and dimethylsulfonium methylide generated from trimethylsulfonium iodide in 50% NaOH with the chiral phase-transfer catalyst (–)-N,N-dimethylephedrinium bromide (1).

$$
\begin{array}{c}
Br^- \quad CH_3 \\
(CH_3)_3\overset{+}{N}\!-\!C\!-\!H \\
| \\
H\!-\!C\!-\!OH \\
| \\
C_6H_5
\end{array}
$$

$$(2, \; \alpha_D - 34°)$$

$$(CH_3)_2\overset{+}{S}CH_3\bar{I} \xrightarrow[77\%]{\substack{(1),\,NaOH \\ C_6H_5CHO}} C_6H_5\overset{*}{HC}\!\!\stackrel{O}{\overbrace{\quad}}\!\!CH_2 \quad \alpha_D + 4.36°$$

$$(CH_3)_2\overset{+}{S}CH_3\bar{I} \xrightarrow[71\%]{\substack{(2),\,NaOH \\ C_6H_5CHO}} C_6H_5\overset{*}{HC}\!\!\stackrel{O}{\overbrace{\quad}}\!\!CH_2 \quad \alpha_D - 2.64°$$

The optical yield of the oxirane obtained using (1) is 67%. If the asymmetric salt (2) derived from ψ-ephedrine is used, the enantiomeric oxirane is formed preferentially. This striking asymmetric synthesis can be achieved with only 0.2 eq. of the catalyst. The choice of solvent is also important. Little asymmetric induction is observed with THF or acetonitrile as solvent; use of benzene, however, results in a high degree of induction.

*Asymmetric additions of dihalocarbenes.*[2] Asymmetric additions of dihalocarbenes generated with (–)-N,N-ethylmethylephedrinium bromide [(3), one methyl group in (1) replaced by ethyl] as the phase-transfer catalyst have been reported. Thus styrene reacts with dichlorocarbene generated with (3) to give an adduct showing some optical activity. The extent of optical induction is low (∼1.0%), but even so, this asymmetric synthesis is unprecedented.[2]

$$\text{CHCl}_3 \xrightarrow[\substack{(3),\ \text{NaOH, H}_2\text{O} \\ \underline{\text{C}_6\text{H}_5\text{CH=CH}_2} \\ 70\%}]{} \quad \text{Cl}_2\text{C} \overset{\text{H} \diagdown \quad \diagup \text{C}_6\text{H}_5}{\underset{\text{H}' \diagdown \quad \diagup \text{H}}{<}}$$

$$\alpha_D + 3.18°$$

Asymmetric syntheses with dibromocarbene generated with (3) as catalyst have also been achieved. In these asymmetric syntheses, the $\beta$-hydroxyethyl unit plays an important role, since use of (+)-trimethyl-$\alpha$-phenethylammonium bromide as catalyst gives a product with only slight optical activity, $\alpha_D$ -0.14°. The $\beta$-hydroxyethyl unit also influences the selectivity of dichlorocarbene. Thus dichlorocarbene generated with (4) as catalyst reacts with several substrates that contain two or three double bonds to give only monoadducts, whereas

$$\underset{\overset{|}{\underset{\text{CH}_2\text{CH}_2\text{OH}}{}}}{\overset{+}{\text{C}_6\text{H}_5\text{CH}_2\text{N}(\text{CH}_3)_2}} \quad \text{OH}^-$$

(4)

dichlorocarbene generated with the usual catalyst cetyltrimethylammonium bromide reacts with all the double bonds in the substrate. In the selective cyclopropanation reaction, the more substituted double bond is the site of attack.

[1] T. Hiyama, T. Mishima, H. Sawada, and H. Nozaki, *Am. Soc.,* **97**, 1626 (1975).
[2] T. Hiyama, H. Sawada, M. Tsukanaka, and H. Nozaki, *Tetrahedron Letters,* 3013 (1975).

### Dimethylformamide–Phosphorus pentachloride.

The Vilsmeier reagent, dimethylchloroformiminium chloride, $(\text{CH}_3)_2\overset{+}{\text{N}}\text{CHClCl}^-$ (1), has generally been prepared by the reaction of dimethylformamide with phosphoryl chloride or thionyl chloride (**1**, 284–285, 286–289; **2**, 154; **3**, 116; **4**, 186). Hepburn and Hudson[1] now report that the crystalline reagent can be prepared in 88% yield by addition of $\text{PCl}_5$ to an excess of DMF (exothermic reaction).

*Alkyl halides.* The Vilsmeier reagent has been used to a limited extent to replace an hydroxyl group by chlorine. Actually (1) is a useful reagent for preparation of alkyl chlorides from alcohols without rearrangement. The reaction is conducted at 75–100° in dioxane, DMF, $\text{CH}_3\text{CN}$, or HMPT. The replacement occurs with inversion. Yields are usually in the range 75–85%.

Reaction of DMF and $\text{PBr}_5$ is complex, but (1) can be converted into $(\text{CH}_3)_2\overset{+}{\text{N}}\text{CHBrBr}^-$ (2) by treatment with anhydrous hydrogen bromide in chloroform at 0°. This reagent (2) converts alcohols into alkyl bromides (dioxane, 100°, 25–60 min.). Rearrangements have been encountered in this reaction in two cases.

[1] D. R. Hepburn and H. R. Hudson, *Chem. Ind.,* 664 (1974).

**N,N-Dimethylformamide dibenzyl acetal,** $(CH_3)_2NCH(OCH_2C_6H_5)_2$. Mol. wt. 271.35.

*Benzylation of nucleosides.*[1]  Uridine (1) is converted into $N^3$-benzyluridine (2) in quantitative yield when heated with DMF dibenzyl acetal in DMF for 3 hrs. at 80°. This derivative is obtained in only low yield using benzyl bromide and sodium hydride. The blocking group of (2) is removed by sodium naphthalenide (THF, 3 hr., 84% yield). Guanosine (3) has also been blocked in the same way.

[1]K. D. Philips and J. P. Horwitz, *J. Org.*, **40**, 1856 (1975).

**N,N-Dimethylformamide dimethyl acetal,** **1**, 281–282; **2**, 154; **3**, 115–116; **4**, 184–185.

*Dehydrative decarboxylation of β-hydroxy carboxylic acids.*[1]  Olefins are formed when β-hydroxy carboxylic acids are refluxed with excess DMF dimethyl acetal in $CHCl_3$ for 5–10 hr. Yields generally are fair to good (50–90%).

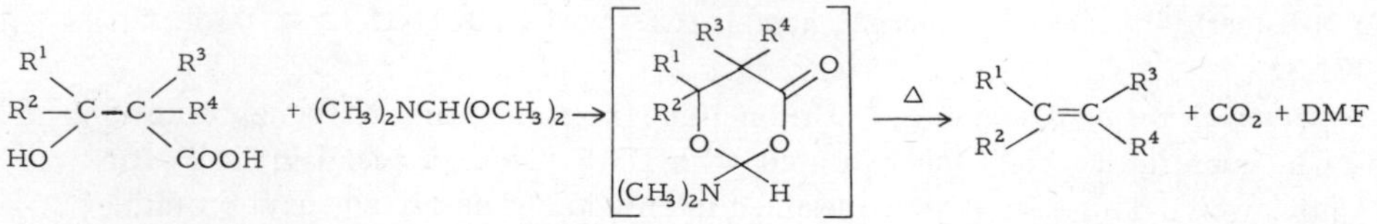

A possible intermediate is formulated.

[1]S. Hara, H. Taguchi, H. Yamamoto, and H. Nozaki, *Tetrahedron Letters,* 1545 (1975).

### N,N-Dimethylformamide dineopentyl acetal, 1, 283.

*Decarboxylative dehydration.*[1] δ-Hydroxy-β,γ-unsaturated cyclohexenecarboxylic acids (1) when treated with this acetal of DMF at room temperature or somewhat higher are converted into 1,3-cyclohexadienes (2) in high yield; co-products are DMF and neopentyl alcohol. α-Methylnaphthalene (b.p. 80°/0.05

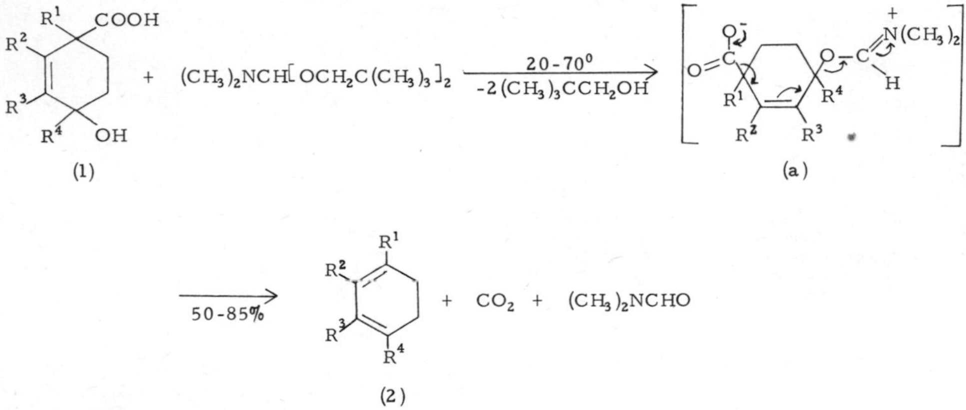

(1)                                                (a)

(2)

mm.) is used as solvent since the high boiling point does not interfere with isolation of the dienes by distillation. For a similar reaction *see* **Dimethylformamide dimethyl acetal,** this volume.

Eschenmoser *et al.* recommend zinc borohydride for reduction of keto esters (3) to the starting hydroxy acids. The Grignard reaction is also applicable.

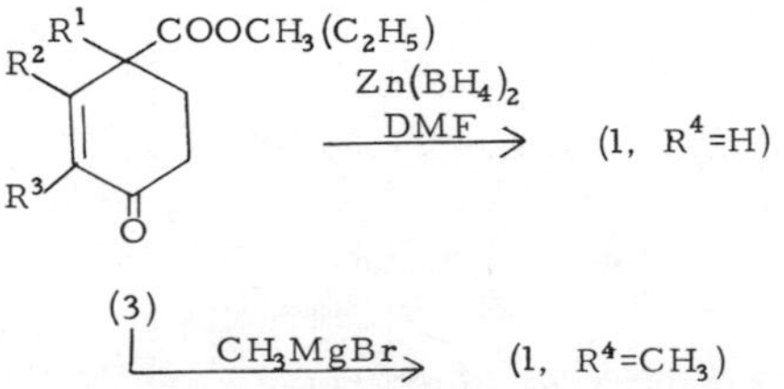

(3)

[1]A. Rüttimann, A. Wick, and A. Eschenmoser, *Helv.,* **58,** 1451 (1975).

**N,N-Dimethylhydrazine, 1,** 289–290; **2,** 154–155; **3,** 117.

*Half acid–esters.*[1] Treatment of the diester (1) with N,N-dimethylhydrazine for 22 hr. at 25° yields the salt of the half acid–ester (2). Treatment of (2) with a mineral acid yields the half acid–ester (3). The method is general, but yields of the intermediate salt vary from 25 to 60%.

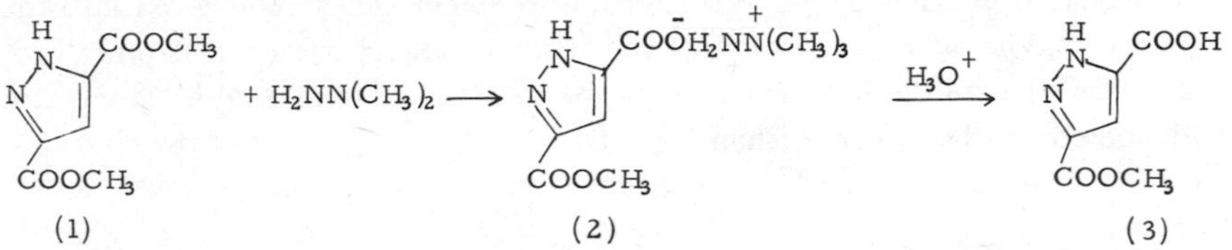

(1)                    (2)                    (3)

[1] J. Nematollahi and S. Kasina, *J.C.S. Chem. Comm.,* 775 (1974).

**Dimethylketene, 1,** 290–292.

*Cycloaddition to allenes.*[1] The thermal cycloaddition of dimethylketene to allenes is regiospecific; the central carbon of the allene is joined in all cases to the carbon of the carbonyl group of the ketene. The products are conjugated methylene- or alkylidenecyclobutanones.

*Examples:*

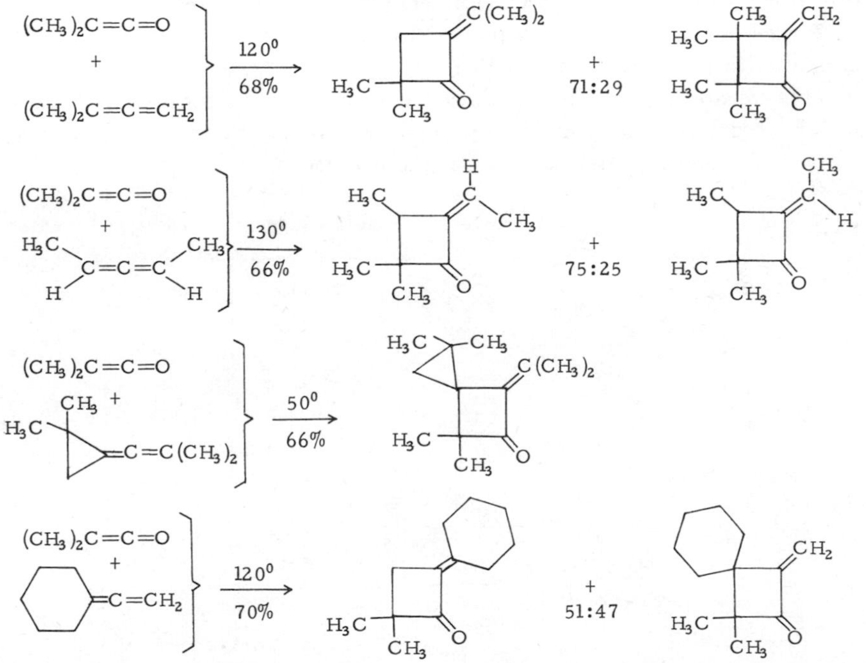

[1] M. Bertrand, R. Maurin, J. L. Gras, and G. Gil, *Tetrahedron,* **31,** 849 (1975); M. Bertrand, J.-L. Gras, and J. Gore, *ibid.,* **31,** 857 (1975).

**2,4-Dimethylpyridine (2,4-Lutidine).** Mol. wt. 107.16, b.p. 159°. Suppliers: Aldrich, Eastman, others.

*Decarboxylations.* Henrick *et al.*[1] report that decarboxylation of the diacid (1) to the lactone (2) at 100° proceeds more rapidly in 2,4-dimethyl-pyridine than in pyridine and that the presence of a copper salt has no detect-able effect. If the reaction is prolonged, the diacid can be converted into the monocarboxylic acid by opening of the lactone ring. However, it is more con-venient to prepare the lactone and then convert the lactone into the mono acid with sodium methoxide in methanol at 70°.

(1)                                                     (2)

[1]C. A. Henrick, W. E. Willy, J. W. Baum, T. A. Baer, B. A. Garcia, T. A. Mastre, and S. M. Chang, *J. Org.,* **40,** 1 (1975).

**Dimethyl selenoxide,** $(CH_3)_2$SeO. Mol. wt. 125.03, m.p. 85–86°.

*Preparation.*[1] The reagent is prepared in 75% yield by reaction of dimethyl selenide with 1 mole of ozone in $CHCl_3$ at –50°.

*Oxyselenation of olefins.*[2] The reagent reacts with olefins in HOAc–CHCl₃ at 60° to give 2-acetoxyalkyl methyl selenides, formed by stereospecific *trans*-addition. Presumably $CH_3$SeOAc is the reactive species.

[1]G. Ayrey, D. Barnard, and D. T. Woodbridge, *J. Chem. Soc.,* 2089 (1962).
[2]N. Miyoshi, S. Furui, S. Murai, and N. Sonoda, *J.C.S. Chem. Comm.,* 293 (1975).

**Dimethyl sulfide–Cuprous bromide,** $(CH_3)_2SCuBr$. Mol. wt. 205.59, white prisms, m.p. 124–129° dec.

*Lithium organocuprates.* House *et al.*[1] have found that certain undesirable side reactions in the preparation of lithium organocuprates can be minimized by use of this complex rather than commercial cuprous bromide itself, which apparently contains some impurities. The complex is readily prepared in 90% yield from $(CH_3)_2S$ and CuBr. It is insoluble in ether, hexane, acetone, methanol, and water, but dissolves in several solvents in the presence of excess $(CH_3)_2S$. Thus a solution of the complex in ether and $(CH_3)_2S$ is used; the excess sulfide is readily separated from reaction products. The soluble copper reagent *t*-BuC≡CCu can also be used instead of CuBr, but the precursor, *t*-butylacetylene, is expensive.[2] The use of the complex was illustrated for reactions of $(CH_3)_2CuLi$ and $(CH_2=CH)_2CuLi$.

[1] H. O. House, C.-Y. Chu, J. M. Wilkins, and M. J. Umen, *J. Org.*, **40**, 1460 (1975).
[2] H. O. House and M. J. Umen, *J. Org.*, **38**, 3893 (1973).

**Dimethyl sulfide ditriflate,** $(CH_3)_2\overset{+}{S}\!-\!OSO_2CF_3 \ \overset{-}{O}SO_2CF_3$ (1). Mol. wt. 360.29, m.p. 85–90° dec.

The complex (1) is obtained by the reaction of dimethyl sulfoxide with trifluoromethanesulfonic anhydride at −78°. It decomposes overnight at room temperature or upon exposure to air or water. For synthetic purposes, it is prepared *in situ.*

*Oxidation of alcohols.*[1] The reagent oxidizes benzoin to benzil at −78° in methylene chloride in 97% yield. However, yields of ketones from most secondary alcohols are in the range of 50–60%. Considerable amounts of methylthiomethyl ethers are usually formed.

[1] J. B. Hendrickson and S. M. Schwartzman, *Tetrahedron Letters*, 273 (1975).

**Dimethyl sulfoxide, 1,** 296–310; **2,** 157–158; **3,** 119–123; **4,** 192–194; **5,** 263–266.

### Solvent effects

*Oxidation with sodium dichromate.* Primary and secondary alcohols are oxidized to aldehydes and ketones in 80–90% yield by $Na_2Cr_2O_7 \cdot 2H_2O$ and sulfuric acid in DMSO at 70° (90 min.). DMSO is not involved and acts as a solvent. In its absence, considerable charring occurs with further oxidation of the carbonyl product. $CrO_3$ can also be used as oxidant.[1a]

*Knorr reaction.* When conducted in DMSO the Knorr reaction[1b] leads principally to oxazolidine-2-thiones (equation I).[1c]

*Reactions*

**Reaction with epoxides** (4, 193; 5, 266).[2] Details concerning the reaction of epoxides with DMSO and strong acids are available. The reaction is usually regio- and stereoselective (equations I and II). The initial salts can be hydrolyzed

(I)     $C_6H_5CH-CH_2$  $\xrightarrow[50-85\%]{\text{DMSO, HA}}$  $C_6H_5CHCH_2OH$
                \__O__/                                        $O-S^+(CH_3)_2$ $A^-$

(II)    $\xrightarrow[38.6\%]{\text{DMSO, TFA}}$  $+$ $CF_3COO^-$ $\xrightarrow[46\%]{H_2O}$

to 1,2-diols. Treatment of the salts with bases leads not only to the expected α-hydroxy ketones, but also to 1,2-diols, even in the absence of water (equation III). Other by-products are also formed. The ratio of ketol to glycol depends on the base used and on the $A^-$ anion of the salt.

(III)     $C_6H_5CHCH_2OH$  $\xrightarrow{N(C_2H_5)_3}$  $C_6H_5CCH_2OH$ $+$ $C_6H_5CHOHCH_2OH$
          $CF_3COO^- \; \overset{+}{O}S(CH_3)_2$                   $\underset{O}{\parallel}$

                                                    (20%)            (43%)

**Oxidation of alcohols** (1, 308). Finch and co-workers[3] report that the Barton oxidation of chloroformates (1, 308–309) was superior to Jones oxidation in the case of aldehyde (1). The *trans*-isomer of (1) was obtained in somewhat higher yield (81%).

(1)

The original procedure was reported to give rather low yields in the case of ketones, for example, 20% for cholestanone. Barton and Forbes[4] have since improved this oxidation method by addition of an acid scavenger, usually 1,2-epoxypropane, to the reaction with DMSO prior to the addition of base. Cholestanone is obtained in 80% yield by this modification. Even so, the yield is low in the case of the hindered ketone 11-ketoprogesterone (24%).

These chemists also reported a related, high-yield oxidation of secondary alcohols with nitrosyl tetrafluoroborate in the presence of hexamethyldisiloxane as acid scavenger.

**α-Diketones from α-methylene ketones.** This transformation can be carried out in two steps.[5] First, the ketone is brominated with cupric bromide (1, 161) in chloroform—ethyl acetate. The second step is a modification of the Kornblum oxidation (1, 303–304). The oxidation of secondary bromides by DMSO alone

$$\underset{RCCH_2R^1}{\overset{O}{\overset{\|}{}}} \xrightarrow[90-97\%]{\underset{CHCl_3,\ C_2H_5OAc}{CuBr_2}} \underset{RCCHBrR^1}{\overset{O}{\overset{\|}{}}} \xrightarrow[65-95\%]{\underset{Na_2CO_3}{DMSO,\ KI}} \underset{RC-CR^1}{\overset{O\ \ O}{\overset{\|\ \ \|}{}}}$$

is generally slow. However, addition of potassium iodide usually permits oxidation at ambient temperature within 3–5 hr. Even with this improvement, the ketone $(CH_3)_3CCOCHBrC(CH_3)_3$ is not oxidized, even after 25 hr at 150°. The ketone can, however, be oxidized by DMSO activated by silver nitrate[6] (47% yield).

*1,2,3-Tricarbonyl compounds.* 1,3-Dicarbonyl compounds can be converted into 1,2,3-tricarbonyl compounds by monobromination (bromine and pyridine or triethylamine) followed by Kornblum oxidation (1, 303) in DMSO at 70–80°:

$$\underset{(1)}{RCOCH_2COR^1} \xrightarrow[80-95\%]{Br_2,\ Py} \underset{(2)}{RCOCHBrCOR^1} \xrightarrow[75-90\%]{DMSO} \underset{(3)}{RCOCOCOR^1}$$

The oxidation can also be conducted in one step, (1)→(3), with selenium dioxide, usually in refluxing dioxane for 12–20 hr. Yields are generally in the range of 40–70%.[7]

*Oxidation of thiophosphoryl and selenophosphoryl groups* (*cf.* 4, 194).[8] These groups are oxidized to phosphoryl groups by DMSO and an acid catalyst ($Cl_3CCOOH$, $H_2SO_4$, $p$-TsOH). The reaction proceeds with inversion at the

$$\underset{R^3}{\overset{R^1}{\underset{R^2\cdots}{}}}P=S(Se) + (CH_3)_2SO \xrightarrow[50-95\%]{H^+,\ 60-80°} O=P\underset{R^3}{\overset{R^1}{\underset{\cdots R^2}{}}} + (CH_3)_2S + S(Se)$$

phosphorus atom, except in the case of cyclic compounds such as (1), which undergo the reaction with retention of configuration.

(1)

*Pfitzner–Moffatt oxidation* (1, 304–307; 2, 162; 3, 121).[9] Improved results have been obtained by substitution of DCC with the water-soluble diimide 1-cyclohexyl-3-(2-morpholinoethyl)-carbodiimide metho-*p*-toluenesulfonate (*see* 1, 181; supplier: Aldrich).[10]

In a total synthesis of the alkaloid *dl*-camptothecin, Rapoport *et al.*[11] investigated the oxidation of the alcohol (1). The Pfitzner–Moffatt reagent proved

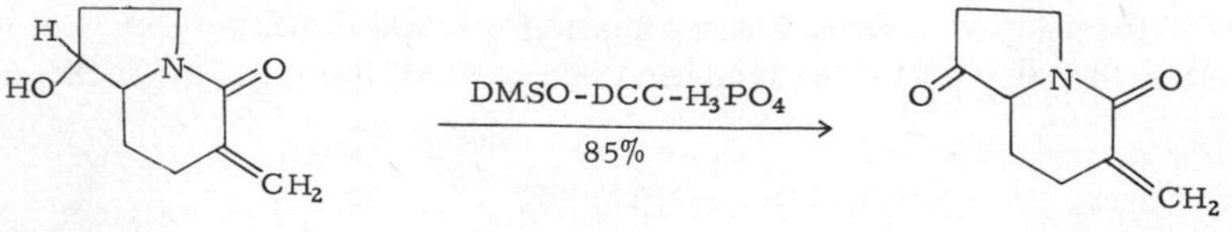

(1)                                        (2)

most satisfactory. Use of DMSO–Ac$_2$O (**1**, 305; **2**, 163–165; **3**, 121–122) gave (**2**) in 57% yield together with the acetate of (**1**). Use of Jones reagent under optimum conditions gave (**2**) in 40% yield. Sarett reagent (**1**, 145–146) gave non-ketonic products and starting alcohol. The alcohol resisted oxidation with ruthenium tetroxide and lead tetraacetate.

The Pfitzner–Moffatt reagent was also shown to be effective for oxidation of the related alcohol (**3**) to the corresponding ketone (76% yield).

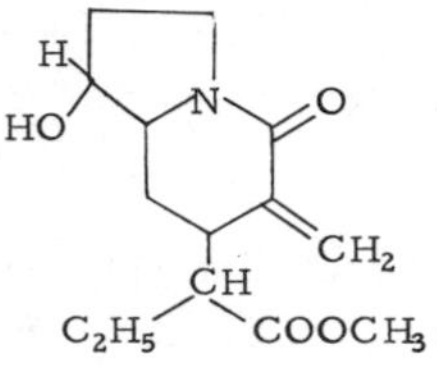

(3)

*Methylation.* The reaction of 7,7-dichlorobicyclo[4.1.0]heptane (**1**) with potassium *t*-butoxide in DMSO unexpectedly gives ethylbenzene (**2**) as a major product. A possible route from (**1**) to (**2**) has been suggested (equation I).[12]

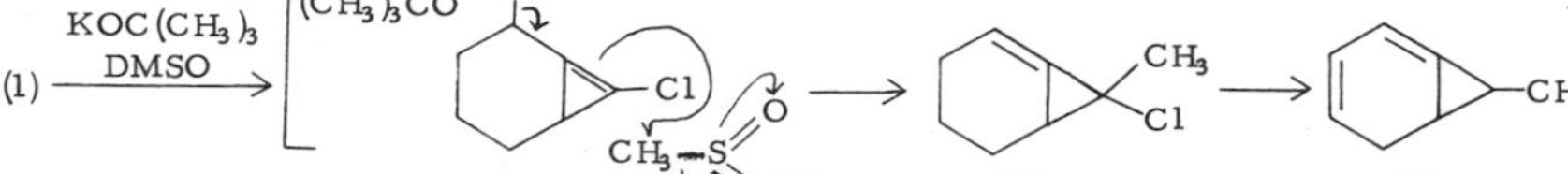

(1)                (2, 25%)        (3, 15%,          (4, ~50%)
                                    both isomers)

(a)

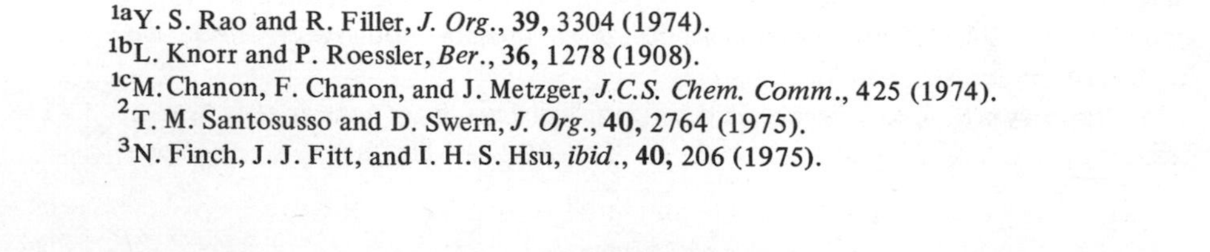

[1a]Y. S. Rao and R. Filler, *J. Org.*, **39**, 3304 (1974).
[1b]L. Knorr and P. Roessler, *Ber.*, **36**, 1278 (1908).
[1c]M. Chanon, F. Chanon, and J. Metzger, *J.C.S. Chem. Comm.*, 425 (1974).
[2]T. M. Santosusso and D. Swern, *J. Org.*, **40**, 2764 (1975).
[3]N. Finch, J. J. Fitt, and I. H. S. Hsu, *ibid.*, **40**, 206 (1975).

[4] D. H. R. Barton and C. P. Forbes, *J.C.S. Perkin I*, 1614 (1975).

[5] D. P. Bauer and R. S. Macomber, *J. Org.*, **40**, 1990 (1975).

[6] N. Kornblum and H. W. Frazier, *Am. Soc.*, **88**, 865 (1966).

[7] F. Dayer, H. L. Dao, H. Gold, H. Rodé-Gowal, and H. Dahn, *Helv.*, **57**, 2201 (1974).

[8] M. Mikolajczyk and J. Luczak, *Chem. Ind.*, 701 (1974).

[9] J. G. Moffatt, *Org. Syn. Coll. Vol.*, **5**, 242 (1973).

[10] N. Finch, J. J. Fitt, and I. H. S. Hsu, *J. Org.*, **40**, 206 (1975).

[11] C. S. F. Tang, C. J. Morrow, and H. Rapoport, *Am. Soc.*, **97**, 159 (1975).

[12] C. J. Ransom and C. B. Reese, *J.C.S. Chem. Comm.*, 970 (1975).

## Dimethyl sulfoxide–Bromine.

*p-Bromoaniline.* Direct bromination of aniline leads to di- and tribromo derivatives. *p*-Bromoaniline can be prepared in high yield by reaction of aniline hydrobromide with DMSO (reflux).[1] Presumably the DMSO–Br$_2$ adduct is the brominating species (*cf.* DMSO–Cl$_2$, 4, 200).

[1] P. A. Zoretic, *J. Org.*, **40**, 1867 (1975).

## Dimethyl sulfoxide–Chlorine, 4, 200.

*Dehydration of aldoximes to nitriles.*[1] This reaction can be carried out in 90–95% yield with the reagent of Corey and Kim in acetonitrile followed by addition of triethylamine.

[1] T.-L. Ho and C. M. Wong, *Syn. Commun.*, **5**, 423 (1975).

## Dimethyl sulfoxide–Sulfur trioxide, 2, 165–166; 4, 200–201.

*Quinones.* Harvey *et al.*[1] were able to oxidize the diol (1) to benzo[*a*]pyrene-4,5-quinone (2) with this reagent under carefully controlled conditions

(exclusion of moisture, pure diol). The related DMSO–Ac$_2$O reagent can also be used, but yields are erratic. Numerous other reagents were found to be valueless.

The diol is sensitive to both dehydration and oxidative cleavage. It can be obtained by osmium tetroxide oxidation of benzo[a] pyrene.

[1] R. G. Harvey, S. H. Goh, and C. Cortez, *Am. Soc.*, **97**, 3468 (1975).

## Dimethyl sulfoxide–Trifluoroacetic anhydride, 5, 267.

*Iminosulfuranes.* Details are available for the preparation of iminosulfuranes [sulfilimines, $(CH_3)_2 \overset{+}{S} \overset{-}{N} Ar$] with this activated form of DMSO. The reaction is general for arylamines, but fails with relatively basic amines such as cyclohexylamine and benzylamine.[1]

[1] A. K. Sharma, T. Ku, A. D. Dawson, and D. Swern, *J. Org.*, **40**, 2758 (1975).

## 6,6-Dimethyl-2-vinyl-5,7-dioxaspiro[2.5]octane-4,8-dione (1). Mol. wt. 184.20, m.p. 51–53°.

*Preparation:*

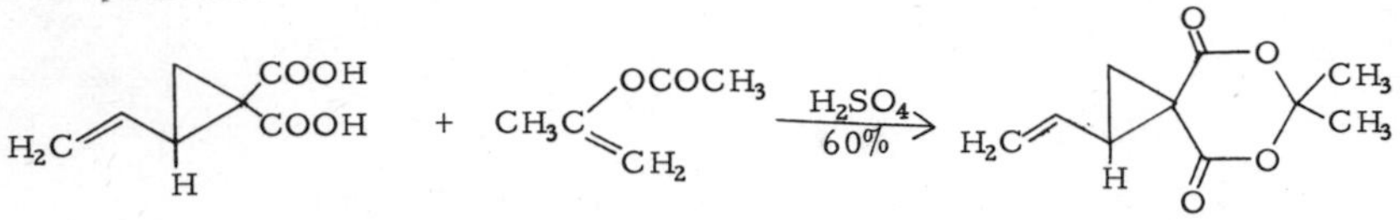

(1)

*Reaction with nucleophiles.* The spiroactivated vinylcyclopropane reacts cleanly with various nucleophiles at $C_2$.[1]

*Examples:*

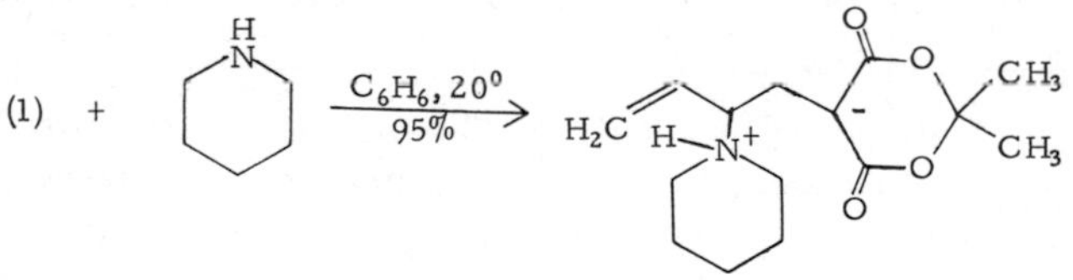

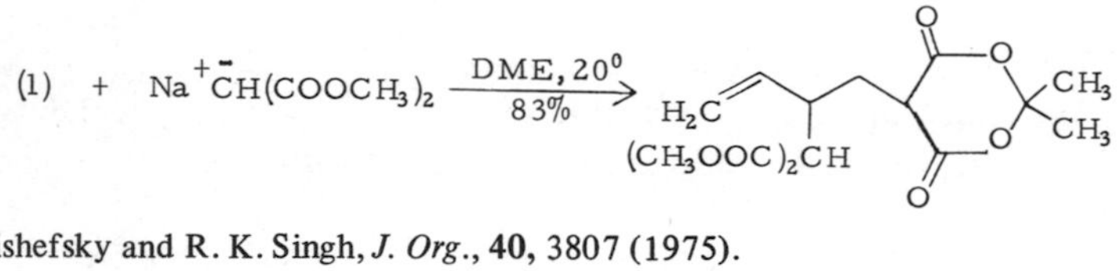

[1] S. Danishefsky and R. K. Singh, *J. Org.*, **40**, 3807 (1975).

## 2,2-Dimethylvinyl triflate, $(CH_3)_2 C = CHOSO_2 CF_3$ (1). Mol. wt. 204.17, b.p. 63–66°/72 mm.

*Preparation*[1]:

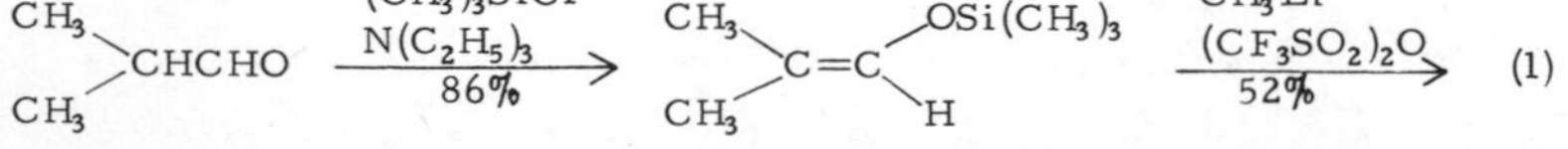

*Isopropylidenecarbene.*[2] Treatment of (1) in glyme with potassium *t*-butoxide in the presence of 2-butyne first at −55° and then at −20° (12–24

hr.) results in formation of an isopropylidenecyclopropene (2). Addition of 70% perchloric acid to the solution of (2) gives a crystalline cyclopropenium perchlorate salt (3); addition of cyclopentadiene gives the Diels–Alder adduct (4). No yields have been given for these products.

When diphenylacetylene is used as the alkyne in this reaction, the $t$-butyl vinyl ether $(CH_3)_2C=CHOC(CH_3)_3$ is the only isolated product; this is probably formed by insertion of the carbene into the O—H bond of $t$-butanol. In the case of phenylacetylene, the only isolated product is (5), the origin of which

is not certain.

Unfortunately, the postulated methylenecyclopropene intermediates polymerize rapidly at room temperature.

[1]P. J. Stang, M. G. Mangum, D. P. Fox, and P. Haak, *Am. Soc.*, **96**, 4562 (1974).
[2]P. J. Stang and M. G. Mangum, *ibid.*, **97**, 3854 (1975).

**2,4-Dinitrobenzenesulfenyl chloride,** 

(1), 1, 319. Mol. wt. 307.03, m.p. 95–96°.

*Preparation.*[1]

*Unsymmetrical disulfides.*[2] Unsymmetrical disulfides can be prepared in 70–95% yield by the procedure shown in equation I. Only traces of symmetrical disulfides are obtained.

*Reaction with norbornene.*[3] The main products of the addition of this reagent to norbornene (2) in HOAc at 25° are the chloride (3) and nortricyclene
(4). In the presence of a strong electrolyte ($LiClO_4$) the reaction is changed

drastically to give the acetates (5) as the major products. The results are inter-
preted as an increase in electrophilicity of (1) by addition of the salt.

[1] N. Kharasch and R. B. Langford, *Org. Syn., Coll. Vol.*, **5**, 474 (1973).
[2] T. Endo, H. Tasai, and T. Ihigami, *Chem. Letters*, 813 (1975).
[3] N. S. Zefirov, N. K. Sadovaja, A. M. Maggerramov, I. V. Bodrikov, and V. R. Karstashov,
*Tetrahedron*, **31**, 2948 (1975).

**2,4-Dinitrobenzenesulfonylhydrazine**, $NO_2$—⟨benzene ring⟩—$SO_2NHNH_2$ (1).    Mol.    wt.

262.04, m.p. 120°.

The hydrazine can be prepared in 70% yield by the reaction of 2,4-dinitro-
benzenesulfonyl chloride with 95% hydrazine in THF at $-78 \rightarrow 65°$.

*Eschenmoser α,β-epoxy ketone cleavage* (**2**, 419–422; **3**, 293). In the synthesis of a prostanoid, Corey and Sachdev[1] encountered a case where the Eschenmoser cleavage reaction with *p*-toluenesulfonylhydrazine gave a very complex mixture from which none of the desired acetylenic aldehyde could be isolated.
They then found that cleavage could be effected with 2,4-dinitrobenzenesulfonyl-
hydrazine in $CH_2Cl_2$ or THF at 0–25°. Pyridine, sodium carbonate, or sodium
bicarbonate can be used as catalysts. The yield can sometimes be improved by
addition of ethyl isocyanate to scavenge the sulfinic acid formed in the fragmentation. This modified procedure was shown to be effective in five cases.

[1] E. J. Corey and H. S. Sachdev, *J. Org.*, **40**, 579 (1975).

## 2,4-Dinitrofluorobenzene, 1, 321–322.

*Primary allylic chlorides.*[1] Primary allylic alcohols can be converted into the allylic chlorides by reaction with 2,4-dinitrofluorobenzene and triethylamine at room temperature. The resulting ether is then treated with lithium chloride in HMPT ($S_N2$ displacement). Yields in both steps are high.

[1] S. Czernecki and C. Georgoulis, *Bull. soc.*, 405 (1975).

**O-2,4-Dinitrophenylhydroxylamine,** $H_2NO$ —⟨benzene ring⟩— $NO_2$ . Mol. wt. 199.13, m.p. $O_2N$ (1) 112–113°.

*Preparation.*[1]

*Nitriles.*[2]   2,4-Dinitrophenyl aldoximes, prepared in high yield from (1), on treatment with base [KOH, $N(C_2H_5)_3$] are converted into nitriles in 80–95% yield.

[1] Y. Tamura, J. Minamikawa, K. Sumoto, S. Fujii, and M. Ikeda, *J. Org.*, **38**, 1239 (1973).
[2] M. J. Miller and G. M. Loudon, *ibid.*, **40**, 126 (1975).

**Diphenylcopperlithium (Lithium diphenyl cuprate)**, 4, 108.

*α-Arylation of carbonyl compounds.* The reaction of α-halo ketones with diphenylcopperlithium is not synthetically useful since mainly products of reduction are formed (equation I). Sacks and Fuchs[1] have solved this problem by

$$(I) \quad \text{[cyclohexanone with Br]} \quad + \quad (C_6H_5)_2CuLi \xrightarrow[-60°]{(C_2H_5)_2O-THF} \text{[cyclohexanone]} + \text{[cyclohexanone with } C_6H_5]$$

80%     4%

use of *p*-toluenesulfonylazocyclohexene-1 (1), prepared as shown. This substance

$$\text{[N-NHSO}_2\text{C}_7\text{H}_7, Br]} \xrightarrow{Na_2CO_3} \text{(1)} \xrightarrow[70-75\%]{(C_6H_5)_2CuLi \text{ or } C_6H_5Cu} \text{(2)} \xrightarrow[95\%]{C_6H_5 \; BF_3 \cdot (C_2H_5)_2O \atop (CH_3)_2CO, H_3O^+} \text{(3)}$$

3.0   C₆H₅Cu
72%

undergoes ready conjugate addition with diphenylcopperlithium or phenyl-copper[2] to yield (2). The desired α-phenyl ketone (3) is obtained by acetone exchange. Actually, (1) can be generated *in situ* with phenylcopper serving as base.

A further use of this new reaction is illustrated in equation II.

[1]C. E. Sacks and P. L. Fuchs, *Am. Soc.*, 97, 7372 (1975).
[2]Phenyllithium is added to a suspension of a slight excess of CuI in ether; THF is added and after 5 min. the suspension is cooled to −60°. The substrate is then added.

**Diphenyldiazomethane**, 1, 338–339; 4, 204.

*Polymeric reagent.* Chapman and Walker[1] have prepared a polymeric form of diphenyldiazomethane[2] from a polystyrene cross-linked with divinylbenzene as shown.

The diazomethylene groups of this polymer react efficiently with carboxylic acids to give "polymer diphenylmethyl esters," which can be used in peptide synthesis.

[1]P. H. Chapman and D. Walker, *J.C.S. Chem. Comm.*, 690 (1975).
[2]G. L. Southard, G. S. Brooke, and J. M. Pettee, *Tetrahedron Letters*, 3505 (1969).

**Diphenyl diselenide, 5,** 272–276.

*Preparation.* Sharpless and Young[1] have published details of a procedure developed by R. F. Lauer for the preparation of this reagent from bromobenzene (77–80% yield).

An alternate preparation has been described in which phenylmagnesium bromide is treated with selenium and then with bromine. Diphenyl diselenide is then obtained (66% yield) by air oxidation of the intermediate phenylselenol.[2]

*trans-Alkenes.* Mitchell has reported a synthesis of alkenes that involves alkylation of benzyl phenyl selenides and selenoxide fragmentation. The process is formulated:

$$RCH_2Br \xrightarrow[\sim 100\%]{\overset{\overset{-}{C_6H_5}\overset{+}{Se}Na}{C_2H_5OH}} \underset{(1)}{RCH_2SeC_6H_5} \xrightarrow[THF]{LDA} \left[\underset{}{R\overset{-}{C}HSeC_6H_5} \overset{Li^+}{}\right] \xrightarrow[THF]{R^1CH_2Br}$$

$$\underset{\underset{(2)}{\overset{|}{CH_2R^1}}}{RCHSeC_6H_5} \xrightarrow[THF]{H_2O_2} \left[\underset{\underset{(b)}{\overset{|}{CH_2R^1}}}{R\overset{|}{C}HSe^+C_6H_5} \overset{O^-}{}\right] \xrightarrow{ca.\ 80\%} \underset{(3)}{\overset{R}{\underset{H}{}}C=C\overset{H}{\underset{R^1}{}}}$$

Only lithium diisopropylamide was found satisfactory for preparation of the anion (a); use of *n*-BuLi resulted in formation of *n*-BuSeC$_6$H$_5$. The disubstituted olefins (3) have the *trans*-configuration because of *syn*-elimination from an intermediate in which the two R groups are staggered.

This synthesis of alkenes is notable in that it involves coupling of two halides.[3]

[1]K. B. Sharpless and M. W. Young, *J. Org.*, **40,** 947 (1975).
[2]H. J. Reich, J. M. Renga, and I. L. Reich, *Am. Soc.*, **97,** 5434 (1975).
[3]R. H. Mitchell, *J.C.S. Chem. Comm.*, 990 (1974).

**Diphenyl disulfide, 5,** 276–277. Additional supplier: Eastman.

*α,α-Dialkyl ketones.* Two laboratories[1,2] have used diphenyl disulfide for sulfenylation of enolates (*see* **Dimethyl disulfide, 5,** 246–247).

Coates *et al.*[3] have extended this reaction to a method for *gem*-dialkylation at a methylene group adjacent to a carbonyl group. α-Phenyl thioketones can be alkylated at the carbon bearing sulfur, and the phenylthio group can be replaced by a second alkyl group by reduction–alkylation. For example, the α-phenyl thioketone (1) is treated with sodium or potassium hydride in THF and then allowed to react with methyl iodide (excess) to give (2). This product is then reduced by a large excess of lithium in liquid ammonia–ether and then alkylated with a large excess of methyl iodide. In principle, any two alkyl groups can be introduced and in either order.

The paper also includes one example of an olefin synthesis involving reductive elimination of a *vic*-hydroxy sulfide:

$$
\underset{\underset{\displaystyle SC_6H_5}{\overset{\displaystyle CHO}{|}}}{n\text{-}C_4H_9\text{—}C\text{—}CH_2R}
\quad
\xrightarrow[\substack{75\%}]{\substack{1)\ NaBH_4,\,CH_3OH \\ 2)\ \underline{n}\text{-}C_4H_9Li,\,THF \\ C_6H_5COCl}}
\quad
\underset{\underset{\displaystyle SC_6H_5}{\overset{\displaystyle CH_2OCOC_6H_5}{|}}}{n\text{—}C_4H_9\text{—}C\text{—}CH_2R}
$$

$$
\xrightarrow[54\%]{Li,\,NH_3,\,(C_2H_5)_2O}
\quad
\underset{}{n\text{-}C_4H_9\overset{\overset{\displaystyle CH_2}{\|}}{—}C\text{—}CH_2R}
$$

*5-Alkyl- $\Delta^2$-cyclopentenones; 6-alkyl- $\Delta^2$-cyclohexenones.* Grieco and Pogonowski[4] have described a route to 5-alkyl-$\Delta^2$-cyclopentenones from cyclopentanone, the first step of which is sulfenylation with diphenyl disulfide and LDA followed by periodate oxidation to the β-keto sulfoxide (1). This product is converted into the dianion with 2 eq. of lithium diisopropylamide; the dianion

(1)                    (2)                    (3)

(4)                              (5)

is alkylated at the 5-position to give (2). Sulfoxide elimination, (2) → (3), completes the sequence. The same method provides a route to 6-alkyl-$\Delta^2$-cyclohexenones (5) from α-phenylsulfinyl cyclohexanone (4).

*1,2-Transpositions of a carbonyl group.* An α-sulfenylated ester (1), prepared as described above, can be reduced by lithium aluminum hydride in THF at 25° in high yield to the alcohol (2). Treatment with thionyl chloride (benzene or ether) yields the primary chloride (3), which can be dehydrohalogenated with potassium *t*-butoxide in DMSO at room temperature. In the case illustrated the initial olefin isomerized to the conjugated isomer (4). Hydrolysis of (4) to the ketone (5) was effected with HgCl$_2$ in refluxing aqueous

$$C_6H_5CH_2\underset{SC_6H_5}{CHCOOC_2H_5} \xrightarrow[\text{quant.}]{\text{LiAlH}_4} C_6H_5CH_2\underset{SC_6H_5}{CHCH_2OH} \xrightarrow[90\%]{\text{SOCl}_2} C_6H_5CH_2\underset{SC_6H_5}{CHCH_2Cl}$$

(1)                    (2)                    (3)

$$\xrightarrow[90\%]{\substack{\text{KOC(CH}_3)_3 \\ \text{DMSO}}} C_6H_5CH=\underset{SC_6H_5}{CCH_3} \xrightarrow[85\%]{\text{HgCl}_2} C_6H_5CH_2\overset{O}{\overset{\|}{C}}CH_3$$

(4)                        (5)

acetonitrile or dioxane. The overall process thus effects the conversion: —CH$_2$COOR → —COCH$_3$. The reactions were applied to two other esters and comparable yields were obtained.

The same process is applicable with slight modifications to transposition of a ketonic group. For example, estrone methyl ether (6) can be converted into the isomeric 16-ketone (11) as formulated. The sequence was shown to be applicable to four-, five-, and six-membered rings.

A variation effects alkylation and transposition of the carbonyl group. For example, methyllithium adds to the sulfenylated ester (12) to give (13). Dehydration, isomerization, and hydrolysis leads to (16).[5]

$$CH_3(CH_2)_7\overset{SC_6H_5}{\underset{}{C}}HCOOC_2H_5 \xrightarrow[\text{quant.}]{CH_3Li} CH_3(CH_2)_7\overset{SC_6H_5}{\underset{OH}{C}}HC(CH_3)_2 \xrightarrow{TsOH} CH_3(CH_2)_7\overset{SC_6H_5}{\underset{}{C}}HC\overset{CH_2}{\underset{CH_3}{}}$$

$$(12) \qquad\qquad (13) \qquad\qquad (14)$$

$$\xrightarrow[94\%]{\underset{DMSO}{KOC(CH_3)_3}} CH_3(CH_2)_7\overset{SC_6H_5}{\underset{}{C}}=C(CH_3)_2 \xrightarrow[80\%]{HgCl_2} CH_3(CH_2)_7\overset{O}{\overset{\|}{C}}CH(CH_3)_2$$

$$(15) \qquad\qquad (16)$$

*α-Keto esters.* Trost and Salzmann[6] have reported that bissulfenylation of esters is possible and provides a route to α-keto esters. Thus treatment of (1) with 2 eq. of lithium N-isopropylcyclohexylamide (LiICA, 4, 306–309) and then with 2 eq. of diphenyl disulfide in THF–HMPT at 0° gives (2) in 79% yield. This is converted into the α-keto ester (4) by transketalization followed

by acid treatment.

[1] B. M. Trost and T. N. Salzmann, *Am. Soc.*, **95**, 6840 (1973).
[2] D. Seebach and M. Teschner, *Tetrahedron Letters*, 5113 (1973).
[3] R. M. Coates, H. D. Pigott, and J. Ollinger, *ibid.*, 3955 (1974).
[4] P. A. Grieco and C. S. Pogonowski, *J.C.S. Chem. Comm.*, 72 (1975).
[5] B. M. Trost, K. Hiroi, and S. Kurozumi, *Am. Soc.*, **97**, 438 (1975).
[6] B. M. Trost and T. N. Salzmann, *J. Org.*, **40**, 149 (1975).

**Diphenyldi(1,1,1,3,3,3-hexafluoro-2-phenyl-2-propoxy)sulfurane**, 4, 205–207; 5, 270–272.

*Cleavage of secondary amides* (4, 206).[1] Details have been published about the cleavage of secondary amides with this sulfurane. Sulfilimines are generally obtained from unhindered amides, whereas hindered amides are generally cleaved to imidates. Both products are obtained in intermediate cases, as in the cleavage N-*n*-butylbenzanilide (equation I). Both products are readily convertible into the free amine.

$$\text{(I)} \quad C_6H_5CONH(C_4H_9\text{-}\underline{n}) + C_6H_5-\underset{\underset{\underset{C_6H_5}{|}}{\underset{|}{F_3C-C-CF_3}}}{\overset{\overset{\overset{C_6H_5}{|}}{\overset{|}{F_3C-C-CF_3}}}{\overset{|}{\underset{|}{O}}}}S-C_6H_5 \longrightarrow C_6H_5CO_2R_F + \underline{n}\text{-}C_4H_9N{=}S(C_6H_5)_2 + \underline{n}\text{-}C_4H_9N{=}C(OR_F)C_6H_5$$

$$\text{(47\%)} \qquad\qquad \text{(47\%)} \qquad\qquad \text{(49\%)}$$

$$[(C_6H_5)_2S(OR_F)_2]$$

This cleavage reaction is potentially useful because the conditions are much milder than those usually required for cleavage of secondary amides. Protection of an amine function as an amide in synthesis is thus feasible.

*Reactions with amines.*[2] The sulfurane (1) reacts with ammonia and primary amines, amides, and sulfonamides to give S,S-diphenylsulfilimines.

$$(C_6H_5)_2\underset{\underset{OR_F}{|}}{\overset{\overset{OR_F}{|}}{S}} + H_2NY \xrightarrow[-R_FOH]{} (C_6H_5)_2\underset{\underset{OR_F}{|}}{\overset{\overset{\overset{Y\diagdown_N\diagup H}{|}}{}}{S}}: \longrightarrow$$

$$(1)$$

$$(C_6H_5)_2\overset{\overset{\overset{Y\diagdown_N\diagup H}{|+}}{}}{S}\,\overset{-}{O}R_F \rightleftharpoons (C_6H_5)_2\overset{+}{S}{-}\overset{-}{N}Y \longleftarrow (C_6H_5)_2S{=}N\diagup^Y + R_FOH$$

$$\text{(50–98\%)}$$

Secondary amines with moderately acidic α-protons are oxidized by (1) to imines.

$$(1) + C_6H_5CH_2NHCH_3 \longrightarrow C_6H_5CH{=}NCH_3 \quad + \quad (C_6H_5)_2S$$
$$\text{(55\%)}$$

$$(1) + (C_6H_5CH_2)_2NH \longrightarrow C_6H_5CH_2N{=}CHC_6H_5 + (C_6H_5)_2S$$
$$\text{(85\%)}$$

Benzylamine is oxidized by 2 eq. of (1) to benzonitrile (89% yield).

[1] J. C. Martin and J. A. Franz, *Am. Soc.*, 97, 6137 (1975).
[2] J. A. Franz and J. C. Martin, *ibid.*, 97, 583 (1975).

**Diphenyl selenide** ("Phenyl selenide"), $(C_6H_5)_2Se$. Mol. wt. 233.17, m.p. 0–3°.
Supplier: Eastman.

*Preparation.*[1]

*Oxiranes.*[2] The reaction of diphenyl selenide (or dimethyl selenide), silver
tetrafluoroborate, and an alkyl iodide or bromide (excess) at 20° under $N_2$
gives alkyldiphenylselenonium tetrafluoroborates (1) in 40–60% yield. These
are converted into ylides (a), which react with nonenolizable aldehydes or

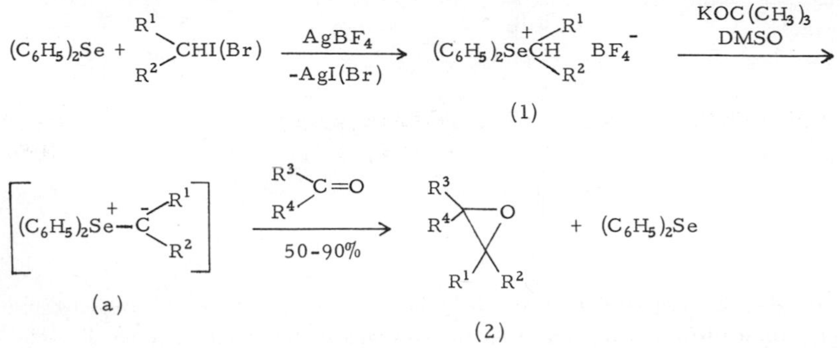

(1)

(a)

(2)

ketones to give oxiranes. Significant amounts of oxiranes are not obtained from
enolizable carbonyl compounds; in this case alkylation products are mainly
formed. Thus dimethylselenonio methylide, $(CH_3)_2\overset{+}{Se}-\overset{-}{C}H_2$, reacts with
acetophenone to give propiophenone (13%) and isobutyrophenone (28%).

[1] A. H. Blatt, *Org. Syn., Coll. Vol.*, **2**, 238 (1943).
[2] W. Dumont, P. Bayet, and A. Krief, *Angew. Chem., internat. Ed.*, **13**, 274 (1974).

**Diphenylseleninic anhydride,** $[C_6H_5\overset{O}{\overset{\|}{Se}}]_2O$.        Mol. wt. 360.12, m.p. 124–126°
and 164–165°.

*Preparation.*[1] The anhydride is prepared by reaction of diphenyl diselenide
with 3 moles of ozone in $CCl_4$ at −5°.

*Oxidation of phenols.*[2] Treatment of phenols with this reagent in $CH_2Cl_2$
at 25° results in *ortho-* and *para-*hydroxylation. Thus oxidation of 2,4,6-tri-
methylphenol (1) results in formation of (2) and (3), the dimer of the *o*-
hydroxycyclohexadienone.

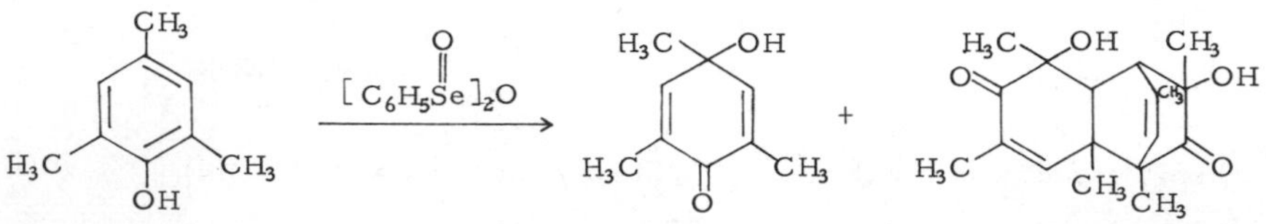

(1)                              (2, 30%)                (3, 48%)

If the oxidation is carried out on the anion of (1), prepared with sodium hydride, (3) is obtained in 55% yield with no evidence of formation of (2).

[1] G. Ayrey, D. Barnard, and D. T. Woodbridge, *J. Chem. Soc.*, 2089 (1962).
[2] D. H. R. Barton, P. D. Magnus, and M. N. Rosenfeld, *J.C.S. Chem. Comm.*, 301 (1975).

**Diphenyl sulfide**, $(C_6H_5)_2S$. Mol. wt. 186.28, b.p. 151–153°/15 mm.

*Singlet oxygen reactions.* Dialkyl sulfides are oxidized to sulfoxides by singlet oxygen. Diphenyl sulfide is practically inert under the same conditions. However, if it is added to the photooxidation of a dialkyl sulfide, it is converted into diphenyl sulfoxide. The interpretation advanced by Foote and Peters[1] is that diphenyl sulfide reacts readily with persulfoxides, formed in the reaction of alkyl sulfides with singlet oxygen:

$$R_2S \xrightarrow{{}^1O_2} R_2\overset{+}{S}O\overset{-}{O} \xrightarrow{R_2S} 2\ R_2\overset{+}{S}-O^-$$

$$\downarrow (C_6H_5)_2S$$

$$(C_6H_5)_2\overset{+}{S}-\overset{-}{O} \ + \ R_2\overset{+}{S}-\overset{-}{O}$$

Wasserman and Saito[2] have used diphenyl sulfide to intercept dioxetanes formed in the sensitized photooxidations of certain unsaturated systems. Thus reaction of *cis*-dimethoxystilbene (1) with singlet oxygen yields only the dioxetane (2) and the product of cleavage, methyl benzoate. However, addition of diphenyl sulfide to the reaction results in formation of benzil dimethyl ketal

(3, 18%) and diphenyl sulfoxide (35%) in addition to methyl benzoate (32%). The formation of (3) is considered to involve reaction of diphenyl sulfide with the dioxetane to give the zwitterion (a), which then undergoes a benzilic acid–like rearrangement to (3). Addition of diphenyl sulfide was shown to alter other dye-sensitized photooxidations known to involve intermediate dioxetanes; it has no effect on singlet oxygen reactions that involve 1,4-transannular peroxides.

[1] C. S. Foote and J. W. Peters, *Am. Soc.*, **93**, 3795 (1971).
[2] H. H. Wasserman and I. Saito, *ibid.*, **97**, 905 (1975).

**Diphenylsulfonium cyclopropylide, 4,** 211–214; **5,** 281.

*Geminal alkylation* (**4,** 212–213). The definitive paper on geminal alkylation of ketones via cyclobutanones has been published.[1] The original method involves dibromination of the cyclobutanone and consequently is not suitable for substrates containing isolated double bonds. In this case an alternative approach is available via α-trimethylenedithiocyclobutanones.[2] Direct condensation of trimethylene dithiotosylate (**4,** 539–540; **5,** 71) with the cyclobutanone enolate fails, but can be accomplished indirectly by conversion of the cyclobutanone into an enamide by reaction with *t*-butoxybis(dimethylamino)methane.[3] The desired dithiane is then obtained by reaction of the enamide and trimethylenedithiotosylate in ethanol buffered with potassium acetate. The sequence is illustrated for 1-tetralone (**1**). The product (**4**) obtained in this way can be cleaved to (5) by methanolic sodium methoxide. The formulation also illustrates conversion of (4) into (9). The overall result is annelation of a cyclopentenone ring onto a carbonyl group.

This new synthetic reaction was used to synthesize methyl desoxypodocarpate (14) from (10). The overall result is conversion of $>C=O$ into $>C\overset{COOCH_3}{\underset{CH_3}{<}}$, a unit encountered in many natural products.

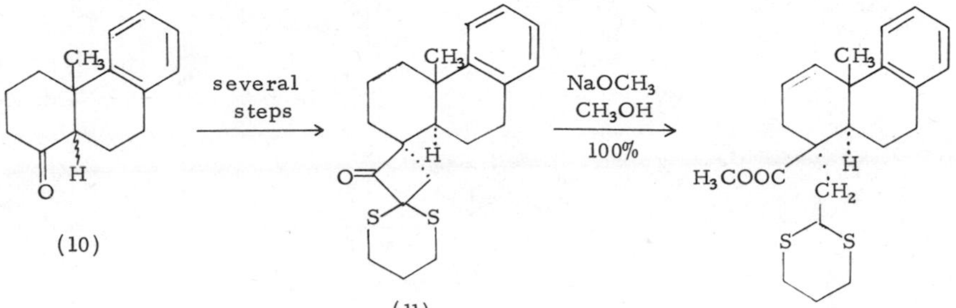

(10)    (11)    (12)

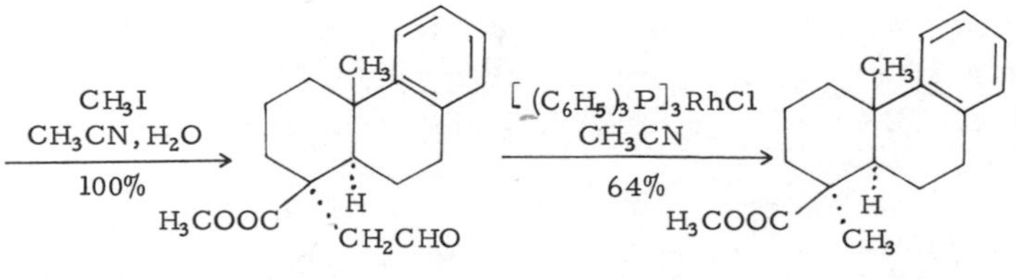

(13)    (14)

[1] B. M. Trost, M. J. Bogdanowicz, and J. Kern, *Am. Soc.*, **97**, 2218 (1975).
[2] B. M. Trost, M. Preckel, and L. M. Leichter, *ibid.*, **97**, 2224 (1975).
[3] H. Bredereck, G. Simchen, S. Rebsdat, W. Kantlehner, P. Horn, R. Wahl, H. Hoffmann, and P. Grieshaber, *Ber.*, **101**, 41 (1968).

**Diphosphorus tetraiodide, 1, 349–350.**

*Improved preparation.* Phosphorus trichloride (0.2 mole) is added cautiously with stirring to a slurry of potassium iodide (0.60 mole) in ether (argon). After reflux of 12 hr., the ether is removed under reduced pressure, and the residue is crystallized from methylene chloride: orange needles, m.p. 123–125°, yield 75–80%.[1]

*Buta-1,2,3-trienes.* Newkome et al.[1] were able to prepare the unstable (Z)- and (E)-1,4-diphenyl-1,4-di(2-pyridyl)butatrienes, (2) and (3), by didehydroxylation of (1) with $P_2I_4$ in anhydrous pyridine. Reduction of (1) with the more usual reagents ($SnCl_2$ or $PBr_3$) resulted mainly in formation of (4).

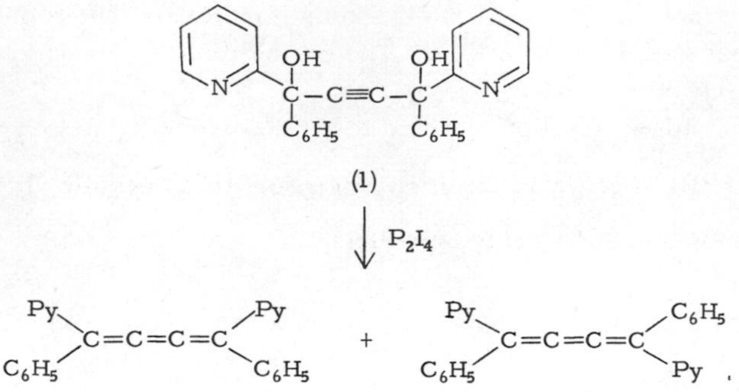

(1)

$$\downarrow P_2I_4$$

(2)          +          (3)

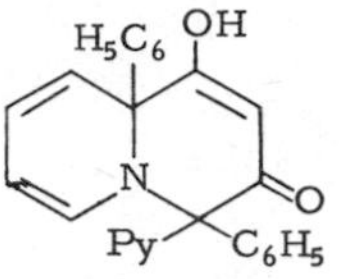

(4)

*1,4-Dienes.* Japanese chemists[2] have reported a synthesis of 1,4-dienes by homo-1,4-elimination of $\alpha,\alpha'$-dihydroxy derivatives of cyclopropanes. For example, treatment of (1)[3] with $P_2I_4$ in refluxing pyridine gave the *trans,trans*-diene (2) in about 50% yield. A complex mixture was obtained when stannous chloride in acid was used.

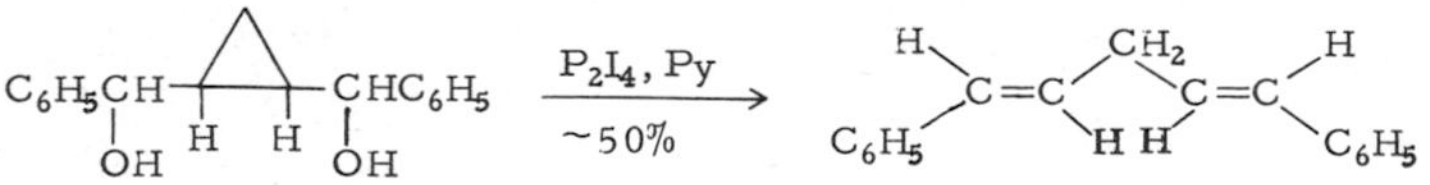

(1)                    (2)

In an analogous reaction, the benzotropylidene (4) was obtained from (3); the yield was not indicated.

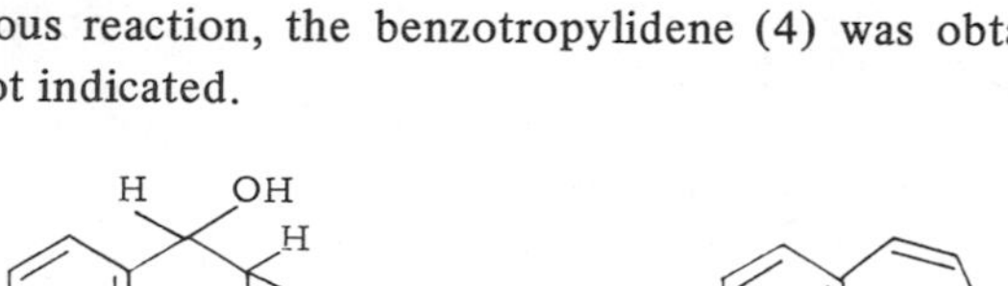

(3)                    (4)

[1] G. R. Newkome, J. D. Sauer, and M. L. Erbland, *J.C.S. Chem. Comm.*, 885 (1975).
[2] T. Hanafusa, S. Imai, K. Ohkata, H. Suzuki, and Y. Suzuki, *J.C.S. Chem. Comm.*, 73 (1974).
[3] Prepared by NaBH$_4$ reduction of *trans*-1,2-dibenzoylcyclopropane: I. Colon, G. W. Griffin, and E. J. O'Connell, Jr., *Org. Syn.*, **52**, 33 (1972).

**Di-*n*-propylcopperlithium (Lithium di-*n*-propyl cuprate), 5, 283–285.**

*Reaction of α,β-epoxysilanes with organocuprates.* Hudrlik *et al.*[1] have shown that the reaction of di-*n*-propylcopperlithium with α,β-epoxysilanes is both regio- and stereospecific. Thus the *cis*-epoxide (1) is converted into the *erythro*-alcohol (2), whereas the *trans*-epoxide (3) is converted into the *threo*-alcohol (4).[2]

Treatment of (2) and (4) with base or acid proceeds with opposite stereochemistry.[3] Thus treatment of (2) with potassium hydride gives *cis*-4-octene (5) in high yield, whereas treatment with $BF_3$ etherate or $H_2SO_4$ gives *trans*-4-octene (6) in high yield. In the same way, (4) can be converted into either (5) or (6). Therefore, the stereochemistry of the base-induced elimination of

$(CH_3)_3SiOH$ is *syn* and that of the acid-catalyzed elimination is *anti*.

The reaction of organocuprates with epoxides is therefore useful for stereospecific synthesis of olefins.

[1]P. F. Hudrlik, D. Peterson, and R. J. Rona, *J. Org.*, **40**, 2263 (1975).
[2]C. R. Johnson, R. W. Herr, and D. M. Wieland, *ibid.*, **38**, 4263 (1973).
[3]P. F. Hudrlik and D. Peterson, *Am. Soc.*, **97**, 1464 (1975).

**2,2'-Dipyridyl disulfide, 5,** 285–286.

*Esters and lactones.* Swiss chemists[1] have used a procedure similar to that of Corey and Nicolaou (**5,** 260) for preparation of esters and lactones, but they carry out the reaction in the presence of 1 eq. of $AgBF_4$ or $AgClO_4$. Coordination of an S-2-pyridylthiocarboxylate with silver ion permits rapid esterification at $20°$ in a few minutes. If the silver salt is omitted in the example cited, the yield of ester is only 5% after 7 days at $20°$.

*Example:*

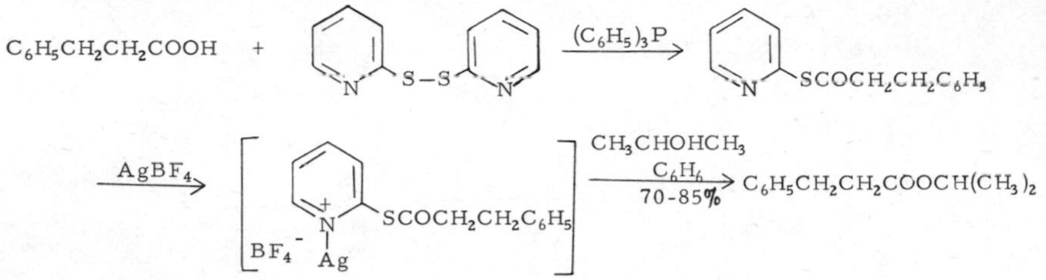

One example of lactonization promoted by $Ag^+$ was reported. Addition of 1 eq. of silver perchlorate to the 2-pyridyl thioester of 15-hydroxypentadecanoic acid in benzene gave a mixture of the monomeric, dimeric, and trimeric lactones in 44% yield within 30 min. at $20°$.

*Macrocyclic lactones.* Corey *et al.*[2] have published several more examples of the conversion of ω-hydroxyalkanoic acids to macrocyclic lactones by use of 2,2'-dipyridyl disulfide and triphenylphosphine. Several lactones in the prostaglandin series have been prepared, for example, (1) and (2). The process has been applied to even more complex natural products and has also been used

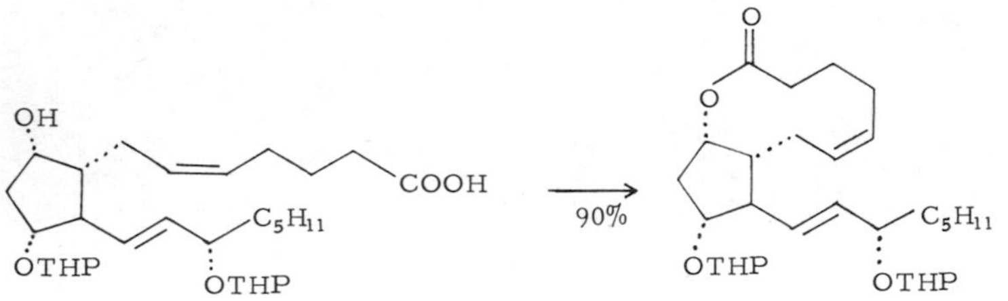

(1)

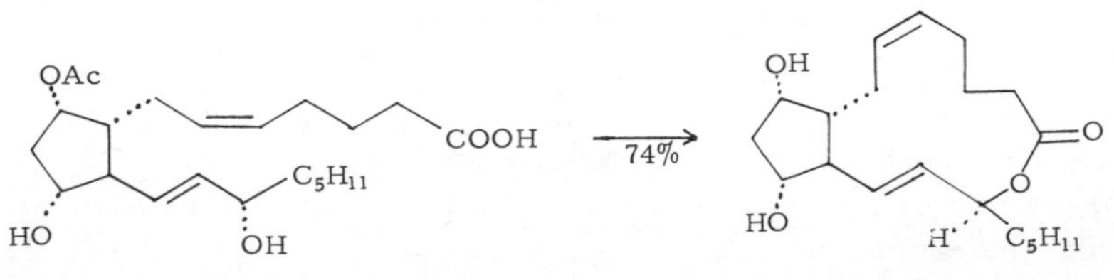

(2)

for the synthesis of several naturally occurring macrocyclic lactones, such as the *Lythracae* alkaloid vertaline (3).

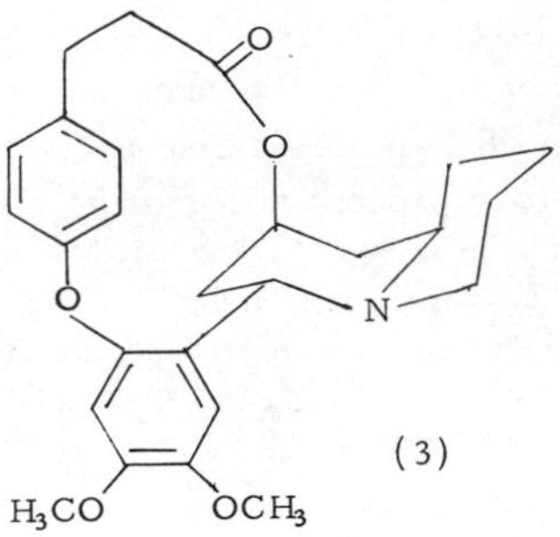

(3)

Application of the cyclization method to N-benzyloxycarbonylcarpamic acid (4) gives the bisbenzyloxycarbonyl derivative of the papaya alkaloid carpaine (5) in >50% yield. Carpaine itself is obtained from (5) by catalytic hydrogenation. In this case there is no evidence for formation of a monolactone in the cyclization.

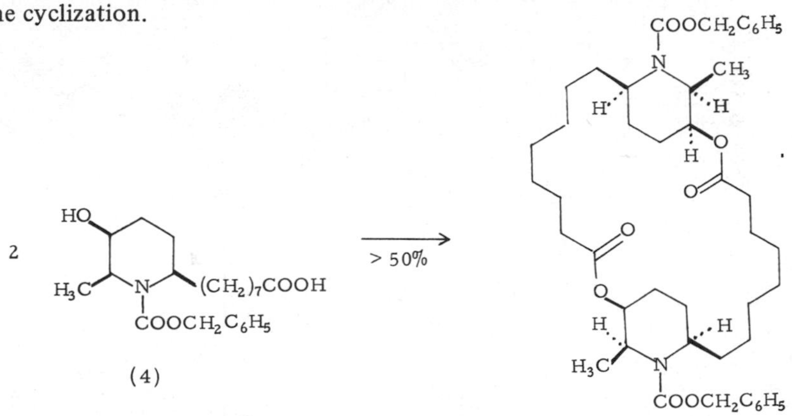

(4)

(5)

This method of lactonization has been used successfully in a total synthesis of the antibiotic (±)-vermiculine (6).[3]

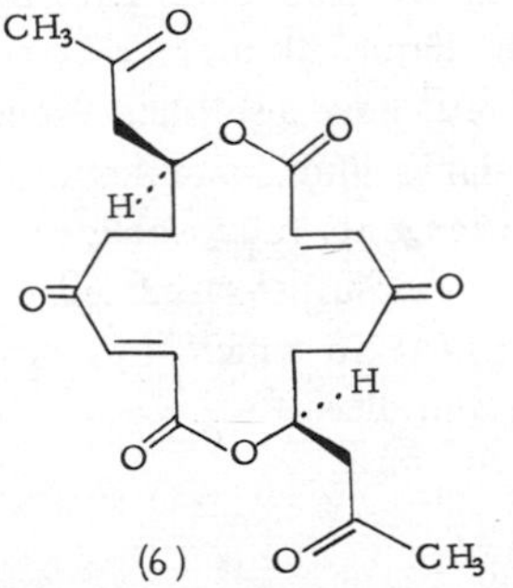

(6)

[1] H. Gerlach and A. Thalmann, *Helv.*, **57**, 2661 (1974).
[2] E. J. Corey, K. C. Nicolaou, and L. S. Melvin, Jr., *Am. Soc.*, **97**, 653, 654 (1975).
[3] E. J. Corey, K. C. Nicolaou, and T. Toru, *Am. Soc.*, **97**, 2287 (1975).

**1,3-Dithiane, 2**, 187; **3**, 135–316; **4**, 216–218.

*Formylation of 2-aminopyridines.* Gassman and Huang[1] have reported the selective *ortho*-formylation of 2-aminopyridine by two methods as shown. The second is based on a previous procedure for *ortho*-alkylation of amines using

dimethyl sulfide (**4**, 190–191). The resulting 3-methylthiomethyl group of (**5**) is oxidized to a formyl group by monochlorination (NCS) followed by hydrolysis of the chlorinated intermediate with mercuric oxide and boron trifluoride.

*Review.* Seebach and Corey[2] have published a general paper on the preparation and metalation of 1,3-dithianes and examples of the reaction of 2-lithio-1,3-dithianes with electrophilic reagents (alkyl halides, carbonyl compounds, acids, and oxides). The value of these sulfur-stabilized anionic reagents is that they are equivalent to acyl anions (a), in which the normal polarity of the carbonyl group is reversed (reversible umpolung).

$$R—\bar{C}=O$$

(a)

[1] P. G. Gassman and C. T. Huang, *J.C.S. Chem. Comm.*, 685 (1974).
[2] D. Seebach and E. J. Corey, *J. Org.*, **40**, 231 (1975).

$D$-(-)-N-Dodecyl-N-methylephedrinium bromide,

$$\begin{array}{c} CH_3 \\ | \\ H-C-N^+(CH_3)_2C_{12}H_{25}\text{-}\underline{n} \\ | \\ H-C-OH \quad Br^- \\ | \\ C_6H_5 \end{array}$$

(1)

$\cdot$ Mol.

wt. 428.40.

The salt is prepared by methylation of D-(-)-ephedrine with formaldehyde followed by quarternization with dodecyl bromide.

*Borohydride reduction of* $>C=O$ *to* $>CHOH$.[1] This salt (1) is a particularly effective catalyst for the reduction of carbonyl groups by potassium or sodium borohydride in a two-phase system (benzene–$H_2O$). Indeed, the rate of reduction is faster than in a homogeneous system. Studies with related salts indicate that the hydroxyl group $\beta$ to the N atom is a contributing factor. (*Cf.* **N,N-Dimethylephedrinium bromide,** this volume.) Even though (1) is optically active, no asymmetric induction was found in these reductions.

In several other typical anion-promoted two-phase reactions, (1) was found to be less effective than hexadecyltributylphosphonium bromide, a typical phase-transfer catalyst.

[1]C. A. Bunton, L. Robinson, and M. F. Stam, *Tetrahedron Letters,* 121 (1971).

# E

**Ethoxycarbonyl isothiocyanate,** $C_2H_5O\overset{\overset{O}{\|}}{C}NCS$. Mol. wt. 131.15, b.p. 44–46°/10 torr.

*Preparation*[1]:

$$C_2H_5OCOCl + KSCN \xrightarrow[60\%]{CH_3CN} C_2H_5O\overset{\overset{O}{\|}}{C}NCS + KCl$$

*Heterocycles.* The use of alkoxycarbonyl isothiocyanates for synthesis of heterocycles has been reviewed.[2]

*Examples:*

[1] R. W. Lamon, *J. Heterocyclic Chem.*, **5**, 837 (1968).
[2] R. Esmail and F. Kurzer, *Synthesis*, 301 (1975).

**N-Ethylacetonitrilium tetrafluoroborate,** $H_3CC\equiv\overset{+}{N}C_2H_5$ $BF_4^-$. Mol. wt. 156.91.

The salt is prepared by alkylation of acetonitrile with triethyloxonium tetrafluoroborate.[1]

*Nitriles.*[2] Aldoximes are dehydrated to nitriles when treated in acetonitrile with N-ethylacetonitrilium tetrafluoroborate (8 hrs. at 20°, then 0.5 hrs. at 80°). The amide coproduct is water soluble and easily separated from the nitrile.

[1] H. Meerwein, P. Laasch, R. Mersch, and J. Spille, *Ber.*, **89**, 209 (1956).
[2] T.-L. Ho, *Synthesis*, 401 (1975).

**Ethyl acrylate,** $CH_2=CHCOOC_2H_5$. Mol. wt. 100.12, b.p. 99°. Suppliers: Aldrich, Eastman, others.

*γ-Keto esters.* Piperidine enamines of aldehydes of type (1) add to ethyl acrylate in refluxing benzene to give enamines (2). These products are converted into γ-keto esters (3) by oxygenation in the presence of cuprous chloride.[1]

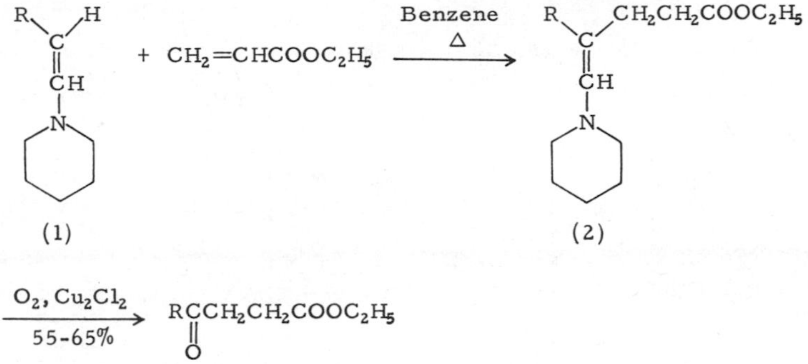

(1)          (2)

( 3 )

[1]T.-L. Ho, *Syn. Commun.*, 135 (1974).

**Ethylaluminum dichloride,** $C_2H_5AlCl_2$. Mol. wt. 126.96; m.p. 32°, b.p. 194°. Supplier: Alfa.

*[2 + 2] Cycloaddition of allenes to alkenes.*[1] This cycloaddition to give methylenecyclobutanes (3) can be realized in the presence of certain Lewis acids as catalysts (listed in approximate order of decreasing reactivity): $C_2H_5AlCl_2$, $GaCl_3$, $AlBr_3$, $AlCl_3$, $FeCl_3$.

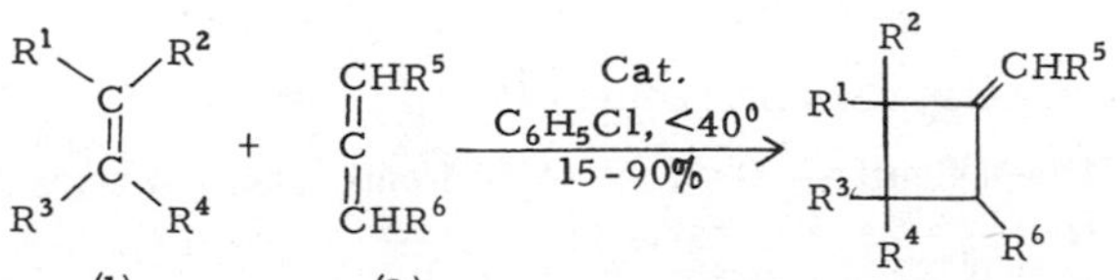

(1)          (2)                    (3)

1-Alkenes (ethylene, propylene) do not react; the hindered 4,4-dimethyl-2-pentene also does not react. Surprisingly, cyclohexene does not undergo this reaction. The yields are also low when both $R^5$ and $R^6$ are alkyl groups. The most serious competing reaction is polymerization of the alkene with deactivation of the catalyst.

*Examples:*

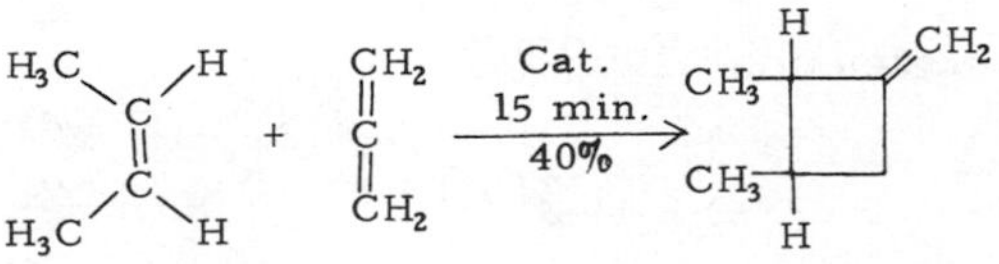

[1] J. H. Lukas, A. P. Kouwenhoven, and F. Baardman, *Angew. Chem. internat. Ed.*, **14**, 709 (1975).

**Ethyl chlorothiolformate (S-Ethyl carbonochloridothioate)**, $ClCOSC_2H_5$, mol. wt. 124.59, b.p. 132°, supplier: Aldrich; **Phenyl chlorothiolformate**, $ClCOSC_6H_5$, mol. wt. 172.63, b.p. 99–101°/10 mm., supplier: Columbia.

*Carbothioate S-esters*, $R\overset{O}{\overset{\|}{C}}SR'$   These esters can be prepared generally in 65–80% yield by reaction of the sodium salt of a carboxylic acid with an ester of chlorothiolformic acid in THF or DMF (pyridine) first at 0° and then at 25°.[1] The conventional method involves reaction of an acid chloride with a mercaptan

and base (Schotten-Baumann procedure).[2]

[1] R. A. Gorski, Dineshkumar, J. Dagli, V. A. Patronik, and J. Wemple, *Synthesis*, 811 (1974).
[2] A. J. Speziale and H. W. Frazier, *J. Org.*, **26**, 3176 (1961).

**Ethyl diazoacetate**, **1**, 367–370; **2**, 193–195; **3**, 138–139; **4**, 228–230; **5**, 295–300.

*Ring expansion* (**1**, 369–370). Ring expansion of unsymmetrical α-mono- and α,α-disubstituted cyclopentanones and cyclohexanones (1) with ethyl diazoacetate in the presence of boron trifluoride etherate gives mixtures of (2) and

(1, n = 1, 2,
R = H, $CH_3$, $C_2H_5$)

(3). The less substituted α-carbon atom migrates preferentially; consequently, (2) predominates. This product (2) is formed exclusively when $R^1 = R^2 = CH_3$.[1]

*A diaziridine.* Ethyl diazoacetate reacts with 4-phenyl-1,2,4-triazoline-3,5-dione to give the 1:1 adduct (1) in nearly quantitative yield. The adduct is

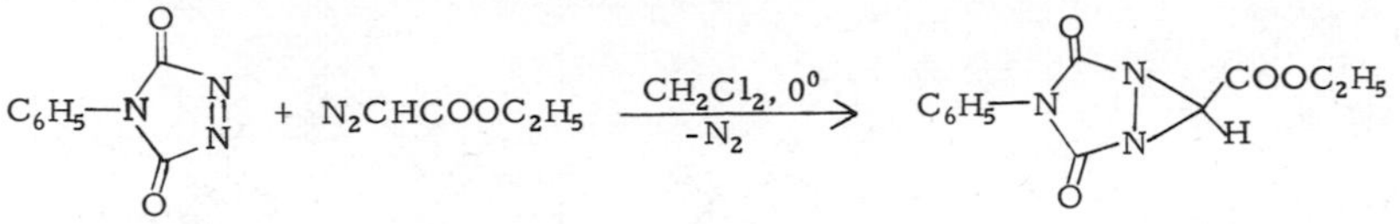

(1)

probably formed by a 1,3-dipolar addition to form a tetrazoline intermediate rather than through a free carbene. The latter reaction requires a higher temperature than necessary in the former case.[2]

[1] H. J. Liu and S. P. Majumdar, *Syn. Commun.*, 5, 125 (1975).
[2] R. A. Izydore and S. McLean, *Am. Soc.*, 97, 5611 (1975).

**Ethyl diethoxyacetate,** $(C_2H_5O)_2CHCOOC_2H_5$. Mol. wt. 176.21, b.p. 199°. Supplier: Aldrich.

*Conjugate addition.* Damon and Schlessinger[1] have found that the anion (1) of this substance can undergo conjugate addition (*cf.* 5, 444–445) and have used this reaction for an improved synthesis of *dl*-4-isoavenaciolide (5), an antifungal metabolite of *Aspergillus avenaceus.*

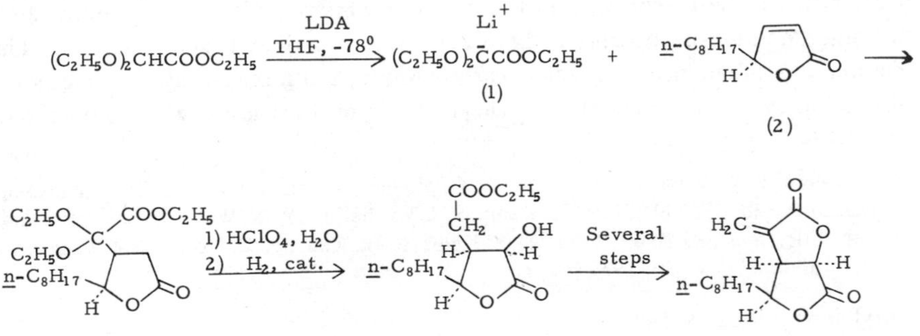

[1] R. E. Damon and R. H. Schlessinger, *Tetrahedron Letters*, 4551 (1975).

**2-Ethyl-7-hydroxybenzisoxazolium tetrafluoroborate** (1), 2, 192. Mol. wt. 251.00, m.p. 136–137°.

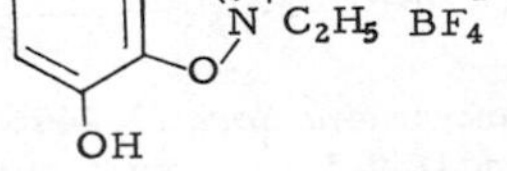

(1)

The salt is available in 78% overall yield from 2,3-dihydroxybenzaldehyde.

*Peptide Synthesis.* The salt reacts with a carboxylic acid (preferably the alkali salt) in an alkaline medium to form an active ester, shown to have structure (2). This structure is more stable than (b) mainly because of the intramolecular hydrogen bond. These active esters react with amines to form amides (3).

RCOONa + (1) $\xrightarrow[90-94\%]{\substack{Py-H_2O \\ pH\ 4.5-5.0}}$

(a)          (b)

$\xrightarrow{R'NH_2}$  $\underset{(3)}{RCNHR'}$

(2)

Kemp *et al.*[1] have now published complete details on the use of this reagent for peptide synthesis. The second step is best carried out with a tetraalkylammonium salt of an amino acid or with an amino acid (or peptide) in DMSO containing 1 eq. of tetramethylguanidine, $[(CH_3)_2N]_2C=NH$. The method has been shown to be useful with all the common amino acids except arginine and histidine. Medium-sized peptides have been prepared. Kemp concludes that this method is comparable to the azide, carbodiimide, and mixed-anhydride methods of peptide synthesis. Difficulties with yield and purification have been observed with fairly large peptides.

[1] D. S. Kemp, S.-W. Wang, J. Rebek, Jr., R. C. Mollan, C. Banquer, and G. Subramanyam, *Tetrahedron*, **30**, 3955 (1974); D. S. Kemp, S. J. Wrobel, Jr., S.-W. Wang, Z. Bernstein, and J. Rebek, Jr., *ibid.*, **30**, 3969 (1974); D. S. Kemp, S.-W. Wang, R. C. Mollan, S.-L. Hsia, and P. N. Confalone, *ibid.*, **30**, 3677 (1974).

### Ethylidene iodide, 3, 141.

*Preparation.* This iodide can be prepared in 60% yield by the reaction of ethylidene chloride (Aldrich) with ethyl iodide and aluminum chloride.[1] It can also be prepared by the reaction of acetaldehyde hydrazone with iodine and triethylamine (4, 260).[2] This method is based on the procedure of Pross and Sternhell.[3]

$$CH_3CHO \xrightarrow[75\%]{H_2NNH_2} CH_3CH=NNH_2 \xrightarrow[34\%\ overall]{I_2,\ N(C_2H_5)_3} CH_3CHI_2$$

[1] R. L. Letsinger and C. W. Kammeyer, *Am. Soc.*, **73**, 4476 (1951).
[2] E. C. Friedrich, S. N. Falling, and D. E. Lyons, *Syn. Commun.*, 33 (1975).
[3] A. Pross and S. Sternhell, *Australian J. Chem.*, **23**, 989 (1970).

**Ethyl lithioacetate, 3,** 172–173.

*dl-Mevalonolactone* (3). A recent simple synthesis of mevalonolactone involved the reaction of 1-acetoxy-3-oxobutane with ethyl lithioacetate (prepared with LDA). The product was then treated with methanolic potassium hydroxide.[1]

$$AcOCH_2CH_2COCH_3 + LiCH_2COOC_2H_5 \xrightarrow{94\%} AcOCH_2CH_2\overset{\overset{\displaystyle OH}{|}}{\underset{\underset{\displaystyle CH_3}{|}}{C}}CH_2COOC_2H_5 \xrightarrow[93\%]{KOH\ CH_3OH}$$

(1)                           (2)                    (3)

[1]R. A. Ellison and P. K. Bhatnagar, *Synthesis*, 719 (1974).

**Ethyl lithiotrimethylsilylacetate, 5,** 373–374. Note corrected nomenclature.

The definitive paper is available.[1]

[1]H. Taguchi, K. Shimoji, H. Yamamoto, and H. Nozaki, *Bull. Chem. Soc. Japan*, **47**, 2529 (1974).

**Ethyl malonate,** $CH_2\begin{smallmatrix} \diagup COOC_2H_5 \\ \diagdown COOH \end{smallmatrix}$ . Mol. wt. 132.11, b.p. 107–109°/24 mm.

This half ester is prepared by partial saponification of diethyl malonate.[1]

*One-pot malonic ester synthesis.*[2] A separate hydrolysis step can be avoided by use of the dilithium salt of ethyl malonate, prepared by reaction of the ester with 2 eq. of lithium isopropylcyclohexylamide (**4,** 306–309) in THF at $-78°$. HMPT and the alkyl halide are then added and the reaction mixture is allowed to come to room temperature for 2 hr. Decarboxylation is then effected by overnight reflux.

$$RX + LiCH(COOLi)COOC_2H_5 \xrightarrow{THF,\ HMPT} RCH_2COOC_2H_5$$

[1]R. E. Strube, *Org. Syn., Coll. Vol.,* **4,** 417 (1963).
[2]J. E. McMurry and J. H. Musser, *J. Org.,* **40,** 2556 (1975).

**Ethyl α-methylsulfinylacetate,** $CH_3SOCH_2COOC_2H_5$. Mol. wt. 150.20.

*Preparation.* Several methods are available for preparation of this α-sulfinyl ester (1); one route is reaction of dimsyllithium in THF with ethyl chloroformate:

$$CH_3SO\overset{-}{C}H_2\ Li^+ + ClCOOC_2H_5 \xrightarrow{\sim 80\%} CH_3SOCH_2COOC_2H_5$$

(1)

*Reactions.*[1] A few of the useful transformations carried out with (1) are formulated. These depend on the fact that (1) is converted into the relatively stable carbanion, $CH_3SO\overset{-}{C}HCOOC_2H_5$, by sodium hydride in DMSO at 25°. The carbanion is readily alkylated by primary alkyl bromides or iodides in satisfactory yields. Although (1) is very stable, the alkylated derivatives on heating are converted into α, β-unsaturated esters. In this way $RCH_2X$ can be converted

(I)    (1)
$$\xrightarrow[\substack{55-80\%}]{\substack{NaH,\ DMSO\\ RCH_2X}} CH_3\overset{O}{\overset{\|}{S}}\underset{CH_2R}{\overset{}{CH}}COOC_2H_5 \xrightarrow[\substack{70-80\%}]{\substack{140^0\\ -CH_3SOH}} \underset{H}{\overset{R}{\diagdown}}C=C\underset{COOC_2H_5}{\overset{H}{\diagup}}$$

(II)    (1)
$$\xrightarrow{\substack{NaH,\ DMSO\\ RX}} CH_3\overset{O}{\overset{\|}{S}}\underset{R}{\overset{}{CH}}COOC_2H_5 \xrightarrow[\substack{\sim 80\%}]{CHCl_3,\ HCl} CH_3\overset{Cl}{\overset{\|}{S}}\underset{R}{\overset{}{C}}COOC_2H_5 \xrightarrow{H_2O} RCOCOOC_2H_5$$

(III)    (1) + $CH_2=CHCOOCH_3$
$$\xrightarrow[\substack{80\%}]{NaH,\ THF} CH_3S\overset{O}{\overset{\|}{-}}\underset{}{\overset{CH_2CH_2COOCH_3}{\overset{|}{CH}}}COOC_2H_5 \xrightarrow{\Delta} C_2H_5OOCCH=CHCH_2COOCH_3$$

into $RCH=CHCOOC_2H_5$ (*trans*). Equation II illustrates the conversion of RX into $RCOCOOC_2H_5$. The carbanion also undergoes Michael addition, as shown by the reaction with methyl acrylate (equation III).

[1] J. J. A. van Asten and R. Louw, *Tetrahedron Letters*, 671 (1975).

**Ethyl α-phenylsulfinylacetate,** $C_6H_5\overset{O}{\overset{\uparrow}{S}}CH_2COOC_2H_5$ (1). Mol. wt. 212.27.
*Preparation:*

$$C_6H_5\overset{O\ Li^+}{\overset{\uparrow}{S}}\bar{C}H_2 + (C_2H_5O)_2C=O \xrightarrow{73\%} (1)$$

**β-Hydroxy esters.** [1] The carbanion of (1) does not react with aldehydes or ketones; however, when (1) is treated with ethylmagnesium iodide in ether at 0° it forms a Grignard reagent (2), which reacts readily with aldehydes or ketones at room temperature to give adduct (3) in 75–95% yield. The adducts are desulfurized by Raney nickel to give β-hydroxy esters (4).

$$(1) + C_2H_5MgI \xrightarrow[\substack{0^0}]{(C_2H_5)_2O} \underset{(2)}{C_6H_5\overset{O\ MgX}{\overset{\uparrow}{S}}\overset{}{-}CHCOOC_2H_5} \xrightarrow[\substack{75-95\%}]{RCOR'}$$

$$\underset{(3)}{C_6H_5\overset{O}{\overset{\uparrow}{S}}\underset{C_2H_5OOC\ \ OH}{\overset{|\ \ \ \ |}{CH}}\diagdown\overset{R}{\underset{R'}{}}} \xrightarrow{Raney\ Ni} \underset{(4)}{\underset{R'}{\overset{R}{\diagup}}C\overset{|}{\underset{OH}{-}}CH_2COOC_2H_5}$$

**β-Keto esters, methyl ketones.** [2] The Grignard reagent derived from (1) can also be used to effect the transformation $RCHO \longrightarrow RCOCH_2COOC_2H_5$. When heated in refluxing benzene the adduct (3) of (2) and an aldehyde eliminates benzenesulfenic acid to give the enol (a) of a β-keto ester (4).

$$\underset{(2)}{C_6H_5\overset{\overset{O}{\uparrow}\,\overset{MgX}{|}}{S}CHCOOC_2H_5} \quad \xrightarrow[85-95\%]{RCHO} \quad \underset{(3)}{R\underset{\underset{H}{|}}{\overset{\overset{OH}{|}}{C}}-\underset{\underset{H}{|}}{\overset{\overset{COOC_2H_5}{|}}{C}}-\underset{\overset{\downarrow}{O}}{S}C_6H_5} \quad \xrightarrow[-C_6H_5SOH]{C_6H_6,\ \Delta}$$

$$\underset{(a)}{\left[ R\overset{\overset{OH}{|}}{C}=CHCOOC_2H_5 \right]} \quad \xrightarrow[75-95\%]{} \quad \underset{(4)}{RCOCH_2COOC_2H_5}$$

Pyrolysis of β-hydroxy sulfoxides (5) results in methyl ketones (6); higher temperatures are required for this elimination.

$$\underset{(5)}{R\underset{\underset{H}{|}}{\overset{\overset{OH}{|}}{C}}-CH_2\underset{\overset{\downarrow}{O}}{S}C_6H_5} \quad \xrightarrow[75-95\%]{150-160^\circ} \quad \underset{(6)}{R\overset{\overset{O}{\|}}{C}CH_3}$$

[1] N. Kunieda, J. Nokami, and M. Kinoshita, *Tetrahedron Letters*, 3997 (1974).
[2] J. Nokami, N. Kunieda, and M. Kinoshita, *ibid.*, 2841 (1975).

**Ethyl phenyl sulfoxide,** $C_2H_5\overset{\overset{O}{\|}}{S}C_6H_5$. Mol. wt. 154.22.

**β-Keto sulfoxides.** A recent short estrone synthesis[1] involves reaction of (1) with the anion of diethyl sulfoxide (**1**, 311) to give the β-keto sulfoxide

(3) in high yield. Heating in diglyme with the dione (4) gives the trione (5) after elimination to an α, β-unsaturated ketone and Michael addition. The trione has been converted into estrone.[2]

When (6), prepared from (1) and the anion of ethyl phenyl sulfoxide, is heated with (4), the cyclized product (7) is obtained in 35% yield. Probably benzene-sulfenic acid, produced by elimination of (6), serves as an acid catalyst for cyclization of (5) to (7).

[1] Y. Oikawa, T. Kurosawa, and O. Yonemitsu, *Chem. Pharm. Bull. Japan*, **23**, 2466 (1975).
[2] G. H. Douglas, J. M. H. Graves, D. Hartley, G. A. Hughes, B. J. McLoughlin, J. Siddall, and H. Smith, *J. Chem. Soc.*, 5072 (1963).

# F

Ferric chloride, 1, 390–392; 2, 199; 3, 145; 4, 236; 5, 307–308.

*Alkenylation of Grignard reagents* (4, 236). Neumann and Kochi[1] have reported extensive studies of the synthesis of olefins by cross-coupling of Grignard reagents with alkenyl halides:

$$RMgX + \ \ \underset{}{>}C=C\underset{}{<}^{X} \ \ \xrightarrow{\text{Fe(III)}} \ \ \underset{}{>}C=C\underset{}{<}^{R} \ + \ MgX_2$$

In the original study, ferric chloride was used as the catalyst. Recent results indicate that other iron complexes are much more effective, particularly Fe(III) complexes containing β-diketonate ligands such as $Fe(CH_3COCHCOCH_3)_3$, $Fe[(CH_3)_3CCOCHCOC(CH_3)_3]_3$, and $Fe(C_6H_5COCHCOC_6H_5)_3$. These are rapidly reduced by the Grignard reagent to a Fe(I) species, which is the actual catalyst. This iron species loses its activity markedly on standing.

R can be primary, secondary, or tertiary. The reaction is stereospecific. From present results, several mechanistic schemes are possible.

*Alkyl chlorides.* Trialkylboranes react with bromine[2] or iodine[3] to give alkyl bromides and iodides, respectively. Alkyl chlorides have been prepared by the reaction of trialkylboranes and cupric chloride[4]:

$$R_3B + 2 \ CuCl_2 \ \xrightarrow[45-90\%]{} \ RCl + R_2BOH + Cu_2Cl_2 + HCl$$

More recently this conversion has been effected with ferric chloride in aqueous THF; yields are high and all three alkyl groups can be utilized.[5]

$$R_3B + 6 \ FeCl_3 + 3 \ H_2O \longrightarrow 3 \ RCl + B(OH)_3 + 6 \ FeCl_2 + 3 \ HCl$$

Similarly, alkyl thiocyanates can be prepared by reaction of trialkylboranes with ferric thiocyanate $[Fe(NH_4)(SO_4)_2 + KSCN]$.

*t-Alkyl iodides.*[6] *t*-Alkyl chlorides are converted into the corresponding iodides on reaction with sodium iodide in benzene with a trace of ferric chloride or mercuric chloride as catalyst. The reaction is slow (2–120 hr.), but yields are almost quantitative. Benzylic chlorides also react under these conditions, but simple primary and secondary chlorides do not react.

[1]S. M. Neumann and J. K. Kochi, *J. Org.*, **40**, 599 (1975).
[2]C. F. Lane and H. C. Brown, *J. Organometal. Chem.*, **26**, C51 (1971).
[3]H. C. Brown, M. W. Rathke, and M. M. Rogić, *Am. Soc.*, **90**, 5038 (1968).
[4]C. F. Lane, *J. Organometal. Chem.*, **31**, 421 (1971).
[5]A. Arase, Y. Masuda, and A. Suzuki, *Bull. Chem. Soc. Japan*, **47**, 2511 (1974).
[6]J. A. Miller and M. J. Nunn, *Tetrahedron Letters*, 2691 (1975).

**Ferric chloride–Acetic anhydride, $FeCl_3$–$(CH_3CO)_2O$.**

*Cleavage of ethers.* Some time ago Knoevenagel[1] reported that diethyl ether is cleaved to ethyl acetate by ferric chloride in acetic anhydride. Actually, the reagent is a worthwhile alternative for cleavage of ethers, now commonly used as protecting groups.[2] For example, *t*-butyldimethylsilyl ethers are cleaved in this way at 0° in 15 min. Simple dialkyl ethers are generally cleaved at steam bath temperatures. Cleavage of (1) to (2) proceeds without isomerization of the double bond. Optically active ethers are usually cleaved with substantial racemization.

(1)                                                          (2)

[1] E. Knoevenagel, *Ann.*, **402**, 111 (1914).
[2] B. Ganem and V. R. Small, Jr., *J. Org.*, **39**, 3728 (1974).

**Ferric chloride–*n*-Butyllithium.**

*Deoxygenation of epoxides.* *n*-Butyllithium (2–3 eq.) reacts with ferric chloride in THF at −78° to form a black, soluble iron species that converts epoxides into olefins in about 60–90% yield. The reaction is not stereospecific; for example, the oxide of *cis*-stilbene is converted into *cis*-stilbene and *trans*-stilbene in the ratio 89 : 11.[1]

[1] T. Fujisawa, K. Sugimoto, and H. Ohta, *Chem. Letters*, 883 (1974).

**Ferrous perchlorate [Iron(II) perchlorate], $Fe(ClO_4)_2 \cdot 6H_2O$. Mol. wt. 362.84. Supplier: Alfa.**

*Reductive decarboxylation of peroxy acids.*[1] Treatment of peroxycyclohexanecarboxylic acid (1) with this reagent gives cyclohexanol (2) in about 25% yield. A similar reaction with a mixture of (3) and (4) gives (5) and (6) with significant loss of configuration. On the other hand, the reaction of (7) and (8) proceeds with small, but significant, retention of configuration. These reactions proceed through free-radical intermediates.

(1)                                          (2)

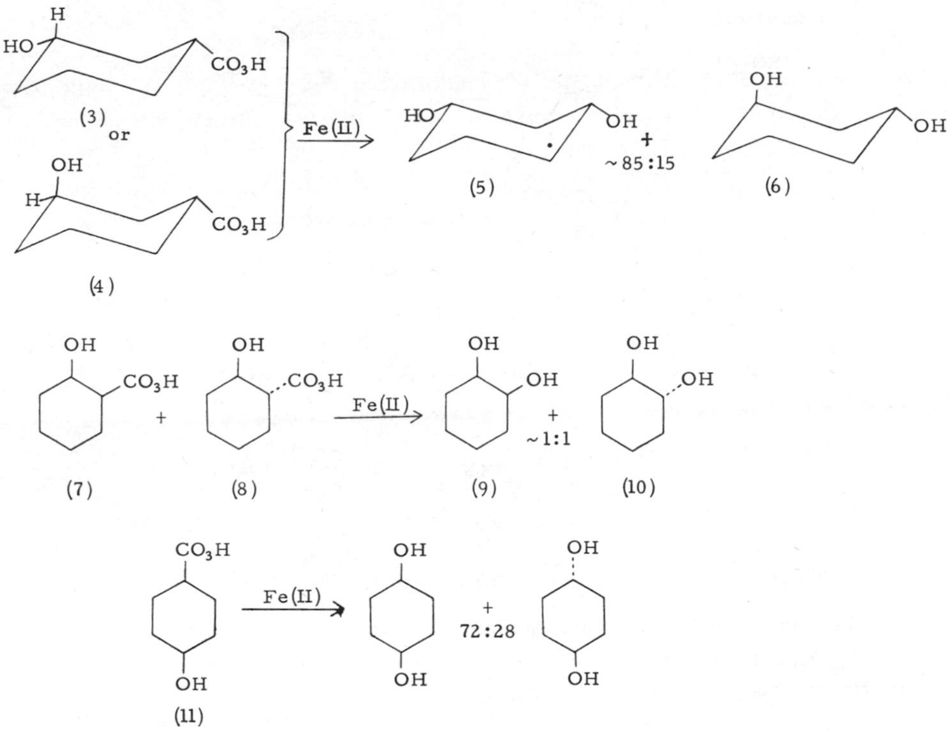

HO — H

CO₃H

(3) or

OH
H

CO₃H

(4)

} Fe(II) →

HO ⟶ OH
(5)

~85:15

+

OH
⟶ OH
(6)

OH
CO₃H

(7)

+

OH
··CO₃H

(8)

Fe(II) →

OH
OH
(9)

~1:1

+

OH
OH
(10)

CO₃H

OH
(11)

Fe(II) →

OH

OH
(9)

+

72:28

OH

OH

[1] J. T. Groves and M. Van Der Puy, *Am. Soc.*, **97**, 7118 (1975).

**Fluorodiiodomethane**, CHFI₂. Mol. wt. 285.84, b.p. 100–101°.

*Preparation.* [1]

*Fluorocarbene.* [2]  Fluorocyclopropanes can be obtained in 15–45% yield by irradiation of fluorodiiodomethane (3500 Å) in the presence of an olefin. The *syn/anti* ratios are approximately one.

[1] J. Hine, R. Butterworth, and P. B. Langford, *Am. Soc.*, **80**, 819 (1958).
[2] J. L. Hahnfeld and D. J. Burton, *Tetrahedron Letters*, 1819 (1975).

**Fluorodimethoxyborane**, FB(OCH₃)₂. Mol. wt. 91.88, b.p. 46–48°.

*Preparation.* The reagent is prepared in 76% yield by reaction of BF₃ etherate (1 eq.) with trimethyl borate (2 eq.).

*Allylic alcohols.* [1]  Alkenes can be converted regio- and stereoselectively into 2-alkene-1-ols by hydrogen—metal exchange to give an alkenylpotassium followed by reaction with this borane. The allylic alcohol is obtained on oxidation of the boronic ester intermediate.

*Examples:*

[1]G. Rauschwalbe and M. Schlosser, *Helv.*, **58**, 1094 (1975).

## Fluoromethylenetriphenylphosphorane, 3, 146.

*Preparation.* Burton and Greenlimb[1] have reported two methods for generation of this ylide (equations I and II).

$$(I) \quad (C_6H_5)_3P + CH_2FI \xrightarrow[80\%]{} [(C_6H_5)_3\overset{+}{P}CH_2F]I^- \xrightarrow[THF, -78^0]{n\text{-}BuLi} (C_6H_5)_3P=CHF$$

$$(II) \quad (C_6H_5)_3P + CHFI_2 \xrightarrow[57.7\%]{} [(C_6H_5)_3\overset{+}{P}CHFI]I^- \xrightarrow[DMF, 0^0]{Zn/Cu} (C_6H_5)_3P=CHF$$

The first method suffers from the disadvantage that fluoroiodomethane is not readily available; furthermore, satisfactory yields of vinyl fluorides are not obtained on reaction with carbonyl compounds unless potassium *t*-butoxide is used to promote decomposition of intermediate betaines.

In the second, preferred method, the ylide is generated *in situ* by dehalogenation of fluoroiodomethyltriphenylphosphonium iodide with zinc–copper couple in DMF at 0°. In general, somewhat higher yields of vinyl fluorides were obtained from reagent generated by this technique. In either case the *cis–trans* ratio is about 50:50.

[1]D. J. Burton and P. E. Greenlimb, *J. Org.*, **40**, 2796 (1975).

## Fluorosulfuric acid, 1, 396–397; 2, 199–200; 5, 310–311.

*Rearrangement of geraniol.*[1] Treatment of geraniol (1), or the *cis*-isomer nerol, with this superacid in $SO_2$–$CS_2$ at −78° followed by careful quenching with potassium carbonate leads to a novel iridoid ether (2) in yields as high as 78%. α-Cyclogeraniol (3) is converted into (2) under the same conditions in somewhat higher yield. One postulated scheme for the rearrangement involves a series of carbonium ions.

[1]D. V. Banthorpe, P. A. Boullier, and W. D. Fordham, *J.C.S. Perkin I*, 1637 (1974).

**1-Fluorovinyl methyl ketone,** $CH_2=CFCOCH_3$. Mol. wt. 76.07, b.p. 71°; polymerizes readily, but can be stabilized with hydroquinone.

*Preparation:*[1,2]

*Annelation.*[2] The reagent reacts with enamines to give fluorinated cyclo-hexenones. Thus the reaction with 2-methyl-1-pyrrolidinopropene (2) gives 6-fluoro-4,4-dimethylcyclohexene-2-one-1 (3) in 65% yield.

[1]F. Nerdel, J. Buddrus, W. Brodowski, P. Hentschel, D. Klamann, and P. Weyerstahl, *Ann.*, **710**, 36 (1967).
[2]H. Molines and C. Wakselman, *J.C.S. Chem. Comm.*, 232 (1975).

**Fluoroxytrifluoromethane,** 2, 200; 3, 146–147; 4, 237–238; 5, 312.

*N,N-Difluoroamines.* Primary amines can be converted into N,N-difluoro-amines in 50–75% yield by conversion into the Schiff base with benzaldehyde or, preferably, the sodium salt of 4-carboxybenzaldehyde, followed by reaction with fluoroxytrifluoromethane in methanol–methylene chloride. In the absence of methanol the alkyl fluoride is formed preferentially.[1]

$$RNH_2 + HCO\!-\!C_6H_4CO\overset{-}{O}Na\overset{+}{-}\underline{p} \longrightarrow RN\!=\!CHC_6H_4CO\overset{-}{O}Na\overset{+}{-}\underline{p} \xrightarrow[CH_3OH]{CF_3OF}$$

$$\underset{|\ \ \ \ |}{\overset{F\ \ \ OCH_3}{R\!-\!\underset{|}{\overset{+}{N}}\!-\!CH\!-\!C_6H_4CO\overset{-}{O}\!-\!\underline{p}}} \xrightarrow{CH_3OH} RNF_2 + (CH_3O)_2CHC_6H_4CO\overset{-}{O}\ \overset{+}{N}a\text{-}\underline{p}$$
$$\overset{}{\underset{F}{}}$$

[1] D. H. R. Barton, R. H. Hesse, T. R. Klose, and M. M. Pechet, *J.C.S. Chem. Comm.*, **97** (1975).

**Formaldehyde, 1**, 397–402; **2**, 200–201; **4**, 238–239; **5**, 312–315.

*α-Hydroxymethyl ketones.* Stork and d'Angelo[1] have discussed methods for preparation of α-hydroxymethyl ketones by the reaction of formaldehyde with regiospecifically generated enolates. For example, reduction of (1) with lithium in liquid ammonia (aniline as proton donor) gives mainly the lithium enolate (a),

which reacts with anhydrous formaldehyde at −78° to give the α-hydroxy-methyl ketone (2) in 60% yield. The yield of (2) is improved considerably by trapping the enol as the trimethylsilyl ether (b), from which the lithium enolate (a) can be regenerated by reaction with methyllithium.

α-Hydroxymethyl ketones are useful for annelation.[2] Thus reaction of (3) with ethyl acetoacetate and base followed by hydrolysis gives (4).[1] Another example of annelation is the reaction of (5) with (6) to give (7), which was con-

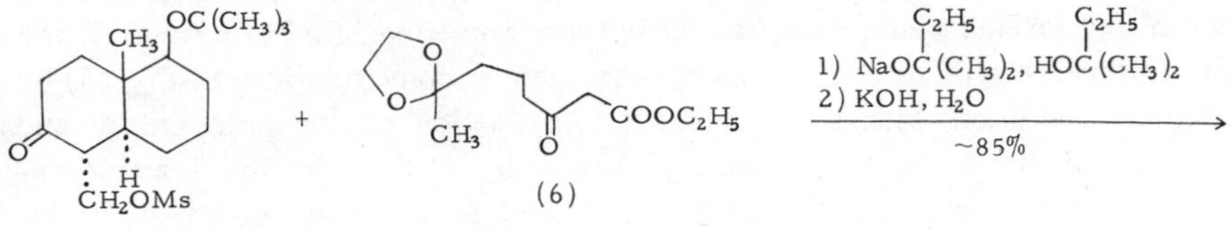

(5)     (6)

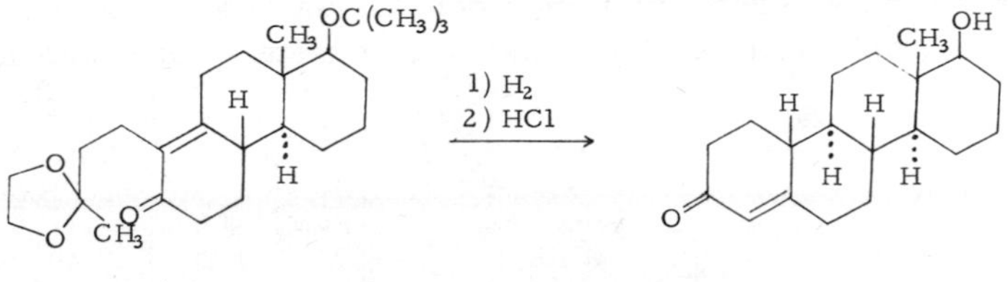

1) NaOC$(CH_3)_2$, HOC$(CH_3)_2$
2) KOH, $H_2O$

~85%

(7)     1) $H_2$   2) HCl     (8)

verted into (±)-D-homo-19-nortestosterone (8).

Stork and Isobe[3,4] have used this hydroxymethylation reaction for two prostaglandin syntheses. In one[3] the starting material is the oxide of cyclopentadiene, which is converted in several steps into (1). Reduction of (1) with methyl diphenylphosphinite[5] gives the enol phosphinate (2), which is cleaved by *t*-

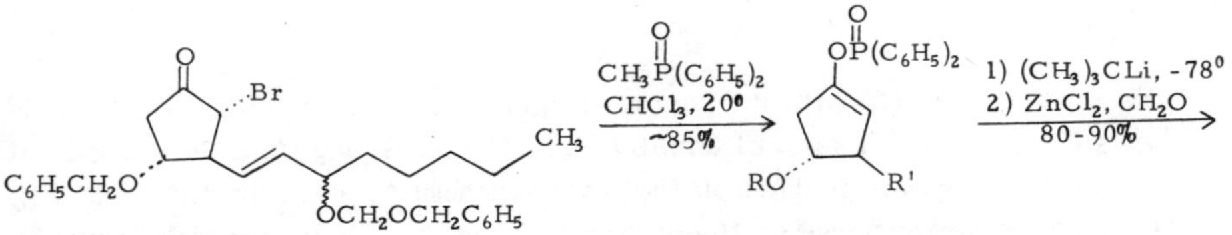

(1)

$CH_3\overset{O}{\overset{\|}{P}}(C_6H_5)_2$
$CHCl_3$, 20°
~85%

(2)

1) $(CH_3)_3CLi$, -78°
2) $ZnCl_2$, $CH_2O$
80-90%

(3)

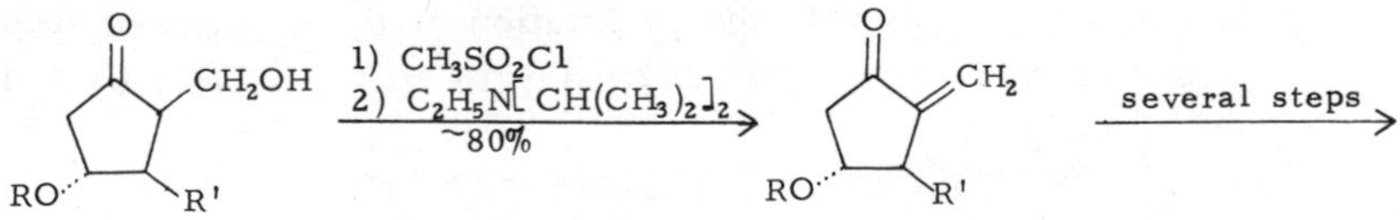

1) $CH_3SO_2Cl$
2) $C_2H_5N[CH(CH_3)_2]_2$
~80%

(4)

several steps

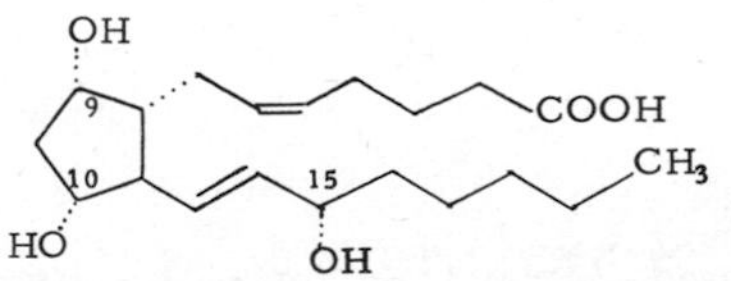

(5)

butyllithium at $-78°$ to the lithium enolate. This enolate reacts with formaldehyde (added zinc chloride, 5, 763) to give the $\alpha$-hydroxymethylcyclopentanone (3) in high overall yield. The product is then converted into (4) by mesylation and treatment with base. This key intermediate is converted into $PGF_{2\alpha}$ (5) by conjugate addition with a divinylcuprate, reduction of the keto group, and cleavage of the hydroxyl-protecting groups by sodium in liquid ammonia-ethanol. $PGF_{2\alpha}$ and the $C_{15}$-epimer obtained in this way are separable by chromatography of the methyl esters.

A second synthesis[4] of $PGF_{2\alpha}$ depends on a combined 1,4-addition of a cuprate with trapping of the intermediate metal enolate with formaldehyde.

Thus reaction of (6) with the cuprate formulated and then with formaldehyde gives (7) and (8) in a ratio of about $1.3:1$. The remaining steps are carried out on the mixture as formulated in the prior synthesis to give a mixture of $PGF_{2\alpha}$ (5) and the mirror image of 15-epi $PGF_{2\alpha}$ in about 17% overall yield from (6). The products are separable by chromatography of the methyl esters.

*α-Methylenation of lactones.* Grieco and Hiroi[6] have applied the $\alpha$-hydroxymethylation procedure to the preparation of the oxygenated system (2). The protective ether group was removed with $HOAc-H_2O-THF$, $5:4:1$; in 71% yield.

Grieco and co-workers[7] have used this method for bis-$\alpha$-methylenation in a total synthesis of desoxyvernolepin (4), an analog of the interesting sesquiterpene vernolepin, which contains a dilactone system. The advantage of this method is that the methylenation reaction is the last step in the synthesis.

(3)                                   (4)

[1] G. Stork and J. d'Angelo, *Am. Soc.*, **96**, 7114 (1974).
[2] Z. G. Hajos and D. R. Parrish, *J. Org.*, **38**, 3244 (1973).
[3] G. Stork and M. Isobe, *Am. Soc.*, **97**, 4745 (1975).
[4] G. Stork and M. Isobe, *Am. Soc.*, **97**, 6260 (1975).
[5] For formation of lithium enolates from vinylphosphinites *see* I. J. Borowitz, E. W. R. Casper, R. K. Crouch, and K. C. Yee, *J. Org.*, **37**, 3873 (1972).
[6] P. A. Grieco and K. Hiroi, *Tetrahedron Letters*, 3467 (1974).
[7] P. A. Greico, J. A. Noguez, and Y. Masaki, *ibid.*, 4213 (1975).

**Formaldehyde diphenyl thioacetal [Bis(phenylthio)methane]**, $H_2C(SC_6H_5)_2$. Mol. wt. 232.37, m.p. 39.5–40°.

*Preparation:*[1]

$$2\ C_6H_5SNa\ +\ CH_2I_2\ \xrightarrow[80\%]{60°}\ H_2C(SC_6H_5)_2$$

*Ketone synthesis.*[2] Metalation of (1) with *n*-butyllithium followed by reaction with an alkyl bromide gives a high yield of the monoalkyl derivative (2), which can be alkylated a second time by treatment with excess sodium amide in

$$(C_6H_5S)_2CH_2\ \xrightarrow[\text{2) RBr}]{\text{1) }\underline{n}\text{-BuLi}}\ (C_6H_5S)_2C\diagup^{R}_{\diagdown H}\ \xrightarrow[70\text{-}80\%]{\substack{\text{1) NaNH}_2,\ \text{THF, HMPT}\\ \text{2) R}^1\text{Br, }20°}}\ (C_6H_5S)_2C\diagup^{R}_{\diagdown R^1}$$

(1)                          (2)                                   (3)

THF and HMPT and an alkyl bromide at 20°. This sequence can be used for synthesis of both symmetrical and unsymmetrical ketones, obtained by hydrolysis of (3).

[1] E. J. Corey and D. Seebach, *J. Org.*, **31**, 4097 (1966).
[2] G. Schill and C. Merkel, *Synthesis*, 387 (1975).

**Formylmethylenetriphenylphosphorane**, $(C_6H_5)_3P=CH-CHO$. Mol. wt. 304.33, m.p. 186–187° dec.

The Wittig reagent is prepared[1] from anhydrous chloroacetaldehyde:

$$ClCH_2CHO\ +\ (C_6H_5)_3P\longrightarrow(C_6H_5)_3\overset{+}{P}CH_2CHOC\overset{-}{l}\ \xrightarrow[C_2H_5OH]{N(C_2H_5)_3}\ (C_6H_5)_3P=CHCHO$$

This stable phosphorane reacts normally with aldehydes, but does not react with ketones.

*Pyridine acrylaldehydes.* These compounds can be prepared[2] conveniently by reaction of pyridinecarboxaldehydes with this Wittig reagent:

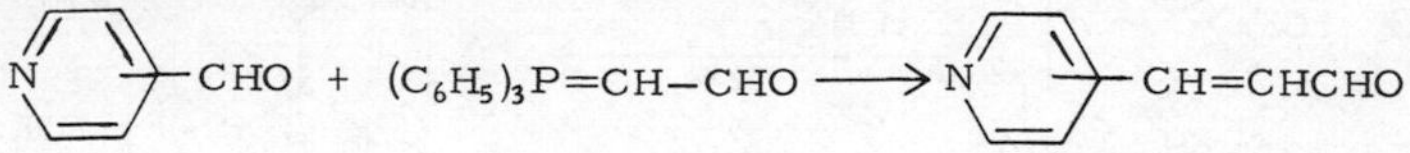

$$\text{N}\!\!\!\diagup\!\!\!\diagdown\!\!\!-\text{CHO} + (\text{C}_6\text{H}_5)_3\text{P}=\text{CH}-\text{CHO} \longrightarrow \text{N}\!\!\!\diagup\!\!\!\diagdown\!\!\!-\text{CH}=\text{CHCHO}$$

(o-, 21%;
m-, 56%;
p-, 62%)

The aldehydes had been prepared by condensation of pyridinecarboxaldehydes with acetaldehyde, but yields are much lower.

[1] S. Trippett and D. M. Walker, *J. Chem. Soc.*, 1266 (1961).
[2] I. Hagedorn and W. Hohler, *Angew. Chem. internat Ed.*, **14**, 486 (1975).

# G

**Graphite bisulfate,** $C_{24}{}^+HSO_4{}^- \cdot 2H_2SO_4$. Hygroscopic.

This lamellar reagent is prepared by electrolysis of 98% $H_2SO_4$ with a graphite anode.[1]

*Esterification.*[2] Graphite bisulfate is a very efficient catalyst for esterification of formic acid and acetic acid. Reactions proceed in high yield (90–98%) with a variety of alcohols at $25°$ in 1–17 hr. Esterification of other acids requires longer periods, but high yields can be realized. However, tertiary alcohols undergo elimination as the main pathway. Benzoic and cinnamic acid are esterified only at higher temperatures. The graphite compound serves both as an acid catalyst and as a dehydrating reagent. It is much more efficient than a combination of graphite and sulfuric acid or a sulfonic acid resin.

[1]G. R. Hennig, *Prog. Inorg. Chem.*, **1**, 125 (1959).
[2]J. Bertin, H. B. Kagan, J.-L. Luche, and R. Setton, *Am. Soc.*, **96**, 8113 (1974).

**Grignard reagents,** **1**, 415–424; **2**, 205; **5**, 321.

*Activation by transition metal compounds.* Reactions of Grignard reagents catalyzed by transition metal compounds have been reviewed, particularly reactions that have been reported during the years 1965–1975.[1]

*α-Allenic alcohols.* These substances (2) can be prepared in over 90% yield by the reaction of alkynyloxiranes (1) with Grignard reagents in THF at $20°$ with cuprous iodide as catalyst.[2]

$$R-C\equiv C-\overset{R^1}{\underset{\triangle}{\diagdown}}O \quad + \; R^2MgBr \quad \xrightarrow{Cu_2I_2} \quad \overset{R}{\underset{R^2}{\diagup}}C=C=C\overset{R^1}{\underset{CH_2OH}{\diagdown}}$$

$$(1) \qquad\qquad\qquad\qquad\qquad (2)$$

*Reaction with allylic ethers.*[3] Alkenes are obtained by reaction of allylic ethers with Grignard reagents catalyzed by cuprous chloride or bromide. Direct replacement of the ether group takes place as the major reaction when the ether is primary. Secondary and tertiary allylic ethers react mainly with rearrangement.

*Examples:*

$$(CH_3)_2C{=}CHCH_2OC_2H_5 \; + \; \underline{n}\text{-}C_7H_{15}MgCl \xrightarrow[83\%]{\substack{Cu_2Cl_2 \\ THF}} (CH_3)_2C{=}CHCH_2{-}C_7H_{15}{-}\underline{n}$$

$$+ \; \underline{n}\text{-}C_7H_{15}{-}\overset{\overset{\textstyle CH_3}{\textstyle |}}{\underset{\underset{\textstyle CH_3}{\textstyle |}}{C}}{-}CH{=}CH_2$$

$$97{:}3$$

$$H_2C=CH\overset{\overset{\displaystyle CH_3}{|}}{\underset{\underset{\displaystyle CH_3}{|}}{C}}OCH_3 \ + \ \underline{n}\text{-}C_7H_{15}MgCl \xrightarrow[70\%]{} \underline{n}\text{-}C_7H_{15}CH_2CH=C(CH_3)_2$$

*1-Alkenes.* Normant *et al.*[4] have developed a method for a regiospecific synthesis of 1-alkenes by addition of organocopper reagents to 1-alkynes. The copper reagents are prepared by the reaction of a Grignard reagent with cuprous bromide in ether at $-40°$. The 1-alkyne is then added to this solution to form a

$$RMgBr \ + \ CuBr \rightarrow RCu \cdot MgBr_2 \xrightarrow{R'C\equiv CH} \overset{R}{\underset{R'}{>}}C=C\overset{H}{\underset{Cu}{<}} \xrightarrow{H_3O^+} \overset{R}{\underset{R'}{>}}C=CH_2$$

vinylcopper derivative, which is hydrolyzed by acid to the 1-alkene. The addition reaction is highly dependent on the presence of magnesium salts that are soluble in the reaction medium. Use of magnesium bromide or magnesium iodide results in high yields. Yields are unsatisfactory when the salt is magnesium chloride or lithium iodide. The beneficial effect is considered to be a result of stabilization of the intermediate complex (a) involved in the reaction.

$$\begin{array}{c} \overset{Br}{\diagup} \overset{\overset{\displaystyle R'}{|}}{\underset{\underset{\displaystyle C}{\parallel \parallel}}{C}} \\ Mg \leftarrow \phantom{|} \leftarrow CuR \\ \diagdown \underset{Br}{} \underset{H}{} \end{array}$$

(a)

*Reaction with $\alpha,\beta$-unsaturated acetals.*[5] Under catalysis with cuprous bromide, Grignard reagents condense with $\alpha,\beta$-unsaturated acetals to give enol ethers (Z-isomers predominating). The products are hydrolyzed to aldehydes by acid. The reaction can be carried out with $\alpha$- or $\beta$-substituted $\alpha,\beta$-unsaturated acetals.

$$\underline{n}\text{-BuMgBr} \ + \ CH_2=CHCH(OC_2H_5)_2 \xrightarrow[75.5\%]{\overset{Cu_2Br_2}{THF,\ 0^0}} \underline{n}\text{-}BuCH_2CH=CHOC_2H_5 \xrightarrow{HCl}$$

$$\xrightarrow[85-90\%]{} \underline{n}\text{-}BuCH_2CH_2CHO$$

[1] H. Felkin and G. Swierczewski, *Tetrahedron*, **31**, 2735 (1975).
[2] P. Vermeer, J. Meijer, C. de Graaf, and H. Schreurs, *Rec. trav.*, **93**, 46 (1974).
[3] A. Commercon, M. Bourgain, M. Delaumeny, J. F. Normant, and J. Villieras, *Tetrahedron Letters*, 3837 (1975).
[4] J.-F. Normant, G. Cahiez, M. Bourgain, C. Chuit, and J. Villieras, *Bull. Soc.*, 1656 (1974).
[5] J. F. Normant, A. Commercon, M. Bourgain, and J. Villieras, *Tetrahedron Letters*, 3833 (1975).

# H

**1,3,4,6-Heptatetraene  (Divinylallene),**

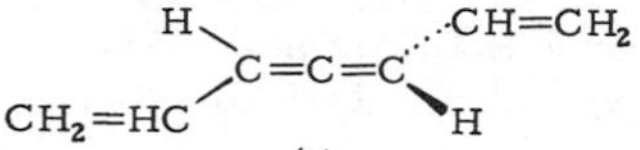

(1)

. Mol. wt.
92.14, b.p. 115°, polymerizes in air.
*Preparation:*

$$CH_2=CH-C\equiv CMgBr \ + \ CH_2=CHCH_2Br \ \xrightarrow[74\%]{\underset{THF}{Cu_2Cl_2}} CH_2=CHC\equiv CCH_2CH=CH_2$$

$$\xrightarrow[\underset{61\%}{C_2H_5OH}]{KOH, Ether} (1)$$

*Diels–Alder reactions.*[1]  This "butadiene dimer" reacts with tetracyanoethylene and maleic anhydride to give only 1:1 adducts, (2) and (3), even under forcing conditions. However, it does give the diadduct (4) with the highly reactive 4-phenyl-1,2,4-triazoline-3,5-dione.

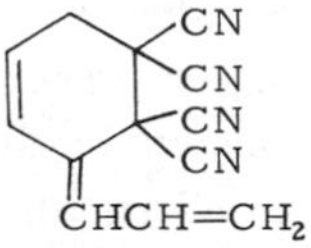

(2) 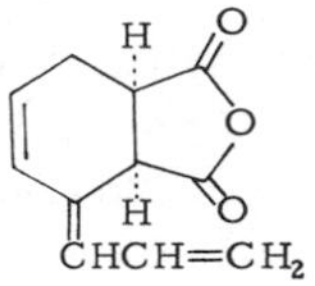　　　　　　(3) 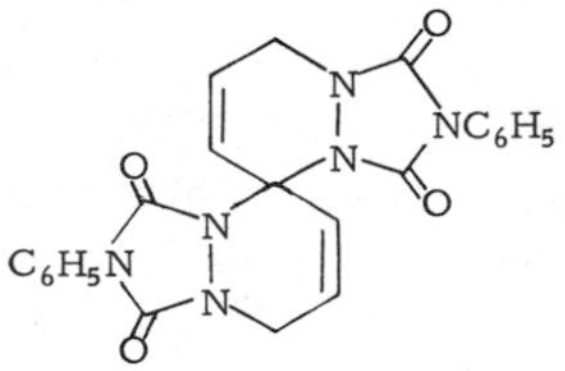　　　　　　(4)

[1] U. Mödlhammer and H. Hopf, *Angew. Chem. internat. Ed.*, **14**, 501 (1975).

## Hexadecyltributylphosphonium bromide, 5, 322–323.

*Sulfides.* Symmetrical sulfides can be prepared in high yield from the reaction of alkyl halides and sodium sulfide in water with this phase-transfer catalyst. Unsymmetrical sulfides are prepared from alkyl halides and sodium mercaptides. Secondary alkyl halides react more slowly than primary halides; bromides are more reactive than chlorides.[1]

$$2 \ RX + Na_2S \ \xrightarrow[\underset{70^0}{cat.}]{} R_2S + 2 \ NaX$$

$$RX + R'SNa \ \xrightarrow[\underset{40-70^0}{cat.}]{} RSR' + NaX$$

Details for the preparation of phenyl neopentyl sulfide are available (equation I).[2]

$$(\text{I}) \quad (CH_3)_3CCH_2Br + C_6H_5SNa \xrightarrow[\substack{H_2O \\ 78\text{-}85\%}]{\overset{+}{C_{16}H_{33}}PBu_3Br^{-}} C_6H_5SCH_2C(CH_3)_3 + NaBr$$

[1] D. Landini and F. Rolla, *Synthesis*, 565 (1974).
[2] *Idem, Org. Syn.*, submitted (1976).

## Hexafluoroantimonic acid, 5, 309–310.

*t-Nitroalkanes.*[1] Primary nitroalkanes can be obtained satisfactorily by reaction of primary iodides or bromides with silver nitrite (Victor Meyer reaction). The reaction is not useful in the case of secondary or tertiary halides. Thus 1-bromoadamantane reacts with silver nitrite in acetonitrile to give 1-nitroadamantane in only 2.5% yield. However, the stable 1-adamantyl hexafluoroantimonate (1, 5, 310) is converted into 1-nitroadamantane (2) in 66% yield. This reaction, however, does not appear to be general; it failed in an

attempted preparation of α-nitrotriphenylmethane and *t*-butyl nitrate.

*Reduction of enones and bicyclic phenols.* French chemists[2] have reported transfer of hydrogen from cyclohexane to enones and phenols in this super acid. The products are the more stable isomers. This method has obvious advantages over Li–NH₃ for reduction of bicyclic phenols.

*Examples:*

(60%)                              (30%)

(97%)

[1]G. A. Olah and H. C. Lin, *Synthesis,* 537 (1975).
[2]J.-M. Coustard, M.-H. Douteau, J.-C. Jacquesy, and R. Jacquesy, *Tetrahedron Letters,* 2029 (1975).

**Hexamethyldisilazane,** 1, 427; 2, 207–208; 5, 323.

*Aryloxysilanes.* Phenols are converted into aryloxytrimethylsilanes (1) when heated slowly with hexamethyldisilazane to about 150° until evolution of ammonia has ceased.[1]

$$\text{ArOH} + (H_3C)_3\text{SiNHSi}(CH_3)_3 \xrightarrow[85-98\%]{20 \to 150^0} \text{ArOSi}(CH_3)_3 \qquad (1)$$

[1]H. Niederpriim, P. Voss, and V. Beyl, *Ann.,* 20 (1973).

**Hexamethylditin,** $(CH_3)_6Sn_2$. Mol. wt. 327.29, m.p. 23°, b.p. 182°. Supplier: Alfa.

*Deoxygenation of nitroarenes.*[1] Aromatic nitro compounds when heated

$$2 \;\text{ArNO}_2 \xrightarrow[150^0,\,16\,hr.]{(CH_3)_6Sn_2} \text{ArN=NAr} + \text{ArN=NAr}$$

$$\overset{\downarrow}{O}$$

(60-75%)          (5-20%)

with hexamethylditin at 150° for 16 hr. are deoxygenated to azoxy compounds and then, more slowly, to azo compounds.

[1]F. P. Tsui and G. Zon, *J. Organometal. Chem.,* **96**, 365 (1975).

**Hexamethylphosphoric triamide (HMPT),** 1, 430–431; 2, 208–210; 3, 149–153; 4, 244–247; 5, 323–325.

**Warning:** A rare form of cancer (squamous cell carcinoma) has been found in rats that had inhaled HMPT in a concentration as low as 400 ppb in a period of 8 months. The incidence was shown to be dose related. Acute toxicity to animals is low to moderate by ingestion, inhalation, and skin absorption; chronic toxicity is more severe.[1]

*Esterification* (**4**, 247). Shaw and Kunerth[2] have reported further studies on the esterification of sodium salts of carboxylic acids with alkyl halides in HMPT at room temperature. The method is applicable to the preparation of ethyl esters of hindered acids; for example, ethyl mesitoate can be obtained in 99% yield. In the esterification of acids that undergo ready decarboxylation, anhydrous potassium carbonate rather than NaOH is used as base (equation I). Diesters can be obtained by reaction of sodium salts of acids with dibromomethane (equation II). Phenols are converted by this method into ethers in

$$(I) \quad CH_2(COOH)_2 \ + \ C_2H_5I \ \xrightarrow[91\%]{K_2CO_3, \, HMPT} \ CH_2(COOC_2H_5)_2$$

$$(II) \quad C_6H_5COOH \ + \ CH_2Br_2 \ \xrightarrow[86\%]{NaOH, \, HMPT} \ (C_6H_5\overset{\overset{O}{\parallel}}{C}\!-\!O)_2CH_2$$

$$(III) \quad C_6H_5OH \ + \ (CH_3)_2CHI \ \xrightarrow[100\%]{NaOH, \, HMPT} \ C_6H_5OCH(CH_3)_2$$

high yield (equation III). Attempts to prepare ethers of alcohols in this way have failed.

*Reductive cleavage of sulfonates.* Sulfonates are cleaved by lithium or potassium in HMPT in the presence of a proton donor, $NH_4Cl$.[3]

$$ROSO_2R' \ \xrightarrow[\substack{NH_4Cl, -30 \text{ to } 0^0}]{Li, \, HMPT} \ \underset{44-76\%}{ROH} \ + \ \underset{10-22\%}{R'H}$$

*Catalysis of an amalgam reaction.* The addition of HMPT as cosolvent with ether markedly accelerates the conversion of *cis*-3,4-dichlorocyclobutene (1) into *syn*-tricyclo [4.2.0.0$^{2.5}$]octa-3,7-diene (2) by 1% sodium amalgam.[4] In the absence of HMPT the reaction is slower, but the yield of (2) is 46–51%.[5]

$$(1) \qquad\qquad\qquad\qquad\qquad (2)$$

*Wittig reactions.*[6] Treatment of absolute[7] HMPT under $N_2$ with a slight excess of potassium gives a mixture of the two bases (1):

$$[(CH_3)_2N]_3P{=}O \ + \ 2 \ K \longrightarrow \underbrace{\overset{+}{K}\overset{-}{O}P[N(CH_3)_2]_2 \ + \ \overset{+}{K}\overset{-}{N}(CH_3)_2}_{(1)}$$

These bases (1) in HMPT convert phosphonium salts into the ylides, which undergo ready Wittig reactions. The reaction with aldehydes occurs stereoselectively to give Z-alkenes. Alkenes are also obtained by oxygenation, and again the Z-isomer is obtained preferentially.

*Examples:*

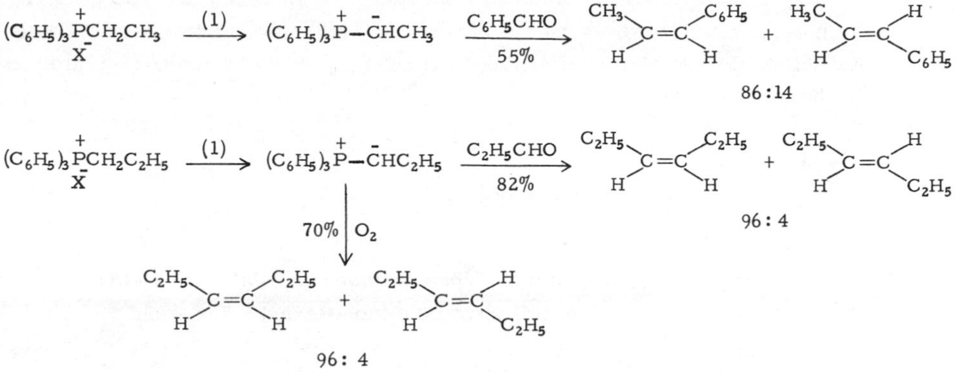

86 : 14

96 : 4

96 : 4

The bases (1) have also been used to effect intramolecular C-alkylation, as formulated for the synthesis of indane.

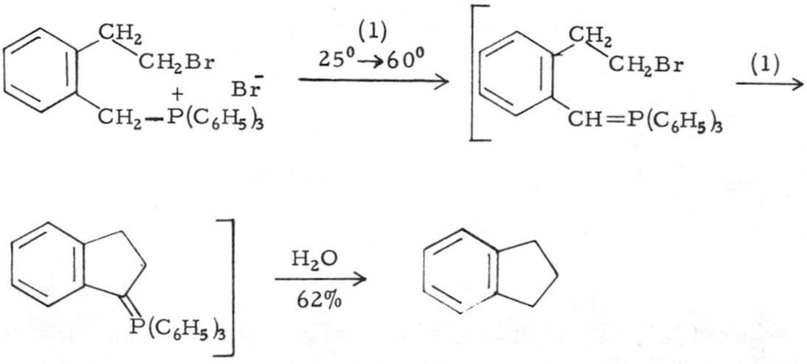

Some insect sex pheromones are derivatives of long-chain alkenes in which a double bond has the Z(*cis*)-configuration. Bestmann *et al.*[8,9] have reported that the Wittig reaction with aldehydes can be modified to give almost entirely the Z-isomer if potassium is used as base and HMPT as solvent.

*Examples:*

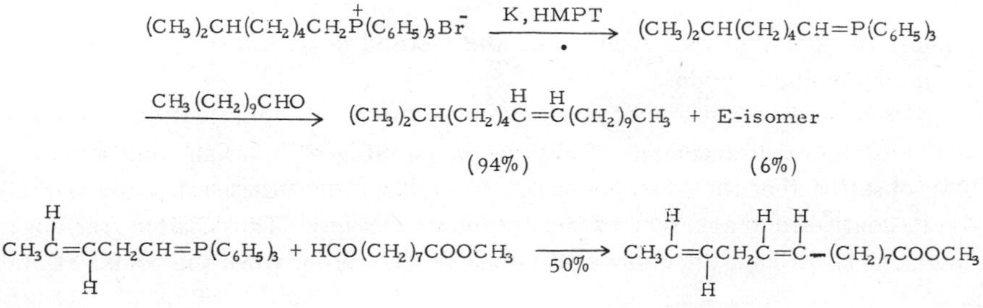

$$(CH_3)_2CH(CH_2)_4CH_2\overset{+}{P}(C_6H_5)_3\overset{-}{Br} \xrightarrow{K, HMPT} (CH_3)_2CH(CH_2)_4CH=P(C_6H_5)_3$$

(94%)                (6%)

Z-Olefins are also obtained in 95% geometrical purity if THF–HMPT (2:1) or THF–DMSO (2:1) is used as solvent. In these systems potassium can be replaced by *n*-butyllithium as base.[10]

*Tetrahydropyranes.* Tetrahydropyranes can be obtained in yields of >70% by heating the monotosylate of a 1,5-diol in HMPT (80°, 6 hr.). Alternatively, the ditosylate is heated in HMPT containing 1 eq. of water to effect hydrolysis to the monotosylate.[11]

*2-Dimethylaminoquinolines.*[12] When refluxed in DMF and HMPT, acetanilides are converted into 2-dimethylaminoquinolines in 40–76% yield.

When refluxed in HMPT with derivatives of acetic acid, acetanilide is converted into 2-dimethylamino-4-methylquinoline:

These one-pot syntheses seem to be the method of choice for preparation of 2-dimethylaminoquinolines.

*Stereochemical control of enolization of esters.* In continuation of studies on the Claisen rearrangement of allyl esters (4, 307–308), Ireland and Willard[13] have observed that the stereochemistry of enolization of these esters (1) and (2) can be controlled to a marked extent by the solvent used. Thus Claisen rearrangement of (1) through the enolate obtained in THF alone gives the acids (3) and

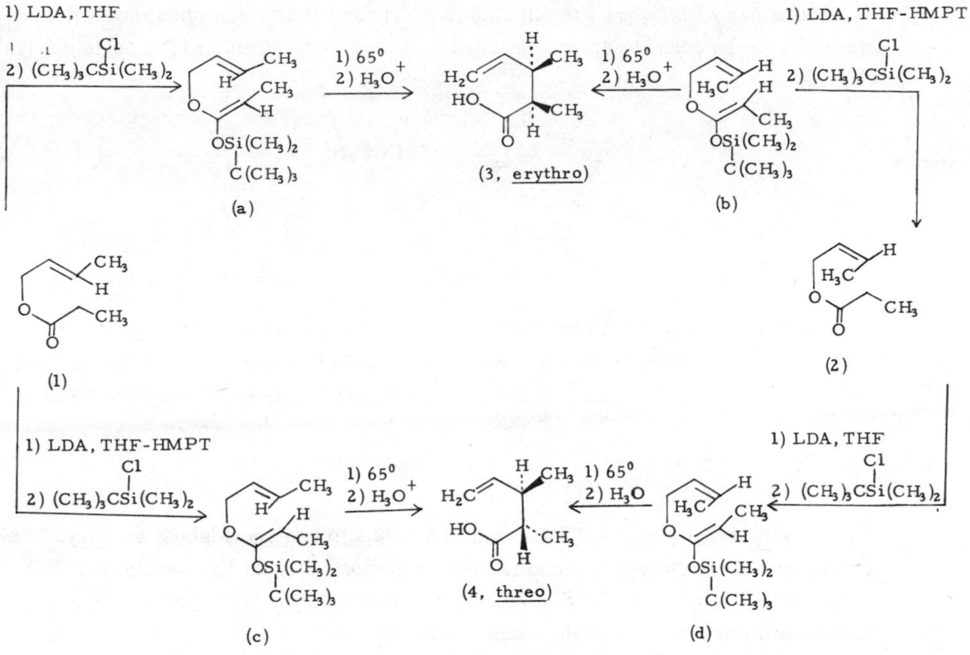

1) LDA, THF

2) (CH$_3$)$_3$CSi(CH$_3$)$_2$

(a)

1) 65°
2) H$_3$O$^+$

(3, erythro)

1) 65°
2) H$_3$O$^+$

(b)

1) LDA, THF-HMPT

2) (CH$_3$)$_3$CSi(CH$_3$)$_2$

(1)

(2)

1) LDA, THF-HMPT

2) (CH$_3$)$_3$CSi(CH$_3$)$_2$

(c)

1) 65°
2) H$_3$O$^+$

(4, threo)

1) 65°
2) H$_3$O

(d)

1) LDA, THF

2) (CH$_3$)$_3$CSi(CH$_3$)$_2$

(4) in the ratio $87:13$. The ratio is reversed to $19:81$ when THF–23% HMPT is used as solvent for generation of the enolates. Similar differences were noted for the rearrangement of the isomeric ester (2). In each case THF alone favors formation of E-enolates (a) and (d), whereas addition of HMPT favors formation of Z-enolates (b) and (c).

Other simple esters and one ketone were shown to be subject to similar control (equations I and II).

(I)　$CH_3CH_2CH_2COOCH_3 \longrightarrow$

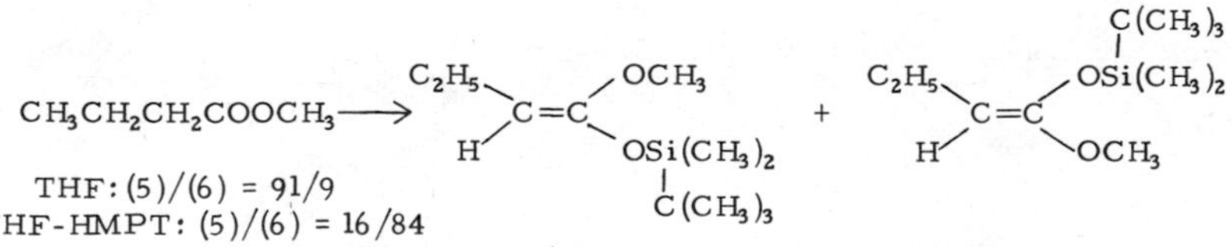

THF: (5)/(6) = 91/9
THF-HMPT: (5)/(6) = 16/84

(5)

(6)

+

(II)　$CH_3CH_2COCH_2CH_3 \longrightarrow$

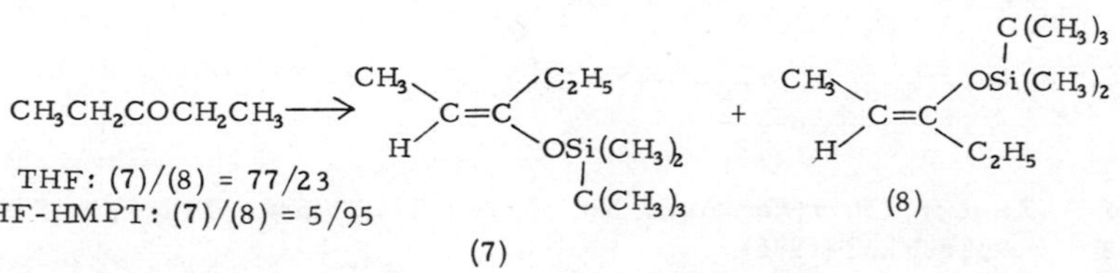

THF: (7)/(8) = 77/23
THF-HMPT: (7)/(8) = 5/95

(7)

(8)

+

*Activation of anions.* French chemists[14] found that some phosphoramidates are also able to activate anions. Actually, the phosphoramide (1) is particularly effective.

(1)

They then turned to polyamines that are known to activate organolithium reagents. And indeed a number of these substances were found to activate acetate anions in the reaction formulated in equation I. Twelve polyamines

$$(I) \quad C_6H_5CH_2Cl + KOAc \xrightarrow[\text{CH}_3\text{CN},\,25^0]{\text{Cat.}} C_6H_5CH_2OAc$$

tested were active, but TMEDA and DABCO showed the highest activity. The polyamines were shown to enhance the nucleophilicity of various anions: $RO^-$, $ArO^-$, $NO_2^-$, $NC^-$, and $NCS^-$. They are comparable to crown ethers. The quaternary ammonium salts of the diamines are inactive.

*Alkyl chlorides.*[15] Alcohols react with hydrogen chloride in HMPT at 50–85° to form alkyl chlorides in 65–90% yield. The reaction of primary alcohols is rapid (30 min.); longer reaction times are required for secondary and tertiary alcohols (about 1 hr.). Neopentyl alcohol reacts without formation of rearranged products. Racemic product was obtained from 2-octanol.

*Decarboalkylation* (5, 445–446). Japanese chemists[16] have used this reaction for the synthesis of certain α-alkylated esters, ketones, and nitriles as illustrated in equations I–III.

$$(I) \quad CH_2(COOC_2H_5)_2 + C_6H_5CH_2Br \xrightarrow[\text{HMPT}]{\text{LiCl}} \underset{(51\%)}{C_6H_5CH_2CH_2COOC_2H_5} + \underset{(26\%)}{(C_6H_5CH_2)_2CHCOOC_2H_5}$$

$$(II) \quad CH_3COCH_2COOC_2H_5 + C_6H_5CH_2Br \xrightarrow[24\%]{} C_6H_5CH_2CH_2COCH_3$$

$$(III) \quad NCCH_2COOC_2H_5 + C_6H_5CH_2Br \longrightarrow \underset{(52\%)}{(C_6H_5CH_2)_2CHCN} + \underset{(15\%)}{C_6H_5CH_2CH_2CN}$$

[1]J. A. Zapp, Jr., *Chem. Eng. News*, No. 39, 17 (1975); *Science*, **190**, 422 (1975); *Chem. Eng. News*, Feb 2, 3 (1976).

[2] J. E. Shaw and D. C. Kunerth, *J. Org.*, **39**, 1968 (1974).

[3] T. Cuvigny and M. Larchevêque, *J. Organometal. Chem.*, **64**, 315 (1974).

[4] M. F. Carmody and L. A. Paquette, *Org. Syn.*, submitted (1976).

[5] M. Avram, I. G. Dinulescu, E. Marica, G. Mateescu, E. Sliam and C. D. Nenitzescu, *Ber.*, **97**, 382 (1964).

[6] H. J. Bestmann and W. Stransky, *Synthesis*, 798 (1974).

[7] HMPT is refluxed for several hours over calcium hydride under vacuum of a water pump and is then distilled through a silver mirror column (b.p. 106°/11 torr).

[8] H. J. Bestmann and O. Vostrowsky, *Tetrahedron Letters*, 207 (1974).

[9] H. J. Bestmann, O. Vostrowsky, and A. Plenchette, *ibid.*, 779 (1974).

[10] P. E. Sonnet, *Org. Prep. Proc. Int.*, 269 (1974).

[11] P. Picard, D. Leclercq, and J. Moulines, *Tetrahedron Letters*, 2731 (1975).

[12] E. B. Pedersen and S.-O. Lawesson, *Acta Chem. Scand.*, **28B**, 1045 (1974).

[13] R. E. Ireland and A. K. Willard, *Tetrahedron Letters*, 3975 (1975).

[14] H. Normant, T. Cuvigny, and P. Savignac, *Synthesis*, 805 (1975).

[15] R. Fuchs and L. L. Cole, *Canad. J. Chem.*, **53**, 3620 (1975).

[16] M. Asaoka, K. Miyake, and H. Takei, *Chem. Letters*, 1149 (1975).

**Hexamethylphosphorous triamide, 1**, 425; **2**, 207; **3**, 148–149; **5**, 242.

*Selective functionalization of primary long-chain diols.* Reaction of diols of type (1) with this phosphine in THF–CCl$_4$ followed by addition of KPF$_6$ gives the mono salt (2) in high yield. These salts react with various nucleophiles (NaN$_3$, KI, C$_6$H$_5$SH) to give monosubstituted derivatives (3) of (1).[1]

$$HO(CH_2)_{\underline{n}}OH \; + \; [(CH_3)_2N]_3P \xrightarrow[\substack{2) \; KPF_6 \\ 70\text{-}94^0}]{1) \; CCl_4, \; THF} HO(CH_2)_{\underline{n}}O\overset{+}{P}[N(CH_3)_2]_3 \; PF_6^-$$

(1)                                                  (2)

$$\xrightarrow[50\text{-}90\%]{} HO(CH_2)_{\underline{n}}X \; + \; OP[N(CH_3)_2]_3$$

(3, X = N$_3$, I, CN, OCH$_3$)

*Formylation of secondary hydroxyl groups.* Treatment of alcohol (1) with hexamethylphosphorous triamide, chloroform, and dimethylformamide gives the corresponding formyl esters (2) in high yield. The reaction is considered to

involve formation of $[(CH_3)_2\overset{+}{N}=C\overset{H}{\underset{Cl}{<}}]Cl^-$, which reacts with the hydroxyl group to give [ROCH=N⁺(CH₃)₂]Cl⁻; hydrolysis then gives the formyl ester, ROCHO.[2]

*Aryl alkyl ethers.* The reaction of an alcohol with hexamethylphosphorous triamide in carbon tetrachloride at low temperatures results in formation of an alkoxytris(dimethylamino)phosphonium chloride (1), which can be converted into the stable hexafluorophosphate (2) by addition of ammonium hexafluorophosphate. The salt (2) reacts with potassium phenolates or aryl thiophenolates

$$[(CH_3)_2N]_3P + Cl-CCl_3 \longrightarrow \overset{+}{[(CH_3)_2N]_3}PCl \ \overset{-}{CCl_3} \xrightarrow{ROH}$$

$$\overset{+}{[(CH_3)_2N]_3}P-O-R \ \overset{-}{Cl} \xrightarrow{NH_4PF_6} \overset{+}{[(CH_3)_2N]_3}POR \ PF_6^{-}$$
$$(1) \hspace{4cm} (2)$$

$$(2) + ArOK(SK) \xrightarrow{DMF, \Delta} ROAr(SAr)$$
$$(3)$$

to give aryl alkyl ethers or thioethers (3). Yields are generally in the range of 65–90%. The reaction proceeds with inversion in the case of a chiral alcohol.[3]

*Desulfurization* (4, 242).[4]  Another example:

$$(C_6H_5CH_2)_2S_2 + [(CH_3)_2N]_3P \xrightarrow[\ \ \ \ \ \ ]{\overset{C_6H_6}{100-105^0}} (C_6H_5CH_2)_2S + [(CH_3)_2N]_3PS$$
$$(\sim 98\%) \hspace{2cm} (97\%)$$

[1] R. Boigegrain, B. Castro, and C. Selve, *Tetrahedron Letters*, 2529 (1975).
[2] S. Czernecki and C. Georgoulis, *Compt. rend.*, **280(C)**, 305 (1975).
[3] I. M. Downie, H. Heaney, and G. Kemp, *Angew. Chem. Internat. Ed.*, **14**, 370 (1975).
[4] D. N. Harpp and R. A. Smith, *Org. Syn.*, submitted (1975).

### Hexamethylphosphorous triamide–Dialkyl azodicarboxylates.

*Mixed carbonates.*[1] Mixed carbonates can be prepared by the reaction of alcohols with dialkyl azodicarboxylates and hexamethylphosphorous triamide:

$$ROH + R'OOCN=NCOOR^1 + [(CH_3)_2N]_3P \xrightarrow[-N_2, -HCOOR^1]{THF} RO-CO-OR^1$$

Yields are in the range of 22–88%.

Note that the related system triphenylphosphine–diethyl azodicarboxylate has been used to convert alcohols into amines (4, 553–554).

[1] G. Grynkiewicz, J. Jurczak, and A. Zamojski, *Tetrahedron*, **31**, 1411 (1975).

### Hydrazine, 1, 434–445; 2, 211; 3, 153; 4, 248; 5, 327–329.

*Cleavage of phthaloyl protective group.* The phthaloyl group is not useful for protection of the amino group of penicillins and cephalosporins because hydrazinolysis (Ing–Manske procedure, 1, 442) disrupts the azetidinone ring. Eli Lilly chemists[1] have reported a new method of dephthaloylation that overcomes this difficulty. The phthalimido compound (1) is hydrolyzed by aqueous sodium sulfide to the corresponding phthalamic acid (2), which on dehydration

(1)                                    (2)

(3)                                    (4, 55-95%)

is converted into the phthalisoimide (3). Dehydration can be effected with DCC, ethyl chloroformate–triethylamine, or trifluoroacetic anhydride–triethylamine. Phthalisoimides (3) undergo hydrazinolysis rapidly (THF, $-20°$, 20–30 min.) and without reaction of the azetidinone carbonyl group. An amine salt of phthalylhydrazide is formed, which on treatment with dilute hydrochloric acid is cleaved to phthalylhydrazide and the hydrochloride of the amine (4). The complex can also be treated with an acyl chloride to give an acyl derivative of (4). Yields are improved when methylhydrazine is used in place of hydrazine.

*Reduction of aromatic nitro compounds.* Aromatic nitro compounds can be reduced to amines by hydrazine in the presence of ferric chloride and an active carbon. The reaction is carried out in refluxing methanol or ethanol (5–26 hr.). Yields are usually greater than 90%.[2]

*Protection of carboxyl groups.* The carboxyl group of N-acyl amino esters can be protected by conversion to the hydrazide by reaction with hydrazine hydrate. Deprotection can be accomplished in quantitative yield by perchloric acid with no evidence of racemization. The method has not been applied as yet to higher peptides.[3]

[1] S. Kukolja and S. R. Lammert, *Am. Soc.*, 97, 5582, 5583 (1975).
[2] T. Hirashima and O. Manabe, *Chem. Letters*, 259 (1975).
[3] J. Schnyder and M. Rottenberg, *Helv.*, 58, 521 (1975).

**Hydriodic acid,** 1, 449–450; 2, 213–214.

*Reduction of alkenylsilanes.* Hydriodic acid (constant boiling) reacts with alkenylsilanes (1) with replacement of $(CH_3)_3 Si$ by H:

(1)                                    (2)

The same replacement can be achieved with iodine and water (or $D_2O$).[1]

[1] K. Utimoto, M. Kitai, and H. Nozaki, *Tetrahedron Letters*, 2825 (1975).

**Hydrobromic acid, 1,** 450–452; **2,** 214–215; **3,** 154; **4,** 249–250; **5,** 332–333.

*Cyclization.* A key step in a recent synthesis[1] of pentazocine (3), a non-narcotic analgesic,[2] involved treatment of the carbinol (1) with concd. hydrobromic acid in HOAc for 38 hr. The reaction involves dehydration, ring closure, and O-demethylation. It is noteworthy that only one (2) of the two possible isomers is formed.

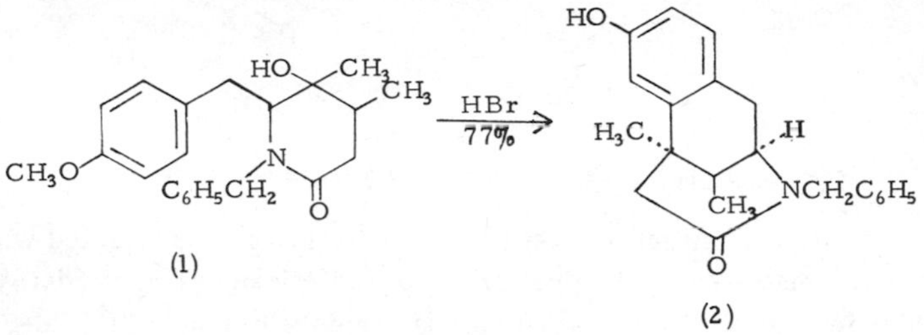

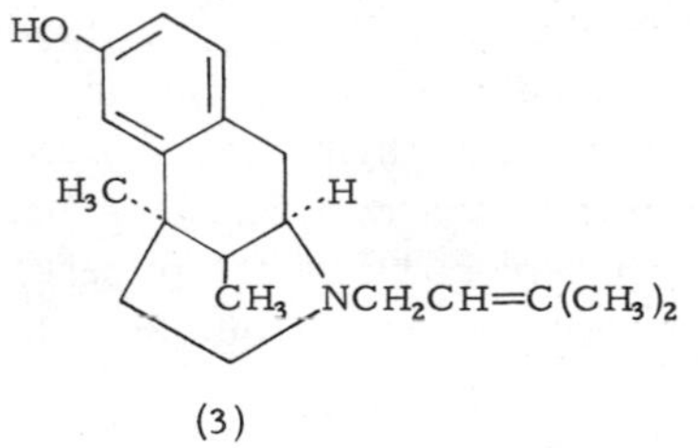

(1)

(2)

(3)

*Rearrangement of 17β-acetoxy-5,6-epoxyandrostane-7-ols.* When treated with hydrobromic acid in glacial acetic acid, the four isomeric 17β-acetoxy-5,6-epoxyandrostane-7-ols (1) rearrange mainly to 17β-acetoxy-4-methylestratriene-1,3,5(10). A skeletal rearrangement is involved rather than migration of the

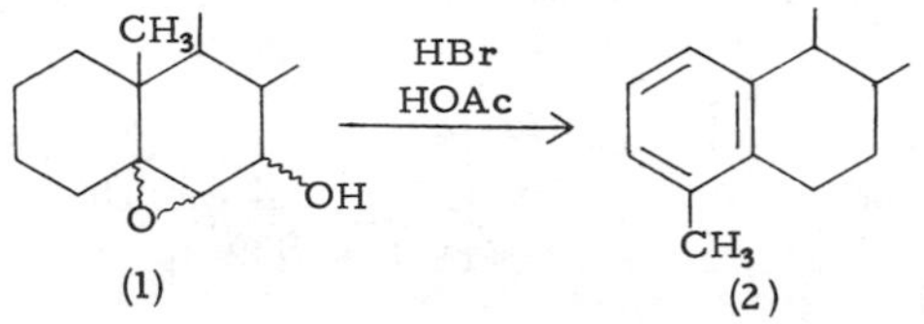

(1)                    (2)

methyl group.[3]

*Indenes.* 2,3-Disubstituted indenes (2) can be obtained by treatment of the diols (1) with 48% hydrobromic acid (6 hr. reflux). Use of concd. hydrochloric acid gives phthalans (3) in high yield.[4]

$$R_3\overset{-}{B}C\equiv CR'Li^+ \xrightarrow[-\text{LiCl}]{\text{HCl, }-78^0} \underset{R^2B}{\overset{R}{\diagdown}}C=CHR' \xrightarrow[H_2O, \Delta]{\text{HCl}} R-B-\underset{\underset{HO}{|}}{\overset{\overset{R}{|}}{C}}-CH_2R'$$

$$\xrightarrow{\text{Oxid.}} HO-\underset{\underset{R}{|}}{\overset{\overset{R}{|}}{C}}-CH_2R'$$

*Phenols.* 1,4-Cyclohexanediones are converted into phenols when refluxed in hydrobromic acid or hydrochloric acid (four examples, yields 70–85%).[5]

[1] T. Kametani, S.-P. Huang, M. Ihara, and K. Fukumoto, *Chem. Pharm. Bull. Japan*, **23**, 2010 (1975).

[2] S. Archer, N. F. Albertson, L. S. Harris, A. K. Pierson, and J. G. Bird, *J. Med. Chem.*, **7**, 123 (1964).

[3] D. Baldwin and J. R. Hanson, *J.C.S. Perkin I*, 1941 (1975).

[4] W. E. Parham and Y. A. Sayed, *Synthesis*, 116 (1976).

[5] C. G. Rao, S. Rengaraju, and M. V. Bhatt, *J.C.S. Chem. Comm.*, 584 (1974).

**Hydrochloric acid, 4,** 450; **5,** 333–334.

*Synthesis of t-alcohols.*[1] Lithium trialkylalkynylborates, when treated in THF with excess hydrochloric acid at room temperature or at reflux, undergo a double migration of alkyl groups from boron to carbon. Complexes containing

highly branched alkyl groups may require a reflux period of 4–8 hr.

The protonation thus can be controlled to give either the alkenylborane or the *t*-alkylborane. This double migration appears to be fairly general, and, when followed by oxidation, provides a route to tertiary alcohols. Yields are in the range of 70–80% when R' is hydrogen; yields are lower when both R and R' are secondary or tertiary alkyl groups.

*Reaction with a β-keto ester.* It was recently reported that the β-keto ester (1) undergoes simple decarboethoxylation when refluxed in 8% HCl.[2] The

(1)                                (a)                                (2)

product is actually a carboxylic acid, (2), formed by aldol condensation of the intermediate dione obtained by decarboethoxylation of (1) followed by cleavage of the intermediate aldol (a).[3]

*2-Alkyl-1-oxo-2-cyclopentenes.* 2-Alkylidene-1-oxocyclopentanes (1), prepared by the reaction of aldehydes with enamines of cyclopentanone, are isomerized to the endocyclic isomers (2) by concentrated hydrochloric acid in butanol (90–100°, 2 hr.).[4]

(1)                                (2)

[1] M. M. Midland and H. C. Brown, *J. Org.*, **40**, 2845 (1975).
[2] V. Dabral, H. Ila, and N. Anand, *Tetrahedron Letters*, 4681 (1975).
[3] V. Dave and E. W. Warnhoff, *ibid.*, 4695 (1976).
[4] M. P. L. Caton, E. C. J. Coffee, T. Parker, and G. L. Watkins, *Syn. Commun.*, 303 (1974).

**Hydrofluoric acid, HF.** Supplier: Alfa.

*Nitration; alkanesulfonyl fluorides.* Swiss chemists[1] have reported two examples in which reactions are carried out successfully in 80–98% aqueous hydrofluoric acid. One is the nitration of anthraquinone with nitric acid or dinitrogen tetroxide in 80–98% aqueous hydrofluoric acid at 30–40°. The other is oxidation of aliphatic thiols to alkanesulfonyl fluorides with nitrogen dioxide in aqueous 80–98% HF.

$$\underline{n}\text{-}C_8H_{17}SH + 3\ NO_2 + HF \xrightarrow[85\%]{} \underline{n}\text{-}C_8H_{17}SO_2F + 3\ NO + H_2O$$

[1] C. Comninellis, P. Javet, and E. Plattner, *Tetrahedron Letters*, 1429 (1975).

**Hydrogen bromide, 1**, 450–453; **4**, 251.

*2-Bromo-1-alkenes.* Hydrogen bromide adds to terminal acetylenes to form 1-bromo-1-alkenes. However, it adds to 1-trimethylsilyl-1-alkynes to form 2-bromo-1-alkenes. Although the reaction is a free radical reaction, a peroxide initiator is not required and may be deleterious. Trimethylsilyl bromide is the

$$RC\equiv CSi(CH_3)_3 \xrightarrow{HBr} R\overset{\overset{\displaystyle Br}{|}}{C}=CHSi(CH_3)_3 \xrightarrow{HBr} R\overset{\overset{\displaystyle Br}{|}}{\underset{\underset{\displaystyle Br}{|}}{C}}-CH_2Si(CH_3)_3 \longrightarrow RCBr=CH_2 + (CH_3)_3SiBr$$

other product formed and can be recycled. The reaction can be carried out neat or in pentane or hexane. No reaction occurs with hydrogen chloride.[1]

[1] R. K. Boeckman, Jr., and D. M. Blum, *J. Org.*, **39**, 3306 (1974).

## Hydrogen chloride, 2, 215; 4, 252; 5, 335–336.

*Thioacyl chlorides.* Aliphatic thioacyl chlorides have been prepared for the first time by addition of hydrogen chloride at $-80°$ to thioketenes in $CFCl_3$.

$$\underset{R^2}{\overset{R^1}{>}}C=C=S \; + \; HCl \; \xrightarrow{-80°} \; \underset{R^2}{\overset{R^1}{>}}CH\overset{\overset{\displaystyle S}{||}}{C}Cl$$

The products are more stable than thioketenes, but generally they decompose at temperatures above $-40°$. They are active thioacylating compounds.[1]

A simpler route to 3,3-dimethylthiobutyryl chloride is shown in equation I.[2]

$$(I) \qquad (CH_3)_3CC\equiv C-S^-K^+ + HCl \xrightarrow[\sim 50\%]{\underset{-80°\to-10°}{n\text{-}C_5H_{12}}} (CH_3)_3CCH_2\overset{\overset{\displaystyle S}{||}}{C}Cl$$

[1] G. Seybold, *Angew. Chem. internat. Ed.*, **14**, 703 (1975).
[2] G. Seybold and C. Heibl, *ibid.*, **14**, 248 (1975).

## Hydrogen fluoride, 1, 455–456; 2, 215–216; 4, 252; 5, 336–337.

*Cyclization.*[1] Treatment of (1) with anhydrous hydrogen fluoride gives the β-diketone (2), formed by a Claisen condensation from (a).

(1)                 (a)                 (2)

[1] V. Askam and T. U. Qazi, *J.C.S. Chem. Comm.*, 798 (1975).

## Hydrogen fluoride—Pyridine (Pyridinium polyhydrogen fluoride).

*Dediazotization.* Aryldiazonium ions react at $80°$ with this sytem to give fluoroarenes. If sodium chloride, bromide, or iodide is added haloarenes are formed in good yield with less than 10% of fluoroarenes. However, the reaction gives unexpected and novel isomer distributions of haloarenes. That is, the halide ion attacks ring positions in addition to the one ($C_1$) from which $N_2$ is lost. The results are rationalized on the basis of the ambient nature of aryldiazonium ions. In addition, the intermediate formation of arynes was also established by experiments carried out in pyridinium polydeuterium fluoride.[1]

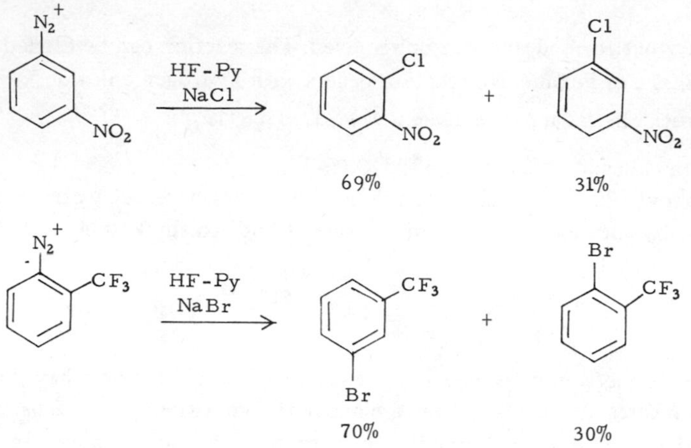

69%                    31%

70%                    30%

[1] G. A. Olah and J. Welch, *Am. Soc.*, **97**, 208 (1975).

**Hydrogen peroxide, 1**, 457–471; **2**, 216–217; **3**, 154–155; **4**, 253–255; **5**, 337–339.

*Hydroxylation of aflatoxin $B_1$.* Büchi *et al.*[1] effected hydroxylation of the carcinogen aflatoxin $B_1$ (1) to the metabolite (2) and the epimeric alcohol by

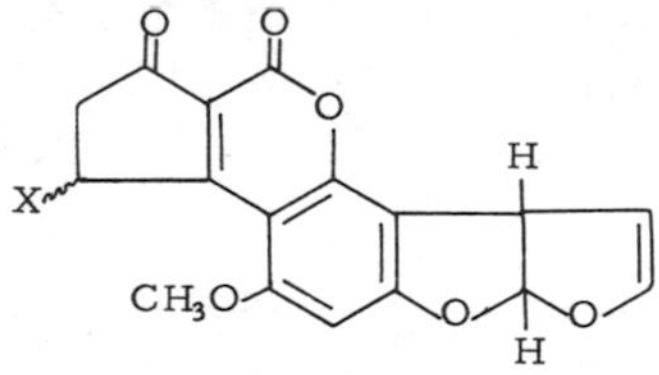

(1, X = H)

(2, X = OH)

oxidation with excess 95% $H_2O_2$ in the presence of 1 eq. of thallium(I) ethoxide at 0° (26% yield). A thallium enolate may be involved. Hydroxylation of (1) was also effected with silver(I) oxide in methanol containing sodium hydroxide (30% yield).

[1] G. Büchi, K.-C. Luk, and P. M. Müller, *J. Org.*, **40**, 3458 (1975).

**Hydrogen peroxide–Acetic acid, 1**, 459–462; **2**, 216; **3**, 155.

*Baeyer–Villiger oxidation.* Peracids are generally used for this reaction. However, Indian chemists[1] have found that the less expensive hydrogen peroxide in acetic acid is satisfactory, particularly for oxidation of rigid polycyclic ketones such as adamantanone and tricyclo[5.2.1.0$^{2,6}$]decane-3-one.

$$\text{(adamantanone)} \xrightarrow[\substack{90-95\%}]{\substack{H_2O_2,\ HOAc \\ 50^\circ,\ 4\ hr.}} \text{(lactone)}$$

$$\text{(bicyclic ketone)} \xrightarrow[\substack{85-90\%}]{\substack{H_2O_2,\ HOAc \\ 50^\circ,\ 2\ hr.}} \text{(lactone)}$$

[1] G. Mehta and P. N. Pandey, *Synthesis*, 404 (1975).

## Hydrogen peroxide—Dibromantin (1, 208).

*1,2-Dioxetanes.*[1] These unstable compounds have been isolated from the reactions of alkenes and singlet oxygen (5, 487–488). Kopecky *et al.*[2] have reported a nonphotochemical route to 1,2-dioxetanes from β-halo hydroperoxides, prepared by reaction in ether of an alkene, 98% $H_2O_2$, and 1,3-dibromo- or 1,3-diiodo-5,5-dimethylhydantoin (1, 208). Treatment of β-halo hydroperoxides derived from trisubstituted olefins with sodium hydroxide in methanol at low temperatures gives 1,2-dioxetanes, as formulated for trimethyl-1,2-dioxetane (1):

$$\underset{CH_3}{\overset{CH_3}{\diagdown}}C=CHCH_3 \xrightarrow[\text{Dibromantin}]{H_2O_2} \underset{\underset{CH_3}{\overset{|}{}}\ \underset{Br}{\overset{|}{}}}{CH_3-\overset{\overset{OOH}{|}}{C}-CHCH_3} \xrightarrow[\substack{\sim 30\%}]{\substack{NaOH \\ CH_3OH}}$$

$$\underset{\overset{|}{CH_3}}{CH_3-\overset{\overset{O-O}{|\ \ \ |}}{C}-CHCH_3}$$

(1)

Treatment of β-halo hydroperoxides of tetrasubstituted olefins with base results exclusively in formation of allylic hydroperoxides. In this case treatment of the β-halo hydroperoxide with silver acetate[3] in $CH_2Cl_2$ proved effective. For example, tetramethyl-1,2-dioxetane (2) was obtained in this way in an optimum yield of ~35%:

$$\underset{\underset{CH_3}{\overset{|}{}}\ \underset{Br}{\overset{|}{}}}{CH_3-\overset{\overset{HOO}{|}}{C}-\overset{\overset{CH_3}{|}}{C}-CH_3} \xrightarrow{AgOAc} \underset{\underset{CH_3}{\overset{|}{}}\ \underset{CH_3}{\overset{|}{}}}{CH_3-\overset{\overset{O-O}{|\ \ |}}{C}-\overset{}{C}-CH_3} + CH_3-\overset{\overset{OOH}{|}}{\underset{\underset{CH_3}{\overset{|}{}}}{C}}-C\overset{\diagup CH_2}{\diagdown CH_3} + (CH_3)_2CCCH_3$$

$$(2,\ \sim 35\%)\qquad\qquad \sim 50\%\qquad\qquad \sim 15\%$$

Several other dioxetanes were prepared in this way. All are yellow and decompose with luminescence, at temperatures between 10 and 70°, to carbonyl cleavage products. They are not converted to allylic hydroperoxides either thermally or by acid or base.

[1] *Caution:* Dioxetanes and hydroperoxides are explosive.
[2] K. R. Kopecky, J. E. Filby, C. Mumford, P. A. Lockwood, and J.-Y. Ding, *Canad. J. Chem.*, **53**, 1103 (1975).
[3] Highest yields were obtained with Baker material.

**Hydrogen selenide,** $H_2Se$. Mol. wt. 80.98, b.p. −41.5°. *Caution:* Poisonous. *Preparation.*[1]

*6-Selenoxo nucleosides and nucleotides.* Treatment of purine nucleosides (1) or nucleotides for several days at 65° with $H_2Se$ in pyridine–water (sealed tube) results in displacement of the amino group to give 6-selenoxo derivatives in 20–75% yield. For example, adenosine (1) is converted into (2) in 56% yield. Adenosine is inert to $H_2S$ under the same conditions.[2]

[1] F. Feher, *Handb. Prep. Inorg. Chem.*, **1**, 418 (1963).
[2] C.-Y. Shiue and S.-H. Chu, *J.C.S. Chem. Comm.*, 319 (1975).

**1-Hydroxybenzotriazole,** 3, 156; 5, 342.

*Depsipeptides.* Klausner and Chorev[1] have prepared depsipeptides[2] in 75–93% yields by alcoholysis of N-protected amino acid active esters (*o*- and *p*-nitrophenyl and 2,4,5-trichlorophenyl) with this catalyst.

*Peptide synthesis.* Coupling reactions in peptide synthesis using active esters are greatly accelerated in the solution phase[3] or the solid phase[4] by addition of 1-hydroxybenzotriazole.

[1] Y. S. Klausner and M. Chorev, *J.C.S. Chem. Comm.*, 973 (1975).
[2] J. ApSimon, Ed., *The Total Synthesis of Natural Products*, Vol. 1, Wiley-Interscience, New York, 1973, pp. 404–426.
[3] W. König and R. Geiger, *Ber.*, **106**, 3626 (1973).
[4] S. A. Khan and K. M. Sivanandaiah, *Tetrahedron Letters*, 199 (1976).

**5-(2′-Hydroxyethyl)-4-methyl-3-benzylthiazolium chloride (1).** Mol. wt. 268.78.

$$\text{HOCH}_2\text{CH}_2 \quad \overset{S}{\diagup} \quad$$

(1)

This thiazolium salt is prepared[1] by quaternization of 5-(2′-hydroxyethyl)-4-methylthiazole[2] with benzyl chloride in acetonitrile.

*Addition of aldehydes to activated double bonds.*[1] Aldehydes add to α,β-unsaturated ketones, esters, and nitriles in the presence of triethylamine and catalytic amounts of this salt. Other thiazolium salts are effective.

*Examples:*

$$\text{CH}_3\text{CHO} + \text{CH}_2{=}\text{CHCOCH}_3 \xrightarrow[50\%]{\overset{\text{cat.}}{\text{N(C}_2\text{H}_5)_3}} \text{CH}_3\text{COCH}_2\text{CH}_2\text{COCH}_3$$

$$\text{CH}_3\text{CHO} + \text{CH}_2{=}\text{CHC}{\equiv}\text{N} \xrightarrow[35\%]{} \text{CH}_3\text{COCH}_2\text{CH}_2\text{C}{\equiv}\text{N}$$

$$\text{C}_6\text{H}_5\text{CHO} + \text{C}_6\text{H}_5\text{CH}{=}\text{CHCOC}_6\text{H}_5 \xrightarrow[75\%]{} \text{C}_6\text{H}_5\text{COCHCH}_2\text{COC}_6\text{H}_5 \atop \qquad\qquad\qquad\quad \text{C}_6\text{H}_5$$

The catalyst is believed to function by reaction with the aldehyde to form an intermediate carbanion (2), which then reacts with the unsaturated compound. Aldehydes in the absence of an unsaturated compound are converted into acyloins in good yield.[3]

$$\text{RCHO} + (1) \longrightarrow$$

(2)

[1] H. Stetter and H. Kuhlmann, *Angew. Chem. internat. Ed.*, **13**, 539 (1974); *idem, Tetrahedron Letters*, 4505 (1974).
[2] Merck AG, Darmstadt, Germany.
[3] R. Breslow, *Am. Soc.*, **80**, 3719 (1958).

**α-Hydroxyhippuric acid,** HOCHCOOH . Mol. wt. 195.18, m.p. 200.5–201.5°.
                                                  $\quad\;$ NHCOC$_6$H$_5$

*Preparation.*[1] The reagent is prepared in 71.8% yield by the reaction of benzamide with glyoxylic acid monohydrate in refluxing acetone (5 hr.).

*α-Amino-γ-keto acids.*[2] The reagent (1) reacts with a 1,3-diketone such as acetylacetone (2) in concd. sulfuric acid to give (3). The reaction of (1) under

(2)                    (3)

(4)
(5)
(6)

similar conditions with a β-keto ester such as methyl acetoacetate (4) gives (5), which can be cyclized to the butenolide (6).

*Amidoalkylation of mercaptans*[1] :

$$RSH \; + \; \underset{\underset{NHCOC_6H_5}{|}}{HOCHCOOH} \; \xrightarrow[\substack{72-78\%}]{\substack{HOAc \\ (H_2SO_4)}} \; \underset{\underset{NHCOC_6H_5}{|}}{RSCHCOOH}$$

[1]U. Zoller and D. Ben-Ishai, *Tetrahedron,* 31, 863 (1975).
[2]D. Ben-Ishai, Z. Berler, and J. Altman, *J.C.S. Chem. Comm.,* 905 (1975).

**Hydroxylamine-O-sulfonic acid,** 1, 481–484; 2, 217–219; 3, 156–157; 4, 256; 5, 343–344.

*Oxidative conversion of aldehydes into nitriles.* Aliphatic aldehydes are converted directly into nitriles when treated with excess hydroxylamine-O-sulfonic acid at 25° (60–90% yield). Aromatic aldehydes under the same conditions are converted into oximino derivatives, $ArCH=NOSO_3H$, which are converted into nitriles either by decomposition at 65° or by cleavage with sodium hydroxide (yields 80–95%).[1]

[1]C. Fizet and J. Streith, *Tetrahedron Letters,* 3187 (1974).

**N-Hydroxy-5-norbornene-*endo*-2,3-dicarboximide,**     (1). Mol. wt.

179.18, m.p. 165–166°, soluble in water and usual organic solvents. Supplier: Fluka.

This reagent is prepared by the reaction of 5-norbornene-*endo*-2,3-dicarboxylic anhydride with hydroxylamine (91% yield).[1]

*Peptide synthesis.*[2] Japanese investigators report that this substance is superior to the related N-hydroxysuccinimide (1, 487) for use in combination with

DCC for peptide synthesis. Yields of peptides are high with only negligible racemization. The luteinizing hormone-releasing hormone, a decapeptide amide, was synthesized by this new active ester method.

[1]L. Bauer and S. V. Miarka, *J. Org.*, **24**, 1293 (1959).
[2]M. Fujino, S. Kobayashi, M. Obayashi, T. Fukuda, S. Shinagawa, and O. Nishimura, *Chem. Pharm. Bull. Japan.* **22**, 1857 (1974).

**2-Hydroxy-5-oxo-5,6-dihydro-2H-pyrane,** (1). Mol. wt. 114.10, b.p. 54–58°.

*Preparation*[1] :

*Diels–Alder reactions.*[2] This compound is a reactive dienophile. For example, it reacts with cyclopentadiene at room temperature to give the adduct (2) in 75% yield. Cleavage of the ring leads to the highly functionalized Diels–Alder adduct (3).

(2)                    (3)

[1]O. Achmatowicz, Jr., P. Bukowski, B. Szechner, Z. Zweirzchowska, and A. Zamojski, *Tetrahedron*, **27**, 1973 (1971).
[2]G. Jones, *Tetrahedron Letters*, 2231 (1974).

**3-Hydroxy-2-pyrone,** (1). Mol. wt. 96.09, m.p. 92°.

*Preparation.* This hydroxypyrone is prepared by dehydration and decarboxylation of mucic acid.[1,2]

*Diels–Alder reactions.*[2] This substance has been used as an equivalent of the unknown vinylketene, $CH_2$=CH—CH=C=O, in Diels–Alder reactions. In general, the reaction is accompanied by decarboxylation to give dihydrophenols or cyclohexenones. The pyrone does not undergo cycloaddition with unactivated olefins.

(1) +   [maleic anhydride structure]   $\xrightarrow[83\%]{C_6H_6,\ 80^0}$   [bicyclic product structure with OH]

(1) +   [fumaric acid structure with H, COOH]   $\xrightarrow[73\%]{C_6H_6,\ 115^0}$   [cyclohexadiene product with OH, COOCH$_3$, COOCH$_3$]

(1) +   [methyl acrylate structure CH$_2$, COOCH$_3$]   $\xrightarrow[56\%]{C_7H_8,\ 170^0}$   [cyclohexadiene product with OH, COOCH$_3$]

(1) +   [benzoquinone structure]   $\xrightarrow{\begin{array}{c}1)\ \Delta\\2)\ \text{Acetylation}\end{array}}$   [naphthalene triacetate product with OAc, AcO, OAc]

[1] R. H. Wiley and C. H. Jarboe, *Am. Soc.*, 78, 2398 (1956).
[2] E. J. Corey and A. P. Kozikowski, *Tetrahedron Letters*, 2389 (1975).

# I

**Indium, In.** At. wt. 114.82.

*Reformatsky reaction.* An active form of indium can be prepared by reduction of $InCl_3$ (Alfa) with potassium metal in refluxing xylene (compare activation of zinc, **5**, 753).

The activated indium can be used to effect the Reformatsky reaction of carbonyl compounds with ethyl bromoacetate. High yields can be realized when an excess of the carbonyl compound is used.[1]

[1] L.-C. Chao and R. D. Rieke, *J. Org.*, **40**, 2253 (1975).

**Iodine, 1**, 495–500; **2**, 220–222; **3**, 159–160; **4**, 258–260; **5**, 346–347.

*Trisubstituted olefins* (**5**, 346). Zweifel and Fisher[1] have extended the synthesis of disubstituted olefins from dialkylalkenylboranes (equation I) to the synthesis of trisubstituted olefins (equation II). Overall yields are in the range

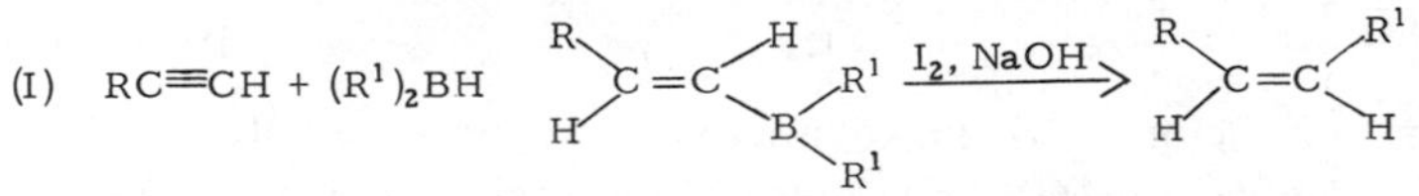

(I) $RC\equiv CH + (R^1)_2BH$

65–80%; the R group migrates as usual with retention of configuration.

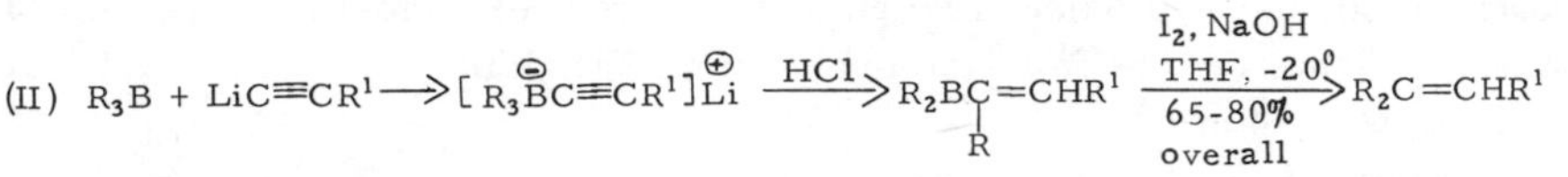

(II) $R_3B + LiC\equiv CR^1 \longrightarrow [R_3\overset{\ominus}{B}C\equiv CR^1]\overset{\oplus}{Li} \xrightarrow{HCl} R_2B\underset{R}{C}=CHR^1 \xrightarrow[\substack{65-80\% \\ overall}]{\substack{I_2, NaOH \\ THF, -20^0}} R_2C=CHR^1$

The reaction has been applied to a synthesis of propylure, the sex attractant of the pink bollworm moth.[2]

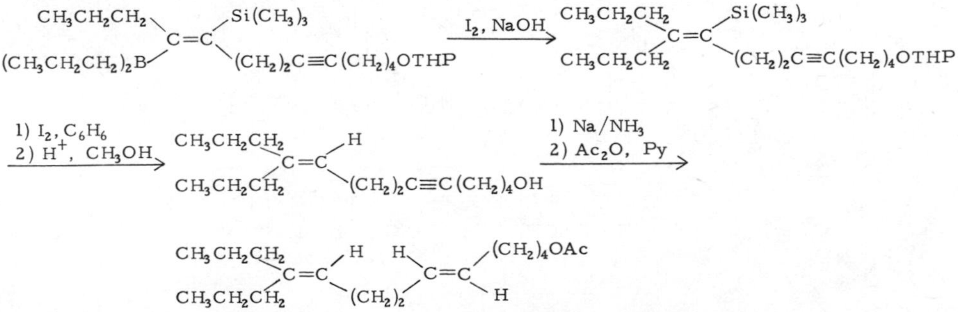

*Symmetrical acetylenes.* Suzuki *et al.*[3] have developed a method for conversion of a trialkylborane into a symmetrical alkyne. Lithium chloroacetylide

*293*

$$\underset{Cl}{\overset{H}{>}}C=C\underset{H}{\overset{Cl}{<}} \quad \xrightarrow[-HCl, -CH_4]{CH_3Li,\ 0^\circ} \quad LiC\equiv CCl \xrightarrow[THF]{(n-C_4H_9)_3B} \quad [(n-C_4H_9)_3\bar{B}C\equiv CCl]Li^+$$

$$(1) \hspace{7cm} (2)$$

$$\xrightarrow[\substack{65\%}]{\substack{1)\ I_2,\ -78^\circ \rightarrow 20^\circ \\ 2)\ H_2O_2,\ OH^-}} \quad n-C_4H_9C\equiv CC_4H_9-n$$

$$(3)$$

(1) is prepared by reaction of *trans*-1,2-dichloroethylene with methyllithium. A trialkylborane, for example tri-*n*-butylborane, in THF is added to form the ate complex (2). Treatment of (2) with iodine followed by the usual alkaline oxidation gives 5-decyne (3) in 65% yield [calculated on the assumption that 1 mole of (3) is formed from 1 mole of (2)]. This reaction is an extension of an earlier synthesis of acetylenes from trialkylboranes (5, 346).

*Symmetrical diynes.*[4] In a new approach to coupling of alkynes, a dialkyl-halogenoborane (1)[5] is treated with an alkynyllithium in THF to form the salt (2); after cooling to -78°, 1 eq. of iodine is added and the reaction is allowed

$$R'_2BX \;+\; 2\,LiC\equiv CR \xrightarrow[-LiX]{THF} \left[R'_2\bar{B}(C\equiv CR)_2\right]Li^+ \xrightarrow[70-95\%]{I_2,\ -78^\circ} RC\equiv CC\equiv CR \;+\; R'C\equiv CR$$

$$(1) \hspace{4cm} (2) \hspace{4cm} (3) \hspace{1cm} (traces)$$

to warm to 0° before work-up. Diynes (3) are obtained, usually in high yield.

*Alkyl iodides* (3, 160). The preparation of alkyl iodides by reaction of tri-alkylboranes with iodine and a base is improved by use of sodium methoxide in place of sodium hydroxide. Two groups on boron react rapidly; in the case of trialkylboranes derived from terminal olefins, the third group also reacts, but more slowly (24 hr.).[6]

*Examples:*

$$CH_3CH_2CH=CH_2 \xrightarrow[\substack{79\%}]{\substack{1)\ BMS,\,THF \\ 2)\ I_2,\ NaOCH_3}} CH_3CH_2CH_2CH_2I$$

$$CH_3CH=CHCH_3 \xrightarrow[66\%]{} CH_3\overset{\overset{\textstyle I}{|}}{C}HCH_2CH_2$$

[1] G. Zweifel and R. P. Fisher, *Synthesis*, 376 (1975).

[2] K. Utimoto, M. Kitai, and H. Nozaki, *Tetrahedron Letters*, 2825 (1975); K. Utimoto, M. Kitai, M. Naruse, and H. Nozaki, *ibid.*, 4233 (1975).

[3] K. Yamada, N. Miyaura, M. Itoh, and A. Suzuki, *ibid.*, 1961 (1975).

[4] A. Pelter, K. Smith, and M. Tabata, *J.C.S. Chem. Comm.*, 857 (1975).

[5] Both bromodicyclohexylborane and chlorobis-(1,2-dimethylpropyl)borane were used; the main requirement is that the alkyl group of the halogenoborane have a low aptitude for migration in the presence of iodine.

[6] N. R. De Lue and H. C. Brown, *Synthesis*, 114 (1976).

## Iodine–Dimethyl sulfoxide, 5, 347.

*Oxidation of thiophosphoryl and thiocarbonyl compounds* (4, 194).[1] Compounds containing group (1) or (2) are oxidized by DMSO containing a catalytic amount of iodine to the oxygen analogs. Oxidation of optically active phosphine

$$
\left.
\begin{array}{c}
\ce{>P=S(Se)} \\
(1) \\[2ex]
\ce{>C=S(Se)} \\
(2)
\end{array}
\right\}
\xrightarrow[80°]{\ce{CH3SCH3}, \ I_2}
\begin{array}{c}
\ce{>P=O} \\
(3) \\[2ex]
\ce{>C=O} \\
(4)
\end{array}
\ + \ \ce{H3CSCH3} \ + \ \ce{S(Se)}
$$

sulfides proceeds with inversion of configuration:

$$
\begin{array}{c}
CH_3 \\
\underline{n}\text{-}C_3H_7\cdots P=S \\
C_6H_5
\end{array}
\xrightarrow{O}
\begin{array}{c}
CH_3 \\
O=P\cdots C_3H_7\text{-}\underline{n} \\
C_6H_5
\end{array}
$$

(20% optical purity)

Yields of the oxo compounds obtained in this way range from 65 to 90%.

$$(\underline{n}\text{-}C_4H_9)_3 P=S(Se) \xrightarrow{85\text{–}89\%} (\underline{n}\text{-}C_4H_9)_3 P=O$$

$$[(CH_3)_2 N]_3 P=S \xrightarrow{68\%} [(CH_3)_2 N]_3 P=O$$

$$(H_2N)_2 C=S \xrightarrow{66\%} (H_2N)_2 C=O$$

80%

[1] M. Mikolajczyk and J. Luczak, *Synthesis*, 114 (1975).

**Iodine–Potassium iodate, $I_2$–$KIO_3$.**

cis-*Hydroxylation of alkenes.*[1] Treatment of alkenes with iodine (0.5 mole) and potassium iodate (0.25 mole) in HOAc ($20°$) and then with potassium acetate under reflux yields cis-1,2-diols after hydrolysis. This method offers an attractive alternative to the Woodward method (**1**, 1002–1003) utilizing the more expensive silver acetate.

$$R^1R^2C=CR^3R^4 \xrightarrow[\substack{2)\ KOAc,\ \Delta \\ 42-90\%}]{1)\ I_2;\ KIO_3,\ HOAc} R^1\underset{R^2}{\overset{OH}{\underset{|}{C}}}-\underset{R^4}{\overset{OH}{\underset{|}{C}}}-R^3$$

[1] L. Mangoni, M. Adinolfi, G. Barone, and M. Parrilli, *Gazz.*, **105**, 377 (1975).

**Iodine–Pyridine–Sulfur dioxide.**

*Deoxygenation of sulfoxides.* Sulfoxides are converted into sulfides in high yield when treated with 1 eq. of iodine and excess pyridine–sulfur dioxide complex in acetonitrile ($80°$). If bromine is used, pyridine is omitted in order to

$$R\overset{O}{\overset{\uparrow}{S}}R \xrightarrow[85-95\%]{\substack{I_2 \\ Py-SO_2}} RSR$$

prevent α-bromination of the sulfoxide.[1]

[1] M. Nojima, T. Nagata, and N. Tokura, *Bull. Chem. Soc. Japan*, **48**, 1343 (1975).

**Iodine–Pyridine N-oxide.**

*Oxidative desulfurization of thioketals and s-trithianes.*[1] Thioketals are converted into the corresponding ketones by treatment with iodine and pyridine N-oxide in DMSO (steam bath, 1 hr.). In the three cases reported, yields were about 80%. Under similar conditions, s-trithianes are converted into aldehydes. The reaction in the latter case may involve the sequence illustrated:

[1] J. B. Chattopadhyaya and A. V. Rama Rao, *Tetrahedron Letters*, 3735 (1973); idem, *Synthesis*, 865 (1974).

**Iodine azide, 1,** 500–501; **2,** 222–223; **3,** 160–161; **4,** 262; **5,** 350–351.

*syn-Addition to a strained olefin.* Two laboratories[1,2] have reported that iodine azide undergoes exclusive *syn*-addition to the strained cyclobutene derivative (1), in contrast to the usual mode of *anti*-addition.

IN$_3$, CH$_3$CN
-5°

COOCH$_3$
COOCH$_3$
(1)

COOCH$_3$
COOCH$_3$
(2)

[1] G. Mehta, P. K. Dutta, and P. N. Pandey, *Tetrahedron Letters,* 445 (1975).
[2] T. Sasaki, K. Kanematsu, and A. Kondo, *Tetrahedron,* **31,** 2215 (1975).

**Iodine bromide, 1,** 501–502; **2,** 224.

*Bromination of methyl 3-oxo-5β-cholanate* (1). Bromination of this 5β-ketosteroid with IBr (2 eq.) in HOAc gives the expected 4β-bromo ketone (2).[1] On standing, (2) rearranges slowly to the 2β-bromo ketone (3) in 50% yield

CH$_3$

2 IBr
2 hr.

(1)

CH$_3$

Br

5 days

Br

(2)

CH$_3$

(3)

from (1). This paper is the first report of bromination of (1) at C$_2$. The rearrangement is facilitated by iodine.

[1] Y. Yanuka and G. Halperin, *J. Org.,* **39,** 3047 (1974).

**Iodobenzene diacetate–Trimethylsilyl azide.**

This combination of reagents reacts with simple acyclic and cyclic olefins to form α-azidoketones (*see* **5,** 354). Cycloalkenes substituted with various groups undergo fragmentation to ω-cyanocarboxylic acid derivatives when $n = 4$.[1]

(CH$_2$)$_n$
CH
CH

35–95%

(CH$_2$)$_n$
C=O
CHN$_3$

(CH$_2$)$_n$
C—X
C—H

40–56%

O=C—X
(CH$_2$)$_n$
C≡N

X = OC$_2$H$_5$, OCOCH$_3$,
Cl, Br, N$_3$

[1] J. Ehrenfreund and E. Zbiral, *Tetrahedron,* **28,** 1697 (1972); *idem, Ann.,* 290 (1973).

**Iodobenzene dichloride (Phenyliodine dichloride),** 1, 505–506; 2, 225–226; 3, 164–165; 4, 264–265; 5, 352–353.

*Remote functionalization of steroids* (4, 264–265; 5, 352–353). Breslow et al.[1] have modified their earlier procedure for chlorination of steroids at $C_9$ or $C_{14}$ by use of an external source of chlorine radicals, iodobenzene dichloride or sulfuryl chloride. Thus irradiation of the m-iodobenzoate of 3α-cholestanol (1) and iodobenzene dichloride in methylene chloride for 28 min. at 25° followed by saponification and acetylation leads to a mixture of 3α-cholestanyl acetate (18.4%) and the acetate of $\Delta^{9(11)}$-cholestene-3α-ol (2, 66%). Application of the

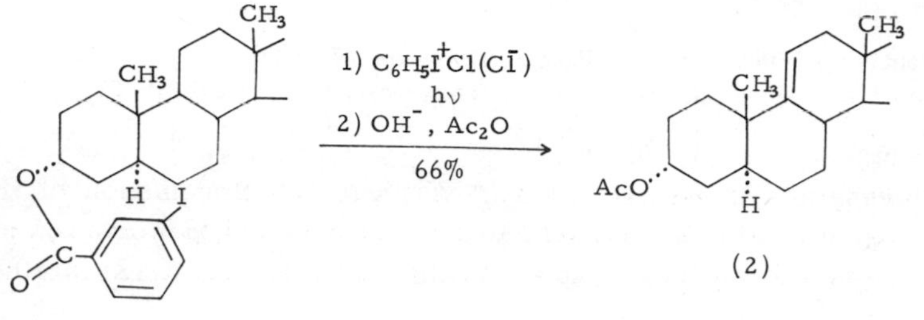

(1)    (2)

same procedure to the ester (3) leads to $\Delta^{14}$-3α-cholestenyl acetate. A radical

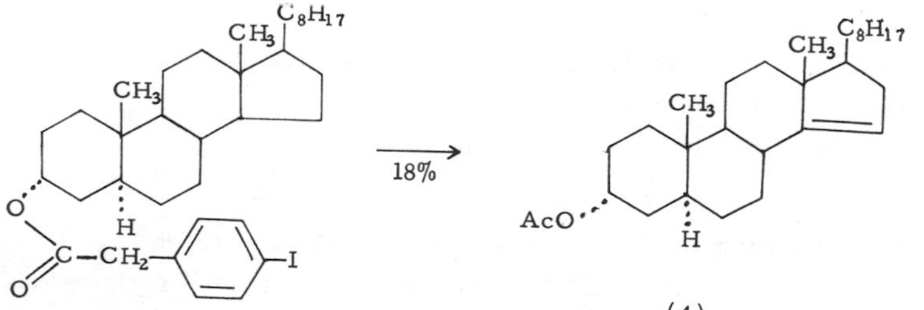

(3)    (4)

relay mechanism is proposed for these reactions in which the aryl iodide dichloride is generated by an external source.

This functionalization has been used for a synthesis of cortisone from cortexolone that is an attractive alternative to biological hydroxylation. Thus irradiation of (5) and iodobenzene dichloride gives, after dehydrohalogenation and saponification, the unsaturated alcohol (6). Hydroboration, oxidation, and deprotection of (6) affords dihydrocortisone, which had been converted previously into cortisone acetate and prednisolone acetate.

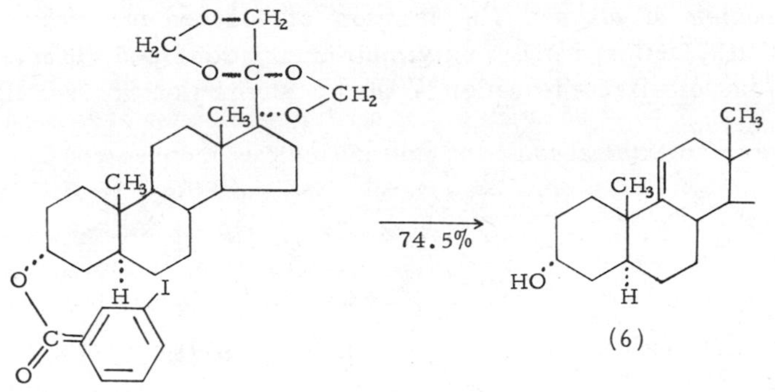

(5)    74.5%    (6)

Breslow *et al.*[2] have since extended this remote halogenation to the conversion of β-cholestanol into androsterone (11) by selective halogenation at $C_{17}$. In order to place the internal chlorinating reagent in proximity to the $C_{17}$-position, 3β-cholestanol is converted, with inversion, into the ester (7). Irradiation of (7) and iodobenzene dichloride gives an intermediate $C_{17}$-chloro compound that is

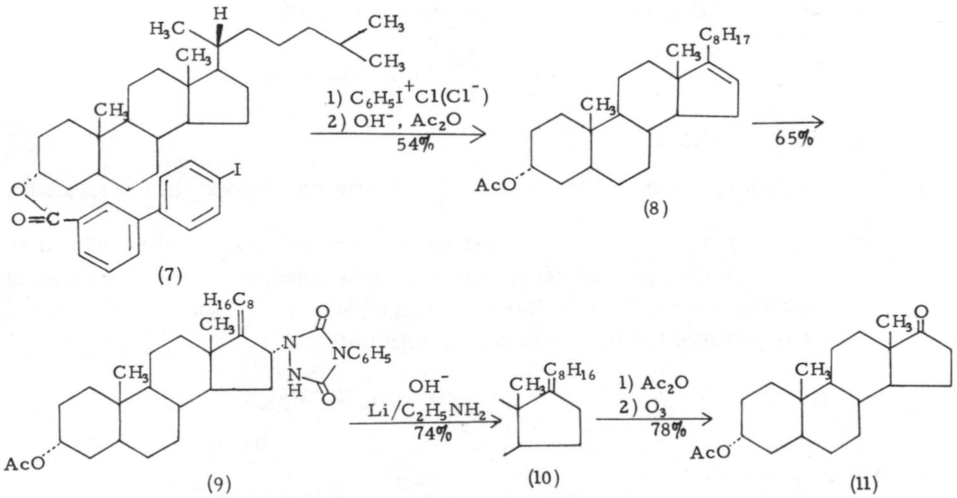

(7)    1) $C_6H_5I^+Cl(Cl^-)$
2) $OH^-$, $Ac_2O$
54%    (8)    65%

(9)    $OH^-$
$Li/C_2H_5NH_2$
74%    (10)    1) $Ac_2O$
2) $O_3$
78%    (11)

readily converted into $\Delta^{16}$-3α-cholestenyl acetate (8) by base. The double bond of (8) is transposed to the 17(20)-position by an ene reaction with N-phenyltriazolinedione; the product (9) on saponification and reduction with lithium in ethylamine gives $\Delta^{17(20)}$-3-cholestenol (10). Androsterone acetate (11) is then obtained by acetylation and ozonolysis of (10).

*Chlorination of alkynes.* The reaction of iodobenzene dichloride with alkynes ($CHCl_3$, reflux, azobisisobutyronitrile initiation) results in *anti*-addition to give E-dichloroalkenes (equation I). Use of chlorine (ionic or radical) leads to complex mixtures.[3]

$$(I)\quad R-C\equiv C-R' \xrightarrow{C_6H_5ICl_2} \underset{Cl}{\overset{R}{>}}C=C\underset{R'}{\overset{Cl}{<}} \;+\; \underset{Cl}{\overset{R}{>}}C=C\underset{Cl}{\overset{R'}{<}}$$

$$(65-95\%) \qquad\qquad (5-35\%)$$

[1] R. Breslow, R. J. Corcoran, and B. B. Snider, *Am. Soc.*, **96**, 6791, 6793 (1974).
[2] *Idem, ibid.*, **97**, 6580 (1975).
[3] A. Debon, S. Masson, and A. Thuillier, *Bull. soc.*, 2493 (1975).

## Iodobenzene difluoride [Phenyliodine(III) difluoride], $C_6H_5IF_2$.

The most convenient preparation involves the reaction shown.[1] Methylene chloride is used as solvent and the solution of the reagent is used directly.

$$C_6H_5ICl_2 + HgO + 2HF \xrightarrow{CH_2Cl_2} C_6H_5IF_2 + H_2O + HgCl_2$$

The reagent has been used for fluorination of olefins. The reaction requires acid catalysis and involves a rearrangement as shown for the reaction with

$$(C_6H_5)_2C=CH_2 \xrightarrow[47\%]{C_6H_5IF_2} C_6H_5CF_2CH_2C_6H_5$$

1,1-diphenylethylene.

*Polymeric reagent,* $(P)\!\!-\!\!\bigcirc\!\!-\!\!I\underset{F}{\overset{F}{<}}$ . Zupan and Pollak[2] have prepared a polymer-supported reagent by iodination of "popcorn"-polystyrene followed by reaction with $XeF_2$. This reagent reacts with phenyl-substituted olefins at 25° to give rearranged *gem*-difluorides in high yield. The polymeric iodobenzene that is also formed is removed simply by filtration.

$$(C_6H_5)_2C=CH_2 \xrightarrow[96\%]{} C_6H_5CF_2CH_2C_6H_5$$

$$(C_6H_5)_2C=CHCH_3 \xrightarrow[95\%]{} C_6H_5CF_2\underset{CH_3}{\overset{}{C}}HC_6H_5$$

$$C_6H_5CH=CH_2 \xrightarrow[86\%]{} C_6H_5CH_2CHF_2$$

[1] W. Carpenter, *J. Org.*, **31**, 2688 (1966).
[2] M. Zupan and A. Pollak, *J.C.S. Chem. Comm.*, 715 (1975).

**Iodobenzene    ditrifluoroacetate    (Phenyliodine(III)    ditrifluoroacetate),** $C_6H_5I(OCOCF_3)_2$. Mol. wt. 430.05, m.p. 112–114° dec.

*Preparation.* The reagent can be prepared conveniently by dissolution of iodobenzene diacetate (formerly named iodosobenzene diacetate) in warm trifluoroacetic acid (53% yield).[1] Other methods have been reported.[2,3]

*Oxidation.*[1] Phenyliodine ditrifluoroacetate is more effective for various dehydrogenations than iodobenzene diacetate. Representative oxidations are formulated. Alcohols are oxidized preferably in the presence of a tertiary amine

$$C_6H_5CH_2OH \xrightarrow[50\%]{C_5H_5N,\ 80°} C_6H_5CHO$$

$$(C_6H_5)_2CHOH \xrightarrow[65\%]{C_5H_5N,\ 80°} (C_6H_5)_2C{=}O$$

$$(C_6H_5)_2\overset{\overset{OH}{|}}{C} - \overset{\overset{OH}{|}}{C}(C_6H_5)_2 \xrightarrow[90\%]{\substack{CH_3COCH_3 \\ 20°}} 2\ (C_6H_5)_2C{=}O$$

$$2\ C_6H_5SH \xrightarrow[92\%]{CH_2Cl_2,\ 20°} C_6H_5S{-}SC_6H_5$$

$$C_6H_5NH{-}NHC_6H_5 \xrightarrow[88\%]{C_2H_5OH,\ 20°} C_6H_5N{=}NC_6H_5$$

and require elevated temperatures. Dehydrogenation of CH—CH groups proceeds in low yield. On the other hand, oxidation of NH—NH groups and of hydroquinones proceeds in high yield at 20°. The reagent also cleaves 1,2-diols in high yield.

[1] S. Spyroudis and A. Varvoglis, *Synthesis*, 445 (1975).
[2] N. W. Alcock and T. C. Waddington, *J. Chem. Soc.*, 4103 (1963).
[3] I. I. Maletina, V. V. Orda, and L. M. Yagupol'skii, *J. Org. Chem. U.S.S.R.*, **10**, 294 (1974).

**Iodomethyl methyl sulfide, 5, 353–354.**

This reagent can be prepared *in situ* by the reaction of chloromethyl methyl sulfide (Aldrich) and sodium iodide in DMF.

*Protection of primary hydroxyl groups.*[1] Sodium alkoxides (from sodium hydride) react with iodomethyl methyl sulfide at $0°$ in DME to form methyl-thiomethyl ethers (MTM ethers) in yields usually $>86\%$. These ethers are stable to bases and nucleophiles (NaH, RLi, NaOR) and fairly stable to acid. Thus acetonide and THP protecting groups can be cleaved more easily by acid than the MTM ether function.

Two methods can be used for deblocking: treatment with mercuric chloride in aqueous acetonitrile at $25°$ or reaction with silver nitrate in aqueous THF buffered with 2,6-lutidine at $25°$. In either case yields are 88–95%.

Only modest yields ($\sim 40\%$) of MTM ethers are obtained from secondary alcohols.

[1] E. J. Corey and M. G. Bock, *Tetrahedron Letters*, 3269 (1975).

**Ion-exchange resins, 1,** 511–519; **2,** 227–228; **4,** 266–268; **5,** 355–356.

*Selective ketalization.* Jones *et al.*[1] have reported selective ketalizations of some diketosteroids with ethylene glycol with an acidic ion-exchange resin (Amberlite IR 120-H) as catalyst. This procedure was found more selective than the usual homogeneous conditions. 3,6-, 3,7-, and 3,17-Diketo-5α-androstanes were converted in this way into the 3-monoketals in very high yield.

*γ-Butyrolactones.*[2] 3-Hydroxy-N,N-dimethyl-1-carboxamides are converted into γ-butyrolactones by treatment in acetone with an acidic ion-exchange resin[3] either for 20 hr. at room temperature or for 2–3 hr. at reflux. The usual acid catalysts (*p*-TsOH, BF$_3$ · etherate) are much less effective.

*Dimerization of vinylarenes.*[4] Amberlyst-15 sulfonic acid resin is superior to PPA for dimerization and cyclization of α-methylstyrene (1) to (2). The reaction is carried out in refluxing cyclohexane (5 hr.). It is also superior to PPA for

(1)                                    (2)

(3)                                                                      (+ Isomer, traces)

(4)

dimerization of $\beta$-vinylnaphthalene (3). The catalyst can be reused and actually appears to become more effective with use.

*Cyclization of costunolide.*[5] The acid form of Amberlite IR-120 resin was used for a cyclization under mild conditions of the costunolide (1), which polymerizes readily in the presence of strong acids, to $\alpha$- and $\beta$-cyclocostunolide, (2)

(1)                          (2, 31%)                          (3, 36%)

and (3), as the major products. The former product is a natural product. Ten-membered carbocyclics have been postulated as biogenetic precursors of bicyclic sesquiterpenes.

*Esters.*[6] Esters can be prepared readily by reaction of carboxylic acids with alkyl iodides or bromides in the presence of Amberlite IRA (OH⁻ form):

$$\text{Resin-}\overset{+}{\text{N}}(\text{CH}_3)_3\text{OH}^- + \text{RCOOH} \xrightarrow[-\text{H}_2\text{O}]{} \text{Resin-}\overset{+}{\text{N}}(\text{CH}_3)_3\text{RCOO}^- \xrightarrow{\text{R'X}}$$

Yields are low when only catalytic amounts of the resin are used. Yields are high ($\sim$90%) with primary alkyl halides, but lower with secondary alkyl halides. Only traces of esters are obtained from *t*-butyl bromide.

*Solid-phase peptide synthesis* (1, 516–517; 2, 227). Originally, Merrifield[7] used stannic chloride and chloromethyl methyl ether for chloromethylation of the polymer beads. More recently, along with Feinberg,[8] he recommended zinc chloride and chloromethyl methyl ether. Sparrow[9] prefers boron trifluoride etherate and chloromethyl ethyl ether (Aldrich). This ether may not be carcinogenic. The resulting resins have desirable physical properties.

[1] E. R. H. Jones, G. D. Meakins, J. Pragnell, W. E. Müller, and A. L. Wilkins, *J.C.S. Perkin I*, 2376 (1974).

[2] W. Sucrow and U. Klein, *Ber.*, **108**, 48 (1975).

[3] E. Merck, Iat I.

[4] W. P. Duncan, E. J. Eisenbraun, A. R. Taylor, and G. W. Keen, *Org. Prep. Proc. Int.*, 7, 225 (1975).

[5] T. C. Jain and J. E. McCloskey, *Tetrahedron*, **31**, 2211 (1975).

[6] G. Cainelli and F. Manescalchi, *Synthesis*, 723 (1975).

[7] R. B. Merrifield, *Biochemistry*, 3, 1385 (1964).

[8] R. S. Feinberg and R. B. Merrifield, *Tetrahedron*, **30**, 3209 (1974).

[9] J. T. Sparrow, *Tetrahedron Letters*, 4637 (1975).

**Iron(III) acetylacetonate, 4,** 268.

*β-Epoxidation of cholesterol.* The β-oxide of cholesterol (2) can be obtained stereoselectively by treatment with a large excess of 30% hydrogen peroxide in acetonitrile containing iron(III) acetylacetonate. $\Delta^4$-Cholestene-3α-ol and

$\Delta^4$-cholestene-3β-ol are also oxidized under these conditions mainly to the corresponding β-oxides.[1]

[1] M. Tohma, T. Tomita, and M. Kimura, *Tetrahedron Letters*, 4359 (1973).

**Iron pentacarbonyl, 1,** 519–520; **2,** 229–230; **3,** 167; **4,** 268–270; **5,** 357–358.

*Benzamidines.*[1] Treatment of oximes of N-alkyl- or N-arylbenzamides (1) with iron pentacarbonyl (1 eq.) in refluxing dry THF gives benzamidines (2) in 70–90% yield.

This reductive dehydroxylation may be involved in deoximation by Fe(CO)$_5$ (**2**, 229–230). Thus treatment of methyl mesityl ketoxime (3) under the conditions used above gives the imine (4). In this case steric hindrance probably inhibits ready hydrolysis to the ketone.

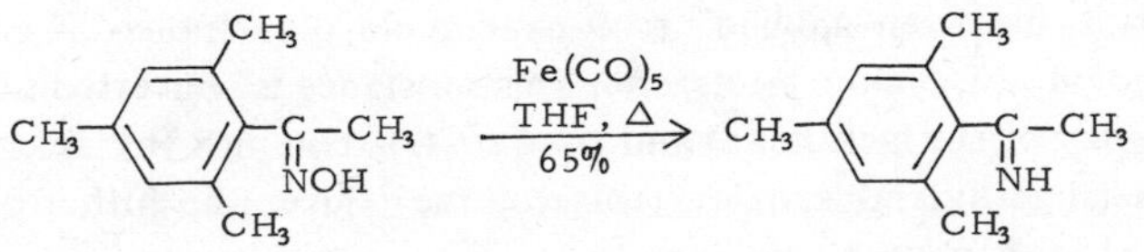

(3)                                    (4)

*Alkali metal iron carbonylates.* An aqueous alcoholic solution of KHFe(CO)$_4$ can be prepared from iron pentacarbonyl (1 mole) and KOH (3 moles). A THF solution of NaHFe(CO)$_4$ can be prepared by the reaction of iron pentacarbonyl with sodium amalgam and then with water (1 mole). Amines can be alkylated by aldehydes in the presence of these reagents. One advantage of this procedure is that primary amines can be converted into either monoalkyl or dialkyl derivatives.[2]

$$C_6H_5NH_2 + HCHO \xrightarrow[65\%]{\substack{KHFe(CO)_4 \\ C_2H_5OH}} C_6H_5NHCH_3$$

$$C_6H_5NH_2 + 2\ HCHO \xrightarrow[90\%]{2\ KHFe(CO)_4} C_6H_5N(CH_3)_2$$

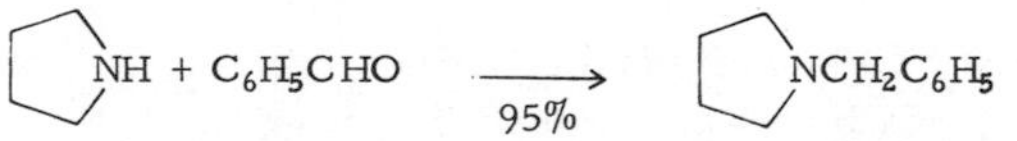

*Reaction with vinyl oxiranes.* α,β-Unsaturated epoxides, for example (1), on irradiation in the presence of iron pentacarbonyl give lactones such as (2). When warmed, (2) loses CO and is converted into the Fe(CO)$_3$ complex of the

(1)                          (2)                              + CO

Fe(CO)$_3$

(3)

hydroxycyclohexadiene (3).[3]

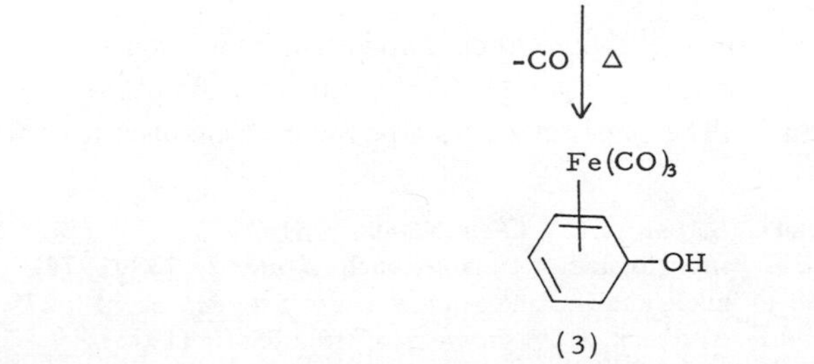

This reaction has been applied[4] to 9-oxabicyclo[6.1.0]nona-2,4,6-triene (4), the monoepoxide of cyclooctatetraene. This substance is converted in 70% yield into (5), by way of the lactone (a) and the $Fe(CO)_3$ complex (b). Decomposition of (5) with trimethylamine oxide (this volume)[5] gives the hitherto unknown 9-oxabicyclo[4.2.1]nona-2,4,7-triene (6).

(4)    + Fe(CO)₅    hν →    (a)    →    (b)

70% →    (5)    (CH₃)₃N→O / 48% →    (6)

***Ring expansion of α- and β-pinene.***[6] When either α- or β-pinene, (1) or (2), is heated neat with an initial pressure of 30 psi CO at 160° for 68 hr. with an equimolar amount of iron pentacarbonyl, two ketones, (3) and (4), are obtained, formed by insertion of CO into the cyclobutane ring together with some α-pinene. Both ketones are optically active. The Cotton effects are of nearly the same

(1)    (2)    Fe(CO)₅ / Δ →    (1) +    (3, 29%, $\phi_D$ -659)    (4, 34% $\phi_D$ + 637)

value, but of opposite sign. Other known rearrangements of pinenes usually involve cleavage of the C—C bond bearing the *gem*-dimethyl group and lead to racemic products. The paper suggests a possible mechanism for this unusual ring expansion.

[1] A. Dondoni and G. Barbaro, *J.C.S. Chem. Comm.*, 761 (1975).
[2] G. P. Boldrini, M. Panunzio, and A. Umani-Ronchi, *Synthesis*, 733 (1974).
[3] R. Aumann, K. Fröhlich, and H. Ring, *Angew. Chem. internat. Ed.*, **13**, 275 (1974).
[4] R. Aumann and H. Averbeck, *J. Organometal. Chem.*, **85**, C4 (1975).
[5] Y. Shvo and E. Hazum, *J.C.S. Chem. Comm.*, 336 (1974).
[6] A. Stockis and E. Weissberger, *Am. Soc.*, **97**, 4288 (1975).

Isoamyl nitrite, 4, 270; 5, 358–359.

*α-Substituted α-diazo esters* (4, 270). The experimental details for the preparation of these compounds have been published.[1]

[1] N. Takamura, T. Mizoguchi, K. Koga, and S. Yamada, *Tetrahedron*, 31, 227 (1975).

Isobutylaluminum dichloride, $i$-$C_4H_9AlCl_2$. Mol. wt. 155.00.

This alkylaluminum dichloride is prepared *in situ* from tri-$i$-butylaluminum and aluminum chloride:

$$(i\text{-}C_4H_9)_3Al + 2\ AlCl_3 \longrightarrow 3\ i\text{-}C_4H_9AlCl_2$$

*Stereospecific reduction.* The usual complex hydrides [*e.g.*, $LiAlH_4$, $HAlCl_2$, $LiAlH(O\text{-}t\text{-}Bu)_3$] reduce (1R)-3-*endo*-aminobornane-2-one (1) to mixtures of (1R)-3-*endo*-aminoborneol (2) and (1R)-3-*endo*-aminoisoborneol (3), with (2) predominating. Stereospecific reduction to (2) can be achieved with alkylaluminum dichlorides.[1]

(1)                    (2)                    (3)

[1] H. Pauling, *Helv.*, 58, 1781 (1975).

Isobutyl chloroformate, $ClCOOCH_2CH(CH_3)_2$. Mol. wt. 136.58, b.p. 128.8°. Supplier: Aldrich.

*Mixed anhydride peptide synthesis.* Lilly chemists[1] have reported the successful synthesis of $BOC\text{-}Gly\text{-}Gly\text{-}OC_2H_5$ and $BOC\text{-}Gly\text{-}Gly\text{-}Gly\text{-}OC_2H_5$ in reasonable yields (77%, 88%) via mixed anhydrides derived from isobutyl chloroformate. Previously, other laboratories[2] reported that considerable quantities of diacylated side products are formed in mixed-anhydride syntheses with glycine when ethyl chloroformate is used.

[1] J. A. Hoffmann and M. A. Tilak, *Org. Prep. Proc. Int.*, 7, 215 (1975).
[2] K. D. Kopple and K. J. Renick, *J. Org.*, 23, 1565 (1958); R. B. Merrifield, A. R. Mitchell, and J. E. Clarke, *ibid.*, 39, 660 (1974).

Isoprene epoxide, (1). Mol. wt. 84.12, b.p. 81° (735 mm.).

*Preparation.*[1]

*Prenyl synthon.*[2] The reagent reacts with phenol (equation I) and phenyl-lithium (equation II) by ring opening with allylic rearrangement.

(I) 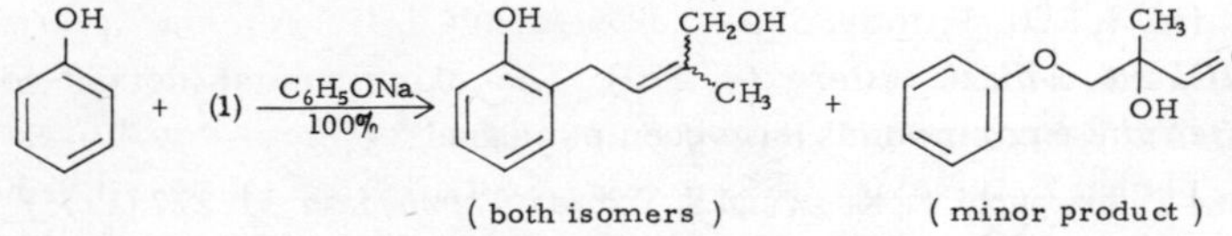    ( both isomers )    ( minor product )

(II) 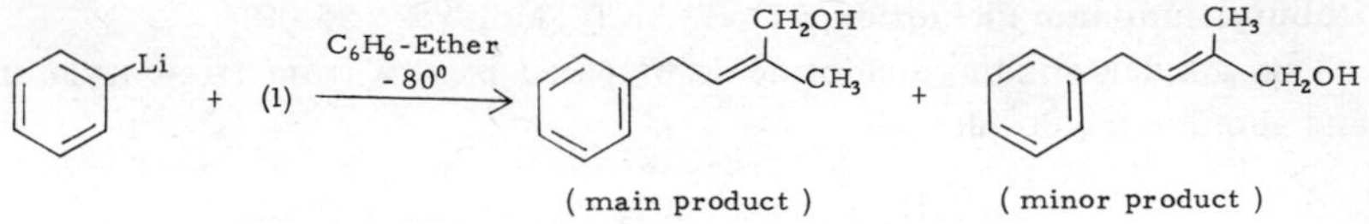    ( main product )    ( minor product )

[1] R. Pummerer and W. Reindel, *Ber.*, **66**, 335 (1933).
[2] G. C. M. Aithie and J. A. Miller, *Tetrahedron Letters*, 4419 (1975).

**Isopropenyl acetate**, 1, 524–526; 2, 191, 230, 239; 3, 103; 4, 273.

$\beta$-*Methoxy ketones.* In the presence of Lewis acids (TiCl$_4$, AlCl$_3$, SnCl$_4$) isopropenyl acetate alkylates dimethyl acetals (1) to form $\beta$-methoxy ketones (2) and traces of $\alpha,\beta$-unsaturated ketones (3).[1]

(1)    (2, 40-90%)    (3, 1-5%)

[1] T. Mukaiyama, T. Izawa, and K. Saigo, *Chem. Letters*, 323 (1974).

**Isopropenyllithium,**

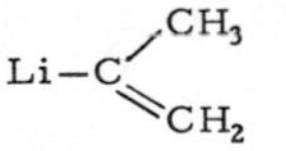

*Isopropenylation of alkenes.*[1] The reaction of trialkylboranes with isopropenyllithium in the presence of iodine followed by oxidative work-up results in isopropenylalkanes.

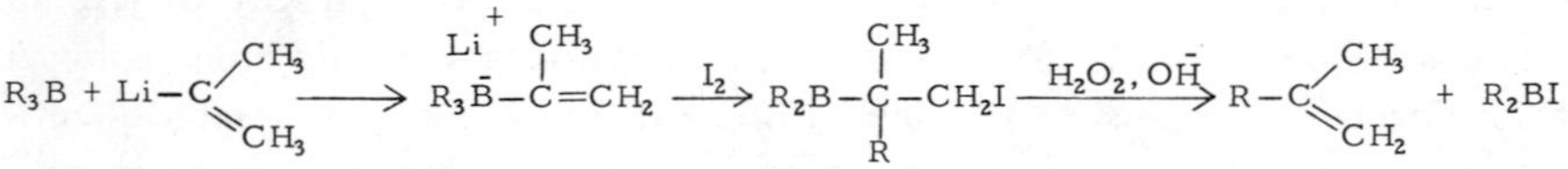

*Examples:*

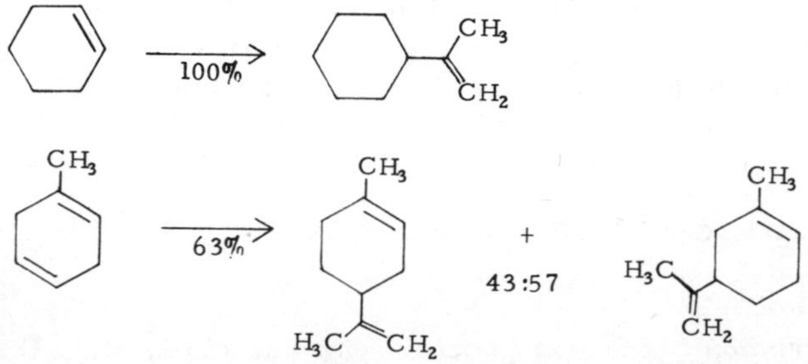

43:57

[1] N. Miyaura, H. Tagami, M. Itoh, and A. Suzuki, *Chem. Letters*, 1411 (1974).

(−)-  and  (+)-2,3-O-Isopropylidene-2,3-dihydroxy-1,4-bis(diphenylphosphino)-butane (DIOP), 4, 273; 5, 360–361.

*Asymmetric homogeneous hydrogenation of a thiophene derivative.*[1] 2-(4-Methoxy-5-phenyl-3-thienyl)acrylic acid (1) on hydrogenation with Kagan's catalyst (4, 273) is converted into (+)-2-(4-methoxy-5-phenyl-3-thienyl)propionic acid (2) in 97% yield and with an optical purity of 88%.

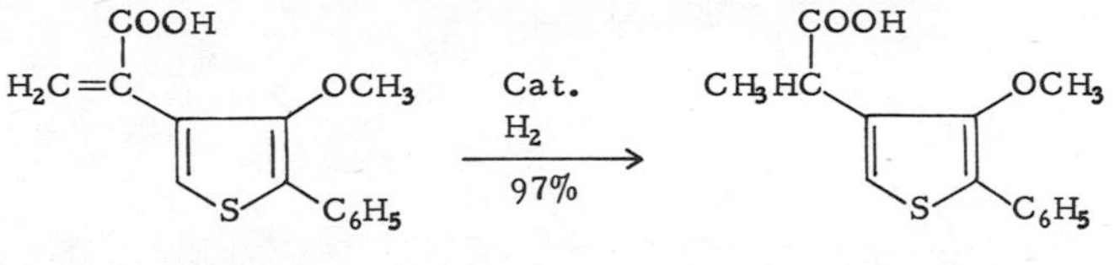

(1)                                                    (2)

[1] A. P. Stoll and R. Süess, *Helv.*, **57**, 2487 (1974).

**Isopropylidene isopropylidenemalonate,**

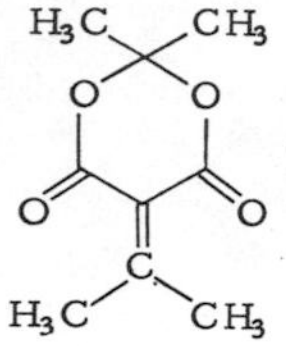

(1). Mol. wt. 184.19, m. p. 74.5–75°.

*Preparation:*

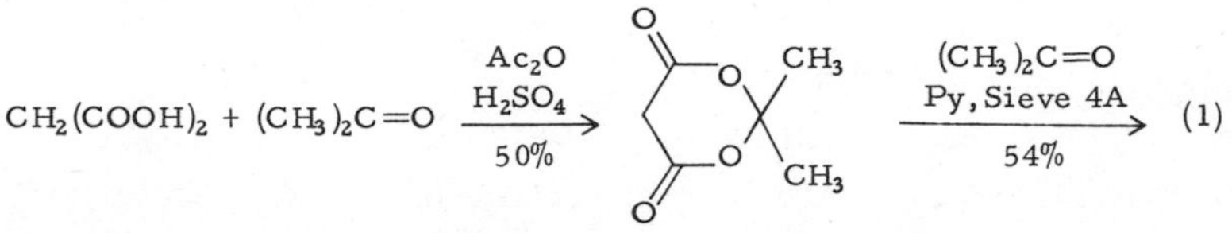

*Diels–Alder dienophile.* Dauben *et al.*[1] have used this ester as a dienophile

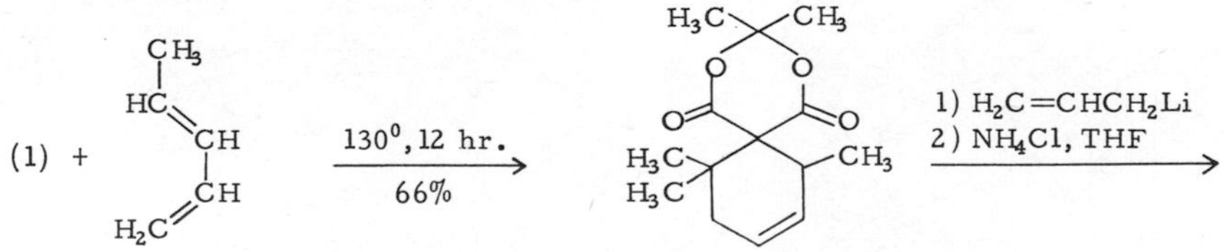

(2)

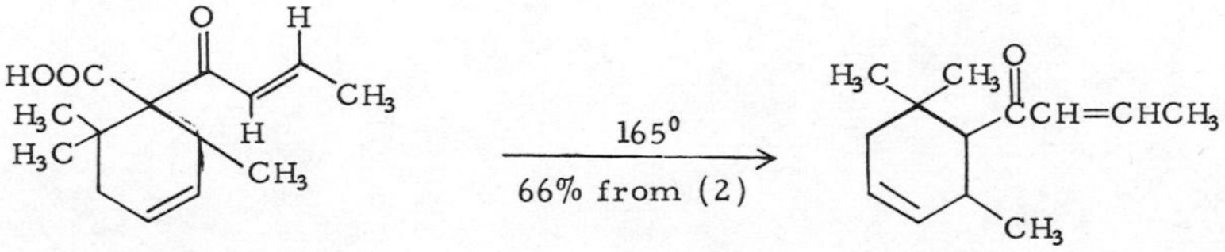

(3)                                                    (4)

in a new synthesis of δ-damascone (4). Cycloaddition of (1) and 1,3-pentadiene gives the adduct (2) in 66% yield. Reaction of (2) with allyllithium gives (3) and the corresponding β,γ-isomer, which is isomerized to (3) by treatment with ammonium chloride. Decarboxylation of (3) at 165° leads to (4).

[1] W. G. Dauben, A. P. Kozikowski, and W. T. Zimmerman, *Tetrahedron Letters*, 515 (1975).

# K

Ketene dimethyl thioacetal monoxide, $H_2C=C\overset{\overset{\displaystyle O}{\uparrow}}{\underset{SCH_3}{<}}^{SCH_3}$ (1). Mol. wt. 104.17.

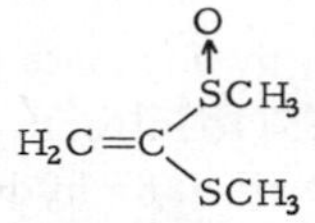

*Preparation.* This reagent is prepared from glyoxylic acid in 77% overall yield[1]:

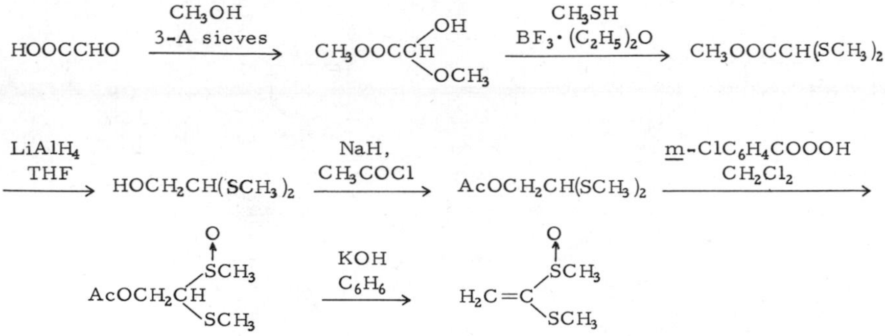

(1)

*Conjugate addition.* Schlessinger *et al.*[1,2] have reported a general synthesis of 1,4-dicarbonyl compounds based on utilization of (1) as a 2-carbon Michael acceptor. Michael additions were realized with enamines, sodium enolates of β-dicarbonyl compounds, and lithium enolates of esters. The products can be converted into aldehydes by hydrolysis in aqueous acetonitrile with perchloric acid.

*Examples:*

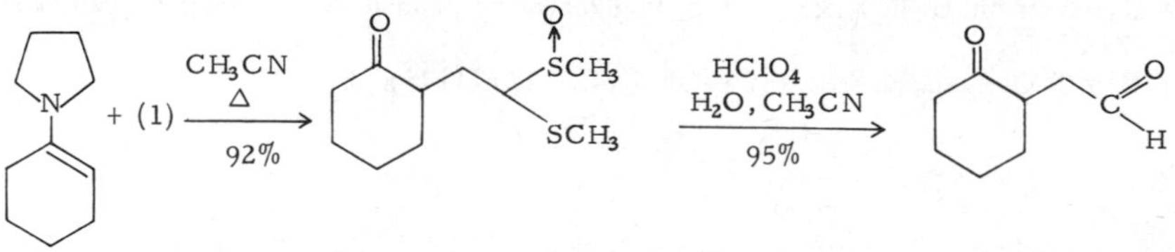

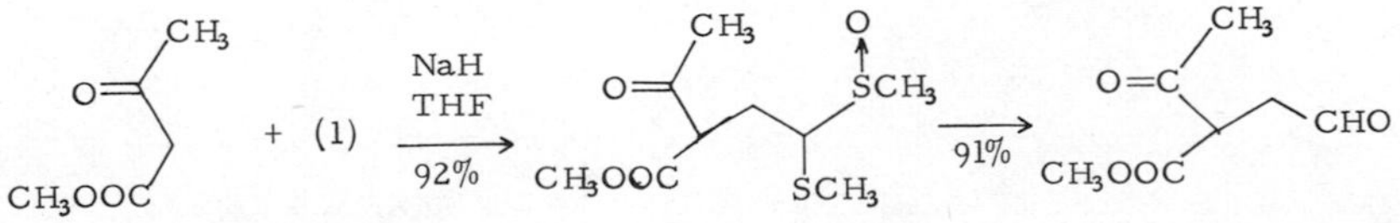

$$\begin{array}{c}\text{COOCH}_3 \\ | \\ \text{CH}_2 \\ | \\ \text{C}_2\text{H}_5\end{array} + (1) \xrightarrow[93\%]{\substack{\text{LDA} \\ \text{THF}}} \text{CH}_3\text{OOC} \diagdown\diagup\diagdown \overset{\overset{\text{O}}{\uparrow}}{\text{SCH}_3} \xrightarrow{97\%} \text{CH}_3\text{OOC}\diagdown\diagup\diagdown \text{CHO}$$

The reagent (1) was used by Japanese chemists[3] in a synthesis of the alkaloid (±)-1-acetylaspidoalbidine (5) to introduce a two-carbon side chain into (2). Two isomers were obtained (3), which on hydrolysis gave the same aldehyde (4). This

(2)         (1) LDA DME 75%         (3, two isomers)    hyd.

(4)       Several steps       (5)

product was converted into the alkaloid (5) in several steps.

[1] J. L. Herrmann, G. R. Kieczykowski, R. F. Romanet, P. L. Wepplo, and R. H. Schlessinger, *Tetrahedron Letters*, 4711 (1973).

[2] J. L. Herrmann, G. R. Kieczykowski, R. F. Romanet, and R. H. Schlessinger, *ibid.*, 4715 (1973).

[3] Y. Ban, T. Ohnuma, K. Seki, and T. Oishi, *ibid.*, 727 (1975).

# L

Lead tetraacetate (LTA), **1**, 537–563; **2**, 234–238; **3**, 168–171; **4**, 278–288; **5**, 365–370.

*Oxidative decarboxylation* (**1**, 554–557; **2**, 235–237; **3**, 168–169; **4**, 280). Russian chemists[1] report that yields are low in the oxidative decarboxylation of ω-alkoxycarbonylalkanoic acids by the standard procedure of Kochi (**2**, 235). They were able to improve the yields of alkyl ω-alkenoates (2) to 25–50% by

$$\text{ROOC}(CH_2)_{\underline{n}}CH_2CH_2COOH + Pb(OAc)_4 \xrightarrow[-CO_2]{Cu(OAc)_2, Py} \text{ROOC}(CH_2)_{\underline{n}}CH=CH_2$$

(1, $\underline{n}$ = 1–6)                                                                 (2)

gradual addition (3–5 hr.) of lead tetraacetate or by preparation of the lead(IV) salt of the acid (1) and gradual addition of this salt to a solution of LTA in benzene containing pyridine and cupric acetate.

*Oxidative ring cleavage.* Trost and Hiroi[2] have developed a method for cleavage of cycloalkanones that involves essentially two steps: selective sulfenylation with diphenyl disulfide (**5**, 276–277; this volume) and oxidation with lead tetraacetate. The method is illustrated for estrone methyl ether as the substrate. Thus this five-membered ring ketone can be converted in high overall yield into

(1)

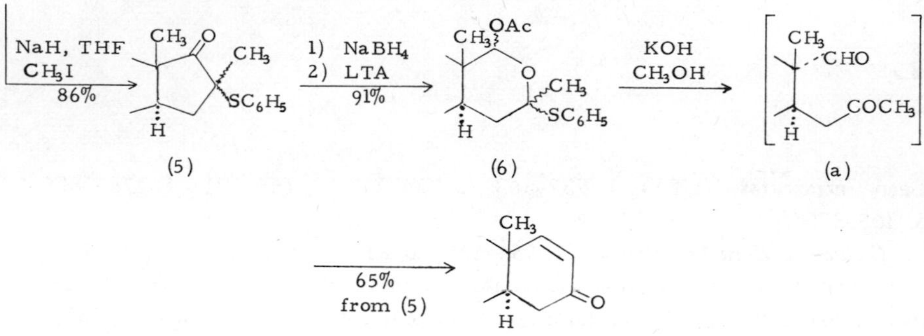

(5)　　　　　　　　(6)　　　　　　　　(a)

$$\xrightarrow[\text{from (5)}]{65\%}$$

(7)

the cyclic hemiacetal (4). A more interesting variant involves alkylation of the intermediate (1), followed by reduction and oxidation, which results in ring enlargement with transposition of the carbonyl group to give (7).

The method is equally applicable to cyclobutanones, but is not successful with cyclohexanones.

*Oxidation of phenols.* Phenols are oxidized by lead tetracetate to *o*- and *p*-quinol acetates or biphenyls depending on reaction conditions. For example, 2,4,6-tri-*t*-butylphenol is oxidized by lead tetraacetate in acetic acid, benzene, or

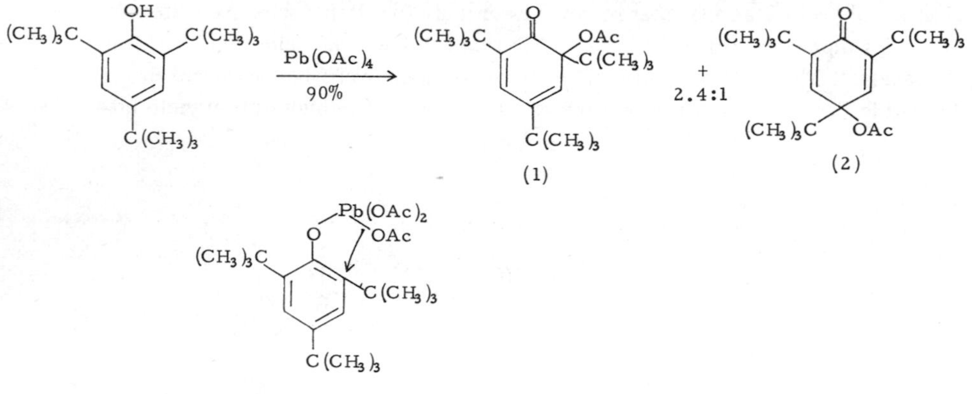

(1)　　　+　　2.4:1　　　(2)

(a)

methylene chloride mainly to (1) and (2). The fact that (1) is formed preferentially suggests that an intramolecular acetoxylation as depicted in (a) may be involved in part.[3]

This oxidation has been used in a recent synthesis of (±)-aeroplysinin-I (7), a metabolite of the marine sponges of the genus *Verongia*.[4] Lead tetraacetate oxidation of (3) gave the acetoxy dienone (4). Direct reduction of (4) with

(3) → Pb(OAc)₄ HOAc 35% → (4) → NaBH₄ → (5)

(4) → CH₃OH TsOH → (6) → NaBH₄ 60% → (7)

sodium borohydride gave (5), an isomer of the desired product, and the corresponding monoacetate. Hydrolysis of the acetate group of (4) followed by reduction gave the desired *trans*-glycol (7) in satisfactory yield. The present synthesis is interesting in that it does not involve an arene oxide and that both *cis*- and *trans*-glycols can be prepared from one precursor.

Oxidation of (±)-thaliporphine (8) with lead tetraacetate in acetic acid gives (9) in quantitative yield. This is probably formed by acetoxylation of the

(8) → Pb(OAc)₄ HOAc, 20° → (a) → quant. → (9)

(9) → 1) $H_3O^+$ 2) $CH_2N_2$ 86.5% overall → (10)

quinone methide (a). The phenol (9) was converted into (±)-cataline (10) as shown.[5]

*Benzpyrene-1-ol.* Reaction of 6-bromobenzpyrene with lead tetraacetate gives the 1-acetoxy derivative (2) as the major product. An unusual product of plumbation (3) is also formed. Reaction of (2) with *n*-butyllithium removes both

the bromine atom and the acetate group to give benzpyrene-1-ol (4), a possible metabolite of benzpyrene.[6]

*21-Acetoxylation of 20-ketosteroids* (1, 541).[7] Oxidation with lead tetraacetate with $BF_3$ catalysis is the best method for accomplishing this transformation in the case of 3β-acetoxypregna-5,14-diene-20-one. The yield is 53% when the temperature is controlled at 5–10°. The yield is surprisingly high since C=C groups are also attacked by the reagent.

*Rearrangement of amides.*[8] Primary amides undergo oxidative rearrangement to isocyanates when treated with lead tetraacetate. The reaction is generally carried out in an alcohol (*t*-butanol generally), in which case the product is isolated as the carbamate. Triethylamine or stannic chloride catalyzes this reaction. This oxidation provides a useful alternative to the classical Hofmann, Schmidt, and Curtius rearrangements.

[1] Y. N. Ogibin, M. I. Katzin, and G. I. Nikishin, *Synthesis*, 889 (1974).

[2] B. M. Trost and K. Hiroi, *Am. Soc.*, 97, 6911 (1975).

[3] M. J. Harrison and R. O. C. Norman, *J. Chem. Soc. (C)*, 728 (1970) and references cited therein.

[4] R. J. Andersen and D. J. Faulkner, *Am. Soc.*, 97, 936 (1975).

[5] O. Hoshino, H. Hara, M. Ogawa, and B. Umezawa, *J.C.S. Chem. Comm.*, 306 (1975); *Chem. Pharm. Bull. Japan*, 23, 2578 (1975).

[6] R. G. Harvey and H. Cho, *J.C.S. Chem. Comm.*, 373 (1975).

[7]E. Yoshi, T. Koizumi, H. Ikeshima, K. Ozaki, and I. Hayashi, *Chem. Pharm. Bull. Japan*, **23**, 2496 (1975).

[8]H. E. Baumgarten, H. L. Smith, and A. Staklis, *J. Org.*, **40**, 3554 (1975).

## Lead tetraacetate—N-Chlorosuccinimide, 5, 370.

*Decarboxylation.*[1] Tetraasterane (3) has been prepared for the first time from the tetracarboxylic acid (1).[2] Degradation of (1) proved difficult, but

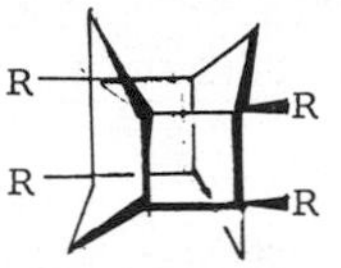

(1, R = COOH)
(2, R = Cl)
(3, R = H)

repeated reaction of (1) with lead tetraacetate and NCS in DMF—HOAc (5:1) according to the procedure of Grob (**5**, 370) gave (2) in about 20% yield. The tetrachloride was reduced to hydrocarbon (3) in 74% yield by sodium in boiling alcohol.

[1]H.-M. Hutmacher, H.-G. Fritz, and H. Musso, *Angew. Chem. internat. Ed.*, **14**, 180 (1975).
[2]Obtained as the dianhydride in 10% yield by irradiation of 3,6-dihydrophthalic anhydride.

## Lead tetraacetate—Trifluoroacetic acid.

*Bridgehead derivatives.*[1] Functional groups can be introduced into saturated hydrocarbons by treatment with lead tetraacetate and trifluoroacetic acid– methylene chloride at 20° followed by addition of a nucleophile.

*Examples:*

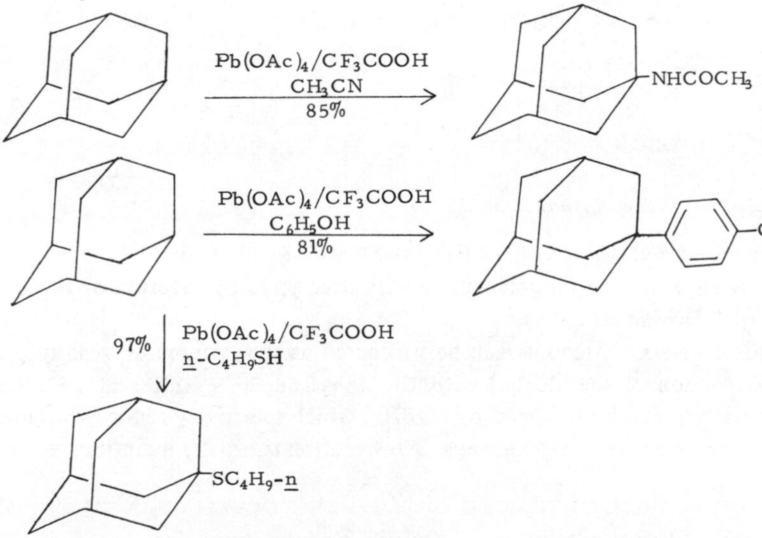

[1]S. R. Jones and J. M. Mellor, *Synthesis*, 32 (1976).

**Lead tetrakis(trifluoroacetate),** 2, 238–239; 4, 282–283; 5, 370–371.

*Reaction with allylic and homoallylic hydroxysteroids.* Westphal and Zbiral[1] have reported the reaction of lead tetrakis(trifluoroacetate) with a number of unsaturated hydroxy steroids.

*Examples:*

[1] D. Westphal and E. Zbiral, *Ann.*, 2038 (1975).

**Levulinic anhydride,** $(CH_3COCH_2CH_2CO)_2O$. Mol. wt. 214.21.

The anhydride can be prepared in quantitative yield by reaction of the acid (1, 564) with DCC in ether (5 hr.).

*Levulinic esters.*[1] Alcohols can be protected as the levulinates, readily prepared by reaction of the alcohol with the anhydride in pyridine at 25°. The protecting group can be removed by $NaBH_4$, which selectively reduces carbonyl groups in the presence of ester groups. This reaction is usually quantitative. This

$$ROH + (CH_3COCH_2CH_2CO)_2O \xrightarrow[65-90\%]{} ROCOCH_2CH_2COCH_3$$

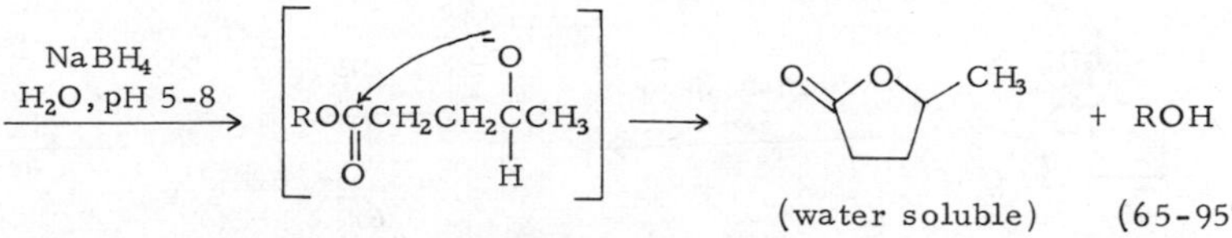

(water soluble)    (65-95%)

method was shown to be applicable to steroids and to several nucleosides. Levulinates are stable for some time to TFA.

[1] A. Hassner, G. Strand, M. Rubenstein, and A. Patchornik, *Am. Soc.*, **97**, 1614 (1975).

**Lindlar catalyst,** 1, 566–567; 3, 171–172; 4, 283.

*Preparation.*[1]

*Reduction of azides to amines.*[2] The reduction of azides to amines is usually carried out by catalytic hydrogenation or with lithium aluminum hydride. Neither method is selective. Corey *et al.* have used Lindlar catalyst for selective reduction of azides in the presence of carbonyl groups, double bonds, and benzyl ether groups. An acetylenic function is also reduced to the *cis*-olefin. Representative amines obtained in this way, with the yields, are formulated.

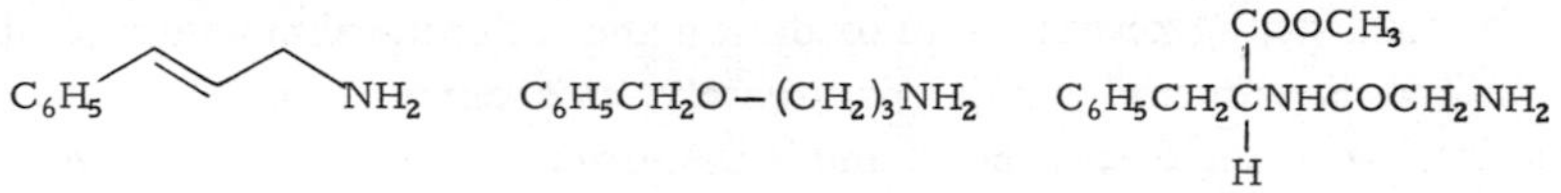

(93%)    (96%)    (97%)

[1] H. Lindlar and R. Dubuis, *Org. Syn.*, *Coll. Vol.* **5**, 880 (1975).
[2] E. J. Corey, K. C. Nicolaou, R. D. Balanson, and Y. Machida, *Synthesis*, 590 (1975).

**1-Lithiocyclopropyl phenyl sulfide,** 5, 372–373.

*Cyclobutyl annelation.* Trost and Keeley[1] have used this reagent for a double cyclobutyl annelation in a stereoselective synthesis of grandisol (8), a constituent of the boll weevil sex pheromone. The starting material (1) is obtained by the conjugate addition of thiophenol to methacrolein. This is converted into the cyclobutanone (2) by reaction with 1-lithiocyclopropyl

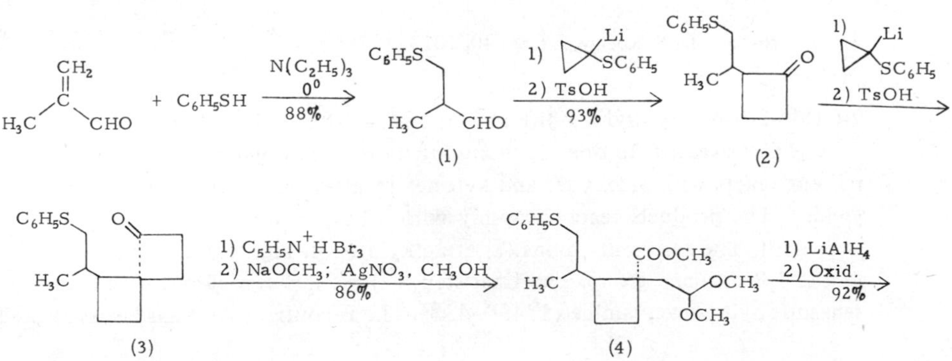

(5) → Wolff-Kishner red. 85% → (6) → 1) $H_3O^+$ 2) $LiAlH_4$ 98%

(7) → 1) $ClC_6H_4COOOH$ 2) 180° 84% → (8)

phenyl sulfide followed by acid-catalyzed rearrangement. The sequence is repeated to give the spiro[3.3]heptanone-1 (3), which is then converted into (4) by a procedure developed earlier by Trost (4, 212–213). The ester group of (4) is converted into a methyl group by standard techniques. The resulting methylcyclobutane (6) is hydrolyzed to the aldehyde and then reduced to the alcohol (7). The final step involves peracid oxidation and sulfoxide elimination to obtain the desired isopropenyl group. Fortunately, this elimination occurs without isomerization of the double bond and is therefore a useful method for introduction of an isopropenyl unit. Although the synthesis is lengthy, only the compounds formulated are purified, and yields in all steps are >80%. However, the NMR spectrum of the saturated methyl region of the final product indicates that it consists of grandisol and the isomeric fragranol (9) in the ratio 4:1.

(9)

[1]B. M. Trost and D. E. Keeley, *J. Org.*, **40**, 2013 (1975).

## 2-Lithio-2-trimethylsilyl-1,3-dithiane, 2, 184; 4, 284–285; 5, 374.

*α,β-Unsaturated ketones by homologation of aldehydes and ketones.*[1] The reagent reacts with aldehydes and ketones to give ketene thioacetals (1) in high yields.[2] The products react sluggishly with *n*-butyllithium in THF unless HMPT is present. The resultant anions (a) are alkylated in high yield at the 2-position of the 1,3-dithiane group. The final step, (2)→(3), is carried out with O-mesitylenesulfonylhydroxylamine (5, 430–433). The E-configuration has been assigned to (3) when the double bond is disubstituted and, tentatively, when the double bond is trisubstituted.

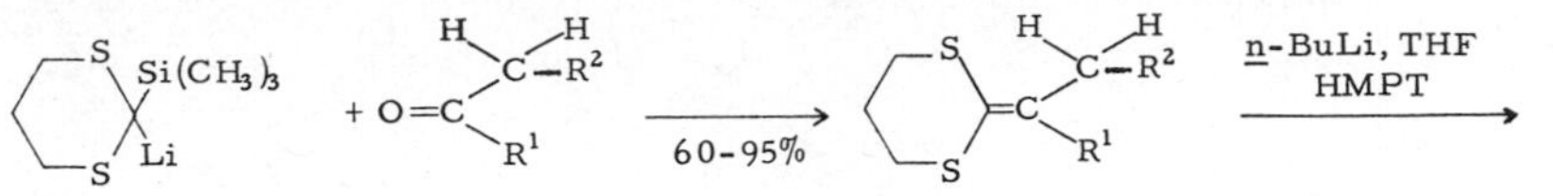

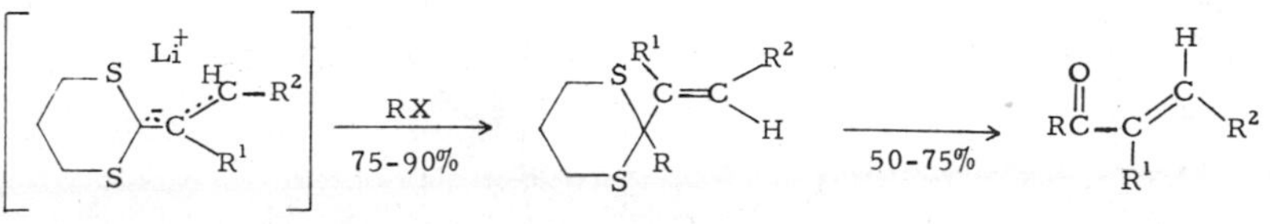

$$\text{(a)} \qquad \text{(2)} \qquad \text{(3)}$$

60-95%

n-BuLi, THF
HMPT

(1)

RX
75-90%

50-75%

*Cationic cyclizations.* Andersen *et al.*[3] have reported some interesting cyclizations by protonation of the ylidene dithiane (1) (equations I and II).

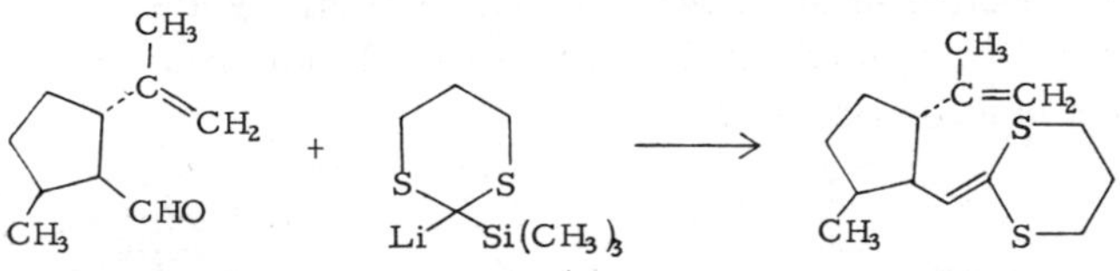

(1)

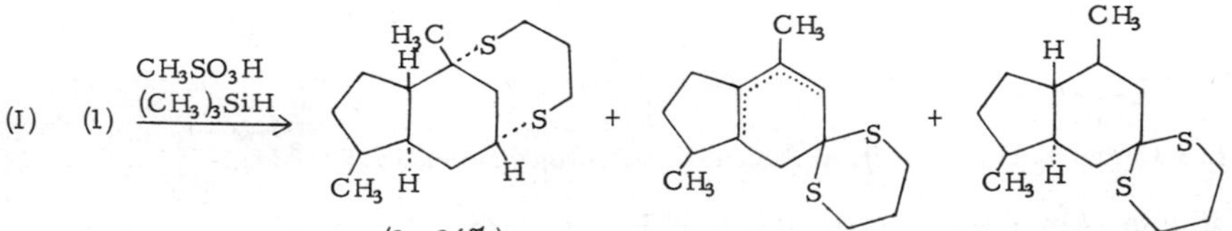

(I)  (1)  $\xrightarrow{\text{CH}_3\text{SO}_3\text{H} \atop (\text{CH}_3)_3\text{SiH}}$

(2, 26%)        (3, 20%)        (4, 15%)

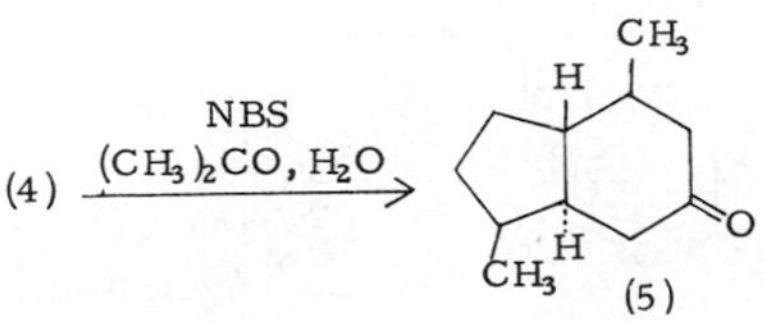

(4)  $\xrightarrow{\text{NBS} \atop (\text{CH}_3)_2\text{CO, H}_2\text{O}}$

(5)

[1] D. Seebach, M. Kolb, and B.-T. Gröbel, *Tetrahedron Letters*, 3171 (1974).
[2] *Idem, Ber.*, **106**, 2277 (1973).
[3] N. H. Andersen, Y. Yamamoto, and A. D. Denniston, *Tetrahedron Letters*, 4547 (1975).

**Lithium–Alkylamines**, 1, 575–581; 2, 241–242; 3, 175; 4, 287–288; 5, 377–379.

*Reductive cleavage of urethanes.* The last step in a synthesis of 2,7-epi-(±)-perhydrohistrionicotoxin (2) involves cleavage of the urethane group of (1) with excess lithium in methylamine at $-78°$.[1]

[1] E. J. Corey, Y. Ueda, and R. A. Ruden, *Tetrahedron Letters*, 4347 (1975).

**Lithium–Ammonia**, 1, 601–603; 2, 205; 3, 179–182; 4, 288–290; 5, 379–381.

*Reduction of dichlorocarbene adducts.*[1] Reduction of the *trans–cis* mixture of the dichlorocarbene adducts (2) of 2-alkoxy-3,4-dihydro-2H-pyranes (1) gives 3-alkoxy-2-oxanorcaranes (3) in yields of 62–78% if sodium benzoate is used as quencher (*see also* **5**, 379). Use of a protonic quencher such as an alcohol or ammonium chloride results in substantially lower yields. Note that compounds of type (3) cannot be prepared by Simmons–Smith cyclopropanation of (1), because these substances are very acid sensitive.

*Reductive alkylation of 2-furoic acid.* 2-Furoic acid (1) is reduced to 2,5-dihydro-2-furoic acid (2, R = H) by lithium (2.5 eq.) in liquid ammonia at

(1)                          (2)

−78° followed by addition of $NH_4Cl$ in 3 min. (80% yield). Addition of an alkyl halide instead of $NH_4Cl$ results in alkylation to give (2). Oxidation of (2) with lead tetraacetate results in loss of $CO_2$ and formation of a 2-alkylfuran.[2]

[1] A. J. Duggan and S. S. Hall, *J. Org.*, **40**, 2238 (1975).
[2] A. J. Birch and J. Slobbe, *Tetrahedron Letters*, 627 (1975).

## Lithium–Hexamethylphosphoric triamide.

*Cleavage of sulfonamides.*[1]  Sulfonamides are reductively cleaved at −30° by lithium in HMPT and a cosolvent such as *t*-butanol.

[1] T. Cuvigny and M. Larchevêque, *J. Organometal. Chem.*, **64**, 315 (1974).

## Lithium–Trimethylchlorosilane.

*Reductive coupling of alkanals.*  Two laboratories[1,2] have reported reductive coupling of alkanals (equation I). A similar reaction of aromatic aldehydes has

also been reported.[3]

[1] J. P. Picard, A. Ekouya, J. Dunogues, N. Duffaut, and R. Calas, *J. Organometal. Chem.*, **93**, 51 (1975).
[2] V. Rautenstrauch, *Synthesis*, 787 (1975).
[3] T. H. Chan and E. Vinokur, *Tetrahedron Letters*, 75 (1972).

**Lithium acetylide, 1**, 573–574; **5**, 382.

*Ethynylcarbinols.* Lithium acetylide can be prepared in high yield as a solution in THF by addition of *n*-butyllithium to acetylene at $-78°$. If the solution is allowed to warm to $0°$, a white precipitate of dilithium acetylide is formed by disporportionation. High yields of ethynylcarbinols can be obtained by reaction of aldehydes or ketones with the solution of lithium acetylide in THF at $-78°$. Lithium acetylide stabilized with ethylenediamine is less reactive than the reagent stabilized with THF.[1]

*Examples:*

$$\underline{n}\text{-}C_5H_{11}CHO \ + \ LiC\equiv CH \xrightarrow[\phantom{xx}]{\substack{THF \\ -78°}} \xrightarrow[98\%]{H_2O,\ 25°} \ \underline{n}\text{-}C_5H_{11}\overset{\overset{\displaystyle OH}{|}}{C}HC\equiv CH$$

$$+ \ LiC\equiv CH \xrightarrow[92\%]{THF \quad H_2O,\ 25°}$$

*Lithium ethynyltrialkylborates; synthesis of methyl ketones.*[2] Lithium acetylide reacts with trialkylboranes at $-78°$ (THF) to give lithium ethynyltrialkylborates (1). On treatment with hydrochloric acid one alkyl group migrates from boron to carbon to give (2). A second migration is prevented by addition of base (also at $-78°$). Oxidation of (2) then gives a methyl ketone (3) in yields of 70–90%. Alternatively, (2) can be converted into a 1,1-disubstituted olefin (4) by treatment with base and iodine (**5**, 346–347).

$$R_3B \ + \ LiC\equiv CH \longrightarrow Li[R_3BC\equiv CH] \xrightarrow[\substack{2)\ NaOH,\ -78°}]{1)\ HCl,\ -78°} \underset{R_2B}{\overset{R}{\diagdown}}C=CH_2$$

$$\underset{(1)}{\phantom{x}} \qquad\qquad \underset{(2)}{\phantom{x}}$$

$$\underset{R}{\overset{R}{\diagdown}}C=CH_2 \xleftarrow{NaOH,\ I_2} \underset{R_2B}{\overset{R}{\diagdown}}C=CH_2 \xrightarrow[70-90\%]{H_2O_2,\ NaOH} R\overset{\overset{\displaystyle O}{\|}}{C}CH_3$$

$$\underset{(4)}{\phantom{x}} \qquad\qquad \underset{(2)}{\phantom{x}} \qquad\qquad \underset{(3)}{\phantom{x}}$$

Note that hydroboration of a terminal alkyne proceeds by anti-Markovnikov addition to give the alkenylborane (5), which can be converted into an aldehyde (6) or a 1,2-disubstituted alkene (7).

$$RC\equiv CH + R'_2BH \longrightarrow \underset{H}{\overset{R}{>}}C=C\underset{BR'_2}{\overset{H}{<}} \xrightarrow[\text{NaOH}]{H_2O_2} RCH_2CHO$$

$$(5) \qquad\qquad\qquad (6)$$

$$\Big| I_2, \text{NaOH}$$

$$\underset{H}{\overset{R}{>}}C=C\underset{H}{\overset{R'}{<}}$$

$$(7)$$

Application of the sequence to vinyllithium results in the Markovnikov secondary alcohol (10):

$$R_3B + LiCH=CH_2 \longrightarrow [R_3BCH=CH_2]Li \xrightarrow[\text{2) NaOH}]{\text{1) HCl}} \underset{R_2B}{\overset{R}{>}}CHCH_3$$

$$(8) \qquad\qquad\qquad\qquad (9)$$

$$\xrightarrow[87-94\%]{\substack{H_2O_2 \\ \text{NaOH}}} \underset{OH}{RCHCH_3}$$

$$(10)$$

[1] M. M. Midland, *J. Org.*, **40**, 2250 (1975).
[2] H. C. Brown, A. B. Levy, and M. M. Midland, *Am. Soc.*, **97**, 5017 (1975).

**Lithium aluminum hydride, 1,** 581–595; **2,** 242; **3,** 176–177; **4,** 291–293; **5,** 382–389.

*Conjugated dienes.* Acetylenes of type (1) can be converted into 1,3-dienes (3) by way of α-allenic alcohols (2). This synthesis is not stereoselective, but yields are satisfactory.[1]

$$R^1-\underset{R^2}{\overset{OR}{C}}-C\equiv C-\underset{R^4}{\overset{OH}{C}}-R^3 \xrightarrow[\text{THF},20-40°]{\text{LiAlH}_4} R^1-\underset{R^2}{\overset{}{C}}=C=CH-\underset{R^4}{\overset{OH}{C}}-R^3 \xrightarrow[\text{THF, reflux}]{\text{LiAlH}_4} R^1-\underset{R^2}{\overset{}{C}}=CH-CH=\underset{R^4}{\overset{}{C}}-R^3$$

$$(1) \qquad\qquad\qquad (2) \qquad\qquad\qquad (3)$$

*N-Alkylation of sec-amines.*[2] The reaction of secondary amines with excess lithium aluminum hydride and an ethyl ester in THF or ether results in N-alkylation:

$$R^3C\underset{OC_2H_5}{\overset{O}{<}} + \text{LiAlH}_4 + HN\underset{R^2}{\overset{R^1}{<}} \longrightarrow \left[R^3\underset{OC_2H_5}{\overset{OAlH_3}{C}}-N\underset{R^2}{\overset{R^1}{<}}\right] \longrightarrow R^3C\underset{N}{\overset{O}{<}}\underset{R^2}{\overset{R^1}{<}}$$

$$\xrightarrow{\text{LiAlH}_4} R^3CH_2N\underset{R^2}{\overset{R^1}{<}}$$

*Examples:*

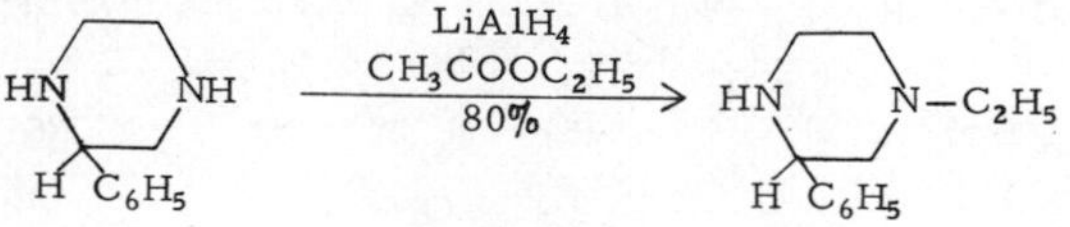

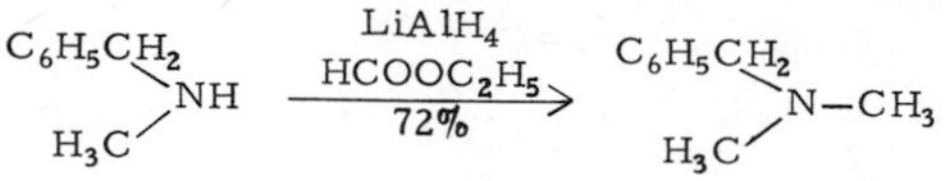

***Reduction of quinones to dihydrodiols.***[3] Reduction of some quinones of polynuclear hydrocarbons with $LiAlH_4$ in refluxing ether or with $NaBH_4$ in methanol leads mainly to *trans*-dihydrodiols rather than hydroquinones. The

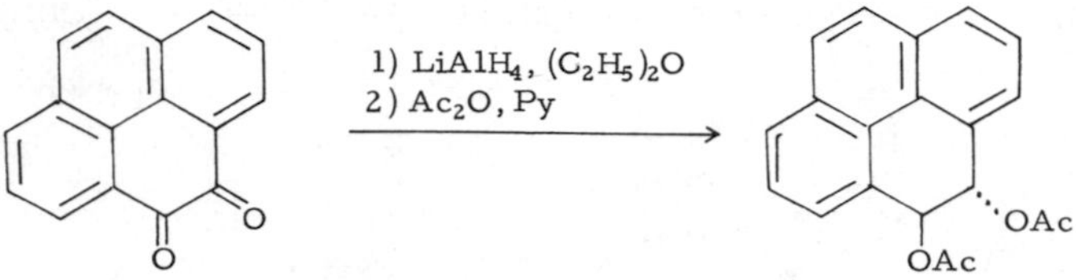

(1, 192 mg.)                    (2, 222 mg.)

factors controlling the course of reduction and the stereochemistry are not clear. The diols are very unstable and are best purified and stored as the diacetates. *See also* **Potassium borohydride**, this volume.

[1] A. Claesson, *Acta Chem. Scand.*, **B29**, 609 (1975).
[2] J. M. Khanna, V. M. Dixit, and N. Anand, *Synthesis*, 607 (1975).
[3] R. G. Harvey, S. H. Goh, and C. Cortez, *Am. Soc.*, **97**, 3468 (1975).

**Lithium aluminum hydride–Cuprous iodide.**
  *Conjugate reduction of enones.*[1] When $LiAlH_4$ in THF is added to a slurry of CuI in THF at $0°$, gas is evolved, and a deep black reagent is formed that is an active reagent for conjugate reduction of enones when the $LiAlH_4$/CuI/enone ratio is $1:4:1$.
  *Examples:*

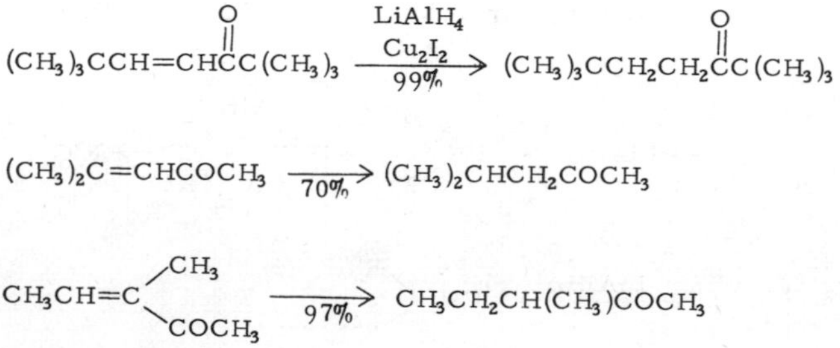

The actual reagent in these reductions is $H_2$ AlI. *cis*-Enones are reduced at a much slower rate than *trans*-enones; in fact, cyclohexenone is not reduced at all.

Lithium aluminum hydride in combination with titanium(III) chloride (**5**, 391–392) also reduces enones by 1,4-addition, but in somewhat lower yields.

*Compare* **Potassium tri-*sec*-butylborohydride—Cuprous iodide**, this volume.

[1] E. C. Ashby and J. J. Lin, *Tetrahedron Letters*, 4453 (1975).

**Lithium azide**, $LiN_3$. Mol. wt. 48.76. Supplier: Fisher.

*Reaction with tertiary bromides.* Edwards and Grieco[1] have presented evidence for $S_N 2$ displacement of tertiary bromine by azide ion in several cases.

[1] O. E. Edwards and C. Grieco, *Canad. J. Chem.*, **52**, 3561 (1974).

**Lithium bis(dialkylamino) cuprates**, $(R_2 N)_2 Cu^- Li^+$ (**1**). Grayish white, stable at 0° and below.

*Preparation.* These complexes are prepared by addition of 0.5 eq. of CuI to a solution of a lithium dialkylamide in ether or THF at −78 to 0°. The complex

of lithium pyrrolidide, , was used for the most part in the reactions described below.

*1,5-Dienes.* Japanese chemists[1] have prepared 1,5-dienes by reductive coupling of allylic halides with these complexes. For example, reaction of cyclohexylidenethyl bromide with the complex of lithium pyrrolidide and cuprous iodide in ether at 0° for 4 hr. gives the three products shown in equation I in essentially quantitative yield. The reaction is very sensitive, however, to the solvent and the temperature and also to the dialkylamine used for preparation of the catalyst.

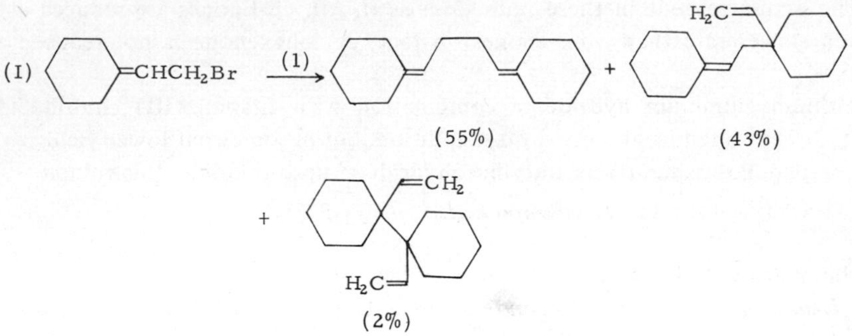

(I) $\bigcirc$=CHCH$_2$Br $\xrightarrow{(1)}$ (55%)  +  (43%)

+ (2%)

Further examples of this procedure are shown in equations II and III. This coupling reaction was used to synthesize all *trans*-squalene (2) in 50% isolated

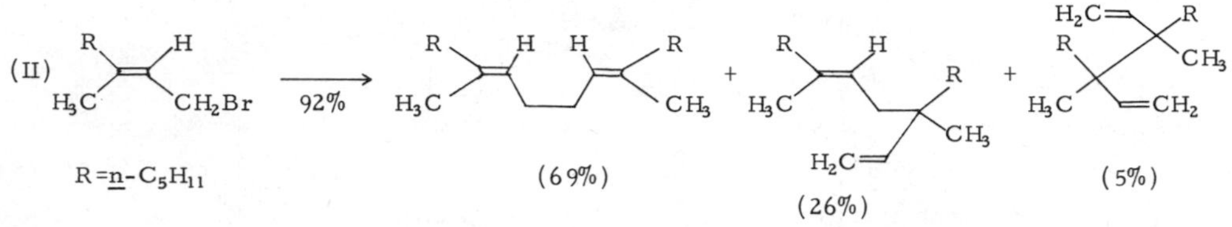

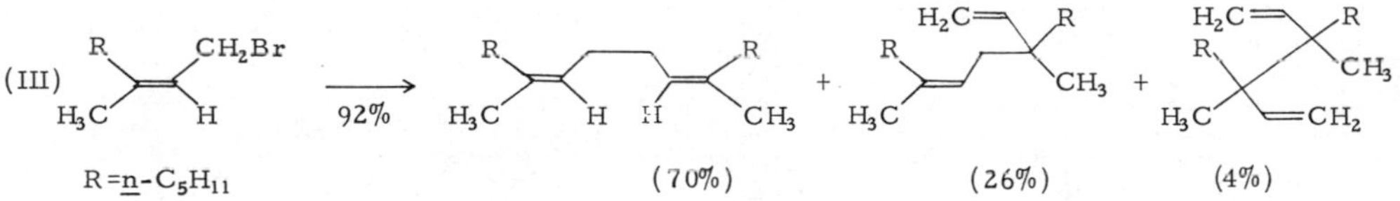

(II) [structure] $\xrightarrow{92\%}$ [structure] (69%) + [structure] (26%) + [structure] (5%)

R = $\underline{n}$-C$_5$H$_{11}$

(III) [structure] $\xrightarrow{92\%}$ [structure] (70%) + [structure] (26%) + [structure] (4%)

R = $\underline{n}$-C$_5$H$_{11}$

yield from *trans,trans*-farnesyl bromide (1).

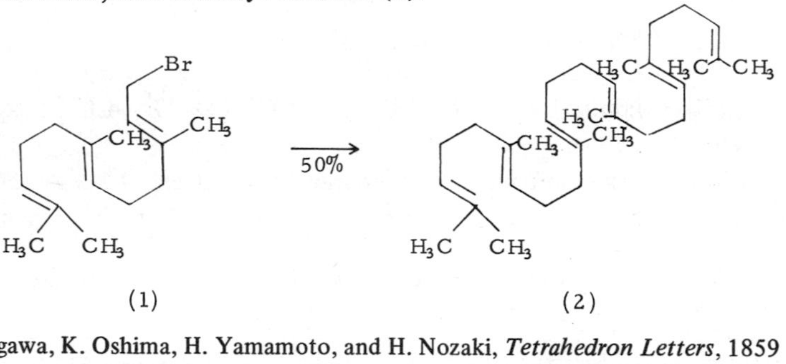

(1)                                           (2)

[1] Y. Kitagawa, K. Oshima, H. Yamamoto, and H. Nozaki, *Tetrahedron Letters*, 1859 (1975).

**Lithium bis(ethylenedioxyboryl)methide, (1).** Mol. wt. 161.69.
*Preparation*[1]:

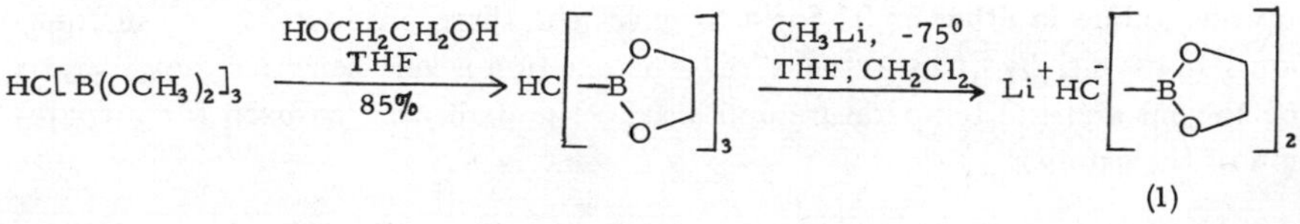

HC[ B(OCH$_3$)$_2$]$_3$ $\xrightarrow[85\%]{\substack{HOCH_2CH_2OH \\ THF}}$ HC$\left[-B\left(\begin{smallmatrix}O\\O\end{smallmatrix}\right)\right]_3$ $\xrightarrow{\substack{CH_3Li, -75^0 \\ THF, CH_2Cl_2}}$ Li$^+$HC$\left[-B\left(\begin{smallmatrix}O\\O\end{smallmatrix}\right)\right]_2^-$

(1)

*Conversion of RCOR' to RR'CHCHO.*[1] The carbanion (1) reacts with carbonyl compounds to give boronic esters (2), which can be oxidized under alkaline conditions to aldehydes (3). However, yields in this oxidation vary

considerably with the pH. Additions of buffers are usually necessary to obtain satisfactory results. However, yields of aldehydes synthesized by this route are often higher than those obtained by Wittig reactions.

Acid hydrolysis of (2) provides a route to boronic acids (4).

One disadvantage of this synthesis is that tris(dimethoxyboryl)methane is not readily available.[2]

[1] D. S. Matteson, R. J. Moody, and P. K. Jesthi, *Am. Soc.*, 97, 5608 (1975).
[2] D. S. Matteson, *Synthesis*, 147 (1975).

**Lithium α-carboethoxyvinyl(1-hexynyl)cuprate,** $Li^+C_4H_9C\equiv CC\bar{u}C\diagup^{CH_2}_{COOC_2H_5}$

(1)

Mol. wt. 250.72.

This organocuprate reagent is derived from ethyl α-bromoacrylate as formulated:

$$CuI + C_4H_9C\equiv CLi \xrightarrow[-LiI]{\underset{0^0}{(C_2H_5)_2O}} CuC\equiv CC_4H_9 \xrightarrow[-30^0]{CH_3Li}$$

$$Li^+CH_3C\bar{u}C\equiv CC_4H_9 + CH_2=C\diagup^{Br}_{COOC_2H_5} \longrightarrow (1)$$

The reagent reacts by transfer of the ethyl acrylate unit, the acetylenic ligand being unreactive.

The reagent reacts specifically with allylic and propargylic halides; alkyl and benzylic halides are unreactive.

$$CH_2=CHCH_2Br + (1) \xrightarrow[70\%]{\underset{overnight}{-78^0}} CH_2=CHCH_2C\diagup^{CH_2}_{COOC_2H_5}$$

$$\text{Br} + (1) \xrightarrow{50\%}$$ (cyclohexenyl)$-\text{C}\overset{\diagup \text{CH}_2}{\diagdown \text{COOC}_2\text{H}_5}$

$$\text{HC}{\equiv}\text{CCH}_2\text{Br} + (1) \xrightarrow{70\%} \text{HC}{\equiv}\text{CCH}_2\text{C}\overset{\diagup \text{CH}_2}{\diagdown \text{COOC}_2\text{H}_5}$$

The reagent has been used in a novel synthesis of α-methylene-γ-butyro-lactones, as shown for the synthesis of the lactone (5) in high overall yield.[1]

(cyclohexenyl)$-\text{C}\overset{\diagup \text{COOC}_2\text{H}_5}{\diagdown}\!\!\!\overset{\|}{\underset{\text{CH}_2}{}}$  $\xrightarrow{\text{OH}^-}$  (cyclohexenyl)$-\text{C}\overset{\diagup \text{COOH}}{\diagdown}\!\!\!\overset{\|}{\underset{\text{CH}_2}{}}$  $\xrightarrow[\text{KI}_3]{\text{NaHCO}_3}$

  (2)        (3)

(4) $\xrightarrow{(\underline{n}\text{-Bu})_3\text{SnH}}$ (5)

*Reaction with carbonyl compounds.* Lithium α-carbomethoxyvinyl(1-hexynyl)cuprate (2) differs from simple divinylcuprates in that it usually reacts with conjugated enones to give mainly, or even exclusively, products of 1,2-addition.[2]

*Examples:*

$$\text{CH}_3\text{COCH}{=}\text{CH}_2 + (2) \xrightarrow[75\%]{-78^0} \text{CH}_3\text{OOC}\overset{}{\underset{\overset{\|}{\text{CH}_2}}{\text{C}}}\text{CH}_2\text{CH}_2\text{COCH}_3$$

$$\text{C}_6\text{H}_5\text{CH}{=}\text{CHCOC}_6\text{H}_5 + (2) \xrightarrow{-78^0} \text{C}_6\text{H}_5\text{CH}{=}\text{CH}\overset{\text{OH}}{\underset{\text{H}_5\text{C}_6}{\overset{|}{\text{C}}}}{-}\underset{\overset{\|}{\text{CH}_2}}{\text{C}}\text{COOCH}_3 + \text{CH}_3\text{OOC}\underset{\overset{\|}{\text{CH}_2}}{\overset{\text{C}_6\text{H}_5}{\overset{|}{\underset{}{\text{C}}}\text{HCH}_2\text{COC}_6\text{H}_5}}$$

      (53%)      (17%)

(cyclohexenone) $+ (2) \xrightarrow[65\%]{-78^0}$ (cyclohexene with OH and $\underset{\overset{\|}{\text{CH}_2}}{\text{C}}\text{COOCH}_3$)

However, the related cuprate reagent (3), prepared from 2-lithio-3,3-diethoxy-propene[3] and the dimethyl sulfide complex of cuprous bromide, undergoes 1,4-addition to cyclic enones even at $-70^\circ$. The adduct of (3) with carvone was converted by hydrolysis and oxidation into the acrylic acid derivative (4), which would have been formed if conjugate addition of (2) were possible.[4]

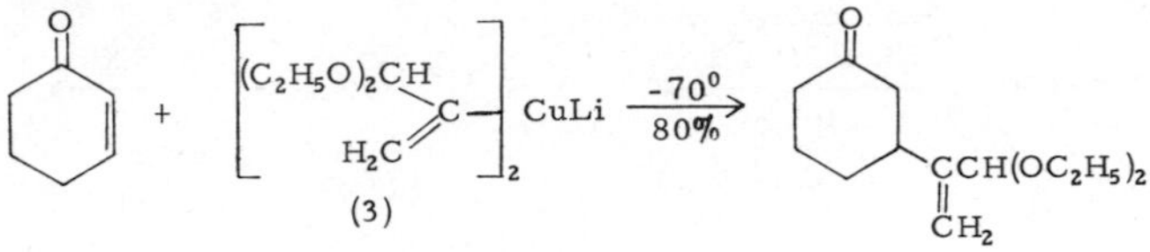

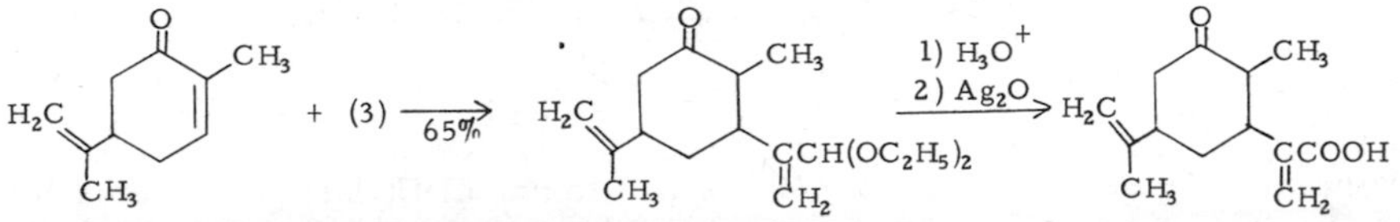

(4)

[1] J. P. Marino and D. M. Floyd, *Am. Soc.*, **96**, 7138 (1974).
[2] *Idem*, *Tetrahedron Letters*, 3897 (1975).
[3] J.-C. Depezay and Y. L. Merrer, *ibid.*, 2751 (1974).
[4] J. P. Marino and J. S. Farina, *Tetrahedron Letters*, 3901 (1975).

**Lithium diethylamide, 1,** 610–611; **2,** 247–248; **4,** 298; **5,** 398–399.

***Alkyl α,α-dichloroalkanoates.***[1] Alkyl dichloracetates can be metallated in THF–HMPT at $-78°$ with this base. The resulting anion (1) is alkylated by primary alkyl bromides at $-40°$ in high yield. Yields are lower with secondary alkyl bromides owing to decomposition of (1). Note that only fair yields are obtained with methyl α-lithiodichloroacetate because of concomitant Claisen self-condensation.

$$\underset{(R^1 = C_2H_5,\ \underline{i}\text{-}C_3H_7)}{CHCl_2COOR^1} + LiN(C_2H_5)_2 \longrightarrow \underset{(1)}{LiCCl_2COOR^1} \xrightarrow[50-95\%]{R^2Br} \underset{(2)}{R^2CCl_2COOR^1}$$

***Isopropyl α-chloroalkanoates.***[2] The products (2) obtained as shown above can be used for the synthesis of α-chlorocarboxylates (4). Treatment of (2) with lithium in THF at $-5$ to $0°$ gives the lithio anion (3), which can be alkylated with primary or secondary alkyl iodides or bromides (HMPT required for secondary alkyl halides). $R^1X$ can be replaced by acid chlorides, aldehydes, or ketones.

$$\underset{(2)}{RCCl_2COOC_3H_7\text{-}\underline{i}} \xrightarrow{Li,\ THF} \underset{(3)}{R\overset{Cl}{\underset{Li}{C}}COOC_3H_7\text{-}\underline{i}} \xrightarrow[70-90\%]{R^1X} \underset{(4)}{R\overset{Cl}{\underset{R^1}{C}}COOC_3H_7\text{-}\underline{i}}$$

[1] J. Villieras, J. R. Disnar, P. Perriot, and J. F. Normant, *Synthesis*, 524 (1975).
[2] J. Villieras, P. Perriot, M. Bourgain, and J. F. Normant, *ibid.*, 533 (1975).

**Lithium diethylamide–Hexamethylphosphoric triamide.**

*Alkylation of ketimines.*[1] The ketimine (1) can be converted into the

$$
\begin{array}{c}
CH_3 \\
\phantom{CH_3}\searrow \\
\phantom{CH_3CH_3}C=N-C_6H_{11} \\
CH_3\nearrow
\end{array}
\quad
\xrightarrow[-40^\circ]{\substack{\text{LiN}(C_2H_5)_2 \\ \text{HMPT}, C_6H_6}}
\quad
\begin{array}{c}
CH_3 \\
\phantom{CH_3}\searrow \\
\phantom{CH_3CH_3}C=N-C_6H_{11} \\
Li^+\bar{C}H_2\nearrow
\end{array}
\quad \xrightarrow{RX}
$$

$$
\text{(1)} \qquad\qquad\qquad\qquad\qquad\qquad \text{(2)}
$$

$$
\begin{array}{c}
CH_3 \\
\phantom{CH_3}\searrow \\
\phantom{CH_3CH_3}C=N-C_6H_{11} \\
RCH_2\nearrow
\end{array}
\quad \xrightarrow{H_3O^+} \quad RCH_2COCH_3
$$

$$
\text{(3)} \qquad\qquad\qquad\qquad \text{(4)}
$$

carbanion (2) by treatment at $-40^\circ$ with lithium diethylamide activated by HMPT. This anion can be alkylated at the same temperature; the products (3) are converted into methyl ketones (4) by acid hydrolysis.

*Examples:*

$$
\text{(1)} + C_7H_{15}Br \xrightarrow[77\%]{} CH_3COC_8H_{17}
$$

$$
\text{(1)} + CH_2=CHCH_2Br \xrightarrow[50\%]{} CH_3COCH_2CH_2CH=CH_2
$$

$$
\text{(1)} + Br(CH_2)_3Br \xrightarrow[62\%]{} CH_3CO(CH_2)_4Br
$$

The ketimine can also be converted into the dicarbanion (5). Dialkylation of this dianion furnishes, after hydrolysis, symmetrical ketones (6).

$$
(Li^+\bar{C}H_2)_2C=N-C_6H_{11} \xrightarrow[20-80\%]{\substack{1)\ 2\ RX \\ 2)\ H_3O^+}} RCH_2COCH_2R
$$

$$
\text{(5)} \qquad\qquad\qquad\qquad \text{(6)}
$$

*1,4-Diketones.*[2] This alkylation of ketimines has been extended to the synthesis of 1,4-diketones by use of a 2,3-dihalopropene as the alkylating reagent. The reaction is carried out with the carbanion, prepared as described above and then diluted with THF at $-60^\circ$. The resultant vinylic halide is then hydrolyzed with sulfuric acid at $-20$ to $0^\circ$.

*Examples:*

$$
\begin{array}{c}
C_6H_5 \\
\phantom{C_6H_5}\searrow \\
\phantom{C_6H_5C_6H_5}C=N-C_6H_{11} \\
Li^+\bar{C}H_2\nearrow
\end{array}
\quad
\xrightarrow[74\%]{\substack{1)\ ClCH_2CCl=CH_2 \\ 2)\ H_3O^+}}
C_6H_5COCH_2CH_2CCl=CH_2
$$

$$
\xrightarrow[75\%]{H_2SO_4,\ 0^\circ} C_6H_5COCH_2CH_2COCH_3
$$

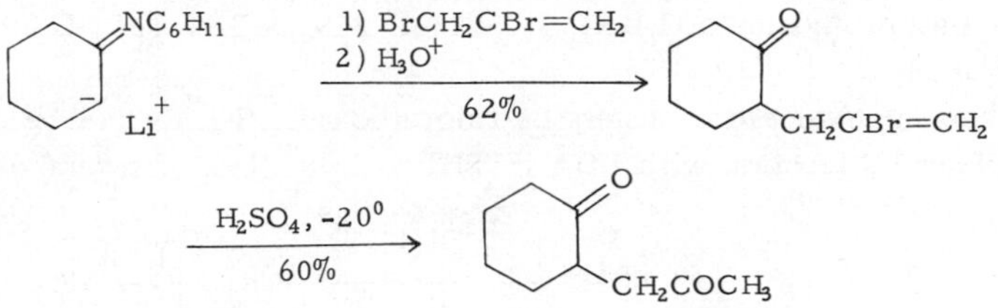

1,4-Diketones can also be prepared by the reaction of lithiated ketimines with epoxides to give, after acid hydrolysis, 1,4-ketols, which are usually not isolated because of their ready dehydration, but are oxidized by Jones reagent to 1,4-diketones.

*Examples:*

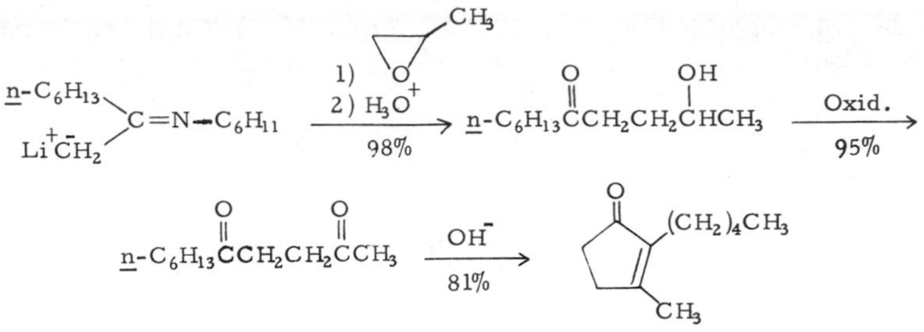

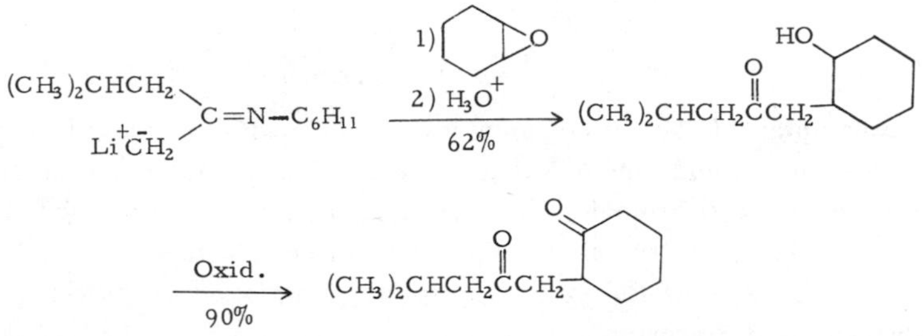

The first example formulates use of this method for the synthesis of dihydro-jasmone in high yield. The second example shows that even a relatively unreactive oxide (cyclohexene oxide) can be used.

Symmetrical 1,4-diketones can be prepared by coupling the anion of the imine by reaction with iodine followed by acid hydrolysis.

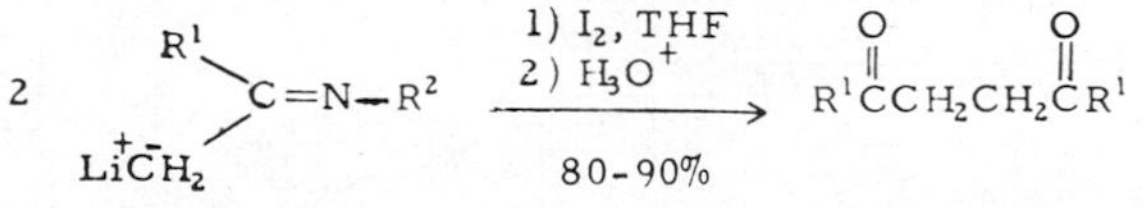

[1] T. Cuvigny, M. Larchevêque, and H. Normant, *Compt. Rend.*, **277**, 511 (1973).
[2] *Idem, Tetrahedron Letters*, 1237 (1974).

**Lithium diisopropylamide (LDA)**, **1**, 611; **2**, 249; **3**, 184–185; **4**, 298–302; **5**, 400–406.

*β-Hydroxy thiol esters.* Esters of thiocarboxylic S-acids can be converted into enolates by reaction with LDA in THF at $-78°$. Use of several other bases

$$R^1_2CHCSR \ + \ R^2CR^3 \ \xrightarrow[\substack{1) \ LDA, \ -78^0 \\ 2) \ H_3O^+ \\ \hline 40-80\%}]{} \ R^2C-C-CSR$$

(potassium *t*-butoxide, triphenylmethyllithium) results in elimination to ketenes. The enolates react with aromatic aldehydes and aliphatic and aromatic ketones to form β-hydroxy thiol esters.[1]

*Isomerization of* cis-*cyclooctene oxide* (**2**, 247; **4**, 309). In the original work on the isomerization of *cis*-cyclooctene oxide (1) with lithium diethylamide, Cope reported that (2) and (3) are formed in a 4 : 1 ratio. In recent work[2]

(1)                (2)                (3)

$LiN(C_2H_5)_2$     (2):(3) = 20:80

$LiN(C_2H_5)_2-LiBr$    (2):(3) = 80:20

$LiN(\underline{i}-C_3H_7)_2$    (2):(3) = 98:2

$Li_3PO_4$      only (3) is formed.

a 1 : 4 ratio was obtained. The discrepancy was then shown to be a result of the presence of lithium bromide in Cope's experiments (his *n*-BuLi was generated from lithium and *n*-butyl bromide). If LDA is used as base, (2) is isolated in 80% yield; less than 2% of the allylic alcohol is formed. Note that the allylic alcohol can be obtained as the exclusive product if potassium *t*-butoxide or lithium phosphate is used as base.

*Cyclization of a δ-chloro ester to a cyclobutane derivative.* A key step in a recent synthesis of (±)-grandisol (5) involves intramolecular alkylation of (1) to give a mixture of the isomeric esters (2) and (3). The desired ester (3) can be obtained as the major product by effecting cyclization with LDA as base at $0°$ in THF containing HMPT.[3]

(1)                (2)        35:65        (3)

(5)                          (4)

*α-Lithio selenoxides and selenides.* Selenides and selenoxides are usually partially or completely cleaved rather than deprotonated by alkyllithium reagents. However, Reich and Shah[4] found that phenyl alkyl selenoxides are deprotonated by LDA at $-78°$ and that the anions formed react with aldehydes, ketones, and the more reactive alkyl halides at this temperature. The products in the case of selenides are then oxidized at low temperature (*m*-chloroperbenzoic acid at $-10°$ or ozone at $-78°$) and are added to refluxing $CH_2Cl_2$ or $CCl_4$ to effect selenoxide elimination. Olefins, dienes, and allylic alcohols can be prepared in this way.

*Examples:*

*Mixed malonates.*[6] Ethyl chloroformate reacts with the α-anion of *t*-butyl acetate to give ethyl *t*-butyl malonate in 70–75% yield.[7] The reaction is a general route to mixed malonates. Ethyl chloroformate can be replaced by other

$$CH_3COOC(CH_3)_3 + ClCOOC_2H_5 \xrightarrow[\text{70-75\%}]{\substack{LDA \\ THF}} CH_2 \begin{smallmatrix} COOC_2H_5 \\ COOC(CH_3)_3 \end{smallmatrix}$$

electrophiles: $(RO)_2POCl$, $C_6H_5SSC_6H_5$, $C_6H_5SeBr$.

*Regiospecific aldol condensations.*[8] Methyl ketones can be converted into the kinetic lithium enolates by lithium diisopropylamide in THF at $-78°$. These can be trapped by aldehydes to give aldols:

$$CH_3CH_2CH_2COCH_3 \xrightarrow[\text{THF}, -78°]{LiN(i\text{-}Pr)_2} CH_3CH_2CH_2\overset{\overset{O^-\ Li^+}{|}}{C}=CH_2 + CH_3CH_2CH_2CHO \xrightarrow{65\%}$$

$$CH_3CH_2CH_2COCH_2\overset{\overset{OH}{|}}{C}HC_3H_7\text{-}\underline{n} \xrightarrow[\text{72\%}]{\substack{p\text{-}TsOH \\ C_6H_6}} CH_3CH_2CH_2COCH=CHC_3H_7\text{-}\underline{n}$$

$\alpha,\beta$-Unsaturated ketones can also be used in regiospecific aldol condensation:

$$CH_3CH=C\overset{\overset{O^-\ Li^+}{|}}{H}C=CH_2 + CH_3CH=CHCHO \xrightarrow{70\%} CH_3CH=CH\overset{\overset{O}{||}}{C}CH_2\overset{\overset{OH}{|}}{C}HCH=CHCH_3$$

Note that alkylation of the kinetic lithium enolates of methyl ketones with alkyl halides leads to mixtures and is generally of little synthetic value.

*Metalation of 3-picoline.* 3-Picoline (1) has been metalated by LDA in ether,[9] but the reaction proceeds more rapidly and in higher yield when LDA in HMPT–THF is used.[10] Reaction of (a) with alkyl bromides also proceeds readily under the same conditions.

(1)                    (a)                              (2)

*Primary and secondary lithium derivatives of nitrosamines.*[11] Primary and secondary lithium derivatives of nitrosamines of types (1) and (2) can be prepared in high yield by metalation with LDA in THF at $-80°$. Metalation is

(1)

(2)

accelerated by addition of HMPT. Use of an alkyllithium or an aryllithium generally gives lower yields owing to competing attack at the nitroso group. The products (1) and (2) undergo electrophilic substitution as expected at temperatures of less than $0°$.

*Examples:*

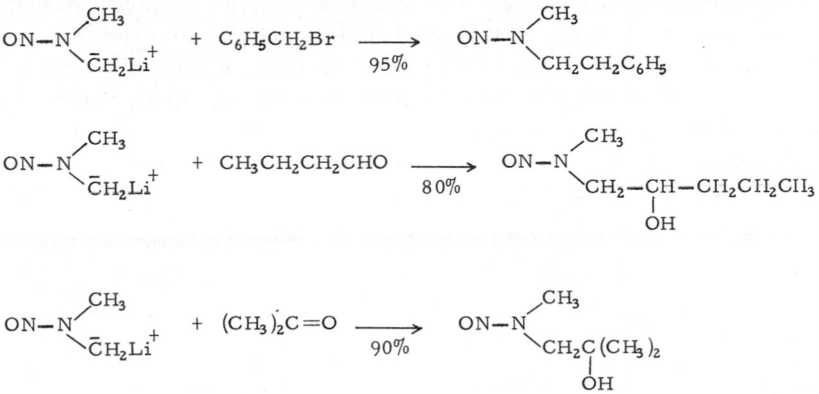

The products can be denitrosated by treatment with gaseous hydrogen chloride or with Raney nickel. The overall consequence then is that secondary amines can be substituted by electrophiles at the $\alpha$-position. Since many nitrosamines are potent carcinogens, Seebach has developed a "one-pot" reaction as shown in equation I. The secondary amine is treated consecutively with the reagents as shown, and the reaction is worked up only after completion of the

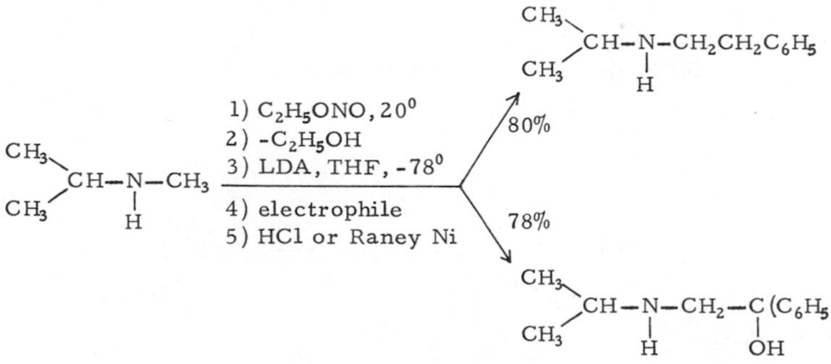

five steps.

At the present time only the nitroso group has been found suitable for conversion of secondary amines into nucleophilic reagents at the $\alpha$-carbon atom. $\alpha$-Amino carbanions ($\underset{}{\overset{R}{\underset{H}{>}C-N<}}$) cannot be prepared directly.

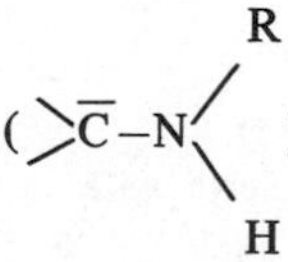

*Allyl selenide anions.*[12] Allyl phenyl selenides, $C_6H_5SeCH_2CH=CR^1R^2$, are converted into α-lithio selenides by LDA in THF at -78 to 0°. The anions are nucleophiles and undergo alkylation predominantly at the α-position to give allyl selenides. When these allyl selenides are oxidized with hydrogen peroxide, allyl selenoxides are formed. These undergo [2.3] sigmatropic rearrangement to give eventually an allylic alcohol. This synthesis of allylic alcohols is related to one developed by Evans *et al.* (5, 400–401) from allylic sulfoxides, but in the latter case a thiophile is required to effect cleavage of the phenyl sulfoxide group. In the selenoxide procedure, the shift occurs more rapidly than selenoxide elimination.

*Examples:*

$$\overset{\overset{\text{Li}}{|}}{C_6H_5Se CHCH=CHCH_3} \ + \ C_6H_5CH_2CH_2Br \ \longrightarrow \ C_6H_5CH_2CH_2\underset{\underset{SeC_6H_5}{|}}{CHCH=CHCH_3}$$

(E and Z isomers)

$$\xrightarrow{H_2O_2} \left[ C_6H_5CH_2CH_2\underset{\underset{O \leftarrow SeC_6H_5}{|}}{CHCH=CHCH_3} \right] \xrightarrow[\text{overall}]{80\%} C_6H_5CH_2CH_2CH=CHC\overset{\diagup OH}{\underset{\diagdown CH_3}{H}} + C_6H_5Se\overset{O}{\overset{\uparrow}{O}}H$$

$$\overset{\overset{\text{Li}}{|}}{C_6H_5Se CHCH=CHC_6H_5} \ + \ CH_3COCH_3 \ \longrightarrow \ HO-\underset{\underset{CH_3}{|}}{\overset{\overset{CH_3}{|}}{C}}-\underset{\underset{SeC_6H_5}{|}}{CHCH=CHC_6H_5}$$

$$\xrightarrow[55\%]{H_2O_2} HO-\underset{\underset{CH_3}{|}}{\overset{\overset{CH_3}{|}}{C}}CH=CHCHOHC_6H_5$$

$$\overset{\overset{\text{Li}}{|}}{C_6H_5Se CHCH=C(CH_3)Cl} \ + \ C_6H_5CH_2Br \ \longrightarrow \ C_6H_5CH_2\underset{\underset{SeC_6H_5}{|}}{CHCH=C(CH_3)Cl}$$

$$\xrightarrow[70\%]{H_2O_2} C_6H_5CH_2CH=CHCOCH_3$$

[1] J. Wemple, *Tetrahedron Letters*, 3255 (1975).
[2] J. K. Whitesell and P. D. White, *Synthesis*, 602 (1975).
[3] J. J. Babler, *Tetrahedron Letters*, 2045 (1975).
[4] H. J. Reich and S. K. Shah, *Am. Soc.*, 97, 3250 (1975).
[5] The selenides are prepared by reaction of $C_6H_5SeNa$ with an appropriate halide or mesylate.

[6]T. J. Brocksom, N. Petragnani, and R. Rodrigues, *J. Org.*, **39**, 2114 (1974).

[7]T. J. Brocksom, N. Petragnani, R. Rodrigues, and H. L. S. Teixeira, *Org. Syn.*, submitted (1975).

[8]G. Stork, G. A. Kraus, and G. A. Garcia, *J. Org.*, **39**, 3459 (1974).

[9]A. D. Miller, C. Osuch, N. N. Goldberg, and R. Levine, *Am. Soc.*, **78**, 674 (1956); W. B. Edwards, III, *J. Heterocyclic Chem.*, **12**, 413 (1975).

[10]E. M. Kaiser and J. D. Petty, *Synthesis*, 705 (1975).

[11]D. Seebach and D. Enders, *Angew. Chem. internat. Ed.*, **14**, 15 (1975).

[12]H. J. Reich, *J. Org.*, **40**, 2570 (1975).

**Lithium dimethoxyphosphinylmethylide,** $LiCH_2\overset{O}{\overset{\|}{P}}(OCH_3)_2$ (1). Mol. wt. 129.99.
*Preparation*[1]:

$$CH_3I + (CH_3O)_3P \xrightarrow[-CH_3I]{reflux} CH_3\overset{O}{\overset{\|}{P}}(OCH_3)_2 \xrightarrow{\substack{n-BuLi \\ THF, -78°}} (1)$$

*Cyclopentenones.* Heathcock *et al.*[2] have developed a cyclopentenone synthesis that involves an intramolecular Wittig–Horner reaction of $\beta,\epsilon$-diketophosphonates. For example, the reaction of 2 eq. of (1) with the ester (2) in THF gives the diketone (3), after hydrolysis of the protective group. When treated with sodium hydride in DME at $20°$, (3) cyclizes to the cyclopentenone (4) in 80% yield.

This route was used for a synthesis of *cis*-jasmone (8) from (5) as shown in the formulation.

[1]E. J. Corey and G. T. Kwiatkowski, *Am. Soc.*, **88**, 5654 (1966).

[2]R. D. Clark, L. G. Kozar, and C. H. Heathcock, *Syn. Commun.*, 1 (1975).

**Lithium diphenylphosphide, 4,** 303–304; **5,** 408–410.

*Allylic phosphine oxides; Wittig–Horner synthesis of 1,3-dienes.*[1] Allylic phosphine oxides can be prepared by reaction of 2,6-dichlorobenzoates[2] of allylic alcohols with lithium diphenylphosphide followed by oxidation. Thus

the pure phosphine oxide (3) can be prepared in >80% yield from (1). The ylide of (3) reacts at $C_\alpha$ with cyclohexanone to give an adduct that eliminates lithium diphenylphosphinate at 25° to give the Z-diene (4) containing less than 3% of the E-isomer.

This modified Wittig–Horner reaction was used for synthesis of 3-desoxyvitamin $D_2$ (5).

Davidson and Warren[3] have used the diphenylphosphinyl group as an activating group in synthesis of 1,3-dienes from an alkyl halide and two carbonyl compounds as outlined in equation I.

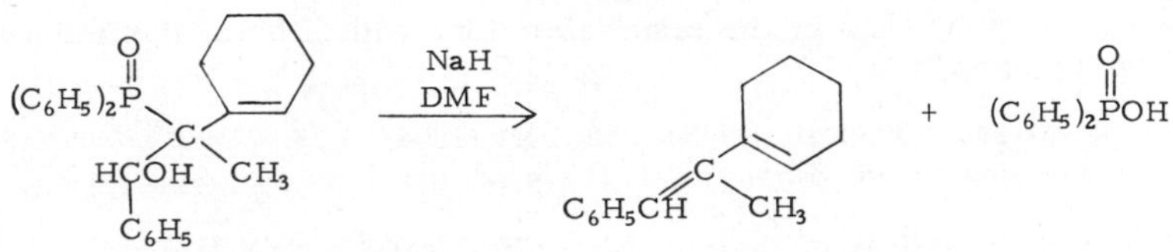

$$(C_6H_5)_2\overset{O}{\underset{}{P}} \text{—} C \underset{\underset{C_6H_5}{|}}{\overset{\overset{}{|}}{\underset{}{}}} \xrightarrow[\text{DMF}]{\text{NaH}}$$

**Debromination.** Reaction of 1,2-dibromoethylene (*cis/trans* = 64:36) in refluxing THF with 2 moles of lithium diphenylphosphide gives acetylene in 80% yield. This debromination also occurs with lithium diphenylarsenide.[4]

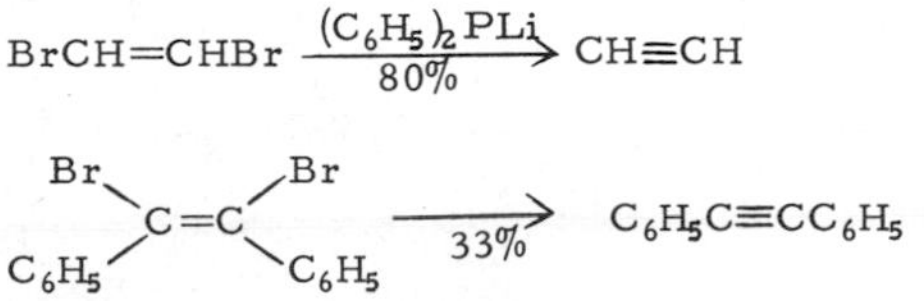

$$\text{BrCH=CHBr} \xrightarrow[80\%]{(C_6H_5)_2PLi} \text{CH≡CH}$$

The reaction of lithium diphenylphosphide with 1,2-dichloroethylene follows a different course[5]:

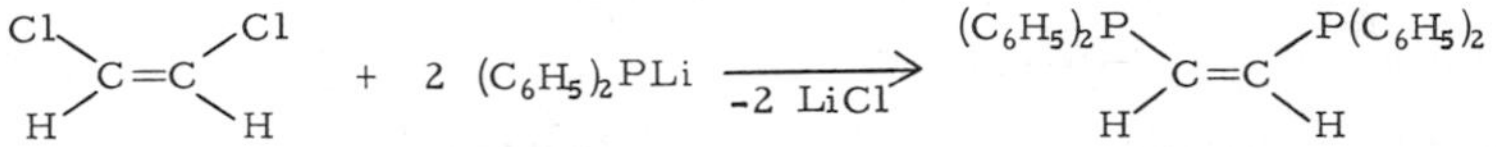

[1] B. Lythgoe, T. A. Moran, M. E. N. Nambudiry, S. Ruston, J. Tideswell, and P. W. Wright, *Tetrahedron Letters*, 3863 (1975).
[2] G. Stork and W. N. White, *Am. Soc.*, 78, 4609 (1956) have used 2,6-dichlorobenzoates of allylic alcohols for displacement reactions without rearrangement.
[3] A. H. Davidson and S. Warren, *J.C.S. Chem. Comm.*, 148 (1975).
[4] D. G. Gillespie and B. J. Walker, *Tetrahedron Letters*, 4709 (1975).
[5] A. M. Aguiar and D. Daigle, *Am. Soc.*, 86, 2299 (1964).

**Lithium methoxyaluminum hydride**, $LiAlH_3OCH_3$. Mol. wt. 67.98.

This reducing agent is mentioned by Brown and Shoaf,[1] but the main interest in this paper is centered on lithium trimethoxyaluminum hydride.

It is prepared by slow addition of anhydrous methanol (0.01 mole) to 20 ml. of an ice-cold 0.5 $M$ solution of lithium aluminum hydride in THF.[2]

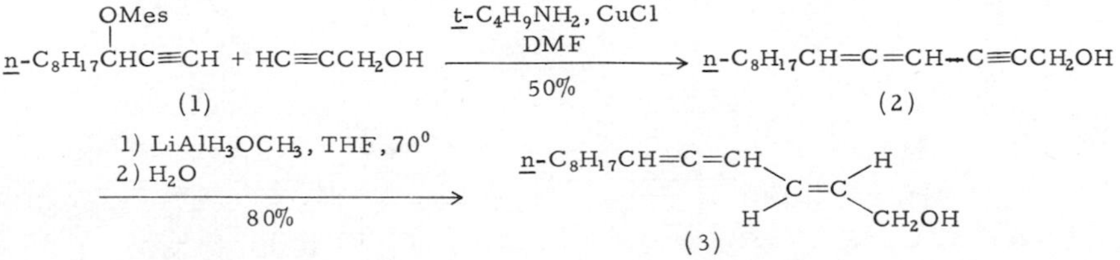

*Selective reduction of an yneallene.* In a recent synthesis of the vinylallene (3), Baudouy and Gore[2] prepared (2) by Cadiot coupling of the acetylenic mesylate (1) with propargyl alcohol. This product was reduced in high yield and stereospecifically to the (E)-eneallene system. This reduction had been carried out previously in the same laboratory with lithium aluminum hydride (yields 50–90%).

[1] H. C. Brown and C. J. Shoaf, *Am. Soc.,* **86,** 1079 (1964).
[2] R. Baudouy and J. Gore, *Synthesis,* 573 (1974).

## Lithium methylsulfinylmethylide (Dimsyllithium), $CH_3SOCH_2^-Li^+$.

$N^8$-*Alkylation of 7,8-dihydropteridines.* Direct monoalkylation at $N^8$ of 2,4-diamino-7,8-dihydropteridines (1) can be accomplished by treatment with 1.1 eq. of *n*-butyllithium in DMSO (the actual reagent is lithium methylsulfinylmethylide) and then with a slight excess of an alkyl halide. No substitution in

(1)       1) n-BuLi, DMSO    2) RX    60–80%       (2)

the pyrimidine ring or amino groups is observed.

[1] M. Chaykovsky, *J. Org.,* **40,** 145 (1975).

## Lithium methyl(vinyl)cuprate, $CH_3(CH_2=CH)CuLi$ (1). Mol. wt. 112.56.

This mixed cuprate is prepared *in situ* by addition of 1 eq. of vinyllithium and then 1 eq. of methyllithium to 1.1 eq. of cuprous iodide in THF at $-35°$.

*trans-2,3-Dialkylcyclopentanones.* Posner *et al.*[1] have developed an efficient method for conversion of 2-cyclopentenone (2) into *trans*-2,3-dialkylcyclo-pentanones, a system present in 11-desoxyprostaglandins. This reagent in THF solution selectively transfers the vinyl group to $C_3$ of (2) in nearly ideal stoichiometry to form the enolate ion (3). This ion undergoes substitution by a variety of electrophiles at $C_2$; *trans*-2,3-dialkylcyclopentanones are formed preferentially. Variations in time and especially temperature are critical factors. Several examples are formulated. By-products include *cis*-2,3-dialkylcyclo-pentanones, 2,4-dialkylcyclopentanones, and 3-vinylcyclopentanone.

It is noteworthy that in diethyl ether both the methyl and vinyl groups of (1) are transferred to $C_3$ of 2-cyclopentenone. Also note that even in THF (1) undergoes coupling reactions with acid chlorides and epoxides with preferential transfer of the methyl group. At the present time the selectivity of organic group transfer of mixed lithium dialkyl cuprates is not well understood.[2]

[1]G. H. Posner, J. J. Sterling, C. E. Whitten, C. M. Lentz, and D. S. Brunelle, *Am. Soc.*, **97**, 107 (1975).

[2]W. H. Mandeville and G. M. Whitesides, *J. Org.*, **39**, 400 (1974).

**Lithium phenylethynolate,** $C_6H_5C\equiv CO^-Li^+ \leftrightarrow C_6H_5\bar{C}=C=OLi^+$. Mol. wt. 124.07.
   *Preparation:*

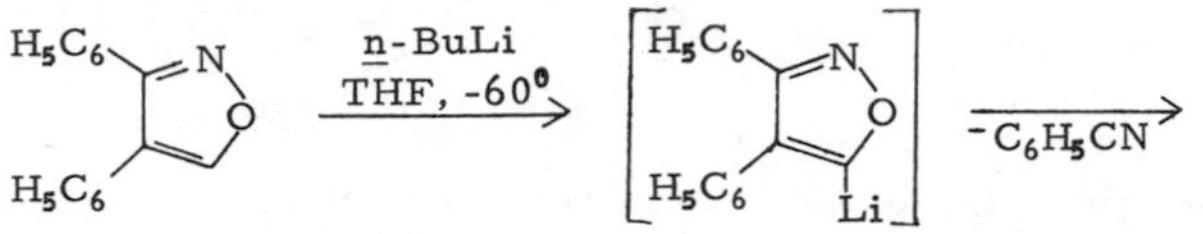

$\beta$-*Lactones*. Highly substituted $\beta$-lactones (2) can be prepared by reaction of (1) with aldehydes and ketones. These undergo quantitative decarboxylation at $140°$ to give olefins (3). The yield in the alkylation step is improved by addition of HMPT.[1]

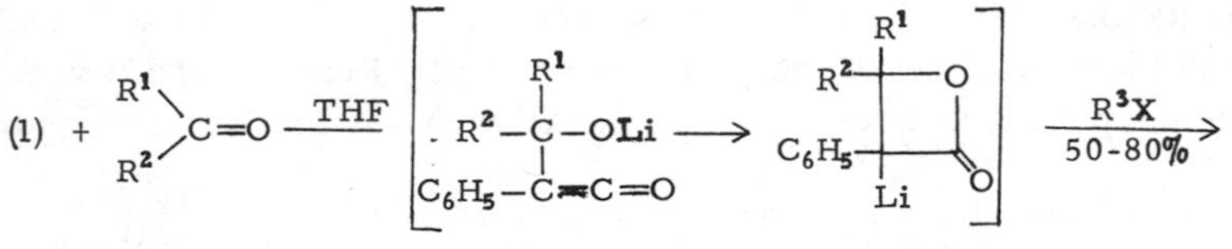

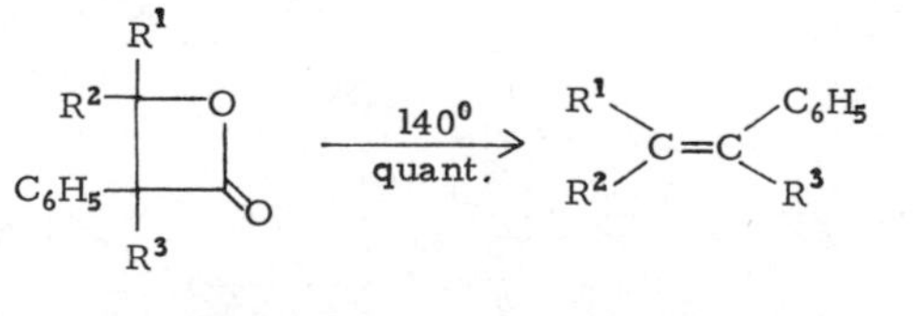

[1]U. Schöllkopf and I. Hoppe, *Angew. Chem. internat. Ed.*, **14**, 765 (1975).

**Lithium phenylthio(alkyl)cuprates, 5,** 414–415.
   $\beta$-*Alkyl-*$\alpha$,$\beta$-*unsaturated ketones*.[1] $\beta$-Bromo-$\alpha$,$\beta$-unsaturated ketones[2] react smoothly with lithium phenylthio(alkyl)cuprates in THF at $-78$ to $0°$ to give $\beta$-alkyl-$\alpha$,$\beta$-unsaturated ketones (70–95% yield). Lithium dialkylcuprates can

be used, but yields are somewhat lower. Products of further addition to the β-alkylenone are not observed with mixed cuprate reagents.

*Examples:*

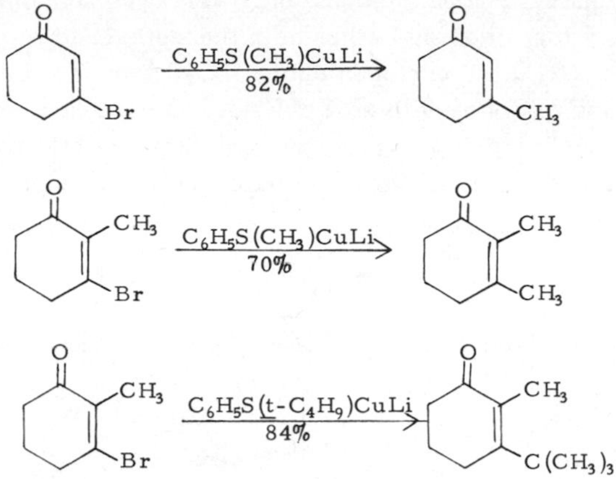

[1] E. Piers and I. Nagakura, *J. Org.,* **40,** 2694 (1975).
[2] Preparation, *see* **Triphenylphosphine dihalides** [*Idem, Syn. Comm.,* 5, 193 (1975)].

**Lithium piperidide,** LiN⟨ ⟩ . Mol. wt. 91.08.

*α-Chlorocycloalkanones.* Cyclic ketones can be converted into the homologous α-chloro ketones by a two-step process illustrated for conversion of cyclohexanone (1) into α-chlorocycloheptanone (3). Treatment with dichloromethyllithium (**4,** 138–139) gives the carbinol (2). Reaction of (2) with 2 eq. of lithium piperidide gives a carbenoid (a), which is unstable and undergoes

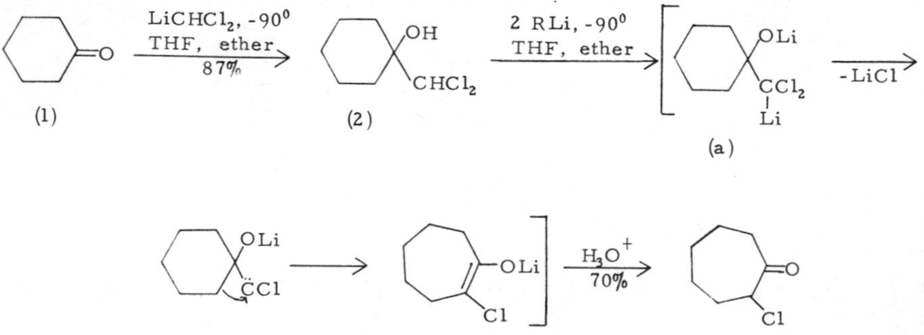

α-elimination to the carbene (b). This intermediate rearranges to the lithium enolate of α-chlorocycloheptanone (c). *n*-Butyllithium is less satisfactory than lithium piperidide in this process.[1]

The reaction is also applicable to acyclic $\alpha,\alpha$-dichloro- and $\alpha,\alpha,\alpha$-trichloro-carbinols; two products can be formed depending on the migratory aptitudes of the groups attached to the carbon bearing the hydroxyl group. The migratory aptitudes have been found to be $C_6H_5 > H > R$.

*Examples:*

$$(CH_3)_2CHCHCHCl_2 \xrightarrow[60\%]{} (CH_3)_2CHCOCH_2Cl$$
$$\overset{|}{O}Li$$

$$C_6H_5CHCHCl_2 \xrightarrow[90\%]{} C_6H_5CHClCHO + C_6H_5COCH_2Cl$$
$$\overset{|}{O}Li \qquad\qquad (65\%) \qquad\qquad (35\%)$$

$$(C_2H_5)_2CCHCl_2 \xrightarrow[58\%]{} (C_2H_5)_2C(OH)CHO + C_2H_5CHClCOC_2H_5$$
$$\overset{|}{O}Li \qquad\qquad (67\%) \qquad\qquad (33\%)$$

Note, however, that reaction of lithium piperidide with (4), prepared by reaction of cyclohexanone with dichloromethyllithium followed by addition of trimethylchlorosilane, results in the more usual $\beta$-elimination with formation of dichloromethylenecyclohexane (5).[2]

$$\xrightarrow[-\,LiOSi(CH_3)_3]{RLi}$$

(4) (5)

[1] J. Villieras, C. Bacquet, and J. F. Normant, *J. Organometal. Chem.*, **97**, 325 (1975).
[2] *Idem, ibid.*, **97**, 355 (1975).

### Lithium *n*-propylmercaptide, 3, 188.

*Dehydration.*[1] Lithium isopropylmercaptide and lithium *t*-butylmercaptide are even more efficient than lithium *n*-propylmercaptide for cleavage of methyl benzoate.

Attempted cleavage of methyl benzylpenicillate (1), however, with lithium *t*-butylmercaptide leads to a thiazole (2), a product of dehydration and cleavage.

$$\xrightarrow[80\%]{\substack{LiSC(CH_3)_3 \\ HMPT}}$$

(1) (2)

[1] S. Wolfe, J.-B. Ducep, and J. D. Greenhorn, *Canad. J. Chem.*, **53**, 3435 (1975).

### Lithium 2,2,6,6-tetramethylpiperidide (LiTMP), 4, 310–311; 5, 417.

*$\gamma$-Alkylation of 2-butynoic acid; isoprenoids.*[1] The first step in a new route to Z-trisubstituted olefins involves conversion of 2-butynoic acid (1)

into the dianion (2) by treatment with 2 eq. of lithium 2,2,6,6-tetramethyl-piperidide in THF—HMPT (7:1) at -100°. Alkylation of (2) with 1-bromo-3-methyl-2-butene followed by methylation gives a mixture of the ene-yne ester

$$CH_3C\equiv CCOOH \xrightarrow[\substack{THF, HMPT \\ -100^0}]{2\ LiTMP} [CH_2 \cdots C \equiv C \cdots COO]^{-2}\ 2\ Li^+ \xrightarrow[\substack{1)\ (CH_3)_2C=CHCH_2Br \\ 2)\ CH_3I,\ DMF}]{}$$

(1)                                                                    (2)

$$(CH_3)_2C=CHCH_2CH_2C\equiv CCOOCH_3 \quad + \quad CH_2=C=C\begin{smallmatrix}COOCH_3 \\ \\ CH_2CH=C(CH_3)_2\end{smallmatrix}$$

(3)                                   2.2:1                     (4)

1) (morpholine)
2) H₃O⁺

$$CH_3COCHCH_2CH=C(CH_3)_2$$
$$\quad\quad |$$
$$\quad\quad COOCH_3$$

(5)

(3) $\xrightarrow[\substack{74\%}]{\substack{(CH_3)_2CuLi \\ CuCH_3}}$ 

(6)                     $\xrightarrow{AlH_3}$                     (7)

(3) formed by γ-alkylation, and the ene-allene ester (4), resulting from α-alkylation. The ratio of (3) to (4) is dependent on the ratio of THF to HMPT and on the temperature. The desired product (3) is separated from (4) by conversion of the latter into (5) by treatment with morpholine followed by acid work-up. Separation of (3) and (5) is then possible by silica gel chromatography. The reaction of (3) with a mixture of dimethylcopperlithium and methylcopper leads almost exclusively to the Z-product (6). Conversion of (6) to nerol (7) has been effected with aluminum hydride (1, 34).

*Diels–Alder reactions with benzyne.* Shepard[2] has recently demonstrated that benzyne generated from chlorobenzene and LiTMP undergoes expected Diels–Alder cycloadditions with reactive dienes. The reaction with furane itself fails because of a base-promoted side reaction.

*Examples:*

Cl + (2,5-dimethylfuran) $\xrightarrow[\substack{24\%}]{\substack{LiTMP \\ THF, \Delta}}$ (product)

*Reaction with benzoate esters.*[3] The reaction of benzyl benzoate (1) with LiTMP in THF at $-78°$ gives benzoin benzoate (2) in 65% yield. The reaction is considered to involve abstraction of a proton to form a dipole-stabilized carbanion (a) followed by reaction of this with the starting ester.

The reaction of alkyl benzoates (3) with this base proceeds in a different way to give *o*-benzoylbenzoates (4) in moderate yields. This reaction is considered to involve *ortho*-lithiation.

*Anthracene synthesis.*[4] Anthracene (1) can be obtained in one step when bromobenzene is refluxed in THF in the presence of this base. The yield is

63% based on bromobenzene. This seemingly simple reaction involves conversion of THF to the enolate ion of acetaldehyde (a) and conversion of bromobenzene to benzyne; these two substances react to form anthracene through intermediates (b), (c), and (d). The reaction is general for synthesis of 9,10-unsubstituted anthracenes.

(a)

(b)

(c)    (d)    (1)

[1] B. S. Pitzele, J. S. Baran, and D. H. Steinman, *J. Org.,* **40,** 269 (1975).
[2] K. L. Shepard, *Tetrahedron Letters,* 3371 (1975).
[3] C. J. Upton and P. Beak, *J. Org.,* **40,** 1094 (1975).
[4] I. Fleming and T. Mah, *J.C.S. Perkin I,* 964 (1975).

## Lithium tri-*sec*-butylborohydride, 4, 312–313. Supplier: Aldrich ("L Selectride").

*Reduction of α,β-unsaturated esters.*[1] α,β-Unsaturated esters undergo 1,4-reduction to the enolate carbanion of the saturated ester (a) when treated with this borohydride in THF at low temperatures. When this reduction is carried out in the absence of a proton donor the carbanion undergoes Claisen condensation to give the dimer of the saturated ester as a coproduct. This side

reaction can be avoided by carrying out the reduction in the presence of *t*-butanol. Under these conditions, reduction can be accomplished in generally good yields. Note that methyl cinnamate is reduced in low yield ($\sim 25\%$).

The intermediate enolates (a) can also be trapped by a reactive electrophile. In this case the enone is added to lithium tri-*sec*-butylborohydride in the absence of an alcohol, and the alkylating agent is added after 20 min. Products of this reductive alkylation are obtained in yields of 50–60%.

[1] B. Ganem and J. M. Fortunato, *J. Org.,* **40,** 2846 (1975).

## Lithium triethylborohydride, 4, 313–314.

*Reductive demethylation of quaternary ammonium salts.*[1] Lithium triethylborohydride is a useful reagent for selective demethylation of quaternary ammonium salts containing at least two methyl groups. Deethylation has been observed in some cases.

*Examples:*

1) $LiHB(C_2H_5)_3$, THF, $25^0$

$C_6H_5\overset{+}{N}(CH_3)_3$ $I^-$  $\xrightarrow[\text{quant.}]{\text{2) } H_3O^+}$  $C_6H_5N(CH_3)_2$

$C_6H_5\overset{+}{N}(C_2H_5)(CH_3)_2$ $I^-$ $\longrightarrow$ $C_6H_5N(C_2H_5)CH_3$ + $C_6H_5N(CH_3)_2$

$\qquad\qquad\qquad\qquad\qquad\qquad\qquad$ 96% $\qquad\qquad$ 4%

$C_6H_5CH_2\overset{+}{N}(CH_3)_3$ $I^-$ $\xrightarrow[\text{quant.}]{}$ $C_6H_5CH_2N(CH_3)_2$

*Ester reductions.* Procedures for the reduction of esters to primary alcohols have been published.[2]

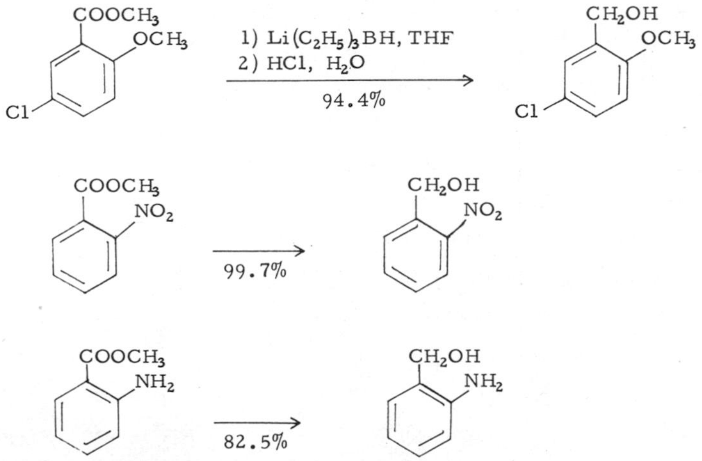

1) $Li(C_2H_5)_3BH$, THF
2) HCl, $H_2O$

94.4%

99.7%

82.5%

*Review.* Reduction with various trialkylborohydrides has been reviewed.[3]

[1]M. P. Cooke, Jr., and R. M. Parlman, *J. Org.,* **40,** 531 (1975).
[2]C. F. Lane, *Aldrichimica Acta,* **7,** 32 (1974).
[3]S. Krishnamurthy, *ibid.,* **7,** 55 (1974).

**Lithium triethylmethoxide, 4,** 314. Soluble in practically all organic solvents.
*Preparation.*[1] The base is prepared by addition of 1 eq. of *n*-butyllithium to triethylmethanol in the solvent to be used.

*Dichlorocarbene.*[2] This base is useful for preparation of dichlorocarbene from chloroform. One advantage is that the reaction can be carried out in a homogeneous system (pentane, hexane, THF). Higher temperatures ($65°$) are necessary than when potassium *t*-butoxide is used, and the reaction is somewhat slower (2–3 hr.), but yields of *gem*-dichlorocyclopropanes are higher, particularly with less reactive alkenes.

$$\text{(cyclohexene)} + \text{CHCl}_3 \xrightarrow[\substack{\text{82\%}}]{\substack{\text{LiOC}(C_2H_5)_3 \\ C_6H_{14}}} \text{(bicyclic product, } Cl, Cl)$$

$$\text{CH}_2{=}\text{CHC}_6\text{H}_{13}\text{-}\underline{n} \xrightarrow[75\%]{} \underset{\text{H}_2\text{C}\text{---}\text{CHC}_6\text{H}_{13}\text{-}\underline{n}}{\overset{\text{Cl} \diagdown \diagup \text{Cl}}{}}$$

[1] H. C. Brown, B. A. Carlson, and R. H. Prager, *Am. Soc.*, **93**, 2070 (1973).
[2] R. H. Prager and H. C. Brown, *Synthesis*, 736 (1974).

# M

Magnesium, 1, 627–629; 2, 254; 3, 189; 4, 315; 5, 419.

*Cyclopentanones; cyclohexanones.* δ- and ε-Iodonitriles undergo intramolecular cyclization to cyclopentanones and cyclohexanones, respectively, when treated with magnesium in ether.[1]

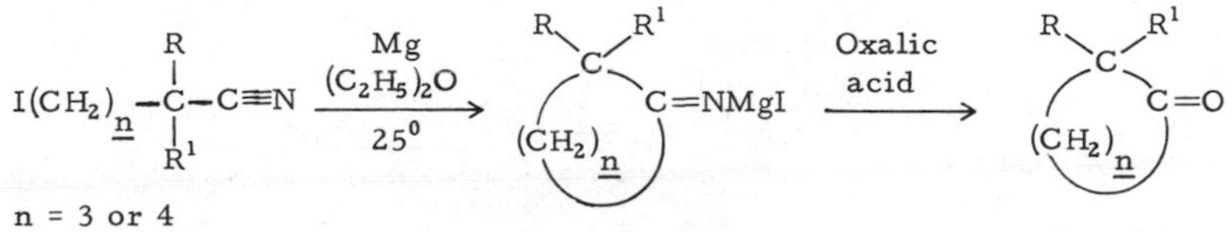

n = 3 or 4

*Examples:*

72%

75%          10%

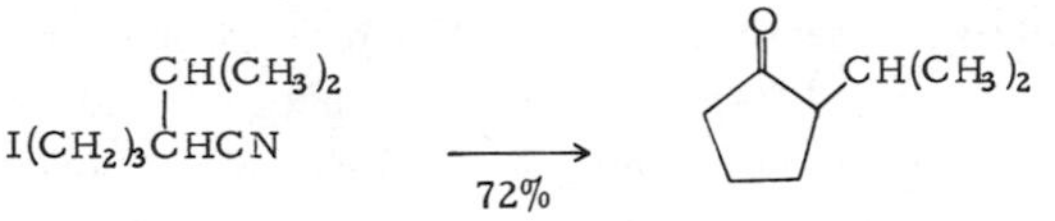

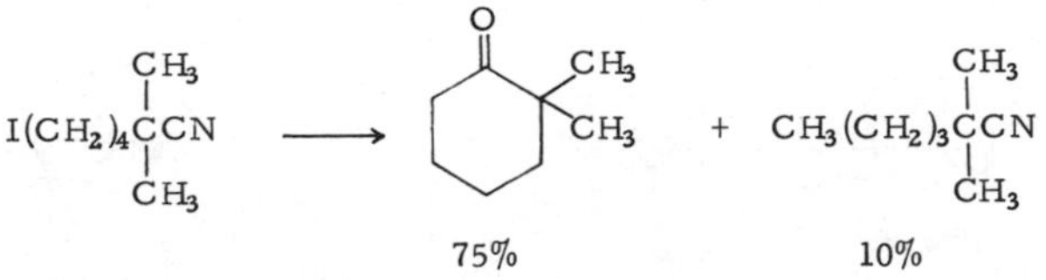

Yields of cyclanones are lower with ω-brominated nitriles or when magnesium is replaced by lithium.

*Reduction of α,β-unsaturated nitriles.*[2] α,β-Unsaturated nitriles can be reduced by magnesium turnings in methanol (exothermic reaction) in high yield. The principal side reaction noted in only a few cases was decyanation rather than reduction. The particular advantage over catalytic hydrogenation is that nonconjugated double bonds are not reduced. The method is compatible with various functional groups: ester, acid, ether, ketal, and nonconjugated C=C bonds.

*Example:*

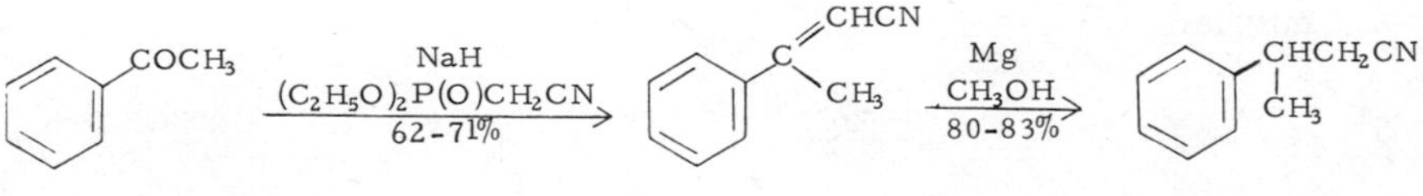

*Stereospecific pinacol reductions.*[3] Pinacol reduction of (1) and (3) in the presence of trimethylchlorosilane (4, 183) gives only one of the three possible

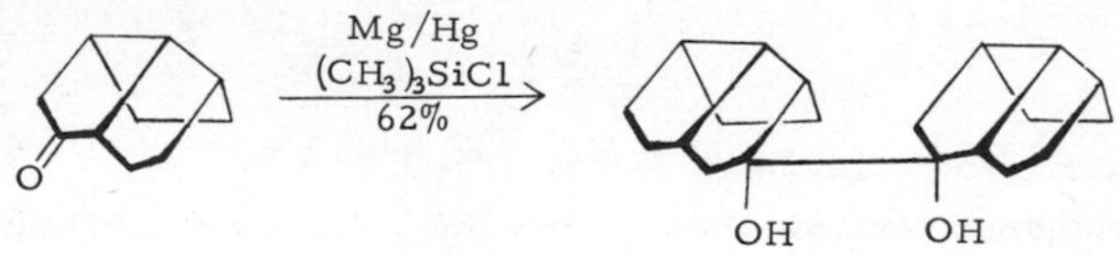

[1, (+)-S]    (2)

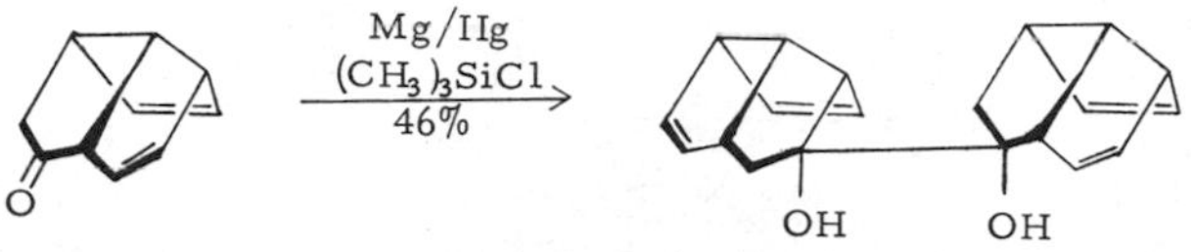

[3, (+)-S]    (4)

diols in each case, (2) and (4). In contrast, pinacol reduction of (±)-(1) and of (±)-(3) gives (2) and (4), respectively, together with the other isomeric diol from *exo, exo* carbon—carbon bond formation. The products, (2) and its isomer, were transformed into *dl*-bivalvane (5) and *meso*-bivalvane (6), respectively. The name bivalvane refers to the physical similarity to mollusks such as clams and oysters.

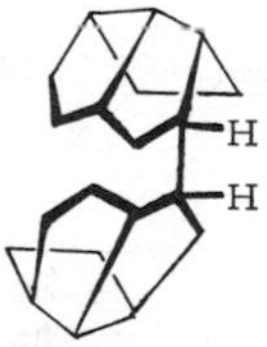

(5) 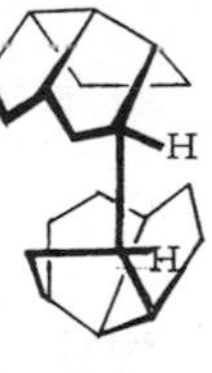    (6)

[1] M. Larcheveque, A. Debal, and T. Cuvigny, *J. Organometal Chem.*, **87**, 25 (1975).
[2] J. A. Profitt, D. S. Watt, and E. J. Corey, *J. Org.*, **40**, 127 (1975); J. A. Profitt, and D. S. Watt, *Org. Syn.*, submitted (1975).
[3] L. A. Paquette, I. Itoh, and W. B. Farnham, *Am. Soc.*, **97**, 7280 (1975).

## Magnesium bromide—Acetic anhydride.

*Ether cleavage.*[1] Cyclic ethers, particularly tetrahydrofuranes, are cleaved by magnesium bromide and acetic anhydride in acetonitrile to ω-bromoacetates. The reaction occurs with inversion (S$_N$2).

*Examples:*

H$_3$C —◁— CH$_3$ (with O in ring)  $\xrightarrow[88\%]{20^0}$  structure with H$_3$C, H, OAc, Br, H, CH$_3$

(tetrahydropyran)  $\xrightarrow[50\%]{85^0}$  AcO—...—Br

(cyclohexene oxide)  $\xrightarrow[80\%]{20^0}$  cyclohexane with OAc and Br

[1] D. J. Goldsmith, E. Kennedy, and R. G. Campbell, *J. Org.,* **40**, 3571 (1975).

**Magnesium iodide,** **5,** 420. Suppliers: K and K, Pfaltz and Bauer, ROC/RIC.

*Homoallylic iodides and bromides.*[1] Homoallylic iodides or bromides (2) can be obtained by reaction of secondary and tertiary cyclopropylcarbinols (1) with magnesium iodide or bromide in refluxing ether (3–90 hr.):

$$\triangleright C \begin{smallmatrix} R^1 \\ R^2 \\ OH \end{smallmatrix} \xrightarrow[\ (C_2H_5)_2O\ ]{MgX_2} X\diagdown\diagup=\begin{smallmatrix} R^1 \\ R^2 \end{smallmatrix}$$

(1)                                    (2)

No reaction is observed when R$^1$ = R$^2$ = H; secondary alcohols (R$^1$ or R$^2$ = H) react more slowly than tertiary alcohols (R$^1$ = R$^2$ = CH$_3$). The reaction is stereoselective; E-isomers of (2) usually predominate.

*α-Keto esters; 1,2-diketones.* French chemists[2] have described the preparation of these compounds from α-chloroglycidic esters (1), available by the Darzens reaction of aldehydes and ketones with alkyl dichloroacetates.[3] The glycidic ester (1) is treated with MgI$_2$ in ether at 35° for 30 min; then an aqueous

$$\begin{smallmatrix} R^2 \\ R^1 \end{smallmatrix} C \underset{O}{-} C \begin{smallmatrix} Cl \\ COOR \end{smallmatrix} \xrightarrow[70-85\%]{\begin{smallmatrix} 1)\ MgI_2 \\ 2)\ NaHSO_3 \end{smallmatrix}} \begin{smallmatrix} R^2 \\ R^1 \end{smallmatrix} CHCCOOR \ (\underset{O}{\parallel})$$

(1)                                    (2)

$\downarrow$ R$^3$MgBr, -78°

$$\begin{smallmatrix} R^2 \\ R^1 \end{smallmatrix} C \underset{O}{-} C \begin{smallmatrix} Cl \\ COR^3 \end{smallmatrix} \xrightarrow[\substack{75-90\% \\ overall}]{\begin{smallmatrix} 1)\ MgI_2 \\ 2)\ NaHSO_3 \end{smallmatrix}} \begin{smallmatrix} R^2 \\ R^1 \end{smallmatrix} CHC-CR^3 \ (\underset{O}{\parallel}\underset{O}{\parallel})$$

(3)                                    (4)

saturated solution of sodium hydrogen sulfite is added. α-Keto esters (2) are obtained in 70–85% yield. Note that both α- and β-keto esters are formed on isomerization of nonhalogenated glycidic esters.

α-Diketones (4) are obtained from (1) by reaction of Grignard reagents with the ester group to form α-chloroepoxy ketones (3). These are converted into α-diketones (4) by sequential treatment with magnesium iodide and sodium hydrogen sulfite. If water is used in place of the reducing agent, α-iodopyruvates are obtained from (1) and α-iododiketones from (3).

[1] J. P. McCormick and D. L. Barton, *J.C.S. Chem. Comm.*, 303 (1975).
[2] P. Coutrot and C. Legris, *Synthesis*, 118 (1975).
[3] J. Villieras *et al.*, *Compt. rend.*, **270**, 1250 (1970); *Bull. soc.*, 1450 (1970).

**Magnesium methyl carbonate (Methyl methoxymagnesium carbonate) (MMC),**
**1**, 631–633; **2**, 256; **3**, 190–191; **5**, 420–421.

*Carboxylation of levulinic acid.* Carboxylation of levulinic acid (1) with MMC in DMF at 135° for 24 hr. followed by esterification affords dimethyl β-ketoadipate (2) in 92% yield.[1]

(a)

(2)

*Cyclodehydration.* Treatment of (1) with MMC in methanol results in formation of (2) rather than the expected product (3). Use of sodium methoxide

(1)                         (2)                         (3)

in methanol also gives (2) rather than (3), but in only 2% yield. Formation of (2) may involve carboxylation of the benzylic $CH_2$ group rather than the methyl group α to the carbonyl group, but evidence for this reaction is lacking.[2]

[1] B. J. Whitlock and H. W. Whitlock, Jr., *J. Org.*, **39**, 3144 (1974).
[2] S. Danishefsky, T. A. Bryson, and J. Puthenpurayil, *J. Org.*, **40**, 1846 (1975).

**Manganese(III) acetate,** $Mn(OCOCH_3)_3$, 2, 263–264; 4, 318.

*Dealkylation of aryldialkylamines.* One alkyl group of amines of this type is replaced by an acetyl group on reaction with metal acetates; cleanest reactions

$$C_6H_5N(CH_3)_2 \xrightarrow[\substack{HCCl_3,\ Ac_2O,\ 25^0 \\ 61\%}]{Mn(OAc)_3} C_6H_5N\begin{smallmatrix}CH_3 \\ COCH_3\end{smallmatrix}$$

$$C_6H_5N\begin{smallmatrix}C_2H_5 \\ CH_3\end{smallmatrix} \longrightarrow C_6H_5N\begin{smallmatrix}CH_3 \\ COCH_3\end{smallmatrix} + C_6H_5N\begin{smallmatrix}C_2H_5 \\ COCH_3\end{smallmatrix}$$

$$61\% \qquad\qquad 8\%$$

$$C_6H_5N\begin{smallmatrix}C_4H_9\text{-}\underline{n} \\ CH_3\end{smallmatrix} \longrightarrow C_6H_5N\begin{smallmatrix}CH_3 \\ COCH_3\end{smallmatrix} + C_6H_5N\begin{smallmatrix}C_4H_9\text{-}\underline{n} \\ COCH_3\end{smallmatrix}$$

$$83\% \qquad\qquad 1\%$$

are obtained with manganese(III) acetate. The formulation indicates that methyl groups are oxidized less readily than ethyl or *n*-butyl groups.[1]

*γ-Lactones* (4, 71). Details of the synthesis of γ-lactones from olefins and carboxylic acids are available.[2] Higher metal salts of manganese, cerium, and

vanadium can be used as catalysts; manganic acetate, $Mn(OAc)_3$, was used most extensively because of availability and solubility. This catalyst is conveniently prepared[3] *in situ* by the reaction of potassium permanganate with manganous acetate. Addition of the potassium salt of the acid component improves the yields. Yields of 60–70% can be obtained with acetic acid; terminal olefins react about five times as fast as internal olefins. Use of other acids results in somewhat lower yields; but even so, this synthesis is useful for preparation of α-substituted γ-lactones.

*Example:*

$$CH_3(CH_2)_4C\equiv CCH_2CH=CH_2 + CH_3COOH \xrightarrow{50\%}$$

The mechanism is considered to involve generation and oxidation of radicals:

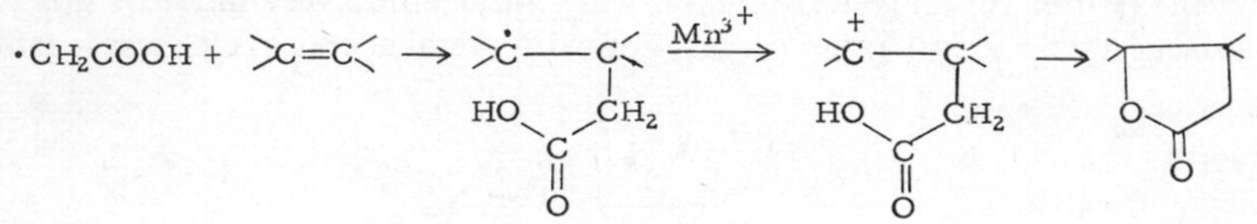

*Dihydrofuranes.*[4] Dihydrofuranes are obtained in fair to high yield by reaction of readily enolizable ketones ($\beta$-diketones, $\beta$-keto esters) with olefins in the presence of 2 eq. of manganic acetate.[5] The reaction is considered to involve generation and oxidation of free radicals.

*Examples:*

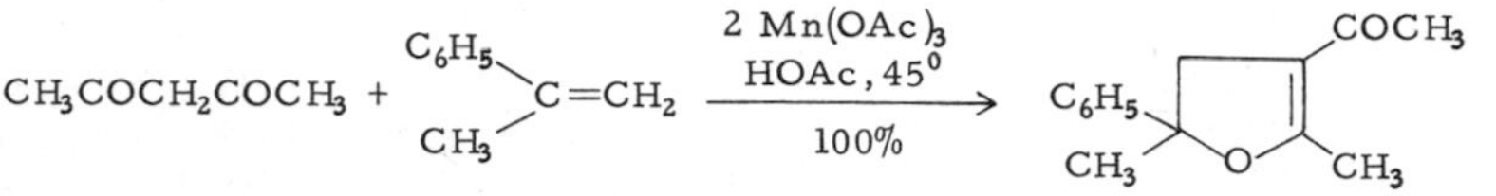

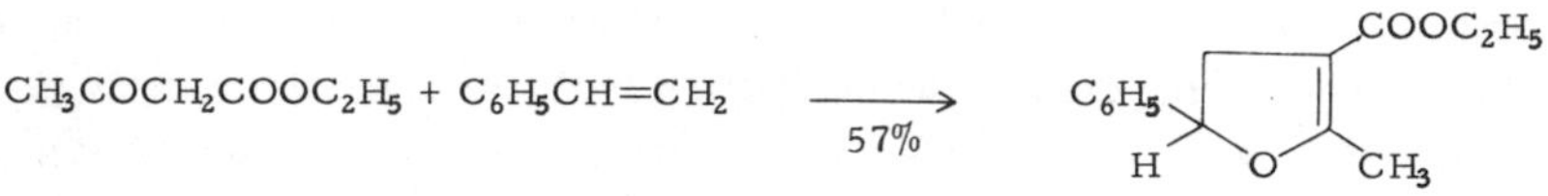

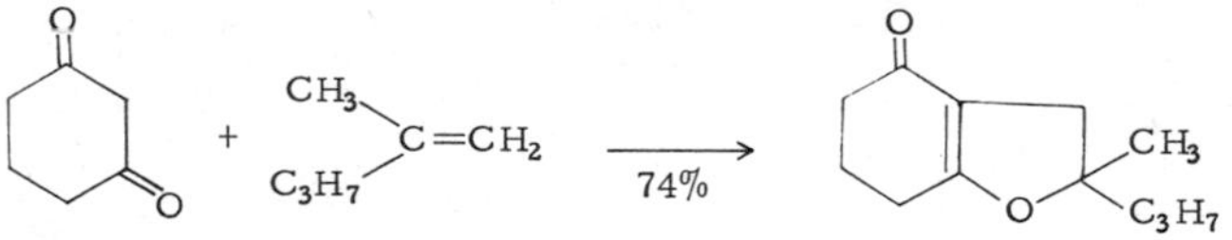

*1,4-Diketones.*[6] The reaction of enol esters with ketones at 50–70° in the presence of manganic acetate (2 eq.) leads to 1,4-diketones in moderate yields:

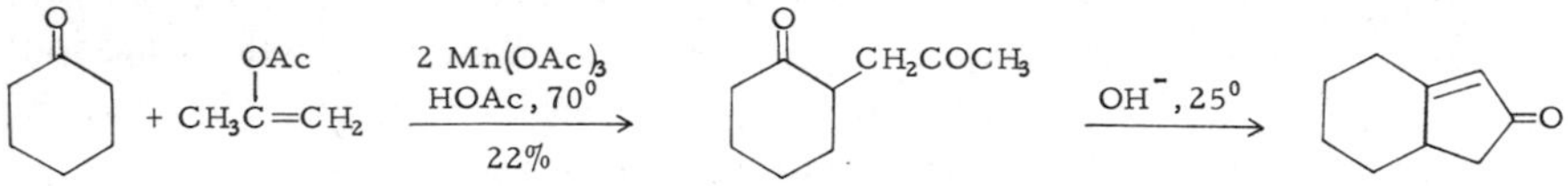

The products are readily cyclized by dilute base to cyclopentenones.
Unsymmetrical ketones give a mixture of two products:

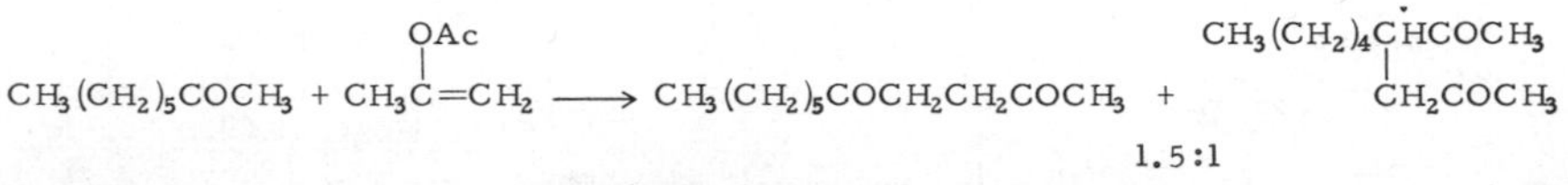

$$1.5:1$$

[1] B. Rindone and C. Scolastico, *Tetrahedron Letters*, 3379 (1974).
[2] E. I. Heiba, R. M. Dessau, and P. G. Rodewald, *Am. Soc.*, **96**, 7977 (1974).
[3] E. I. Heiba, R. M. Dessau, and W. I. Koehl, Jr., *Am. Soc.*, **91**, 138 (1969).
[4] E. I. Heiba and R. M. Dessau, *J. Org.*, **39**, 3456 (1974).

[5] Prepared *in situ* by addition of potassium permanganate to a solution of manganous acetate in acetic acid.
[6] R. M. Dessau and E. I. Heiba, *J. Org.*, **39**, 3457 (1974).

**Manganese dioxide,** 1, 637–643; 2, 257–263; 3, 191–194; 4, 317–318; 5, 422–424.

*2,2′-Biimidazole* (3). 2,2′-Bi(2-imidazoline) (1) can be oxidized to 2,2′-biimidazole (3) by conversion to the bistrimethylsilyl derivative (2), which is soluble in organic solvents, and oxidation of (2) in CCl₄ with the activated manganese dioxide of Goldman (2, 257). The procedure can also be used to oxidize 2-methyl-2-imidazoline to 2-methylimidazole (67% overall yield).[1]

*Cyclohexylideneacetaldehydes.* Brink[2] has reported a two-step synthesis of these aldehydes (3) from esters of cyclohexylideneacetic acids (1).[3] These are reduced in absolute ether with lithium aluminum hydride at room temperature. The alcohols are oxidized with Attenburrow manganese dioxide

in petroleum ether (1–5 hr.) to the aldehydes (3). These aldehydes are unstable to air, acids, and bases, and are best isolated as the bisulfite-addition compound.

*Review.* Oxidations with activated manganese dioxide have been reviewed.[4,5]

[1] E. Duranti and C. Balsamini, *Synthesis*, 815 (1974).
[2] M. Brink, *ibid.*, 253 (1975).
[3] Prepared by zinc-promoted condensation of cyclohexanones with esters of bromoacetic acid in yields of 60–80%; O. Wallach, *Ann.*, **365**, 255 (1909).
[4] A. J. Fatiadi, *Synthesis*, 65, 133 (1976).
[5] J. S. Pizey, *Synthetic Reagents*, Vol. 2, Halsted Press, New York, New York, 1974, pp. 143–174.

**(−)-Menthoxyacetyl chloride,** (1). Mol. wt. 232.75.

Aldrich supplies the free acid.

*Optically active arene oxides.* Arene oxides have been the subject of recent investigation because they are primary intermediates in the metabolism of

arenes. Akhtar and Boyd[1] have recently reported the synthesis of optically active 1,2-oxides of naphthalene and anthracene based on a method of chromato-
graphic separation. The method is outlined for preparation of the former oxide.
The bromohydrin (2) was converted into a diastereomeric mixture of bromo-
menthoxyacetates (3), which were separable by short column chromatography

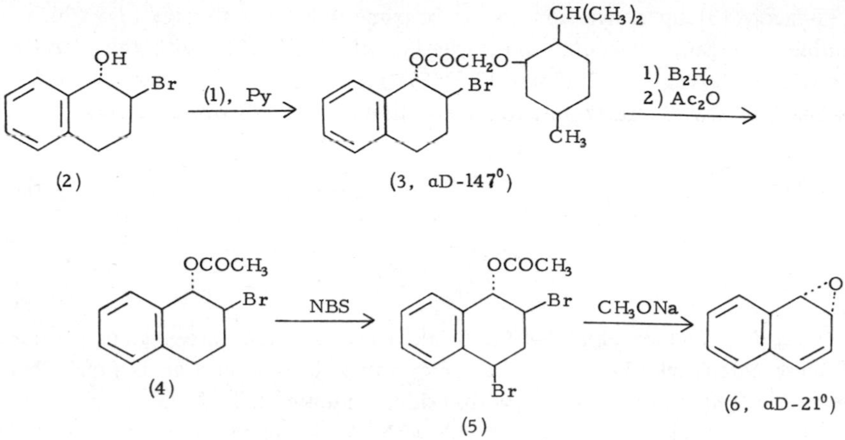

(2)     (3, αD-147°)

(4)     (5)     (6, αD-21°)

on Kieselgel G (Merck). Ester interchange of chiral (3) was then effected; remaining steps to naphthalene 1,2-oxide (6) followed the known synthesis.[2]

[1] M. N. Akhtar and D. R. Boyd, *J.C.S. Chem. Comm.*, 916 (1975).
[2] H. Yagi and D. M. Jerina, *Am. Soc.*, **97**, 3185 (1975).

**Mercuric acetate, 1,** 644–652; **2,** 264–267; **3,** 194–196; **4,** 319–323; **5,** 424–427.

*Reactions in methanol.* The reaction of mercuric acetate in refluxing methanol for 12 hr. with β-pinene (1) leads to a mixture of the methyl ethers of

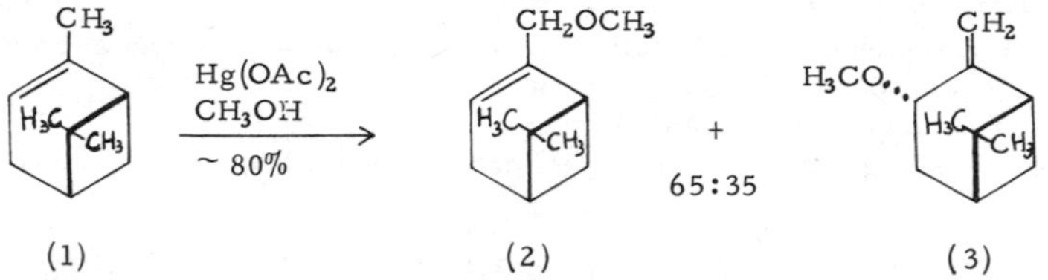

(1)     (2)     (3)
     65:35

myrtenol (2) and of *trans*-pinocarveol (3). The same products are obtained
from α-pinene, but in lower yields.[1]

*Oxidative cyclization.* Treatment of 2′-hydroxychalcones (1) with mercuric
acetate in DMSO gives cyclized adducts (2) and aurones (3); the adducts are converted into (3) by further treatment with CaO in DMSO.[2]

(1)              (2)              (3)

*Oxymercuration–demercuration* (**2,** 265–267; **3,** 194–195; **4,** 319–320; **5,** 425–427).[3] The oxymercuration of *endo*-tricyclo[5.2.2.0$^{2,6}$]undeca-3,8–diene (1) is highly regioselective and stereospecific. In contrast, hydroboration

(1)           (2, 92%)           (3, 8%)

of (1) is stereospecific, but not regiospecific, and results in three products.

[1] A. Kergomard, J.-C. Tardivat, and J.-P. Vuillerme, *Bull. soc.,* 2572 (1974).
[2] M. F. Grundon, D. Stewart, and W. E. Watts, *J.C.S. Chem. Comm.,* 722 (1975).
[3] N. Takaishi, Y. Fujikura, and Y. Inamoto, *J. Org.,* **40,** 3767 (1975).

**Mercuric chloride, 1,** 652–654; **5,** 427–428.

*Hydrolysis of vinyl sulfides to carbonyl compounds.* Corey and Shulman (**3,** 97) in 1970 first reported that vinyl sulfides could be hydrolyzed to ketones with mercuric chloride in aqueous acetonitrile. This method, sometimes in combination with mercuric oxide or calcium carbonate (**4,** 38, 39),[1] has been widely used. However, recently, Grieco, Cohen, and collaborators[2] have reported that mercuric chloride hydrolysis of vinyl sulfides of type (1) to aldehydes is not generally successful. They have found a very simple solution illustrated for hydrolysis of (2) to *n*-heptanal (4).

(1)

$CH_3(CH_2)_3CH_2CH=CHSC_6H_5$ $\xrightarrow[\text{quant.}]{HCl, C_6H_6}$ $CH_3(CH_2)_3CH_2CH_2\overset{\displaystyle Cl}{\underset{\displaystyle SC_6H_5}{C}H}$ $\xrightarrow[\text{quant.}]{HgCl_2, H_2O}$ $CH_3(CH_2)_5CHO$

(2)                         (3)                      (4)

Another expedient is to add thiophenol to the double bond and then hydrolyze the resultant thioacetal with mercuric chloride (*see* **5,** 85).

[1] H. J. Bestmann and J. Angerer, *Tetrahedron Letters,* 3665 (1969).
[2] A. J. Mura, Jr., G. Majetich, P. A. Grieco, and T. Cohen, *ibid.,* 4437 (1975).

**Mercuric oxide,** 1, 655–658; 2, 267–268; 4, 323–324; 5, 428.

*Hydration of acetylenes* (1, 656). Newman and Lee[1] have effected hydration of 1-ethynylcyclohexanol to 1-acetylcyclohexanol by slow addition to a warm stirred solution of yellow mercuric oxide in dilute sulfuric acid; after

which the mixture is heated to 70° for 30 min.

[1]M. S. Newman and V. Lee, *J. Org.*, **40**, 381 (1975).

**Mercuric oxide—Mercuric bromide.**

*Koenigs–Knorr synthesis.* This catalyst system can be used in place of the usual silver oxide or silver carbonate in synthesis of alkyl β-D-glucopyranosides, β-D-galactopyranosides, and β-D-xylopyranosides. The β-anomers are formed almost exclusively in yields of 60–80%. The reaction is conducted in absolute chloroform in the presence of Drierite; reactions are complete in 1–2 hr. at 25°.[1]

[1]L. R. Schroeder and J. W. Green, *J. Chem. Soc.*, 530 (1966); L. R. Schroeder, K. M. Counts, and F. C. Haigh, *Carbohydrate Res.*, **37**, 368 (1974).

**Mercury(I) nitrate,** $Hg_2(NO_3)_2$. Mol. wt. 525.24. Suppliers: Alfa, J. T. Baker, Fisher, ROC/RIC.

*Benzaldehydes.* Treatment of a benzyl bromide (1) with mercurous nitrate in DME gives the corresponding nitrate (2), which is converted into the aldehyde (3) when refluxed in ethanol containing aqueous sodium hydroxide.[1]

[1]A. McKillop and M. E. Ford, *Syn. Commun.*, **4**, 45 (1974).

**Mercury(II) trifluoroacetate,** $Hg(OOCCF_3)_2$. Mol. wt. 426.65. Supplier: Alfa.

The salt can be prepared *in situ* from mercury(II) oxide and trifluoroacetic anhydride (equimolar amounts).

*Allylic trifluoroacetoxylation.* This salt is more reactive than mercury(II) acetate for allylic hydroxylation. Thus cholesterol esters are converted, after hydrolysis, into $\Delta^4$-3β,6β-dihydroxy steroids, even at 0°. Yields are about 50%. $\Delta^5$-3β-Aminosteroids are converted into mixtures of $\Delta^5$-3β-amino-4β-hydroxysteroids and $\Delta^4$-3$\beta$-amino-6$\beta$-hydroxysteroids. $\Delta^5$-Cholestene is converted into a 1 : 1 mixture of $\Delta^{3,5}$- and $\Delta^{4,6}$-cholestadiene.[1]

[1] G. Massiot, H.-P. Husson, and P. Potier, *Synthesis*, 723 (1974).

**1-(Mesitylenesulfonyl)-1,2,4-triazole, 5,** 434.

*Oligonucleotides.* The full paper on the use of arylsulfonyltriazoles for synthesis of oligonucleotides has been published.[1]

[1] N. Katagiri, K. Itakura, and S. A. Narang, *Am. Soc.,* **97,** 7332 (1975).

$\pi$-**2-Methallylnickel bromide,** $CH_3-C$ $G-CH_3$ . Mol. wt. 298.04.

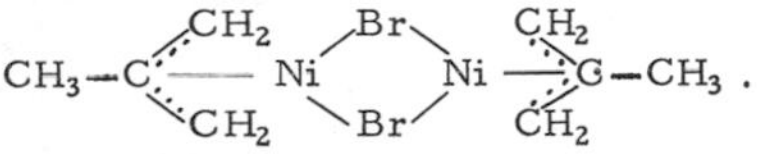

(1)

*Preparation.* The complex is prepared from 2-methallyl bromide and nickel carbonyl in 85% yield by the method of Semmelhack and Helquist (**4,** 353).[1]

*Reaction with carbonyl compounds.*[2] The complex reacts with aldehydes and ketones to produce homoallylic alcohols:

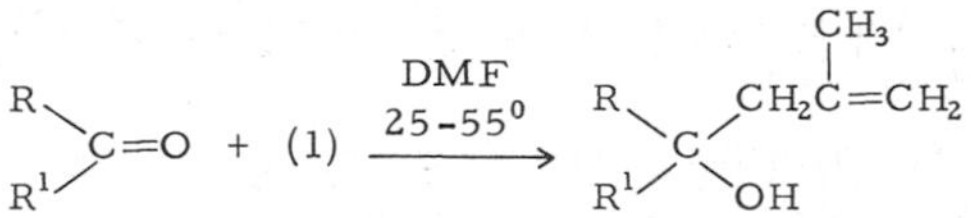

Aliphatic ketones are relatively inert, but aldehydes are generally reactive. $\alpha$-Diketones are the most reactive substrates and undergo exclusive attack at only one carbonyl group. $\alpha,\beta$-Unsaturated ketones undergo only 1,2-attack, even in the presence of added cuprous iodide. Esters and acid chlorides do not react.

The related $\pi$-allyl nickel bromide complex [$\pi$-**(2-carboethoxyallyl)nickel bromide,** this volume] is useful for synthesis of $\alpha$-methylene-$\gamma$-butyrolactones.

[1] M. F. Semmelhack and P. M. Helquist, *Org. Syn.,* **52,** 115 (1972).
[2] L. S. Hegedus, S. D. Wagner, E. L. Waterman, and K. Siirala-Hansen, *J. Org.,* **40,** 593 (1975).

**Methaneselenol,** $CH_3SeH$. Mol. wt. 94.17, b.p. 25.5°, strong odor.

*Preparation.*[1] The reagent is prepared in nearly quantitative yield by reduction of dimethyl diselenide with sodium in liquid ammonia.

*Epoxides via methyl selenoacetals.* Krief *et al.*[2] have reported the synthesis of epoxides from two carbonyl compounds. The first step involves preparation of a dimethyl selenoacetal (1), followed by conversion to an $\alpha$-methyl seleno-carbanion (a). These highly reactive carbanions react with even hindered carbonyl compounds to give $\beta$-hydroxy methyl selenides (2), which are converted into selenonium salts by reaction with methyl iodide or dimethyl sulfate.

$$R^1_{\phantom{1}}\!\!\diagdown\!\!C{=}O \;+\; 2\,CH_3SeH \xrightarrow[70\text{-}90\%]{HCl,\,0^\circ} CH_3Se{-}\underset{R^2}{\overset{R^1}{\underset{|}{\overset{|}{C}}}}{-}SeCH_3 \xrightarrow[-n\text{-}BuSeCH_3]{\substack{n\text{-}BuLi\\ THF,\,-78^\circ}} \left[\,CH_3Se\underset{R^2}{\overset{R^1}{\underset{|}{\overset{|}{C}}}}{}^{-}\;Li^+\right]$$

$$(1) \hspace{5cm} (a)$$

$$\underset{R^4}{\overset{R^3}{\underset{|}{\overset{|}{\diagdown}}}}\!\!C{=}O \xrightarrow{60\text{-}95\%} CH_3Se{-}\underset{R^2\;R^4}{\overset{R^1\;R^3}{\underset{|\;\;|}{\overset{|\;\;|}{C{-}C}}}}{-}OH \xrightarrow{CH_3I}{70\text{-}100\%} (CH_3)_2\overset{+}{Se}{-}\underset{\underset{I^-}{R^2}\;R^4}{\overset{R^1\;R^3}{\underset{|\;\;|}{\overset{|\;\;|}{C{-}C}}}}{-}OH \xrightarrow[DMSO]{KOC(CH_3)_3}$$

$$(2) \hspace{5cm} (3)$$

$$\left[(CH_3)_2\overset{+}{Se}{-}\underset{\underset{I^-}{R^2}\;R^4}{\overset{R^1\;R^3}{\underset{|\;\;|}{\overset{|\;\;|}{C{-}C}}}}{-}O^-K^+\right] \xrightarrow{65\text{-}98\%} \underset{R^2}{\overset{R^1}{\diagdown}}\!\!\overset{\diagup R^3}{\underset{\diagdown}{\triangle}}\!\!\underset{R^4}{\diagup O} \;+\; (CH_3)_2Se$$

$$(b) \hspace{5cm} (4)$$

The final step involves reaction with potassium $t$-butoxide in DMSO at $25^\circ$ for 30 min. Although several steps are involved, the intermediates (2) and (3) need not be purified, and yields of epoxides (4) based on $R^3COR^4$ are generally in the range of 60–80% overall.

This reaction was originally carried out with phenylselenol, $C_6H_5SeH$,[3] but yields are generally higher in the newer procedure. Moreover, the by-products, $n$-BuSeCH$_3$, b.p. $0^\circ/1$ mm., and $(CH_3)_2Se$, b.p. $58^\circ$, are volatile and easily eliminated.

*Caution*: All reactions should be conducted in a hood. Selenium compounds, particularly volatile ones, are toxic.

[1] G. E. Coates, *J. Chem. Soc.*, 2839 (1953).
[2] D. Van Ende, W. Dumont, and A. Krief, *Angew. Chem. internat. Ed.*, **14**, 700 (1975).
[3] W. Dumont and A. Krief, *ibid.*, **14**, 350 (1975).

**Methanesulfonyl chloride, 1**, 662–664; **2**, 268–269; **4**, 326–327; **5**, 435–436.

*Protection of hydroxyl groups as mesylates.* In a synthesis of decinine (1), Lantos and Loev[1] protected the hydroxyl group of one of the intermediates as

$$(1)$$

the mesylate. This derivative did not interfere with an Ullman coupling reaction used during the synthesis and was sufficiently stable to survive a later condensation reaction utilizing calcium hydroxide as base.

[1]I. Lantos and B. Loev, *Tetrahedron Letters*, 2011 (1975).

**4-Methoxy-5-acetoxymethyl-*o*-benzoquinone**, (1) mol. wt. 210.18, m.p. 128–130°; **4-Methoxy-5-methyl-*o*-benzoquinone** (2), mol. wt. 152.14, m.p. 110–120° dec.

(1)          (2)

Both quinones have been prepared[1] from 2-chloromethyl-4-nitroanisole.

*Diels–Alder reactions.*[1] Unlike *p*-benzoquinones, which are potent dienophiles in Diels–Alder reactions, *o*-benzoquinones can function as either dienes or dienophiles. Danishefsky *et al.* reasoned that 4-methoxy-*o*-benzoquinones should serve predominantly as dienophiles. Indeed when heated in a sealed tube at 105° for 5 hr. with 1,3-butadiene, the quinone (1) gives (3) in 63% yield. This product is evidently formed by cycloaddition to the 5,6-double bond followed by enolization to the diosphenol.

$$(1) + \quad \overset{105^\circ}{\underset{63\%}{\longrightarrow}} \quad (3)$$

$$(2) \rightleftharpoons \quad (2a) \quad + \quad \overset{105^\circ}{\underset{48\%}{\longrightarrow}} \quad (4)$$

The *o*-quinone (2) reacts with butadiene to give (4). This product can be considered to be formed by tautomerization to (2a) followed by cycloaddition at the exocyclic methylene with formation of a spiro system.

[1] S. Mazza, S. Danishefsky, and P. McCurry, *J. Org.*, **39**, 3610 (1974).

**2-Methoxyallyl bromide,** $\overset{OCH_3}{\underset{H_2C}{\underset{\diagup}{C}}\diagdown CH_2Br}$ , **4,** 327–328.

*Cycloaddition to arenes.*[1] Benzene reacts with the 2-methoxyallyl cation (a), formed from 2-methoxyallyl bromide by reaction with silver trifluoroacetate, to give bicyclo[3.2.2]nona-6,8-diene-3-one (1) at room temperature (6% yield). Adducts are also obtained from *p*-xylene (3.5% yield) and toluene. The

$$\text{(a)} \quad \overset{25^\circ}{\underset{6\%}{\longrightarrow}} \quad \text{(1)} \quad + \quad CH_3COCH_2C_6H_5 \quad \text{(2)}$$

bicyclic ketone (1) is sensitive to acids and is easily converted into (2); this cleavage may be responsible for the low yield. In spite of the low yields, the reaction provides a simple route to a new bicyclic system.

[1] H. M. R. Hoffmann and A. E. Hill, *Angew. Chem. internat. Ed.*, **13,** 136 (1974).

**S-(2-Methoxyallyl)-N,N-dimethyldithiocarbamate** (1). Mol. wt. 175.32.
*Preparation:*

$$(CH_3)_2N\overset{S}{\overset{\|}{C}}SNa \quad + \quad CH_2{=}\overset{OCH_3}{\overset{|}{C}}{-}CH_2Br \quad \xrightarrow{88\%} \quad CH_2{=}\overset{OCH_3}{\overset{|}{C}}{-}CH_2S\overset{S}{\overset{\|}{C}}N(CH_3)_2$$
$$(1)$$

*Methyl ketones, aldols, 1,4-diketones.* Japanese chemists[1] have used this allyl dithiocarbamate for introduction of the $CH_3COCH_2$ group into various electrophiles: alkyl halides, carbonyl compounds, and epoxides. Typical results are summarized in equations I–III. The reagent is first converted into the lithium salt (2) by LDA in THF at $-78°$ in essentially quantitative yield.

(I) 

$$CH_2=C \underset{\underset{Li \leftarrow N(CH_3)_2}{\overset{OCH_3}{|}}{\overset{\gamma}{\underset{\alpha}{CH}}} \overset{S}{\underset{}{C}} \xrightarrow[95-98\%]{RX} CH_2=C \underset{\overset{}{R}}{\overset{OCH_3}{|}} \overset{S}{\underset{}{SCN(CH_3)_2}} \xrightarrow{H_3O^+} CH_3 \overset{O}{\overset{||}{C}} CHS \overset{S}{\overset{||}{C}} N(CH_3)_2$$

(2)        ( + traces of γ-adduct)

$$\xrightarrow[70-80\%]{\underset{Ni}{Raney}} CH_3 \overset{O}{\overset{||}{C}} CH_2R$$

(II)    (2)  +  $\underset{R^2}{\overset{R^1}{}} C=O \longrightarrow \underset{R^2}{\overset{R^1}{}} C \underset{\overset{}{H_2}}{\overset{OH}{\underset{}{C}}} \overset{OCH_3}{\underset{\overset{}{H}}{C}} \overset{S}{\underset{}{SCN(CH_3)_2}} \xrightarrow{H_3O^+} \underset{R^2}{\overset{R^1}{}} C \underset{CH_2COCH_2SCN(CH_3)_2}{\overset{OH}{}}$

( + ~20% of α-adduct)

$$\xrightarrow[65-85\%]{\underset{Ni}{Raney}} \underset{R^2}{\overset{R^1}{}} C \underset{CHCOCH_3}{\overset{OH}{}}$$

(III)    (2)  +  $C_6H_{13}$ $\longrightarrow C_6H_{13}CHOHCH_2 \overset{OCH_3}{\underset{\overset{}{H}}{C}} = \overset{S}{\underset{}{SCN(CH_3)_2}} \xrightarrow{H_3O^+}$

( + some α-adduct)

$$C_6H_{13}CHOHCH_2CH_2 \overset{O}{\overset{||}{C}} CH_2S \overset{S}{\overset{||}{C}} N(CH_3)_2 \xrightarrow[64\%]{Raney\ Ni}$$

$$C_6H_{13}CHOHCH_2CH_2 \overset{O}{\overset{||}{C}} CH_3 \xrightarrow[oxid.]{Jones} C_6H_{13} \overset{O}{\overset{||}{C}} CH_2CH_2 \overset{O}{\overset{||}{C}} CH_3$$

The anion (2) reacts with alkyl halides almost exclusively at the α-position; it reacts with carbonyl compounds and epoxides preferentially at the γ-position.

[1] T. Nakai, H. Shiono, and M. Okawara, *Tetrahedron Letters*, 4027 (1975).

**p-Methoxybenzyl  itaconate,**  $\underset{}{\text{p-}CH_3OC_6H_4CH_2OOCCH_2 \overset{\overset{CH_2}{||}}{C} COOH}$ (1).  Mol.  wt. 240.26, m.p. 86.8–87.2°.

The ester is prepared by the reaction of *p*-methoxybenzyl alcohol with itaconic anhydride (30 hr., 55–60°); yield, 68%.

**α-Methylene-γ-butyrolactones.**[1] The dianion of (1), generated with LDA in THF at $-78°$, reacts with aldehydes and ketones to give, after treatment with acid, α-methylene-γ-butyrolactones (2). The ester group can be hydrolyzed by trifluoroacetic acid. Use of itaconic acid itself leads to low yields of ketones. Use

(2)                                                        (3)

of the dianion of methyl itaconate gives lactones in acceptable yields, but hydrolysis of the ester group is attended with isomerization of the α-methylene double bond. Yields are low to moderate, but the reaction provides a very direct synthesis of products of type (3). It has been used for synthesis of *dl*-protolichesterinic acid $(R_2 = CH_3(CH_2)_{12}-)$, *dl*-"*trans*-nephrosterinic acid" $(R_2 = CH_3(CH_2)_{10}-)$, and *dl*-"*cis*-nephrosterinic acid," $(R_1 = CH_3(CH_2)_{10}-)$.

[1] R. M. Carlson and A. R. Oyler, *Tetrahedron Letters*, 4099 (1975).

**cis- and *trans*-2-Methoxycyclopropyllithium,** (1) and (2).
  *Preparation:*

(44%)                                           (1)

(2)

**β,γ-Unsaturated aldehydes.**[1] These reagents serve as the equivalent to the unknown Wittig reagent $(C_6H_5)_3P=CHCH_2CHO$. Thus aldehydes react with

(3)                          RCHO                          (4)

$$(3) \text{ or } (4) \xrightarrow[\substack{2) \ CH_3OH, \ -40^\circ \\ 85-95\%}]{1) \ CH_3SO_2Cl} \quad \begin{matrix} R \\ \diagdown \\ H \diagup \end{matrix} C = CHCH_2CH(OCH_3)_2$$

(5)

60-80% | HSCH$_2$CH$_2$OH
BF$_3$ · (C$_2$H$_5$)$_2$O

95% | (HOOC)$_2$, H$_2$O
CH$_3$COCH$_3$

$$RCH = CHCH_2C\overset{S}{\underset{O}{H}} \xrightarrow[\substack{\text{or 1) } FSO_3CH_3, \ CH_3CN \\ 2) \ CaCO_3, \ H_2O}]{\substack{HgCl_2, \ CaCO_3, \\ CH_3CN, \ H_2O,}} RCH = CHCH_2CHO$$

(6)                                                                (7)

(1) and (2) in ether at $-78^\circ$ to give the epimeric carbinols (3) and (4), respectively. These can be converted into $\beta,\gamma$-unsaturated aldehydes (7) via the dimethyl acetals (5) or the ethylene hemithioacetals (6). In most cases examined to date, this synthesis leads stereoselectively to the *trans*-isomer of (7). Differences in the *cis/trans* ratio are observed depending on whether the precursor is (3) or (4), but the stereoselectivity is controlled also by the nature of the R group. From (4), the ratio of *trans* to *cis* products is about 5:1 when R is a primary alkyl group, but is much higher with (3) or (4) when R is cyclohexyl, isopropyl, phenyl, or 1-methyl-*trans*-propenyl. The nature of the leaving group is also important; arylsulfonates give somewhat higher *trans* to *cis* ratios than mesylates.

One example of the reaction of (1) with a ketone, cyclohexanone, was reported.[1]

[1] E. J. Corey and P. Ulrich, *Tetrahedron Letters*, 3685 (1975).

## 4-Methoxy-5,6-dihydro-2H-pyrane, 2, 271; 3, 197–198; 5, 438.

*Improved preparation.* Reese et al.[1] have reported an improved preparation by which this reagent (1) for the protection of hydroxyl groups has been obtained for the first time as an almost pure (>98%) liquid, b.p. $50.5^\circ/10$–11 torr, stable under N$_2$ at $20^\circ$ for several months.

$$\text{(Cl—CO—CH}_2\text{—CH}_2\text{—CH}_2\text{—Cl)} \xrightarrow[\substack{H_2O, \ 100^\circ \\ 46-54\%}]{NaH_2PO_4, \ H_3PO_4} \text{(tetrahydropyran-4-one)} \xrightarrow[\substack{p\text{-}TsOH, \ CH_3OH \\ 92\%}]{HC(OCH_3)_3}$$

$$\text{(4,4-dimethoxytetrahydropyran, } H_3CO \ OCH_3) \xrightarrow[\substack{\Delta \\ 71\%}]{HC(OCH_3)_3, \ p\text{-}TsOH} \text{(4-methoxy-5,6-dihydro-2H-pyrane, } OCH_3)$$

(1)

[1] R. Arentzen, Y. T. Yan Kui, and C. B. Reese, *Synthesis*, 509 (1975).

**Methoxymethylenetriphenylphosphorane, 1, 671.**

*Conversion of 17-ketosteroids to 20-ketosteroids.* Danishefsky et al.[1] have

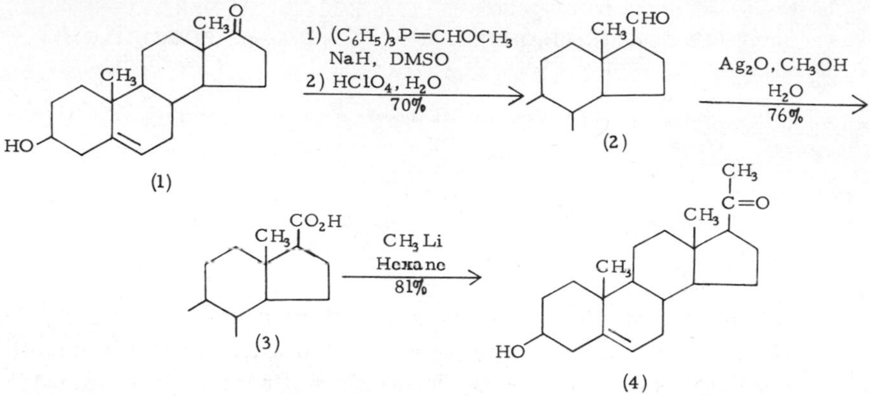

(1)

1) $(C_6H_5)_3P=CHOCH_3$
   NaH, DMSO
2) $HClO_4$, $H_2O$
   70%

(2)

$Ag_2O$, $CH_3OH$
$H_2O$
76%

(3)

$CH_3Li$
Hexane
81%

(4)

developed a simple method for conversion of androstenolone (1) into preg-nenolone (4). The final step, addition of an alkyllithium to an etianic acid,[2] should be useful for elaboration of various side chains.

[1] S. Danishefsky, K. Nagasawa, and N. Wang, *J. Org.*, **40**, 1989, (1975).
[2] Review: M. J. Jorgenson, *Org. Reac.*, **18**, 1 (1970).

**2-(o-Methoxyphenyl)-4,4-dimethyl-2-oxazoline,** 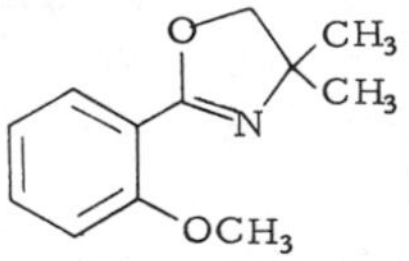 (1). Mol. wt. 193.25.

The reagent is prepared in satisfactory yield from o-methoxybenzoic acid and 2-amino-2-methyl-1-propanol.

*Diphenic acids and o-alkylbenzoic acids.*[1] These substances are readily available from (1).

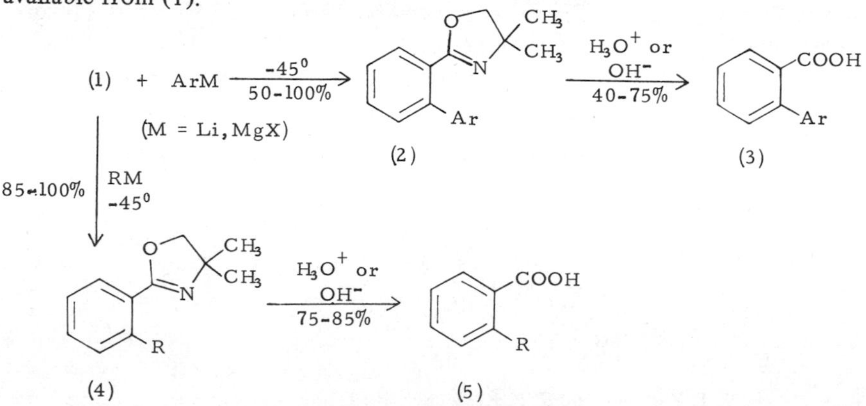

(1) + ArM  $\xrightarrow[50-100\%]{-45^0}$  (2)  $\xrightarrow[40-75\%]{H_3O^+ \text{ or } OH^-}$  (3)

(M = Li, MgX)

85-100%  RM  $-45^0$

(4)  $\xrightarrow[75-85\%]{H_3O^+ \text{ or } OH^-}$  (5)

[1] A. I. Meyers and E. D. Mihelich, *Am. Soc.*, **97**, 7383 (1975).

**Methoxyphenylthiomethyllithium,** $\begin{matrix} C_6H_5S \\ \\ CH_3O \end{matrix}$ CHLi. Mol. wt. 160.16.

The reagent is prepared by treatment of methoxymethyl phenyl thioether[1] with $n$-butyllithium in THF at $-30°$.

**$\alpha$-Methylene-$\delta$-lactones.**[2] Trost and Miller have described a ring expansion

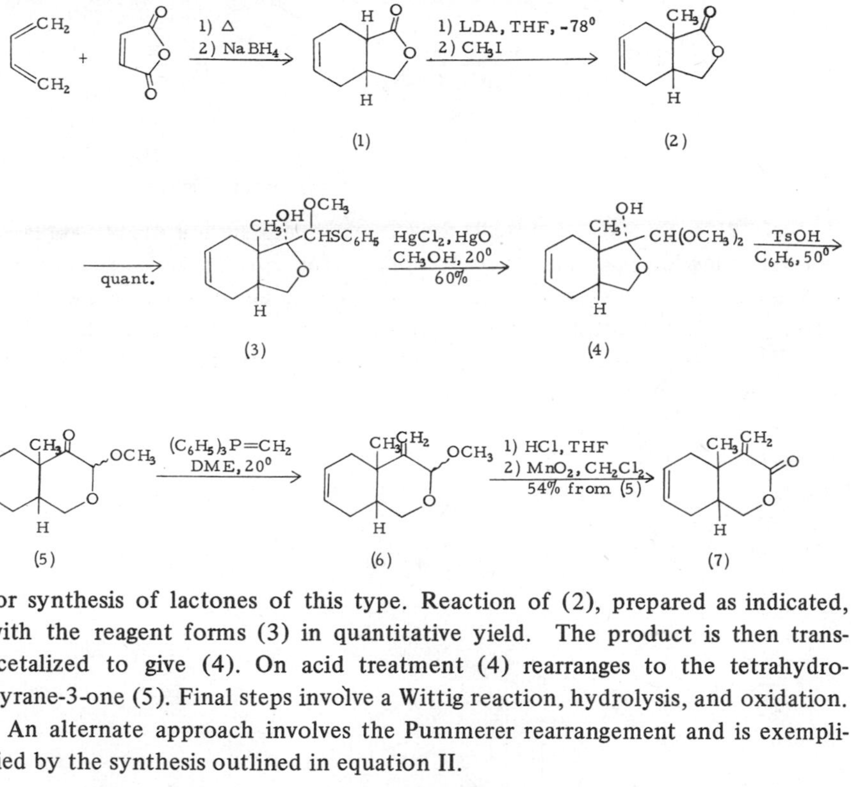

(1)          (2)

(3)          (4)

(5)          (6)          (7)

for synthesis of lactones of this type. Reaction of (2), prepared as indicated, with the reagent forms (3) in quantitative yield. The product is then trans-acetalized to give (4). On acid treatment (4) rearranges to the tetrahydro-pyrane-3-one (5). Final steps involve a Wittig reaction, hydrolysis, and oxidation.

An alternate approach involves the Pummerer rearrangement and is exemplified by the synthesis outlined in equation II.

(II)

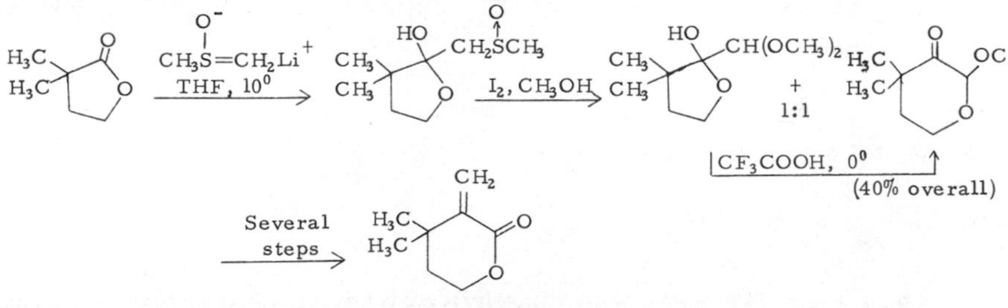

[1] Obtained by base-catalyzed condensation of thiophenol and chloromethyl methyl ether: J. de Lattre, *Bull. Soc. Chem. Belg.*, **26**, 323 (1912); S. Oae, T. Masuda, K. Tsujihara, and N. Furukawa, *Bull. Chem. Soc. Japan*, **45**, 3586 (1972).
[2] B. M. Trost and C. H. Miller, *Am. Soc.*, **97**, 7182 (1975).

***trans*-1-Methoxy-3-trimethylsilyloxy-1,3-butadiene** (1). Mol. wt. 134.25, b.p. 54–55°/5 mm.

*Preparation.* This diene (1) is prepared[1] in 68% yield by the reaction of *trans*-4-methoxybutene-2-one in benzene with trimethylchlorosilane in the presence of anhydrous zinc chloride as catalyst. The reaction is carried out at 40° with stirring overnight.

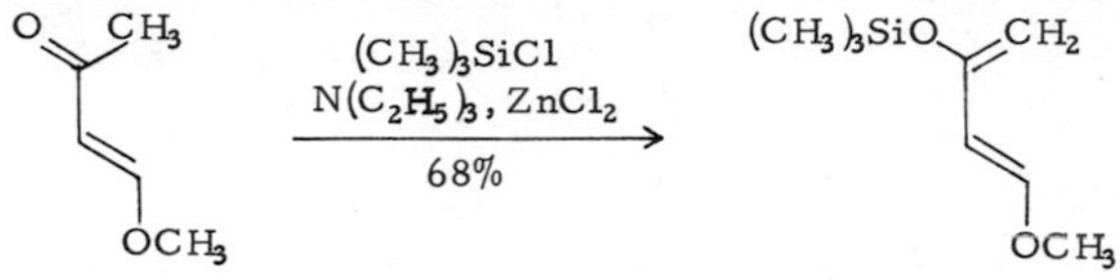

(1)

*Diels–Alder reaction.* 1,3-Dialkoxy-1,3-butadienes are not readily available; however, they are potentially useful dienes for [4 + 2] cycloadditions. The siloxydiene, on the other hand, is readily available and has been shown to be useful for organic syntheses.[1] For example, (1) reacts with maleic anhydride at room temperature to give the adduct (2); work-up with HCl in THF gives the methoxyketo anhydride (3) in 93% yield:

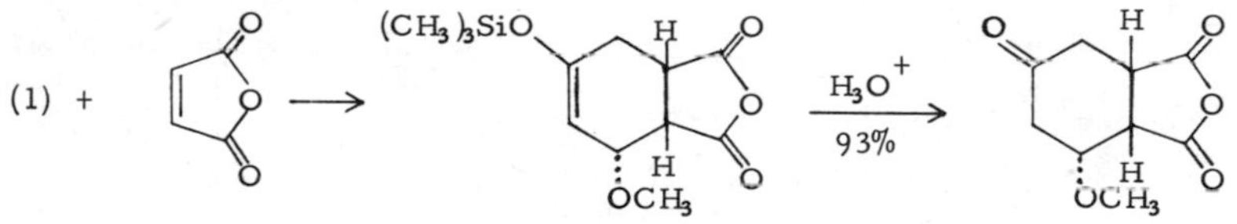

(2)            (3)

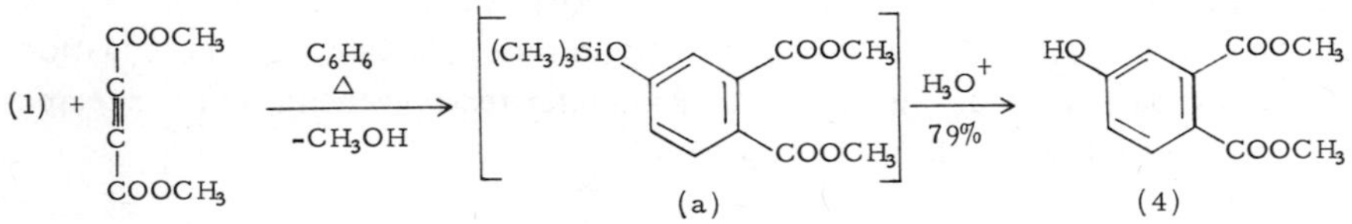

(a)           (4)

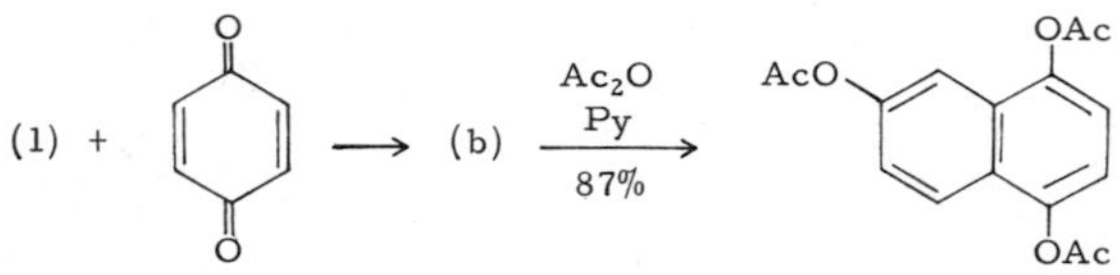

(5)

The reagent (1) reacts with dimethyl acetylenedicarboxylate in refluxing benzene with loss of methanol to give an adduct (a) that gives dimethyl 4-hydroxy-*o*-phthalate (4) on hydrolysis. Cycloaddition of (1) and *p*-benzoquinone gives, after acetylation, 1,4,6-triacetoxynaphthalene (5).

The reaction of (1) with methyl vinyl ketone (95°, 20 hr.) followed by rapid acidic work-up gives a mixture of (6) and (7). Treatment of the mixture with HCl–THF results in conversion of (6) to (7), obtained in 86% overall yield.

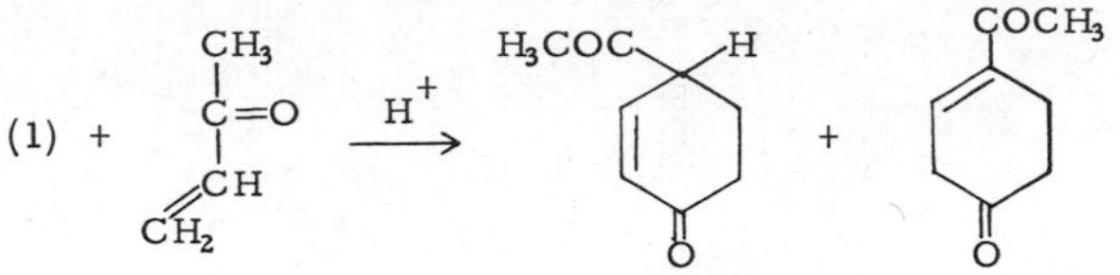

$$(6) \xrightarrow[\substack{86\% \\ \text{overall}}]{H^+} (7)$$

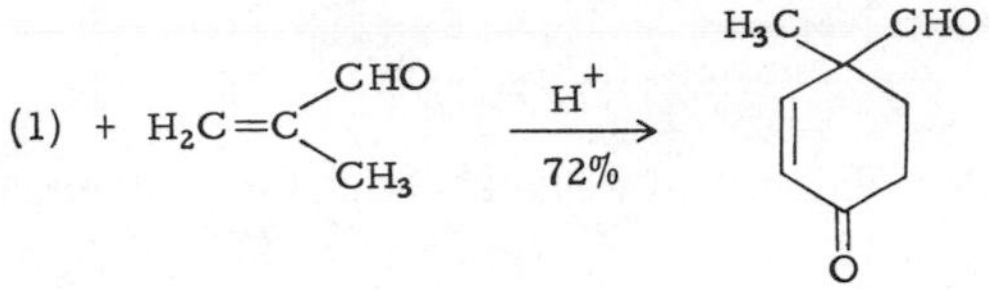

(8)

Reaction of (1) with methyl vinyl ketone to give (7) and with methacrolein to give (8) illustrates the value of (1) in the synthesis of cyclohexenones. The formation of (6) may involve (c) as an intermediate:

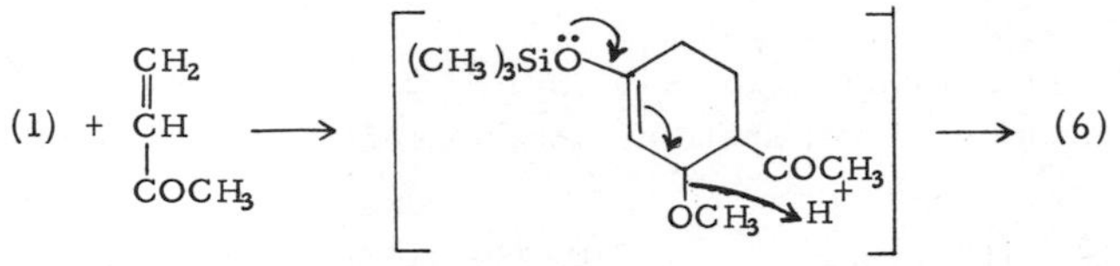

(c)

**cis-$\Delta^1$-3-Octalones.**[2] This substituted diene (1) undergoes cycloaddition with relatively unreactive dienophiles. The reaction has been used as a route to *cis*-$\Delta^1$-3-octalones. Thus (1) reacts with (2) at 190° (30 hr.) to give (3) as the major product after acidic work-up.

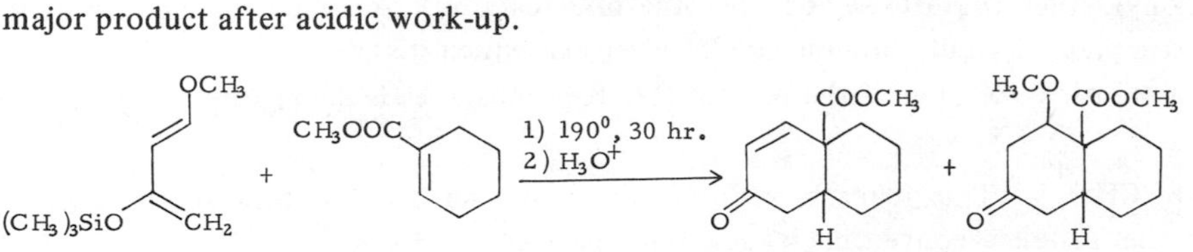

This report also includes an example of sequential Diels–Alder reactions with (1). Thus reaction of (1) with methyl acrylate followed by ketalization leads to the ketal ester (5), which can be used in a second Diels–Alder reaction with (1) resulting in (6) as the major product.

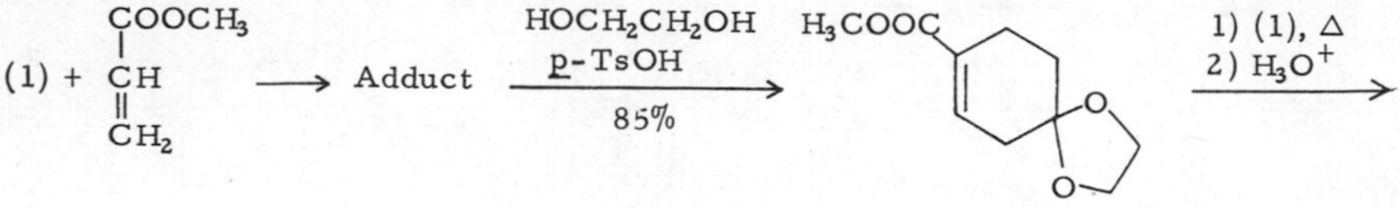

(5)

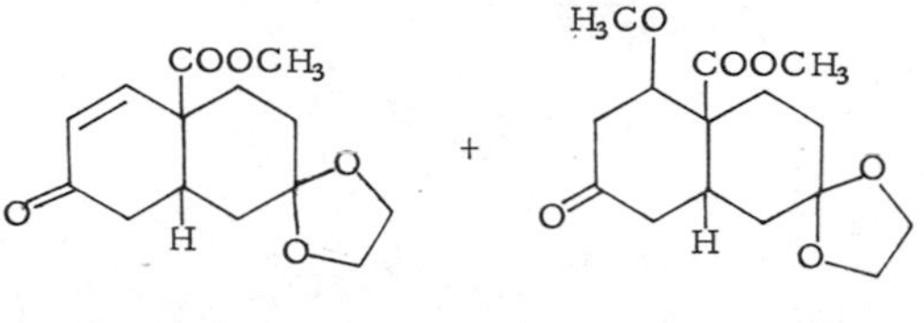

(6, 73%)          (7, 2%)

A third example of synthesis of a cis-$\Delta^1$-3-octalone is the reaction of (1) with 2-methylcylohexenone (8).

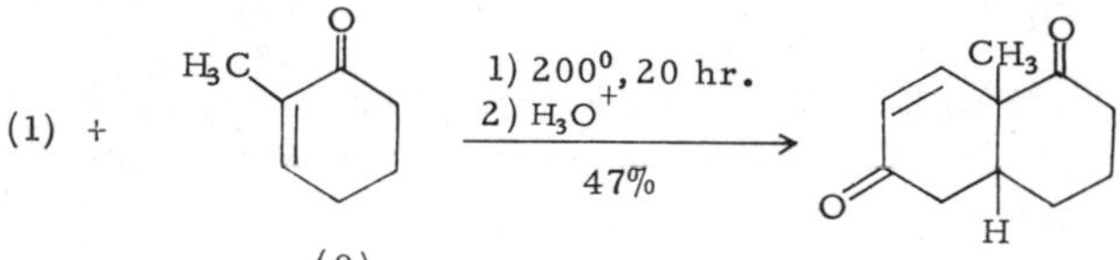

(8)

(9)

Note that the conventional route to trans-$\Delta^1$-3-octalones involves bromination and debromination of corresponding decalones. This method is not suitable in the case of cis-$\Delta^1$-3-octalones.

[1] S. Danishefsky and T. Kitahara, Am. Soc., 96, 7807 (1974).
[2] Idem, J. Org., 40, 538 (1975).

**α-Methoxyvinyllithium**, $\overset{\overset{\displaystyle CH_2}{\|}}{Li\overset{}{C}}-OCH_3$. Mol. wt. 64.01.

This metalated enol ether is prepared by reaction of t-butyllithium with methyl vinyl ether in THF at $-65°$ (exothermic reaction). A yellow precipitate of a complex of t-butyllithium and THF forms, which dissolves at $-5$ to $0°$ to give a colorless solution of the reagent (1). This substance is the synthetic equivalent of $CH_3\overset{\overset{\displaystyle O}{\|}}{C}{}^-$.[1] Thus it reacts with an electrophile such as a ketone to give a vinyl ether product, convertible by acid into the methyl ketone:

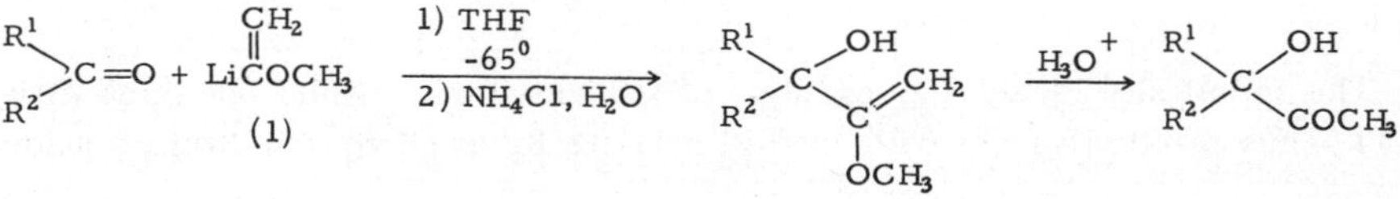

*Examples:*

$$C_6H_5CHO \xrightarrow{78\%} C_6H_5\overset{\displaystyle OH}{\underset{\displaystyle OCH_3}{CH}}\!\!\!-C\!\!=\!\!CH_2 \xrightarrow{91\%} C_6H_5\overset{\displaystyle OH}{CH}\diagdown COCH_3$$

$$\xrightarrow{90\%}$$

$$\xrightarrow{75\%} \quad \xrightarrow{61\%}$$

$$C_6H_5COOCH_3 \xrightarrow[74\%]{2\;(1)} CH_3O\diagdown \underset{CH_2}{C}\diagdown \overset{C_6H_5}{\underset{CH_2}{C}} \diagup \overset{OH}{\underset{\phantom{x}}{C}} \diagdown OCH_3 \xrightarrow{64\%} CH_3 \diagdown \underset{O}{C} \diagdown \overset{C_6H_5}{C} \diagup \overset{OH}{C} \diagdown \underset{O}{C} \diagdown CH_3$$

$$C_6H_5COOH \longrightarrow C_6H_5\overset{O}{\overset{\|}{C}}\!\!-\!C\diagup\overset{OCH_3}{\diagdown CH_2} \xrightarrow{77\%} C_6H_5\overset{O}{\overset{\|}{C}}\!\!-\!\overset{O}{\overset{\|}{C}}\!\!-\!CH_3$$

$$\underline{n}\text{-}C_8H_{17}I \longrightarrow \underline{n}\text{-}C_8H_{17}C\diagup\overset{OCH_3}{\diagdown CH_2} \longrightarrow \underline{n}\text{-}C_8H_{17}\overset{O}{\overset{\|}{C}}CH_3$$

$$C_6H_5COCH_2Br \longrightarrow CH_3O\diagdown C \diagdown \overset{C_6H_5}{\diagup}\overset{O}{\diagdown CH_2} \longrightarrow \overset{C_6H_5}{\diagdown}\!\!C\!\!\diagup\overset{O}{\overset{\|}{\diagdown}CCH_3} \diagdown CH_2OH$$

The third example shows that the reagent undergoes 1,2-addition with α,β-
unsaturated carbonyl compounds. The reaction of esters with 2 eq. of the reagent
results in 3-hydroxy-3-substituted pentanediones. The reaction with acids leads
ultimately to α-diketones.

[1] J. E. Baldwin, G. A. Höfle, and O. W. Lever, Jr., *Am. Soc.*, **96**, 7125 (1974); see also,
U. Schöllkopf and P. Hänssle, *Ann.*, **763**, 208 (1972).

## Methylal (Dimethoxymethane), 1, 671–672.

*Methoxymethyl ethers.* Methoxymethylation of alcohols can be carried out in high yield by reaction with methylal at $25°$ in chloroform catalyzed by phosphorus pentoxide.[1] These ethers are preferable to tetrahydropyranyl ethers

$$ROH + CH_3OCH_2OCH_3 \xrightarrow{P_2O_5} ROCH_2OCH_3$$

for protection of hydroxyl groups because they do not introduce an asymmetric center.

[1] K. Fuji, S. Nakano, and F. Fujite, *Synthesis*, 276 (1975).

## Methyl benzenesulfinate, $C_6H_5S(O)OCH_3$. Mol. wt. 156.20, b.p. $76–78°/0.45$ mm. *Caution:* Skin irritant.

*Preparation*[1]:

$$(C_6H_5S)_2 + 3\ Pb(OAc)_4 + 4\ CH_3OH \xrightarrow[62-68\%]{} 2\ C_6H_5S(O)OCH_3 + 3\ Pb(OAc)_2$$
$$+ 4\ HOAc + 2\ CH_3OAc$$

*β-Keto phenyl sulfoxides.*[2] Treatment of a carbonyl compound with an activated methylene group with this reagent and sodium hydride in anhydrous ether gives β-keto phenyl sulfoxides in 50–85% yield. Methyl *p*-toluenesulfinate can be used in the same way.

*Examples:*

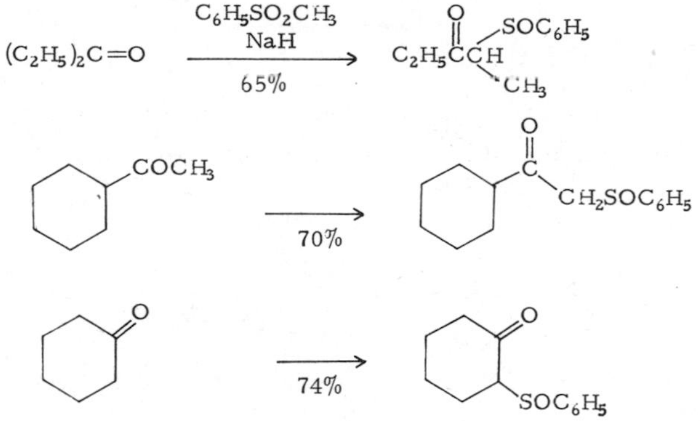

[1] L. Field and J. M. Locke, *Org. Syn., Coll. Vol.*, **5**, 723 (1973).
[2] H. J. Monteiro and J. P. de Souza, *Tetrahedron Letters*, 921 (1975).

## 3-Methylbenzothiazole-2-thione, (1). Mol. wt. 181.20, m.p. $92°$.

*Preparation*[1]:

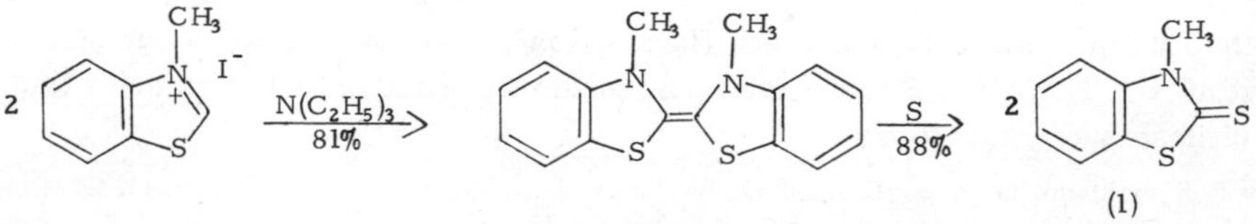

(1)

*Thiiranes.*[2]  Oxiranes react with this reagent and TFA in $CH_2Cl_2$ at $0°$ to give thiiranes, usually in quantitative yields. The reaction is stereospecific: *cis*-stilbene oxide is converted exclusively into *cis*-stilbene episulfide.

[1]H.- W. Wanzlick, H.- J. Kleiner, I. Lasch, H. W. Füldner, and H. Steinmaus, *Ann.*, **708**, 155 (1967).
[2]V. Calò, L. Lopez, L. Marchese, and G. Pesce, *J.C.S. Chem. Comm.*, 621 (1975).

**Methyl(bismethylthio)sulfonium hexachloroantimonate,** $(CH_3S)_2\overset{+}{S}CH_3 SbCl_6^-$ (1). Mol. wt. 475.76.

*Preparation.*[1]

*Thiirenium salts.*[2]  The thiirenium salt (2) is formed quantitatively (NMR spectroscopy) by reaction of (1) with 2-butyne at $-80°$ in liquid $SO_2$. The salt is stable for hours at $-70$ to $-50°$. It is also formed as a transient species by the

(2)

reaction of methanesulfenyl chloride with 2-butyne.

*Episulfonium salts.*[3]  The reagent reacts with alkenes in $CH_2Cl_2$ at $0°$ or in $SO_2$ at $-60°$ to give thiiranium salts (2, episulfonium salts) in 85–90% yield. These products are stable for some time at $-10°$.

(2)

[1]G. Capozzi, V. Lucchini, G. Modena, and F. Rivetti, *J.C.S. Perkin II*, (1975).
[2]G. Capozzi, O. De Lucchi, V. Lucchini, and G. Modena, *J.C.S. Chem. Comm.*, 248 (1975).
[3]*Idem, Tetrahedron Letters*, 2603 (1975).

**1-Methyl-2-bromopyridinium iodide,** (1).

*Esterification.*[1]  The reaction of carboxylic acids and alcohols in the presence of 1.2 eq. of this pyridinium salt and 2.4 eq. of tri-*n*-butylamine affords esters in 60–90% yield. Optimum yields are obtained in refluxing toluene or $CH_2Cl_2$. The method is useful for synthesis of sterically hindered esters.

$$(1) + RCOOH \xrightarrow{\text{Base}} \left[ \begin{array}{c} \text{RCO} \end{array} \right] \xrightarrow[\text{base}]{R'OH} RCOR' + \quad + \text{Base·HI}$$

[1] T. Mukaiyama, M. Usui, E. Shimada, and K. Saigo, *Chem. Letters*, 1045 (1975).

**Methyl chloroformate,** $ClCOOCH_3$. Mol. wt. 94.50, b.p. 68–71°. Suppliers: Aldrich, Eastman, others.

*1,2- and 1,4-Dihydropyridines.* N-Carbomethoxy-1,2- and N-carbomethoxy-1,4-dihydropyridine, (1) and (2), can be prepared by treatment of a mixture of pyridine and sodium borohydride in THF with methyl chloroformate at 0–10°. The mixture contains 35–40% of (2), which can be isolated easily by

$$\text{pyridine} + NaBH_4 \xrightarrow[\text{THF, 0–10}^0]{ClCOOCH_3} \underset{(1)}{\text{COOCH}_3} + \underset{(2)}{\text{COOCH}_3}$$

treatment of the reaction mixture with maleic anhydride, which undergoes a Diels–Alder reaction with (1). The 1,2-isomer is obtained almost exclusively when the reaction is carried out in methanol at −70°. These dihydropyridines are fairly stable to oxygen at 25°. The carbomethoxy group can be converted into an N-methyl group by reduction with $LiAlH_4$. It can be replaced by hydrogen by reaction with methyllithium and then water.[1]

1-Carbomethoxy-4-alkyl(aryl)-1,4-dihydropyridines can be obtained in good yield by the reaction of a dialkyl(aryl)copperlithium with pyridine and methyl chloroformate in ether at −78° and then at 0°. Only 2–10% of the corresponding 1,2-dihydropyridine derivative is formed. No reaction takes place in the absence of methyl chloroformate.[2]

$$\text{pyridine} + R_2CuLi \xrightarrow[-78 \to 0^0]{ClCOOCH_3} \underset{\substack{COOCH_3 \\ (3, \text{ main product})}}{R} + \underset{\substack{COOCH_3 \\ (4, \text{ minor product})}}{R}$$

[1] F. W. Fowler, *J. Org.*, **37**, 1321 (1972).
[2] E. Piers and M. Soucy, *Canad. J. Chem.*, **52**, 3563 (1974).

**Methyl chlorosulfinate—Dimethyl sulfoxide.**

*Alkoxysulfonium salts.* Methyl chlorosulfinate and dimethyl sulfoxide form the adduct (a), isolated as the salt (1). This product (1) reacts with various

(a)                              (1)

substrates to form sulfonium salts.[1]

*Examples:*

[1]Y. Hara and M. Matsuda, *J.C.S. Chem. Comm.*, 919 (1974).

**Methylcopper, 4**, 334–335; **5**, 148.

*(E,E)-1,3-Dienes.* Dialkenylchloroboranes (1), obtained by the reaction of alkynes with chloroborane etherate (**4**, 346–347), react with methylcopper (3 eq. are necessary for satisfactory results) at 0° in ether to give (E,E)-1,3-dienes (2) in 65–100% yield (glpc). Only small amounts of coupling products

(1)                              (2)

are obtained with other copper derivatives such as dimethylcopperlithium or cuprous or cupric salts.[1]

*Addition to α,β-acetylenic sulfoxides.*[2] Methylcopper reacts with α,β-acetylenic sulfoxides (1) by *cis*-addition to give β-methyl-α,β-ethylenic sulfoxides (2) in high yield:

(1)                              (2)

[1]Y. Yamamoto, H. Yatagai, and I. Moritani, *Am. Soc.*, **97**, 5606 (1975).
[2]W. E. Truce and M. J. Lusch, *J. Org.*, **39**, 3174 (1974).

**2-Methylcyclopentenone-3-dimethylsulfoxonium methylide** (1). Mol. wt. 174.26, m.p. 170–173°.

*Preparation:*

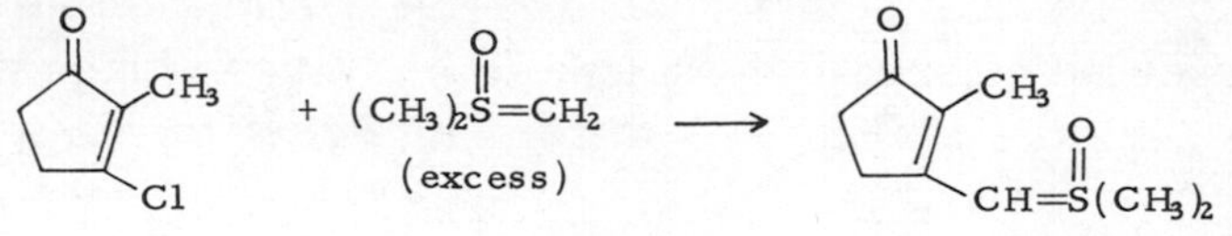

(1)

*Hydroazulenes.* The reagent (1) reacts with various Michael acceptors to give vinylcyclopropanes, for example (2), obtained as a 7:1 mixture of *trans/cis* isomers. The mixture reacts with Wittig reagents to form a *trans*-divinylcyclopropane (3) and a hydroazulene (4), formed directly from the *cis*-isomer of (2).

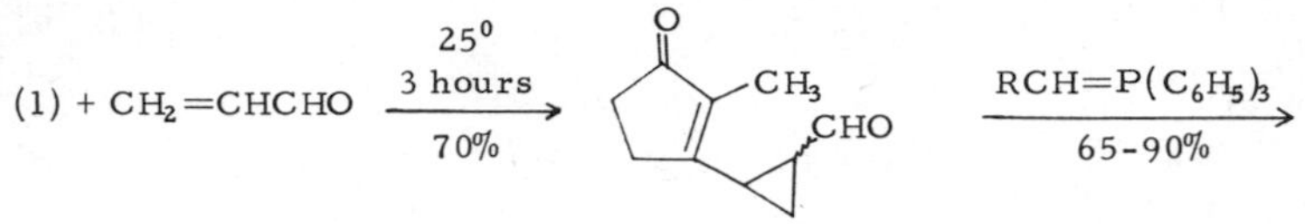

(2, two isomers)

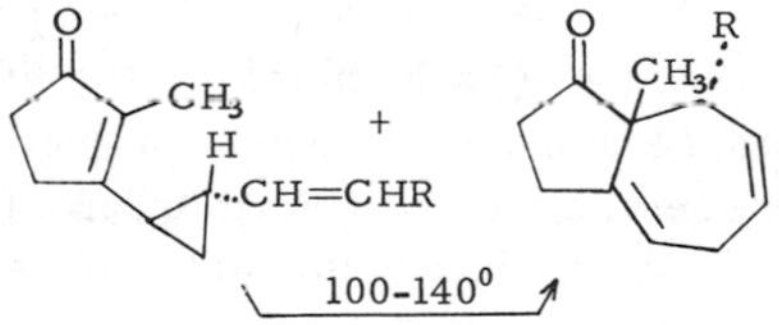

(3)                      (4)

Pyrolysis of (3) results in rearrangement to (4). Overall yields of hydroazulenes are satisfactory.[1]

[1] J. P. Marino and T. Kaneko, *J. Org.*, **39**, 3175 (1974).

**N-Methyl-N,N′-di-*t*-butylcarbodiimium tetrafluoroborate,**

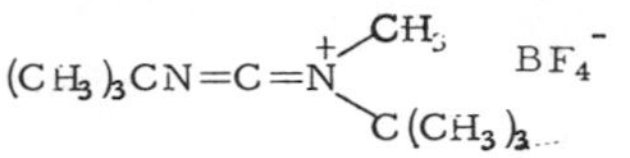

$$(CH_3)_3CN=C=N\overset{+}{\underset{C(CH_3)_3}{\overset{CH_3}{\big|}}} \quad BF_4^- \quad (1)$$

Mol. wt. 256.11, m.p. 85–86°, hygroscopic.

The salt is prepared in 62% yield by alkylation of di-*t*-butylcarbodiimide with trimethyloxonium tetrafluoroborate.[1]

*Dehydration.* Schnur and van Tamelen[2] have investigated use of this reagent for dehydration. Diethylene glycol (2) is converted, in $CHCl_3$–$C_6H_5NO_2$, into *p*-dioxane (3) when heated to 130° with 1 eq. of (1) and a slight excess of triethylamine. Similarly, 1,4-butanediol is converted into tetrahydrofurane (64%),

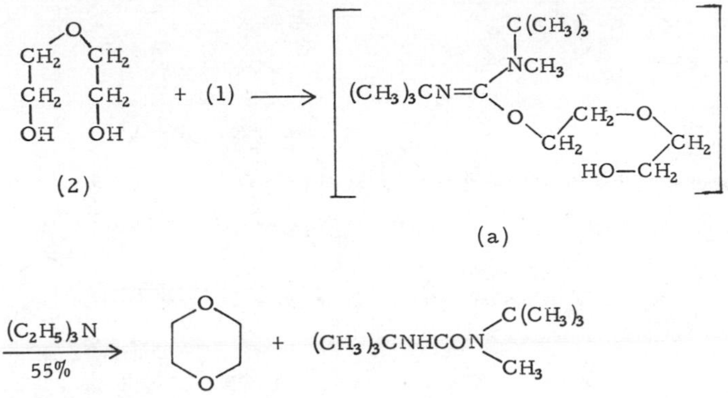

(a)

(3)

and 1,6-hexanediol into oxepane (22%).

The reagent (1) also effects synthesis of dipeptides from an N-acylamino acid and an amino acid ester hydrochloride. Yields are in the range of 70–80%. The optical yield in the one case reported was 86%.

Nucleotides are converted into symmetrical pyrophosphates by the reagent. However, only low yields of dinucleotides are obtained from suitably protected nucleotides and nucleosides with this reagent.

[1] K. Hartke, F. Rossbach, and M. Radau, *Ann.*, **762**, 167 (1972).
[2] R. C. Schnur and E. E. van Tamelen, *Am. Soc.*, **97**, 464 (1975).

**2-Methylene-1,3-propanediol**, $H_2C=C(CH_2OH)_2$. Mol. wt. 58.10, b.p. 127–129°/32 mm.

The reagent is prepared from 5-norbornene-2-carboxaldehyde (Aldrich) by a Cannizzaro reaction with formaldehyde followed by a retro Diels–Alder reaction at 520°.

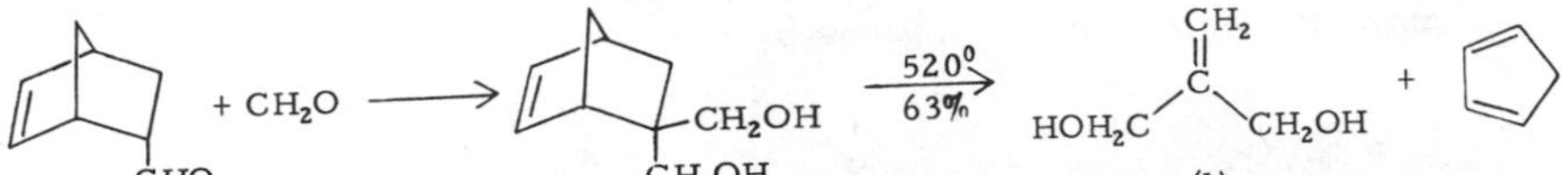

*Protection of carbonyl groups.*[1]  Aldehydes and ketones are converted into acetals and ketals in high yield by reaction with (1) in refluxing benzene (TsOH catalysis) with azeotropic removal of water. Deprotection has been carried out by

five methods, as illustrated for the case of the ketal of 4-*t*-butylcyclohexanone
(2).

[1]E. J. Corey and J. W. Suggs, *Tetrahedron Letters*, 3775 (1975).

## Methylenetriphenylphosphorane, 1, 678.

*Isopropenyl compounds.* The reaction of methylenetriphenylphosphorane with esters in ether solution leads to the corresponding methyl ketones:
$RCOOR' \longrightarrow RCOCH_3$.[1] The reaction follows a different course in DMSO
to give, after hydrolysis, isopropenyl compounds. Three equivalents of the

ylide are consumed for each mole of ester. Yields range from 30 to 75%.[2]
    *Examples:*

$$CH_3(CH_2)_6COOC_2H_5 \longrightarrow CH_3(CH_2)_6C\underset{CH_2}{\overset{CH_3}{\diagup}} + CH_3(CH_2)_6COCH_3$$

$$(40\%) \qquad\qquad (17\%)$$

[1]F. Ramirez and S. Dershowitz, *J. Org.*, **22**, 41 (1957); H. J. Bestmann and B. Arnason, *Ber.*, **95**, 1513 (1962).

[2]A. P. Uijttewaal, F. L. Jonkers, and A. van der Gen, *Tetrahedron Letters*, 1439 (1975).

## Methyl fluoride—Antimony pentafluoride—Sulfur dioxide, 3, 201-202.

*Ene reactions.*[1] The antimony pentafluoride—methyl fluoride complex in $SO_2$ undergoes the ene reaction with vinyl halides (1) at $-65°$ to give $\beta,\gamma$-unsaturated methyl sulfinates (2), isolated after addition of methanol. Yields

$$R = H, CH_2CH_2OCOCH_3$$

are high (70-80%). The reaction presumably involves methylated sulfur dioxide, $[CH_3O{-}S{=}O]^+SbF_6^-$, as the enophile. The ene reaction is reversed by base (60% yield).

[1]P. E. Peterson, R. Brockington, and M. Dunham, *Am. Soc.*, **97**, 3517 (1975).

## Methyl fluorosulfonate, 3, 202; 4, 339-340; 5, 445-446.

*Thioacetal—hemithioacetal—acetal interchange.* Corey and Hase[1] have disclosed a method for direct interchange of these functional groups under neutral conditions. The method is formulated for acetal interchange of heptanal.

This acetal interchange is applicable also to trimethylene dithioacetals, as shown by conversion of the 1,3-dithane derivative (1) into the hemithioacetal (2).

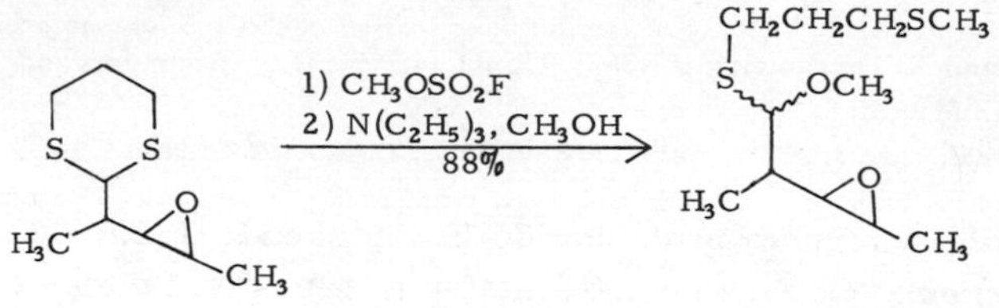

1) $CH_3OSO_2F$
2) $N(C_2H_5)_3$, $CH_3OH$
88%

The present method has several attractive features: It does not go through the free carbonyl system (which could be sensitive to acidic or basic conditions); replacement of thioacetal by acetal is desirable in reactions involving oxidations; and hemithioacetal groups are cleaved under neutral conditions by $HgCl_2$.

*Alkylation of ambident nucleophiles.* Beak and Lee[2] have reported several examples of alkylation of ambident compounds in which the product results from methylation at the hetero atom that does not bear the proton in the major tautomer. Intermediate salts such as (a) have been isolated and characterized in these reactions. The reaction is significant in that, at least in some cases, the

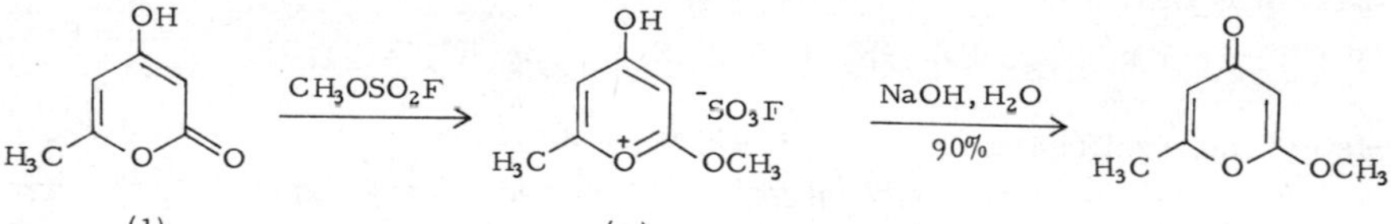

$CH_3OSO_2F$

$^-SO_3F$

$NaOH$, $H_2O$
90%

(1)          (a)          (2)

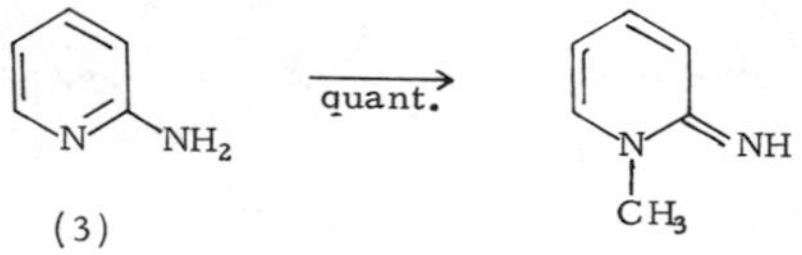

quant.

(3)          (4)

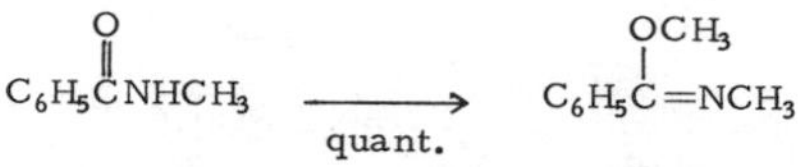

quant.

(5)          (6)

less stable alkylated derivative is formed.

*Cyclobutenones.*[3] 4,4-Dimethylcyclobutenone (3) has been prepared by alkylation of (1) with methyl fluorosulfonate. The resulting sulfonium salt undergoes elimination of dimethyl sulfide when treated with quinoline in DMSO at 25° for 2–3 hr.

(1)

(2)                                                                (3)

This cyclobutenone was shown to be a reactive dienophile.

*Alkylation of a cephalosporin.*[4] Reaction of the ester (1) with methyl fluorosulfonate gives the optically active sulfonium salt (2) formed with inversion at $C_6$. This is the first report of an S-alkylated penam or cepham. The

(1)                                                                (2)

reaction is believed to involve alkylation of sulfur followed by cleavage and reclosure of the $C_6$–S bond.

**trans-α-*Methoxystilbene.*[5]** The enol of desoxybenzoin (1) is predominately O-methylated by methyl fluorosulfonate in HMPT:

(1)                                                                (2)

The yield of (2) using dimethyl sulfate in HMPT is 43%.

[1] E. J. Corey and T. Hase, *Tetrahedron Letters,* 3267 (1975).
[2] P. Beak and J.-K. Lee, *J. Org.,* **40,** 147 (1975).
[3] T. R. Kelly and R. W. McNutt, *Tetrahedron Letters,* 285 (1975).
[4] D. K. Herron, *Tetrahedron Letters,* 2145 (1975).
[5] G. R. Krow and E. Michener, *Synthesis,* 572 (1974).

**Methyl iodide, 1**, 682–685; **2**, 274; **3**, 202; **4**, 341.

*Caution:* Methyl iodide can cause serious toxic effects in the central nervous system. Since it is an alkylating agent, it should be handled as a possible carcinogen.[1]

*Corey-Winter olefin synthesis* (**1**, 1233–1234; **2**, 439–441; **3**, 315–316; **4**, 269–270, 541–542; **5**, 34, 661). Vedejs and Wu[2] converted thionocarbonates into olefins by alkylation with methyl iodide (90°, DME, sealed tube) followed by reduction (zinc dust—ethanol or magnesium amalgam—THF). This two-step procedure was used for preparation of the cyclobutene (2); in this case the reaction with triethyl phosphite was immeasurably slow.

(1)    (2)

This variation of the Corey-Winter reaction is not stereoselective; thus the thionocarbonates of either *cis-* or *trans*-cyclooctane-1,2-diol give only *cis*-cyclooctene.

[1]*Documentation of the Threshold Limit Values,* 3rd ed., American Conference of Governmental Industrial Hygienists, (1971), p. 166.
[2]E. Vedejs and E. S. C. Wu, *J. Org.,* **39**, 3641 (1974).

**Methyllithium, 1**, 686–689; **2**, 274-278; **3**, 202-204; **5**, 448-454.

*Reaction with β,γ-unsaturated carboxylic acids.* A few years ago Dalton and Chan[1] reported that the β,γ-unsaturated carboxylic acid (1) reacts with methyllithium (2 eq.) in hexane to give the expected methyl ketone (2) and that a product (3) of rearrangement is obtained if the reaction is conducted in ether. This reaction is not connected with the cyclic structure of (1), since a

(2)    (1)    (3)

(5)    (4)    (6)

similar result has now been obtained with the β,γ-unsaturated carboxylic acid (4).[2] Reaction of (4) with methyllithium in ether gives the expected methyl ketone (5); reaction in THF–ether (4:2) leads to the rearranged methyl ketone (6). Thus, by proper choice of solvent, it is apparently possible to conduct this reaction to give either the normal or the rearranged product without formation of mixtures.

*β-Methyl-α,β-unsaturated ketones.* Hiyama et al.[3] have reported a one-carbon homologation via the dichlorocarbene adducts of enol ethers, illustrated for the adduct of 1-ethoxycyclododecene (1). Thus treatment of (1) with 2 eq.

$$\text{(1)} \quad \xrightarrow[-95 \to 25^0]{2\text{CH}_3\text{Li, THF/HMPT}} \quad \left[ \text{(a)} \to \text{(b)} \right]$$

$$\xrightarrow[75\%]{\text{H}_3\text{O}^+} \quad \text{(2, E/Z = 3:1)}$$

of methyllithium in THF–HMPT at low temperatures and then at room temperature followed by treatment with dilute sulfuric acid gives (2) in 75% yield. The yield is lower (47%) when HMPT is omitted. Addition of TMEDA also improves the yield.

The reaction is believed to involve loss of hydrogen chloride (a), and addition of methyllithium to give the ethoxyallene (b).

The sequence was used for a synthesis of muscone from 1-ethoxycyclotetra-decene in about 45% overall yield.

*6-Oxatricyclo[3.2.1.0²,⁷]octane.*[4] Treatment of the *gem*-dibromocyclo-propane (1) with methyllithium in ether for 10 min. gives a mixture of (2) and (3). The origin of (3) is not understood, but evidence has been obtained that (2) arises by an unusual intramolecular carbenoid insertion into the O–H bond.

$$\text{(1)} \longrightarrow \text{(2)} + \text{(3)}$$

[1] J. C. Dalton and H.-F. Chan, *Tetrahedron Letters*, 3145 (1973).
[2] J. C. Dalton and B. G. Stokes, *ibid.*, 3179 (1975).
[3] T. Hiyama, T. Mishima, K. Kitatani, and H. Nozaki, *Tetrahedron Letters*, 3297 (1974).
[4] A. R. Allan and M. S. Baird, *J.C.S. Chem. Comm.*, 172 (1975).

## Methyllithium—Dimethylcopperlithium.

*Equatorial methylation of cyclohexanones.* MacDonald and Still[1] report that a 3:2 mixture of methyllithium and dimethylcopperlithium reacts with 4-*t*-butylcyclohexanone (1) at $-70°$ to give almost exclusively *trans*-4-*t*-butyl-1-methylcyclohexanol (2).

Mixtures of (2) and (3) formed with other reagents are as follows:

|              | Percent (2) | Percent (3) |
|--------------|-------------|-------------|
| $CH_3MgI$    | 51          | 49          |
| $(CH_3)_2Mg$ | 70          | 30          |
| $CH_3Li, -78°$ | 79        | 21          |

Thus $CH_3Li$ in combination with $(CH_3)_2CuLi$ shows the highest stereochemical control reported to date for this methylation. Another example of this enhanced stereoselectivity was reported in the same paper[1] (equation I).

The structure of this new reagent is uncertain; both $(CH_3)_3CuLi_2$ and $(CH_3)_4CuLi_3$ are suggested as possibilities for this new cuprate.

[1] T. L. Macdonald and W. C. Still, *Am. Soc.*, **97**, 5280 (1975).

**2-Methyl-4-methoxymethyl-5-phenyl-2-oxazoline,** (1).

Mol. wt. 189.25. Supplier: Aldrich.

*Preparation.*[1] The 4S,5S-(−)-isomer of this oxazole can be prepared from 1S,2S-(+)-1-phenyl-2-amino-1,3-propanediol (Aldrich, Strem).

$$\text{(structure: amino alcohol with HO, C}_6\text{H}_5\text{, H}_2\text{N, CH}_2\text{OH)} \quad + \quad CH_3C(OC_2H_5)=NH \cdot HCl \longrightarrow CH_3-C \text{(oxazoline, C}_6\text{H}_5, \text{CH}_2\text{OH)} \xrightarrow[\;CH_3I\;]{NaH} \quad (1)$$

*Asymmetric synthesis of dialkylacetic acids.* (Compare **3**, 314.) Meyers and Knaus[1] have reported the asymmetric synthesis of R- or S-dialkylacetic acids from (1) by metalation with 1 eq. of lithium diisopropylamide and addition of

$$(-)\text{-}(1) \quad \begin{array}{l} \text{1) LDA, R}'X \\ \text{2) LDA, RX} \end{array} \Bigg\} \longrightarrow \quad \text{(oxazoline (2), R, R}', C_6H_5, CH_2OCH_3)$$

$$\xrightarrow{\;H_3O^+\;} \quad \text{(S)}\text{-}(3)\ \text{[R}', H, R, COOH]}$$

$$(-)\text{-}(1) \quad \begin{array}{l} \text{1) LDA, RX} \\ \text{2) LDA, R}'X \end{array} \Bigg\} \nearrow \quad (2) \qquad \xrightarrow{\;H_3O^+\;} \quad R\text{-}(3)\ \text{[R, COOH, H, R}']}$$

an alkyl halide, followed by a second alkylation in the same way. If R' has a higher priority than R (Cahn–Prelog–Ingold sequence), the first sequence leads to the (S)-dialkylacetic acid, and the second sequence to the (R)-acid. Overall yields are 65–75%; optical yields of 75% can be obtained, particularly if the temperature is low (about −100°).

The related oxazoline (4) has been used in the same way for synthesis of (S)-2-methylalkanoic acids (5).[2]

$$\text{(4) [H, H}_3C, O, C_6H_5, N, CH_2OCH_3]} \xrightarrow[\;\;2)\ H_3O^+\;\;]{\begin{array}{c}\text{1) LDA, −78°}\\ RI\end{array}} \text{(S)-(5) [H, R, H}_3C, COOH]}$$

*γ-Butyrolactones; 1,4-butanediols.* Meyers and Mihelich[3] have developed a synthesis of (R)-γ-butyrolactones and (R)-1,4-butanediols from this chiral oxazoline (1). The sequence is shown in the formulation. The lactones (5) obtained

(1)                                                    (a)

(2)                                                    (3)

(4)                    (R-5)                    (R-6)

in this way have the R-configuration. Yields of (5) from (2) are in the range of 60–75%. Note that the amino alcohol (4) can be recycled to (1).

*3-Alkylalkanoic acids.*[4] This chiral oxazoline (1) can be converted into an

(1)                                            (2)

(3)                                            (4)

optically active 3-substituted alkanoic acid (4) by condensation of the lithium salt with an aldehyde followed by dehydration to an alkenyloxazoline (2). This product undergoes 1,4-addition with an organolithium derivative to give, after hydrolysis, the carboxylic acid (4) with an optical purity of >90%. The enantiomeric acid can be obtained by simply reversing the order of introduction of the R and R' groups.

[1] A. I. Meyers and G. Knaus, *Am. Soc.*, **96**, 6508 (1974); *idem, ibid*, **98**, 567 (1976).
[2] A. I. Meyers, G. Knaus, and K. Kamata, *ibid.*, **96**, 268 (1974).
[3] A. I. Meyers and E. D. Mihelich, *J. Org.*, **40**, 1186 (1975).
[4] A. I. Meyers and C. E. Whitten, *Am. Soc.*, **97**, 6266 (1975).

**Methyl γ-methylthiocrotonate,**

$$\underset{H}{\overset{CH_3SCH_2}{>}}C=C\underset{COOCH_3}{\overset{H}{<}}$$

(1). Mol. wt. 146.21, b.p. 95–97°/12 torr.

*Preparation.*[1]

$$\underset{H}{\overset{BrCH_2}{>}}C=C\underset{COOCH_3}{\overset{H}{<}} + CH_3SH \xrightarrow[71\%]{\underset{(C_2H_5)_2O}{N(C_2H_5)_3}} \underset{H}{\overset{CH_3SCH_2}{>}}C=C\underset{COOCH_3}{\overset{H}{<}}$$

(1)

*Synthetic applications.*[2] This α,β-unsaturated ester (1) is converted by lithium diisopropylamide (LDA) into the extended enolate (a), which undergoes alkylation at the α-carbon atom to give a product with a potential carbonyl group on the γ-carbon atom. Under the same conditions a second alkyl group

$$(1) \xrightarrow[THF, -78°]{LDA} \left[ \overset{Li^+}{\underset{CH_3SCH\!\cdots\!CH\!\cdots\!CHCOOCH_3}{}} \right] \xrightarrow{RX} \underset{H}{\overset{CH_3S}{>}}C=CHC\underset{COOCH_3}{\overset{R}{H}} \xrightarrow[2)\ R'X]{1)\ LDA, -78°}$$

(a)                  (2)

$$\underset{H}{\overset{CH_3S}{>}}C=CHC\underset{COOCH_3}{\overset{R}{\underset{|}{\overset{|}{C}}}}-R'$$

(3)

can be introduced at the α-position of (2) to give (3).

Direct hydrolysis of the enol thioether function in (2) and (3) proved unexpectedly difficult; however, it was possible to hydrolyze the free acid (4) by 1:1 HCl–HOAc at 90° to give (5), which contains a masked aldehyde group. This substance undergoes a normal Wittig reaction with carbomethoxymethylenetriphenylphosphorane to give (6).

$$\underset{H_3C}{\overset{H_3C}{>}}C\underset{\overset{|}{CH}}{\overset{COOH}{<}} \xrightarrow[67\%]{HCl/HOAc} \quad (5) \quad \xrightarrow{(C_6H_5)_3P=CHCOOCH_3} \quad (6)$$

(4)                  (5)                  (6)

Treatment of the ester (7) with HCl–HOAc leads to the cyclohexenone (8) in high yield. This reaction probably involves hydration of the alkyne bond.

The ester (9) undergoes Michael addition to methyl vinyl ketone to give the adduct (10), which cyclizes readily to the acylcyclopentenecarboxylic ester (11).

The allyl derivative (12) undergoes Cope rearrangement at 200° (8 hr.) to give (13) in 70% yield. This can be converted into the triene (14) by sulfoxide elimination.

[1] L. Birkofer and I. Hartwig, *Ber.*, **87**, 1189 (1954).
[2] A. S. Kende, D. Constantinides, S. J. Lee, and L. Liebeskind, *Tetrahedron Letters*, 405 (1975).

**Methyl methylthiomethyl sulfoxide**, 4, 341–342; 5, 456–457.

*Cyclobutanones.* Reaction of methyl methylthiomethyl sulfoxide with 1,3-dibromopropane in the presence of 2 eq. of potassium hydride leads to cyclo-butanone dimethyl dithioacetal S-oxide (2) in high yield.[1] Intermediates (a)

(1)    + Br(CH₂)₃Br   →[2 KH][THF]   (a)   →   (b)

→[97%] (2)   →[H₃O⁺][79%]   (3)

and (b) are suggested. Hydrolysis of (2) to cyclobutanone (3) is carried out with ethanol containing a few drops of sulfuric acid (18 hr. at 25° or 5 hr. at 45°).

The method was used to synthesize 2- and 3-substituted cyclobutanones.

*Cycloalkanones.*[2] Five- and six-membered cyclic ketones can also be prepared by reaction of methyl methylthiomethyl sulfoxide (1) and a 1,ω-dihalo-( or ditosyloxy)alkane (2) in the presence of 2 eq. of base (potassium hydride or *n*-butyllithium) in THF at -70 to 20°. The product (3) is hydrolyzed to the ketone (4) by acid. Ogura *et al.* suggest that 2 eq. of base are required because

(1)   +   (2, $\underline{n}$=3,4,5)   →[2 n-BuLi][THF]   (a)   →[70-90%]

(3)   →[H₂SO₄][C₂H₅OH][80-90%]   (4)

a Stevens-type rearrangement is involved in the formation of (3). The ready formation of a four-membered ring is also accounted for on the basis of an intermediate such as (a), which contains one more atom in the ring than the final ketone. The reaction fails when *n* = 6.

This method was also used for the synthesis of tetrahydro-γ-pyrone (equation I) and of 3-cyclopentenone (equation II).

[1] K. Ogura, M. Yamashita, M. Suzuki, and G. Tsuchihashi, *Tetrahedron Letters*, 3653 (1974).
[2] K. Ogura, M. Yamashita, S. Furukawa, M. Suzuki, and G. Tsuchihashi, *ibid.*, 2767 (1975).

## N,N-Methylphenylaminotriphenylphosphonium iodide, $(C_6H_5)_3P^+N(CH_3)C_6H_5I^-$ (1).

*Preparation.*[1]

*Unsymmetrical amines.*[2] Alcohols can be converted into unsymmetrical secondary and tertiary amines by treatment of the alkoxide (sodium hydride, DMF) with an amine and this phosphonium salt.

*Unsymmetrical sulfides.*[3] Unsymmetrical sulfides can be prepared in 65–85% yield by reaction of sulfides with the sodium salts of alcohols in the presence of this aminophosphonium salt (equation I). The sulfuration proceeds with inversion of configuration (equation II).

[1] H. Zimmer and G. Singh, *J. Org.*, **28**, 483 (1963).
[2] Y. Tanigawa, S.-I. Murahashi, and I. Moritani, *Tetrahedron Letters*, 471 (1975).
[3] Y. Tanigawa, H. Kanamaru, and S.-I. Murahashi, *ibid.*, 4655 (1975).

**3-Methyl-1-phenyl-2-phospholene,** (1). Mol. wt. 164.18.

*Preparation.* The phospholene is prepared from isoprene and dichlorophenylphosphine.[1]

*Aromatic aldehydes.*[2] The reagent reacts with aromatic acid chlorides in the presence of triethylamine in an inert solvent to form an acylphospholenium salt (2), which is usually not isolated, but is hydrolyzed to the aldehyde (3) and the phospholene oxide (4). The oxide can be reduced to (1) by polymethylhydrogen siloxane (DC 1107, Dow).

$$(1) \xrightarrow[\text{N}(C_2H_5)_3]{\text{ArCOCl}} \quad (2) \xrightarrow{H_2O} ArCHO + (4)$$

(2)                              (3)                 (4)

Aliphatic acid chlorides under these conditions are merely hydrolyzed to the acid. The method also fails with aromatic acid chlorides containing nitro and hydroxy groups. In general, yields of aldehydes are in the range of 65–90%.

This synthesis provides an alternative to Rosenmund reduction, which is sometimes difficult.

[1] W. B. McCormack, *Org. Syn.*, **43**, 73 (1963).
[2] D. G. Smith and D. J. H. Smith, *J.C.S. Chem. Comm.*, 459 (1975); *cf.* F. Mathey, *Tetrahedron*, **29**, 707 (1973).

**Methyl 2-phenylsulfinylacetate**, $C_6H_5\overset{O}{\overset{\uparrow}{S}}CH_2COOCH_3$ (1). Mol. wt. 198.24, m.p. 46–48°.

*Preparation:*

$$C_6H_5SH + BrCH_2COOCH_3 \longrightarrow C_6H_5SCH_2COOCH_3 \xrightarrow{NaIO_4} (1)$$

*α,β-Unsaturated esters.* Trost et al.[1] have developed a synthesis of olefins based on the fact that sulfoxides with an α electronegative substituent undergo thermal elimination rather easily to give olefins. For example, the enolate anion (2) of (1), obtained by reaction with NaH or LDA, reacts with alkyl halides at room temperature in HMPT to give (3); the temperature is then raised to ~80° and β-elimination is allowed to proceed at this temperature for 10–15 hr. α,β-Unsaturated esters (4) are obtained in about 80% yield. Alkylation proceeds rather sluggishly except for benzyl or allyl bromides and alkyl iodides. All disubstituted olefins obtained in this way have the E-configuration.

$$RCH_2X + C_6H_5\overset{O}{\overset{\uparrow}{S}}CH=\overset{O^-}{\overset{|}{C}}OCH_3 \longrightarrow RCH_2\overset{\overset{O}{\overset{\uparrow}{SC_6H_5}}}{\underset{COOCH_3}{CH}} \xrightarrow{\Delta} \overset{R}{\underset{H}{}}C=C\overset{H}{\underset{COOCH_3}{}} + C_6H_5SOH$$

(2)                              (3)                              (4)

*Examples:*

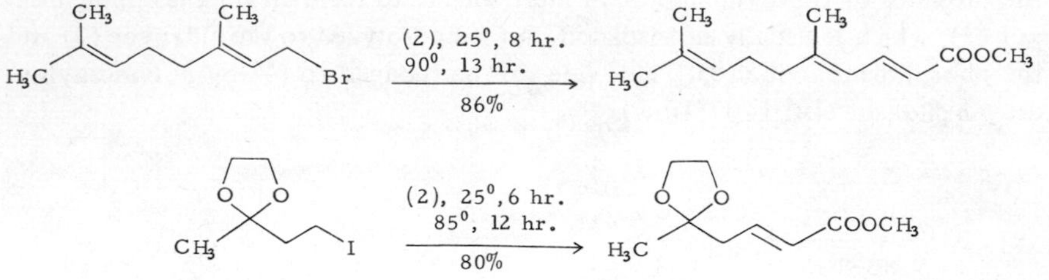

$\alpha,\beta$-Unsaturated esters can also be synthesized by use of the $\pi$-allylpalladium complexes of olefins and a phosphine ligand. Highest yields are obtained in DMSO with HMP[2] as ligand:

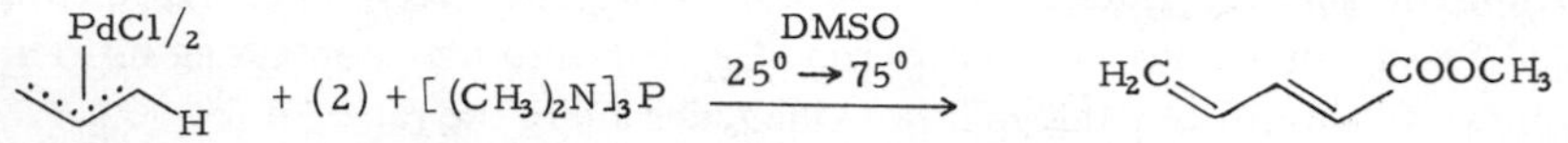

Trisubstituted olefins obtained in this procedure are usually mixtures of E- and Z-isomers, but disubstituted olefins have the E-configuration.

*Other activated sulfoxides.* This alkylative elimination reaction has been extended by Trost and Bridges[3] to a one-pot synthesis of alkenes, vinyl sulfides, $\alpha,\beta$-unsaturated sulfoxides, and $\alpha,\beta$-unsaturated nitriles. The sulfoxides (1–4) are converted into the anions by lithium N-isopropylcyclohexylamide or sodium hydride and are then alkylated at 20° in THF or DME; elimination is then effected by raising the temperature to reflux. In some cases trimethyl phosphite is added as a scavenger for phenylsulfenic acid. Typical results are formulated in the equations. The elimination reaction is facilitated by an aryl, thioaryl, or

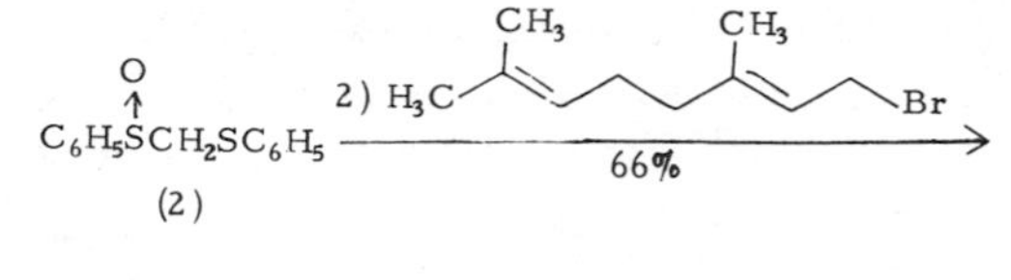

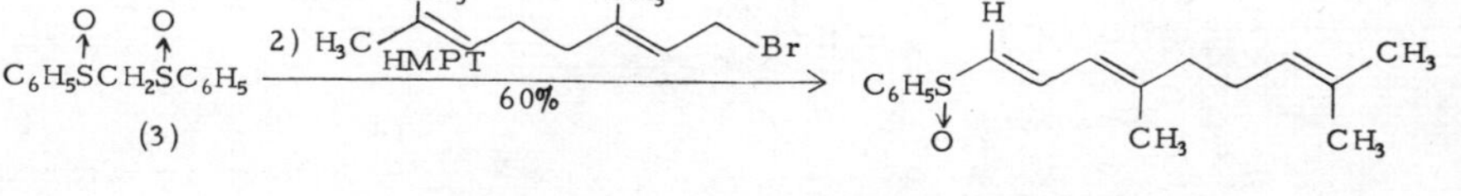

$$\begin{array}{c} O \\ \uparrow \\ C_6H_5SCH_2CN \end{array} \xrightarrow[\substack{2)\ H_3C \\ 49\%}]{1)\ base} \quad \text{(see scheme)}$$

(4)

cyano group on the α-carbon atom, but is decreased by a phenylsulfinyl group (sulfoxide 3). In this case HMPT is used as solvent. In the case of (1) and (3) the introduced double bond has the E-stereochemistry. In the two other cases both possible isomers are formed.

[1] B. M. Trost, W. P. Conway, P. E. Strege, and T. J. Dietsche, *Am. Soc.*, **96**, 7165 (1974).
[2] $[(CH_3)_2N]_3P$.
[3] B. M. Trost and A. J. Bridges, *J. Org.*, **40**, 2014 (1975).

**N-Methylphenylsulfonimidoylmethyllithium,** $C_6H_5\overset{O}{\underset{NCH_3}{\overset{\|}{S}}}CH_2Li$ . Mol. wt. 165.17.

(1)

*Preparation:*

$$C_6H_5\overset{O}{\overset{\|}{S}}CH_3 \xrightarrow[\substack{NaN_3 \\ H_2SO_4 \\ CHCl_3}]{} C_6H_5\overset{O}{\underset{NH}{\overset{\|}{S}}}CH_3 \xrightarrow[\substack{CH_2O \\ HCOOH}]{} C_6H_5\overset{O}{\underset{NCH_3}{\overset{\|}{S}}}CH_3 \xrightarrow[\substack{n\text{-}BuLi \\ THF}]{} (1)$$

*Methylenation of carbonyl compounds.*[1]  The reagent reacts with aldehydes and ketones in THF at 0° to give β-hydroxy sulfoximines (2). On reduction of (2) with aluminum amalgam in aqueous THF containing acetic acid, reductive elimination occurs to give an olefin (3).

$$\begin{array}{c} R^1 \\ R^2 \end{array}\!\!\!C=O + (1) \longrightarrow \begin{array}{c} R^1 \\ R^2 \end{array}\!\!\!\underset{OH}{\overset{}{C}}-CH_2\overset{O}{\underset{NCH_3}{\overset{\|}{S}}}C_6H_5 \xrightarrow[\substack{Al/Hg,\ HOAc \\ THF,\ H_2O}]{} \begin{array}{c} R^1 \\ R^2 \end{array}\!\!\!C=CH_2 + C_6H_5\overset{O}{\overset{\|}{S}}NHCH_3$$

(2)  (3)

*Examples:*

$$CH_3(CH_2)_6CHO \xrightarrow{60\%} CH_3(CH_2)_6CH=CH_2$$

$$(CH_3)_3C\!\!-\!\!\bigcirc\!\!=O \xrightarrow{73\%} (CH_3)_3C\!\!-\!\!\bigcirc\!\!=CH_2$$

[1] C. R. Johnson, J. R. Shanklin, and R. A. Kirchhoff, *Am. Soc.*, **95**, 6462 (1973).

**Methylthioacetic acid,** $CH_3SCH_2COOH$ (1). Mol. wt. 104.14.

This reagent is prepared by reaction of methylmercaptan with chloroacetic acid.

*Acyl anion equivalent.*[1] This acid can be converted into the dianion with LDA in THF–HMPT at `0°`. Alkylation followed by oxidative decarboxylation (**N-Bromosuccinimide**, this volume) with sodium metaperiodate results in an aldehyde, as shown in equation I. A ketone is obtained by bis-alkylation followed by oxidative decarboxylation (equation II). Note that the carbonyl group is derived from the reagent, which is thus equivalent to the dianion $\bar{\bar{C}}=O$.

(I) $\quad CH_3\bar{S}CHCOO^- 2\,Li^+ + Cl(CH_2)_{10}CH_2I \xrightarrow[69\%]{} Cl(CH_2)_{10}CH_2\underset{\underset{SCH_3}{|}}{C}HCOOH \xrightarrow[51\%]{NaIO_4 \atop CH_3OH}$

$$Cl(CH_2)_{10}CH_2CH(OCH_3)_2$$

(II) $\quad CH_3\bar{S}CHCOO^- 2\,Li^+ + (CH_3)_2CHI \xrightarrow[59\%]{} (CH_3)_2\underset{\underset{SCH_3}{|}}{C}H\overset{\overset{COOH}{|}}{C}H \xrightarrow[63\%]{1)\ 2\ LDA \atop 2)\ C_6H_5CH_2Br}$

$$(CH_3)_2CH-\underset{\underset{SCH_3}{|}}{\overset{\overset{COOH}{|}}{C}}CH_2C_6H_5 \xrightarrow[80\%]{1)\ NCS,\ NaHCO_3 \atop 2)\ H_3O^+} (CH_3)_2CH\overset{\overset{O}{\|}}{C}CH_2C_6H_5$$

[1] B. M. Trost and Y. Tamaru, *Tetrahedron Letters*, 3797 (1975).

## 3-Methylthio-1,4-diphenyl-*s*-triazolium iodide (1). Mol. wt. 395.26, m.p. 244–245° dec.

*Preparation.* The heterocycle is readily available by cyclization of 1,4-diphenylthiosemicarbazide and methylation of the intermediate 1,4-diphenyl-*s*-triazoline-3-thione (equation I).

(I) $\quad C_6H_5HNNH-\overset{\overset{S}{\|}}{C}NHC_6H_5 \xrightarrow[71\%]{CH(OC_2H_5)_3,\ CH_3I \atop DMF}$

(1)

*Aldehyde synthesis.*[1] Treatment of (1) with excess sodium hydride in DMF at 0° produces a "nucleophilic carbene" (2); treatment of the solution of (2)

(2)                             (3)

(a)

with an alkyl halide and aqueous potassium iodide gives a 5-alkyl derivative
(3). Reduction of these products (3) with sodium borohydride followed by
hydrolysis gives aldehydes (4) in yields of 30–80%.

[1] G. Doleschall, *Tetrahedron Letters*, 1889 (1975).

## 1-Methylthio-3-methoxypropyne, $CH_3SC \equiv CCH_2OCH_3$ (1). Mol. wt. 116.18.

*Preparation.*[1] This reagent is prepared by addition of 1 eq. of dimethyl
disulfide to 1-lithio-3-methoxypropyne[2] at $-30°$; the reaction mixture is then
allowed to stand at room temperature for 3 days.

*β-Keto aldehydes; 3-furanones.*[1] The carbanion of (1) reacts entirely in the
allenic form (2b)[3] on alkylation to give the corresponding substituted allene (3).
These products can be hydrolyzed selectively either to dimethyl acetals of β-keto
aldehydes (4) or to alkylthioacrolein derivatives (5).

The alkylated allene (3) can also be converted into a carbanion (6). Reaction of this carbanion with an aldehyde or a ketone gives the product (7), which has not been isolated, but which has been hydrolyzed to an intermediate that cyclizes to a 3-furanone (8). Preparation of the furanones (8) from (1) can be carried out without isolation of any intermediates.

[1] R. M. Carlson, R. W. Jones, and A. S. Hatcher, *Tetrahedron Letters*, 1741 (1975).
[2] L. Brandsma, *Preparative Acetylenic Chemistry*, Elsevier, Amsterdam, 1971, p. 172.
[3] Compare 3,3-diethoxy-l-methylthiopropyne, **5**, 207–208.

**Methylthiomethyl-N,N-dimethyldithiocarbamate,** $CH_3SCH_2\overset{\displaystyle S}{\overset{\|}{C}}-N(CH_3)_2$ (1). Mol. wt. 181.35, b.p. 135–137°/3.5 torr.

*Preparation:*

$$CS_2 + HN(CH_3)_2 \longrightarrow HS\overset{\displaystyle S}{\overset{\|}{C}}-N(CH_3)_2 \xrightarrow[76\%]{CH_3SCH_2Cl} CH_3SCH_2S\overset{\displaystyle S}{\overset{\|}{C}}-N(CH_3)_2$$

(1)

*Aldehyde synthesis.* Japanese chemists[1] have reported a synthesis of dimethyl acetals of aldehydes (3) by alkylation of the lithium salt of (1) with an alkyl iodide followed by hydrolysis of the resulting 1-methylthioalkyl N,N-dimethyldithiocarbamate (2) with mercuric oxide and mercuric chloride in methanol.

$$(1) \xrightarrow[\text{THF}, -55°]{\text{n-BuLi}} LiC\overset{SCH_3}{\underset{SCN(CH_3)_2}{\overset{|}{\underset{\|}{H}}}}\;\; \xrightarrow{RX}\;\; R-C\overset{SCH_3}{\underset{SCN(CH_3)_2}{\overset{|}{\underset{\|}{H}}}}\;\; \xrightarrow[\substack{65-70\% \\ \text{overall}}]{\substack{HgO, HgCl_2 \\ CH_3OH, \text{reflux}}}\;\; R-CH(OCH_3)_2$$

(2)                                                                   (3)

[1] I. Hori, T. Hayashi, and H. Midorikawa, *Synthesis*, 705 (1974).

**Methylthiomethyllithium,** $LiCH_2SCH_3$. Mol. wt. 68.07.

*Preparation* (**2**, 403).

*One-carbon homologation of organoboranes.*[1] The reaction of trialkylboranes in THF with this α-lithio derivative of dimethyl sulfide in TMEDA results in formation of an α-thioorganoborate anion, detectable by NMR. Addition of methyl iodide (or methyl fluorosulfonate) results in homologation.

$$(n\text{-}C_4H_9)_3B + Li^+\bar{C}H_2SCH_3 \xrightarrow[\text{TMEDA}]{\text{THF}} (n\text{-}C_4H_9)_3\bar{B}CH_2SCH_3\,Li^+ \xrightarrow{CH_3I}$$

$$\left[ (n\text{-}C_4H_9)_2\overset{C_4H_9\text{-}n}{\overset{|}{B}}-CH_2-\overset{+}{S}(CH_3)_2 \right] \xrightarrow{93\%} (n\text{-}C_4H_9)_2BCH_2C_4H_9\text{-}n + S(CH_3)_2$$

Two useful applications of the homologation are conversion of arylboranes into benzylboranes (equation I) and of alkenylboranes into allylboranes (equation II).

(I)   $CH_3O$—〈benzene ring〉—$B$〈〉   $\xrightarrow[84\%]{\begin{array}{l}1)\ LiCH_2SCH_3\text{-}TMEDA\\2)\ CH_3I\end{array}}$   $CH_3O$—〈benzene ring〉—$CH_2$—$B$〈〉

(II)   $\begin{array}{c}n\text{-}C_4H_9\\ \\H\end{array}C=C\begin{array}{c}H\\ \\B(Sia)_2\end{array}$   $\xrightarrow{72\%}$   $\begin{array}{c}n\text{-}C_4H_9\\ \\H\end{array}C=C\begin{array}{c}H\\ \\CH_2B(Sia)_2\end{array}$

2-Lithiothiomethoxy-1,3-thiazoline (1) is also satisfactory as the homologation reagent.[2]

(1)

[1] E. Negishi, T. Yoshida, A. Silveira, Jr., and B. L. Chiou, *J. Org.*, **40**, 814 (1975).
[2] K. Hirai, H. Matsuda, and Y. Kishida, *Tetrahedron Letters*, 4359 (1971).

**Methylthiotrimethylsilane,** $CH_3SSi(CH_3)_3$. Mol. wt. 120.29, b.p. 110–111°; very sensitive to moisture.

*Preparation.*[1,2]

$$CH_3SSCH_3 \xrightarrow{LiAlH_4} CH_3SLi \xrightarrow[65\text{-}70\%]{(CH_3)_3SiCl} CH_3SSi(CH_3)_3$$

*Thioketalization.*[2]  This silane converts aldehydes and ketones into dimethyl thioacetals and thioketals in high yield at 20° without acid catalysis. The rate

$$\begin{array}{c}R^1\\ \\R^2\end{array}C=O\ +\ 2\ CH_3SSi(CH_3)_3 \longrightarrow \begin{array}{c}R^1\\ \\R^2\end{array}C\begin{array}{c}SCH_3\\ \\SCH_3\end{array}\ +\ O[Si(CH_3)_3]_2$$

appears to be proportional to solvent polarity.

The silane undergoes 1,4-addition with $\alpha,\beta$-unsaturated carbonyl compounds:

$$CH_2=\overset{\overset{\displaystyle CH_3}{|}}{C}CHO\ +\ CH_3SSi(CH_3)_3 \xrightarrow[82\%]{CH_2Cl_2,\ 0^0} CH_3SCH_2\overset{\overset{\displaystyle CH_3}{|}}{C}=CHOSi(CH_3)_3$$

(E and Z isomers)

Other alkyl- or arylthiotrimethylsilanes are much less reactive and require prolonged reaction at elevated temperatures for thioketalization. However, the reaction of phenylthiotrimethylsilane can be conducted at 25° with anionic catalysts (potassium cyanide–18-crown-6[3]) or with triphenylphosphine.

[1] K. A. Hooten and A. L. Allred, *Inorg. Chem.*, **4**, 671 (1965).
[2] D. A. Evans, E. G. Grimm, and L. K. Truesdale, *Am. Soc.*, **97**, 3229 (1975).
[3] D. A. Evans and L. K. Truesdale, *Tetrahedron Letters*, **49**, 4929 (1973).

**Methyl p-toluenesulfinate**, $(\underline{p})\text{-}CH_3C_6H_4\overset{\overset{O}{\|}}{S}OCH_3$ . Mol. wt. 170.23, b.p. 129–130°/ 14 mm. *Caution:* The reagent can cause a persistent skin rash.

The ester can be prepared in 65% yield by the reaction of p-toluenesulfinyl chloride and methanol in ether–pyridine.[1]

*α-p-Toluenesulfinyl ketones.*[2] These useful intermediates can be prepared in 50–75% yield by the reaction in DME of sodium enolates with the ester:

$$R^1\overset{\overset{O}{\|}}{C}CH_2R^2 \ + \ CH_3\overset{\overset{O}{\|}}{S}C_6H_4CH_3 \xrightarrow[50-75\%]{NaH,\ DME} R^1\overset{\overset{O}{\|}}{C}\underset{\underset{R^2}{|}}{C}H\overset{\overset{O}{\|}}{S}C_6H_4CH_3$$

[1] J. W. Wilt, R. G. Stein, and W. J. Wagner, *J. Org.*, **32**, 2097 (1967).
[2] R. M. Coates and H. D. Pigott, *Synthesis*, 319 (1975).

## S-Methyl p-toluenethiosulfonate (Methyl thiotosylate), 5, 460.

*Beckmann fragmentation.* Autrey and Scullard[1] have shown that Beckmann fragmentation[2] rather than Beckmann rearrangement is favored by introduction of a methylsulfenyl group α to the oxime group; in addition, the termini of the bond that is cleaved are obtained in different oxidation states. Thus cleavage of the oxime of 2-methylthio-7-methoxytetralone-1 (4), prepared as shown, leads to approximately 1 : 1 mixtures of the enol thioethers (5) and (6).

Autrey and Scullard[3] used this fragmentation in a synthesis of corynantheine (11) from yohimban-17-one (7). The synthesis also involved a novel desulfurization of the enol thioether (9) without reduction of the vinyl group. This step

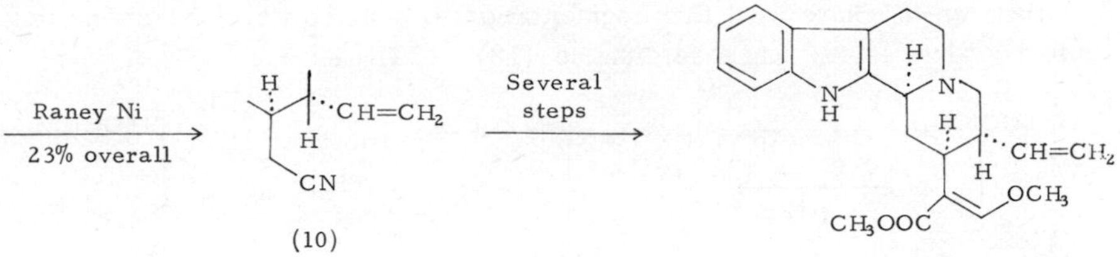

(10)

(11)

was carried out with W-2 Raney nickel deactivated under carefully controlled conditions.[4]

Shimizu[5] used this version of the Beckmann fragmentation to construct the characteristic A ring of the steroidal alkaloids of salamanders from an androstane-3-one. Thus treatment of (12), prepared as described above from 5$\beta$-androstane-17$\beta$-ol-3-one, with tosyl chloride followed by desulfurization gave (14). This intermediate was converted to (15) by epoxidation followed by reaction with

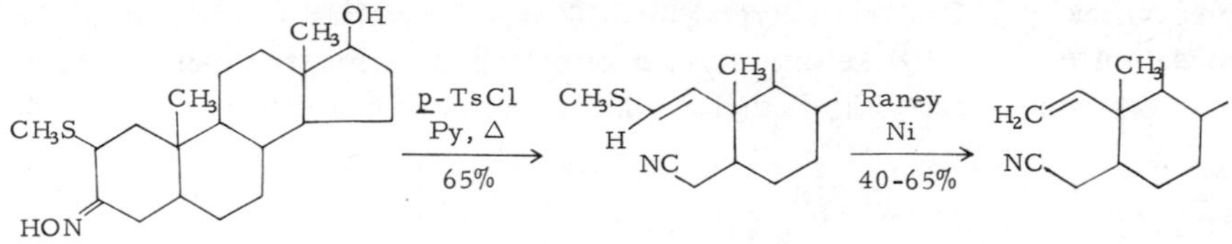

(12)

(13)

(14)

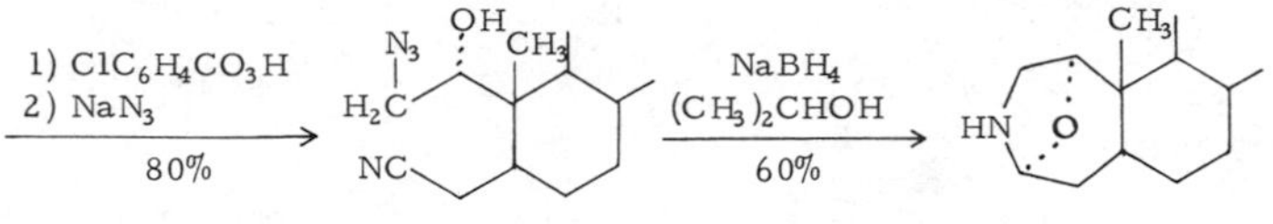

(15)

(16)

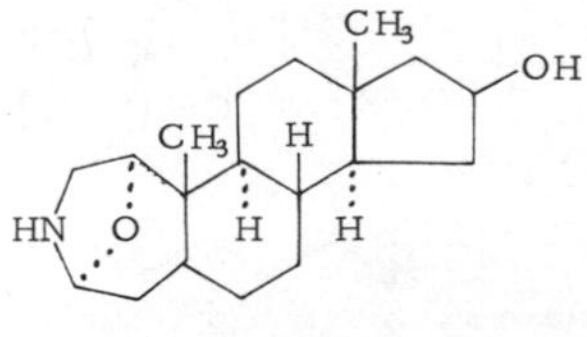

(17)

sodium azide. Treatment of (15) with sodium borohydride reduced the azide group to $-NH_2$ and the nitrile group to $-CHO$, and this intermediate cyclized to (16). The final step in the synthesis of samandarine (17) involved transposition of the $C_{17}$-hydroxyl group to the 16-position.[6]

Grieco *et al.*[7] have used this fragmentation to convert a cyclohexanone ring into a δ-valerolactone ring as formulated; (18) → (22).

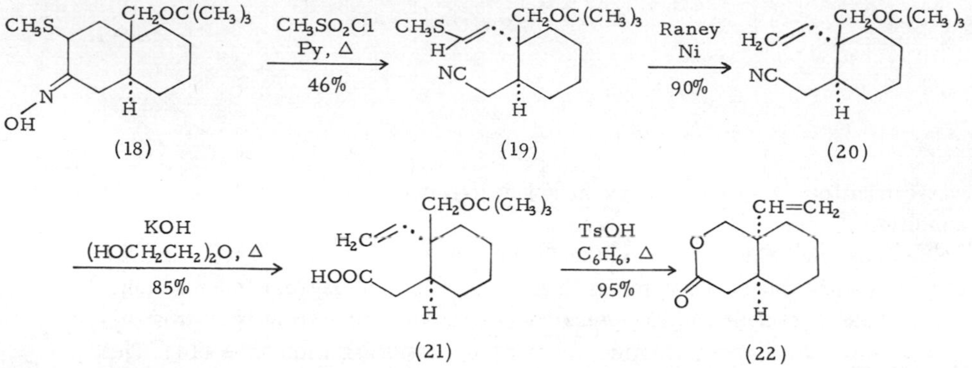

(18)  →  (19)  →  (20)

CH$_3$SO$_2$Cl
Py, Δ
46%

Raney
Ni
90%

KOH
(HOCH$_2$CH$_2$)$_2$O, Δ
85%

(21)

TsOH
C$_6$H$_6$, Δ
95%

(22)

*Oxidation of a methyl ketone to an α-ketoaldehyde.*[8]   A key step in a recent preparation of a 21-dimethoxypregnene-14-one-20 involves a new method for oxidation of a methyl ketone to an α-ketoaldehyde. Thus the methyl ketone (1) was condensed with diethyl oxalate (sodium ethoxide), and the crude

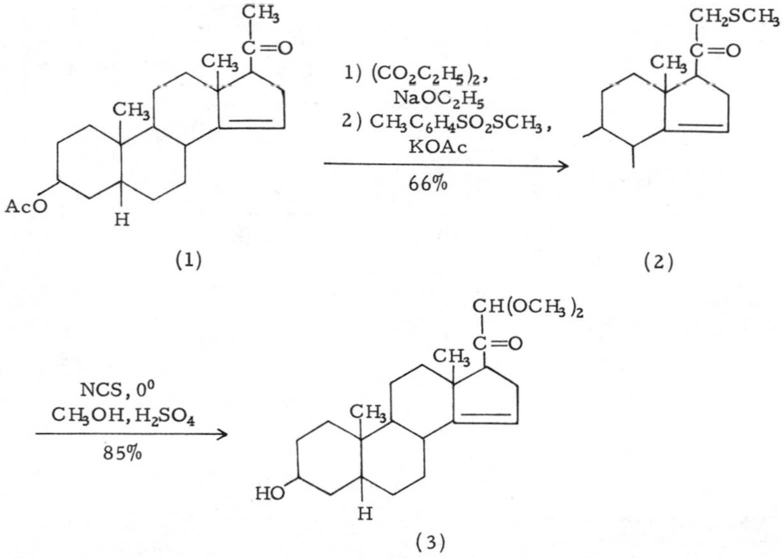

(1)

1) (CO$_2$C$_2$H$_5$)$_2$,
NaOC$_2$H$_5$
2) CH$_3$C$_6$H$_4$SO$_2$SCH$_3$,
KOAc
66%

(2)

NCS, 0°
CH$_3$OH, H$_2$SO$_4$
85%

(3)

oxalyl ketone was then allowed to react with methyl thiotosylate (1.1 eq.) in refluxing ethanol containing excess potassium acetate. The monosulfenylated product (2) was then oxidized with NCS (2.2 eq.) in 2% methanolic sulfuric acid at 0°. The dimethyl acetal (3) was obtained in high yield.

*Cardenolides.* A new, efficient synthesis of cardenolides is shown in the formulation.[9]

[1]R. L. Autrey and P. W. Scullard, *Am. Soc.*, **90**, 4924 (1968).
[2]C. A. Grob, H. P. Fischer, W. Raudenbusch, and J. Zergenyi, *Helv.*, **47**, 1003 (1964).
[3]R. L. Autrey and P. W. Scullard, *Am. Soc.*, **90**, 4917 (1968).
[4]The catalyst was washed four times with acetone, then refluxed with acetone for 3 hr. The catalyst was then washed several times with water, thrice with 95% ethanol, and thrice with methanol. The catalyst was stored in a sealed bottle completely filled with methanol.
[5]Y. Shimizu, *Tetrahedron Letters*, 2919 (1972).
[6]S. Hara and K. Oka, *Am. Soc.*, **89**, 1041 (1967).
[7]P. A. Grieco, K. Hiroi, J. J. Reap, and J. A. Noguez, *J. Org.*, **40**, 1450 (1975).
[8]E. Yoshii, T. Miwa, T. Koizumi, and E. Kitatsuji, *Chem. Pharm. Bull. Japan*, **23**, 462 (1975).
[9]E. Yoshii, T. Koizumi, H. Ikeshima, K. Ozaki, and I. Hayashi, *ibid.*, **23**, 2496 (1975).

**2-Methyl-2-thiazoline**      (1), **4**, 344–345.

*Aldehydes, β-hydroxy aldehydes, homoallylic alcohols.* Meyers and Durandetta[1] have published details for the synthesis of aldehydes from this heterocycle. A typical synthesis is formulated for the preparation of 5-phenylvaleraldehyde (4).

$$(1) \xrightarrow[\substack{2)\ C_6H_5CH_2CH_2CH_2I \\ 74\%}]{1)\ \underline{n}\text{-BuLi, THF}} (2) \xrightarrow[\substack{(C_2H_5)_2O,\ H_2O \\ 90\text{-}95\%}]{Al-Hg}$$

$$(3) \xrightarrow[\substack{CH_3CN,\ H_2O \\ 71.1\%}]{HgCl_2} O{=}CHCH_2(CH_2)_3C_6H_5 \quad (4)$$

If the carbanion of (1) is allowed to react with an aldehyde or ketone, β-hydroxy aldehydes can be prepared.[2] Since β-hydroxy aldehydes are usually labile, it is advantageous to trap the adduct (a) with chloromethyl methyl ether as shown for (5). The protected hydroxy aldehydes (6) so obtained can be converted into homoallylic alcohols by a Wittig reaction.

(6) $\xrightarrow[\substack{78\%}]{C_6H_5CH{=}P(C_6H_5)_3}$ (7)

$\xrightarrow[HCl]{THF}$ (8)

[1] A. I. Meyers and J. L. Durandetta, *J. Org.*, **40**, 2021 (1975).
[2] A. I. Meyers, J. L. Durandetta, and R. Munavu, *ibid.*, **40**, 2025 (1975).

**Methyltricaprylylammonium chloride** (Aliquat 336), 4, 28, 30; 5, 460.

*Evaluation of phase-transfer catalysts.* Herriott and Picker[1] have examined the catalytic effects of 21 quaternary ammonium and phosphonium salts on the $S_N2$ reaction of sodium thiophenoxide with 1-bromooctane in a two-phase system of aqueous sodium hydroxide and an organic solvent. They report the following conclusions: The catalytic efficiency increases with the length of the long chain in the salt, and the more symmetrical salts are superior to those with only one long chain. Phosphonium salts are somewhat more effective than the corresponding ammonium salts, but are more prone to decomposition. Surprisingly, benzyltriethylammonium chloride (3, 19; 4, 27–31) is relatively ineffective. The authors conclude that methyltricaprylylammonium chloride is one of the most effective catalysts; the availability and low cost are additional favorable factors to be considered.

Herriott and Picker also note that the crown ether dicyclohexyl-18-crown-6 functions as an efficient phase-transfer catalyst in the benzene–water system. They suggest that some of the catalytic effects reported for this crown ether may result from phase-transfer catalysis rather than complexation of cations.

In the same paper they also report that the rate of the $S_N2$ reaction can also be increased by use of a more polar organic solvent. Thus rates are increased by change of solvent from benzene to $o$-dichlorobenzene, methylene chloride, or 1,2-dichloroethane.

The present evidence supports the mechanism of Starks,[2] in which the primary role of the catalyst is to transport an anion into the organic phase.

*Acetoacetic ester condensation.* Durst and Liebeskind[3] have used this phase-transfer catalyst in a solid–liquid phase acetoacetic ester condensation. In the

$$CH_3COCH_2COOCH_3 \xrightarrow[85\%]{\substack{1)\ NaH,\ C_6H_6 \\ 2)\ Cat.,C_6H_5CH_2Cl}} CH_3COCHCOOCH_3 \ (CH_2C_6H_5)$$

general procedure methyl acetoacetate is treated with sodium hydride, benzene being added as solvent after the mineral oil is removed. Aliquat 336 is added and when the solution is refluxing, the alkyl halide is added. Reflux is continued for 8 hr. Note that only the product of C-alkylation is obtained, a result that is favored in protic solvents under usual conditions. Benzene or other common solvents can be used in the phase-transfer method. A variety of alkylating reagents can be used.

*Optically active secondary alkyl halides.* Landini et al.[4] have extended their study of the conversion of methanesulfonates to alkyl fluorides to the reaction of the methanesulfonate (1) of (–)-(R)-2-octanol with KF, KCl, KBr, or KI in the presence of a phase-transfer catalyst in a two-phase system. In the case of

$$\underset{(1)}{MsOCHC_6H_{13}\text{-}\underline{n}\ (CH_3)} \xrightarrow[Cat.,H_2O]{KX} \underset{(2)}{X-CHC_6H_{13}\text{-}\underline{n}\ (CH_3)}$$

potassium chloride and bromide, the reaction proceeds with 95 and 86% inversion, respectively ($S_N2$ reaction). In the case of potassium iodide the product, 2-iodooctane, is almost completely racemized owing to halogen–halogen exchange.

*Sulfides and dithioacetals.* Sulfides can be prepared in high yield (85–95%) by reaction of thiols with an alkyl halide using the phase-transfer technique (equation I). The reaction is complete in about 15 min.

$$\text{(I)} \quad R^1SH + R^2X \xrightarrow[\underset{85-95\%}{}]{C_6H_6,\ H_2O,\ NaOH \atop CH_3N^+(C_8H_{17})_3\ Cl^-} R^1SR^2$$

$$\text{(II)} \quad 2\ RSH + CH_2Cl_2 \xrightarrow[\underset{90-96\%}{}]{H_2O,\ NaOH, \atop CH_3N^+(C_8H_{17})_3\ Cl^-} H_2C{\overset{\displaystyle SR}{\underset{\displaystyle SR}{\big<}}}$$

The reaction of thiols with methylene chloride (cosolvent) in aqueous sodium hydroxide with the catalyst gives high yields of symmetrical dithioacetals (equation II). This reaction gives only low yields (5%) of 1,3-dithiane from 1,3-dithiopropane and methylene chloride.[5]

*Nitriles.* Newman et al.[6] have converted mesylates into nitriles in about 70% yield by reaction with KCN, water, and benzene at reflux (5–6 hr.) with this phase-transfer catalyst. Reaction with KCN in aqueous DMSO is less satisfactory.

[1] A. W. Herriott and D. Picker, *Am. Soc.*, 97, 2345 (1975).
[2] C. M. Starks, *ibid.*, 93, 195 (1971); C. M. Starks and R. M. Owens, *ibid.*, 95, 3613 (1973).
[3] H. D. Durst and L. Liebeskind, *J. Org.*, 39, 3271 (1974).
[4] D. Landini, S. Quici, and F. Rolla, *Synthesis*, 430 (1975).
[5] A. W. Herriott and D. Picker, *ibid.*, 447 (1975).
[6] M. S. Newman et al., *J. Org.*, 40, 2863 (1975).

**Methyl trifluoromethanesulfonate**, $CH_3OSO_2CF_3$. Mol. wt. 164.1, $n_D$ 1.3244, corrosive. Supplier: Aldrich.

*Methylation of carbohydrates.*[1] Carbohydrates can be methylated in >80% yield by methyl triflate and a sterically hindered, weak base such as 2,6-di-*t*-butylpyridine and 2,6-di-*t*-butyl-4-methylpyridine. The method is useful for sugars that are sensitive to acids or bases.

[1] J. Arnarp, L. Kenne, B. Lindberg, and J. Lönngren, *Carbohydrate Res.*, 44, C5 (1975).

**Methyl α-trimethylsilylvinyl ketone (3-Trimethylsilyl-3-butene-2-one)**, 5, 461–463.

*Preparation*[1]:

$$CH_2{=}CHBr \xrightarrow[\underset{77-93\%}{}]{1)\ Mg,\ THF \atop 2)\ (CH_3)_3SiCl} CH_2{=}CHSi(CH_3)_3 \xrightarrow[\underset{70-71\%}{}]{1)\ Br_2,\ -78^0 \atop 2)\ HN(C_2H_5)_2} CH_2{=}C{\overset{\displaystyle Si(CH_3)_3}{\underset{\displaystyle Br}{\big<}}}$$

$$\xrightarrow[\underset{}{}]{1)\ Mg,\ THF \atop 2)\ CH_3CHO} \left[ CH_2{=}C{\overset{\displaystyle Si(CH_3)_3}{\underset{\displaystyle \underset{\displaystyle OH}{\overset{|}{CHCH_3}}}{\big<}}} \right] \xrightarrow[\underset{45-56\%}{}]{Jones \atop reagent} CH_2{=}C{\overset{\displaystyle Si(CH_3)_3}{\underset{\displaystyle COCH_3}{\big<}}}$$

*Annelation.*[2] Details are available for the preparation of *cis*-5,6-dimethyl-$\Delta^{1(10)}$-decalone-2 (2) from 2-methyl-2-cyclohexene-1-one (1) using this reagent. Direct annelation of 2,3-dimethylcyclohexanone gives a 3:2 mixture of the *cis*- and *trans*-isomers of (2) in 15% yields.[3]

[1] R. K. Boeckman, Jr., D. M. Blum, B. Ganem, and N. Halvey, *Org. Syn.*, submitted (1975).
[2] R. K. Boeckman, Jr., D. M. Blum, and B. Ganem, *Org. Syn.*, submitted (1975).
[3] G. Berger, M. Franck-Neumann, and G. Ourisson, *Tetrahedron Letters*, 3451 (1968).

## Methyl vinyl ketone, 1, 697–703; 2, 283–285; 5, 464.

*Annelation of pyridinium rings.* Protonated nitrogen heterocycles form adducts with methyl vinyl ketone that can be converted into fused pyridinium salts.[1] For example, (1) reacts with the reagent at 120° in acetonitrile or with dimethylacetamide to give the adduct (2), which when heated cyclizes to (3). The corresponding reaction with (4) gives an adduct that undergoes spontaneous cyclization and aromatization to form (5).

This annelation reaction has been extended to mesityl oxide and related ketones.[2] Thus (6), isoquinolinium perchlorate, reacts with mesityl oxide at 120° (several hours) to form (7) by a spontaneous oxidative aromatization.

(6)                                                          (a)

(b)                                          (7)

*Synthesis of phenols from α-ketols.* Hardegger *et al.*[3a] in 1974 synthesized orchinol (2) in low yield by the base-catalyzed condensation of the hydroxy tetralone (1) with methyl vinyl ketone. Later studies[3b] indicate that this method

(1)                        (a)                        (2)

is often useful for the preparation of phenols from benzoins. The reaction involves Michael addition and aldol condensation to give a six-membered cyclic α,β-unsaturated γ-hydroxy ketone, which can be isolated. Dehydration of these products leads to phenols, often in satisfactory overall yield. The method is useful for preparation of 3,4-disubstituted phenols. For example, 3,4-diphenylphenol (3) can be prepared in good yield from benzoin and methyl vinyl ketone.

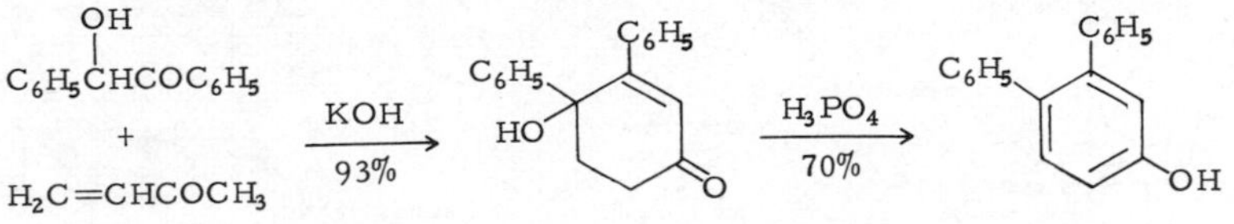

(3)

The reaction of propionoin and methyl vinyl ketone takes a different course and leads to 2,3-dimethyl-6-ethylphenol (4).

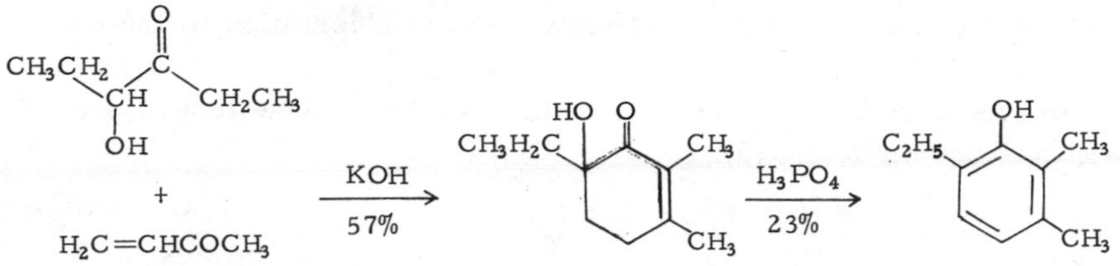

(4)

[1] D. D. Chapman, J. K. Elwood, D. W. Heseltine, H. M. Hess, and D. W. Kurtz, *J.C.S. Chem. Comm.*, 647 (1974).

[2] D. D. Chapman, *ibid.*, 489 (1975).

[3] (a) K. Steiner, Ch. Egli, N. Rigassi, S. E. Helali, and E. Hardegger, *Helv.*, 57, 1137 (1974); (b) Ch. Egli, S. E. Helali, and E. Hardegger, *Helv.*, 58, 104 (1975).

**2-Methyl-6-vinylpyridine,** $CH_2$ (1). Mol. wt. 119.16, b.p. 65–70°/12 mm.

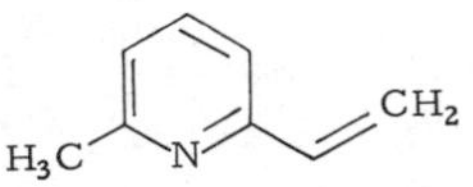

This pyridine derivative can be prepared from 2,6-lutidine by hydroxymethylation and dehydration (27% yield).[1]

*Bis-annelations.* Danishefsky *et al.*[2] have shown that this reagent can be used to effect bis-annelation. The product of addition of (1) to an enamine of a ketone, for example cyclohexanone, gives, after ketalization, the product (2). The key step involves Birch reduction of the pyridine ring to a dihydro product

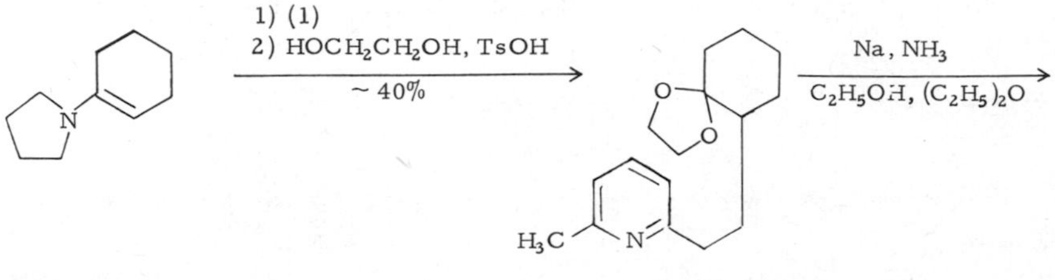

(2)

(a), which contains two enamine linkages. Hydrolysis of (a) followed by cycliza-tion gives (3). The final step involves vinylogous aldolization to the tricyclic ketone (4).

This procedure has been used in a synthesis of *dl*-D-homoestrone (6) from the Wieland–Miescher ketone (5).

*Optically active estrone.* Danishefsky and Cain[3,4] have reported a synthesis of optically active estrone that utilizes the tris-annelating pyridine reagent (1).

Michael addition of (2) to (1) gives the trione (3) in almost quantitative yield. The next step, conversion of (3) to optically active (4), is based on the procedure of Eder *et al.*[5] and Hajos and Parrish[6] for asymmetrical aldolization. These chemists found that asymmetric cyclization of (I) to optically active (II) or the

(I)                    (II)                    (III)

dehydration product (III) could be effected in high yield if carried out in the presence of an optically active amino acid, particularly proline. The asymmetric induction is strongly dependent on the structure of (I), the solvent, and the amine component, but optical yields of 93.4% were realized in the conversion of (I) to (II). Use of L-proline leads to 7aS absolute configuration in (II) and (III); use of D-proline leads to the enantiomeric 7aR-isomer. Hajos suggests that this asymmetric synthesis is a result of the "meso"-like character of the prochiral carbon atom: It bears two dissimilar groups and two identical groups ($-COCH_2$), and an optically active amino acid is then able to differentiate between the two identical groups.

This efficient asymmetric synthesis was used to convert (3) into (4) with an optical purity of 86%. In this case L-phenylalanine proved to be much more efficient than L-proline. The optically active product (4) was converted into optically active estrone (6) in overall yield of 13% from (2). The intermediate (5) was also converted into several optically active 19-norsteroids such as (7).

[1] K. S. N. Prasad and R. Raper, *J. Chem. Soc.*, 217 (1956).
[2] S. Danishefsky, P. Cain, and A. Nagel, *Am. Soc.*, 97, 380 (1975).
[3] S. Danishefsky and P. Cain, *ibid.*, 97, 5282 (1975).
[4] *Idem, J. Org.*, 39, 2925 (1974).
[5] U. Eder, G. Sauer, and R. Wiechert, *Angew. Chem. internat. Ed.*, 10, 496 (1971).
[6] Z. G. Hajos and D. R. Parrish, *J. Org.*, 39, 1615 (1974).

**Molecular sieves, 1**, 703–705; **2**, 286–287; **3**, 206; **4**, 345; **5**, 465.

*Polyenals.* δ-Alkoxy-α,β-unsaturated aldehydes are converted into polyenals by treatment with a tertiary amine (DBU or DBN) in $CH_2Cl_2$ in the presence of

molecular sieves 3A or 4A.[1] This reaction was used in one step of a synthesis of vitamin A (equation I).[2]

*Scavenger for hydrogen chloride.*[3] The acylation of sensitive secondary amides with acid chlorides generally proceeds in low yield because hydrogen chloride is formed and can result in cleavage of the desired products. Merck chemists have found two solutions to this problem. One involves absorption of the HCl by powdered 4A molecular sieves (600 mesh). Molecular sieves have previously been used for removal of water and methanol, but, apparently, never for absorption of HCl.

The second solution is to carry out acylations in the presence of trimethysilyl trifluoroacetamide or trimethylsilyl urcthane, $(CH_3)_3SiNHCOOC_2H_5$, both of which are readily hydrolyzed by HCl to an amide and trimethylchlorosilane. Collidine is a poor acceptor of HCl in these acylations.

These methods were originally investigated for use in acylations of β-lactam antibiotics with acid chlorides.

[1] A. Ishida and T. Mukaiyama, *Chem. Letters.*, 1167 (1975).
[2] T. Muakiayama and A. Ishida, *ibid.*, 1201 (1975).
[3] L. M. Weinstock, S. Karady, F. E. Roberts, A. M. Hoinowski, G. S. Brenner, T. B. K. Lee, W. C. Lumma, and M. Sletzinger, *Tetrahedron Letters*, 3979 (1975).

**Molybdenum hexafluoride,** $MoF_6$. Mol. wt. 209.95. Suppliers: Alfa, ROC/RIC.

*gem-Difluoro compounds.*[1] With catalysis by $BF_3$ this reagent converts aromatic aldehydes and ketones into *gem*-difluoro compounds. Yields are modest, however, unless the aromatic ring is substituted by electron-attracting substituents, particularly in the *para*-position. Yields are low if the carbonyl group is hindered.

[1] F. Mathey and J. Bensoam, *Tetrahedron*, 31, 391 (1975).

**Molybdenum pentachloride,** $MoCl_5$. Mol. wt. 273.21. Suppliers: Alfa, ROC/RIC.
*Preparation.*[1]
*Halogen exchange.*[2] Secondary and tertiary alkyl iodides, bromides, and fluorides are converted into the corresponding alkyl chlorides in fair to good yields by reaction with $MoCl_5$ in $CH_2Cl_2$ at 20°.

**vic-Dichlorides.**[3] Alkyl chlorides and haloalkanes are converted into *vic*-dichlorides by reaction with $MoCl_5$ in $CH_2Cl_2$ at $20°$ (8 hr.).

*Examples:*

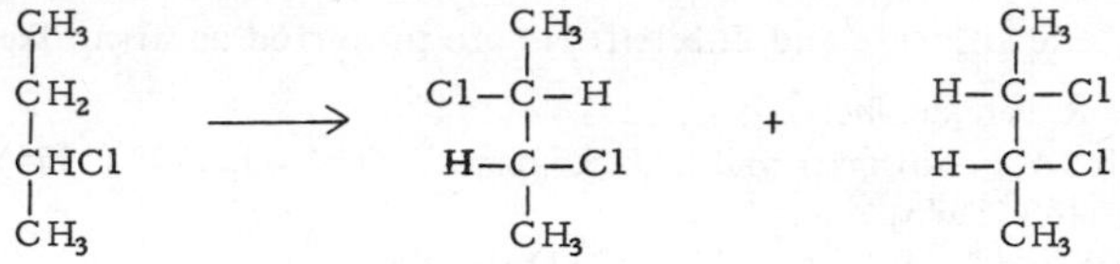

$$
\begin{array}{c}
CH_3 \\
| \\
CH_2 \\
| \\
CHCl \\
| \\
CH_3
\end{array}
\longrightarrow
\begin{array}{c}
CH_3 \\
| \\
Cl-C-H \\
| \\
H-C-Cl \\
| \\
CH_3
\end{array}
\quad + \quad
\begin{array}{c}
CH_3 \\
| \\
H-C-Cl \\
| \\
H-C-Cl \\
| \\
CH_3
\end{array}
$$

(dl, 58%)              (meso, 38%)

$$
\begin{array}{c}
CH_3 \\
| \\
CH_2 \\
| \\
CHBr \\
| \\
CH_3
\end{array}
\longrightarrow
\begin{array}{c}
CH_3 \\
| \\
Cl-C-H \\
| \\
H-C-Cl \\
| \\
CH_3
\end{array}
\quad + \quad
\begin{array}{c}
CH_3 \\
| \\
H-C-Cl \\
| \\
H-C-Cl \\
| \\
CH_3
\end{array}
$$

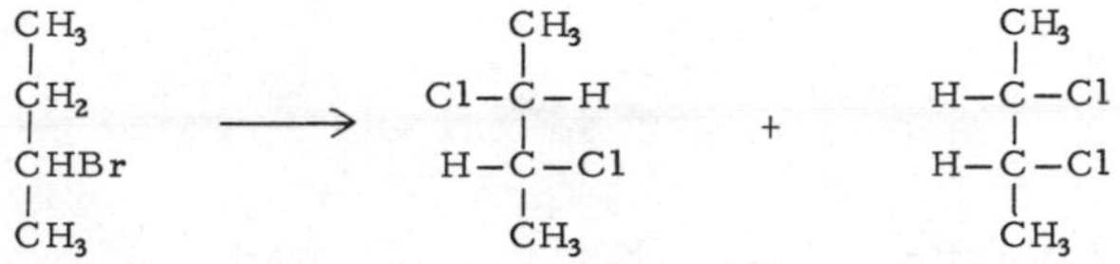

(54%)                   (38%)

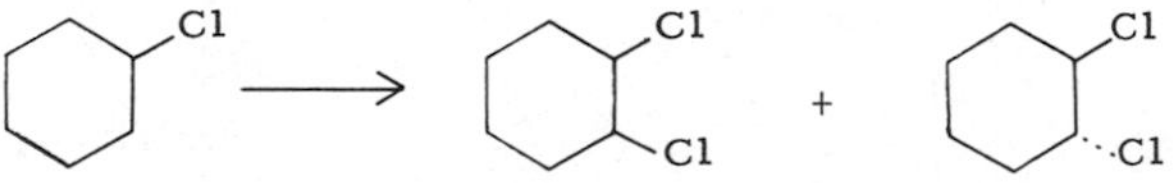

(75%)                (< 1%)

*Chlorination of alkenes and alkynes.* McCann and Brown[4] have noted that under irradiation tetrachloroethylene reacts with molybdenum(V) chloride to form hexachloroethane and molybdenum(IV) chloride. The yield in this reaction is approximately quantitative.

$$Cl_2C=CCl_2 + 2\ MoCl_5 \longrightarrow Cl_3C-CCl_3 + 2\ MoCl_4$$

San Filippo *et al.*[5] have extended this reaction to alkenes and alkynes to give dichloroalkanes and dichloroalkenes in fair to good yield. Irradiation is not necessary. In both cases the reaction involves *cis*-addition to the unsaturated linkage as shown in the examples:

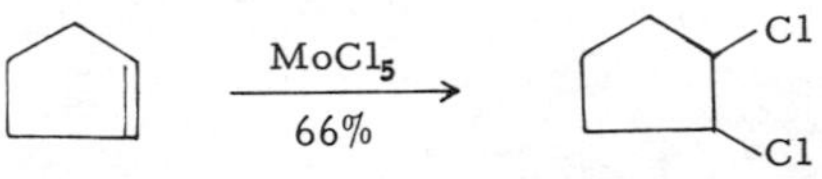

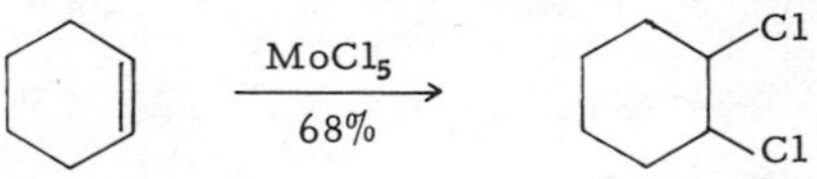

$$CH_3(CH_2)_2C \equiv C(CH_2)_2CH_3 \quad \xrightarrow[36\%]{MoCl_5} \quad \underset{Cl}{\overset{CH_3(CH_2)_2}{>}}C = C\underset{Cl}{\overset{(CH_2)_2CH_3}{<}}$$

Yields are lower with terminal, tri- and tetrasubstituted olefins, and terminal alkynes. Methylene chloride and chloroform are preferred as solvents.

[1] A. J. Leffler and R. Penque, *Inorg. Syn.*, 12, 187 (1970).
[2] J. San Filippo, Jr., A. F. Sowinski, and L. J. Romano, *J. Org.*, 40, 3295 (1975).
[3] *Idem, ibid.*, 40, 3463 (1975).
[4] E. L. McCann, III, and T. M. Brown, *ibid.*, 12, 181 (1970).
[5] J. San Filippo, Jr., A. F. Sowinski, and L. J. Romano, *Am. Soc.*, 97, 1599 (1975).

# N

**Naphthalene–Lithium (Lithium naphthalenide), 2**, 288–289; **3**, 208; **4**, 348–349; **5**, 468.

*Deoxygenation of 1,4-epoxy-1,4-dihydronaphthalenes.* The Diels–Alder adducts of benzynes with furanes can be aromatized by this radical anion in

clean conversions in 50–60% yield. Extrusion of oxygen can be effected photochemically in triethylamine, but in low yield.[1]

*γ-Butyrolactones.*[2] Lithium naphthalenide in THF in the presence of diethylamine reacts with carboxylic acids to form the α-anions of lithium carboxylates. Addition of epoxides gives the corresponding γ-hydroxy acids. These are cyclized to γ-butyrolactones when refluxed in benzene for 5 hr. The reaction is useful for synthesis of substituted lactones.

[1] S. B. Polovsky and R. W. Franck, *J. Org.*, **39**, 3010 (1974).
[2] T. Fujita, S. Watanabe, and K. Suga, *Australian J. Chem.*, **27**, 2205 (1974).

**α-Naphthyldiphenylmethyl chloride,** (1). Mol. wt. 316.82.

*Preparation.*[1]
*Protection of primary hydroxyl groups.*[2] The 5′-hydroxyl groups of nucleosides can be protected as the α-naphthyldiphenylmethyl ether. The group is selectively cleaved reductively in the presence of a *p*-methoxytrityl protecting group by sodium anthracenide in THF (HMPT). Sodium naphthalenide is less selective. The α-naphthyldiphenylmethyl group is relatively stable toward the benzophenone radical anion, which readily cleaves β,β,β-trichloroethyl groups.

[1] G. H. Holmberg, *Acta Acad. Aboensis, Ser. B.*, **16**, 138 (1948); *C.A.*, **45**, 560b (1951).
[2] R. L. Letsinger and J. L. Finnan, *Am. Soc.*, **97**, 7197 (1975).

**(R)-(-)-1-(1-Naphthyl)ethyl isocyanate** (1). Mol. wt. 197.23, b.p. 106–108°/ 0.16 mm., $\alpha_D$ -50.5°.

This isocyanate can be prepared[1] from the commercially available (R)-(+)-1-(1-naphthyl)ethylamine (2), $\alpha_D$ +50.5°.[2]

[2, R-(+)]                                            [1, R-(-)]

*Resolution of alcohols.* About 20 alcohols have been resolved by reaction with (1) to form diastereomeric carbamates that can be resolved by crystallization or preparative liquid chromatography.[3]

[1] W. H. Pirkle and M. S. Hoekstra, *J. Org.*, **39**, 3904 (1974).
[2] Norse Chemical Co.
[3] W. H. Pirkle and R. W. Anderson, *J. Org.*, **39**, 3901 (1974).

**Neomenthyldiphenylphosphine,** (1). Mol. wt. 324.45,

m.p. 96–99°, $\alpha_D$ +94.4°. Supplier: Strem.

*Preparation.*[1] This phosphine is prepared in ~35% yield by the reaction of sodium diphenylphosphine with (-)-menthyl chloride[2] in THF. This material contains about 5% of neomenthyldiphenylphosphine oxide as a tenacious impurity.

*Asymmetric hydrogenation.* Morrison et al.[1,3] have reported on asymmetric hydrogenations catalyzed by rhodium(I) complexes of the Wilkinson type containing chiral ligands. This type of asymmetric synthesis had been carried out previously with relatively inaccessible phosphine ligands that are asymmetric at phosphorus. Phosphines that are asymmetric at carbon are more readily available and appear to be more efficient. Thus reduction of (E)-β-methylcinnamic acid with prereduced tris(neomenthyldiphenylphosphine)chlororhodium in the presence of triethylamine leads to 3-phenylbutanoic acid, $\alpha_D$ +34.5°, which contains 61% enantiomeric excess of the S-isomer. Hydrogenations of olefins exhibit a lower degree of asymmetric bias.

[1] J. D. Morrison and W. F. Masler, *J. Org.*, **39**, 270 (1974).
[2] J. G. Smith and G. F. Wright, *J. Org.*, **17**, 1116 (1952).
[3] J. D. Morrison, R. E. Burnett, A. M. Aguiar, C. J. Morrow, and C. Phillips, *Am. Soc.*, **93**, 1301 (1971).

**Nickel(II) acetate,** $Ni(OOCCH_3)_2 \cdot 4\ H_2O$. Mol. wt. 248.86. Suppliers: Alfa, ROC/RIC.

*Tetracyclic metal complexes.*[1] The 14-membered macrocycle (3) can be prepared by the condensation of ethylenediamine (2) with phenylazomalondialdehyde (1) in the presence of this nickel salt. Divalent copper salts can also be used.

$$(1) + NH_2CH_2CH_2NH_2\ (2) \xrightarrow[\begin{array}{c}C_2H_5OH,\,60^0\\59.9\%\end{array}]{Ni(II)} (3)$$

[1] F. A. L'Eplattenier and A. Pugin, *Helv.*, **58**, 2283 (1975).

**Nickel(II) acetylacetonate** [$Ni(acac)_2$],

(1). Mol. wt. 208.86.

Supplier: ROC/RIC.

*C-Alkylation.*[1] This chelated β-diketone reacts with primary alkyl halides in DMF to give mainly C-monoalkylation products. C-Dialkylation products are obtained in addition from alkylation with benzylic or allylic halides. O-Alkylation has not been observed.

$$(1) + 2\ RX \xrightarrow{DMF} 2\ CH_3COCHCOCH_3$$
$$\overset{|}{R}$$

Use of $Cu(acac)_2$, $Fe(acac)_3$, or $Cr(acac)_3$ gives much lower yields.

C-Alkylation has also been observed with the nickel(II) chelate of ethyl acetoacetate.

[1] M. Boya, M. Moreno-Mañas, and M. Prior, *Tetrahedron Letters*, 1727 (1975).

**Nickel carbonyl (Nickel tetracarbonyl),** **1**, 720–723; **2**, 290–293; **3**, 210–212; **4**, 353–355; **5**, 472–474.

*Cyclization of allylic dibromides* (**1**, 722–723; **2**, 290–292; **3**, 211; **4**, 355). Corey and Helquist[1] have extended the synthesis of large-ring cycloalkenes by cyclization of allylic dibromides to a synthesis of large-ring methylenecyclo-alkanes. Thus cyclization of the allylic dibromides (1), prepared as shown in equation I, was carried out by slow addition of dilute solutions in DMF to a solution of excess nickel carbonyl in DMF at temperatures of 50–70°. The

(2)

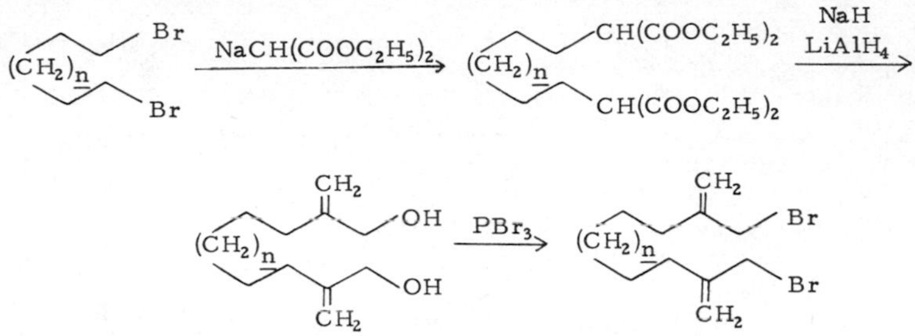

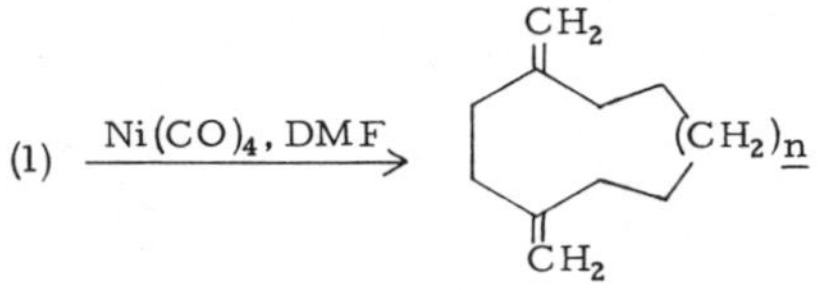

Yield:
$n = 6$, 15%
$n = 8$, 81%
$n = 12$, 40%

cyclization was successful when $n = 6$, 8, and 12, but failed when $n = 4$. The cyclization of allylic dibromides of type (1) is definitely less favorable than that of allylic dibromides of type (3), in which case yields of cyclic 1,5-dienes of 60–85% are obtainable when $n = 4, 6$, and 10.

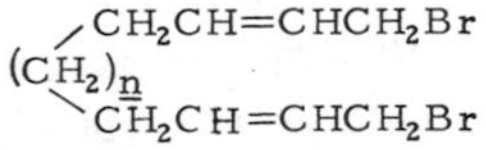

(3)

Corey and Helquist also report a further transformation of (2, $n = 8$) to a fused-ring cyclopentenone (5).

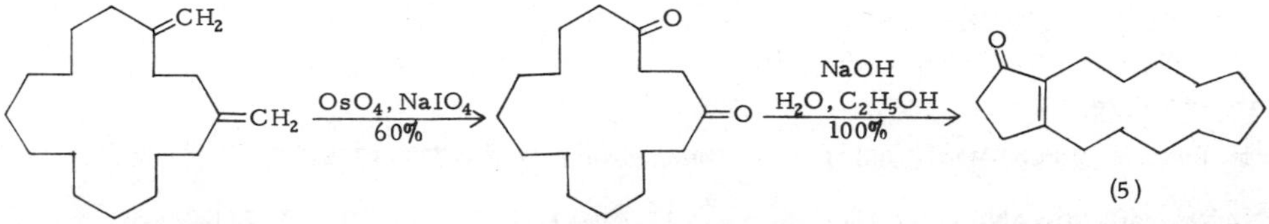

(2, $\underline{n} = 8$)                    (4)                                                                (5)

*Reaction of π-allylnickel halides with organic halides* (2, 291–292; 4, 353). Hegedus and Miller[2] have presented evidence that this reaction involves a radical chain mechanism. Thus the reaction is completely inhibited by a trace of *m*-dinitrobenzene, a radical anion scavenger.[3] Moreover, the reaction proceeds with complete racemization in the case of an optically active halide.

*Cembrene synthesis* (5, 473). The complete report of the synthesis of cembrene by coupling of two terminal allylic bromides is available.[4]

[1] E. J. Corey and P. Helquist, *Tetrahedron Letters*, 4091 (1975).

[2] L. S. Hegedus and L. L. Miller, *Am. Soc.*, 97, 459 (1975).

[3] N. Kornblum, R. E. Michel, and R. C. Kerber, *Am. Soc.*, 88, 5662 (1966); G. A. Russell and W. C. Danen, *ibid.*, 88, 5663 (1966).

[4] W. G. Dauben, G. H. Beasley, M. D. Broadhurst, B. Muller, D. J. Peppard, P. Pesnelle, and C. Suter, *Am. Soc.*, 97, 4973 (1975).

## *o*-Nitrobenzenesulfenyl chloride, 1, 745.

*Novel heterocycles.*[1] Some novel heterocycles have been prepared by deoxygenation[2] of indolyl *o*-nitrophenyl sulfides, prepared by reaction of indoles with 2-nitrobenzenesulfenyl chloride.[3] The reaction involves intermediate spirocyclic compounds.

*Examples:*

*Deblocking cysteine protecting groups.*[4] The common S-protecting groups can be cleaved by reaction with 2-nitrophenylsulfenyl chloride in acetic acid followed by reduction of the S-(2-nitrophenylsulfenyl)cysteine residues (equation I).

[1] A. H. Jackson, D. N. Johnston, and P. V. R. Shannon, *J.C.S. Chem. Comm.*, 911 (1975).

[2] J. I. G. Cadogan and S. Kulik, *J. Chem. Soc.*, 2621 (1971); J. I. G. Cadogan, *Accts. Chem. Res.*, 5, 303 (1972).

[3] A. Fontana, F. Marchiori, R. Rocchi, and P. Pajetta, *Gazzetta*, 96, 1301 (1966).

[4] A. Fontana, *J.C.S. Chem. Comm.*, 976 (1975).

## *o*-Nitrobenzenesulfonyl chloride, $O_2NC_6H_4SO_2Cl$. Mol. wt. 221.62, m.p. 65–67°. Supplier: Aldrich.

*Elimination of 3β-hydroxy and 11α-hydroxy steroids.*[1] *o*-Nitrobenzene-sulfonyloxy groups undergo elimination in HMPT much more readily than *p*-toluenesulfonate groups.

*Examples:*

The corresponding reactions with *p*-toluenesulfonates proceed in low yield.

[1] U. Zehavi, *J. Org.*, 40, 3870 (1975).

**Nitronium trifluoromethanesulfonate**, $NO_2^+F_3CSO_3^-$. Mol. wt. 195.09, m.p. 198–200° dec.

This salt is prepared[1] in 97% yield by the reaction of dinitrogen pentoxide with trifluoromethanesulfonic acid:

$$N_2O_5 + F_3CSO_3H \xrightarrow[20-35^0]{ClCH_2CH_2Cl} NO_2^+ \ F_3CSO_3^- + HNO_3$$

The related nitronium salt, $NO_2^+(F_3CSO_3^-)_2H_3O^+$, has also been prepared.[2]

*Nitration.*[1] Limited experimental data suggest that nitronium trifluoromethanesulfonate is more effective than nitronium tetrafluoroborate (1, 742–743) under heterogenous conditions (1,2-dichloroethane), but less effective under homogenous conditions (sulfolane and 1,2-dichloroethane).

[1] F. Effenberger and J. Geke, *Synthesis*, 40 (1975).
[2] C. L. Coon, W. G. Blucher, and M. E. Hill, *J. Org.*, 38, 4243 (1973).

*o*-**Nitrophenyl selenocyanate**, —SeCN (1). Mol. wt. 227.08, m.p. 139–141°.

*Preparation.*[1,2] The reagent can be prepared in 66% yield by diazotization of *o*-nitroaniline followed by treatment of the diazonium salt with potassium selenocyanate.[3]

*Olefin synthesis.* The preparation of olefins by decomposition of alkyl phenyl selenoxides (**Diphenyl diselenide, 5,** 272–276) is very useful, except in the case of primary alkyl phenyl selenoxides, which usually give low yields of terminal olefins on decomposition. However, the presence of electron-withdrawing substituents on the benzene ring increases both the rate of elimination and the yield of olefins. In instances where use of diphenyl diselenide results in low yields, Sharpless and Young[2] recommend use of *o*-nitrophenyl selenocyanate (1) or 4,4'-dichlorodiphenyl diselenide, both of which are converted into the corresponding $ArSe^-Na^+$ reagents on reduction with sodium borohydride in ethanol.

For example, the selenides (2) and (3) both give methylenecyclohexane (4) on decomposition, but the yield of (4) is much higher from (3).

The reagent $Na^+Se^-C_6H_4Cl(p)$ is nearly equivalent to $Na^+Se^-C_6H_4NO_2(o)$ and has the advantage that 4,4'-dichlorodiphenyl diselenide is as easy to prepare as diphenyl diselenide (75% yield from *p*-bromochlorobenzene).

*Moenocinol.*[4] A recent synthesis of the sesterterpene ($C_{25}$) moenocinol (5) involved coupling of nerol and geraniol derivatives to give (1), which contains the correct carbon skeleton of the natural product. The $-COOCH_3$ group was then converted into $-CH_2OMs$. Treatment of the mesylate with the

$$\xrightarrow[\text{CH}_3\text{OH},\ 0^\circ]{\text{TsOH}}$$

(5)

o-nitrophenylselenium anion then gave (3) in high overall yield. This was oxidized to the o-nitrophenyl selenoxide, which underwent elimination to give the olefin (4) in 76% yield. The final step involved acid hydrolysis of the tetrahydropyranyl ether group to give the free alcohol (5). Attempts to use the phenyl selenoxide derivative corresponding to (3) gave less than 4% of (4) on decomposition.

[1] H. Bauer, *Ber.*, **46**, 92 (1913).
[2] K. B. Sharpless and M. W. Young, *J. Org.*, **40**, 947 (1975).
[3] G. R. Waitkins and R. Shutt, *Inorg. Syn.*, **2**, 186 (1946); Supplier: Strem.
[4] P. A. Grieco, Y. Masaki, and D. Boxler, *Am. Soc.*, **97**, 1597 (1975).

**β-Nitrostyrene,** $C_6H_5CH{=}CHNO_2$. Mol. wt. 149.15, m.p. 56–58°, lachrymator. Supplier: Aldrich.

*Protection of thiol groups.*[1]  Under catalysis by a base (N-methylmorpholine) the reagent reacts with cysteine groups:

$$C_6H_5CH{=}CHNO_2\ +\ {-}NH{-}CH{-}CO{-} \rightleftharpoons$$

(1)

(2)

The adducts (2) are stable to acid conditions; the thiol group of (2) is liberated by mild alkaline conditions. The protective group was used in two new syntheses of glutathione. The adducts (2) contain a new asymmetric center.

[1] G. Jung, H. Fouad, and G. Heusel, *Angew. Chem. internat. Ed.*, **14**, 817 (1975).

**Nitrosyl chloride, 1,** 748–755; **2,** 298–299.

*Nitrosolysis.* Nitrosation of medium-sized ring ketones under usual conditions gives $\alpha,\alpha'$-dioximino ketones.[1]

Mononitrosation[2] of cyclohexanone can be effected by reaction with slightly more than 1 eq. of nitrosyl chloride in liquid sulfur dioxide in the presence of an alcohol and 1 eq. of hydrochloric acid at about $-10^\circ$ (equation I). The product (1) is a 2-alkoxy-3-oximinocycloalkene, the enol ether of an α-oximino ketone. These products can be obtained in yields as high as 40%.

(I)

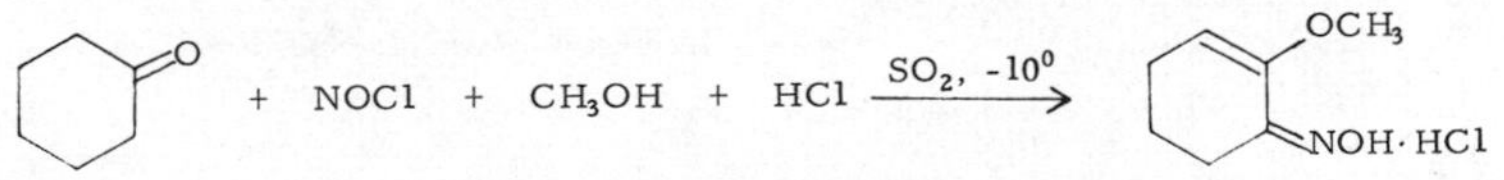

(1)

If the reaction formulated above is conducted in the absence of additional acid, the ring is cleaved to give an ester of $\omega$-oximinocaproic acid (2). Intermediates such as (a) and (b) are considered as precursors of (2) (equation I).

(II)

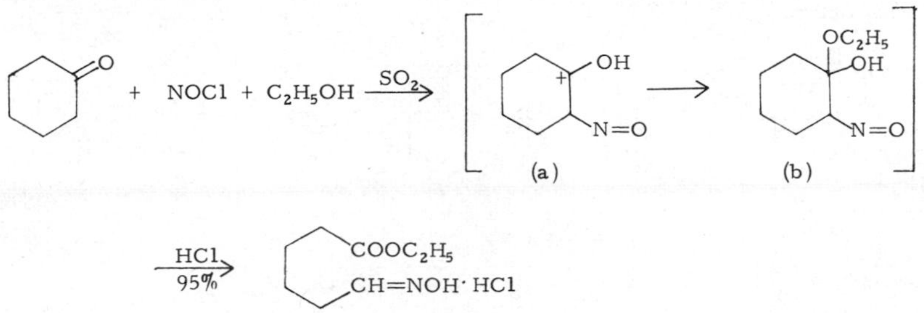

(a)          (b)

(2)

The nitrosolysis reaction is observed only in polar nonbasic solvents: sulfur dioxide, nitromethane, or sulfolane.

[1] O. Touster, *Org. Reac.*, **7**, 327 (1953); A. F. Ferris, G. S. Johnson, and F. E. Gould, *J. Org.*, **25**, 496 (1960).
[2] M. M. Rogić, J. Vitrone, and M. D. Swerdloff, *Am. Soc.*, **97**, 3848 (1975).

# O

**Osmium tetroxide–Sodium chlorate,** 1, 764; 2, 301; 4, 361.

*Reaction with benzoquinone.*[1] Apinol tetramethyl ether (1,2,3,4-tetra-methoxybenzene (3) can be prepared readily from hydroquinone (1) as formulated.

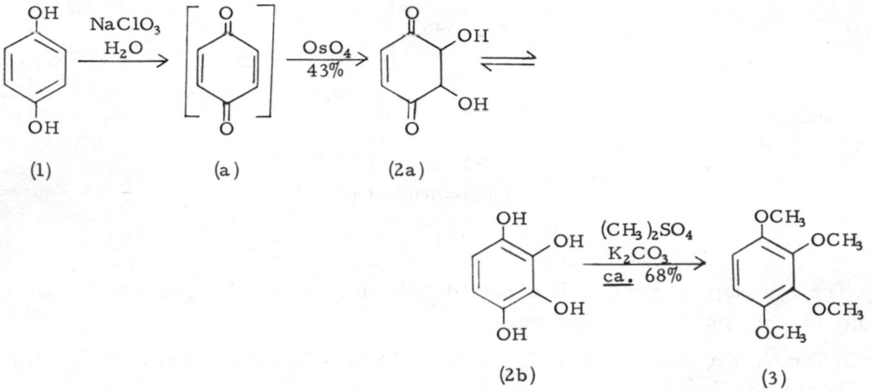

(1)               (a)              (2a)

(2b)            (3)

[1] A. J. Quillinan and F. Scheinmann, *J.C.S. Perkin I.*, 1329 (1973); A. J. Quillinan, *Org. Syn.*, submitted (1975).

**Oxalyl chloride,** 1, 767–772; 2, 301–302; 3, 216–217; 4, 361; 5, 481–482.

*2,3-Furanediones.* Alkenyloxysilanes (1) react with oxalyl chloride in ether at 20° to give yellow furanediones (2) in yields generally in the range 65–85%. When heated under reduced pressure the diones lose carbon monoxide to give α-oxoketenes (a), which cyclodimerize to pyranediones (3). When $R^2$ = H, the dimers exist as 4-hydroxy-2-pyrones (4).[1]

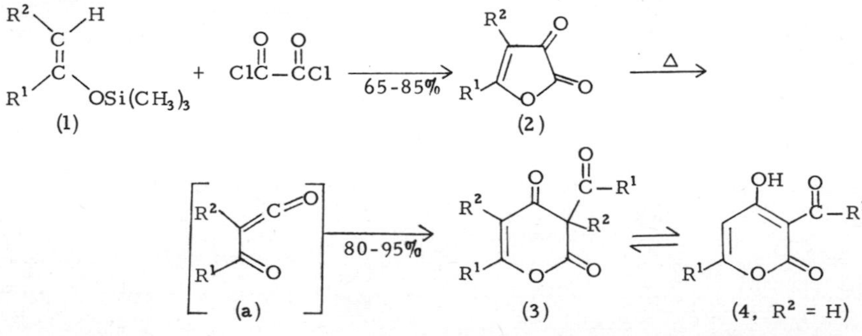

(1)                    (2)

(a)           (3)           (4, $R^2$ = H)

[1] S. Murai, K. Hasegawa, and N. Sonoda, *Angew. Chem. internat. Ed.*, 14, 636 (1975).

**$\mu$-Oxo-bis[tris(dimethylamino)phosphonium] bistetrafluoroborate,** $[(CH_3)_2N]_3$-
$\overset{+}{P}OP\overset{+}{[}N(CH_3)_2]_3$ $2BF_4^-$ **(1).** Mol. wt. 516, m.p. 194–204°; insol. in ether, THF, $CH_2Cl_2$; sol. in $CH_3CN$, DMF, HMPT. Supplier: Fluka ("Bates' Reagent").
   *Preparation:*

$$\left(CH_3-\!\!\!\left\langle\bigcirc\right\rangle\!\!\!-SO_2\right)_2\!\!O \ + \ [(CH_3)_2N]_3P{=}O \ \xrightarrow{60\%} \ [(CH_3)_2N]_3\overset{+}{P}O\overset{+}{P}[N(CH_3)_2]_3 \ \underset{80\%}{\overset{NaBF_4}{\xrightarrow{CH_3CN}}} \ (1)$$
$$2\ TsO^-$$

   *Polypeptide synthesis.* This reagent has been used for polypeptide synthesis, possibly by the mechanism formulated in scheme I, although there is no conclusive evidence of the involvement of an acyloxyphosphonium salt. This synthesis has the advantage that the coproduct HMPT is water-soluble and easily

$$\text{(I)} \quad \text{(1)} \ \xrightarrow[-[(CH_3)_2N]_3P=O]{RCOO^-} \ RCO{-}O{-}\overset{+}{P}[N(CH_3)_2]_3 \ \xrightarrow{R'-NH_2} $$
$$BF_4^-$$

$$RCO{-}NHR' + [(CH_3)_2N]_3P{=}O$$

removed from the peptide. Triethylamine or N-methylmorpholine is used as base. One disadvantage is that racemization in some cases is considerable, but it can be decreased markedly by addition of 1-hydroxybenzotriazole (**3**, 156) or N-hydroxysuccinimide (**1**, 487).[1]

[1] A. J. Bates, I. J. Galpin, A. Hallett, D. Hudson, G. W. Kenner, R. Ramage, R. C. Sheppard, *Helv.*, **58**, 688 (1975).

**Oxotriruthenium   acetate   complex   [Tris(aquo)hexa-$\mu$-acetato-$\mu_3$-oxotriruthenium acetate],** $[Ru_3O(OCOCH_3)_6(H_2O)_3]OCOCH_3$. Mol. wt. 585.06, green powder. Soluble in water, lower alcohols, and DMF.
   *Preparation.*[1] This oxotriruthenium acetate complex is obtained by treating ruthenium trichloride hydrate with acetic acid and sodium acetate in ethanol (1 hr. reflux). It can be purified by crystallization from methanol–acetone.
   *Isomerization of allylic alcohols to ketones.*[2] 1-Alkene-3-ols in organic solvents are isomerized to saturated ketones in high yield by this ruthenium complex by intramolecular transfer of hydrogen:

$$CH_3(CH_2)_2CH(OH)CH{=}CH_2 \ \underset{98\%}{\overset{cat.,85°}{\xrightarrow{\hspace{1.5cm}}}} \ CH_3(CH_2)_2COCH_2CH_3$$

Hydridochlorotris(triphenylphosphine)ruthenium(II), $RuHCl[P(C_6H_5)_3]_3$,[3] is also effective, but this substance is fairly unstable to air and is difficult to recover.
   If a vinyl ketone (*e.g.*, methyl vinyl ketone) is added to the reaction of 1-hexene-3-ol formulated above, then 1-hexene-3-one is obtained in about 90%

yield by intermolecular transfer of hydrogen to methyl vinyl ketone to give 2-butanone[4]:

$$CH_3(CH_2)_2CH(OH)CH=CH_2 + CH_2=CHCOCH_3 \xrightarrow{cat.} CH_3(CH_2)_2COCH=CH_2 + CH_3CH_2COCH_3$$

[1] A. Spencer and G. Wilkinson, *J.C.S. Dalton*, 1570 (1972).
[2] Y. Sasson and G. L. Rempel, *Tetrahedron Letters*, 4133 (1974).
[3] B. Hudson, D. E. Webster, and P. B. Wells, *J.C.S. Dalton*, 1204 (1972).
[4] Y. Sasson and G. L. Rempel, *Canad. J. Chem.*, **52**, 3825 (1974).

**Oxygen, 4,** 362; **5,** 482–486.

*Degradation of the bile acid side chain.* Fetizon *et al.*[1] have described a convenient new procedure for degradation of the side chains of bile acids and lanosterol. The method is illustrated for methyl cholanate (1). The first steps[2]

involve conversion to (3) by classical steps. The olefin (3) is then converted into ketone (4) by bromination followed by treatment with silver nitrate (5, 59).[3] The enolizable ketone (4) is cleaved in base by oxygen[4] to the bisnorcholanic acid (5). The product (5) can be converted into a 20-ketosteroid (7) by lead tetraacetate oxidation[5] to $C_{20}$-acetates (6, diastereoisomers) followed by hydrolysis and oxidation. The overall yield of (7) from (1) is 32%. The degradation is equally successful with the acetate of methyl lithocholate.

*Oxidative decarboxylation.*[6] Dianions of carboxylic acids (prepared with LDA) react with $O_2$ in ether at $-78°$ to form α-hydroperoxy acids. These need not be isolated; when DMF dimethyl acetal or an acid is added at room temperature, carbon dioxide is lost with formation of a carbonyl compound.

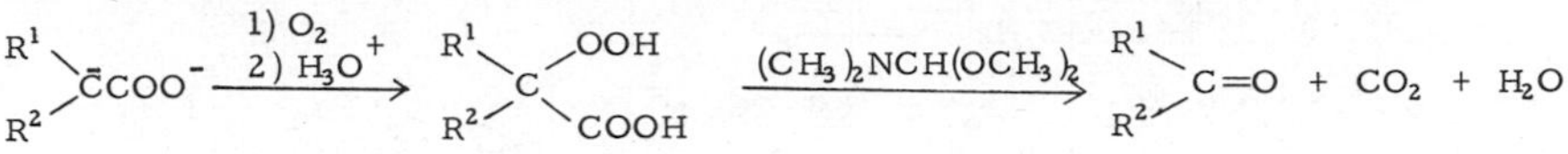

*Examples:*

$$(C_6H_5)_2CHCOOH \xrightarrow[80\%]{} (C_6H_5)_2C=O$$

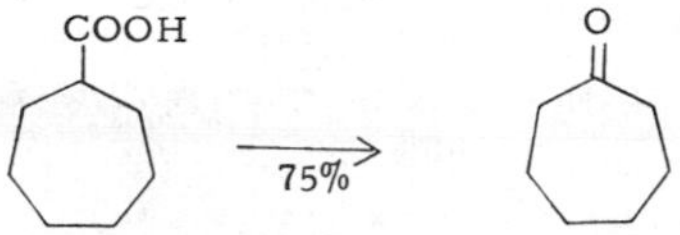

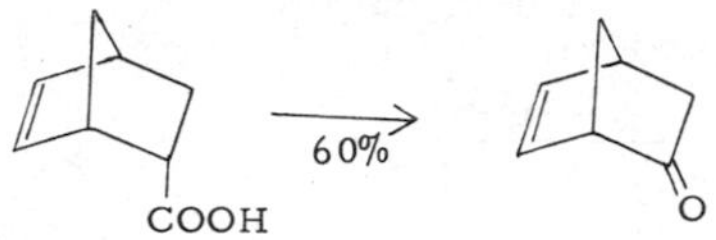

*α-Hydroxylation of amides and esters.*[7] The lithium carbanions of N,N-dialkylamides undergo rapid oxidation by molecular oxygen at $0°$. The α-hydroperoxides are reduced by sodium sulfite to α-hydroxy derivatives. Yields are in the range of 70–85%.

*Examples:*

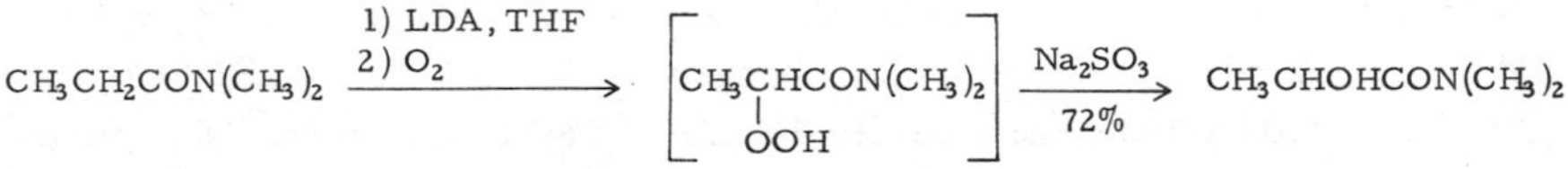

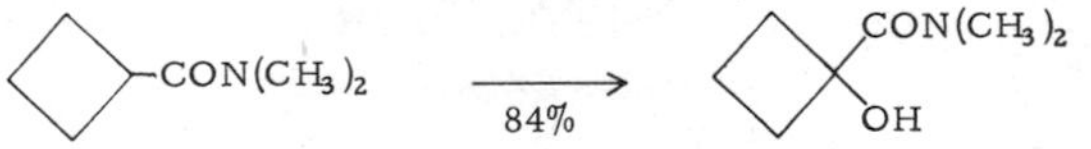

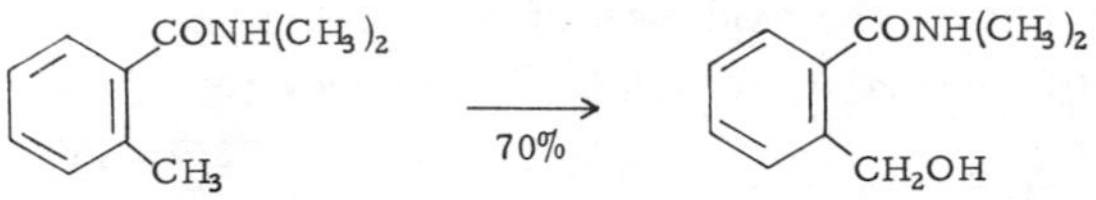

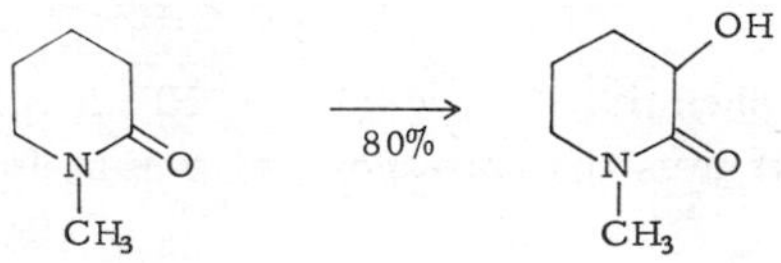

$$CH_3CH_2CON(CH_3)_2 \xrightarrow[2)\ O_2]{1)\ LDA,\ THF} \left[ CH_3CHCON(CH_3)_2 \atop \qquad | \atop \qquad OOH \right] \xrightarrow[72\%]{Na_2SO_3} CH_3CHOHCON(CH_3)_2$$

Lithium enolates of esters containing an α-tertiary carbon atom also undergo this α-hydroxylation reaction. One example of successful oxygenation of an ester with an α-methylene group was reported by Wasserman and Lipshutz.[7]

$$(C_6H_5)_2CHCH_2COOC_2H_5 \xrightarrow{90\%} (C_6H_5)_2CHCH(OH)COOC_2H_5$$

**Hydroxylation of hydrocarbons.** Mimoun and De Roch[8] have described several systems that convert oxygen into a polarized peroxidic form that is active in hydroxylation of hydrocarbons at ambient temperatures. The most active system is composed of oxygen, ferrous chloride, hydrazobenzene (or *o*-phenylenediamine), and a carboxylic acid, usually benzoic acid. The overall stoichiometry of the hydroxylation is represented in equation I, but the reaction is actually more complex. This system converts cyclohexane into cyclohexanol in 20–25% yield based on hydrazobenzene or in 35% yield based on the absorbed oxygen. Cyclohexene is oxidized mainly to cyclohexene-3-ol (no epoxide

$$(\text{I}) \quad RH + C_6H_5-NH-NH-C_6H_5 + O_2 \xrightarrow[\text{acetone}]{\overset{\text{FeCl}_2}{C_6H_5COOH}} ROH + C_6H_5N{=}NC_6H_5 + H_2O$$

is formed). Toluene is converted mainly into benzyl alcohol; the three cresols are formed to a minor extent. The oxidation of 2-methylbutane indicates preferential hydroxylation of a tertiary C—H bond.

**p-Quinols.** Oxygenation of phenols of type (1) at 20° in diethylamine containing sodium amide (excess) gives the corresponding *p*-quinols (2) in high yield.[9]

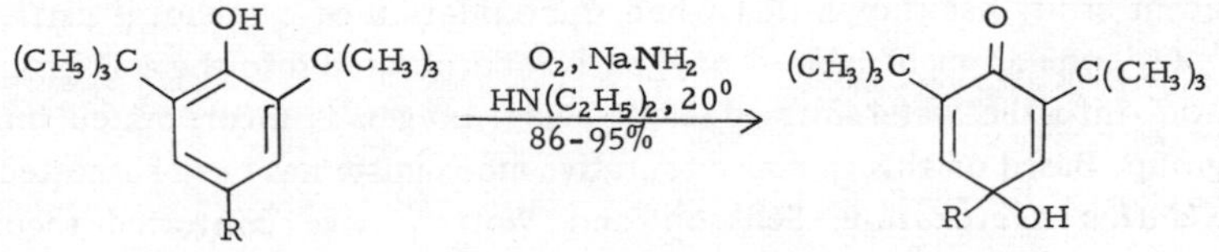

(1, R = CH₃, C₂H₅
 i-C₃H₇, t-C₄H₉)

(2)

*Oxidative cleavage of acid hydrazides.* Oxygen activated by copper salts converts acid hydrazides into carboxylic acids (equation I), esters (equation II), or amides (equation III). All reactions occur at $20°$ without oxidation of other sensitive groups.[10]

$$(I) \quad RCONHNH_2 \xrightarrow[\substack{CH_3OH, 20° \\ 78-95\%}]{O_2, Cu(OAc)_2} RCOOH + N_2 + H_2O$$

$$(II) \quad RCONHNH_2 \xrightarrow[\substack{CH_3OH, 20° \\ \sim 80\%}]{O_2, CuCl, NaOCH_3} RCOOCH_3$$

(III)

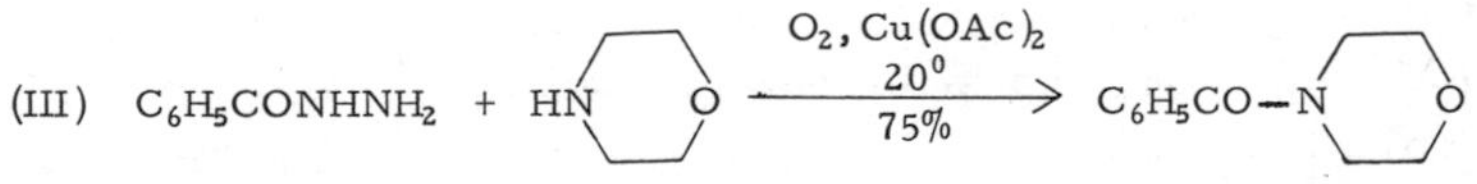

*Oxidation of catechol.* Molecular oxygen activated by cuprous chloride in pyridine and in the presence of methanol oxidizes catechol (1) to the mono-methyl ester of *cis,cis*-muconic acid (2) in 70% yield. Of several catalysts, cuprous chloride is the most efficient. In the absence of methanol, only a trace of *cis,cis*-muconic acid is obtained. High concentrations of methanol result

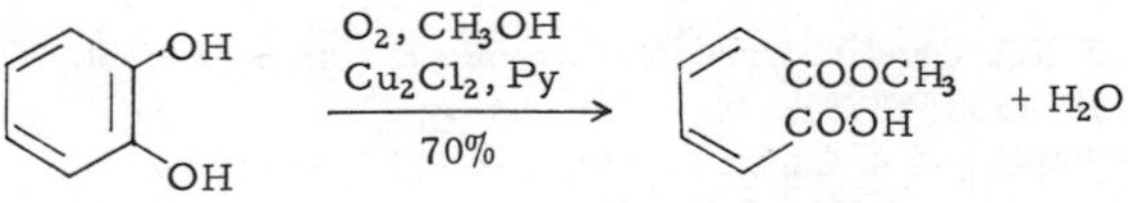

(1)                              (2)

in a decreased yield of (2).[11] The reaction is of interest as a model reaction for enzymatic cleavage of catechol by pyrocatechase, an oxygenase that requires $Cu^+$ for maximum activity. In addition, it has synthetic value. The same catalyst system has been used for oxidation of α-dihydrazones to acetylenes in high yield[12] and for oxidation of *o*-phenylenediamine to *cis,cis*-muconitrile.[13]

A recent study has shown that when the oxidation of catechol is carried out with $^{18}O_2$, one atom of labeled oxygen is incorporated into the acid group and one atom into the water formed; no labeled oxygen is incorporated into the ester group. Based on this finding a tentative mechanism has been suggested.[14]

*Oxidative decyanation.* Selikson and Watt[15] have converted secondary nitriles, for example (1), into ketones (4) by the following sequence. The anion (a) is generated with lithium diisopropylamide in THF at $-78°$ and then oxygen is bubbled into the solution of (a). A lithium $\alpha$-cyanohydroperoxide (b) is formed. Quenching with aqueous acid or acetyl chloride provides the isolable $\alpha$-hydroperoxynitrile (2) or the corresponding acetate. Reduction of (2) leads

$$
\begin{array}{ccccc}
\underset{C_6H_5CH_2}{\overset{H}{\diagdown}}\underset{CH_3}{\overset{CN}{C}} & \xrightarrow{\ LiN(\underline{i}\text{-}Pr)_2\ } & \left[\ \underset{C_6H_5CH_2}{\overset{Li^+}{\diagdown}}\underset{CH_3}{\overset{CN}{\bar{C}}} \xrightarrow{\ O_2\ } \underset{C_6H_5CH_2}{\overset{Li^+\ OO}{\diagdown}}\underset{CH_3}{\overset{CN}{C}}\ \right] & \xrightarrow[92\%]{\ H^+\ } \\
(1) & & (a) \hspace{4em} (b) &
\end{array}
$$

$$
\underset{C_6H_5CH_2}{\overset{HOO}{\diagdown}}\underset{CH_3}{\overset{CN}{C}} \xrightarrow[89\%]{\ SnCl_2\ } \underset{C_6H_5CH_2}{\overset{HO}{\diagdown}}\underset{CH_3}{\overset{CN}{C}} \xrightarrow[98\%]{\ 1\underline{M}\ NaOH\ } \underset{C_6H_5CH_2}{\overset{O}{\diagdown}}\underset{CH_3}{\overset{\|}{C}}
$$

$$\ \ \ (2) \hspace{8em} (3) \hspace{8em} (4)$$

to the cyanohydrin (3). The ketone (4) is then obtained by alkaline hydrolysis of (3). The reaction can be conducted in 82% yield without isolation of intermediates. In 16 reported cases of this procedure overall yields of ketones ranged from 65 to 92%.

[1] M. Fetizon, F. J. Kakis, and V. Ignatiadou-Ragoussis, *Tetrahedron*, **30**, 3981 (1974).
[2] *Idem, J. Org.*, **38**, 4308 (1973).
[3] F. J. Kakis, D. Brase, and A. Oshima, *J. Org.*, **36**, 4117 (1971).
[4] T. J. Wallace, H. Pobiner, and A. Schriesheim, *J. Org.*, **30**, 3768 (1965).
[5] P. Rosen and G. Oliva, *J. Org.*, **38**, 3040 (1973).
[6] H. H. Wasserman and B. H. Lipshutz, *Tetrahedron Letters*, 4611 (1975).
[7] *Idem, ibid., Tetrahedron Letters*, 1731 (1975).
[8] H. Mimoun and I. S. De Roch, *Tetrahedron*, **31**, 777 (1975).
[9] A. Nishinaga, T. Itahara, and T. Matsuura, *Bull. Chem. Soc. Japan*, **48**, 1683 (1975).
[10] T. Tsuji, S. Hayakawa, and H. Takayanagi, *Chem. Letters*, 437 (1975).
[11] J. Tsuji and H. Takayanagi, *Am. Soc.*, **96**, 7349 (1974).
[12] J. Tsuji, H. Takahashi, and T. Kajimoto, *Tetrahedron Letters*, 4573 (1973).
[13] H. Takahashi, T. Kajimoto, and J. Tsuji, *Syn. Commun.*, **2**, 181 (1972).
[14] J. Tsuji, H. Takayanagi, and I. Sakai, *Tetrahedron Letters*, 1245 (1975).
[15] S. J. Selikson and D. S. Watt, *J. Org.*, **40**, 267 (1975).

**Oxygen, singlet, 4**, 362–363; **5**, 486–491.

*1,4-Endoperoxides from acylic 1,3-dienes.* 1,4-Endoperoxides have been obtained in the photooxygenation of acyclic 1,3-dienes (**5**, 488). These are the usual products of oxygenation of cyclic 1,3-dienes. In fact, it was originally suggested that an acyclic 1,3-diene is more reactive to $^1O_2$ than an isolated double bond. However, more recent results suggest the following order of reactivity of carbon—carbon double bonds toward $^1O_2$: trisubstituted monoalkene > 1,3-diene > 1,1-disubstituted monoalkene.[1]

*Examples:*

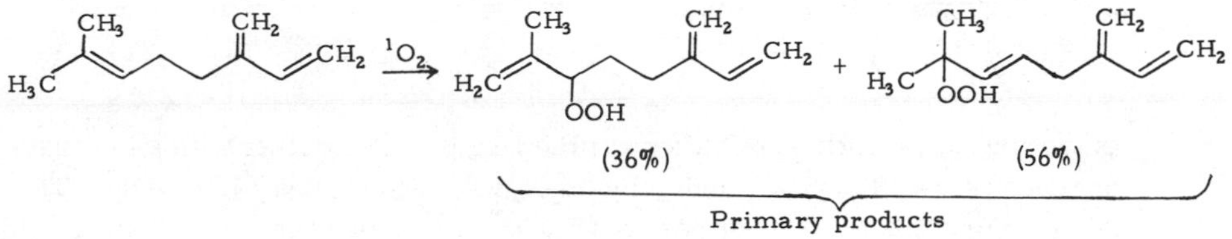

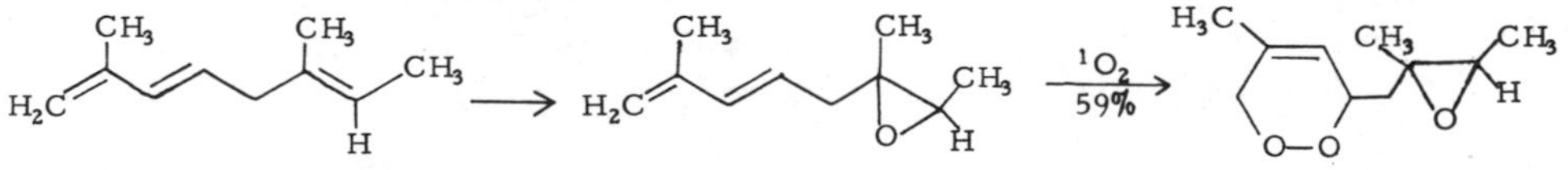

(36%)                    (56%)

Primary products

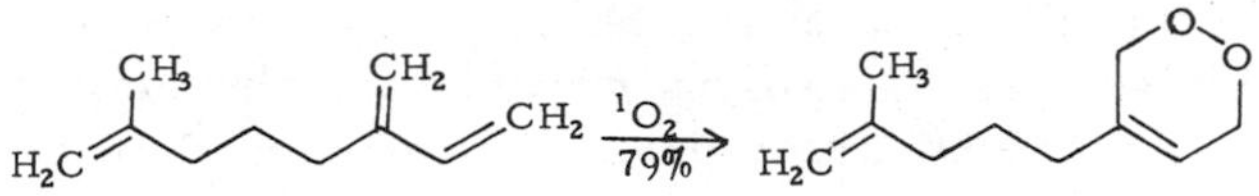

Epoxidation can be used to protect the more reactive isolated double bond (third example).

*Homogeneous photosensitization.*[2] Rose Bengal and Eosin-Y, anionic dyes used for generation of singlet oxygen, are not soluble in aprotic solvents. They can be solubilized in $CS_2$ and $CH_2Cl_2$ by complexation with 18-crown-6 or with Aliquat 336.

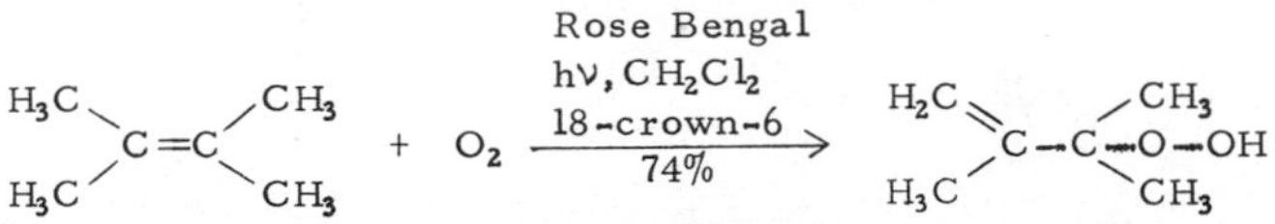

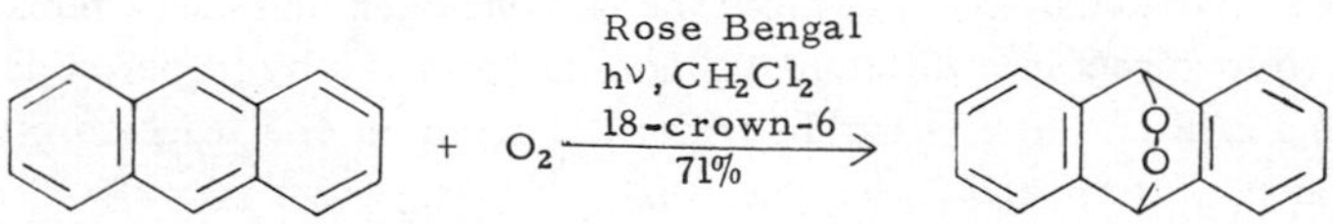

*Biogenetic-type oxidation of p-hydroxyphenylpyruvic acid to homogentisic acid.*[3,4]   Photooxidation of *p*-hydroxyphenylpyruvic acid (1) with Rose Bengal

(1)                              (a)                                   (b)

(2)                                          (3)

as sensitizer in a phosphate buffer at pH>10 gives (2) together with *p*-hydroxy-benzaldehyde (12% yield) and *p*-hydroxyphenylacetic acid (15% yield). This latter compound was shown not to be a precursor of quinol (2). Saito *et al.* propose that (2) is formed via (a) and (b). In biological systems the conversion of (1) to (3) is catalyzed by the enzyme *p*-hydroxyphenylpyruvate hydroxylase, a monooxygenase. Possibly singlet oxygen is also involved in the enzymatic system.

*Ene reaction.*[5]  1-Methyl-4*a*,5,6,7,8,8*a-trans*-hexahydronaphthalene (1) undergoes the expected Diels–Alder reaction with singlet oxygen to give at least an 80% yield of the expected peroxide (2) (equation I).

(1)                              (2, 70%)

(3)                    (4, 85–90%)              (5, 10–15%)

However, the diene (3) in which the 4*a*-hydrogen has been replaced by a methyl group reacts in a different fashion to form the hydroperoxides (4) and (5). In this case products (4) and (5) derived from an ene reaction obtain, possibly because of steric effects (equation II).

The diene (3) reacts with the more potent dienophile 4-phenyl-1,2,4-triazo-line-3,5-dione (**1**, 849–850; **2**, 324–326; **3**, 223–224; **4**, 381–383; **5**, 528–530) to give two major products, the normal (2 + 4) adduct (6) and the ene product (7). The diene (1) reacts with this dieneophile to form only the normal adduct (quantitative yield).

(6, 65%)                    (7, 35%)

*Cycloaddition and ene reactions.* Japanese chemists[6] have considered theoretical aspects of the mechanism of ene, [2 + 2], and [6 + 2] cycloaddition reactions of singlet oxygen and believe that perepoxides are involved as quasiintermediates.

*Photooxygenations of enamines.*[7] Photooxygenation of enamines of cyclic ketones at room temperature in methanol leads to α-diketones. If the reaction is conducted at −78° intermediate dioxetanes can be isolated. Thus photo-oxygenation of N-cyclohexenylmorpholine (1) at 25° gives cyclohexane-1,2-dione (3), isolated as the bisphenylhydrazone, in 75% yield. The intermediate

(1)                    (2)                    (3)

dioxetane (2) can be isolated in 55% yield by photooxidation in dimethyl ether at −78°.

Photooxidation of 3-morpholino-2-cholestene (4) at −78° give a 1:1 mixture of 3-morpholino-3-cholestene-2-one (5) and cholestane-2,3-dione (6).

(4)                    (5)                    (6)

Photooxidation of enamines of open-chain ketones under these conditions results mainly in cleavage to the amide and ketone.

*Photooxygenation of indoles.* Photooxygenation (Rose Bengal) of N-methyltryptophol (1) in methanol at room temperature gives the expected

(1)                              (a)                              (2)

95% | $^1O_2$, $CH_3OH$, $-70°$

(3)                    (4)                    (5)

cleavage product (2). However, if the oxygenation is conducted at $-70°$, 3-hydroperoxyindoline (3) is formed in high yield. This product is converted into the 2,3-dihydro-1,4-benzoxazine (5) by reduction and mild treatment with acid. Such a transformation may be involved in biological systems.[8]

*Reaction with norbornene.* Norbornene (1) has been reported to be inert toward singlet oxygen. However, Jefford and Boschung[9] find that it does react, although slowly, to give (2) and (3) as major products. One is *cis*-cyclopentane-1,3-dicarboxaldehyde (2), probably formed via the dioxetane (a). The other

(1)                    (a)                    (2)

(3)

product is *exo*-norbornene epoxide (3), the origin of which is not clear. In acetonitrile with Methylene Blue as sensitizer, (2) is formed in 61% yield and (3) in 39% yield.

*Regiospecific hydroperoxidation.* Chung and Scott report that photosensitized oxygenation of (1) gives the hydroperoxide (2). They did not isolate any other hydroperoxides or products of transannular cycloaddition.[10]

*Addition to arene oxides* (5, 480).[11] Photosensitized oxygenation of benzene oxide (1) gives mainly phenol. However, singlet oxygen generated from

NaOCl (Clorox) and 30% $H_2O_2$[12] reacts with (1) at −5 to −15° to form the *endo*-peroxide (2) in 37% yield. This substance readily rearranges at 45° to *trans*-benzene trioxide (3). Singlet oxygen generated from triphenyl phosphite ozonide (3, 323–324) reacts with (1) to form (2), but purification by sublimation results in conversion to (3). This trioxide also results from photolysis of (2) (27% yield).

Photosensitized (Rose Bengal) oxygenation of indane 8,9-oxide (5) is facile, perhaps because (5) exists entirely in the arene oxide form. The product (6) rearranges to (7) very readily.

(5)             (6)             (7)

$$85\% \downarrow \quad P(OC_6H_5)_3$$

(8)

*Oxidation of phosphites to phosphates.* Phosphites are converted into phosphates by photooxidation in acetone in the presence of Rose Bengal or Methylene Blue; yield 65–85%.[13]

[1] M. Matsumoto and K. Kondo, *J. Org.*, **40**, 2259 (1975).

[2] R. M. Boden, *Synthesis*, 783 (1975).

[3] I. Saito, M. Yamane, H. Shimazu, T. Matsuura, and H. J. Cahnmann, *Tetrahedron Letters*, 641 (1975).

[4] I. Saito, Y. Chujo, H. Shimazu, M. Yamane, T. Matsuura, and H. J. Cahnman, *Am. Soc.*, **97**, 5272 (1975).

[5] I. Sasson and J. Labovitz, *J. Org.*, **40**, 3670 (1975).

[6] S. Inagaki and K. Fukui, *Am. Soc.*, **97**, 7480 (1975).

[7] H. H. Wasserman and S. Terao, *Tetrahedron Letters*, 1735 (1975).

[8] I. Saito, M. Imuta, S. Matsugo, and T. Matsuura, *Am. Soc.*, **97**, 7191 (1975).

[9] C. W. Jefford and A. F. Boschung, *Helv.*, **57**, 2257 (1974).

[10] S.-K. Chung and A. I. Scott, *J. Org.*, **40**, 1652 (1975).

[11] C. H. Foster and G. A. Berchtold, *J. Org.*, **40**, 3743 (1975).

[12] C. S. Foote, S. Wexler, W. Ando, and R. Higgins, *Am. Soc.*, **90**, 975 (1968).

[13] P. R. Bolduc and G. L. Goe, *J. Org.*, **39**, 3178 (1974).

**Ozone**, 1, 773–777; 4, 363–364, 5, 491–495.

*Oxidation of 2,3-polymethylenebenzo- and naphthofuranes.* Substances of type (1) and (2) are resistant to peracids and to osmium tetroxide—sodium periodate. However, the large-ring ketolactones (3) and (4) are obtained readily

(1, $\underline{n}$ = 4–6, 10)                    (3)

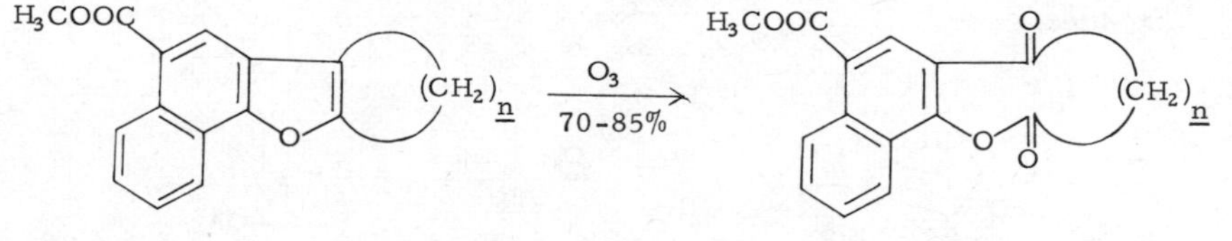

$$(2, \underline{n} = 4\text{-}6, 10)$$    $$(4)$$

by ozonization. Chromic anhydride also effects conversion of (1) to (3), but in lower yield; this reagent converts (2) only into intractable products.[1]

*Oxidation of acetals to esters* (4, 364). The details for oxidation of acetals to esters by ozone have been published.[2] The reaction is general and the yields are essentially quantitative. However, cyclic acetals react much faster than acyclic ones (examples I and II).

(I)    $\underline{n}\text{-}C_6H_{13}CH(OCH_3)_2$  $\xrightarrow[91\%]{\substack{O_3, \ -78^0 \\ 15 \ hr.}}$  $\underline{n}\text{-}C_6H_{13}COOCH_3$

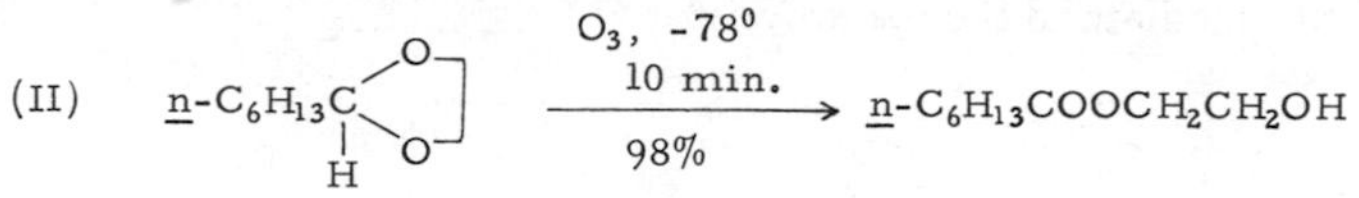

(II)     $\xrightarrow[98\%]{\substack{O_3, \ -78^0 \\ 10 \ min.}}$  $\underline{n}\text{-}C_6H_{13}COOCH_2CH_2OH$

Tetrahydropyranyl ethers (1, unsymmetrical acetals) are oxidized exclusively to the hydroxy ester (2); on heating, this is converted into δ-valerolactone (3) with liberation of the alcohol. The reaction thus constitutes a useful method for cleavage of tetrahydropyranyl ethers under neutral conditions.

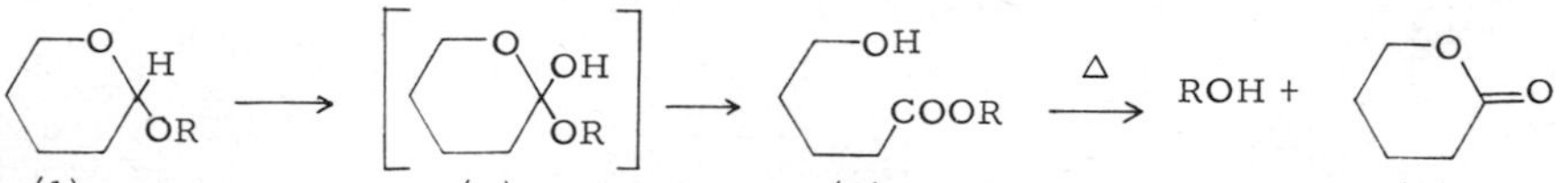

(1)    (a)    (2)    (3)

Tetrahydrofuranyl ethers (4) are also smoothly oxidized to γ-hydroxy esters (5).

(4)    (5)

Ozonization of α- and β-methyl glucopyranosides has also been examined. The latter are converted into the corresponding 5-acetoxyaldonic acid methyl esters, whereas the former are recovered unchanged. The polyacetyl derivatives are used to protect hydroxyl functions. The difference in reactivity of α- and β-anomers can be used to purify α-anomers.

*Example:*

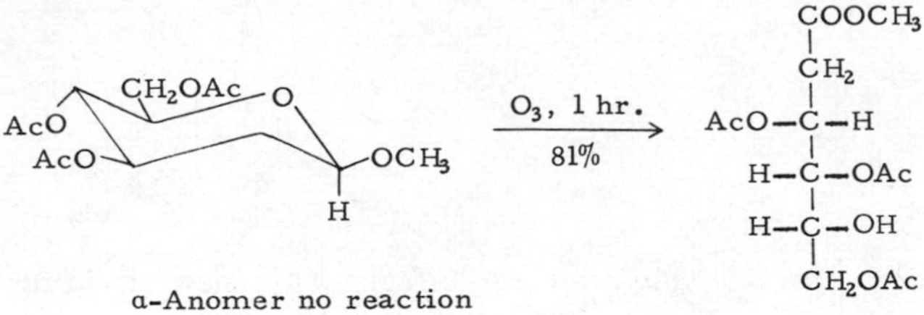

α-Anomer no reaction

From these and other results, Deslongchamps concludes that the reactivity to ozone is related to the conformation of the acetal group.

*Oxidation of benzylidene acetals.* Deslongchamps[3] has extended his studies on ozonization of acetals to benzylidene acetals. Ozonization of the benzylidene derivatives (1) of 9β,10α-decalin-2β,3β-diol in acetic anhydride containing sodium acetate results in formation of a single product (2) with an axial benzoate group and an equatorial acetate group. This is the same isomer that is obtained from the hemiortho ester (3); thus (3) or an equivalent substance is a probable intermediate in the ozonolysis of (1).

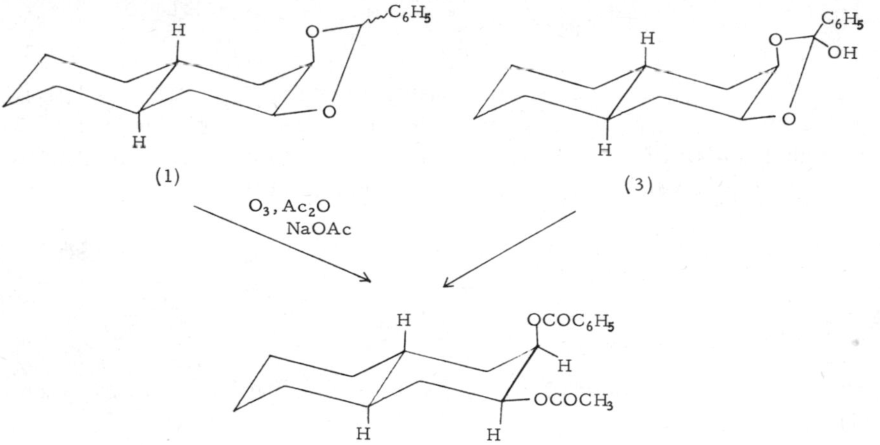

(1)

(3)

$O_3, Ac_2O$
NaOAc

(2)

Ozonolysis of the cholestane benzylidene derivative (4) similarly leads to (5) as the major product.

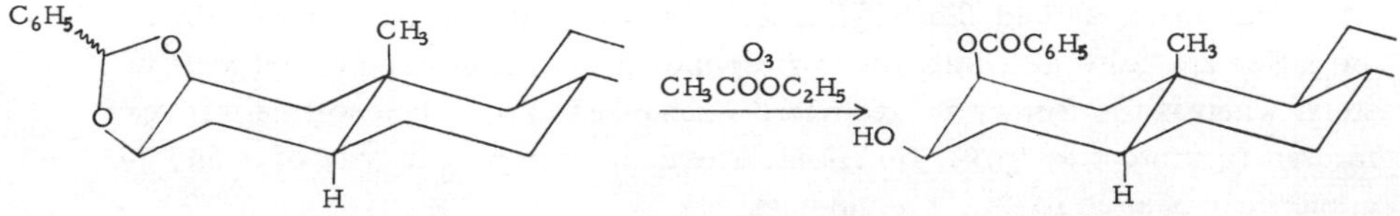

(4)

$O_3$
$CH_3COOC_2H_5$

(5, 92%)

The reaction was also applied to the 4,6-O-benzylidene derivative of methyl α-D-glucopyranosides. The method is useful for preparation of partially esterified sugars.

*Examples:*

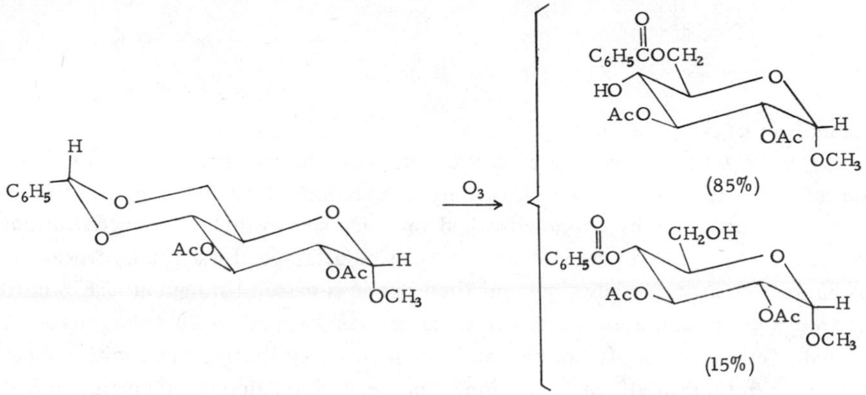

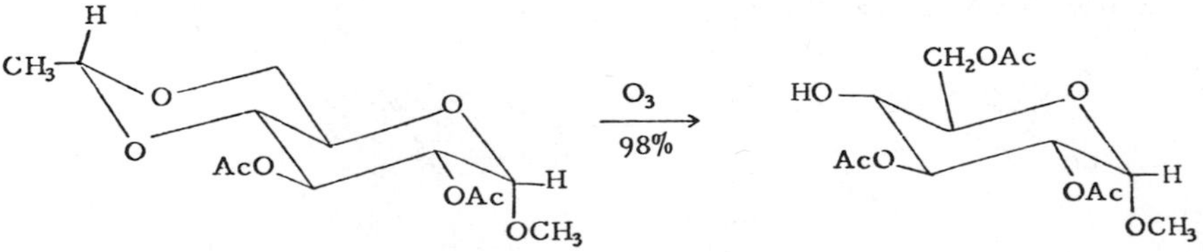

$$C_6H_5\overset{\overset{\displaystyle O}{\|}}{C}OCH_2$$

(85%)

(15%)

98%

*Cleavage of* —CH=CCl—.[4] Substrates containing the vinyl chloride unit are oxidized by ozone in methanol at $-78°$ to methyl esters:

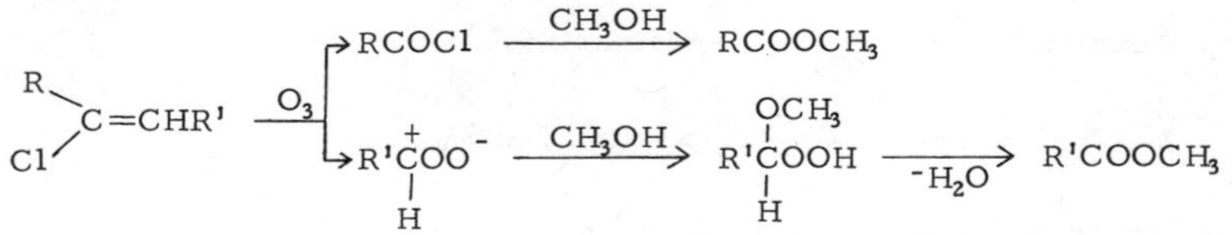

$$\begin{array}{c} RCOCl \xrightarrow{CH_3OH} RCOOCH_3 \\ R'\overset{+}{C}OO^- \xrightarrow{CH_3OH} R'\underset{\underset{\displaystyle H}{|}}{\overset{\overset{\displaystyle OCH_3}{|}}{C}}OOH \xrightarrow{-H_2O} R'COOCH_3 \end{array}$$

*Examples:*

$$CH_3CCl{=}CHCH_3 \xrightarrow{95\%} 2\ CH_3COOCH_3$$

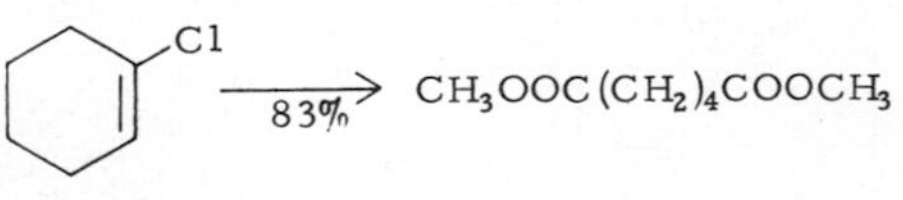

$$\xrightarrow{83\%} CH_3OOC(CH_2)_4COOCH_3$$

**β-Cyclocitral.** This substance (2) can be obtained in high yield by partial

(1)                                                                    (2)

ozonolysis of β-ionone (1).[5]

*Hydroxylation of tertiary carbon atoms on silica gel.* Israeli chemists[6] have devised an experimental method for hydroxylation of saturated hydrocarbons at tertiary positions by ozone adsorbed on silica gel; in this way concentrations of ozone of ~4.5% by weight at −78° can be attained. First the hydrocarbon is impregnated on the silica gel and then ozone is passed through at −78°[7] until the silica gel is saturated. After the silica gel has warmed to 20°, the product is eluted. Tertiary alcohols can be obtained in this way in high yield with almost complete retention of configuration. The secondary alcohol adamantane-2-ol was oxidized to the ketone by this method.

*Examples:*

Adamantane $\xrightarrow{99\%}$ 1-Adamantanol

2-Adamantanol $\xrightarrow{>99.5\%}$ 2-Adamantanone

**δ-Lactones.** δ-Lactones can be prepared conveniently from cyclohexenones by ozonolysis at −60° followed by reduction with sodium borohydride at 0°. Yields are only moderate (25–60%).[8]

(1)                                 (a)                                 (2)

*Oxidation of arenes.* Under the conditions of Mazur,[6] arenes are oxidized to carboxylic acids.[9]

*Examples:*

*Cyclic ketoimides.*[10] Ozonization of cyclic enamine lactams (1) is a route to medium and macrocyclic ketoimides (2). Yields are lower if *m*-chloroperbenzoic acid is used. However, this peracid is preferable to ozone for oxidation of N-acetyl enamines of type (3).

*Review.* The mechanism of ozonolysis has been reviewed.[11]

[1] J. R. Mahajan and H. C. Araújo, *Synthesis*, 54 (1975).

[2] P. Deslongchamps, P. Atlani, D. Fréhel, A. Malaval, and C. Moreau, *Canad. J. Chem.*, **52**, 3651 (1974).

[3] P. Deslongchamps, C. Moreau, D. Fréhel, and R. Chênevert, *ibid.*, **53**, 1204 (1975).

[4] K. Griesbaum and H. Keul, *Angew. Chem. internat. Ed.*, **14**, 716 (1975).

[5] N. Müller and W. Hoffmann, *Synthesis*, 781 (1975).

[6] Z. Cohen, E. Keinan, Y. Mazur, and T. H. Varkony, *J. Org.*, **40**, 2142 (1975).

[7] *Caution*: Ozonizations at temperature lower than $-100°$ can lead to explosions.

[8] C. G. Chavdarian and C. H. Heathcock, *J. Org.*, **40**, 2970 (1975).

[9] H. Klein and A. Steinmetz, *Tetrahedron Letters*, 4249 (1975).

[10] J. R. Mahajan, G. A. L. Ferreira, H. C. Araújo, and B. J. Nunes, *Synthesis*, 112 (1976).

[11] R. Criegee, *Angew. Chem. internat. Ed.*, **14**, 745 (1975).

# P

**Palladium(II) acetate, 1**, 778; **2**, 303; **4**, 365; **5**, 496–497.

*Cyclopropanation* (5, 496). Vorbrüggen *et al.*[1] have extended the cyclopropanation procedure of styrene by Paulissen *et al.* to α,β-unsaturated ketones and esters. The reaction proceeds stereospecifically *cis* to α,α- or α,β-disubstituted ketones and esters in excellent yield, but fails with trisubstituted carbonyl compounds. The reaction evidently involves a carbenoid, since there is no evidence for formation of pyrazolines.

*Examples:*

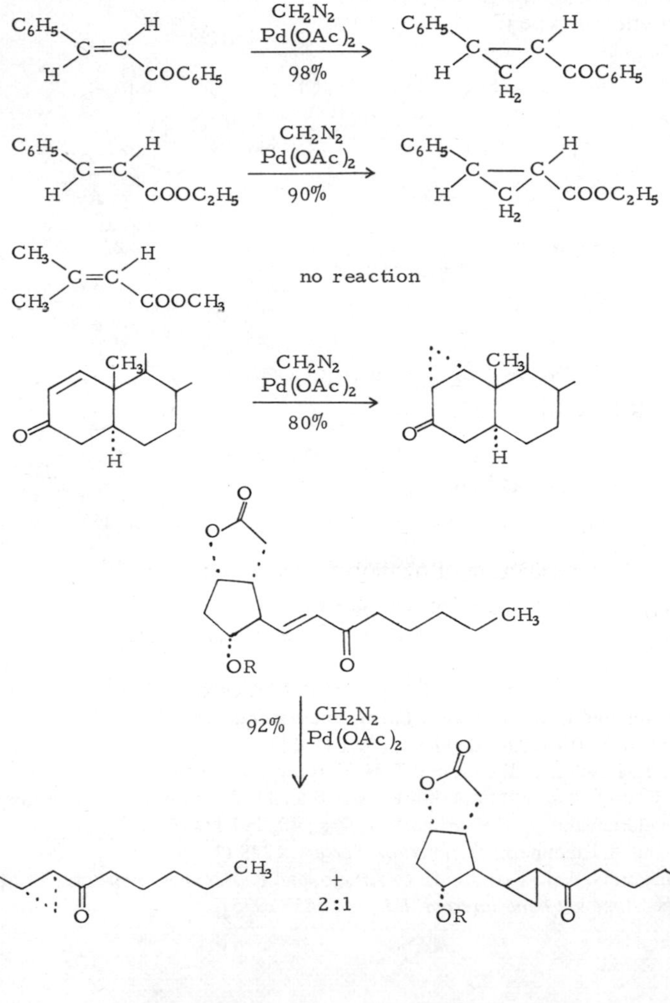

442

The last example shows an application of the cyclopropanation reaction in one step of a synthesis of 13,14-dihydro-13,14-methylene derivatives of $PGF_{2\alpha}$ and of $PGE_2$. Cyclopropanation with the Simmons–Smith reagent failed in this case.

Although simple alkenes such as cyclohexene do not react with diazomethane $-Pd(OAc)_2$, strained alkenes such as norbornadiene and hexamethyl-Dewar benzene readily react at room temperature.[2]

$$\overset{\displaystyle CH_2N_2}{\underset{\displaystyle 63\%}{\overset{\displaystyle Pd(OAc)_2}{\xrightarrow{\hspace{1.5cm}}}}}$$

*ArI → Ar—Ar.* Iodobenzene and *para*-disubstituted derivatives are converted into biaryls in about 50% yield when treated with a catalytic quantity of palladium(II) acetate in triethylamine at $100°$. Presence of water results in replacement of iodine by hydrogen. The reaction fails or proceeds in low yield with *ortho*-substituted aryl iodides. During the reaction Pd(0) is deposited.[3]

[1] U. Mende, B. Radüchel, W. Skuballa, and H. Vorbrüggen, *Tetrahedron Letters*, 629 (1975); B. Radüchel, U. Mende, G. Cleve, G.-A. Hoyer, and H. Vorbrüggen, *ibid.*, 633 (1975).
[2] J. Kottwitz and H. Vorbrüggen, *Synthesis*, 636 (1975).
[3] F. R. S. Clark, R. O. C. Norman, and C. B. Thomas, *J.C.S. Perkin I*, 121 (1975).

**Palladium acetate—Sodium chloride.**

*o-Alkylbenzaldehydes.* Aromatic aldehydes can be alkylated in the *ortho*-position in 60–95% yield on conversion to a Schiff base (1) and then into the σ-bonded palladium complex (2) by reaction with palladium acetate and sodium chloride. Addition of a ligand (usually triphenylphosphine) and then an alkyllithium or Grignard reagent results in *ortho*-alkylation with elimination of Pd(0).[1]

[1] S.-I. Murahashi, Y. Tanba, M. Yamamura, and I. Moritani, *Tetrahedron Letters*, 3749 (1974).

**Palladium black,** 1, 778–782; 2, 203; 4, 365–366; 5, 498–499.

*Amine-exchange reaction.*[1] Unsymmetrical secondary and tertiary amines can be synthesized by dehydrogenation of primary or secondary amines by palladium black (25–200°, 3–20 hr.). The initial step is considered to involve dehydrogenation to an imine followed by addition of the amine and elimination:

*Examples:*

$$C_6H_5CH_2NH_2 \xrightarrow[\text{}]{\overset{Pd}{80^0, 5 \text{ hr.}}} (C_6H_5CH_2)_2NH + C_6H_5CH=NCH_2C_6H_5$$
$$45\% \qquad\qquad\qquad 45\%$$

$$CH_3CH_2NHC_6H_5 \xrightarrow[\substack{98\%}]{\overset{Pd}{150^0, 48 \text{ hr.}}} (CH_3CH_2)_2NC_6H_5$$

Two different amines can be used; for example, a second amine can react with the intermediate imine postulated above as shown:

$$\lfloor R^1CH=NR^2 \rfloor + R^3NHR^4 \rightarrow \begin{bmatrix} R^1CHNHR^2 \\ | \\ NR^3R^4 \end{bmatrix} \xrightarrow[-R^2NH_2]{} R^1CH_2NR^3R^4$$

*Example:*

$$CH_3CH_2NHC_6H_5 + \underline{n}\text{-}C_6H_{13}NH_2 \xrightarrow[\substack{95\%}]{\overset{Pd}{120^0, 40 \text{ hr.}}} CH_3CH_2NHC_6H_{13}\text{-}\underline{n}$$

**sec-** *or* **t-Amines; pyrroles.**[2]   Benzyl or allyl alcohols react with primary or secondary amines in the presence of palladium black at 80–120° for 6–26 hr. to give *sec-* or *t*-amines.

*Examples:*

$$C_6H_5CH_2OH + \underline{n}\text{-}C_6H_{13}NH_2 \xrightarrow[\substack{98\%}]{Pd} C_6H_5CH_2NHC_6H_{13}\text{-}\underline{n}$$

$$\begin{matrix} C_6H_5 \\ \diagdown \\ \diagup \\ CH_3 \end{matrix}CHOH + \underline{n}\text{-}C_6H_{13}NH_2 \xrightarrow[\substack{83\%}]{Pd} \begin{matrix} C_6H_5 \\ \diagdown \\ \diagup \\ CH_3 \end{matrix}CHNHC_6H_{13}\text{-}\underline{n}$$

$$CH_2=CHCH_2OH + \underline{n}\text{-}C_6H_{13}NH_2 \xrightarrow[\substack{87\%}]{Pd} CH_2=CHCH_2NHC_6H_{13}\text{-}\underline{n}$$

N-Substituted pyrroles can be synthesized by a similar reaction:

$$HOCH_2CH=CHCH_2OH + RNH_2 \xrightarrow[\substack{50-95\%}]{Pd, 120^0} \underset{\substack{| \\ R}}{\boxed{\phantom{xx}}}$$
$$(\underline{cis} \text{ or } \underline{trans})$$

*Hydrogen transfer from diamines* (5, 498–499).   In a continuation of a study of hydrogen transfer from amines, Murahashi *et al.*[3] have found that dienes are converted into enes by catalyzed transfer of hydrogen from diamines such as

1,3-propanediamine (1). Thus when 1,5-cyclooctadiene and (1) are stirred at 140° for 2 hr. in the presence of palladium black, cyclooctene is obtained in 85% yield together with (2). Presumably allylamine is formed initially and reacts

with (1) to give 2-ethylhexahydropyrimidine (3), which is dehydrogenated to (2). Actually, (3) is very efficient for selective hydrogenation of dienes such as dicyclopentadiene (4).

Note that structures containing a 1,3-diaminopropane unit have been implicated in biological hydrogen transfer.

*Cleavage of protective groups of amino acids.* Some amino acid protective groups, for example CBZ and benzyl esters and ethers, are cleaved by palladium black[4]; BOC, *t*-butyl esters, and ethers are stable. If liquid ammonia[5] is used as solvent, the catalyst is not poisoned by sulfur-containing amino acid residues.[6]

[1] N. Yoshimura, I. Moritani, T. Shimamura, and S.-I. Murahashi, *Am. Soc.*, **95**, 3038 (1973).
[2] S.-I. Murahashi, T. Shimamura, and I. Moritani, *J.C.S. Chem. Comm.*, 931 (1974).
[3] S.-I. Murahashi, T. Yano, and K. Hino, *Tetrahedron Letters*, 4235 (1975).
[4] R. Willstätter and E. Waldschmidt-Leitz, *Ber.*, **54**, 113 (1921).
[5] V. du Vigneaud *et al.*, *Am. Soc.*, **52**, 4500 (1930); **75**, 4879 (1953).
[6] J. Meienhofer and K. Kuromizu, *Tetrahedron Letters*, 3259 (1974).

**Palladium catalysts, 1,** 778–782; **2,** 203; **4,** 368–369; **5,** 499.

*Transfer hydrogenation of aromatic nitro compounds.*[1] Aromatic nitro compounds are reduced to anilines when refluxed in excess cyclohexene in the presence of ordinary commerical 10% Pd/C catalyst. The method is very useful for selective reduction of polynitrobenzenes. Halogen, if present, is eliminated. Cyclohexene is superior to cyclohexa-1,3-diene as hydrogen donor. The reaction is usually successful, but slower, with sulfur-containing substrates. 4-Methoxy-2,5-dinitroanisole is reduced to 2,5-dimethoxy-4-nitroaniline in 10 min. in the steam bath under these conditions.

*Borohydride-reduced palladium.*[2] Reduction of palladium chloride in methanol with sodium borohydride until evolution of a gas ceases leads to a black material, which is not particularly sensitive to air and is not pyrophoric. The material is useful for selective hydrogenations. It catalyzes rapid hydrogenation of bonds of the type C=C, N=N, and N=O, but not the type C=N and C=O. No hydrogenolysis of nitrogen or oxygen functions is observed in alcohols, amines, amides, esters, ethers, or lactones. Epoxides are opened to alcohols very slowly.

*Deoxygenation of sulfoxides.*[3] Palladium (5%) on charcoal is an effective catalyst for hydrogenation of alkyl and aryl sulfoxides to sulfides (ethanol, autoclave, 80–90°, 1 atm. of hydrogen). Carbonyl groups are not reduced, but a carbon–carbon double bond is reduced, although more slowly than the S=O group.

This reaction was used in a procedure for homologation of aromatic aldehydes, as illustrated for conversion of benzaldehyde into phenylacetaldehyde (5). The first step is a Knoevenagel-like reaction with methyl methylthiomethyl sulfoxide

$$C_6H_5CHO + CH_3SOCH_2SCH_3 \xrightarrow[\underline{ca.~80\%}]{Triton~B} C_6H_5CH=C\begin{smallmatrix} SCH_3 \\ \\ SOCH_3 \end{smallmatrix} \xrightarrow[80\%]{H_2 \atop Pd/C}$$

(1)                                                                          (2)

$$C_6H_5CH_2CH\begin{smallmatrix} SCH_3 \\ \\ SCH_3 \end{smallmatrix} \xrightarrow[77\%]{H_2O_2,~HOAc} C_6H_5CH_2CH\begin{smallmatrix} SCH_3 \\ \\ SOCH_3 \end{smallmatrix} \xrightarrow[88\%]{HC(OC_2H_5)_3 \atop H_2SO_4} C_6H_5CH_2CHO$$

(3)                                              (4)                                              (5)

(1, 4, 341–342) with Triton B as catalyst. The product (2) is reduced to (3). Oxidation of (3) with hydrogen peroxide gives the dimethyl dithioacetal S-oxide (4), which is hydrolyzed by acid to (5).

*Selective reduction of polynitro compounds.* 2,6-Dinitroanilines can be reduced to nitrophenylenediamines by partial hydrogenation with 20% Pd on

charcoal. The method is superior to reduction with sulfides (3, 270–271).[4]

[1] I. D. Entwistle, R. A. W. Johnstone, and T. J. Povall, *J.C.S. Perkin I*, 1300 (1975).
[2] T. W. Russell and D. M. Duncan, *J. Org.*, **39**, 3050 (1974).
[3] K. Ogura, M. Yamashita, and G. Tsuchihashi, *Synthesis*, 385 (1975).
[4] R. E. Lyle and J. L. LaMattina, *Synthesis*, 726 (1974).

**Palladium(II) chloride, 1**, 782; **2**, 303–305; **4**, 369–370; **5**, 500–503.

$\pi$-*Allylpalladium chloride dimers* (**4**, 369; **5**, 500–501). Trost and Weber[1] have developed a short, stereoselective synthesis of acyclic terpenes based on the fact that $\pi$-allylpalladium chloride dimers are formed selectively with the nonconjugated double bond of substances such as methyl geraniate (1). This effect was used for a simple synthesis of the dimethyl ester of a pheromone of

the Monarch butterfly (4). When activated by 1,2-bis(diphenylphosphine)ethane (diphos), the complex can be alkylated by the anion of dimethyl malonate to give (3), which is then decarbomethoxylated by the method of McMurry and Wong.[2]

All-*trans*-farnesol (9) was synthesized from (2) by isoprenylation with the anion of the sulfone ester (5), prepared as shown in equation I. Reaction of this

anion with (2) in THF leads to formation of (6), which is converted into (9)

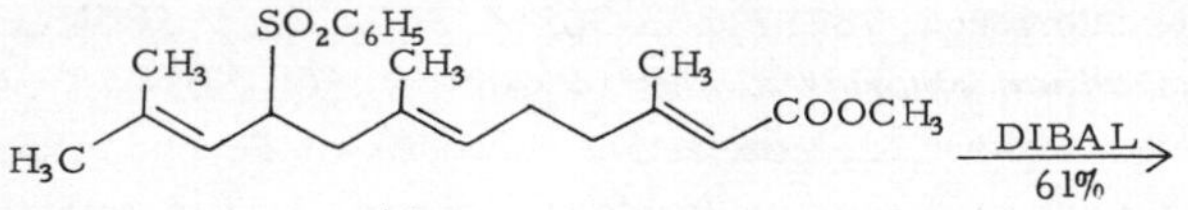

(7)

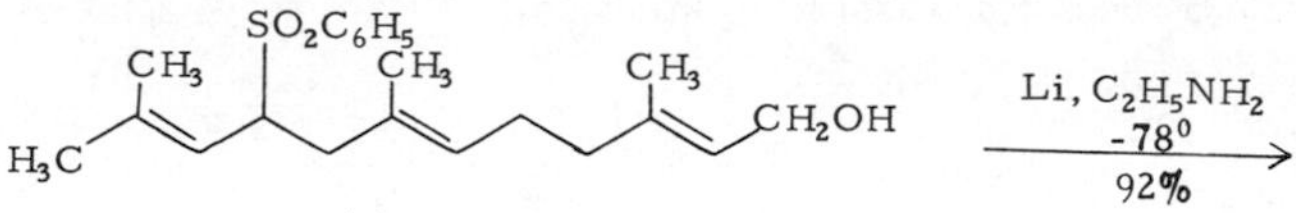

(8)

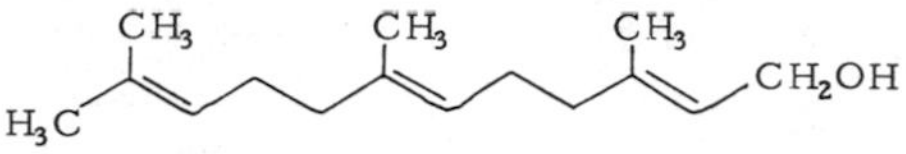

(9)

by known methods. The same sequence was used to convert methyl farnesoate (10) into geranylgeraniol (11).

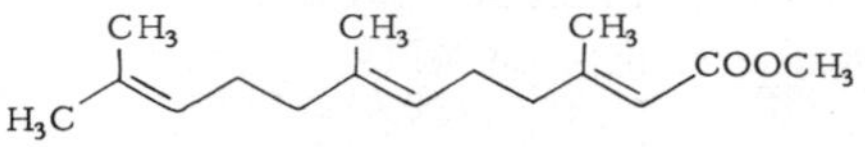

(10) 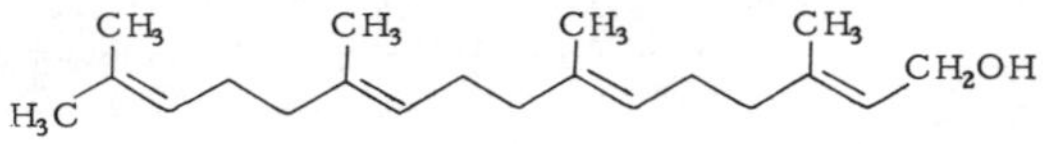                                          (11)

*Alkylation of π-allylpalladium complexes* (**4**, 369; **5**, 500). Trost and Strege[3] have reported regioselectivity and stereoselectivity in allylic alkylation via the π-allypalladium complex (1) derived from methylenecyclohexane. The site of alkylation of (1) is strikingly influenced by the ligand used. Thus reaction

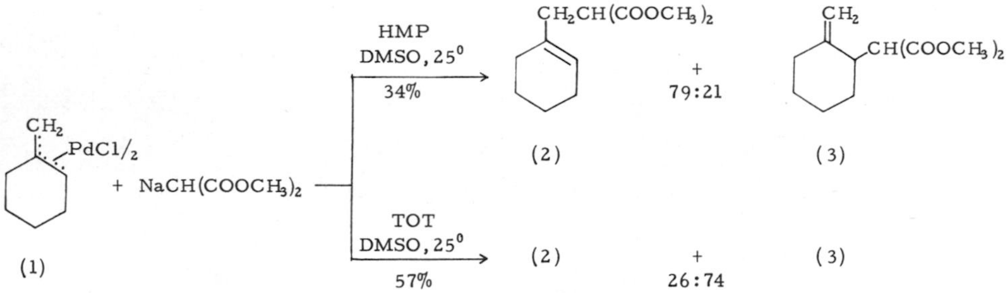

of (1) with the anion of dimethyl malonate in the presence of hexamethylphosphorous triamide (HMP) results in substitution mainly at the primary carbon atom to give (2). However, when a more bulky ligand such as tri-*o*-tolylphosphine (TOT) is used, then substitution at the secondary carbon atom to give (3) is favored.

The stereochemical course of alkylation was investigated by reaction of the 4-*t*-butyl derivative (4) of (1) with the anion of dimethyl malonate with TOT as the ligand. The initial product was decarbomethoxylated to give (5) and (6) in the ratio 18:1. The assignment of configuration based on NMR spectra was

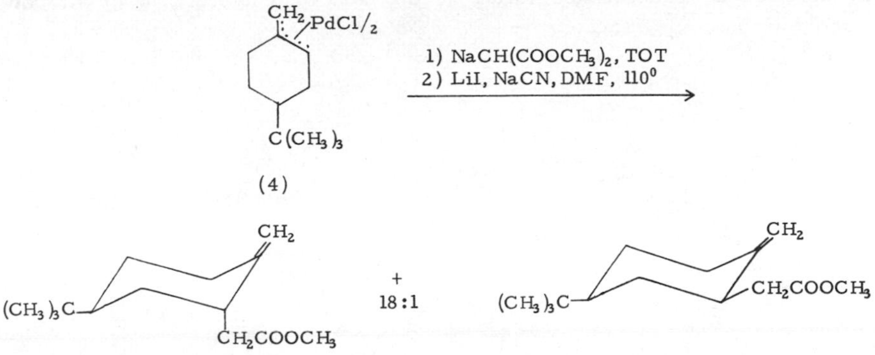

(4)

1) NaCH(COOCH$_3$)$_2$, TOT
2) LiI, NaCN, DMF, 110°

(5)

+
18:1

(6)

confirmed by synthesis of (6). Thus allylic alkylation via π-allyl palladium complexes shows a marked preference for formation of the less stable axial isomer.

**π-Allylpalladium chloride dimers of α,β-unsaturated ketones.**[4] The π-allylpalladium complex of ethyl crotonate (2) reacts with the anion of diethyl malonate in DMSO and THF to give exclusively the product (3) of γ-functionalization of (1) with retention of configuration of the double bond. On the

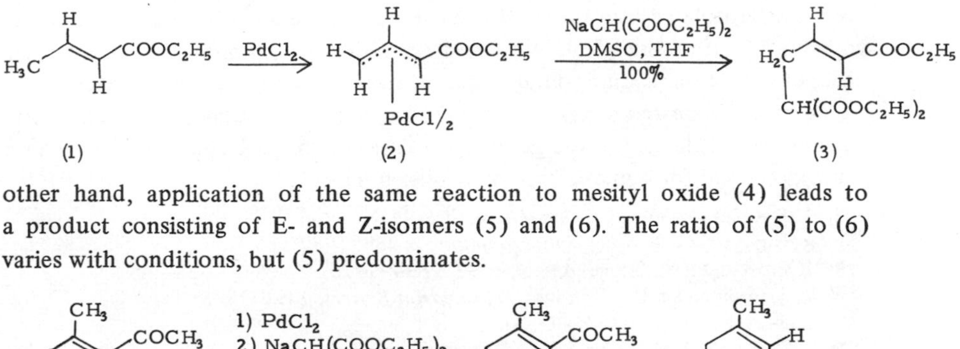

(1)                    (2)                    (3)

other hand, application of the same reaction to mesityl oxide (4) leads to a product consisting of E- and Z-isomers (5) and (6). The ratio of (5) to (6) varies with conditions, but (5) predominates.

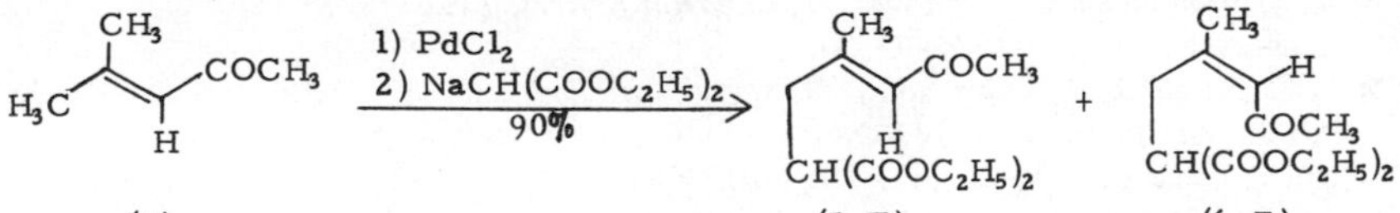

(4)              (5, E)              (6, Z)

**Hydrolysis of nitriles to amides.**[5] Nitriles form complexes with palladium chloride. When heated with water, these complexes are partially hydrolyzed to amides with liberation of PdCl$_2$.[6] The hydrolysis can also be carried out *in situ* without isolation of the complex. Yields of amides are low in the case of CH$_3$C≡N or C$_6$H$_5$C≡N (20–30%); yields are in the range of 70–85% in the case of XCH$_2$C≡N.

*Pyridine- and quinolinecarboxylic acids.*[7] Methylpyridines and methyl-quinolines form complexes (1) with palladium chloride. These complexes can be oxidized by concentrated $H_2O_2$ at 80° and then hydrolyzed to the corresponding carboxylic acid (3).

(1)                    (2)                              (3)

*Carbodiimides.*[8] Palladium(II) chloride, primary amines, and isonitriles form a carbene—metal complex (1),[9] which on treatment with silver oxide is converted into a carbodiimide (2) in yields of 75-95%. Actually, the complex

(1)                                    (2)

need not be isolated, and the synthesis can be conducted with catalytic amounts of $PdCl_2$. In this case yields are generally improved by addition of molecular sieves or anhydrous sodium sulfate, since water is one product of the reaction.

*Polymer-supported reagent.*[10] Palladium(II) chloride supported on polymeric diphenylbenzylphosphine is a very active catalyst for hydrogenation of alkenes and alkynes and for isomerization of double bonds.

[1] B. M. Trost and L. Weber, *J. Org.*, **40**, 3617 (1975).
[2] J. McMurry and G. B. Wong, *Syn. Commun.*, **2**, 389 (1972).
[3] B. M. Trost and P. E. Strege, *Am. Soc.*, **97**, 2534 (1975).
[4] W. R. Jackson and J. U. G. Strauss, *Tetrahedron Letters*, 2591 (1975).
[5] S. Paraskewas, *Synthesis*, 574 (1974).
[6] M. S. Kharasch, R. C. Seyler, and F. R. Mayo, *Am. Soc.*, **60**, 882 (1938).
[7] S. Paraskewas, *Synthesis*, 819 (1974).
[8] Y. Ito, T. Hirao and T. Saegusa, *J. Org.*, **40**, 2981 (1975).
[9] B. Crociani, T. Boschi, and U. Belluco, *Inorg. Chem.*, **9**, 2021 (1970).
[10] K. Kaneda, M. Terasawa, T. Imanaka, and S. Teranishi, *Chem. Letters*, 1005 (1975).

## Palladium(II) chloride—Thiourea.

*Hydrocarboxylation of acetylene.*[1] Italian chemists have prepared a very active catalyst for hydrocarboxylation reactions from palladium(II) chloride and 2 eq. of thiourea. When acetylene, carbon monoxide, and oxygen are passed into a solution of the catalyst in methanol at room temperature, dimethyl maleate is

$$HC{\equiv}CH + 2\ CO + 2\ CH_3OH \xrightarrow[> 90\%]{Cat.} \underset{H_3COOC\qquad\quad COOCH_3}{\overset{H\qquad\quad H}{C{=}C}}$$

formed in more than 90% yield. The oxygen is added to remove the hydrogen evolved in the reaction. Dimethyl fumarate and isomers of dimethyl muconate are formed in trace amounts.

*α-Methylene-γ-butyrolactones.* This catalyst system was used by Norton *et al.*[2] for carbonylation of the acetylenic alcohol (2) to give the bicyclic α-methylene-γ-butyrolactone (3) in high yield. The reaction was suggested by an

(1)                   (2)                              (3)

early synthesis of α-methylene-γ-lactone (5) from 3-butyne-1-ol (4) and nickel carbonyl.[3] In this new method $PdCl_2$ and thiourea were used in equimolar amounts. Only traces of product are formed in the absence of thiourea.

(4)                                    (5)

[1] G. P. Chiusoli, C. Venturello, and S. Merzoni, *Chem. Ind.*, 977 (1968); L. Casser, G. P. Chiusoli, and F. Guerrieri, *Synthesis*, 509 (1973).
[2] J. R. Norton, K. E. Shenton, and J. Schwartz, *Tetrahedron Letters*, 51 (1975).
[3] E. R. H. Jones, T. Y. Shen, and M. C. Whiting, *J. Chem. Soc.*, 230 (1950).

**Palladium nitrate,** $Pd(NO_3)_2$. Mol. wt. 230.72. Suppliers: Alfa, Fisher.

*Piperidines.* Palladium nitrate and triphenylphosphine (ratio 1:3) catalyze

the reaction of butadiene with Schiff bases to form isomeric 2,3,6-trisubstituted piperidines. Palladium chloride and palladium acetylacetonate are ineffective as catalysts.[1]

[1] J. Kiji, K. Yamamoto, H. Tomita, and J. Furukawa, *J.C.S. Chem. Comm.*, 506 (1974).

**Pentafluorophenylcopper,** 5, 504–505. Supplier: Peninsular Chem. Co.

*Ullmann diphenyl ether synthesis.* Copper or one of its salts or oxides has been used to effect this reaction between an aromatic halide and a phenol. Cava and Afzali[1] have found that yields are considerably improved by use of the soluble pentafluorophenylcopper and of pyridine as solvent. For example 6'-phenoxylaudanosine $(2, X = OC_6H_5)$ was obtained in 52% yield in this way

(1, X=Br)
(2, X=OC$_6$H$_5$)

from $(1, X = Br)$. The highest yield obtained by the classical method was 13% $(K_2CO_3, CuO, Py)$. The improved procedure was used for condensation of (1) with several phenolic alkaloids to prepare natural bisbenzylisoquinolines; yields were in the range of 42–54% based on (1).

[1] M. P. Cava and A. Afzali, *J. Org.*, **40**, 1553 (1975).

**Peracetic acid, 1**, 785–791; **2**, 307–309; **3**, 219; **4**, 372; **5**, 505–506.

*Esterification.*[1,2] Carboxylic acids can be esterified by oxidation of a hydrazone with peracetic acid with a trace of iodine as catalyst (equation 1). This

$$(I) \quad RCOOH + (C_6H_5)_2C{=}NNH_2 \xrightarrow{HOOAc, I_2} RCOOCH(C_6H_5)_2$$

reaction is especially useful for preparation of diphenylmethyl esters. It was developed particularly for esterification of the sensitive penam S-oxides (1). Esters were obtained from (1) in yields of 75–90%.

(1)

Apparently diphenyldiazomethane is the actual reagent, and indeed this substance can be prepared in good yield by oxidation of benzophenone hydrazone with $CH_3COOOH$ in $CH_2Cl_2$ in the presence of iodine and a base, particularly 1,1,3,3-tetramethylguanidine. The oxidation can also be effected in $CH_2Cl_2{-}H_2O$ in the presence of a phase-transfer catalyst such as trioctylpropylammonium chloride (85% yield).

A variety of aryldiazoalkanes were obtained in this way.

[1] R. Bywood, G. Gallagher, G. K. Sharma, and D. Walker, *J.C.S. Perkin I*, 2019 (1975).
[2] J. R. Adamson, R. Bywood, D. T. Eastlick, G. Gallagher, D. Walker, and E. M. Wilson, *ibid.*, 2030 (1975).

**Perbenzoic acid, 1, 791–796; 3, 219.**

*Polymer-supported reagent.* A polymeric peracid reagent (1) has been prepared from a polystyrene resin cross-linked with 1 or 2% of divinylbenzene by chloromethylation and oxidation to give a resin substituted by carboxyl groups. The —COOH groups are then converted into —COOOH groups by treatment

$$\cdots CH-CH_2 \cdots$$
(1)

with 85% $H_2O_2$ in methanesulfonic acid. This material is stable on storage at $-20°$, but decomposes slowly at $20°$. It epoxidizes olefins in THF at $40°$. Good yields are generally obtained from disubstituted olefins. The advantage is that the spent reagent is removable by filtration or centrifugation.[1]

A similar reagent has been reported by Japanese chemists.[2]

[1] C. R. Harrison and P. Hodge, *J.C.S. Chem. Comm.*, 1009 (1974).
[2] K. Hirao, O. Setoyama, T. Saito, and O. Yonemitsu, *Chem. Pharm. Bull. Japan*, 22, 2757 (1974).

**Perchloric acid, 1, 796–802; 2, 309–310; 3, 220; 5, 506–507.**

*Bicyclic γ-butyrolactones.* Cyclohexanols and cyclopentanols fused to a cyclopropane ring at the 2,3-position and substituted at the 4-position by an acetic acid group rearrange when treated with 7% aqeuous perchloric acid (methanol as cosolvent) at room temperature to bicyclic γ-butyrolactones.[1]

*Examples:*

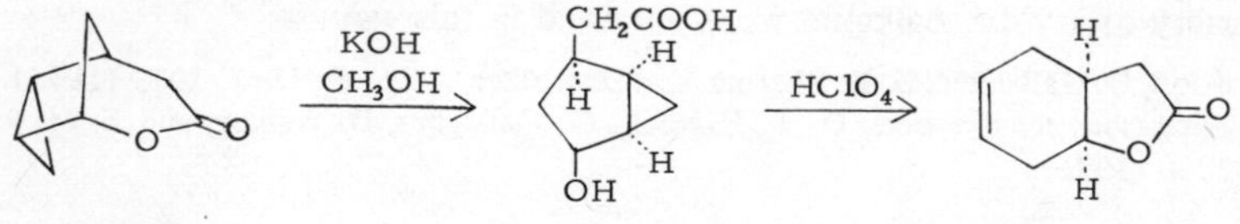

(5)                    (6)                    (7)

The butyrolactones can also be obtained from bicyclic lactones. For example, (2) can be obtained on solvolysis of either (1) or (3). This synthesis is stereoselective; the stereochemistry of the fused ring system is apparently controlled by the $C_3/C_4$ steric arrangement and not by the configuration of the hydroxyl group.

[1] J. A. Marshall and R. H. Ellison, *J. Org.*, **40**, 2070 (1975).

## Perchloryl fluoride, 1, 802–808.

*Caution*: Two laboratories[1] have reported serious explosions in reactions with perchloryl fluoride.

[1] J. H. J. Peet and B. W. Rocket, *J. Organometal. Chem.*, **82**, C57 (1974); W. Adcock and T. C. Khor, *ibid.*, **91**, C20 (1975).

**Perfluorotetramethylcyclopentadienone,** 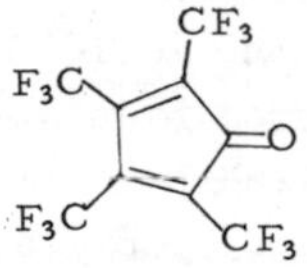 (1). Mol. wt. 352.08, m.p. 44–45°, yellow, extremely volatile.

*Preparation.*[1]

*Diels–Alder reactions.*[2] Unlike cyclopentadienone, this substance is stable as the monomer, probably because of the bulky substitutents. However, this dienone undergoes Diels–Alder reactions with alkenes, sometimes at room temperature or below. In reactions with 1,3-dienes, it functions as the diene rather than the dienophile, probably because of steric effects.

*Examples:*

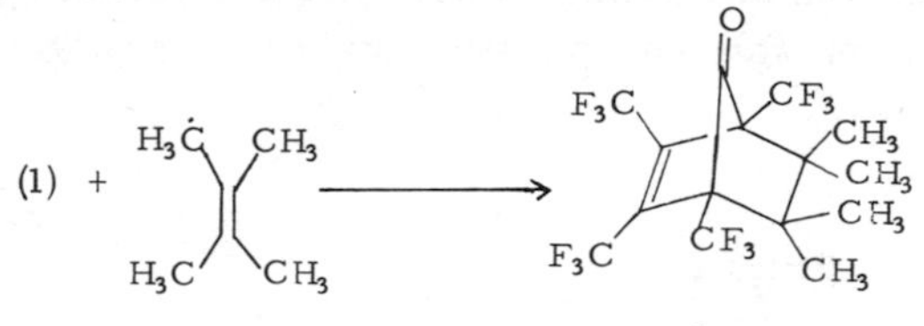

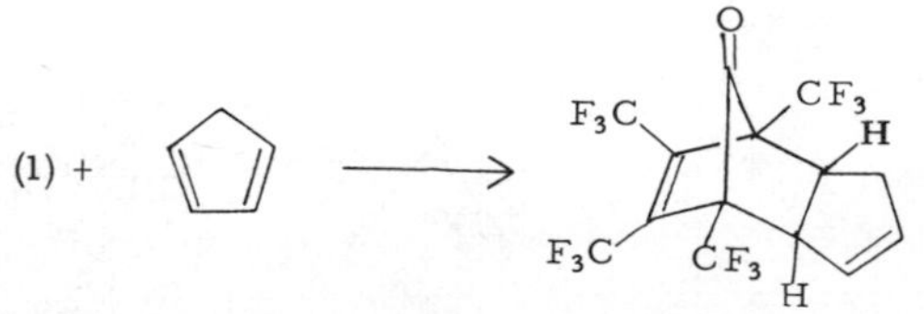

[1] R. S. Dickson and G. Wilkinson, *J. Chem. Soc.*, 2699 (1964).
[2] S. Szilagyi, J. A. Ross, and D. M. Lemal, *Am. Soc.*, **97**, 5586 (1975).

**Periodates, 1,** 809–815; **2,** 311–313; **4,** 373–374; **5,** 507–508.

*Lemieux–Rudloff oxidation* (**1,** 810–812). Bernassau and Fétizon[1] have reported a short and efficient degradation of the side chain of lanosterol (**1**). The first step involves periodate–permanganate oxidation; the last step, irradiation

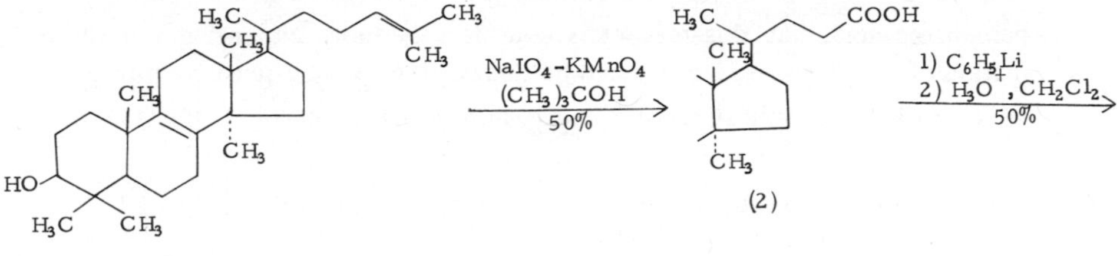

(1)

$$\xrightarrow[\substack{(CH_3)_3COH \\ 50\%}]{NaIO_4\text{-}KMnO_4}$$

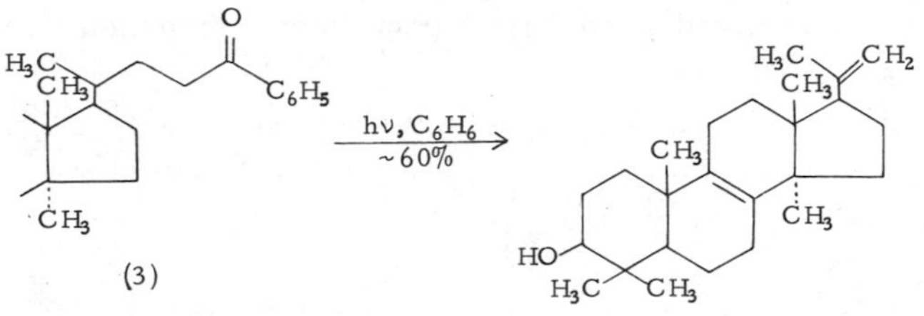

(2)

$$\xrightarrow[\substack{2)\ H_3O^+,\ CH_2Cl_2 \\ 50\%}]{1)\ C_6H_5Li}$$

(3)

$$\xrightarrow[\sim 60\%]{h\nu,\ C_6H_6}$$

(4)

of (3), has been described earlier.[2] The overall yield of (4) is about 12%.

Oxidation of β-pinene with aqueous alkaline potassium permanganate and catalytic amounts of potassium periodate leads to the expected product, norpinone (5), and the unusual β-ketol (6), formed by a new type of rearrangement of the pinene skeleton.[3]

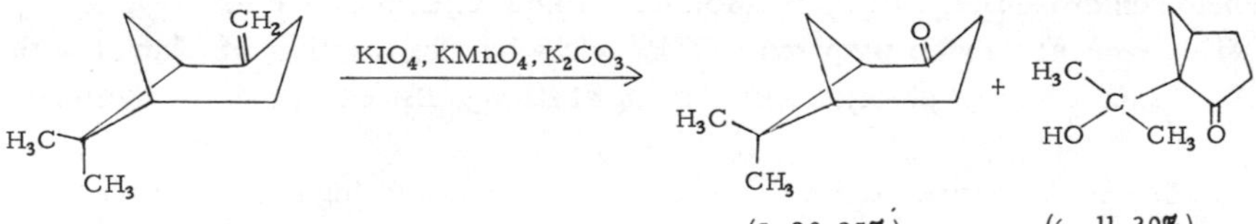

$$\xrightarrow{KIO_4,\ KMnO_4,\ K_2CO_3}$$

(5, 20-25%)        (6, 11-20%)

[1] J.-M. Bernassau and M. Fétizon, *Synthesis*, 795 (1975).
[2] B. Ganem and M. S. Kellog, *J. Org.*, **39**, 575 (1974).
[3] C. W. Jefford, A. Roussel, and S. M. Evans, *Helv.*, **58**, 2151 (1975).

**Peroxybenzimidic acid (Payne's reagent), 1,** 469–470; **4,** 375; **5,** 511.

*Nitroxides.* Payne's reagent is the most satisfactory oxidant for conversion of substituted piperidines to their nitroxides in a nonpolar solvent.[1] The reaction is catalyzed by sodium tungstate. Yields are equally high with *m*-chloroperbenzoic acid, but this reagent is considerably more expensive and the formation of $m$-chlorobenzoic acid can be undesirable for acid-sensitive nitroxides.

[1] E. J. Rauckman, G. M. Rosen, and M. B. Abou-Donia, *Syn. Commun.*, **5**, 409 (1975).

**Phenacylidinedimethylsulfurane,**

(1). Mol. wt. 180.27.

*Preparation.*[1]

*2-Phenylindoles.*[2] The reagent reacts with aniline hydrobromide in the presence of N,N-diethylaniline to form 2-phenylindole (equation I). Actually, the

precursor of (1), phenacyldimethylsulfonium bromide, can be used equally well.

[1] A. W. Johnson and R. T. Amel, *Tetrahedron Letters*, 819 (1966).
[2] H. Junjappa, *Synthesis*, 798 (1975).

**Phenyl chlorosulfite,** $C_6H_5O\overset{O}{\overset{\|}{S}}Cl$. Mol. wt. 176.62, b.p. 95°/10 mm.

This reagent can be prepared in 71% yield by the reaction of phenol with 300% excess thionyl chloride with stirring at strong reflux to expel the hydrogen chloride formed.[1]

*Nitriles.* Aldoximes can be converted into nitriles in high yield by reaction with phenyl chlorosulfite in ether containing 1 eq. of pyridine at 5°. On standing at 25° the phenyl oxime-O-sulfonate decomposes to a nitrile.[2] Carbonates of oximes also decompose to nitriles, but only at temperatures above 100°.[3]

$$RCH{=}NOH \ + \ Cl\overset{O}{\overset{\|}{S}}OC_6H_5 \xrightarrow[\text{}]{Py, 5^\circ} RCH{=}NO\overset{O}{\overset{\|}{S}}OC_6H_5 \xrightarrow[90-98\%]{25^\circ} RC{\equiv}N \ + \ SO_2 \ + \ C_6H_5OH$$

[1] W. E. Bissinger and F. E. Kung, *Am. Soc.*, **70**, 2664 (1948).
[2] J. G. Krause and S. Shaikh, *Synthesis*, 502 (1975).
[3] J. M. Prokipcak and P. A. Forte, *Canad. J. Chem.*, **49**, 1321 (1971).

Phenyl diazomethyl sulfoxide (Phenylsulfinyldiazomethane), $C_6H_5\overset{O}{\underset{\uparrow}{S}}CHN_2$. Mol. wt. 166.20.

*Preparation:*

$$CH_2N_2 + C_6H_5\overset{O}{\underset{\uparrow}{S}}Cl \longrightarrow C_6H_5\overset{O}{\underset{\uparrow}{S}}CHN_2 \quad \underset{2:1}{+} \quad C_6H_5\overset{O}{\underset{\uparrow}{S}}CH_2Cl$$

*Phenylsulfinylcarbene,* $C_6H_5\overset{O}{\underset{\uparrow}{S}}CH:$. The diazo compound reacts with olefins at 20° to form phenylsulfinylcyclopropanes.[1] The reaction is highly stereoselective. Thus E-2-butene reacts to give cyclopropanes with >97% of the methyl groups *trans* to each other; Z-2-butene is converted into cyclopropanes with >99% of the methyl groups in the *cis*-orientation. Moreover, the *anti*-adduct is formed almost exclusively in the case of the latter butene.

[1]C. G. Venier, H. J. Barager, III, and M. A. Ward, *Am. Soc.*, 97, 3238 (1975).

(R)-(+)-α-Phenylethylamine (α-Methylbenzylamine), 1, 838; 2, 272–273; 3, 199–200.

*Oxaziridines.* In the past optically active oxaziridines were prepared by oxidation of imines with optically active peracids. Polish chemists[1] have now prepared these heterocycles by peracid oxidation of imines formed from carbonyl compounds and (R)-(+)-α-phenylethylamine. Two diastereomers are formed in high optical yield. These are separable by high-pressure liquid chromatography.

*Example:*

$$\alpha_{436} + 118.5° \qquad\qquad \alpha_{436} + 205.4°$$

[1]C. Belzecki and D. Mostowicz, *J. Org.*, 40, 3878 (1975).

Phenylhydrazine, 1, 838–842; 2, 322.

*2,3-Dialkylindoles.* The Fischer indole synthesis can be conducted in one flask by heating a ketone and phenylhydrazine hydrochloride in acetic acid at 90°.[1]

*Examples:*

$$CH_3CC_2H_5 + C_6H_5NHNH_2 \cdot HCl \xrightarrow[-H_2O]{CH_3COOH} \left[ CH_3CC_2H_5 + HCl \right]$$
(with $O$ above first $CC_2H_5$, and $NNHC_6H_5$ above the bracketed $CH_3CC_2H_5$)

$$\xrightarrow[\text{pure}]{63\% \mid 90^0}$$

$$n\text{-}C_5H_{11}CCH_3 + C_6H_5NHNH_2 \cdot HCl \xrightarrow[75-80\%]{}$$
(with $O$ above $CCH_3$)

[1]V. Dave, *Org. Syn.*, submitted (1975).

**Phenyl isocyanide dichloride**, $C_6H_5N{=}CCl_2$, **1**, 843–845.

*Dealkylation of cyclic t-amines.* Cyclic *t*-amines react with the reagent to give, after hydrolysis, urea derivatives in which the ring has been opened or the alkyl group has been replaced.[1]

$$C_6H_5N{=}CCl_2 + \underset{}{\overset{CH_3}{\triangle}} \xrightarrow[87\%]{H_2O} C_6H_5NHCONCH_2CH_2Cl$$
(with $CH_3$ below the N in product)

$$C_6H_5N{=}CCl_2 + CH_3N\bigcirc \xrightarrow[73\%]{H_2O} C_6H_5NHCO{-}N\bigcirc$$

[1]G. Leclerc, B. Rouot, and C. G. Wermuth, *Tetrahedron Letters*, 3765 (1974).

**Phenylmercuric acetate**, $C_6H_5HgOOCCH_3$. Mol. wt. 336.75, m.p. 150°. Suppliers: Alfa, ROC/RIC.

Organolead acetates can be prepared in high yield by reaction of organolead halides with phenylmercuric acetate.[1] The reagent is less toxic than thallium(I)

$$R_3PbX + C_6H_5HgOAc \xrightarrow[80-100\%]{CHCl_3, 20^0} R_3PbOAc + C_6H_5HgX$$

acetate; it can be recovered by reaction of $C_6H_5HgX$ with acetic acid.

[1]D. de Vos, D. C. van Beelen, H. O. van der Kooi, and J. Wolters, *Rec. trav.*, **94**, 100 (1975).

**Phenylselenenyl bromide and chloride, 5,** 518–522.

*α,β-Unsaturated ketones,* (5, 518–519). Reich *et al.*[1] have published complete details on the conversion of ketones to enones by selenenylation followed by selenoxide elimination. Several procedures are available for preparation of α-phenylseleno ketones.

The most useful method is reaction of ketone (and ester) lithium enolates, usually prepared by deprotonation of ketones with LDA, with either $C_6H_5SeBr$ or $C_6H_5SeCl$. Enol acetates can be converted into α-phenylseleno ketones by reaction with phenylselenenyl trifluoroacetate, prepared *in situ* by treatment of $C_6H_5SeCl$ or $C_6H_5SeBr$ with silver trifluoroacetate or by conversion to the lithium enolate and reaction with $C_6H_5SeBr$. It is sometimes possible to obtain isomeric α-phenylseleno ketones by use of these two methods (equations I and II).

Several oxidants can be used for oxidation to the selenoxides. Reich considers hydrogen peroxide the oxidant of choice under aqueous conditions; this oxidation can be carried out directly without isolation of the selenide. The oxidation can also be carried out in a two-phase system ($H_2O$–$CH_2Cl_2$); in this case addition of pyridine as buffer is usually advantageous. Ozonization in $CH_2Cl_2$ is useful where the presence of water is undesirable. *m*-Chloroperbenzoic acid has been used to some extent in the case of unsaturated substrates, since selenoxides are formed more readily than epoxides. Reich considers sodium metaperiodate the reagent of last resort because of expense and necessity for an aqueous methanolic medium.

The selenoxide elimination reaction proceeds in satisfactory yield in the case of acylic α-phenylseleno ketones and of cyclic α-phenylseleno ketones in which the carbon bearing the selenium is fully substituted.

The chief advantage of selenium reagents over corresponding sulfur reagents is that selenoxide elimination occurs at a temperature about 100° lower than that required for sulfoxide elimination. However, selenium compounds are toxic and they are more expensive than sulfur compounds.

Reich *et al.* have explored the use of phenylseleninyl chloride, $C_6H_5SeOCl$, m.p. 56–64°. This reagent can be prepared in 92% yield by ozonization of

$C_6H_5SeCl$ in $CH_2Cl_2$. This reagent reacts with lithium enolates of ketones to give enones directly (equation III). However, yields are variable and often lower than those obtained by the two-step procedure. Moreover, the reagent is very hygroscopic. Reich recommends use of this compound only where normal selenide oxidation is not successful.

$$\text{(III)} \quad C_6H_5\overset{\overset{O}{\|}}{C}CH_2CH_2CH_3 \xrightarrow[\substack{80\%}]{\substack{1)\ LDA,\ -78^0 \\ 2)\ C_6H_5SeOCl \\ 3)\ HOAc,\ 25^0}} C_6H_5\overset{\overset{O}{\|}}{C}CH=CHCH_3$$

*Furanes.* Grieco et al.[2] have developed a general synthesis of 2,4- and 2,3,4-substituted furanes from substituted $\gamma$-lactones, as illustrated for the conversion of (1), $\gamma$-decalactone, into 2-hexyl-4-benzylfurane (5). The lactone enolate of (1) is alkylated to give (2). Selenenylation of the enolate of (2) with phenylselenenyl chloride leads to (3). Selenoxide fragmentation of (3) results almost

exclusively in formation of the endocyclic $\alpha,\beta$-unsaturated butenolide (4). The final step involves reduction with diisobutylaluminum hydride (1, 262).

The same paper discloses one example of another route to $\alpha$-alkylated $\gamma$-lactones: conjugate addition of a dialkylcopperlithium to an $\alpha$-methylene-$\gamma$-lactone, as shown in the formulation of (6) $\rightarrow$ (10).

[1] H. J. Reich, J. M. Renga, and I. L. Reich, *Am. Soc.*, **97**, 5434 (1975).
[2] P. A. Grieco, C. S. Pogonowski, and S. Burke, *J. Org.*, **40**, 542 (1975).

**β-(Phenylsulfonyl)propionaldehyde ethylene acetal,** (1). Mol. wt. 242.29, m.p. 67–68°.

*Preparation:*

$$CH_2=CHCHO \ + \ C_6H_5SH \longrightarrow C_6H_5SCH_2CH_2CHO \longrightarrow$$

$$\text{(1)}$$

*α,β-Unsaturated aldehydes.*[1]  The sulfone (1) is converted into the α-carbanion by *n*-butyllithium at −75°; this anion can be alkylated to give (2) in good yield. The free aldehyde (3) is liberated by acid treatment. When (3) is treated with base (1% aqueous NaOH; $N(C_2H_5)_3$–THF–$H_2O$; 10% aqueous Triton B),

benzenesulfinic acid is eliminated in about 30 min. at 20° to give the (E)-α,β-unsaturated aldehyde (4). The method thus effects β-alkylation of acrolein, from which (1) is prepared.

Application of the same sequence to the sulfone acetal (5) provides a route to (E)-α,β-unsaturated aldehydes of type (6). This sequence was used in a synthesis of (±)-nuciferal (7).

*δ-Lactols.*[2] The anion (1) also reacts with epoxides to give addition products (2) in about 60–80% yield. The phenylsulfonyl group is cleaved reductively by

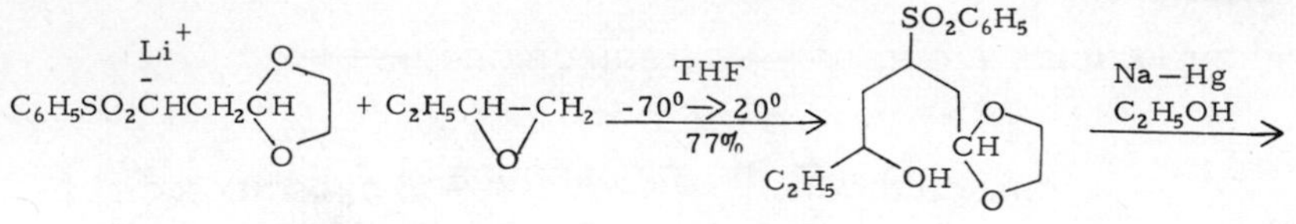

(1)                                                   (2)

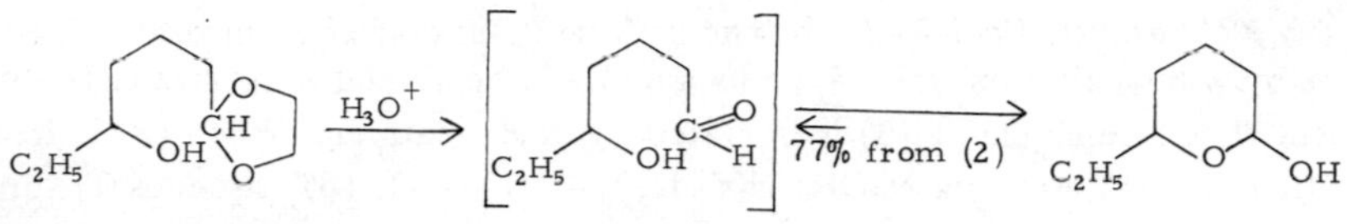

(3)                        (4a)                            (4b)

6% sodium amalgam at 20°. Treatment of the products (3) with acid in aqueous acetone affords the δ-hydroxy aldehydes (4a), which exist in the δ-lactol form (4b). The lactols can be oxidized readily to lactones.

This sequence has been used for the synthesis of jasmine lactone (7) as formulated.

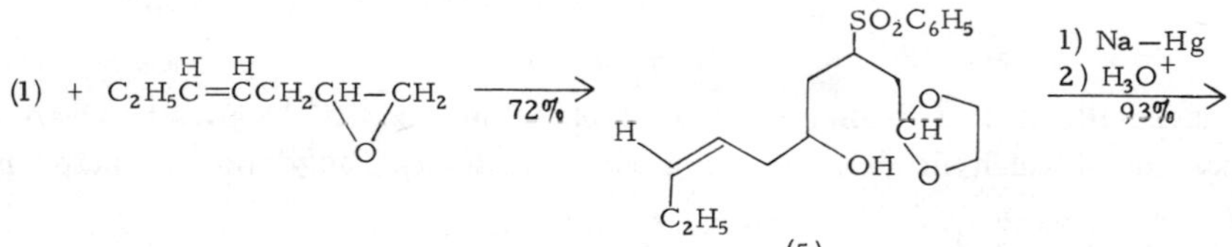

(5)

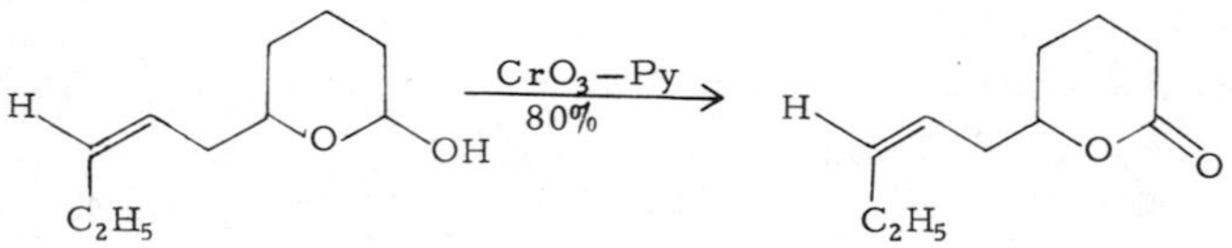

(6)                                              (7)

The related sulfone ketal (8) derived from methyl vinyl ketone has also been used in these reactions.

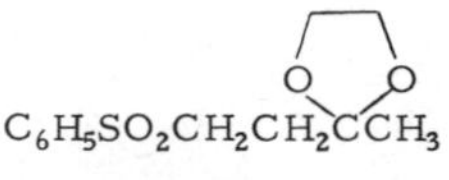

(8)

[1] K. Kondo and D. Tunemoto, *Tetrahedron Letters,* 1007 (1975).
[2] K. Kondo, E. Saito, and D. Tunemoto, *ibid.,* 2275 (1975).

**Phenylthioacetic acid,** $C_6H_5SCH_2COOH$. Mol. wt. 168.21, m.p. 64–66°. Supplier: Aldrich.

*β-Hydroxy sulfides.*[1] Phenylthioacetic acid can be converted into the dianion (1) by treatment with 2 eq. of lithium diisopropylamide in THF at 0°. The dianion can be alkylated in high yield. The monoalkylated product can be converted into the fully substituted acid (3) by repetition of the process—conversion to the dianion followed by alkylation. Reduction of (3) with lithium aluminum hydride then gives a β-hydroxy sulfide (4) in overall yields of 50–90%.

$$\underset{(1)}{\overset{\overset{\displaystyle Li}{|}}{C_6H_5SCHCOOLi}} \xrightarrow[0°\rightarrow 20°]{R^1X,\ THF} \underset{(2)}{\overset{\overset{\displaystyle R^1}{|}}{C_6H_5SCHCOOH}} \xrightarrow[2)\ R^2X]{1)\ 2\ LDA} \underset{(3)}{\overset{\overset{\displaystyle R^1}{|}}{\underset{\underset{\displaystyle R^2}{|}}{C_6H_5SCCOOH}}}$$

$$\xrightarrow[\substack{50-90\% \\ Overall}]{\substack{LiAlH_4 \\ THF,\ \triangle}} \underset{(4)}{\overset{\overset{\displaystyle R^1}{|}}{\underset{\underset{\displaystyle R^2}{|}}{C_6H_5SCCH_2OH}}}$$

β-Hydroxy sulfides (4) can be converted into olefins of the type $\overset{R^1}{\underset{R^2}{>}}C=CH_2$ by lithium–ammonia reduction of the corresponding benzoates (procedure of R. L. Sowerby and R. M. Coates, **4**, 379–380). The method provides an alternative to the more common methylenation of aldehydes or ketones with methylenetriphenylphosphorane or phenylthiomethyllithium.

*Examples:*

$$\underset{\overset{|}{C_6H_5SCHCH_2OH}}{\overset{CH_3(CH_2)_{10}CH_2}{}} \xrightarrow[86\%]{\substack{1)\ \underline{n}\text{-BuLi} \\ 2)\ C_6H_5COCl}} \underset{C_6H_5SCHCH_2OCOC_6H_5}{\overset{CH_3(CH_2)_{10}CH_2}{}} \xrightarrow[65\%]{Li/NH_3} CH_3(CH_2)_{11}CH=CH_2$$

$$\underset{91\%}{\xrightarrow{}}$$

$$\underset{CH_3(CH_2)_{10}CH_2}{\overset{(CH_3)_2CH}{|}}C_6H_5SCCH_2OH \xrightarrow[91\%]{} \underset{CH_3(CH_2)_{10}CH_2}{\overset{(CH_3)_2CH}{|}}C_6H_5SCCH_2OCOC_6H_5 \xrightarrow[64\%]{} \overset{(CH_3)_2CH}{\underset{CH_3(CH_2)_{11}}{>}}C=CH_2$$

The β-hydroxy sulfides (4) can also be converted into terminal oxiranes by the method of Shanklin *et al.* (5, 528):

$$C_6H_5S-\underset{\underset{CH_2(CH_2)_2CH_3}{|}}{\overset{\overset{CH_2(CH_2)_{10}CH_3}{|}}{C}}-CH_2OH \xrightarrow[\text{CH}_2\text{Cl}_2]{(CH_3)_3O^+BF_4^-} \underset{CH_3}{\overset{C_6H_5}{\phantom{x}}}\overset{+}{S}-\underset{\underset{CH_2(CH_2)_2CH_3}{|}}{\overset{\overset{CH_2(CH_2)_{10}CH_3}{|}}{C}}CH_2OH \xrightarrow[\text{overall}]{\substack{\text{NaOH}\\70\%}}$$

$$\underset{CH_3(CH_2)_{11}}{\overset{CH_3(CH_2)_3}{>}}C\overset{O}{\underset{CH_2}{\diagdown\diagup}}CH_2 \quad + \quad C_6H_5SCH_3$$

Grieco and Wang report one example of the conversion of β-hydroxy sulfides of type (5) into a methyl ketone by the method of Trost and co-workers.[2]

$$\underset{C_6H_5S\overset{|}{C}HCH_2OH}{\overset{CH_3(CH_2)_{10}\overset{|}{C}H_2}{\phantom{x}}} \xrightarrow[20^0]{SOCl_2} \underset{C_6H_5S\overset{|}{C}HCH_2Cl}{\overset{CH_3(CH_2)_{10}\overset{|}{C}H_2}{\phantom{x}}} \xrightarrow[\text{DMSO},20^0]{KOC(CH_3)_3}$$

(5)

$$\underset{C_6H_5S\overset{|}{C}=CH_2}{\overset{CH_3(CH_2)_{10}\overset{|}{C}H_2}{\phantom{x}}} \xrightarrow[\substack{CH_3CN,\,H_2O\\>85\%}]{HgCl_2} C_{12}H_{25}\overset{\overset{O}{\|}}{C}CH_3$$

overall from (5)

**γ-Substituted Δ^α,β-butenolides.[3]** The dianion (1) reacts with terminal oxiranes at −60 to 20° to give a γ-hydroxy acid (a), which is not isolated, but is converted into the isomeric lactones (2). Oxidation (NaIO$_4$, H$_2$O$_2$, peracid)

$$(1) \quad + \quad \underset{O}{RCH-CH_2} \longrightarrow \left[ \begin{array}{l} CH_2-\overset{|}{C}HSC_6H_5 \\ R\overset{|}{C}H \quad COOH \\ \phantom{RCH}\diagdown OH \end{array} \right] \xrightarrow[70-95\%]{\substack{C_6H_6,\Delta\\+\\H_3O^+}} \underset{R}{\phantom{x}}\diagup\!\diagdown_{O}\diagdown\!\diagup_{O}^{SC_6H_5}$$

(a)                                    (2)

$$\xrightarrow[\text{quant.}]{\text{Oxid.}} \underset{R}{\phantom{x}}\diagup\!\diagdown_{O}\diagdown\!\diagup_{O}^{\overset{\overset{O}{\uparrow}}{SC_6H_5}} \xrightarrow[75-85\%]{\sim110^0} \underset{R}{\phantom{x}}\diagup\!\diagdown_{O}\diagdown\!\diagup_{O}$$

(3)                                    (4)

to the sulfoxide (3) followed by pyrolysis gives the butenolide (4) in moderate overall yields. Alkylation of (1) and then application of this sequence results in α,γ-disubstituted Δ^α,β-butenolides.

[1] P. A. Grieco and C.-L. J. Wang, *J.C.S. Chem. Comm.*, 714 (1975).
[2] B. M. Trost, K. Hiroi, and S. Kurozumi, *Am. Soc.*, **97**, 438 (1975).
[3] K. Iwai, M. Kawai, H. Kosugi, and H. Uda, *Chem. Letters*, 385 (1974).

**Phenylthiocopper,** $C_6H_5SCu$. Mol. wt. 172.71. Stable to air for prolonged periods.

*Preparation.*[1]  The reagent is prepared by the reaction of thiophenol and red copper(I) oxide in refluxing ethanol with stirring for 19 hr.

$$2\ C_6H_5SH + Cu_2O \xrightarrow[91\%]{-H_2O} 2\ C_6H_5SCu$$

*Lithium phenylthio(alkyl)cuprates,* $C_6H_5SCu{<}^{R}_{Li}$ . These useful mixed cuprates (**5**, 414) can be prepared directly from phenylthiocopper and alkyllithium reagents[2]:

$$C_6H_5SCu + RLi \xrightarrow{THF,\,-20^0} \underset{R}{\overset{C_6H_5S}{>}}CuLi$$

The cuprates are particularly useful for transfer of tertiary alkyl groups to organic substrates:

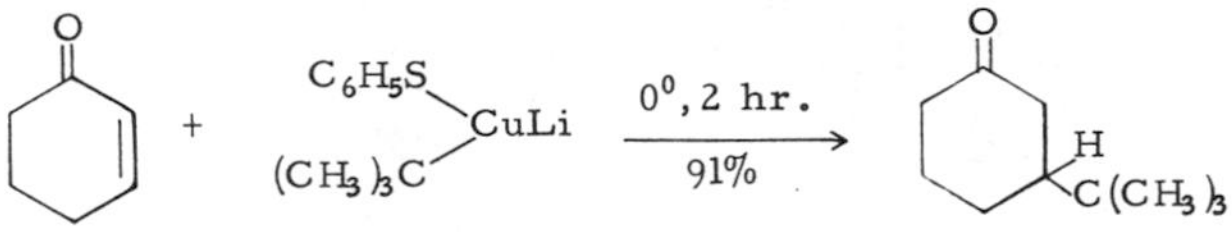

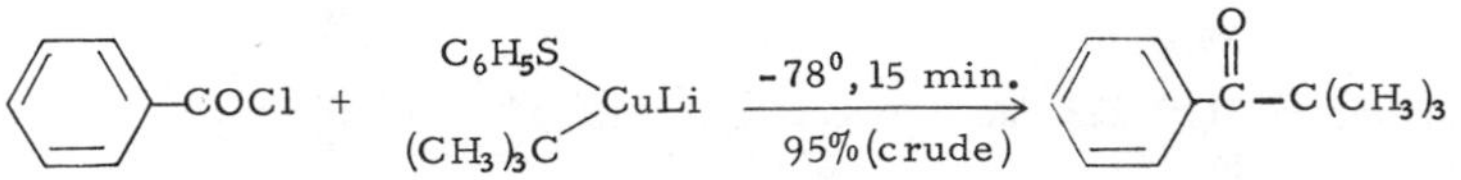

[1] R. Adams, W. Reifschneider, and A. Ferretti, *Org. Syn. Coll. Vol.*, **V**, 107 (1973).
[2] G. H. Posner, D. J. Brunelle, and L. Sinoway, *Synthesis*, 662 (1974).

**1-Phenylthiocyclopropyltriphenylphosphonium tetrafluoroborate** (1). Mol. wt. 498.32, m.p. 201.5–202.5°.

*Preparation:*

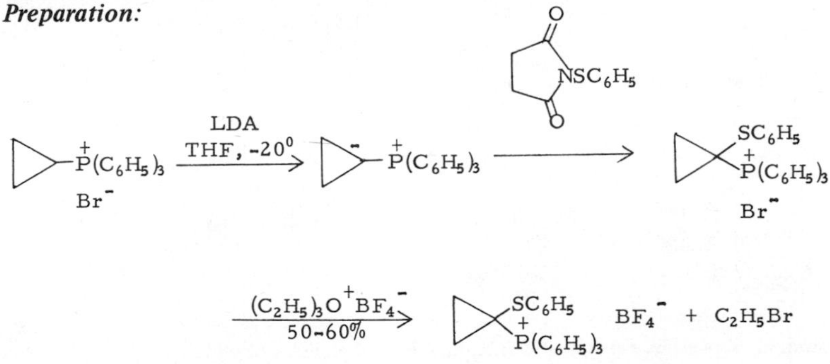

(1)

*Cyclopentanones.*[1] The synthon reacts with sodium enolates of acetoacetic esters (2) in THF (65°) to form cyclopentenyl sulfides (3), which are hydrolyzed by 20% HCl–dioxane to keto acids (4).

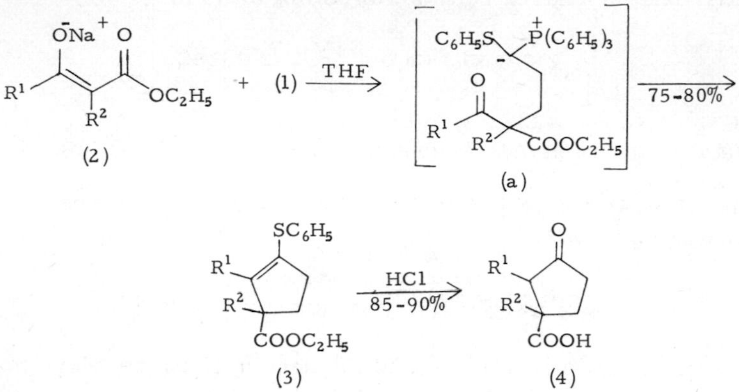

(2)    + (1) $\xrightarrow{THF}$    (a)    $\xrightarrow{75-80\%}$

(3)    $\xrightarrow[85-90\%]{HCl}$    (4)

The reagent also reacts with enolates of cyclic β-dicarbonyl compounds to give annelated products (equation I).

(I).

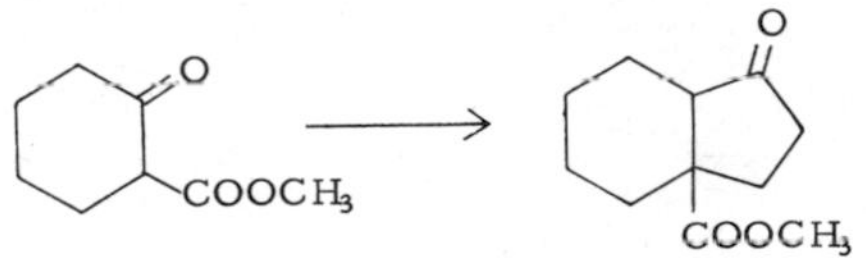

[1] J. P. Marino and R. C. Landick, *Tetrahedron Letters,* 4531 (1975).

**2-Phenylthiopropionic acid,** $C_6H_5S\overset{\underset{CH_3}{|}}{C}HCOOH$ . Mol. wt. 182.24.

   *Preparation.*[1]

   *α-Methylenelactones.* Trost and Leung[2] have disclosed a simple route to these compounds as shown in equation I.

(I)

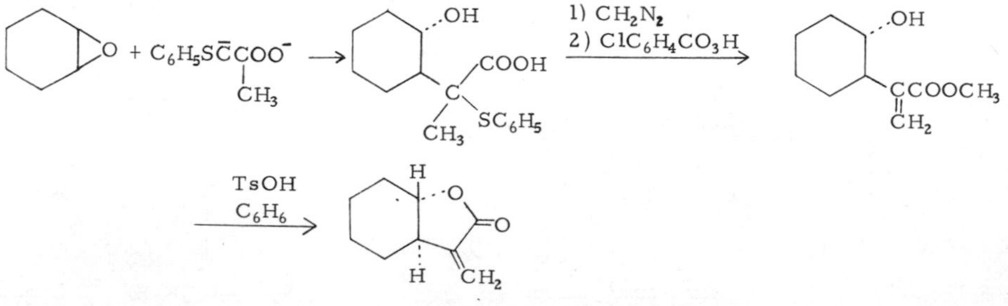

$\xrightarrow[C_6H_6]{TsOH}$

[1] K. Iwai, M. Kawai, H. Kosugi, and H. Uda, *Chem. Letters,* 385 (1974).
[2] B. M. Trost and K. K. Leung, *Tetrahedron Letters,* 4197 (1975).

**4-Phenyl-1,2,4-triazoline-3,5-dione** (PTAD), **1**, 849–850; **2**, 324–326; **3**, 223–224; **4**, 381–383; **5**, 528–530.

**in situ** *Generation.*[1] The reagent can be generated in high yield by oxidation of 4-phenylurazole with DMSO and *p*-toluenesulfonyl isocyanate (Upjohn Chem. Co.) at 0 to 25°. Isolation from the reaction mixture is difficult, but it is not necessary for successful Diels–Alder reactions. The oxidation is considered to be related to the Pfitzner–Moffatt oxidation.

*Adducts of steroidal 5,7-dienes* (**3**, 223–224).[2] The adducts of steroidal 3-keto-5,7-dienes (1) are decomposed at 20° in benzene containing BF₃ etherate to 4,6,8(14)-triene-3-ones (2) in unspecified yield.

Steroidal 3-keto-4,4-dimethyl-5,7-dienes (3) also form 1 : 1 adducts (4) with PTAD, but of a different type. These are decomposed by ethanolic HCl or by BF₃ etherate to 5,7,14-triene-3-ones (5) and tetrahydro-4-phenyltriazole-3,5-dione.

[1] J. A. Moore, R. Muth, and R. Sorace, *J. Org.*, **39**, 3799 (1974).
[2] J. Brynjolffssen, A. Emke, D. Hands, J. M. Midgley, and W. B. Whalley, *J.C.S. Chem. Comm.*, 633 (1975).

**Phenyl(tribromomethyl)mercury, 1,** 851–854; **2,** 326–328; **3,** 225; **4,** 385.

*Ring expansion of 1-methylindenes.* Gillespie *et al.*[1] have described a convenient synthesis of phenyl(tribromomethyl)mercury by mixing phenylmercuric chloride, sodium hydride, and bromoform in benzene with methanol as initiator. They have converted 1-methylindenes into 3-bromo-1-methylnaphthalenes by reaction with dibromocarbene generated from this precursor. Reaction of 1-methylindenes with dibromocarbene generated from bromoform with base

(X = 6-Cl)

leads to 2-bromo-1-methylnaphthalenes (low yield).

[1] J. S. Gillespie, Jr., S. P. Acharya, and D. A. Shamblee, *J. Org.,* **40,** 1838 (1975).

**Phenyl vinyl sulfoxide,** $CH_2=CH-\overset{O}{\overset{\|}{S}}C_6H_5$ . Mol. wt. 152.21.

The sulfoxide is prepared by treatment of diphenyl disulfide with vinyllithium followed by peracid oxidation.

*Vinyl carbonium ion equivalent.*[1] This sulfoxide has been used as an equivalent of the vinyl carbonium ion as illustrated for the synthesis of (3) from (1).

[1] G. A. Koppel and M. D. Kinnick, *J.C.S. Chem Comm.,* 473 (1975).

**1-Phospha-2,8,9-trioxaadamantane ozonide (2). Mol. wt. 208.11.**

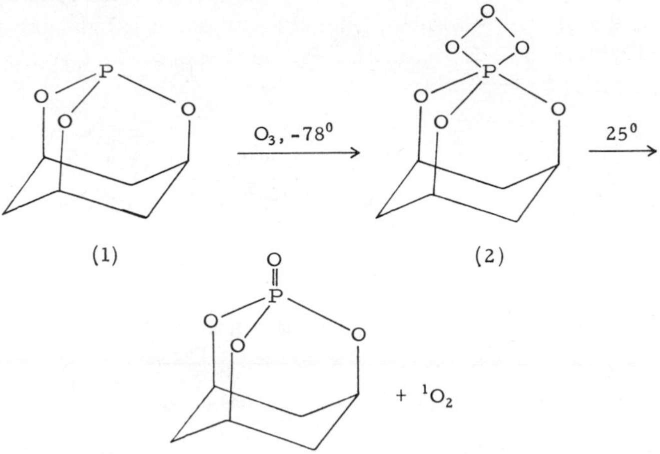

(1)    $O_3, -78°$ →    (2)    $25°$ →

+ $^1O_2$

(3)

This relatively stable ozonide is prepared[1] by ozonization of 1-phospha-2,8,9-trioxaadamantane[2] in methylene chloride at $-78°$. It decomposes at room temperature to singlet oxygen and the phosphate (3).

[1] A. P. Schaap, K. Kees, and A. L. Thayer, *J. Org.*, **40**, 1185 (1975).
[2] J. G. Verkade, T. J. Huttemann, M. K. Fung, and R. W. King, *Inorg. Chem.*, **4**, 83 (1965).

**Phosphonitrilic chloride, 2, 206–207; 4, 380–387.**

*Cleavage of α-hydroxy ketoximes.*[1] α-Hydroxy ketoximes when treated with phosphonitrilic chloride (1) and pyridine undergo fragmentation to aldehydes or ketones and nitriles in high yield:

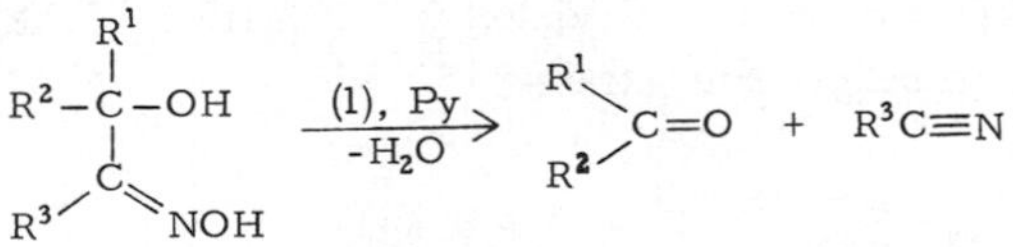

$$R^2-\underset{\underset{C}{\overset{|}{\underset{\|}{C}}}}{\overset{R^1}{\overset{|}{C}}}-OH \xrightarrow[-H_2O]{(1),\ Py} \overset{R^1}{\underset{R^2}{}}C=O\ +\ R^3C\equiv N$$

*Examples:*

$$\xrightarrow[75\%]{(1),\ Py\ \ THF} OHC-(CH_2)_4-CN$$

85%

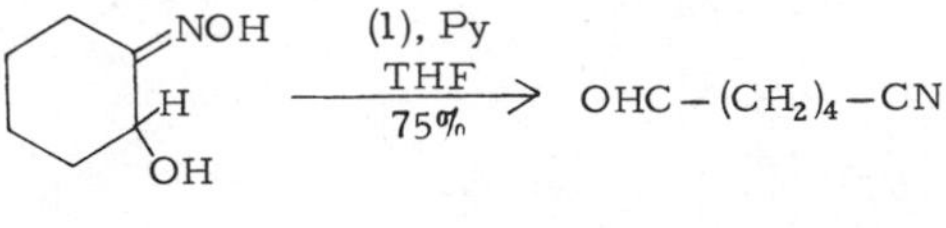

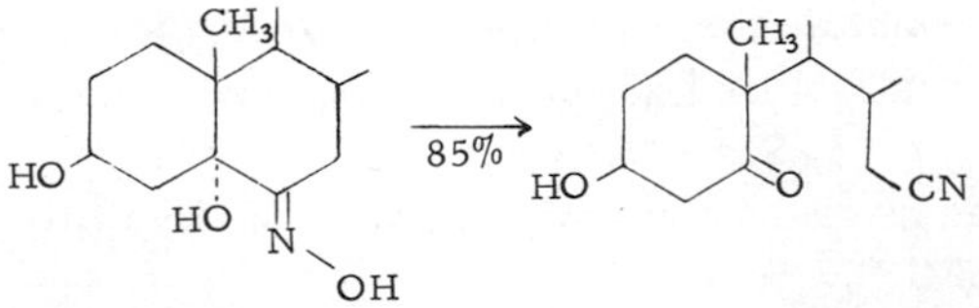

*Active esters of amino acids.* *o*-and *p*-Nitrophenyl esters of amino acids (**1**, 743; **5**, 477) can be prepared in about 45–95% yield by reaction of an N-protected amino acid with phosphonitrilic chloride and triethylamine in chloroform, ethyl acetate, or THF. After 15 min. the phenol is added and the reaction is stirred for about 12 hr.[2]

$$ \underset{\text{(phosphonitrilic chloride)}}{\text{P}_3\text{N}_3\text{Cl}_6} \;+\; \text{HOOCCHNHBOC} \xrightarrow{\text{N(C}_2\text{H}_5\text{)}_3} [\text{P}_3\text{N}_3\text{Cl}_5-\text{OCOCHNHBOC}] $$

$$ \xrightarrow{\text{ArOH}} \text{ArOCOCHNHBOC} $$

[1]G. Rosini, A. Medici, and S. Cacchi, *Synthesis*, 665 (1975).
[2]J. Martinez and F. Winternitz, *Tetrahedron Letters*, 2631 (1975).

## Phosphorus (red)–Iodine, 1, 862.

*Desoxybenzoins.* Benzoins are reduced to desoxybenzoins by treatment with red phosphorus and iodine in carbon disulfide containing pyridine (3.5 hr., 25°).[1] Presumably an α-iodo ketone is formed as an intermediate. This deoxygenation has been carried out also with hydriodic acid (**1**, 449).

$$ \underset{}{\text{ArC}-\text{CHAr}} \xrightarrow[80-90\%]{\text{Py, I}_2} \text{ArC}-\text{CH}_2\text{Ar} $$

[1]T.-L. Ho and C. M. Wong, *Synthesis*, 161 (1975).

## Phosphorus pentasulfide, 1, 870–871; 3, 226–228; 4, 389; 5, 534–535.

*3-Thiabicyclo[3.2.0]hepta-1,4-diene system.* This strained heterocyclic system (**2**) has been formed in low yield by the reaction of the 1,4-diketone (**1**) with phosphorus pentasulfide in tetralin (150°).[1]

(1)                    (2)

*Phosphorus pentachloride, sulfur monochloride.* The reaction of chlorine with phosphorus pentasulfide leads mainly to these two products:

$$ \text{P}_4\text{S}_{10} + 15\,\text{Cl}_2 \rightarrow 5\,\text{S}_2\text{Cl}_2 + 4\,\text{PCl}_5 $$

The reaction can be used for *in situ* preparation of these two reagents as shown in equations I and II.[2]

(I)

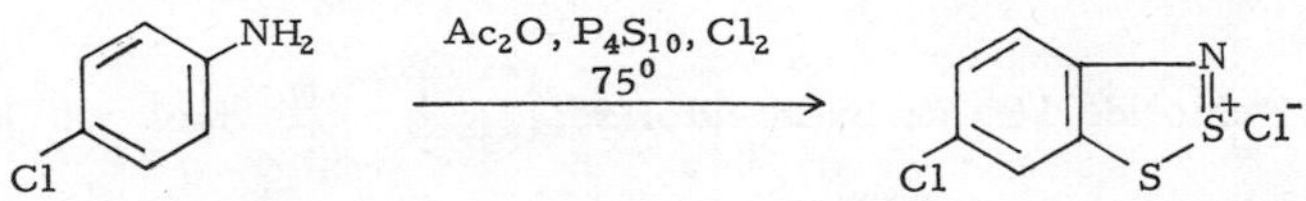

$Ac_2O, P_4S_{10}, Cl_2$
$75^0$

(II)

$P_4S_{10}, Cl_2$
$CCl_4, 20^0$

quant.

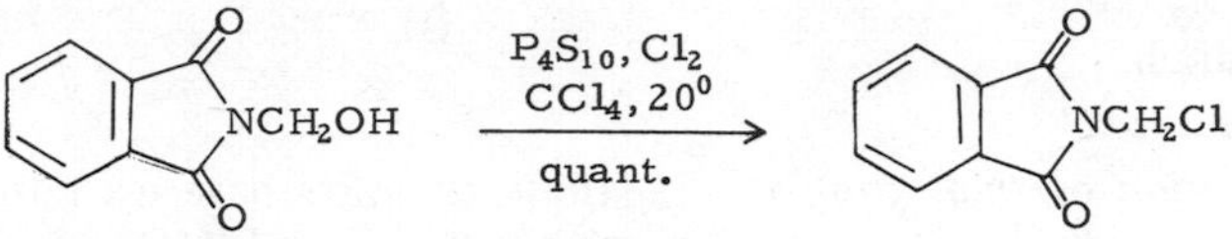

[1] P. J. Garratt and S. B. Neoh, *J. Org.*, **40**, 970 (1975).
[2] Z. Zur and E. Dykman, *Chem. Ind.*, 436 (1975).

**Phthaloyl peroxide,** 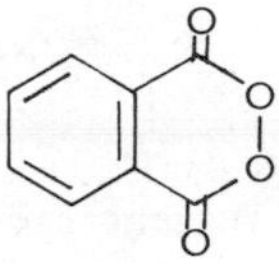 (1). Mol. wt. 164.12, m.p. 126° dec.

The peroxide can be obtained in 55% yield by reaction of *o*-phthaloyl dichloride with sodium peroxide.[1]

*Singlet oxygen.*[2] When (2) or (4) is shaken in benzene solution with (1), the known photoxidation products (3) and (5), respectively, are formed.

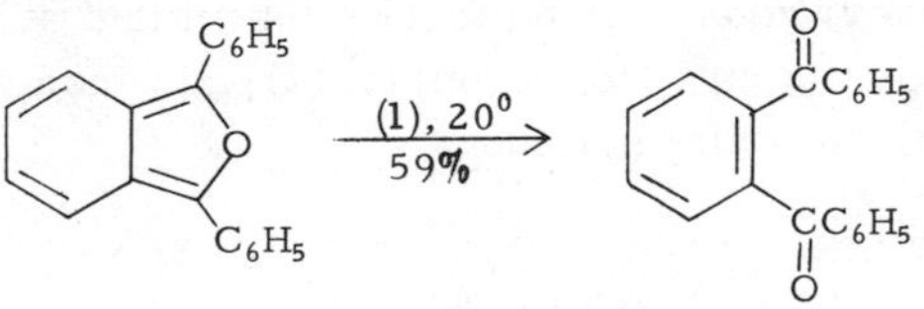

(2)

(1), 20°
59%

(3)

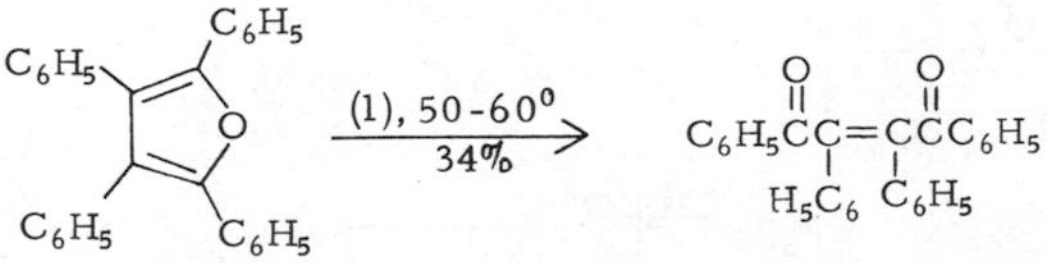

(4)

(1), 50-60°
34%

$C_6H_5CC=CCC_6H_5$

(5, cis and trans)

The reaction of (1) with 9,10-diphenylanthracene is chemiluminescent; the diol (7) is a major product. The expected *endo*-peroxide has not been detected.

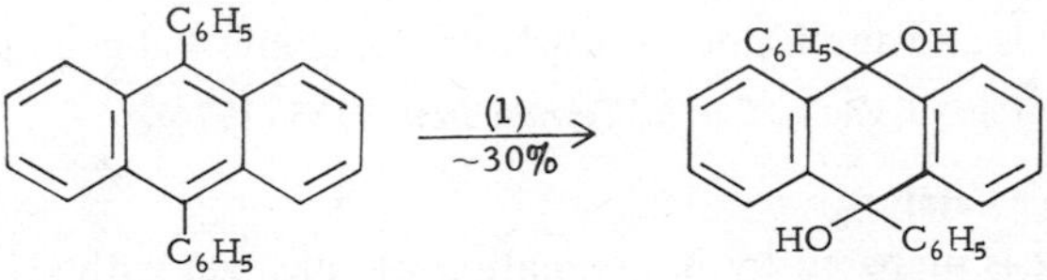

(6)

(1)
~30%

(7)

[1] K. E. Russel, *Am. Soc.*, **77**, 4814 (1955).
[2] K.-D. Gundermann and M. Steinfatt, *Angew. Chem. internat. Ed.*, **14**, 560 (1975).

**2-Picolyl chloride 1-oxide hydrochloride,** [structure] ·HCl Mol. wt. 180.03. Sup-

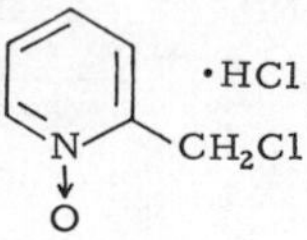

(1)

plier: Aldrich.

*Preparation.*[1]

*Protection of thiol groups.*[2] Japanese chemists have used the 2-picolyl 1-oxide group for protection of thiols. The protective group is removable by acetic

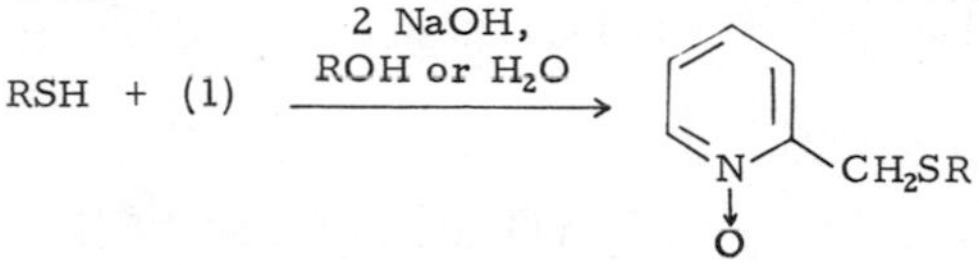

$$RSH + (1) \xrightarrow[\text{ROH or } H_2O]{2\ NaOH,} [structure]\ CH_2SR$$

anhydride (25°). One advantage is that the reagent can be used in aqueous solution.

[1] P. T. Sullivan, M. Kester, and S. T. Norton, *J. Med. Chem.*, **11**, 1172 (1968).
[2] Y. Mizuno and K. Ikeda, *Chem. Pharm. Bull. Japan*, **22**, 2889 (1974).

**Piperidine,** 1, 886–890; 2, 332; 4, 393.

*Decarbomethoxylation.*[1] Heterocycles substituted at one carbon atom by two carbomethoxy groups [for example, (1)] are monodecarbomethoxylated when heated with piperidine in toluene:

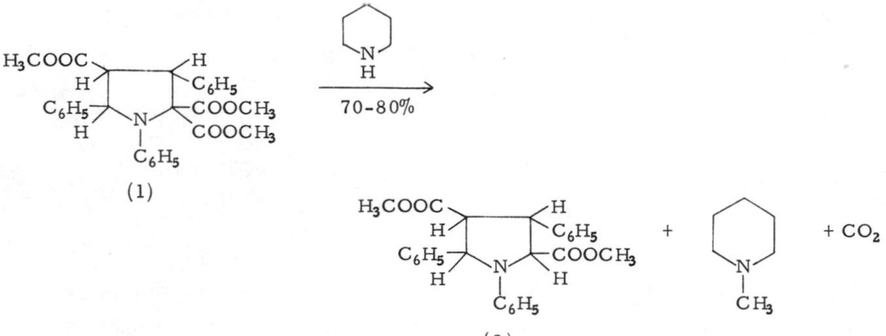

(1)

$$\xrightarrow{70-80\%}$$

(2)

The reaction is also applicable to substituted malonic and cyanoacetic esters.

[1] F. Texier, E. Marchand, and R. Carríe, *Tetrahedron*, 3185 (1974).

**Platinum—Silica catalyst.**

This catalyst is prepared by impregnation of silica gel with $H_2PtCl_4$, which is then reduced to the metal by hydrogenation at 500°.

*Triamantane.* This diamantoid (2) has been prepared by a multiple rearrangement of (1) in hydrogen in the gas phase on the platinum—silica catalyst.[1]

(1)                                    (2)

[1]W. Burns, M. A. McKervey, and J. J. Rooney, *J.C.S. Chem. Comm.*, 965 (1975).

**Polyhydrogen fluoride—Pyridine, 5, 538–539.**

*α-Fluorocarboxylic acids.*[1] α-Amino acids are converted into α-fluorocarboxylic acids by diazotization with sodium nitrite in 70% polyhydrogen fluoride—pyridine at ambient temperatures. For the most part, yields are in the range

$$\underset{\underset{NH_2}{|}}{RCHCOOH} \xrightarrow{\underset{(HF)_x\,F^-\,HN\overset{+}{\diagdown}}{NaNO_2}} \underset{\underset{F}{|}}{RCHCOOH}$$

60–95%; however, yields are low in the case of glycine (38%) and glutamic acid (12%).

*Alkyl fluoroformates.*[2] These substances can be obtained in 30–75% yield by diazotization of alkyl carbamates in polyhydrogen fluoride—pyridine.

$$\underset{RO-\overset{\overset{O}{\|}}{C}-NH_2}{} \xrightarrow{\underset{(HF)_x\,F^-\,HN\overset{+}{\diagdown}}{NaNO_2}} RO-\overset{\overset{O}{\|}}{C}-F$$

*Alkyl halides.*[3] Secondary and tertiary alcohols react with the reagent to form alkyl fluorides. If an alkali halide (NaF, NaCl, $NH_4Br$, KI) is added to the reaction, alkyl halides, even primary, are formed corresponding to the added salt. The displacement proceeds with inversion.

*Reaction of diazoalkanes and α-diazoketones.*[4] Phenyldiazomethane reacts with the reagent to give benzyl fluoride in 70% yield:

$$C_6H_5CHN_2 \xrightarrow[-N_2]{H^+} \left[C_6H_5\overset{+}{C}H_2\right] \xrightarrow[70\%]{F^-} C_6H_5CH_2F$$

α-Fluorinated ketones are obtained in the same way from α-diazo ketones:

$$C_6H_5COCHN_2 \xrightarrow[32\%]{HF-Py} C_6H_5COCH_2F$$

$$C_6H_{11}COCHN_2 \xrightarrow[50\%]{HF-Py} C_6H_{11}COCH_2F$$

Addition of alkali halides or N-halosuccinimides results in substitution by halide ion:

$$C_2H_5OCOCHN_2 + NBS \xrightarrow[50\%]{HF-Py} C_2H_5OCOCHBrF$$

[1] G. A. Olah and J. Welch, *Synthesis*, 652 (1974).
[2] *Idem, ibid.*, 654 (1974).
[3] *Idem, ibid.*, 653 (1974).
[4] *Idem, ibid.*, 896 (1974).

**Polyphosphate ester, 1**, 892–894; **2**, 333–334; **3**, 229–231; **4**, 394–395; **5**, 539–540.

*N-Alkylation.*[1] Primary and secondary amino groups are alkylated by polyphosphoric acid esters of methanol or ethanol at 130–160° (4–8 hr.). Concomitant cyclodehydrations can be effected.

*Examples:*

[1] M. Oklobdžija, V. Šunjić, F. Kajfež, V. Čaplar, and D. Kolbah, *Synthesis*, 596–597 (1975).

**Polyphosphoric acid (PPA), 1**, 894–895; **2**, 334–336; **3**, 231–233; **4**, 395–397; **5**, 540–542.

*Anthracene.* Israeli chemists[1] have modified the Haworth succinic anhydride synthesis of phenanthrenes[2] to a synthesis of anthracene. Thus treatment of the ketone (1) with PPA at 140° for an extended period results in formation of the ketone (2) by reversal of the Friedel–Crafts acylation reaction. This ketone is aromatized to anthracene (3) by treatment with potassium hydroxide–sodium hydroxide (Birch aromatization reaction).[3]

*Tetrahydroisophosphinolinium and tetrahydrophosphinolinium salts.*[4] These rare heterocyclic systems have been prepared as shown in equations I and II. Cyclization of the intermediate salts was carried out with 115% polyphosphoric acid.[5]

*Intramolecular ester condensation.* Indian chemists[6] have synthesized 2-(6'-methoxycarbonylhexyl)cyclopentene-2-one-1 (4) from methyl 10-undecenoate (1). The enone has been used for synthesis of prostaglandins.

[1] I. Agranat and Y.-S. Shih, *Synthesis*, 865 (1974).
[2] E. Berliner, *Org. React.*, 5, 229 (1949).
[3] A. J. Birch and D. A. White, *J. Chem. Soc.*, 4086 (1964).
[4] G. A. Dilbeck, D. L. Morris, and K. D. Berlin, *J. Org.*, 40, 1150 (1975).
[5] This material contains a minimum of 83.2% $P_2O_5$; it was obtained from FMC Corp.
[6] P. D. Gokhale, V. S. Dalavoy, A. S. C. Prakasa Rao, U. R. Nayak, S. Dev, *Synthesis*, 718 (1974).

**Potassium acetate, 1, 206–207.**

*Dehydration.* Methyl 3,7-diacetylcholate (1) can be converted into methyl $\Delta^{11}$-3,7-diacetylcholenate (3) by mesylation followed by dehydromesylation with potassium acetate in HMPT (100°, 2 days).[1] Potassium acetate presumably serves as buffer and as a weak base. The temperature is critical for high yields. Note that (1) does not form a tosylate, probably for steric reasons.

[1]C. H. Chen, *Synthesis*, 125 (1976).

## Potassium 3-aminopropylamide, $K^+ H\bar{N} (CH_2)_3 NH_2$ (1). Mol. wt. 112.22.

This base is formed quantitatively by reaction of potassium hydride with excess 3-aminopropylamine (trimethylenediamine). The base is fairly soluble in the amine ($\geqslant 1.5\ M$); the solutions are stable at room temperature for at least 8 hr.

*Isomerization of alkynes.*[1] This difunctional base induces rapid migration of triple bonds from the interior of the chain to the terminus in seconds at 0°:

$$H(CH_2)_6C\equiv C(CH_2)_6H \xrightarrow[\quad]{\overset{(1)}{H_2N(CH_2)_3NH_2}} H(CH_2)_{12}C\equiv\bar{C}K^+ \xrightarrow[89\%]{H_2O} H(CH_2)_{12}C\equiv CH$$

Formerly, only very strong bases (Na, $NaNH_2$, $KNH_2$) were found to effect isomerization of alkynes, and then only at elevated temperatures and only of 2-alkynes to 1-alkynes. Since internal alkynes are more stable than terminal alkynes, the isomerization is evidently possible because of formation of a metal acetylide. Migration is blocked by a branch in the chain. The base is also highly active in the exchange of benzene C—D bonds. It converts limonene to *p*-cymene with evolution of hydrogen at room temperature in high yield. Brown notes that the base can exist as a chelate (a), a structure that may contribute to the high basicity.

(a)

[1]C. A. Brown, *J.C.S. Chem. Comm.*, 22 (1975); C. A. Brown and A. Yamashita, *Am. Soc.,* **97**, 891 (1975).

## Potassium borohydride, 4, 398–399.

*Reduction of 9,10-phenanthrenequinones.* Dey and Neumeyer[1] have reported that some 9,10-phenanthrenequinones are reduced to 9,10-dihydro-

9,10-*trans*-diols (2) in higher yield with $KBH_4$ in absolute alcohol than with $LiAlH_4$ in THF. Thus (1) is reduced to (2) in 53% yield with $KBH_4$; when

$$KBH_4 \atop C_2H_5OH \atop 53\%$$

(1)                              (2)

$LiAlH_4$ is used, the yield of (2) is 26% and the product is contaminated with some of the quinone. This borohydride has also been found to be superior to $LiAlH_4$ for reduction of (3) to (4).[2]

(3)                              (4)

[1] A. S. Dey and J. L. Neumeyer, *J. Med. Chem.*, **17**, 1095 (1974).
[2] R. M. Moriarty, P. Dansette, and D. M. Jerina, *Tetrahedron Letters*, 2587 (1975).

**Potassium *t*-butoxide, 1,** 911–916; **2,** 338–339; **3,** 233–234; **4,** 399–405; **5,** 544–553.

*Oxygenation of* **p**-*cresols.*[1] In the presence of excess potassium *t*-butoxide, the methyl group of 2,6-disubstituted *p*-cresols is oxygenated exclusively to an

$$\xrightarrow[\text{90\%}]{O_2 \atop KOC(CH_3)_3, \, DMF}$$

aldehyde group. DMF is the most satisfactory solvent. The reactivity of the *p*-methyl group depends on the *ortho*-substituents and decreases in the order:

$$C(CH_3)_3 > OCH_3 > CH_3 > H > Br.$$

*Dehydrohalogenation.* Billups and Chow[2] have prepared naphtho[*b*]-cyclopropene (2) by the method used previously to prepare benzocyclopropene (**4,** 402). Other base–solvent systems were less successful.

(1)                                                              (2)

Ippen and Vogel[3] have also used this method to prepare the dicyclopropene derivative of naphthalene (2) from either the *cis*- or *trans*-isomers (1). This naphthalene derivative is sensitive to shock and requires cautious handling.

(1, cis or trans)                                                (2)

The preparation of tri-*t*-butylcyclopropenyl fluoroborate (4, 399–400) has been published.[4]

*t*-**Butylacetylene.** Kocienski[5] has published a convenient synthesis of *t*-butylacetylene from pinacolone; potassium *t*-butoxide in DMSO is used for dehydrochlorination of the *gem*-dichloro intermediate.

$$(CH_3)_3CCOCH_3 \xrightarrow{PCl_5} (CH_3)_3C\underset{\underset{Cl}{|}}{\overset{\overset{Cl}{|}}{C}}CH_3 \xrightarrow[95\%]{\underset{DMSO}{KOC(CH_3)_3}} (CH_3)_3CC\equiv CH$$

*Intramolecular Diels–Alder reaction.* Näf and Ohloff[6] have reported a short stereoselective synthesis of racemic patchouli alcohol (3) that involves the intra-molecular Diels–Alder reaction of (1) to give (2). This reaction was carried out in

(1)                                   (2)        (3)

(4)

decalin in a sealed tube at 280° in the presence of 5% of potassium *t*-butoxide. In the absence of this base, only (4) was obtained. The base may act as a stabilizer for (1) or it may be a catalyst.

*2-Pyridones.* 2-Halopyridines are converted into 2-pyridones when refluxed with potassium *t*-butoxide in *t*-butanol. The reaction presumably proceeds through the *t*-butyl ether. 3- and 4-Halopyridines do not undergo this transformation.[7]

[1] A. Nishinaga, T. Itahara, and T. Matsuura, *Angew. Chem. internat. Ed.*, **14**, 356 (1975).
[2] W. E. Billups and W. Y. Chow, *Am. Soc.*, **95**, 4099 (1973).
[3] J. Ippen and E. Vogel, *Angew. Chem. internat. Ed.*, **13**, 736 (1974).
[4] J. Ciabattoni, E. C. Nathan, A. E. Feiring, and P. J. Kocienski, *Org. Syn.*, **54**, 97 (1974).
[5] P. J. Kocienski, *J. Org.*, **39**, 3285 (1974).
[6] F. Näf and G. Ohloff, *Helv.*, **57**, 1868 (1974).
[7] G. R. Newkome, J. Broussard, S. K. Staires, J. D. Sauer, *Synthesis*, 707 (1974).

## Potassium *t*-butoxide–Dimethyl sulfoxide.

*Dicyclopropylidenemethane* (5).[1] This hydrocarbon has been obtained recently by the simple synthesis formulated.

[1] R. Kopp and M. Hanack, *Angew. Chem. internat. Ed.*, **14**, 821 (1975).

## Potassium carbonate–Dimethyl sulfoxide.

*Macrolides.* Macrolides can be prepared in high yield by dehydrobromination of $\omega$-bromocarboxylic acids with $K_2CO_3$ in DMSO at $100°$.[1] The reaction

does not require high dilution. 11-Undecanolide was obtained from 11-bromo-undecanoic acid in about 80% yield. Yields are nearly quantitative in the formation of 16- and 18-membered rings.

[1] C. Galli and L. Mandolini, *Org. Syn.*, submitted (1975).

**Potassium chloride (fluoride),** KCl, KF.

*Chlorodecarbomethoxylation; dechlorocarbomethoxylation.*[1] Dimethyl 2-chloroethylene-1,1-dicarboxylate (1) reacts with potassium fluoride in the

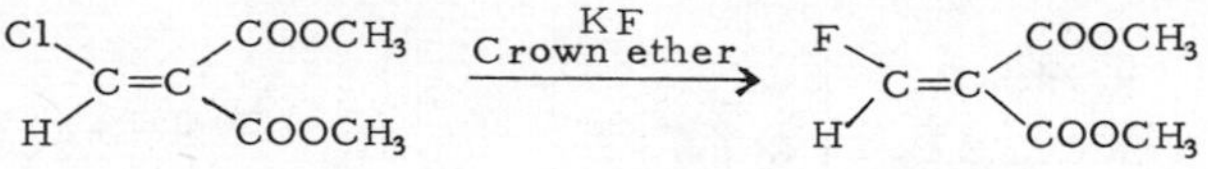

(1)                                                          (2)

presence of dicyclohexyl-18-crown-6 with replacement of Cl by F to give (2).

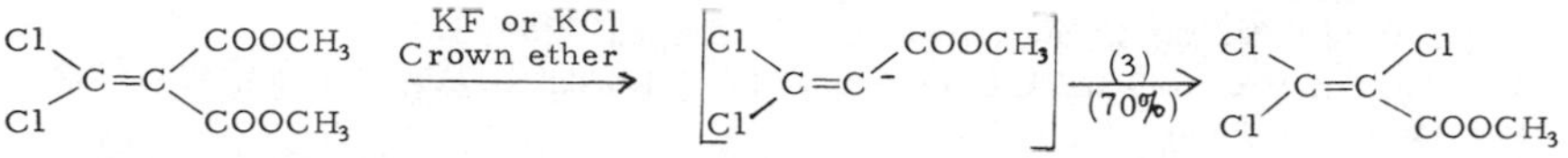

(3)                                    (a)                                    (4)

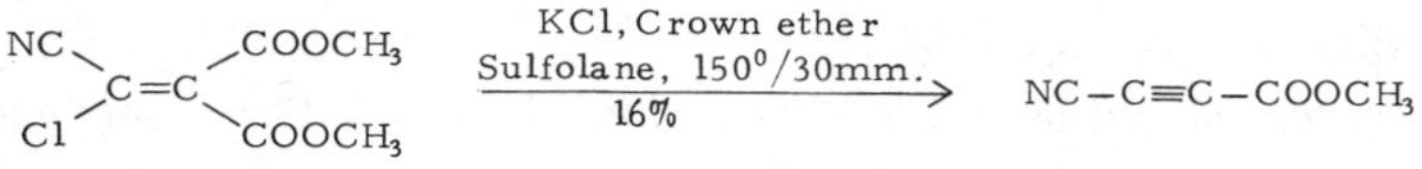

(5)                                                          (6)

Surprisingly, a similar reaction with (3) affords (4) in high yield [if it is assumed that (4) is formed from two molecules of (3)]. Ykman and Hall suggest that this unusual chlorodecarbomethoxylation involves a vinyl carbanion (a). Such a anion could also lose Cl⁻ to form an acetylene, and indeed (5) can be converted into the acetylene (6), but only under forcing conditions.

However, this dechlorocarbomethoxylation reaction was realized under milder conditions with saturated esters using tetraethylammonium chloride (equations I and II).

(I)

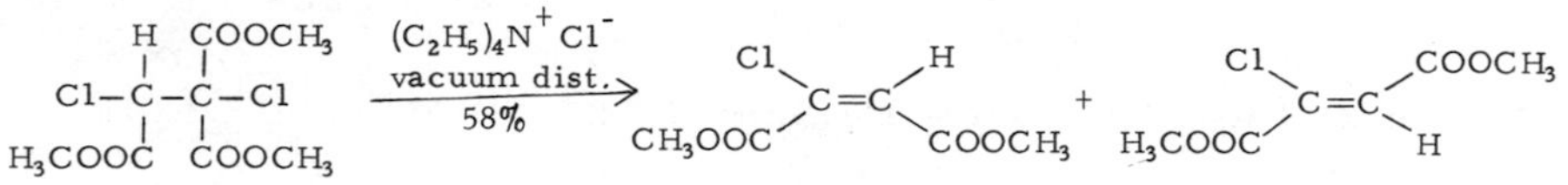

(II)

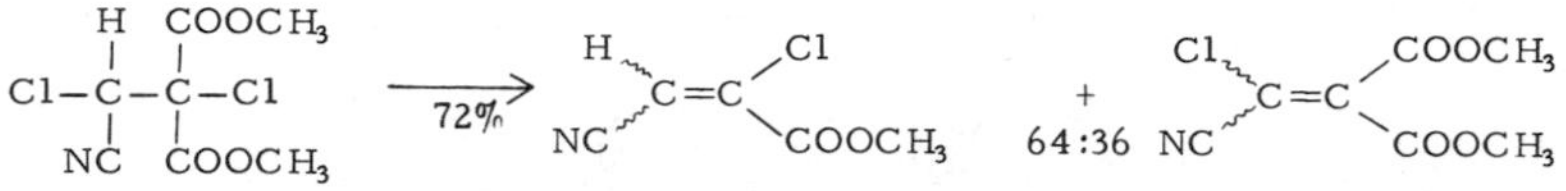

[1]P. Ykman and H. K. Hall, Jr., *Tetrahedron Letters*, 2429 (1975).

**Potassium ferricyanide, 1,** 929–933; **2,** 345; **4,** 406–407; **5,** 554–555.

*Oxidative phenol coupling* (**4,** 406–407). Ferricyanide oxidation of the amide (1) gives the dienone (2) in remarkably high yield (67%). When the

C=O group of (1) is replaced by $CH_2$, a complex mixture is obtained on oxidation.[1]

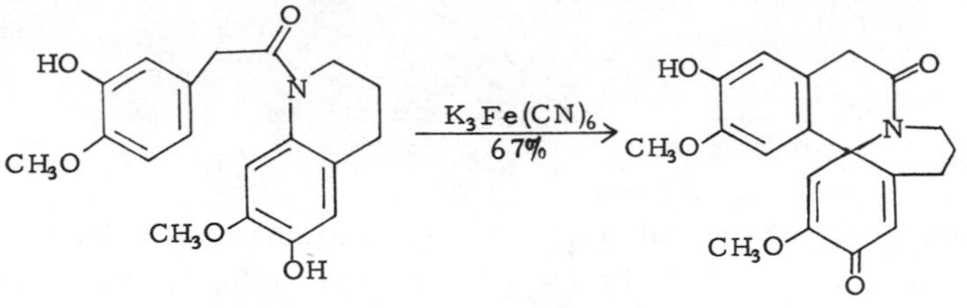

(1)                                    (2)

Ferricyanide oxidizes the dibenzazecine (3) to the dienone (4) in 60% yield.[2]

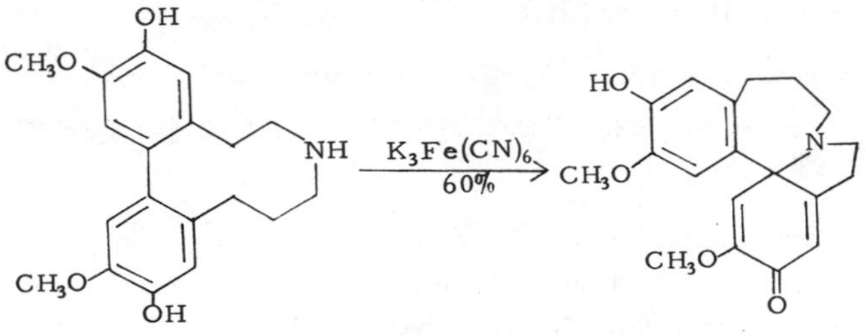

(3)                                    (4)

Oxidations of this type may mimic biosynthesis of various alkaloids.

[1] E. McDonald and A. Suksamrarn, *Tetrahedron Letters*, 4421 (1975).
[2] *Idem, ibid.*, 4425 (1975).

**Potassium fluoride, 1**, 933–935; **2**, 346; **5**, 555–556.

*Michael additions of nitro olefins.*[1] Potassium fluoride catalyzes the Michael addition of 1,3-diketones and simple nitro olefins (equation I). Products of further transformation of the initial adduct are formed in some cases (equation II).

(I)

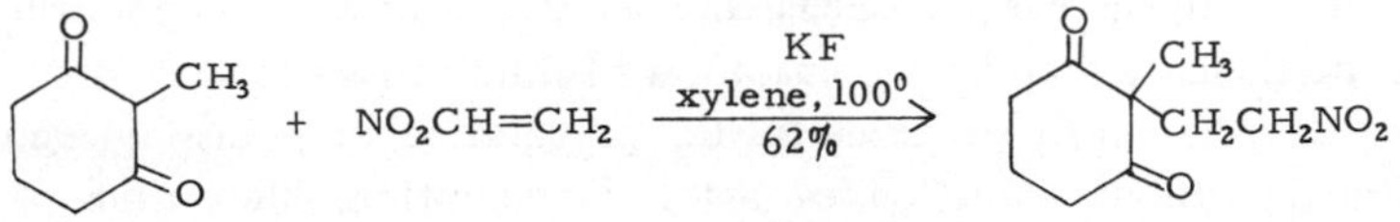

(II)

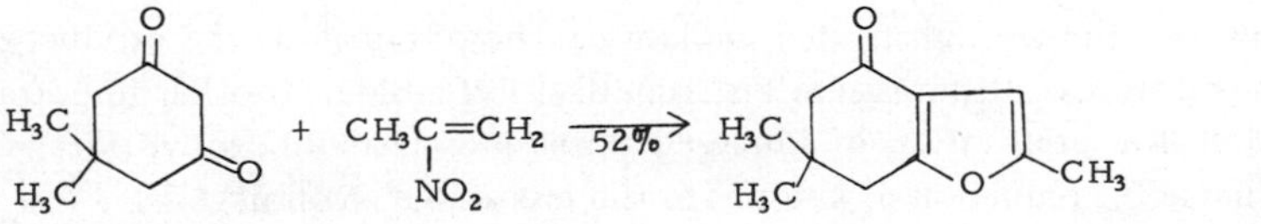

*Allene oxide–cyclopropanone systems.* Treatment of the oxide of 3-chloro-2-triphenylsilyl-l-phenylpropene (1) leads to β-elimination of $(C_6H_5)_3SiF$ and

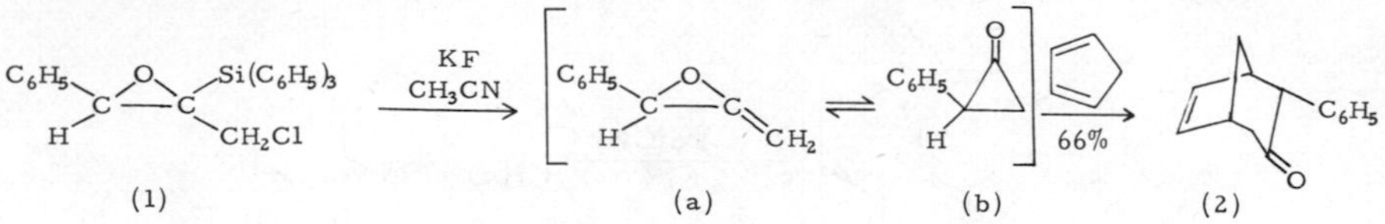

(1)                     (a)              (b)           (2)

generation of an allene oxide (a), which is in equilibrium with the cyclopropanone (b). That (b) is indeed present is shown by cycloaddition to cyclopentadiene to give (2). A similar adduct was formed with furane.[2] Reaction of conjugated dienes with cyclopropanones is a known route to seven-membered ring compounds.[3]

*Cleavage of vinylsilanes.* Vinylsilanes bearing a β-hydroxyl group are cleaved by fluoride ion in DMSO or $CH_3CN$:

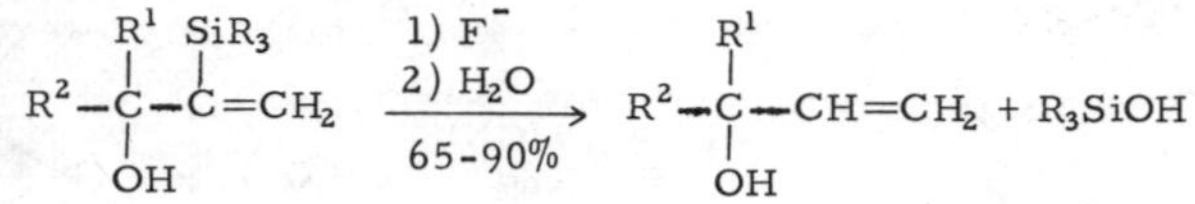

Tetraethylammonium fluoride or potassium fluoride can be used; the corresponding bromides, chlorides, and iodides are inactive. The OH group is essential; vinyltrimethylsilane itself is not cleaved.[4]

[1] T. Yanami, M. Kato, and A. Yoshikoshi, *J.C.S. Chem. Comm.*, 727 (1975).
[2] T. H. Chan, M. P. Li, W. Mychajlowskij, and D. N. Harpp, *Tetrahedron Letters*, 3511 (1974).
[3] N. J. Turro, *Accts. Chem. Res.*, **2**, 25 (1969); H. M. R. Hoffmann, *Angew. Chem. internat. Ed.*, **12**, 819 (1973).
[4] T. H. Chan and W. Mychajlowskij, *Tetrahedron Letters*, 3479 (1974).

**Potassium hydride, 1,** 935; **2,** 346; **4,** 409; **5,** 557. Suppliers: Pressure Chem. Co., Pittsburgh, Pa., and Alfa Inorganics supply dispersions of KH in mineral oil (20–35% by weight). *Caution*: Potassium hydride should be handled with the same precautions as sodium hydride. In the absence of the protective oil, it should be protected from oxygen and moisture.

*Uses.* Brown[1] has published details of his recent work on potassium hydride. It is particularly useful for preparation of hindered bases such as potassium bis-(trimethylsilyl)amide. It is useful for preparation of potassium enolates from ketones, without reduction or aldol condensation. Metalation of 2-methyl-cyclohexanone with KH gives an equilibrium mixture of potassium enolates. However, the less substituted enolate can be obtained as the product of kinetic control by use of potassium bis(trimethylsilyl)amide. Another important use of KH is for preparation of hindered potassium trialkylborohydrides, which are valuable for reduction of ketones to the less stable alcohols.

$$67\% \qquad 33\%$$
$$5\% \qquad 95\%$$

*Condensation of esters and nitriles.* Potassium hydride reacts rapidly with esters or nitriles in THF at 20°. Hydrogen is evolved generally within a minute. After evolution is complete, the reaction mixture is hydrolyzed at 0° (foaming). This condensation requires 1 molar eq. of base. Yields are high.[2]

*Examples:*

[1]C. A. Brown, *J. Org.*, **39**, 3913 (1974).
[2]*Idem, Synthesis*, 326 (1975).

**Potassium hydridotetracarbonylferrate, K⁺HFe(CO)₄⁻, 4, 268–269; 5, 357–358.**

*Dehalogenation.* This hydride dehalogenates alkyl halides at 20° in DME containing a trace of water. Yields are good to excellent. Aryl halides are not reduced. 1-Bromoadamantane is the only unreactive nonaromatic halide reported.[1]

$$(\underline{p})\text{-}BrC_6H_4COCH_2Br \xrightarrow[66\%]{K^+HFe(CO)_4^-} (\underline{p})\text{-}BrC_6H_4COCH_3$$

The reaction occurs with inversion of configuration as shown by reduction of $\alpha$-bromocamphor with $K^+DFe(CO_4)^-$:

*Desulfurization.* Thioketones and thioamides are converted by this reagent in refluxing DME into hydrocarbons and amines, respectively.[2]

*Examples:*

$$(C_6H_5)_2C{=}S \xrightarrow[60\%]{KHFe(CO)_4} (C_6H_5)_2CH_2$$

$$(\underline{p}\text{-}CH_3OC_6H_4)_2C{=}S \xrightarrow{77\%} (\underline{p}\text{-}CH_3OC_6H_4)_2CH_2$$

$$\text{Adamantanethione} \xrightarrow{74\%} \text{Adamantane}$$

$$C_6H_5CSNHC_6H_5 \xrightarrow{38\%} C_6H_5CH_2NHC_6H_5$$

$$CH_3CSNHC_6H_5 \xrightarrow{51\%} C_2H_5NHC_6H_5$$

*Methyl ketones.* The products of Knoevenagel condensation of aldehydes and 2,4-pentanedione are reduced to methyl ketones by this reagent in 60–80% yield.[3]

$$RCHO + CH_2(COCH_3)_2 \xrightarrow{60-75\%} RCH{=}C(COCH_3)_2 \xrightarrow[60-80\%]{KHFe(CO)_4} RCH_2CH_2COCH_3$$

*N,N-Dialkylation of primary amines.*[4] Primary amines can be dialkylated by the reaction with an aldehyde and potassium hydridotetracarbonylferrate in ethanol under carbon monoxide at 20°.

$$C_6H_5NH_2 + 2\ HCHO \xrightarrow[\sim 100\%]{\substack{KHFe(CO)_4 \\ CO}} C_6H_5N(CH_3)_2$$

$$C_6H_5NH_2 + 2\ CH_3CHO \xrightarrow{\sim 100\%} C_6H_5N(C_2H_5)_2$$

This method can also be used for N-monoalkylation by use of a 1:1 mole ratio of amine and aldehyde.

*Hydroacylation of $\alpha,\beta$-unsaturated carboxylic esters.*[5] $\alpha,\beta$-Unsaturated esters (1) can be hydroacylated with sodium hydridotetracarbonylferrate and an alkyl iodide in THF at ambient temperatures. The reaction involves an alkylcarbonylferrate (2), one of which has been isolated as the bis(triphenylphosphine)iminium salt.

$$R^1C \underset{H}{\overset{R^2}{=}} CCOOR \ + \ NaFeH(CO)_4 \xrightarrow{THF} \left[ R^1CH_2 \underset{Fe(CO)_4}{\overset{R^2}{C}} COOR \right]^{-} Na^+ \xrightarrow[45-75\%]{CH_3I} R^1CH_2 \underset{O=CCH_3}{\overset{R^2}{C}} COOR$$

$$(1) \qquad\qquad\qquad (2) \qquad\qquad\qquad (3)$$

*N-Substituted glycines.* These can be synthesized as formulated in fair to good yields from glyoxylic acid and primary aliphatic and aromatic amines under carbon monoxide. The reaction probably involves reduction of an intermediate $RN{=}CHCOOH$.[6]

$$OHCCOOH \ + \ RNH_2 \xrightarrow[35-80\%]{\underset{C_2H_5OH}{KHFe(CO)_4}} RNHCH_2COOH$$

*Alkylation of carbonyl compounds.*[7] The reaction of a carbonyl compound, an aldehyde, and potassium or sodium hydridotetracarbonylferrate in a $1:1:1$ molar ratio results in alkylation of the carbonyl compound (equation I).

$$\text{I} \qquad R^1COCH_2R^2 \ + \ R^3CHO \xrightarrow[H_2, -H_2O]{MHFe(CO)_4} R^1COCHR^2 \atop \underset{CH_2R^3}{|}$$

$$\text{II} \qquad CH_2(COOC_2H_5)_2 \ + \ HCHO \xrightarrow[85\%]{MHFe(CO)_4} CH_3CH(COOC_2H_5)_2$$

$$\text{III} \qquad C_6H_5CH_2CN \ + \ C_6H_5CHO \xrightarrow[75\%]{MHFe(CO)_4} C_6H_5CH_2CHCN \atop \underset{C_6H_5}{|}$$

Yields are usually in the range 40–70%. The reaction is general for alkyl, alicyclic, and aralkyl ketones and for aldehydes as the carbonyl component. Yields are generally higher with aryl aldehydes than with alkyl aldehydes. A methyl or methylene group α to the carbonyl group is essential. Undoubtedly, a reduction of an unsaturated intermediate is involved.

The hydridoiron complex has also been used in the alkylation of active methylene compounds by formaldehyde or benzaldehyde (equations II and III).

[1]H. Alper, *Tetrahedron Letters,* 2257 (1975).
[2]*Idem, J. Org.,* **40,** 2694 (1975).
[3]M. Yamashita, Y. Watanabe, T. Mitsudo, and Y. Takegami, *Tetrahedron Letters,* 1867 (1975).
[4]Y. Watanabe, T. Mitsudo, M. Yamashita, S. C. Shim, and Y. Takegami, *Chem. Letters,* 1265 (1974).
[5]T. Mitsudo, Y. Watanabe, M. Yamashita, and Y. Takegami, *ibid.,* 1385 (1974).

[6] Y. Watanabe, S. C. Shim, T. Mitsudo, M. Yamashita, and Y. Takegami, *ibid.*, 699 (1975).
[7] G. Cainelli, M. Panunzio, and A. Umani-Ronchi, *J.C.S. Perkin I,* 1273 (1975).

**Potassium hydroxide, 5,** 557–560.

*Alkynes by dehydrohalogenation.* Alkynes can be prepared from *vic*-dibromides by treatment at elevated temperatures with KOH dissolved in DMSO.[1]

$$(CH_3)_3CCl \ + \ BrCH=CH_2 \xrightarrow[\substack{-20^0 \\ 98\%}]{AlCl_3} (CH_3)_3CCH_2CHBrCl \xrightarrow[\substack{110-130^0 \\ 89\%}]{KOH, DMSO} (CH_3)_3CC \equiv CH$$

$$C_8H_{17}CHBrCH_2Br \xrightarrow[\substack{130-140^0 \\ 65.7\%}]{KOH, DMSO} C_8H_{17}C \equiv CH$$

*Synthesis of an anthraquinone from a β-polycarbonyl compound.*[2] Oxalyl-diacetone (1) when heated in 30% KOH undergoes intermolecular condensation to the anthraquinone 3,7-dimethylanthrarufin (2). The reaction is of interest

(1)                                                    (2)

because it represents a possible biosynthetic route to anthraquinones.

[1] J. R. Sowa, E. J. Lamby, E. C. Calamai, D. A. Benko, and A. Gordinier, *Org. Prep. Proc. Int.,* 7, 137 (1975).
[2] K. Balenović and M. Poje, *Tetrahedron Letters,* 3427 (1975).

**Potassium hypochlorite, 1,** 938.

*Epoxidation of α,β-unsaturated sulfones.* Epoxidation of *cis*-1-phenyl-2-(*p*-tolylsulfonyl)ethene (1) with alkaline hydrogen peroxide in aqueous acetone

(1)                                                    (2)

(3)

at 45° gives mainly the *trans*-epoxysulfone (2). Similar epoxidation of α,β-unsaturated carbonyl compounds is also nonstereospecific and leads to the more stable epoxides. However, epoxidation of (1) with alkaline potassium hypochlorite gives only the epoxide (3). Epoxidation of (1) with alkaline *m*-chloroperbenzoic acid gives (3) predominately along with about 5% of (2).[1]

[1] R. Curci and F. DiFuria, *Tetrahedron Letters*, 4085 (1974).

**Potassium perchromate, 4, 412.**

*Singlet oxygen.* The detailed paper on the aqueous decomposition of potassium perchromate to generate singlet oxygen has been published by Pitts *et al.*[1] However, they point out that unidentified oxygenation products that do not involve singlet oxygen have been isolated and suggest that extreme caution be used before results are attributed to singlet oxygen. Thermal decomposition of the solid salt gives rise to potassium chromate, potassium superoxide, and oxygen.

[1] J. W. Peters, P. J. Bekowies, A. M. Winer, and J. N. Pitts, Jr., *Am. Soc.*, 97, 3299 (1975).

**Potassium permanganate—Acetic anhydride, 4, 412–413; 5, 563.**

*Oxidation of olefins.*[1] The relative rate of oxidation of a double bond with this reagent decreases with increasing alkyl substitution. Such selectivity is unusual. Thus the rate of oxidation of a double bond with osmium tetroxide or ruthenium tetroxide increases with alkyl substitution.

[1] K. B. Sharpless and D. R. Williams, *Tetrahedron Letters*, 3045 (1975).

**Potassium selenocyanate, KSeCN. Mol. wt. 144.08, m.p. 100° dec. Supplier: Alfa.**

*Isomerization of disubstituted alkenes.*[1] A new method for isomerization of olefins (1) involves the well-known conversion to the bromohydrin (2) by *trans*-addition of OH and Br. This product is then treated with KSeCN in DMF

(60°, 24–48 hr.) to give a β-hydroxyselenocyanate (3) in high yield. On treatment with base a selenirane (4) is formed, but it loses selenium under the experimental conditions to give as the major product an alkene (5) that is isomeric with the starting alkene. Potassium thiocyanate can be used in place of potassium selenocyanate. In this case the intermediate thiirane is stable and can be isolated, but the isomerization is less stereoselective. The conversion of (2) to (3) evidently proceeds predominately with inversion, but the fact that (5) is accompanied by the original olefin (1) suggests some participation of the neighboring hydroxyl group.

*Conversion of epoxides into olefins.*[2] Epoxides are converted into olefins with retention of configuration by treatment with KSeCN in $H_2O-CH_3OH$ at 25–65°. Selenium is deposited. The reaction fails in aprotic solvents (DMSO, DMF). Yields are high in the case of straight-chain compounds, but epoxides

of cyclic compounds vary considerably in reactivity. Epoxides of cyclopentene, cyclooctene, and cyclododecene are unchanged after 3 days even at 65°. Cyclohexene oxide is converted into cyclohexene in quantitative yield at 25°.

[1] D. Van Ende and A. Krief, *Tetrahedron Letters,* 2709 (1975).
[2] J. M. Behan, R. A. W. Johnstone, and M. J. Wright, J.C.S. Perkin I, 1216 (1975).

**Potassium superoxide, $KO_2$.** Mol. wt. 71.10. Suppliers: K and K, Ventron. Protect from water, with which $KO_2$ reacts rapidly.

*Reaction with alkyl halides and sulfonates.* Previously this reagent found little use in organic chemistry because it is unstable in protic solvents and insoluble in aprotic solvents. However, three laboratories[1-3] have now reported that addition of crown ethers, particularly 18-crown-6, permits preparation of solutions of $KO_2$ in various solvents: DMSO, DMF, DME, and even benzene and ether. In DMSO, a crown ether is unnecessary.[3]

The reaction of primary alkyl bromides or sulfonate esters with potassium superoxide solubilized in benzene gives dialkyl peroxides, ROOR, as the major product. Significant amounts of alcohols and olefins are also formed.[1] If the reaction is carried out in DMSO or DMSO–DMF, primary alcohols are formed very rapidly with inversion of configuration.[2,3] Catalytic amounts of a crown ether are sufficient, but use of 1–2 eq. leads to a more rapid reaction. Secondary alcohols can be obtained, often in high yield, by this displacement reaction, but elimination predominates with tertiary substrates. This $S_N2$ reaction has been

*Examples:*

$$\text{Benzyl bromide} \xrightarrow[\text{DMSO, DMF}]{\text{KO}_2} \text{Benzyl alcohol (75\%)}$$

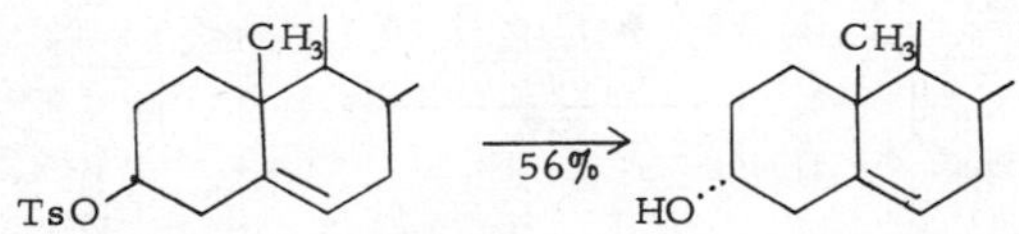

$$\xrightarrow{56\%}$$

used by Corey et al.[3,4] for the synthesis of the $C_9$- and $C_{11}$-epimers of prosta-glandin $F_{2\alpha}$ and to convert the 15-R-prostanoid (1) into the 15-S-prostanoid (2). This transformation had previously failed with a number of other reagents.

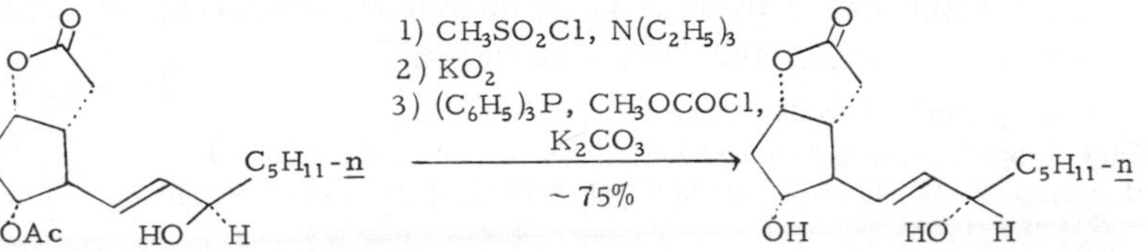

1) $CH_3SO_2Cl$, $N(C_2H_5)_3$
2) $KO_2$
3) $(C_6H_5)_3P$, $CH_3OCOCl$,
   $K_2CO_3$
$$\xrightarrow{\sim 75\%}$$

(1)                                      (2)

A number of interesting elimination reactions have been reported.[3] t-Butyl bromide is converted by $KO_2$ into t-butanol and isobutylene. Other useful reactions are formulated in equations I and II.

(I)

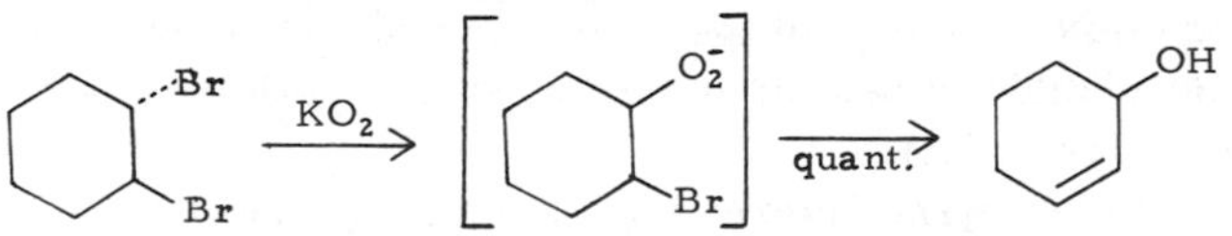

$$\xrightarrow{KO_2} \qquad \xrightarrow[\text{quant.}]{}$$

(II)

+

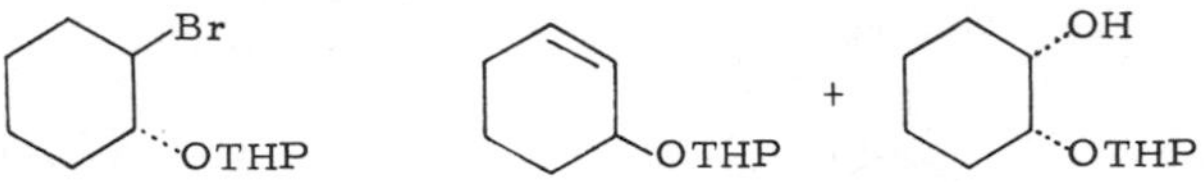

(67%)                          (25%)

Corey's group has also reported the preparation of a cyclic peroxide (4) from a 1,3-dimesylate (3):

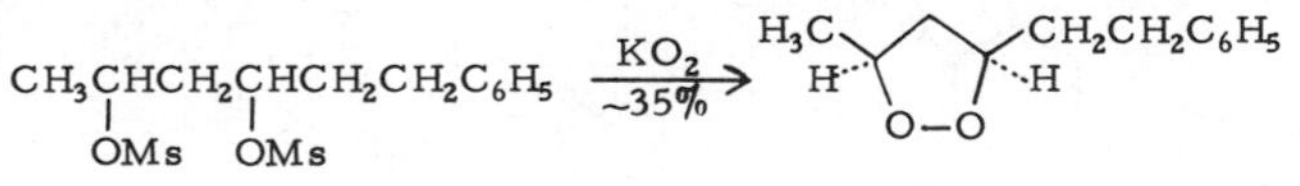

$$CH_3CHCH_2CHCH_2CH_2C_6H_5 \xrightarrow[\sim35\%]{KO_2}$$

OMs  OMs

(3)

(4)

Valentine and Curtis[5] have reported a preliminary study of the reaction of $KO_2$ with a Cu(II) complex, which results in formation of a Cu(I) complex and liberation of molecular oxygen. This reaction is of interest in connection with superoxide dismutase enzymes, which contain transition metals.

*Oxidation of tetracyclone.*[6] Tetracyclone solubilized with dicyclohexyl-18-crown-6 in benzene is oxidized to 2-hydroxy-2,4,5-triphenylfuranone-3 (2)

(1)                                             (2)

in about 80% yield. Neither atmospheric oxygen nor water is involved; apparently the hydroxyl group of (2) is derived from the crown ether. This unusual reaction does not appear to be general; fluorenone, tetraphenylethylene, and 2,5-diphenylfurane do not undergo this oxidation.

[1] R. A. Johnson and E. G. Nidy, *J. Org.*, **40**, 1680 (1975).
[2] J. San Filippo, Jr., C.-I. Chern, and J. S. Valentine, *ibid.*, **40**, 1678 (1975).
[3] E. J. Corey, K. C. Nicolaou, M. Shibasaki, Y. Machida, and C. S. Shiner, *Tetrahedron Letters*, 3183 (1975).
[4] E. J. Corey, K. C. Nicolaou, and M. Shibasaki, *J.C.S. Chem. Comm.*, 658 (1975).
[5] J. S. Valentine and A. B. Curtis, *Am. Soc.*, **97**, 224 (1975).
[6] I. Rosenthal and A. Frimer, *Tetrahedron Letters*, 3731 (1975).

**Potassium tri-sec-butylborohydride,** $K(sec\text{-}C_4H_9)_3BH$. Mol. wt. 222.27. Supplier: Aldrich (K Selectride).

*Preparation.* The reagent is prepared in quantitative yield by the reaction of potassium hydride suspended in THF at 20–22° with tri-sec-butylborane (complete in less than 1 hr.).[1]

*Reduction of cyclic ketones.* Potassium tri-sec-butylborohydride is equal or superior to lithium tri-sec-butylborohydride (4, 312–313) for stereoselective reduction of cyclic ketones.

*γ-Butyrolactones.* In a total synthesis of (±)-isoalantolactone (3, an eudesmane sesquiterpene), Miller and Nash[2] prepared the *cis*-lactone (2) in 96% yield by reduction of the keto ester (1) with this borohydride in THF at −78° followed by treatment with base. Use of sodium borohydride gave (2) in only 66% yield, presumably because the equatorial alcohol at $C_2$ is also formed. The lactone (2) is not formed by reduction of the keto acid itself. It was converted into (3) by a method developed previously by Behare and Miller (3, 190–191).

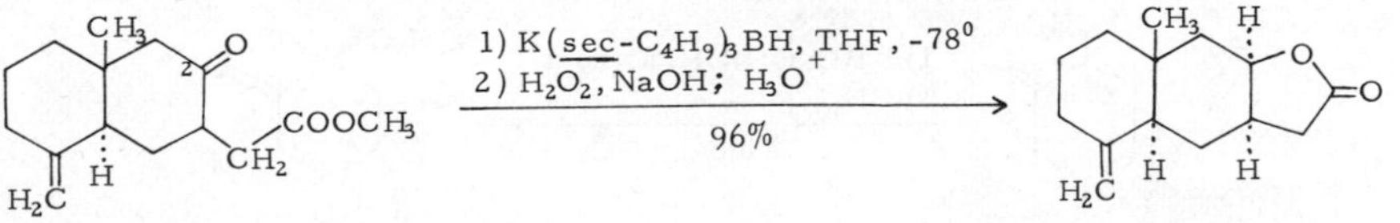

(1)

1) K($\underline{\text{sec}}$-C$_4$H$_9$)$_3$BH, THF, -78$^0$
2) H$_2$O$_2$, NaOH; H$_3$O$^+$

96%

(2)

~45%

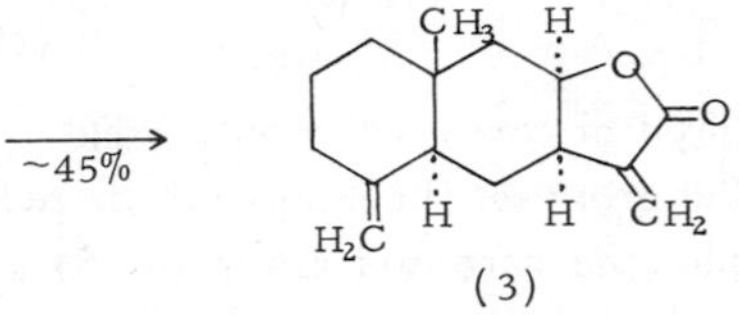

(3)

This lactonization reaction was also used in a synthesis of (±)-yomogin (6).[3] Thus reduction of the keto ester (4) led to the lactone (5) in about 60% yield. The product was converted into the α-methylene derivative by reaction with formaldehyde (4, 298–299); the final step in the synthesis of (6) involved dehydrogenation with DDQ.

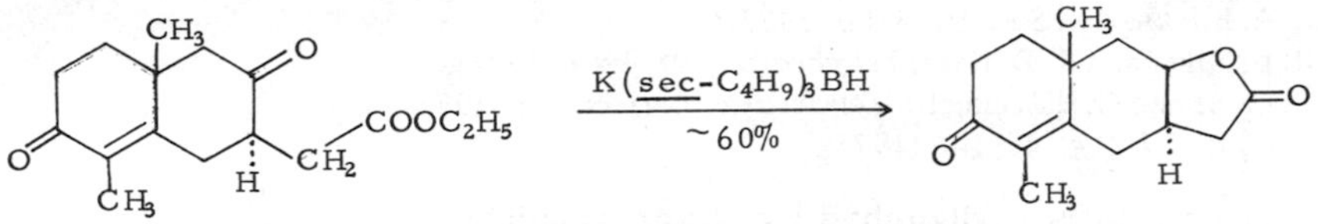

(4)

K($\underline{\text{sec}}$-C$_4$H$_9$)$_3$BH

~60%

(5)

several steps

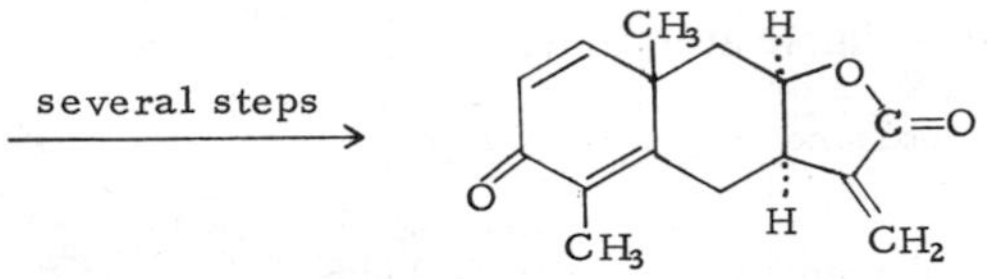

(6)

*Conjugate reduction.* Cyclohexenones unsubstituted at a β-position are reduced by potassium tri-*sec*-butylborohydride to the corresponding saturated ketones in nearly quantitative yield. The reaction can be conducted in THF or ether–THF and is rapid even at -78°. An example is the reduction of (1) to (2). Combined reduction and alkylation is also possible. Thus treatment of carvone (3) with 1 eq. of the reducing agent and then with a slight excess of methyl

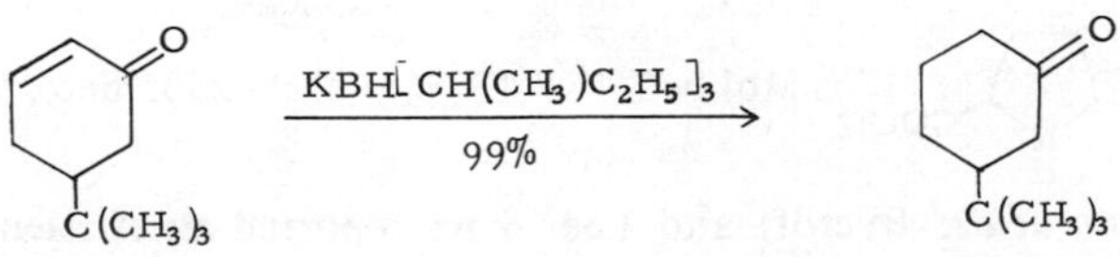

KBH[ CH(CH$_3$)C$_2$H$_5$ ]$_3$

99%

(1)                    (2)

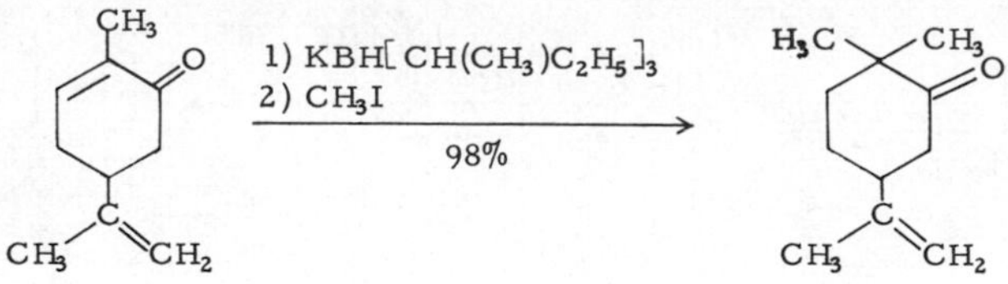

(3)     (4)

iodide leads to 1-methyl-1,6-dihydrocarvone (4) in 98% yield.

Cyclohexenones with an alkyl group at the $\beta$-position are reduced to mixtures of allylic alcohols; for example, the carbonyl group of (5) is reduced in 83% yield.[4]

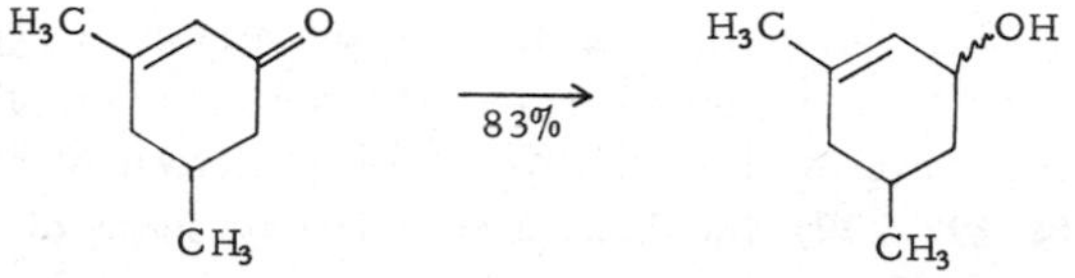

(5)     (6)

[1] C. A. Brown, *Am. Soc.*, **95**, 4100 (1973).
[2] R. B. Miller and R. D. Nash, *Tetrahedron*, **30**, 2961 (1974).
[3] D. Caine and G. Hasenhuettl, *Tetrahedron Letters*, 743 (1975).
[4] B. Ganem, *J. Org.*, **40**, 146 (1975).

**Potassium tri-*sec*-butylborohydride—Cuprous iodide.**

Addition of 2 moles of the borohydride reagent in THF to 1 mole of CuI produces a reagent (2:1 reagent) as a dark grey powder that reduces aliphatic and aromatic halides as well as simple acetylenes:

1-Iodooctane $\xrightarrow[100\%]{}$ Octane

Benzyl chloride $\xrightarrow[98\%]{}$ Toluene

1-Bromonaphthalene $\xrightarrow[95\%]{}$ Naphthalene

5-Decyne $\xrightarrow[94\%]{}$ Z-5-Decene

Ketones are also reduced. The 2:1 reagent may have the formula $K(CuH_2)_n$. In contrast, the 1:1 reagent does not reduce organic halides.[1]

[1] T. Yoshida and E. Negishi, *J.C.S. Chem. Comm.*, 762 (1974).

**(L)-Proline,** 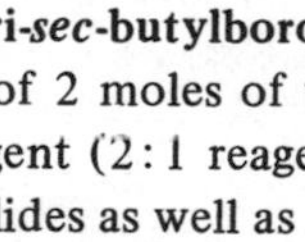 . Mol. wt. 115.13, m.p. 228–233° dec., $\alpha_D$ −8°.

**L-$\alpha$-Amino acids.** Bycroft and Lee[1] have reported an efficient asymmetric synthesis of L-amino acids from $\alpha$-keto acids and ammonia with L-proline as the chiral reagent (recoverable). The asymmetric induction in the hydrogenation step is >90%. Yields are satisfactory: pyruvic acid ⟶ L-alanine (in 60% yield).

(S, S)

L-Proline is more efficient in this synthesis than L-phenylalanine or L-alanine.

[1] B. W. Bycroft and G. R. Lee, *J.C.S. Chem. Comm.*, 988 (1975).

**1,3-Propanedithiol, 1, 956; 4, 413.**

*1,4-Diketones.* The first step in a recent synthesis[1] of 1,4-diketones of the type $RCOCH_2CH_2COR$ involves thioacetalization of succindialdehyde with 1,3-propanedithiol to give 2,2'-ethylenebis(m-dithiane) (1).[2] In some cases

$$OHC(CH_2)_2CHO + HS(CH_2)_3SH \xrightarrow{\text{80-90\%}}$$

(1)

dialkylation of (1) can be accomplished in one step by treatment of (1) with 2 eq. of n-butyllithium and 2 eq. of the alkyl halide. In general, however, dialkylation is best carried out in two steps. The 1,4-diketone is then obtained as usual by hydrolysis with $HgCl_2$ and HgO in aqueous methanol. This method has the limitation that it fails with secondary and tertiary halides, with the exception of isopropyl iodide, and with aryl halides.

[1] W. B. Sudweeks and H. S. Broadbent, *J. Org.*, **40**, 1131 (1975).
[2] D. Seebach, N. R. Jones, and E. J. Corey, *ibid.*, **33**, 300 (1968).

**Propargyl bromide,** $HC{\equiv}CCH_2Br$. Mol. wt. 118.97, b.p. 88–90°. Suppliers: Aldrich, Fluka, K and K, Polysciences.

*Reaction with lithium trialkylalkynylborates.* Alkylation of borates (1), prepared as shown, with this bromide proceeds with migration of one alkyl group to give an intermediate (a), which is not isolated, but is oxidized to a

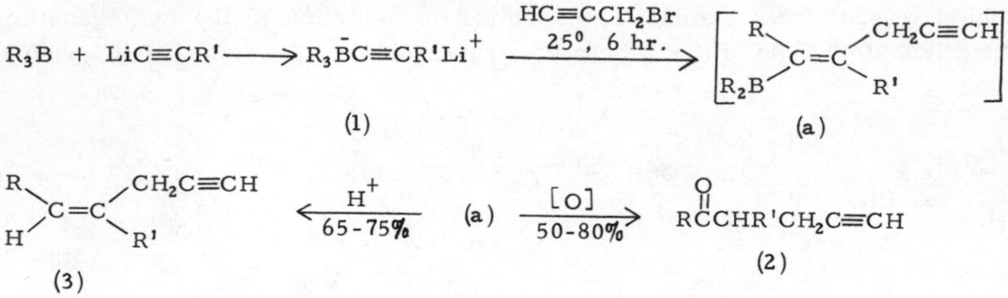

$$R_3B \ + \ LiC{\equiv}CR' \longrightarrow R_3\bar{B}C{\equiv}CR'Li^+ \xrightarrow[25^0, \ 6 \ hr.]{HC{\equiv}CCH_2Br} \left[ \begin{array}{c} R \\ \\ R_2B \end{array} C{=}C \begin{array}{c} CH_2C{\equiv}CH \\ \\ R' \end{array} \right]$$

(1)                                        (a)

$$\begin{array}{c} R \\ \\ H \end{array} C{=}C \begin{array}{c} CH_2C{\equiv}CH \\ \\ R' \end{array} \xleftarrow[65-75\%]{H^+} \ (a) \ \xrightarrow[50-80\%]{[O]} \ \underset{O}{\overset{O}{R\overset{\|}{C}CHR'CH_2C{\equiv}CH}}$$

(3)                                        (2)

γ-ketoalkyne (2) or hydrolyzed with isobutyric acid to a trisubstituted olefin (3). The sequence is stereospecific; only the Z-isomer is formed.[1]

Similar results have been obtained using iodoacetonitrile, $ICH_2C{\equiv}N$.

[1] A. Pelter, K. J. Gould, and C. R. Harrison, *Tetrahedron Letters*, 3327 (1975).

**Propargyltriphenylphosphonium bromide** [Triphenyl(prop-2-ynyl)phosphonium bromide], $HC{\equiv}CCH_2\overset{+}{P}(C_6H_5)_3Br^-$ (1). Mol. wt. 381.26. Supplier: Fluka.

*Heterocycles.* Schweizer et al.[1] have reported the synthesis of several heterocyclic systems using this reagent, which undergoes addition with primary amines to give β-aminopropenyltriphenylphosphonium bromides. On treatment with bases these products form heterocyclic systems with extrusion of triphenylphosphine oxide or methylenetriphenylphosphorane.

*Examples:*

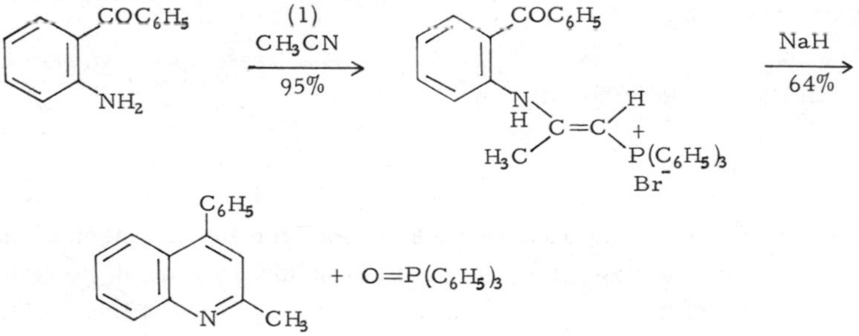

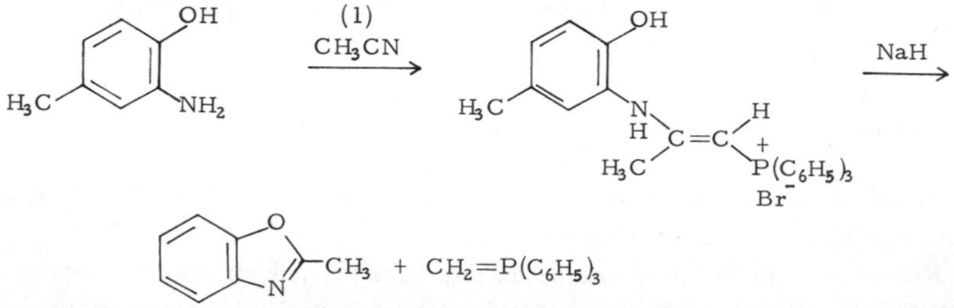

[1] E. E. Schweizer, C. S. Kim, C. S. Labaw, and W. P. Murray, *J.C.S. Chem. Comm.*, 7 (1973); E. E. Schweizer and S. V. De Voe, *J. Org.*, **40**, 144 (1975).

*trans*-1-Propenyllithium. $\overset{CH_3}{\underset{H}{\diagdown}}C=C\overset{H}{\underset{Li}{\diagup}}$ . Mol. wt. 48.01.

*Preparation.*[1] The reagent is prepared by the reaction of *trans*-1-chloro-1-propene (99% purity) with lithium in ether at ice-water bath temperature.

$$\overset{CH_3}{\underset{H}{\diagdown}}C=C\overset{H}{\underset{Cl}{\diagup}} \quad \xrightarrow[\;(C_2H_5)_2O\;]{Li\,(1\%\ Na)} \quad \overset{CH_3}{\underset{H}{\diagdown}}C=C\overset{H}{\underset{Li}{\diagup}}$$

*Synthesis of olefins.*[2] This organolithium reagent undergoes coupling with 1-iodooctane or 1-bromooctane in THF at 25° (1.7 hr.) to give (E)-2-undecene in high yield:

$$\overset{CH_3}{\underset{H}{\diagdown}}C=C\overset{H}{\underset{Li}{\diagup}} \quad + \quad C_7H_{15}CH_2Br(I) \quad \xrightarrow{89.8\%} \quad \overset{CH_3}{\underset{H}{\diagdown}}C=C\overset{H}{\underset{CH_2C_7H_{15}}{\diagup}}$$

The order of reactivity is I > Br > Cl. Thus the yield of (E)-2-undecene using 1-chlorooctane in DME is only 50%. Tosylates do not enter into the reaction. This reaction has also been carried out with the cuprate reagent $\left(\overset{CH_3}{\underset{H}{\diagdown}}C=C\overset{H}{\diagup}\right)_2$ CuLi,[3] but in this case only one alkenyl group is utilized.

Elimination is the predominant reaction with secondary halides.

[1] G. M. Whitesides, C. P. Casey, and J. K. Krieger, *Am. Soc.*, **93**, 1379 (1971).
[2] G. Linstrumelle, *Tetrahedron Letters*, 3809 (1974).
[3] G. Linstrumelle, J. K. Krieger, and G. M. Whitesides, *Org. Syn.*, **53**, 165 (1974).

**Propiolic acid**, HC≡CCOOH. Mol. wt. 70.05, b.p. 102°/200 mm. Suppliers: Aldrich, Farchan.

*δ-Hydroxyacetylenic acids.*[1] The dianion of propiolic acid (1) can be generated at –45° with LDA in 1:1 THF–HMPT solution. The dianion reacts with epoxides to form δ-hydroxyacetylenic acids (2) in moderate yields. The products can be converted into substituted 5,6-dihydro-2H-pyrane-2-ones, a ring

$$\underset{(1)}{LiC\equiv CCOOLi} \quad + \quad R^4CH\overset{O}{\overbrace{\qquad}}CH_2 \quad \xrightarrow{30-50\%} \quad \underset{(2)}{R^4\underset{OH}{\underset{|}{C}}HCH_2C\equiv CCOOH}$$

system that is encountered in various natural products known as polyketides. Routes to four types of polyketides are summarized in Scheme I.

Scheme I

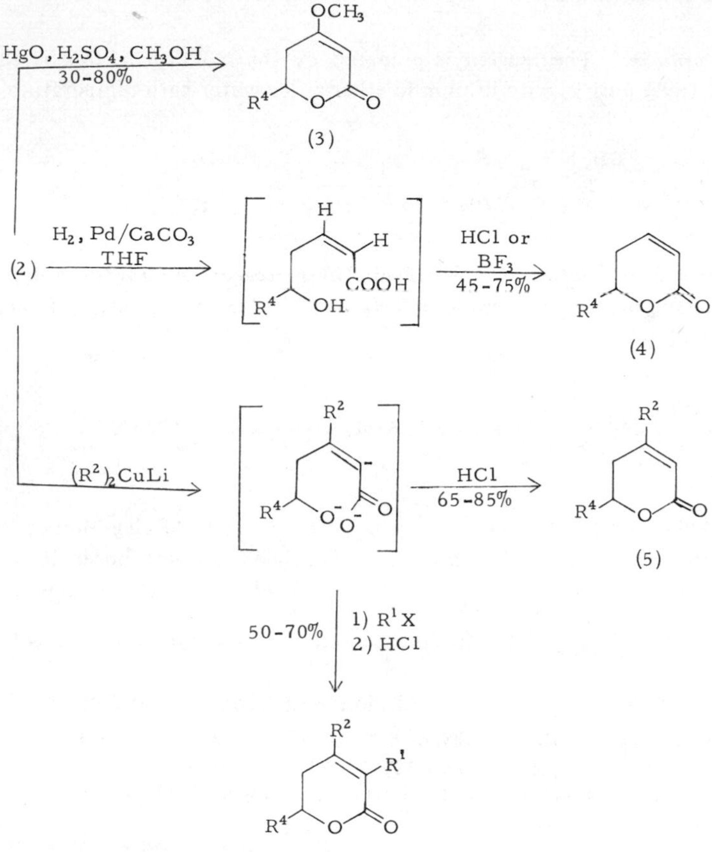

[1] R. M. Carlson and A. R. Oyler, *Tetrahedron Letters*, 2615 (1974); R. M. Carlson, A. R. Oyler, and J. R. Peterson, *J. Org.*, **40**, 1610 (1975).

*n*-Propylthiolithium, $n\text{-}C_3H_7S^-Li^+$.

*Selective debenzylation.* Benzyl groups of quaternary ammonium salts can be removed selectively by treatment with *n*-propylthiolithium in HMPT at 0–5°.[1]

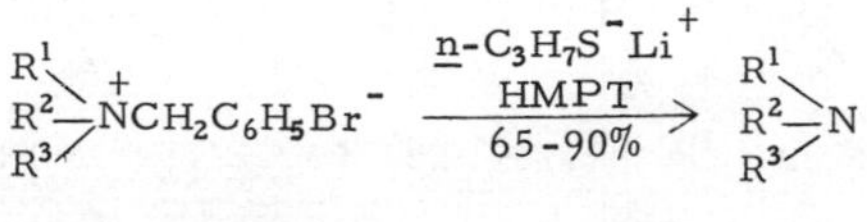

[1] J. P. Kutney, G. B. Fuller, J. Greenhouse, and I. Itoh, *Syn. Commun.*, **4**, 183 (1974).

**Pyridine, 1**, 958–963; **2**, 349–351; **4**, 414–415.

*Sulfonyl chlorides.* Sulfonic acids can be converted in satisfactory yield into sulfonyl chlorides via pyridinium sulfonates. Aminobenzenesulfonic acids can be converted quantitatively into pyridinium acetamidobenzenesulfonates by

$$RSO_3H + \underset{N}{\text{(pyridine)}} \longrightarrow RSO_3^- H\overset{+}{N}\text{(pyridinium)} \quad \xrightarrow[55\text{-}60\%]{PCl_5 \text{ or } SOCl_2} \quad RSO_2Cl$$

treatment with acetic anhydride and pyridine. This method is therefore recommended mainly for preparation of acetamidobenzenesulfonyl chlorides.[1]

*Dehalogenation of α-halo ketones.* α-Halo ketones can be dehalogenated by treatment with pyridine in acetone followed by addition of sodium hydrosulfite; yield 50–75%. The method involves formation of a pyridinium salt, which is then reduced in acetic acid to a 1,4-dihydropyridine derivative; this fragments spontaneously to the ketone and pyridine.[2]

*Dihydroheptalenes.*[3] Treatment of a mixture of isomeric dibromides (1), obtained by carbenoid addition to 2,7-dimethoxy-1,4,5,8-tetrahydronaphthalene, with pyridine at 100° for 1 hr. leads to ring enlargement to 2,9-dimethoxy-1,8-dihydroheptalene (2) in 28% yield. Treatment of (1) with silver salts (**4**, 432–433) results in decomposition to a large number of products.

(1)                    (2)

[1] A. Barco, S. Benetti, G. P. Pollini, and R. Taddia, *Synthesis*, 877 (1974).
[2] T.-L. Ho and C. M. Wong, *J. Org.*, **39**, 562 (1974).
[3] J. D. White and L. G. Wade, Jr., *J. Org.*, **40**, 118 (1975).

**Pyridine hydrochloride, 1**, 964–966; **2**, 352–353; **3**, 239–240; **4**, 415–418; **5**, 566–567.

*Cleavage of a tricyclic ether.*[1] The tricyclic ether (1) is cleaved by pyridine hydrochloride mainly to 9α-acetoxypinene (2); fenchyl chloroacetates, (3a) and

(1)                          (2)

(3a, $R^1$ = OAc, $R^2$ = Cl)
(3b, $R^1$ = Cl, $R^2$ = OAc)

(3b), are formed as minor products. The reaction provides a route to α-pinenes substituted at $C_9$.

*Cleavage of methyl ethers.* Methyl ethers of methoxyformylbenzofuranes such as (1) are cleaved to the corresponding phenol in higher yield by pyridine hydrochloride in quinoline at the boiling point than by neat reagent.[2]

(1)

*Demethylation–cyclodehydration.* Two cases where demethylation of phenolic ethers with pyridine hydrochloride was accompanied by cyclodehydration have been reported.[3]

[1] Y. Bessière-Chrétien and C. Crison, *Bull. Soc.*, 2499 (1975).
[2] L. René, J.-P. Buisson, and R. Royer, *ibid.*, 2763, 2703 (1975).
[3] R. B. Mane and G. S. Krishna Rao, *Chem. Ind.*, 279 (1975).

**Pyridinium chlorochromate,** $C_5H_5\overset{+}{N}HCrO_3Cl^-$. Mol. wt. 215.56, m.p. 205° dec., yellow-orange, air stable. Suppliers: Aldrich, Alfa.

*Preparation.* This reagent is prepared by addition of pyridine (1 mole) to a solution of $CrO_3$ (1 mole) in 6 M HCl (1.1 moles); 84% yield.

*Oxidation.* This compound has been in the literature for some time, but was recognized as a useful oxidation reagent only recently by Corey and Suggs.[1] It oxidizes primary and secondary alcohols in yields equal to or greater than those obtained with Collins reagent (**4,** 215–216) and has the advantage that a large excess is not necessary. The oxidations are usually conducted in methylene chloride at room temperature (1–2 hr.). With acid-sensitive substrates, the reaction can be buffered with sodium acetate as in the last example.

*Examples:*

      1-Decanol $\longrightarrow$ Decanal (92%)

      Oct-2-yn-1-ol $\longrightarrow$ Oct-2-ynal (84%)

      Benzhydrol $\longrightarrow$ Benzophenone (100%)

      Citronellol $\longrightarrow$ Citronellal (82%)

      $\overset{c}{\text{HOCH}_2\text{CH}}$=CHCH$_2$OTHP $\longrightarrow$ $\overset{t}{\text{OCHCH}}$=CHCH$_2$OTHP (81%)

[1] E. J. Corey and J. W. Suggs, *Tetrahedron Letters,* 2647 (1975).

## Pyridinium hydrobromide perbromide, 1, 967–970; 5, 568.

*Dehydrogenation.* The enediones (1a) and (1b) are conveniently aromatized to DL-sugiol (2a) and DL-nimbiol (2b), respectively, by treatment with pyridinium hydrobromide perbromide (HOAc, 20°, 0.5 hr.).[1] Aromatization with DDQ or palladium results in disappointing yields. Presumably, intermediate bromo compounds are formed.

    [ **1a,** R = CH(CH$_3$)$_2$]          (**2a,** 93%)
    [ **1b,** R = CH$_3$]               (**2b,** 45%)

*Polynuclear arenes from metacyclophanes.* A few years ago, Japanese chemists[2] reported that [2.2]metacyclophane (1) is converted into tetrahydropyrene (2) in ~100% yield by treatment with bromine and iron. In an extension of this reaction to (4), pyridinium hydrobromide perbromide was used to obtain

       (1)               (2)               (3)

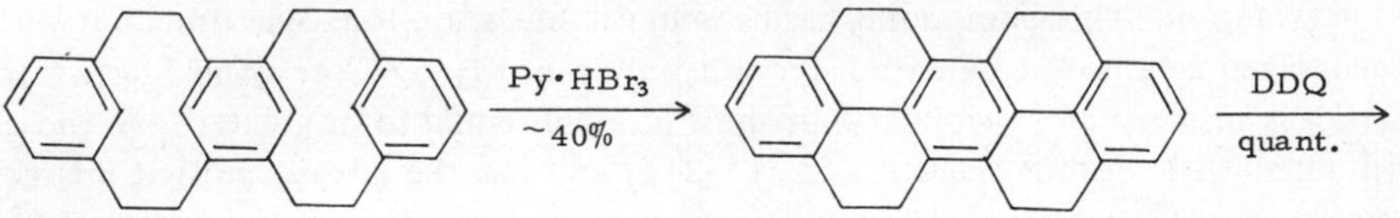

(4)                                    (5)

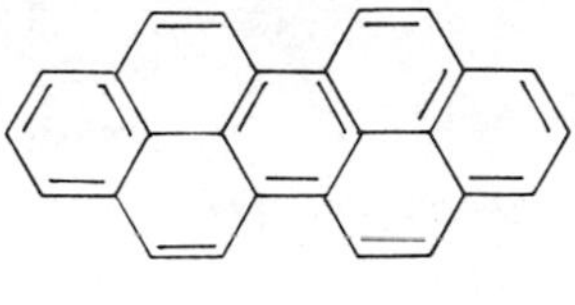

(6)

(5). Dehydrogenation of (2) and of (5) with DDQ in refluxing benzene gave pyrene (3) and peropyrene (6), respectively, in quantitative yield.[3]

[1] W. L. Meyer, G. B. Clemans, and R. A. Manning, *J. Org.*, **40**, 3686 (1975).
[2] T. Sato, M. Wakabayashi, Y. Okamura, T. Amada, and K. Hata, *Bull. Chem. Soc. Japan*, **40**, 2363 (1967).
[3] T. Umemoto, T. Kawashima, Y. Sakata, and S. Misumi, *Tetrahedron Letters*, 1005 (1975).

**Pyruvaldehyde 2-phenylhydrazone (2-Phenylhydrazonopropanal)**, $CH_3 \overset{\displaystyle ||}{\underset{\displaystyle NNHC_6H_5}{C}}CHO$
(1). Mol. wt. 162.19, m.p. 125°.

*Preparation:*

$$CH_3COCH(OCH_3)_2 + C_6H_5NHNH_2 \xrightarrow[92\%]{H_3O^+} \quad (1)$$

*1,4-Enediones.*[1]   $\alpha$-Methyl or $\alpha$-methylene ketones condense with (1) in the presence of potassium ethoxide to give 2-phenylhydrazonopropylidene derivatives (2). Hydrolysis of (2) leads to enediones (3), reducible by zinc and hydrochloric acid to the corresponding saturated 1,4-diketones.

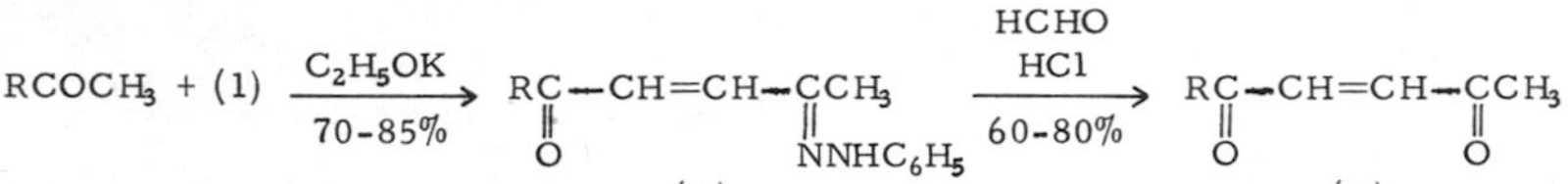

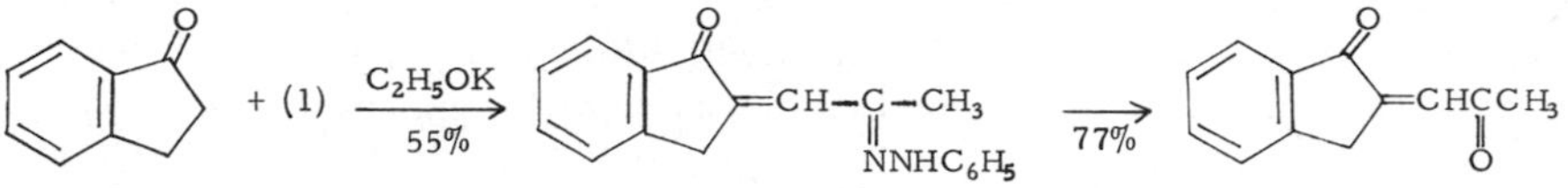

[1] T. Severin and R. Adam, *Ber.*, **108**, 88 (1975).

# Q

**Quinine (1); cinchonine (2).**

$$(1, a_{578} - 172°)$$

$$(2, a_D + 228°)$$

*Asymmetric induction in Michael reactions.* Wynberg and Helder[1] have reported asymmetric syntheses of Michael adducts from inactive donors and methyl vinyl ketone in the presence of catalytic amounts of quinine and cinchonine. Enantiomeric ratios were affected by the solvent; highest inductions were obtained in toluene and tetrachloromethane. Quinine and cinchonine[2] favored different enantiomers of the adduct. The enantiomeric excess was determined in one case from PMR spectroscopy to be 68%.

[1]H. Wynberg and R. Helder, *Tetrahedron Letters*, 4057 (1975).
[2]These alkaloids have the same configuration at $C_3$ and $C_4$, but differ at $C_8$ and $C_9$.

## Quinoline—Acetic acid.

*Hydrolysis of hindered esters.* Aranda and Fétizon[1] observed hydrolysis of a hindered ester group accompanying deacetylation with boiling quinoline. They then found that some hindered esters are hydrolyzed when refluxed in quinoline and acetic acid. This reaction can be accompanied by acetylation of hydroxyl groups, if present. Thus methyl 18α-glycyrrhetate is converted by this procedure into the acid (43% yield), the 3β-acetate (26% yield), and the 3α-acetate (29% yield).

[1]G. Aranda and M. Fétizon, *Synthesis*, 330 (1975).

**3-Quinuclidinol,** (1). Mol. wt. 127.19, m.p. 221–223°. Supplier: Aldrich.

*Cleavage of β-keto esters.* β-Keto and vinylogous β-keto esters are cleaved in high yield (~95%) when refluxed with 5 eq. of this base in *o*-xylene for 6 hr.[1]

[1]E. J. Parish, B.-S. Huang, and D. H. Miles, *Syn. Commun.*, 341 (1975).

# R

Raney nickel, **1**, 723–731; **2**, 293–294; **5**, 570–571.

*Reduction of α,β-unsaturated ketones.* Reduction of an α,β-unsaturated ketone in a fused ring system such as (1) with Raney nickel in 2 $N$ NaOH and

ethanol proceeds stereoselectively to the saturated alcohol (2),[1] which has the configuration opposite to that of reduction of (1) with lithium in liquid ammonia.[2]

*ArOH ⟶ ArH.* A new method for removal of phenolic hydroxyl groups involves conversion to the potassium aryl sulfate and reduction of the salt in aqueous KOH with Raney nickel. Yields are equal to those obtained by other methods.[3]

$$2 \; Py \; + \; ClSO_3H \; \longrightarrow \; Py \cdot HCl \; + \; Py \cdot SO_3$$

$$ArOH + \; Py \cdot SO_3 \; \xrightarrow[80-99\%]{KOH} \; ArOSO_3^- K^+ \; \xrightarrow[50-95\%]{\substack{Raney \; Ni \\ KOH, H_2O}} \; ArH \; + \; K_2SO_4$$

*Hydrogenation of ergosterol.* Hydrogenation of ergosterol in pure ethyl acetate or dioxane with Raney nickel gives $\Delta^7$-5α-ergostene-3β-ol in almost quantitative yield. Addition of dimethylaniline or 4-dimethylaminobenzaldehyde retards hydrogenation of the 22,23-bond; in this way selective reduction to $\Delta^{7,22}$-ergostadiene-3β-ol can be achieved in high yield.[4]

*Hydrogenolysis of aryl halides.* Aryl halides are converted into arenes by hydrogenation in a 0.2 $M$ solution of KOH in ethanol at room temperature and atmospheric pressure in the presence of Raney nickel. The reaction is subject to steric hindrance.[5]

[1] N. F. Hayes, *Synthesis*, 702 (1975).
[2] R. Muneyuki and H. Tanida, *Am. Soc.*, **90**, 656 (1968).

[3]W. Lonsky, H. Traitler, and K. Kratzl, *J.C.S. Perkin I*, 169 (1975).
[4]W. Tadros and A. L. Boulos, *Helv.*, **58**, 668 (1975).
[5]A. J. de Koning, *Org. Prep. Proc. Int.*, **7**, 31 (1975).

## Rhodium (5%) on alumina, 4, 418.

*α-Amino alcohols.* 2-Hydroxy oximes have been reduced to α-hydroxy amines in high yield by hydrogenation over 5% rhodium on alumina.[1] Palladium and platinum catalysts were unsuccessful. Reduction with sodium in liquid ammonia or with lithium aluminum hydride gave amino alcohols in low yield.

The products were converted into the N-monoacetates in >90% yield by acylation with 1 eq. of acetic anhydride in absolute ethanol.[2]

[1]M. S. Newman and V. Lee, *J. Org.*, **40**, 381 (1975).
[2]M. S. Newman and Z. ud Din, *ibid.*, **38**, 547 (1973).

## Rhodium on carbon, 1, 982–983; 4, 418–419.

*Hydrogenation of naphthols and tetralones.*[1] Use of Rh/C as catalyst permits hydrogenation of naphthols to decalols and of tetralones to decalols without hydrogenolysis of alkoxy substituents. One example of complete hydrogenation of a 4-hydroxyquinoline is reported.

(25%)                    (65%)

90%

[1]K. Chebaane, M. Guyot, and D. Molko, *Bull. soc.*, 244 (1975).

**Rhodium carbonyl**, $Rh_6(CO)_{16}$. Mol. wt. 1065.62. Supplier: ROC/RIC.

*Oxidation of ketones to carboxylic acids.* This transition metal cluster compound catalyzes the oxidation of CO to $CO_2$ by molecular oxygen. When acetone is used as solvent, acetic acid is identified as another product of oxidation. When cyclohexanone is used as solvent, adipic acid is formed in high yield.

The oxidation cannot be carried out in pure $O_2$, however, because of decomposition of $Rh_6(CO)_{16}$ in the absence of CO. This example of homogeneous catalysis of C–C bond cleavage of ketones is apparently unique.[1]

[1]G. D. Mercer, J. S. Shu, T. B. Rauchfuss, and D. M. Roundhill, *Am. Soc.*, **97**, 1967 (1975).

**Rhodium trichloride–Silica**, $RhCl_3$–$SiO_2$.

*Dimerization of ethylene.* Rhodium(III) chloride supported on silica gel is an active heterogeneous catalyst for dimerization of ethylene to 1-butene and then to 2-butene; the *trans/cis* ratio is 2.9. The active catalyst may have the structure depicted in (1).[1]

(1).

[1]N. Takahashi, I. Okura, and T. Keii, *Am. Soc.*, **97**, 7489 (1975).

**Rhodium trichloride hydrate, 1**, 986–989; **2**, 357–359; **3**, 243–244; **4**, 420–421.

*Deuterium exchange.* Garnett *et al.*[1] have found that rhodium, in addition to platinum and iridium (3, 134), catalyzes deuterium exchange. $RhCl_3$ functions as a homogeneous catalyst, whereas $RhCl_3$ reduced with aqueous $NaBH_4$ functions as a heterogeneous catalyst. Rhodium has several advantages, one of which is that deuteration of saturated hydrocarbons (*e.g.*, cyclohexane) is possible at reasonable rates.

[1]M. R. Blake, J. L. Garnett, I. K. Gregor, W. Hannan, K. Hoa, and M. A. Long, *J.C.S. Chem. Comm.*, 930 (1975).

**Ruthenium tetroxide, 1**, 986–989; **2**, 357–359; **3**, 243–244; **4**, 420–421.

*Oxidation of cyclic amines.* Sheehan and Tulis[1] have examined the oxidation of acylated cyclic amines with ruthenium tetroxide in a chlorinated solvent (single-phase system) or with ruthenium dioxide–sodium metaperiodate in $CHCl_3$–$H_2O$ (two-phase system). Lactams can be obtained in good yield by oxidation of 1-methyloxalylpiperidine or 1-methyloxalylpyrrolidine (1) with $RuO_4$. The protective group is cleaved with sodium methoxide in methanol. The N-carboethoxy derivative of azetidine ($n = 0$) is oxidized in low yield

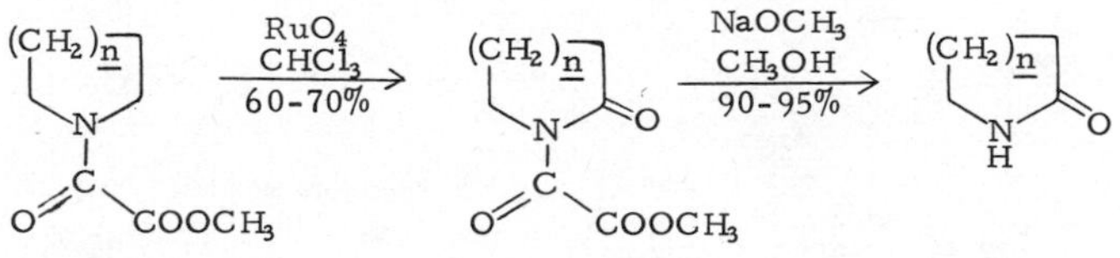

(1, n = 1, 2)                    (2)                         (3)

(22%). Two-phase oxidation of (1) (n = 1, 2) yields cyclic imides (4).

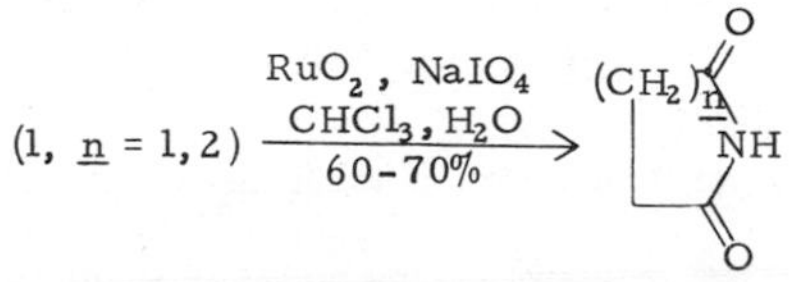

(4)

2-Substituted N-acetyl-pyrrolidines and -piperidines (5) are oxidized to the corresponding N-acetyllactam (6) in about 60% yield. Chiral substrates are oxidized with retention of configuration. 3-Substituted N-acetyl-pyrrolidines

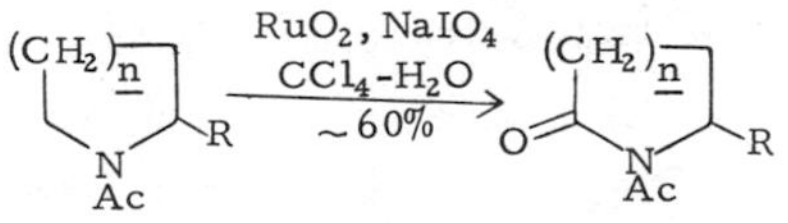

(5, n = 1, 2)                    (6)

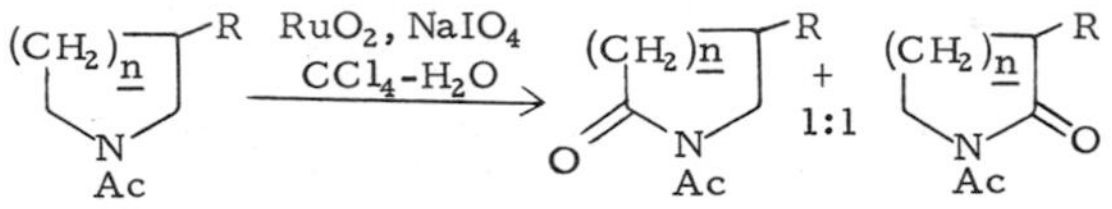

(7)                    (8)              (9)

and -piperidines (7) are oxidized also to lactams (8) and (9) in a 1:1 ratio. No imides have been detected.[2]

*Amino acids.* Oxidation of aliphatic primary amines with ruthenium tetroxide leads to complex products. However, oxidation of aralkylamines at pH 3.0 with periodate (12 eq.) and $RuCl_3 \cdot 3H_2O$ (0.02 eq.) converts the aromatic ring into a carboxyl group to give amino acids. Cleavage is facilitated by a 4-methoxy or hydroxy substituent in the ring.[3]

*Examples:*

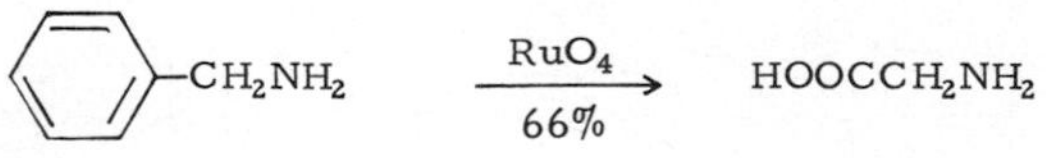

$CH_3O$—⟨aryl⟩—$\underset{\underset{CH_3}{|}}{C}HNH_2$ $\xrightarrow{50\%}$ $HOOC\underset{\underset{CH_3}{|}}{C}HNH_2$

$HO$—⟨aryl⟩—$CH_2CH_2NH_2$ $\xrightarrow{86\%}$ $HOOCCH_2CH_2NH_2$

$HO$—⟨aryl⟩—$CH_2\underset{\underset{NH_2}{|}}{C}HCOOH$ $\xrightarrow{60\%}$ $HOOCCH_2\underset{\underset{NH_2}{|}}{C}HCOOH$

*Phthalic acid from naphthalene.* Naphthalene can be oxidized to phthalic

$$\xrightarrow[\substack{62-65\%}]{\substack{RuO_2, NaOCl \\ CCl_4, H_2O}}$$

acid by ruthenium dioxide and household bleach in a two-phase system, water and carbon tetrachloride. In the case of substituted naphthalenes mixtures are obtained unless the group is either strongly electron attracting or donating. In the former case the substituted ring is protected; in the latter case the substituted ring is oxidized.[4]

[1] J. C. Sheehan and R. W. Tulis, *J. Org.*, **39**, 2264 (1974).
[2] N. Tangari and V. Tortorella, *J.C.S. Chem. Comm.*, 71 (1975).
[3] D. C. Ayres, *J.C.S. Chem. Comm.*, 440 (1975).
[4] U. A. Spitzer and D. G. Lee, *J. Org.*, **39**, 2468 (1974).

# S

**Salcomine [Bis(salicylidene)ethylenediiminocobalt(II)]**, 2, 360; 3, 245.

*Preparation.* Details for the preparation of salcomine (1) in high yield have been presented.[1]

(1)

*2,6-Disubstituted-1,4-benzoquinones.* Salcomine is a useful catalyst for oxygenation of 2,6-disubstituted phenols to the corresponding *p*-benzoquinones.

DMF is the most satisfactory solvent; polyphenylene ethers and/or diphenoquinones are also formed when chloroform or methanol is used as solvent.

[1]C. R. H. I. de Jonge, H. J. Hageman, G. Hoentjen and W. J. Mijs, *Org. Syn.*, submitted (1975).

**Selenium**, 1, 990–992; 4, 222; 5, 575.

*Heterocycles.* Two laboratories[1,2] have reported the syntheses of five-membered heterocycles by the reaction of 1,2-amino alcohols or thiols with carbon monoxide and oxygen catalyzed by selenium (*cf.* 5, 575). Selenium is required in only catalytic amounts because it is converted during the reaction to $H_2Se$, which is then oxidized to Se. The reaction is not useful for preparation of six-membered heterocycles because ureas are the major products in the case of 1,3-amino alcohols and thiols.

*Examples:*

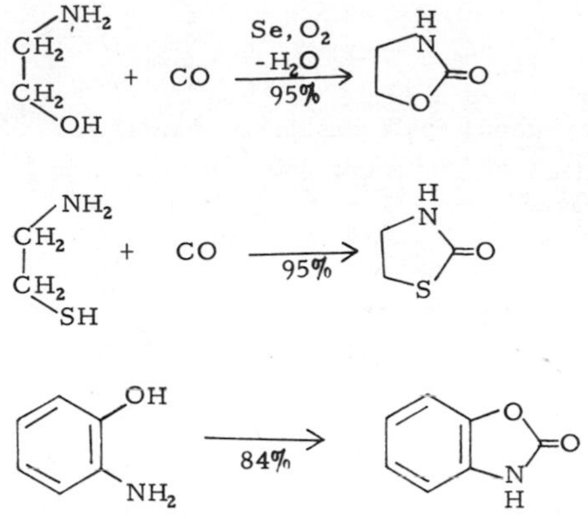

**Selenoketones.** Barton *et al.*[3] have converted ketones into the corresponding selenoketones by treatment of the hydrazone with triphenylphosphine dibromide to give the ketone triphenylphosphazine,[4] for example (1). This derivative (1 eq.), when heated with selenium (2 eq.) and a trace of triethylamine at 120° for 20 hr. under $N_2$, gives the blue selenoketone (2). On reaction with *m*-chloroperbenzoic acid, (2) is converted into di-*t*-butyl ketone and selenium.

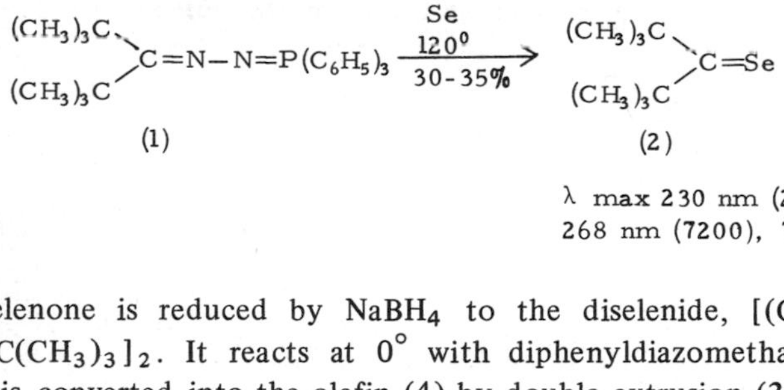

$\lambda$ max 230 nm (2800),
268 nm (7200), 710 nm (21)

The selenone is reduced by $NaBH_4$ to the diselenide, $[(CH_3)_3C]_2CHSe-SeCH[C(CH_3)_3]_2$. It reacts at 0° with diphenyldiazomethane to form (3), which is converted into the olefin (4) by double extrusion (**3**, 319–320) when heated.

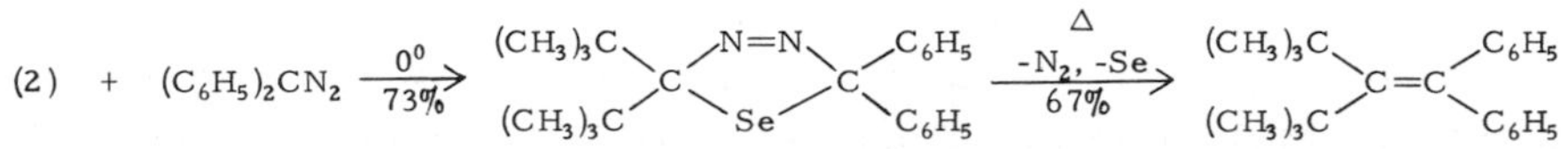

(−)-Selenofenchone (5) was also prepared; when heated with (−)-thiofenchone (6), fenchylidenefenchane (7) was obtained. This product is probably the most hindered known alkene.

(5)                (6)                          (7)

**Thiocarbamates.** Thiocarbamates can be prepared in fair to high yield by reaction of amines and disulfides with carbon monoxide with selenium as catalyst. If triethylamine is added as cocatalyst, aromatic amines can also be used.[5]

$$RNH_2 + R'SSR' + CO \xrightarrow[50-90\%]{\substack{Se \\ 20-60^0}} RNHCOSR' + HSR'$$

[1]P. Koch and E. Perrotti, *Tetrahedron Letters*, 2899 (1974).
[2]N. Sonoda, G. Yamamoto, K. Natsukawa, K. Kondo, and S. Murai, *ibid.*, 1969 (1975).
[3]T. G. Back, D. H. R. Barton, M. R. Britten-Kelly, and F. S. Guziec, Jr., *J. C. S. Chem. Comm.*, 539 (1975).
[4]H. J. Bestmann and H. Fritzsche, *Ber.*, **94**, 2477 (1961).
[5]P. Koch, *Tetrahedron Letters*, 2087 (1975).

**Selenium dioxide, 1**, 992–1000; **2**, 360–362; **3**, 245–247; **4**, 422–424; **5**, 575–576.

*Oxidation of triacetic lactone methyl ether.*[1] The allylic 6-methyl group of triacetic lactone methyl ether (1) is oxidized to a formyl group in high yield by

(1)                          (2)

(2) +                                                    (4)

(3)

selenium dioxide. This reaction to give (2) was useful for the synthesis of natural pyrones such as yangonin (4) by a Wittig reaction of (2) with (3). Yangonin had been synthesized earlier in 18% yield by the condensation of (1) with *p*-methoxy-benzaldehyde (2, 256).

*Dialkyl selenides.* The reaction of trialkylboranes with an equimolar amount of selenium dioxide suspended in THF leads to dialkyl selenides as the major

$$R_3B \xrightarrow[\quad O_2 \quad]{\text{SeO}_2, \text{THF}} \text{RSeR} + \text{RSeSeR}$$

(main product)     (trace)

product together with traces of dialkyl diselenides. The reaction is catalyzed by oxygen and inhibited by galvinoxyl, facts that point to a radical mechanism.[2]

These substances can also be prepared, in somewhat low yield, by reaction of Grignard reagents or alkyllithiums with selenium dioxide.[3]

$$\begin{array}{c} 2\text{ RMgBr} \\ \text{or} \\ 2\text{ RLi} \end{array} \xrightarrow[33\text{-}62\%]{\text{SeO}_2 \atop \text{THF}} \text{RSeR}$$

[1] E. Suzuki, R. Hamajima, and S. Inoue, *Synthesis*, 192 (1975).
[2] A. Arase and Y. Masuda, *Chem. Letters*, 419 (1975).
[3] *Idem, ibid.*, 1331 (1975).

**Silicic acid**, $H_2SiO_3 \cdot n\,H_2O$. Suppliers: Fisher, ROC/RIC.

*Nuclear halogenation of alkylarenes.* Halogenation of alkylarenes absorbed in silicic acid leads to substitution predominately in the aromatic ring. The same result is obtained with NBS or NCS.[1] (*cf.* **N-Bromopolymaleimide, 4**, 49, and **N-Chloropolymaleimide, 4**, 87).

Sulfuryl chloride also chlorinates arenes in the nucleus if silica gel is present in small amounts.[2] Silica gel is also a satisfactory catalyst for condensation of arenes with *p*-toluenesulfenyl chloride, TsSCl, and $C_6H_5SCH_2Cl$.[3]

[1] C. Yaroslavsky, *Tetrahedron Letters*, 3395 (1974).
[2] M. Hojo and R. Masuda, *Syn. Commun.*, 5, 169 (1975).
[3] *Idem, ibid.*, 5, 173 (1975).

**Siloxene**, $(Si_6O_3H_5)_n$.

*Preparation.*[1,2] Calcium silicide (2.5 g.) is added to *n*-propanol (300 ml.), water (55 ml.), and concd. hydrochloric acid at 0°; the mixture is agitated under nitrogen for 50–70 hr., filtered, and washed with *n*-propanol followed by ether at 0°. The product is then heated under vacuum at 200° for 5 hr.

*Isomerization of butene-2.* Siloxene catalyzes the *cis–trans* isomerization of butene-2, without isomerization of the double bond to form butene-1. A free-radical mechanism has been suggested.[2]

[1] H. Kautsky and H. Pfleger, *Z. Anorg. Allg. Chem.*, **295** 206 (1958).
[2] Y. Ono, Y. Sendoda, and T. Keii, *Am. Soc.*, **97**, 5284 (1975).

**Silver acetate**, **1**, 1002–1004; **2**, 362–363.

*Fluoro allyl alcohols.* Schlosser and Chau[1] have described a new route to these alcohols. Chlorofluorocarbene is allowed to react with an olefin in a two-phase system; the resulting chlorofluorocyclopropanes are then opened with silver acetate in acetic acid. The final step involves alkaline hydrolysis.

[1] M. Schlosser and L. V. Chau, *Helv.*, **58**, 2595 (1975).

**Silver carbonate**, **1**, 1005; **2**, 363; **4**, 425.

*Königs–Knorr synthesis* (**1**, 1005; **4**, 67). The Königs–Knorr synthesis of β-glucosides is improved with respect to reaction conditions and to yield by use of silver salts of dicarboxylic acids (*e.g.*, oxalic acid)[1] or of hydroxy carboxylic acids (*e.g.*, 4-hydroxyvaleric acid).[2] Use of the silver salt of a polymeric reagent offers the further advantages of other polymeric reagents.[3]

[1] B. Helferich and W. M. Müller, *Ber.*, **103**, 3350 (1970).
[2] G. Wulff, G. Röhle, and W. Krüger, *Angew. Chem. internat. Ed.*, **9**, 455 (1970).
[3] V. Eschenfelder, R. Brossmer, and M. Wachter, *ibid.*, **14**, 517 (1975).

**Silver carbonate–Celite**, **2**, 363; **3**, 247–249; **4**, 425–428; **5**, 577–580.

*Selective oxidation of dihydroxy steroids.* Jones *et al.*[1] effected selective oxidations of 5α-androstanediols with Fetizon's reagent, usually in refluxing benzene for 4 hr. Thus the 3β-hydroxy group can be selectively oxidized in the presence of 6α-, 7α-, 7β-, 11α-, 11β-, 12α-, 12β-, and 15α-hydroxyl groups. The hydroxyl groups of 3β,6β-diols differ only slightly in the rate of oxidation under the standard conditions. However, selective oxidation of the 3-hydroxyl group can be achieved by using a mixture of chloroform and benzene, the composition of which is changed during the oxidation.

*Oxidation of primary allylic alcohols.* In the synthesis of several isomeric phytoenes ($C_{40}$-conjugated trienes), Weedon *et al.*[2] used Fetizon's reagent for oxidation of primary allylic alcohols to allylic aldehydes. Yields of >85% were reported. Use of manganese dioxide led to much lower yields (40–50%). 2,3-Dichloro-5,6-dicyano-1,4-benzoquinone also gave high yields, but mixtures of stereoisomers were obtained.

*Oxidation of vic-diols.* The course of oxidation of diols of type (1) and (2) depends on the stereochemistry: *threo*-glycols undergo cleavage; *erythro*-glycols are oxidized to both possible ketols. The ratio of the ketols formed depends

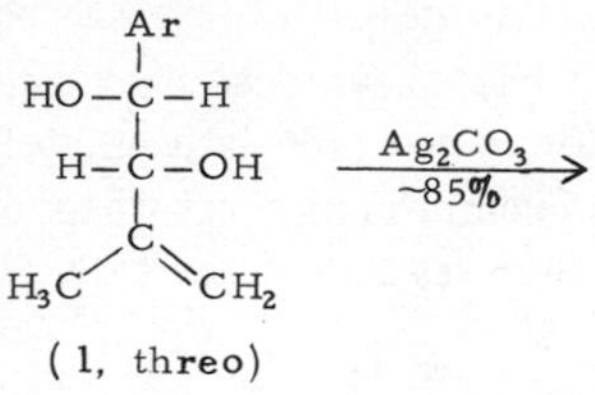

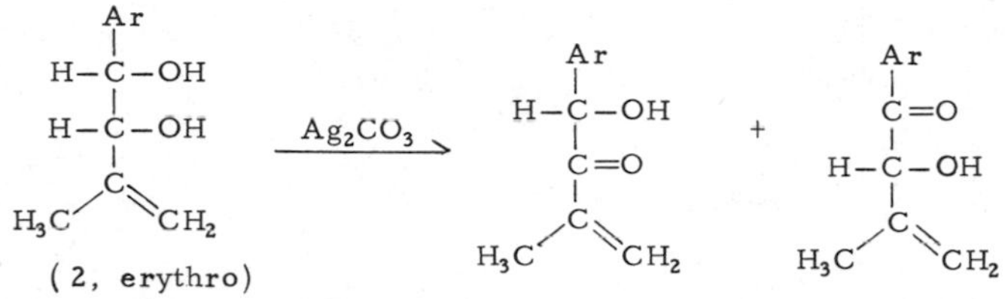

(1, threo)

$Ar-C-H$ etc. → ArCHO (Ag$_2$CO$_3$, ~85%)

(2, erythro)

on the electronic effects of substituents in the Ar group.[3]

*Lactones.* Primary 1,4-, 1,5-, and 1,6-diols are oxidized to lactones in satisfactory yield by silver carbonate on Celite (20–25 eq.) in refluxing benzene (1–30 hr.)[4]

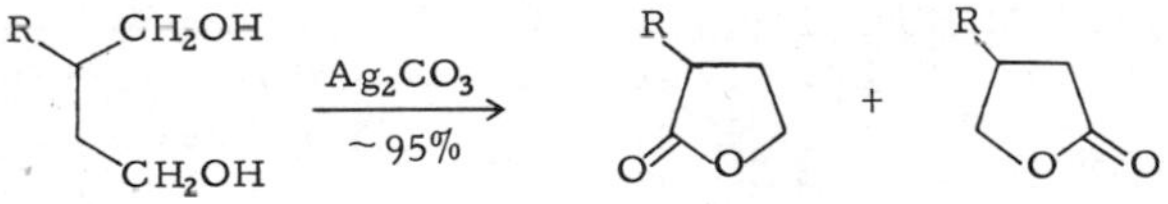

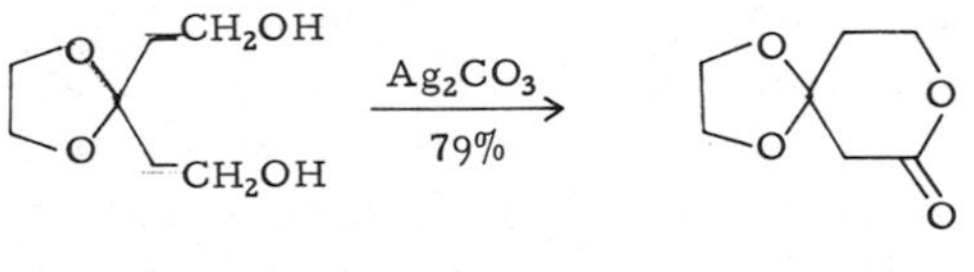

The oxidation was used in a new synthesis of (±)-mevalonolactone (3) from 4-methyl-1,6-heptadiene-4-ol (1).

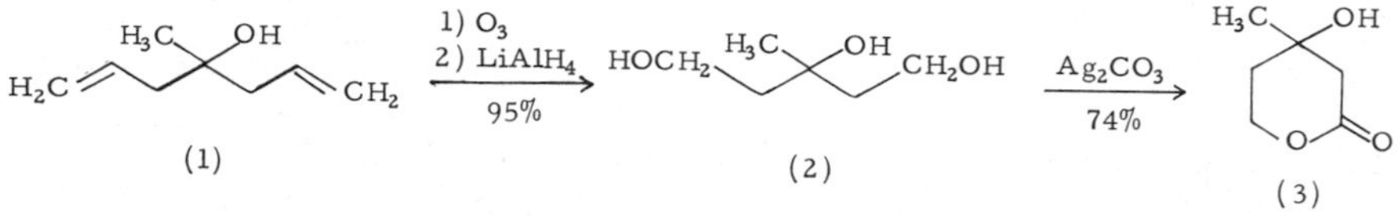

(1)                (2)                (3)

*5-Aryl-γ-lactones.* Dutch chemists[5] have reported a method for synthesis of lactones of type (4) with an imidazolyl, pyridyl, or aryl group at the 5-position. The process involves reaction of an aromatic aldehyde or ketone with the Grignard derivative of 2-(2-bromoethyl)dioxolane-1,3 (1).[6] The product (2) is cyclized to a lactol (3) by dilute sulfuric acid (reflux). The final step involves

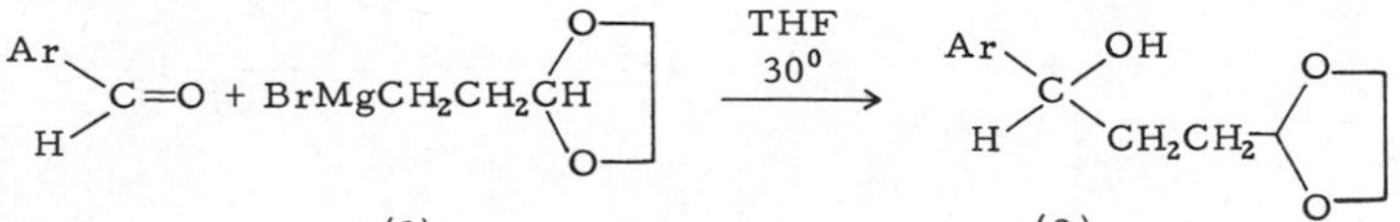

(1)    (2)

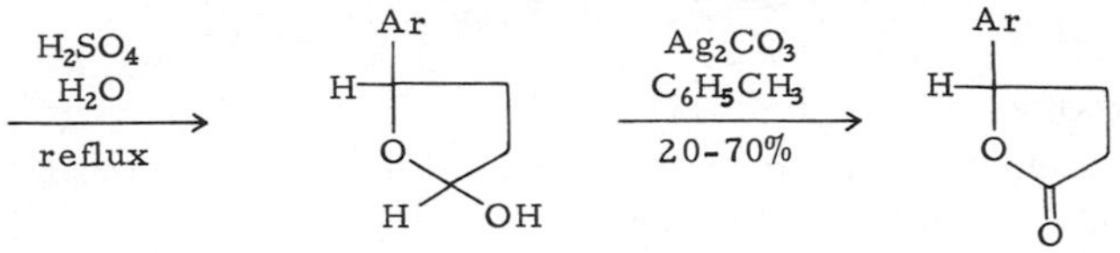

(3)    (4)

oxidation to a lactone (4). Silver carbonate on Celite was used for this purpose. Conventional oxidants (potassium permanganate, chromic acid, manganese dioxide, silver oxide) were ineffective.

*Oxidation of anilines.*[7] Oxidation of anilines with this reagent gives azo-benzenes (N—N coupling), quinonimines (N—C coupling), and phenazines. No products of C—C coupling have been isolated.

*Examples:*

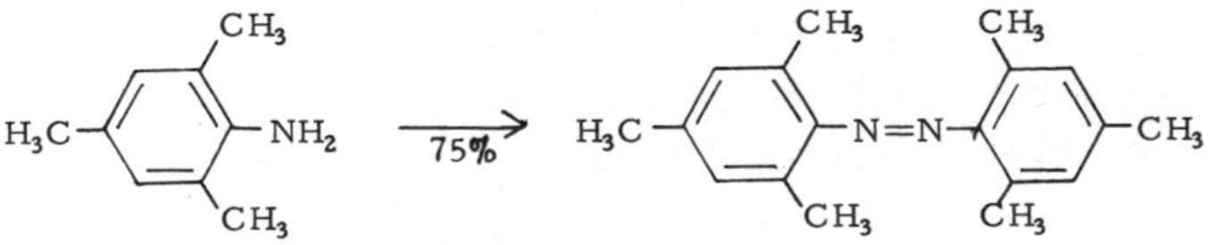

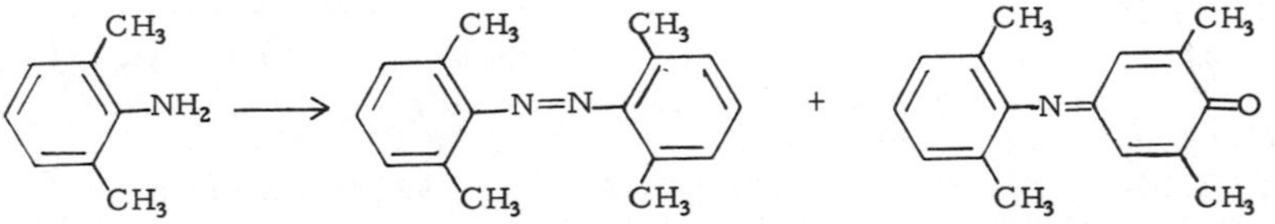

(35%)    (10%)

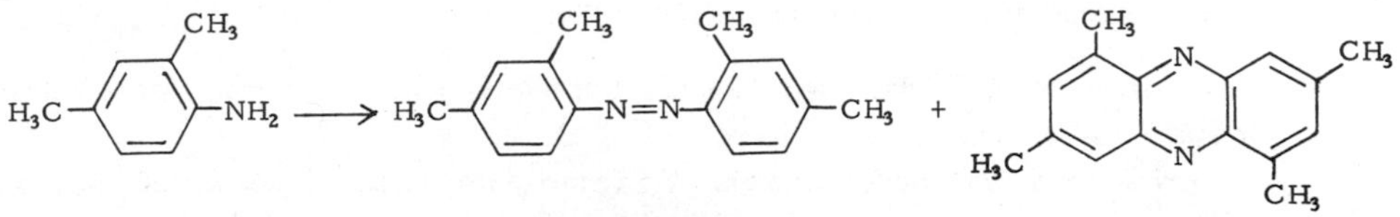

(30%)    (45%)

*Oxidation of various nitrogen compounds.*[8] Primary aromatic amines are oxidized slowly to azobenzenes in yields on the order of 40–50%. Esters of hydrazodicarboxylic acid are oxidized rapidly and in high yield in refluxing

$$2\ \mathrm{ArNH_2} \xrightarrow{\mathrm{Ag_2CO_3}} [2\ \mathrm{Ar\overset{.}{N}H} \longrightarrow \mathrm{ArNH\!-\!NHAr}] \xrightarrow[40\text{-}50\%]{} \mathrm{ArN}\!\!=\!\!\mathrm{NAr}$$

benzene to esters of azodicarboxylic acid.

Hydrazones are rapidly oxidized to diazoalkanes, which on prolonged oxidation are converted into ketones and azines:

$$\mathrm{Ar_2C}\!\!=\!\!\mathrm{NNH_2} \xrightarrow[\sim90\%]{\substack{\mathrm{Ag_2CO_3}\\ 5\ \min.}} \mathrm{Ar_2C}\!\!=\!\!\mathrm{N_2} \xrightarrow{\mathrm{Ag_2CO_3}} \mathrm{Ar_2C}\!\!=\!\!\mathrm{O} + \mathrm{Ar_2C}\!\!=\!\!\mathrm{N\!-\!N}\!\!=\!\!\mathrm{CAr_2}$$
$$(\sim50\%) \qquad (\sim50\%)$$

*Cardenolide glycosides.* Glycosides of various cardenolides can be prepared by the Königs–Knorr reaction in yields of 60–75% using the Fetizon reagent.[9] Yields using ordinary silver carbonate are much lower (0–45%). A further advantage is that elimination of tertiary hydroxyl groups in the aglycone is not observed. Oxidation of the secondary hydroxyl group at $C_3$ is not a competing reaction.

[1] E. R. H. Jones, G. D. Meakins, J. Pragnell, W. E. Müller, and A. L. Wilkins, *J.C.S. Perkin I,* 2376 (1974).

[2] N. Khan, D. E. Loeber, T. P. Toube, and B. C. L. Weedon, *J.C.S. Perkin I,* 1457 (1975).

[3] S. L. T. Thuan and M. P. Maitte, *Tetrahedron Letters,* 2027 (1975).

[4] M. Fetizon, M. Golfier, and J.-M. Louis, *Tetrahedron,* **31,** 171 (1975).

[5] H. J. J. Loozen, E. F. Godefroi, and J. S. M. Besters, *J. Org.,* **40,** 892 (1975).

[6] G. Büchi and H. Wüest, *J. Org.,* **34,** 1122 (1969).

[7] M. Hedayatullah, J. P. Dechatre, and L. Denivelle, *Tetrahedron Letters,* 2039 (1975).

[8] M. Fetizon, M. Golfier, R. Milcent, and I. Papadakis, *Tetrahedron,* **31,** 165 (1975).

[9] J. Hartenstein and G. Satzinger, *Ann.,* 1763 (1974).

**Silver fluoride,** AgF. Mol. wt. 126.87. Suppliers: Alfa, Columbia Org. Chem., K and K, MCB, Schuckardt.

*Bridgehead fluoroadamantanes.* 1-Bromoadamantane is converted by anhydrous zinc fluoride in refluxing cyclohexane into 1-fluoroadamantane in 61% yield. However, the yield is sensitive to the heat treatment required to obtain anhydrous salt from the tetrahydrate and the reaction becomes increasingly slow as the number of bridgehead bromine atoms is increased. In this case the more expensive silver fluoride is used. Even 1,3,5,7-tetrabromoadamantane can be converted into the corresponding tetrafluoroadamantane in 76% yield with silver fluoride.[1]

[1] K. S. Bhandari and R. E. Pincock, *Synthesis*, 655 (1974).

**Silver heptafluorobutanoate,** $CF_3(CF_2)_2COOAg$. Mol. wt. 194.97. Supplier: ROC/RIC.

*Ketocarbenoids.* On reaction of 3-diazocamphor (1) in THF with this silver salt at 45°, nitrogen is evolved and tricyclanone (2) is obtained in about 97% yield. The reaction probably involves an intermediate ketocarbene (a).[1] *See Diethylzinc,* **5,** 219–220 for a related, but less efficient, synthesis of (2).

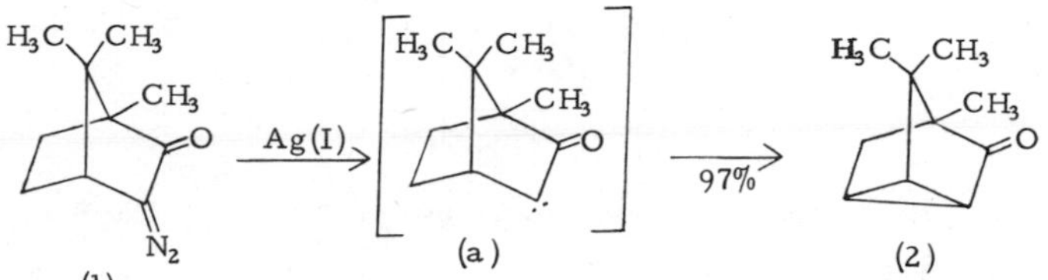

[1] F. C. Brown, D. G. Morris, and A. M. Murray, *Syn. Commun.*, **5,** 477 (1975).

**Silver nitrite, 1,** 1011.

*N-Demethylation of tertiary amines.* Tertiary amines are converted into N-demethyl-N-nitrosamines (40–80% yield) by treatment with $AgNO_2$ (4–8 molar eq.) in DMF (70°, 2–24 hr.). N-Nitrosamines are readily hydrolyzed to secondary amines by acid.[1]

[1] L. Bernardi and G. Bosisio, *J.C.S. Chem. Comm.*, 690 (1974).

**Silver oxide, 1,** 1011–1015; **2,** 368; **3,** 252–254; **4,** 430–431; **5,** 583–585.

*1,4-Diketones.* The reaction of trimethylsilyl enol ethers with $Ag_2O$ in DMSO at 65–100° for 2–5 hr. results in formation of 1,4-diketones in yields of 25–80%. The reaction is considered to proceed through a silver(I) enolate:

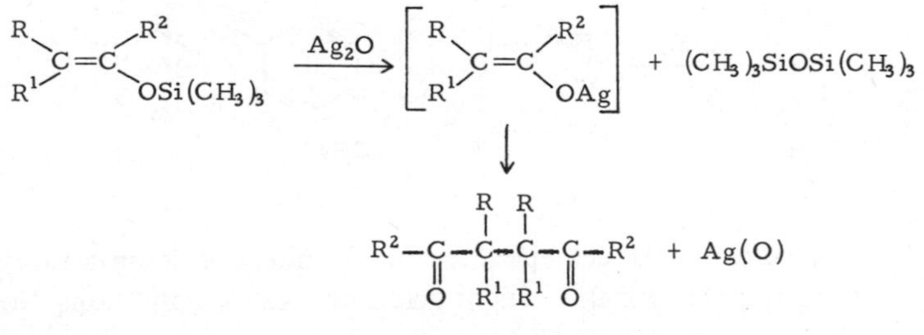

The present method is useful for synthesis of symmetrical 1,4-diketones. The reaction is highly dependent on the solvent. Aprotic polar solvents such as HMPT, DMF, and acetonitrile can be used; no reaction occurs in toluene or diglyme.

The paper also reports two successful cross-coupling reactions leading to unsymmetrical 1,4-diketones.[1]

*Intermolecular oxidative coupling of phenols.* Italian chemists[2] have reported a synthesis of the unusual neolignan (±)-eusiderin (4) patterned on a

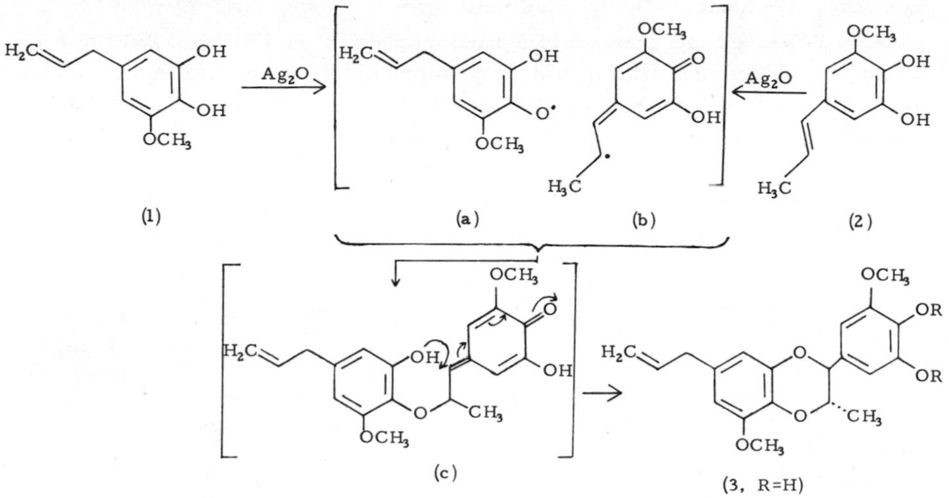

(1)          (a)          (b)          (2)

(c)

(3, R=H)
(4, R=CH₃)

possible biosynthesis suggested by H. Erdtman. Oxidation of equimolar amounts of the phenols (1) and (2) with silver oxide gave, in unspecified yield, the product (3), which is a result of coupling of the radicals (a) and (b). The natural product was obtained by methylation (diazomethane) of (3).

Oxidation of (2) alone gave (5) "in good yield."

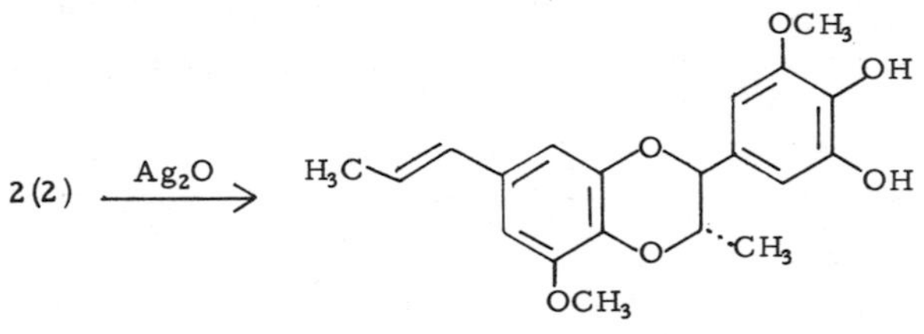

$$2\,(2) \xrightarrow{\text{Ag}_2\text{O}}$$

(5)

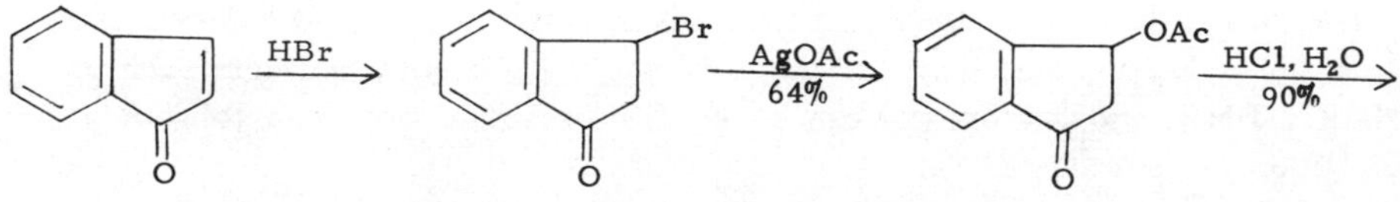

*2,3-Epoxyindanone.* Direct epoxidation of indenone is unsatisfactory because of extensive polymerization under acidic or alkaline conditions. Norwegian chemists[3] have prepared the oxide by an indirect method using freshly prepared silver oxide.[4]

*C-Alkylation.* Grifolin (2) has been synthesized in 15% yield by the reaction of orcinol (1, excess) with farnesyl bromide in dry dioxane containing silver oxide. Some of the 4-farnesyl isomer is formed as well. The reaction of orcinol

with farnesyl bromide catalyzed by acids (*p*-TsOH, oxalic acid) results in a lower yield of C-alkylated products.[5] Reaction between orcinol and farnesol in aqueous oxalic acid also gives grifolin, but O-alkylation occurs to a greater extent than in the silver oxide method.[6]

The reaction of 5,7-dihydroxy-4-methylphthalide (3) with an allylic bromide to form mycophenolic acid (4) is also best conducted in the presence of silver oxide.[7]

[1] Y. Ito, T. Konoike, and T. Saegusa, *Am. Soc.*, **97**, 649 (1975).
[2] L. Merlini and A. Zanarotti, *Tetrahedron Letters*, 3621 (1975).
[3] K. Undheim and B. P. Nilsen, *Acta Chem. Scand.*, **29B**, 503 (1975).
[4] D. Y. Curtin and R. J. Harder, *Am. Soc.*, **82**, 2357 (1960).
[5] J. H. P. Tyman, W. A. Baldwin, and C. J. Strawson, *Chem. Ind.*, 41 (1975).

[6]G. D. Manners, L. Jurd, and K. L. Stevens, *ibid.*, 616 (1974).
[7]L. Canonica, B. Rindone, E. Santamello, and C. Scolastico, *Tetrahedron Letters*, 2691 (1971).

**Silver(II) oxide (Argentic oxide), 2, 369; 4, 431–432.**

*Addition of acetone to terminal alkenes.* Acetone undergoes anti-Markownikoff addition to terminal olefins in the presence of argentic oxide. Straight-chain alkyl methyl ketones are formed in 73–83% yields. Internal olefins undergo

$$RCH{=}CH_2 + CH_3COCH_3 \xrightarrow[73-83\%]{AgO,\,55^0} RCH_2CH_2CH_2COCH_3$$

this reaction less readily; cyclohexene is converted into acetonylcyclohexane in 54% yield (33 hr.). In the absence of an olefin, acetonylacetone is formed. The addition reaction appears to involve a radical chain reaction.[1]

[1]M. Hájek, P. Šilhavý, and J. Málek, *Tetrahedron Letters*, 3193 (1974).

**Silver perchlorate, 2, 369–370; 4, 432–435; 5, 585–587.**

*Rearrangement of polycyclic systems.* (4, 432–435; 5, 585–587). Paquette[1] has reviewed silver ion-catalyzed rearrangements of cyclic systems, particularly those that are useful synthetically. A few of these are formulated. One is the preparation of homotropilidene (2) from 3-norcarene (1) shown in equation I. Note that rearrangement of (3), with a methyl group at a bridgehead position,

$$\text{(I)} \qquad \xrightarrow[42\%]{\substack{1)\ :CBr_2 \\ 2)\ CH_3Li}} \qquad \xrightarrow[\text{high}]{\substack{AgClO_4 \\ C_6H_6,\,40^0}}$$

(1)                                                                 (2)

$$\text{(II)} \qquad \xrightarrow[44\%]{AgClO_4}$$

(3)                              (4)

leads to a completely different isomerization (equation II).

Another useful rearrangement is that of bicyclopropenyl compounds, which provides an attractive route to Dewar benzene derivatives as formulated in equations III and IV. The ratio of the products (9) and (10) formed in the latter reaction is highly dependent on the polarity of the solvent.

$$\text{(III)} \qquad \xrightarrow[45\%]{Ag(I)} \qquad \xrightarrow{\text{slow}}$$

(5)                              (6)                              (7)

(IV) 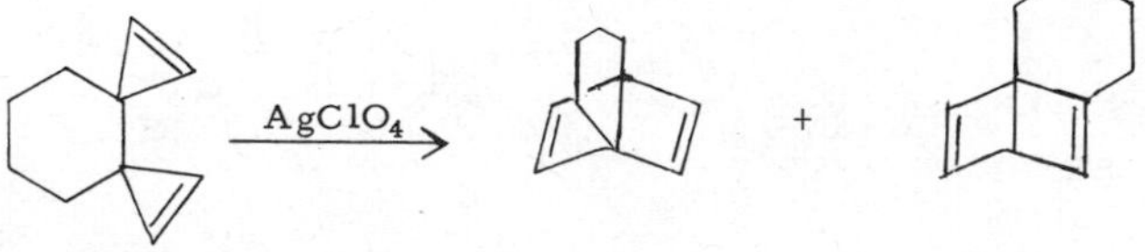     (8)          (9)          (10)

Another useful isomerization is that of the 1,8-bishomocubane derivative (11), which has been used to obtain the $(CH)_{10}$ hydrocarbon (13) with the trivial name snoutene (equation V).

(V) 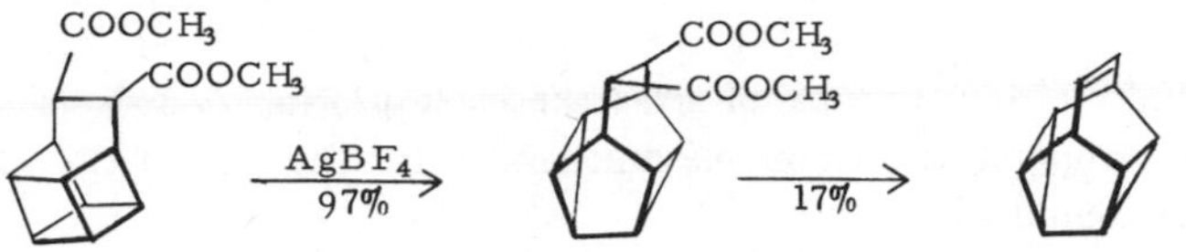     (11)          (12)          (13)

A two-step ring contraction of cyclooctatetraene oxide (14) has been effected with Ag(I) (equation VI).

(VI) 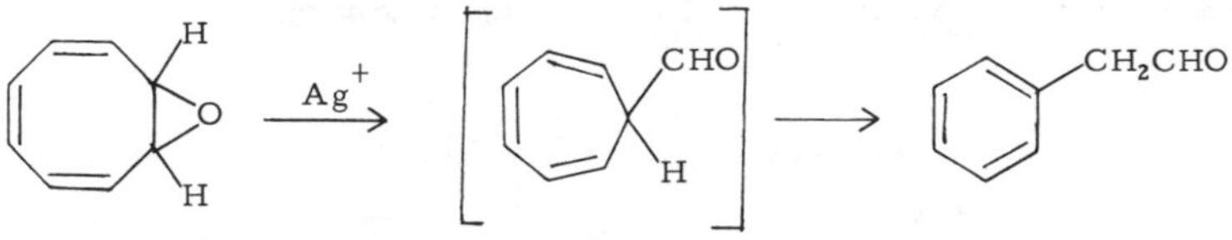     (14)          (15)

*Methanolysis of halogenocarbene adducts* (4, 432–433). Complete details for this solvolytic ring-expansion reaction have been published. The reaction is particularly useful for preparation of medium-sized *trans*-cycloalkene derivatives.[2]

[1]L. A. Paquette, *Synthesis*, 347 (1975).
[2]C. B. Reese and A. Shaw, *J.C.S. Perkin I*, 2422 (1975).

**Silver tetrafluoroborate, 1,** 1015–1018; **2,** 365–366; **3,** 250–251; **4,** 428–429; **5,** 587–588.

*Esters and lactones.*[1] Treatment of the 2-pyridyl ester of a thiocarboxylic S-acid such as (1), prepared as shown, with either silver tetrafluoroborate or silver

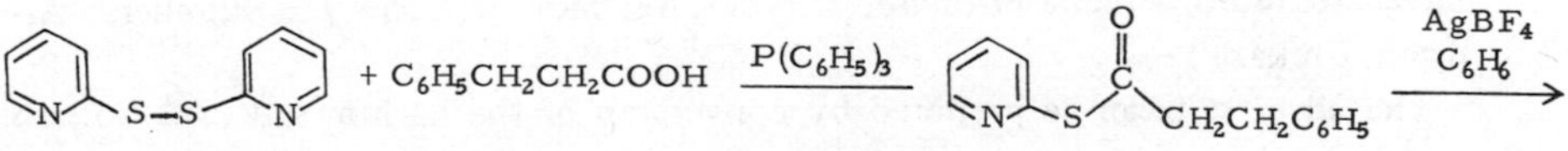

(1)

(a)                                                                                          (b)                                                  70–85%

(2)

perchlorate and 2-propanol gives the ester (2) in 70–85% yield in 10 min. at room temperature. A possible mechanism for the role of the silver-ion catalysis is formulated.

Similarly, treatment of the 2-pyridyl thiol ester of 15-hydroxypentadecanoic acid in benzene with 1 eq. of silver perchlorate leads to rapid formation of the corresponding monomeric, dimeric, and trimeric lactones.

*Cyclopentenones.*[2] Russian chemists report a new synthesis of 2-cyclo-pentenones by the reaction of alkynes with an acyl tetrafluoroborate, formed *in situ* from an acyl chloride and silver tetrafluoroborate:

The original paper suggests a possible route for this reaction.

[1] H. Gerlach and A. Thalmann, *Helv.*, **57**, 2661 (1974).

[2] A. A. Schegolev, W. A. Smit, G. V. Roitburd, and V. F. Kucherov, *Tetrahedron Letters*, 3373 (1974).

**Silver trifluoromethanesulfonate**, $CF_3SO_3Ag$. Mol. wt. 256.94. Suppliers: Aldrich, Fluka.

The silver salt can be prepared by conversion of the barium salt (3M Co.) to the acid, neutralization of the aqueous solution of the acid with silver carbonate, and evaporation (yield 90–95%).[1]

*Diaryl sulfones.*[2] Aryl sulfonyl bromides (but not sulfonyl chlorides) react with silver trifluoromethanesulfonate in nitromethane at 0° to form unstable mixed sulfonic anhydrides (*cf.* **4**, 533–534). These react with arenes (threefold excess) at the same temperature to give diaryl sulfones usually in yields of 80–100%.

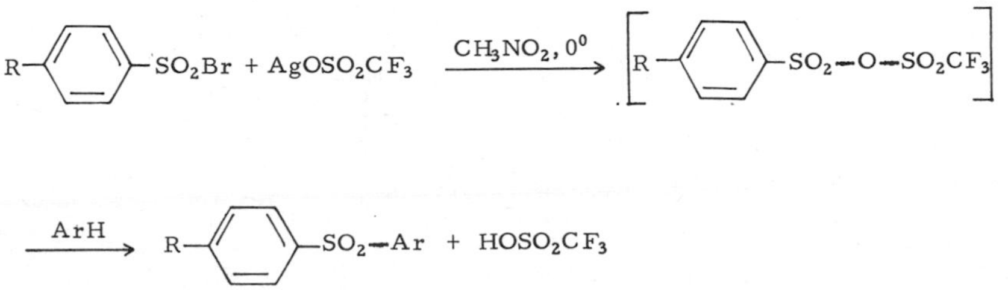

[1] T. Gramstad and R. N. Haszeldine, *J. Chem. Soc.*, 173 (1956); A. Streitwieser, Jr., C. L. Wilkins, and E. Kiehlmann, *Am. Soc.*, **90**, 1598 (1968).
[2] F. Effenberger and K. Huthmacher, *Angew. Chem. internat. Ed.*, **13**, 409 (1974).

**Simmons–Smith reagent, 1**, 1019–1022; **2**, 371–372; **3**, 255–258; **4**, 436–437; **5**, 588–589.

*Monomethylation of cycloalkenones.* Girard and Conia[1] have reported this conversion, using the improved modification of the Simmons–Smith reaction (**4**, 436). α′-Methylation is illustrated for the case of cyclohexenone. The trimethylsilyl ether (2) is prepared under conditions of kinetic control (**3**, 310–311); this undergoes monocyclopropanation almost exclusively in the 1,2-

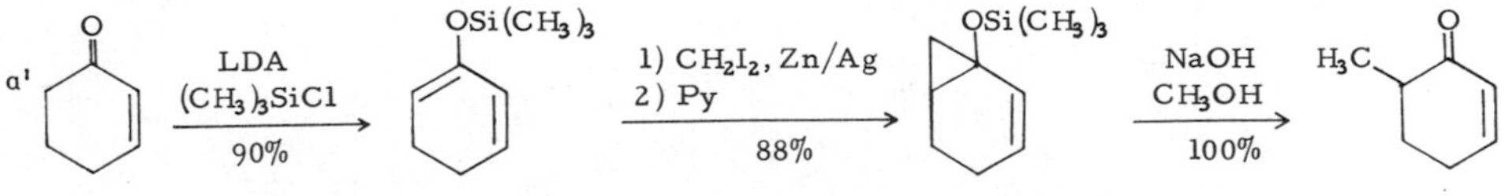

position to give (3). Alkaline or acidic hydrolysis of (3) gives the α′-methyl derivative (4) of (1).

Actually, either α- or α′-monomethylation of a cycloalkenone can be effected in high yield as illustrated for testosterone (5). If the enol ether is generated under thermodynamic control, 4-methyltestosterone (8) is obtained by the sequence illustrated above. 2α-Methyltestosterone (11) can be prepared readily if the enol is generated under kinetic control.

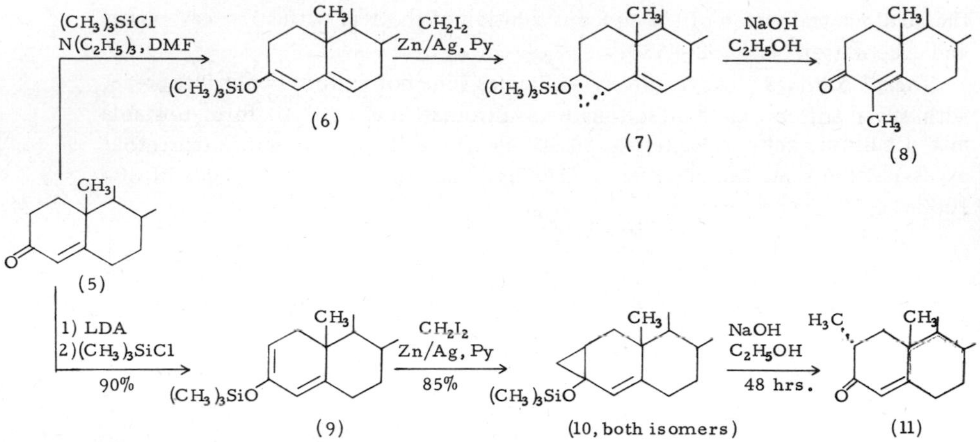

(CH$_3$)$_3$SiCl
N(C$_2$H$_5$)$_3$, DMF

(CH$_3$)$_3$SiO

(6)

CH$_2$I$_2$
Zn/Ag, Py

(CH$_3$)$_3$SiO

(7)

NaOH
C$_2$H$_5$OH

CH$_3$

(8)

CH$_3$

(5)

1) LDA
2)(CH$_3$)$_3$SiCl
90%

(CH$_3$)$_3$SiO

(9)

CH$_2$I$_2$
Zn/Ag, Py
85%

(CH$_3$)$_3$SiO

(10, both isomers)

NaOH
C$_2$H$_5$OH
48 hrs.

H$_3$C

CH$_3$

(11)

*Reaction with $\alpha,\beta$-unsaturated ketones.* 1-Siloxy-1-vinylcyclopropanes can be prepared easily by cyclopropanation of cisoid or labile $\alpha,\beta$-ethylenic ketones. It has already been shown that 1-vinylcyclopropanols undergo ring expansion to cyclobutanones on treatment with acid and to cyclopentanones on thermolysis.[2] Two examples of this useful new cyclopropanation sequence are shown in

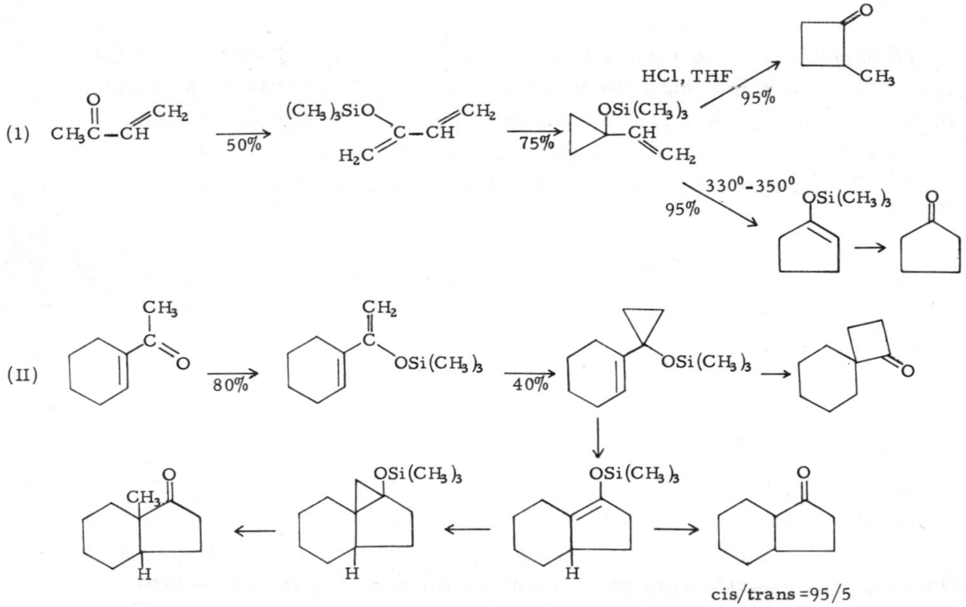

(1)  CH$_3$C—CH
       ‖   ∕CH$_2$
       O

50% →

(CH$_3$)$_3$SiO
H$_2$C
C—CH ∕CH$_2$

75% →

OSi(CH$_3$)$_3$
—CH
∕CH$_2$

HCl, THF
95%

O
CH$_3$

330°–350°
95%

OSi(CH$_3$)$_3$

→

O

(II)

CH$_3$
C=O

80% →

CH$_2$
C
OSi(CH$_3$)$_3$

40% →

OSi(CH$_3$)$_3$

→

O

CH$_3$
O

←

OSi(CH$_3$)$_3$

←

OSi(CH$_3$)$_3$

→

O

cis/trans = 95/5

equations I and II. Note that only the double bond bearing the OSi(CH$_3$)$_3$ group is attacked by 1 eq. of reagent. If, however, 3 eq. are used, double cyclopropanation can be achieved in 80–90% yield. The products can be converted to 1-cyclo-propylcyclopropanols and to cyclopropyl ketones.[3] Two examples are shown in equations III and IV.

(III)  $CH_3\overset{O}{\overset{\|}{C}}-\overset{CH_2}{\underset{}{\overset{\diagup}{CH}}}$  $\xrightarrow{50\%}$  $\underset{H_2C}{(CH_3)_3SiO}\overset{CH_2}{\underset{}{\overset{\diagup}{C-CH}}}$  $\xrightarrow{90\%}$  [bicyclopropyl with OSi(CH$_3$)$_3$ and H]

$\xrightarrow[\Delta]{CH_3OH,}$ 95%  [cyclopropylcyclopropanol with OH]

$\xrightarrow[95\%]{NaOH \atop CH_3OH, \Delta}$  $CH_3CH_2\overset{O}{\overset{\|}{C}}$—[cyclopropyl]—H

(IV)  [cyclohexenone]  $\xrightarrow{90\%}$  [cyclohexadiene with OSi(CH$_3$)]  $\xrightarrow{90\%}$  [bicyclic with OSi(CH$_3$)]

→ [bicyclic alcohol with OH]

→ $H_3C$—[bicyclic ketone]

[1]C. Girard and J. M. Conia, *Tetrahedron Letters*, 3327 (1974).
[2]C. Girard, P. Amice, J. P. Barnier, and J. M. Conia, *ibid.*, 3329 (1974).
[3]C. Girard and J. M. Conia, *ibid.*, 3333 (1974).

**Sodium–Ammonia, 1,** 1041; **2,** 374–376; **3,** 259; **4,** 438; **5,** 589–591.

*Reduction of allenes* (**2,** 374–376; **4,** 438). Indian chemists[1] have extended the reduction of allenes to a general method for synthesis of cyclic trisubstituted *cis*-enes. An example is the synthesis of *cis*-1-methylcyclononene (3) from 1-methylcyclooctene (1).

[cyclooctene with CH$_3$ and H] (1)  $\xrightarrow[65\%]{CBr_4, CH_3Li \atop -65^0}$  [cyclononadiene with CH$_3$, C=C, H] (2)  $\xrightarrow[95\%]{Na/NH_3}$  [cyclononene with CH$_3$ and H] (3)

[1]S. N. Moorthy, R. Vaidyanathaswamy, and D. Devaprabhakara, *Synthesis*, 194 (1975).

**Sodium–*t*-Butanol–Tetrahydrofurane, 1,** 1056; **3,** 378–379.

*Dechlorination.* Benzobarrelene (3) is readily available by the reaction of tetrachlorobenzyne (a) with benzene to give the adduct (2), which is then dechlorinated with sodium in THF containing *t*-butanol. The best weight ratio of THF/*t*-butanol/sodium/(2) is 40:3:1:1. Note that reaction of benzyne with benzene gives only about 2% of (3).[1]

[1]N. J. Hales, H. Heany, and J. H. Hollinshead, *Synthesis*, 707 (1975).

## Sodium–Potassium alloy, 1, 1102–1103.

*Fragmentation of γ-haloketones.* Treatment of a γ-haloketone with sodium–potassium alloy in ether results in an exothermic reaction accompanied by a characteristic deep blue color. Protonation then gives a γ,δ-unsaturated alcohol in which the C—X bond has been reduced and the original α,β-C—C bond has been cleaved. An example is the fragmentation of (1) to (2). In the second example the crude product is oxidized to (4) before work-up.[1]

[1]D. P. G. Hamon, G. F. Taylor, and R. N. Young, *Synthesis*, 428 (1975).

## Sodium aluminum chloride, 1, 1027–1029; 2, 372; 4, 438.

*Cyclodehydrogenation of heterohelicenes.* Several syntheses of coronene and related polynuclear hydrocarbons involve cyclodehydrogenation with aluminum chloride or sodium aluminum chloride to form aryl–aryl bonds[1] (Scholl reaction).[2] In their studies on the synthesis of heterohelicenes, Wynberg and collaborators[3] noted that the heterohexahelicene (1) underwent ring closure with aluminum chloride in benzene under unusually mild conditions (20°). The yield of (2), however, was only about 10%. In later work[4] the yield of (2) was

(1)                                    (2)

raised to 69% (pure) by use of aluminum chloride and sodium chloride (about 2:1 molar ratio) at 140°. Yields of other dehydrohelicenes were sometimes as high as 95%. A number of acids were found to be ineffective for this reaction: $FSO_3H$, $H_2SO_4$, 40% HF, $CF_3COOH$, and HCl. The reaction is also limited to hetero[5]- and hetero[6]helicenes.

[1] L. F. Fieser and M. Fieser, Reinhold, New York, *Topics in Organic Chemistry*, 38–42 (1963).
[2] R. Scholl and C. Seer, *Ann.*, **394**, 111 (1912).
[3] M. B. Groen, H. Schadenberg, and H. Wynberg, *J. Org.*, **36**, 2797 (1971).
[4] J. H. Dopper, D. Oudman, and H. Wynberg, *ibid.*, **40**, 3398 (1975).

**Sodium amide, 1**, 1034–1041; **2**, 373–374; **4**, 439; **5**, 591–593.

*Carboxamides.* The cleavage of nonenolizable ketones with sodium amide to form carboxamides (Haller–Bauer reaction[1]) has been reported to require freshly prepared reagent. Kaiser and Warner,[2] however, report that commercial sodium amide is satisfactory if an equimolar amount of 1,4-diazabicyclo[2.2.2]-octane (DABCO) is added. Benzamide is formed from benzophenone in this way in 73% yield. In the absence of DABCO, the yield is only 18%.

$$C_6H_5COC_6H_5 \xrightarrow[\substack{DABCO \\ 73\%}]{NaNH_2, C_6H_6, \Delta} C_6H_5CONH_2$$

*A bridgehead alkene.* Japanese chemists[3] have reported the first example of formation of a bridgehead alkene by dehydrohalogenation. Thus the bromide (2), obtained by bromination of 4-homoisotwistane (1), when refluxed in

(1)                        (2)                                    (3)

toluene with sodium amide gives (3) in 52% yield with no trace of the 3(4)- or 3(8)-ene. The ready formation of (3), which violates Bredt's rule, is considered to involve a planar *cis*-elimination of bromine and the 2-*exo* hydrogen atom. The olefin is fairly stable at 20° in the absence of air.

[1] K. E. Hamlin and A. W. Weston, *Org. React.*, 9, 1 (1957).
[2] E. M. Kaiser and C. D. Warner, *Synthesis*, 395 (1975).
[3] N. Takaishi, Y. Fujikura, Y. Inamoto, H. Ikeda, and K. Aigami, *J.C.S. Chem. Comm.*, 372 (1975).

**Sodium amide–Sodium *t*-butoxide**, 4, 439–440; 5, 593.

*Review.*[1]

*Indole synthesis.* Lalloz and Caubère[2] have reported a new synthesis of indoles from a halogenated aryl anil of type (1), prepared by condensation of a haloaniline with a ketone.[3] Treatment of the anils with this complex base

(1, X or X′ = Cl or Br)

(a)

(b)

(2)

in THF or THF–HMPT at room temperature leads to formation of indoles (2) in satisfactory yields. The synthesis is believed to involve enolization and dehydrohalogenation to an aryne (a) followed by cyclization ($S_N$Ar mechanism). Indoles substituted at the 5- and 6-positions can be prepared by the method.

[1] P. Caubère, *Accts. Chem. Res.*, 7, 301 (1974).
[2] L. Lalloz and P. Caubère, *J.C.S. Chem. Comm.*, 745 (1975).
[3] M. P. Grammaticakis, *Bull. soc.*, 768 (1949).

**Sodium benzenesulfinate**, $C_6H_5SO_2Na$. Mol. wt. 164.16. Suppliers: Aldrich, Fluka.

*Synthesis of carotinoids via sulfones.* Fischli and Mayer[1] have developed a synthesis of carotinoids based on the fact that allylic aryl sulfones in the presence of base readily eliminate arylsulfinic acid with formation of a conjugated double bond. An example is the synthesis of 15,15′-dehydro-β-carotene (4). β-Ionol (1) is converted into β-ionyl phenyl sulfone (2) by reaction with sodium benzenesulfinate. When this sulfone is treated with finely powdered sodium hydroxide it is deprotonated to the sulfonyl ylide, which is alkylated by the bis allylic chloride (3) to give isomeric 15,15′-dehydro-β-carotenes by way of the intermediate bis sulfone (a). Irradiation with an IR lamp effects isomerization to the all-*trans*-15,15′-dehydro-β-carotene (4).

The same method has been used for synthesis of the apocarotinoid (5) from vitamin A alcohol.

(5)

[1] A. Fischli and H. Mayer, *Helv.*, 58, 1492, 1584 (1975).

**Sodium bis-(2-methoxyethoxy)aluminum hydride,** 3, 260–261; 4, 441–442; 5, 596.

*Dehydrogenation of equilins to equilenins.*[1] The equilin derivative (1) is dehydrogenated to the equilenin derivative (2) when refluxed in toluene with sodium bis-(2-methoxyethoxy)aluminum hydride. If lithium aluminum hydride

(1)                                                                    (2)

is used, the yield is 4.6%. The reaction may involve deprotonation at $C_9$ and then elimination of a hydride ion.

*Reductive methylation* (4, 441–442). Further examples of reductive methylation by this complex metal hydride have been reported.[2] In all cases methylation takes place exclusively at a benzylic carbon atom activated by a phenyl group as shown in the examples.

*Reduction of lactones to lactols.* This reaction has generally been conducted with diisobutylaluminum hydride (1, 261; 2, 140). Japanese chemists[3] effected selective reduction of the lactone (1) to the corresponding lactol (2) with sodium bis-(2-methoxyethoxy)aluminum hydride and 1 eq. of ethanol without reduction of the angular carbomethoxy group.

(1)                                    (2)

*Reduction of sulfoxides.* Sulfoxides are reduced rapidly and generally in high yield to sulfides by this reagent. Sulfones require prolonged reactions at elevated temperatures.[4]

[1] J.-C. Hilscher, *Ber.*, **108**, 727 (1975).
[2] J. Málek and M. Černý, *J. Organometal. Chem.*, **84**, 139 (1975).
[3] R. Kanazawa, H. Kotsuki, and T. Tokoroyama, *Tetrahedron Letters*, 3651 (1975).
[4] T.-L. Ho and C. M. Wong, *Org. Prep. Proc. Int.*, **7**, 163 (1975).

**Sodium bistrimethylsilylamide, 1,** 1046–1047; **3,** 261–262; **4,** 442–443.

*Cyclopropanone and cyclobutanone cyanohydrins.* Stork has extended his synthesis of ketones[1] from protected cyanohydrins of aldehydes (equation I) to a synthesis of cyclic ketones, in particular, cyclopropanone and cyclobu-

tanone cyanohydrins,[2] from protected cyanohydrins of haloaldehydes. The synthesis of cyclopropanone cyanohydrin (5) from the chloro ether (1) is illustrated. It was converted into the cyanohydrin of 3-chloropropionaldehyde

(2) by hydrolysis in the presence of sodium cyanide. The hydroxyl group was protected by reaction of (2) with ethyl vinyl ether (1, 387–388). Cyclization of (3) to (4) was carried out with sodium bistrimethylsilylamide (1 eq., 3 hr., 95°). The protective group was removed by brief treatment with dilute sulfuric acid in methanol.

Cyclobutanone cyanohydrin (8) was prepared in the same way from the protected cyanohydrin of 4-chlorobutyraldehyde (6). Treatment of (8) in pentane with 1 N sodium hydroxide solution saturated with sodium chloride for 5 min.

at 0° gave cyclobutanone (9).

*Intramolecular Claisen condensation.*[3] This base was found to be by far the most satisfactory base for conversion of the keto ester (1) to (2). The reaction was effected in benzene at 79–80° (the temperature is critical). The product (2) was used in the synthesis of several ylango sesquiterpenes.[4]

*Methylenecyclopropanes.* Arora and Binger[5] have used both sodium bis(trimethylsilyl)amide and *n*-butyllithium in ether to generate chloromethylcarbene from 1,1-dichloroethane. The carbene reacts with alkenes to form 1-chloro-1-methylcyclopropanes in 40–80% yields. These products are converted into methylenecyclopropanes in 60–100% yield by potassium *t*-butoxide in DMSO.

[1] G. Stork and L. Maldonado, *Am. Soc.*, **93**, 5286 (1971); *idem, ibid.*, **96**, 5272 (1974).
[2] G. Stork, J. C. Depezay, and J. d'Angelo, *Tetrahedron Letters*, 389 (1975).
[3] E. Piers, R. W. Britton, M. B. Geraghty, R. T. Keziere, and R. D. Smillie, *Canad. J. Chem.*, **53**, 2827 (1975).
[4] E. Piers, R. W. Britton, M. B. Geraghty, R. J. Keziere, and F. Kido, *ibid.*, **53**, 2838 (1975).
[5] S. Arora and P. Binger, *Synthesis*, 801 (1974).

**Sodium borohydride, 1,** 1049–1055; **2,** 377–378; **3,** 262–264; **4,** 443–444; **5,** 597–601.

*Deamination of primary amines.* The conversion of primary amines into N-sulfonamides under Schotten–Baumann conditions is the basis for the classical Hinsberg test. Baumgarten *et al.*[1] have found that N-sulfonamides can be converted into N,N-disulfonimides, usually in high yield, by conversion into the sodium salt (NaH) and subsequent reaction with a sulfonyl chloride in DMF as solvent:

$$\text{RNHSO}_2\text{R}' \xrightarrow[-\text{H}_2]{\substack{\text{NaH} \\ \text{DMF}}} [\text{RN̄SO}_2\text{R}']\overset{\text{Na}^+}{} \xrightarrow[-\text{NaCl}]{\text{ClSO}_2\text{R}'} \text{RN(SO}_2\text{R}')_2$$

The diarylsulfonimides in general are not particularly susceptible to solvolysis, but they are susceptible to nucleophilic displacement at the carbon atom bearing nitrogen[1]:

$$(\underline{p}\text{-NO}_2\text{C}_6\text{H}_4\text{SO}_2)_2\text{N(CH}_2)_5\text{CH}_3 + \text{KI} \xrightarrow{\text{DMF, }110^0} \text{CH}_3(\text{CH}_2)_5\text{I} + (\underline{p}\text{-NO}_2\text{C}_6\text{H}_4\text{SO}_2)_2\text{NK}$$

Hutchins *et al.*[2] have since reported that the disulfonimides undergo nucleophilic displacement with borohydride anion in HMPT to give hydrocarbons. In the case of di-*p*-toluenesulfonimides temperatures of 150–175° and a twofold excess of borohydride are used in order to obtain reasonable reaction times.

$$\text{RN(SO}_2\text{R}')_2 \xrightarrow{\text{NaBH}_4, \text{HMPT}} \text{RH} + \text{NaN(SO}_2\text{R}')_2$$

Yields of hydrocarbons are good to excellent when R is a benzyl or a primary alkyl group, or even an unhindered secondary alkyl group. However, cyclododecyl-N(Ts)$_2$ is converted mainly into cyclododecyl-*p*-toluenesulfonamide by attack at nitrogen rather than carbon.

*Alkylation of aromatic amines.* Reduction of indole (1) with sodium borohydride in acetic acid results in reduction of the indole double bond and alkylation of the nitrogen atom to give N-ethylindoline. The reaction can be carried out in two steps: (1) → (3) → (2).

Primary and secondary aromatic amines can be alkylated by this same procedure, and the reaction can be controlled to result in mono- or dialkylation of primary amines. This amine alkylation is considered to involve reduction of the carboxylic acid to the aldehyde, formation of an iminium ion, and then reduction to the amine product. Actually, aldehydes or ketones can be used in place of a carboxylic acid.[3]

Similar alkylations have been reported by Italian chemists.[4]

*Reduction of tosylhydrazones* (**1**, 1051; **2**, 377–378). The Caglioti reduction of tosylhydrazones has been extended to a synthesis of dimethyl alkanephosphonates (2) from the tosylhydrazones (1). Reduction with sodium borohydride in THF results in formation of (2); reduction in methanol leads to dimethyl 1-diazoalkanephosphonates (3).[5]

*Reduction of carboxylic acid esters.* Esters are not reduced by sodium borohydride. However, if ethanedithiol is added (excess), most benzoate esters are reduced to benzyl alcohols in high yield. Thiophenol and ethylmercaptan do not share this property. Several aliphatic esters are also reduced by $NaBH_4$ activated by $HSCH_2CH_2SH$.[6]

Sodium borohydride in combination with acetanilide or benzanilide also is effective for selective reduction of methyl esters; other functional groups (amide, nitrile, isopropyl esters) are not reduced.[7]

*Reduction of nitro compounds.*[8] In the presence of thiols (usually $C_6H_5SH$), sodium borohydride reduces nitro groups to amino, hydroxylamino, azo, and azoxy groups.

*Dehalogenations.*[9] The $\alpha$-bromocyclopropyl trifluoroacetate (1) is reduced by sodium borohydride to (2) in about 60% yield.

*gem*-Dibromocyclopropanes are also appreciably reactive to NaBH₄ and afford the corresponding monobromides in high yield. Thus 7,7-dibromonorcarane is reduced to a mixture of *cis*- and *trans*-7-bromonorcarane in 79% yield by NaBH₄ in DMF or CH₃CN. The reaction shows characteristics of a radical reaction: inhibition by oxygen, incorporation of deuterium from sodium borodeuteride, product distribution similar to that obtained with tri-*n*-butyltin hydride.

*α,β-Unsaturated ketones.* Japanese chemists[10] have reported a new method for conversion of 1,3-diketones into α,β-unsaturated ketones. 3,5-Dimethylisoxazole, prepared from 2,4-pentanedione and hydroxylamine, can be alkylated

regiospecifically at the 5-methyl group to give (1). Büchi and Vederas (4, 260) had previously shown that isoxazoles of type (1) are reduced to α,β-unsaturated ketones (2) by sodium and *t*-butanol in liquid ammonia. Hydrogenation of (1) follows a different course and results in formation of a 2-amino-2-alkene-3-one (3). The carbonyl group of these products is inert to sodium borohydride; however, the corresponding N-acyl derivatives (4) are reducible, and the products (5) on treatment with acid yield α,β-unsaturated ketones (6), isomeric with (2). Yields in the different steps vary considerably.

*Photo-Birch reductions.* Irradiation of aromatic hydrocarbons in aqueous acetonitrile in the presence of sodium borohydride and 1,2-, 1,3-, or 1,4-dicyanobenzene with a high-pressure mercury arc results in reduction to dihydro derivatives. Phenanthrene is reduced to 9,10-dihydrophenanthrene (71%), anthracene to 9,10-dihydroanthracene (70%), and naphthalene to 1,4-

dihydronaphthalene (30%). Photoreduction is slow in the absence of a dicyano-benzene; water is involved, since deuterium is incorporated if $D_2O$ replaces $H_2O$. The exact mechanism is uncertain.[11]

[1]P. J. DeChristopher, J. P. Adamek, G. D. Lyon, J. J. Galante, H. E. Haffner, R. J. Bŏggio, and R. J. Baumgarten *Am. Soc.*, **91**, 2384 (1969); P. J. DeChristopher, J. P. Adamak, G. D. Lyon, S. A. Klein, and R. J. Baumgarten, *J. Org.*, **39**, 3525 (1974).

[2]R. O. Hutchins, F. Cistone, B. Goldsmith, and P. Heuman, *J. Org.*, **40**, 2018 (1975).

[3]G. W. Gribble, P. D. Lord, J. Skotnicki, S. E. Dietz, J. T. Eaton, and J. L. Johnson, *Am. Soc.*, **96**, 7812 (1974).

[4]P. Marchini, G. Liso, A. Reho, F. Liberatore, and F. M. Moracci, *J. Org.*, **40**, 3453 (1975).

[5]G. Rosini, G. Baccolini, and S. Cacchi, *Synthesis*, 44 (1975).

[6]Y. Maki and K. Kikuchi, *Tetrahedron Letters*, 3295 (1975).

[7]Y. Kikugawa, *Chem. Letters*, 1029 (1975).

[8]Y. Maki, A. Sugiyama, K. Kikuchi, and S. Seto, *ibid.*, 1093 (1975).

[9]J. T. Groves and K. W. Ma, *Tetrahedron Letters*, 909 (1974); *idem, Am. Soc.*, **96**, 6527 (1974).

[10]C. Kashima, Y. Yamamoto, and Y. Tsuda, *J. Org.*, **40**, 526 (1975).

[11]K. Mizuno, H. Okamoto, C. Pac, and H. Sakurai, *J.C.S. Chem. Comm.*, 839 (1975).

**Sodium borohydride, sulfurated, 3**, 264; **4**, 444; **5**, 601.

*Thiols and sulfides.* Benzylic halides can be converted into either a thiol or a sulfide by reduction with sulfurated sodium borohydride depending on conditions of hydrolysis. Aryl halides do not react with the reagent. Aliphatic

halides react in the same way as benzylic halides to give mainly thiols on acid hydrolysis.[1]

[1]J.-R. Brindle and J.-L. Liard, *Canad. J. Chem.*, **53**, 1480 (1975).

**Sodium chloride–Dimethyl sulfoxide, 4**, 445.

*Cyclopropyl ketones.* Cyclopropyl ketones (2) are obtained in 50–100% yield when $\alpha$-acyl-$\gamma$-butyrolactones (1) are heated in DMSO or DMF with catalytic amounts of alkali metal halides (particularly NaCl or NaBr), quaternary ammonium halides [$(CH_3)_4NBr$], or DABCO. Water is not necessary for this decarboxylative ring contraction.[1]

[1]S. Takei and Y. Kawano, *Tetrahedron Letters*, 4389 (1975).

**Sodium cyanide, 4, 446–447; 5, 606–607.**

*Addition of aldehydes to activated double bonds* (4, 447). γ-Ketonitriles can be obtained by addition of aromatic and heterocyclic aldehydes to α,β-unsaturated nitriles under catalysis by cyanide ions in DMSO.[1]

$$\text{ArCHO} \xrightarrow{\text{CN}^-} \underset{\underset{CN}{|}}{\overset{\overset{OH}{|}}{Ar\overset{|}{C}}} + R^1CH{=}\overset{\overset{R^2}{|}}{C}{-}CN \longrightarrow \left[ Ar\overset{\overset{O^-}{|}}{\underset{\underset{CN}{|}}{C}}{-}\overset{\overset{R^1}{|}}{CH}{-}\overset{\overset{R^2}{|}}{CH}CN \right] \xrightarrow[70\text{–}90\%]{-CN^-} Ar\overset{\overset{O}{\|}}{C}\overset{\overset{R^1}{|}}{CH}{-}\overset{\overset{R^2}{|}}{CH}CN$$

*Cyanoboration* (4, 445–447; 5, 446–447); *ketone synthesis.* Symmetrical ketones can be prepared in high yield by conversion of trialkylboranes into sodium trialkylcyanoborates (1) by reaction with sodium cyanide. Reaction of (1) with an electrophile (an imidoyl chloride, benzoyl chloride, or especially,

$$R_3\bar{B}CNNa^+ \xrightarrow{(CF_3CO)_2O} \left[ R_3\bar{B}{-}C{\equiv}\overset{+}{N}{-}COCF_3 \rightarrow R{-}B\overset{R}{\underset{O=C}{\diagdown}}\overset{\underset{\diagup}{C}}{\underset{\diagdown CF_3}{N}} \longrightarrow R{-}B\overset{R\diagup R}{\underset{O}{\diagdown}}\overset{}{\underset{\underset{CF_3}{|}}{N}} \right] \xrightarrow[85\text{–}100\%]{H_2O_2,\ NaOH} R_2C{=}O$$

$$(1) \qquad\qquad\qquad (a) \qquad\qquad\qquad (b) \qquad\qquad (c) \qquad\qquad\qquad (2)$$

trifluoroacetic anhydride) is accompanied by migration of two alkyl groups from boron to carbon to give an intermediate (c), which is oxidized to a symmetrical ketone (2) in high overall yield.

Unsymmetrical ketones (4) can be synthesized by use of dialkylcyanothexylborates (3), since thexyl groups are known to migrate more slowly than other alkyl groups.

$$\underset{R^2}{\overset{R^1}{\diagdown}}\overset{}{\underset{\big|}{B}}{-}CNNa^+ \xrightarrow[75\text{–}85\%]{\substack{1)\ (CF_3CO)_2O,\ -78°\rightarrow0° \\ 2)\ H_2O_2,\ NaOH}} \underset{R^2}{\overset{R^1}{\diagdown}}C{=}O$$

$$(3) \qquad\qquad\qquad\qquad\qquad (4)$$

The synthesis has been used to convert (+)-limonene (5) into the bridged ring system (6a and 6b) and the diene (7) into the fused perhydrophenanthrene-9-one (8).

$$(5) \qquad\qquad (6a) \qquad\qquad (6b) \qquad\qquad (7) \qquad\qquad (8)$$

This cyanoborate process represents an alternative to the carbonylation of trialkylboranes (2, 60).[2]

[1]H. Stetter and M. Schreckenberg, *Ber.*, **107**, 210 (1974).

[2]A. Pelter, K. Smith, M. G. Hutchings, and K. Rowe, *J.C.S. Perkin I*, 129 (1975); A. Pelter, M. G. Hutchings, K. Rowe, and K. Smith, *ibid.*, 138 (1975); A. Pelter, M. G. Hutchings, and K. Smith, *ibid.*, 142 (1975); A. Pelter, M. G. Hutchings, K. Smith, and D. J. Williams, *ibid.*, 145 (1975).

## Sodium cyanide–Hexamethylphosphoric triamide.

*Decarboalkoxylation* (5, 323). Greene and Crabbé[1] have reported a novel prostanoid synthesis that employs α-tropolone methyl ether (1) as starting material. Irradiation of (1) leads to the bicycloheptadienone (2), which is con-

verted in two steps into the keto ester (4), which is then alkylated to give (5). The key step in the synthesis is decarbomethoxylation of (5) to give (6). This reaction was accomplished in high yield with NaCN in HMPT at 70° under dry argon (1 hr.). The aldehyde group was then liberated by acid-catalyzed exchange of the acetal group with acetone to give (7). This was converted by known reactions into *dl*-11-deoxy-PGE₁ (8).

[1]A. Greene and P. Crabbé, *Tetrahedron Letters*, 2215 (1975).

**Sodium cyanoborohydride,** 4, 448–451; 5, 607–609.

*Review.*[1]

*α-Methylenelactones* (5, 608–609). Complete details for reductive amination of α-formyllactones are available.[2]

*Deoxygenation of α,β-unsaturated carbonyl compounds.* Tosylhydrazones of aliphatic aldehydes and ketones are reduced to hydrocarbons by $NaBH_3CN$ in acidic 1:1 DMF–sulfolane (5, 607).[3] Reduction of α,β-unsaturated tosylhydrazones leads to alkenes in which the double bond has migrated to the position formerly occupied by the carbonyl group. One advantage of the procedure is that the less stable positional alkene can be obtained. Thus it is possible to convert endocyclic to exocyclic alkenes and to move a double bond out of conjugation with an aromatic ring or another double bond. The migration leads preferentially to the more stable E-isomer. However, the method does not appear to be useful with cyclohexenones. Thus the tosylhydrazone of $\Delta^4$-cholestenone-3 gives a complex mixture of saturated and unsaturated products.

*Examples:*

$$C_6H_5CH=CHCH=NNHTs \xrightarrow[98\%]{NaBH_3CN} C_6H_5CH_2CH=CH_2$$

$$C_6H_5CH=CH\overset{\overset{\displaystyle NNHTs}{\|}}{C}C_6H_5 \xrightarrow{60\%} C_6H_5CH_2CH=CHC_6H_5$$

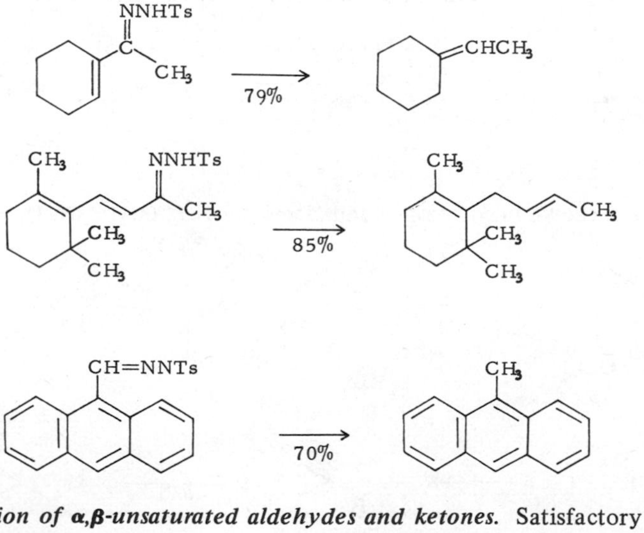

*Reduction of α,β-unsaturated aldehydes and ketones.* Satisfactory yields of allylic alcohols can be realized by reduction with this reagent of acyclic α,β-unsaturated aldehydes and ketones in acidic methanol or HMPT. Cyclic enones, however, are reduced mainly to the corresponding saturated alcohols. Reduction of these compounds to allylic alcohols is best effected with diisobutylaluminum hydride (3, 101–102).[4]

*Reductive alkylation of hydroxylamines.* Oximes and nitrones are reduced readily by sodium cyanoborohydride to N-monoalkylhydroxylamines and N,N-dialkylhydroxylamines, respectively. The oximes or nitrones need not be isolated, but can be generated at the proper pH in the presence of the reducing agent.[5]

$$R^1_{\phantom{1}}\!\!\diagdown C=O \; + \; H_2NOH \longrightarrow \; R^1_{\phantom{1}}\!\!\diagdown C=NOH \; \xrightarrow[\text{NaBH}_3\text{CN}]{\text{pH 6-8}} \; R^1_{\phantom{1}}\!\!\diagdown CH-NHOH$$

$$R^1_{\phantom{1}}\!\!\diagdown C=O \; + \; RHNOH \longrightarrow \; R^1_{\phantom{1}}\!\!\diagdown C=\overset{O^-}{\underset{}{N^+}}R \; \xrightarrow[\text{NaBH}_3\text{CN}]{\text{pH 5-6}} \; R^1_{\phantom{1}}\!\!\diagdown CH-N\!\!\diagup^{R}_{OH}$$

[1]C. F. Lane, *Synthesis*, 135 (1975).
[2]A. D. Harmon and C. R. Hutchinson, *J. Org.*, **40**, 3474 (1975).
[3]R. O. Hutchins, M. Kacher, and L. Rua, *ibid.*, **40**, 923 (1975).
[4]R. O. Hutchins and D. Kandasamy, *ibid.*, **40**, 2531 (1975).
[5]P. H. Morgan and A. H. Beckett, *Tetrahedron*, **31**, 2595 (1975).

## Sodium cyclopentadienyldicarbonylferrate, 5, 610.

*Michael-type reactions.* Complexation of methyl vinyl ketone with $C_5H_5Fe(CO)_2^+$ as shown in (1) permits conjugate addition reactions under mild

conditions as shown in equations I and II[1] [Fp = $Fe(CO)_2(C_5H_5)$].

*Deoxygenation of epoxides.* The deoxygenation of epoxides (1) to olefins (2) with retention of stereochemistry using this reagent has already been mentioned (5, 610).[2] Recently, the reaction has been modified to give the olefin

(3) with inversion of stereochemistry. Thus thermal decomposition of the intermediate alkoxide (a) results in (3), with only traces of the isomeric olefin (2). This sequence is particularly well suited to diaryl and dialkyl epoxides; it is less effective for epoxides of conjugated enones.[3]

[1] A. Rosan and M. Rosenblum, *J. Org.*, **40**, 3621 (1975).
[2] W. P. Giering, M. Rosenblum, and J. Tancrede, *Am. Soc.*, **94**, 7170 (1972).
[3] M. Rosenblum, M. R. Saidi, and M. Madhavarao, *Tetrahedron Letters*, 4009 (1975).

**Sodium N,N-dimethyldithiocarbamate,** $(CH_3)_2NC\overset{S}{\underset{SNa}{}}$ . Mol. wt. 143.21. Supplier: Eastman.

*α,β-Unsaturated aldehydes.* The first step in a synthesis of α,β-unsaturated aldehydes from alkyl halides involves the reaction of sodium N,N-dimethyldithiocarbamate with allyl chloride and methallyl bromide to form the S-allyl dithiocarbamates (1a) and (1b), respectively. The product is sulfenylated to give, after work-up, the rearranged S-γ-methylthioallyl dithiocarbamate (2a) and S-γ-methylthiomethallyl dithiocarbamate (2b). The next steps are metalation and reaction with an alkyl halide at $-78°$ to give (3), which is hydrolyzed readily to the $\alpha,\beta$-unsaturated aldehyde (4). Only the E-isomer is obtained; presumably the less stable Z-isomer isomerizes readily under hydrolytic conditions.

It is also possible to carry out the alkylation step before the sulfenylation step. In this case prolonged heat treatment is required to effect the thio-Claisen rearrangement.[1]

[1] T. Nakai, H. Shiono, and M. Okawara, *Tetrahedron Letters*, 3625 (1974).

**Sodium ethanedithiolate,** $NaSCH_2CH_2SNa$. Mol. wt. 138.16.

*Cleavage of 2-haloethyl esters.*[1] This reagent, prepared from sodium hydride and ethanedithiol, cleaves 2-haloethyl esters selectively in 80–95% yield (four examples). *Compare* **Sodium trithiocarbonate,** this volume.

[1] T.-L. Ho, *Synthesis*, 510 (1975).

**Sodium ethoxide,** 1, 1065–1073; 4, 451–452.

*Naphthpyridines.* Treatment of the dinitriles (1) and (3) with sodium ethoxide results in cyclization to aminoalkoxynaphpyridines (2) and (4).[1] Other sodium alkoxides can be used; however, yields of cyclized products decrease with sodium *n*- and *i*-propoxides and the reaction fails completely with sodium *t*-butoxide. Treatment of 2-cyanobenzyl cyanide results in dimerization.[2]

[1] F. Alhaique, F. M. Riccieri, and E. Santucci, *Tetrahedron Letters*, 173 (1975).
[2] F. Johnson and W. A. Nasutavicus, *J. Org.*, 27, 3953 (1962).

**Sodium hydride, 1**, 1075–1081; **2**, 380–383; **4**, 452–455; **5**, 610–614.

*Methylation of carotenoids.* Norwegian chemists[1] report that O-methylation with methyl iodide and sodium hydride (**4**, 455) appears to be superior to the Kuhn procedure (**2**, 274) in the case of carotenoids. They achieved methylation of phenolic carotenoids in this way for the first time. Enolic hydroxyl groups are methylated, but in this case iodinated by-products are formed. The method is the most satisfactory one for methylation of nonallylic and allylic secondary hydroxyl groups, even sterically hindered ones. Tertiary hydroxyl groups appear to be rather unreactive. Carboxylic acids are esterified.

*O-Methylation of ethynylcarbinols.*[2] Regioselective O-alkylation of even hindered acetylenic carbinols in the presence of other hydroxyl groups can be achieved by conversion to the anion with 2 eq. of sodium hydride in DMF at 20° followed by addition of excess methyl iodide.

*Arene imines.* The N-acetyl imine of the so-called K region of phenanthrene has been prepared (equation I).[3]

*Reverse Dieckmann reaction.* 1,3-Diketones of type (1), prepared as shown, can be converted into 11-membered lactones (2) by an intramolecular reverse Dieckmann reaction.[4]

(1)                                    (2)

[1] G. Nybraaten and S. Liaaen-Jensen, *Acta. Chem. Scand.*, **28 B**, 584 (1974).
[2] Z. G. Hajos and G. R. Duncan, *Canad. J. Chem.*, **53**, 2971 (1975).
[3] J. Blum, Y. Ittah, and I. Shahak, *Tetrahedron Letters*, 4607 (1975).
[4] J. R. Mahajan, *Synthesis*, 110 (1976).

## Sodium hydride—Dimethylformamide.

*Darzens synthesis of glycidic thiol esters.* Standard conditions for the Darzens reaction of $\alpha$-halo esters use sodium alkoxides as the base and the corresponding alcohols as solvent. These conditions fail with S-thioic esters, $-\overset{\text{O}}{\overset{\|}{\text{C}}}-\text{SR}$. Dagli and Wemple[1] report that glycidic thiol esters can be prepared by the Darzens method if sodium hydride or lithium bis(trimethylsilyl)amide (nonnucleophilic bases) is used as base and DMF or THF (polar, aprotic) is used as solvent. It is also important to use $\alpha$-bromo thiol esters rather than $\alpha$-chloro thiol esters.[2] The products are obtained as mixtures of the *cis* and *trans* isomers; the factors governing the stereochemical outcome are not clear.

cis and trans

[1] D. J. Dagli and J. Wemple, *J. Org.*, **39**, 2938 (1974).
[2] R. F. Borch [*Tetrahedron Letters*, 3761 (1972)] also reported that $\alpha$-bromo esters are superior to $\alpha$-chloro esters in Darzens reactions with low molecular weight aldehydes.

## Sodium hydrogen selenide, NaHSe. Mol. wt. 102.97, readily oxidized by air.

*Preparation.*[1] The reagent can be prepared in aqueous or alcoholic solution from selenium (1 eq.) and sodium borohydride (2 eq.). Addition of another equivalent of selenium results in formation of sodium diselenide ($Na_2Se_2$).

*Reduction of disulfides to thiols.*[2] Disulfides are reduced to the corresponding thiols in high yields under mild conditions. The selenium can be reused, so that essentially only sodium borohydride is consumed.

$$RSSR + 2\ HSe^- \xrightarrow{\ H^+\ } 2\ RSH + Se$$

[1] D. L. Klayman and T. S. Griffin, *Am. Soc.*, **95**, 197 (1973).
[2] T. S. Woods and D. L. Klayman, *J. Org.*, **39**, 3716 (1974).

**Sodium hypochlorite, 1,** 1084–1087; **2,** 67; **3,** 45, 243; **4,** 456; **5,** 617.

*Conversion of tosylhydrazones to ketones.*[1] Ketones can be recovered from the tosylhydrazones in 60–85% yield by treatment in $CHCl_3$ with commercial bleach (5% NaOCl) for 5 min. The reaction gives low yields in the cleavage of tosylhydrazones of aldehydes.

*Chlorination of indoles.* The reaction of an aqueous solution of sodium hypochlorite with indole in pentane (or chloroform) at $0°$ (3 hr.) gives N-chloroindole (1) in 90–92% yield. This substance rearranges to 3-chloroindole (2) when heated in *n*-butanol containing potassium carbonate, possibly through (a).[2]

(1)                     (a)                     (2)

[1] T.-L. Ho and C. M. Wong, *J. Org.*, **39**, 3453 (1974).
[2] M. De Rosa, *J.C.S. Chem. Comm.*, 482 (1975).

**Sodium iodide, 1,** 1087–1090; **2,** 384; **3,** 267; **4,** 456–457.

*Dehalogenation* (**1,** 1089). Debromination of *vic*-dibromides can be carried out conveniently in a two-phase system ($H_2O$–toluene) using hexadecyltributyl-phosphonium bromide (**5,** 322–323) as phase-transfer catalyst. The combination of sodium iodide and sodium thiosulfate (**4,** 466) is used as the reagent; sodium iodide is needed in catalytic amounts only, since it is regenerated by the thiosulfate. *meso*-Dibromides react faster than the *dl*-isomers and afford only *trans*-alkenes; the *dl*-isomers give mixtures of *cis*- and *trans*-alkenes.[1]

[1] D. Landini, S. Quici, and F. Rolla, *Synthesis*, 397 (1975).

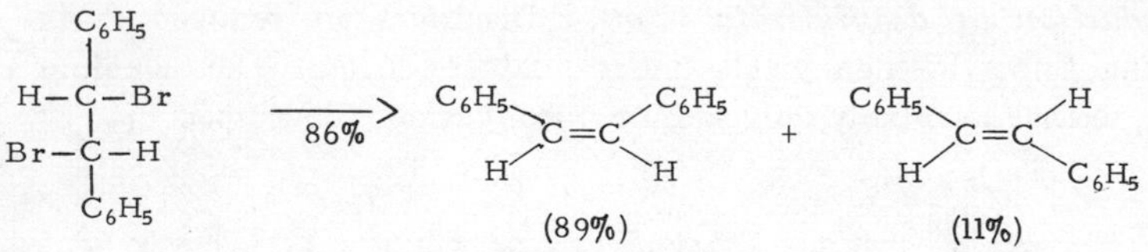

$$C_6H_5-\overset{\displaystyle H}{\underset{\displaystyle Br}{C}}-Br \quad Br-\overset{\displaystyle C}{\underset{\displaystyle C_6H_5}{\underset{|}{C}}}-H \quad \xrightarrow{86\%}$$

(89%)     +     (11%)

## Sodium Iodide–Copper.

*6, 7-Dehydrotropinones.* Reductive dehalogenation of $\alpha,\alpha'$-dibromoketones (1) in the presence of a pyrrole (2) and copper powder in acetonitrile at 25° results in formation of 6,7-dehydrotropinones (3) in 50–90% yield. The products are sensitive, but can be stored as the hydrochlorides at 0°. Tropinones can be obtained by hydrogenation of the olefinic double bond.[1]

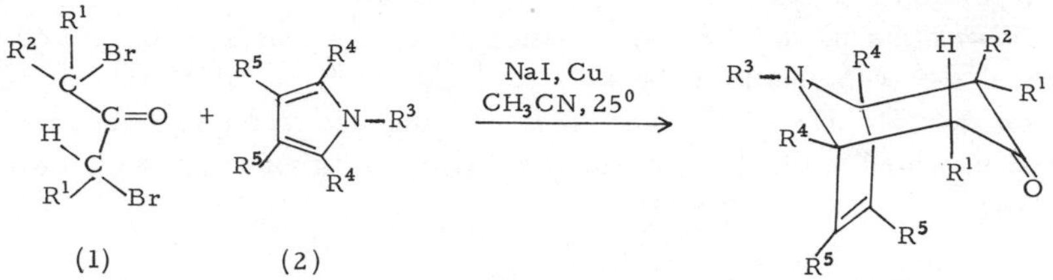

$$\text{(1)} + \text{(2)} \xrightarrow[\text{CH}_3\text{CN, 25}^0]{\text{NaI, Cu}} \text{(3)}$$

$$R^1 \rightarrow R^5 = \text{H or CH}_3$$

Hoffmann[2] has suggested that the reaction involves transformation of the dibromoketone into an allyl cation (b), which reacts by cycloaddition to (2) to give (3).

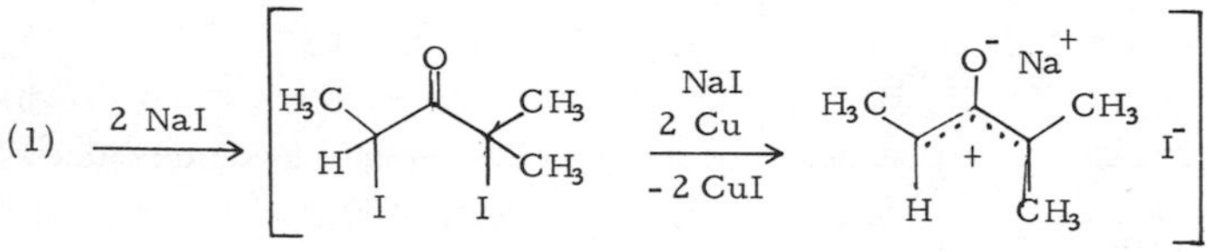

$$\text{(1)} \xrightarrow{\text{2 NaI}} \text{(a)} \xrightarrow[-\text{2 CuI}]{\substack{\text{NaI}\\\text{2 Cu}}} \text{(b)}$$

(a)                    (b)

[1] G. Fierz, R. Chidgey, and H. M. R. Hoffmann, *Angew. Chem. internat. Ed.*, **13**, 410 (1974).
[2] H. M. R. Hoffmann, *ibid.*, **12**, 819 (1973).

## Sodium iodide–Sodium acetate.

*Desoxyketoses.*[1] Epoxy ketoses are reduced by sodium iodide in acetone in the presence of acetic acid and sodium acetate at 20° to desoxyketoses in high yield:

$$C_6H_5CHBrCHBrC_6H_5 + NaOCH_3 \xrightarrow[93\%]{DMSO} C_6H_5C{\equiv}CC_6H_5$$

The starting materials are prepared by oxidation of selectively blocked epoxy sugars with $RuO_2$–$NaIO_4$ (3, 243).

[1]H. Paulsen, K. Eberstein, and W. Koebernick, *Tetrahedron Letters*, 4377 (1974).

**Sodium methoxide, 1,** 1091–1094; **2,** 385–386; **3,** 259–260; **4,** 457–459; **5,** 617–620.

*Diphenylacetylene* is prepared conveniently by dehydrobromination of dibromostilbene with sodium methoxide in DMSO[1] :

*Conversion of primary nitro compounds to aldehydes.* $RCH_2NO_2 \longrightarrow RCH$-$(OCH_3)_2$. Jacobson[2] reports a modified Nef reaction in which a solution of the primary nitroalkane in methanolic sodium methoxide is added slowly to a solution of sulfuric acid in methanol cooled to $-35°$. The acetal of the aldehyde is formed, usually in good yield. Yields are considerably lower in the conventional (aqueous) Nef reaction,[3] resulting in carbonyl compounds directly.

[1]J. R. Sowa and E. G. Calamai, *Org. Prep. Proc. Int.*, **6,** 183 (1974).
[2]R. M. Jacobson, *Tetrahedron Letters*, 3215 (1974).
[3]W. E. Noland, *Chem. Rev.*, **55,** 137 (1955).

**Sodium N-methylanilide, 1,** 1095–1096; **4,** 459.

*Ziegler–Thorpe cyclization.* A recent paper of Newman *et al.*[1] outlines several useful modifications of this reaction. Sodium N-methylanilide was prepared by the reaction of sodium sand with N-methylaniline in dry ether with isoprene rather than styrene or naphthalene as assistant. Potassium N-methylanilide (prepared from KH and N-methylaniline without isoprene) was also used, but did not prove more convenient. Sodium bis(trimethylsilyl)amide, $[(CH_3)_3Si]_2NNa$, was used to a limited extent. The paper also describes a simplified apparatus for carrying out reactions at high dilution.

[1]M. S. Newman, T. G. Barbee, Jr., C. N. Blakesley, Z. ud Din, S. Gromelski, Jr., V. K. Khanna, L.-F. Lee, J. Radhakrishnan, R. L. Robey, V. Sankaran, S. K. Sankarappa, and J. M. Springer, *J. Org.*, **40**, 2863 (1975).

**Sodium 2-methyl-2-butoxide, 1**, 1096.

*Ramberg–Bäcklund rearrangement.* This base was used in the preparation of bicyclo[2.1.1]hexene (2) by Ramberg–Bäcklund rearrangement of the α-chloro sulfone (1), available from 1,3-cyclobutanedicarboxylic acid (several steps).[1]

$$\xrightarrow[68\%]{CH_3CH_2C(ONa)(CH_3)_2}$$

(1)                                                                    (2)

[1]R. G. Carlson and K. D. May, *Tetrahedron Letters*, 947 (1975).

**Sodium methylsulfinylmethylide (Dimsylsodium), 1**, 310–313; **2**, 166–169; **3**, 123–124; **4**, 195–196; **5**, 621.

*Stereoselective Wittig reaction.* A stereoselective synthesis has been reported of the sex pheromone of the Egyptian cotton leafworm, (Z)-9,(E)-11-tetradecadien-l-yl acetate (3), by the Wittig reaction of (E)-2-pentenal (1) with (2) in dry dimethyl sulfoxide with dimsylsodium as base followed by deprotection and acetylation. The 9Z,11E- and 9E,11E-isomers were obtained in a

$H_3C$ ⟍⟋⟍$CHO$  +  $THPO(CH_2)_8CH_2\overset{+}{P}(C_6H_5)_3Br^-$

(1)                                   (2)

$$\xrightarrow[53\%]{\substack{1)\ CH_3SO\overset{-}{C}H_2Na\ \overset{+}{} \\ 2)\ H^+;\ Ac_2O,\ Py}}$$

(3)                          80:20                          (4)

ratio of 4 : 1. The isomers were separated by selective reaction of (4) with tetracyanoethylene. The overall yield of (3) was 38%.[1]

*Stevens rearrangement.* Rearrangement of the N-methyltetrahydroprotoberberinium iodide (1) to the 1-spiroisoquinoline (2) has been effected with dimsylsodium in DMSO in 80% yield.[2]

(1)                                        (2)

[1]D. R. Hall, P. S. Beevor, R. Lester, R. G. Poppi, and B. F. Nesbitt, *Chem. Ind.*, 216 (1975).
[2]S. Kano, T. Yokomatsu, E. Komiyama, S. Tokita, Y. Takahagi, and S. Shibuya, *Chem. Pharm. Bull. Japan*, **23**, 1171 (1975).

**Sodium nitrite, 1,** 1097–1101; **2,** 386–387; **4,** 459–460.

*Nitrosative cyclization.* Toxoflavin (2), an antibiotic isolated from *Psudomonas cocovenenans*, has been synthesized recently by reaction of the aldehyde hydrazone (1) with a slight excess of sodium nitrite in acetic acid at 5°.[1] This reaction was discovered by Goldner *et al.*[2] and has been used for synthesis

(1)                    (2, 35%)                    (3)

of fervenulin.[3] Toxoflavin is also obtained (35% yield) when sodium nitrite is replaced with isoamyl nitrite in $C_2H_5OH$–HOAc (3:2).

When (2) is refluxed in DMF or DMA, the 1-methyl group is lost to form (4), 8-demethylfervenulin.[4] A possible mechanism for this novel demethylation has been offered.

(4)

[1]F. Yoneda and T. Nagamatsu, *Chem. Pharm. Bull. Japan*, **23**, 2001 (1975).
[2]H. Goldner, G. Dietz, and E. Carstens, *Ann.*, **694**, 142 (1966).
[3]G. Blankenhorn and W. Pfleiderer, *Ber.*, **105**, 3334 (1972).
[4]F. Yoneda and T. Nagamatsu, *Am. Soc.*, **95**, 5735 (1973).

**Sodium nitrite–Acetic acid.**

cis-*Hydroxylation.* Treatment of tetracyclone (1) with sodium nitrite ($NaNO_2$) in aqueous acetic acid gives the cis-diol (3) in 85% yield. Study of this

(1)                          (2)                                      (3)

unexpected reaction revealed that the actual reagent is $N_2O_3$ and that the reaction proceeds through the [4 + 2] adduct (2), which can be prepared under anhydrous conditions. The adduct (2) is converted into (3) easily by aqueous solvents.[1]

An example of this cis-hydroxylation had been reported earlier by Criegee and Schröder[2]: the conversion of the nickel chloride adduct of tetramethyl-cyclobutadiene (4) into cis-dihydroxytetramethylcyclobutene (5).

(4)                                      (5)

A variety of olefins were found to be unchanged by treatment with $NaNO_2$ and aqueous acetic acid. Thus it appears that this method of hydroxylation is applicable only to reactive dienes.

[1] S. Ranganathan and S. K. Kar, Tetrahedron, **31**, 1391 (1975).
[2] R. Criegee and G. Schröder, Ann., **623**, 1 (1959).

**Sodium peroxide,** $Na_2O_2$. Mol wt. 78.00. Suppliers: Baker, Fisher, Fluka, ROC/RIC.

*Caution*: Utmost safety precautions should be observed in the reaction of peroxides with organic compounds. The reaction of sodium peroxide with water is exothermic.

*Hydrolysis of amides.*[1] Treatment of an aqueous suspension of an amide with 1 eq. of sodium peroxide at $50°$ (or on a steam bath) added gradually results in hydrolysis within 1 hr. Yields are usually greater than 85%. Extremely water-insoluble amides fail to react. The reaction is particularly valuable for heterocyclic carboxamides when the derived acid is prone to decarboxylation.

[1] H. L. Vaughn and M. D. Robbins, J. Org., **40**, 1187 (1975).

**Sodium selenophenolate,** $C_6H_5SeNa$, **5**, 273.

*Olefin synthesis.* Reich and Chow[1] have developed a method for conversion of β-hydroxy selenides (1) into olefins. The starting materials are generally prepared from alkyl phenyl selenides, readily available by reaction of tosylates,

$$C_6H_5SeCH\underset{R^2}{\overset{R^1}{|}}\xrightarrow{ClC_6H_4CO_3H} C_6H_5\overset{O}{\overset{||}{Se}}CH\underset{R^2}{\overset{R^1}{|}}\xrightarrow{LDA} C_6H_5\overset{O}{\overset{||}{Se}}\overset{R^1}{\underset{R^2}{C^-}} Li^+ + \overset{R^3}{\underset{R^4}{}}C{=}O \longrightarrow$$

(1)                    (2)                         (3)

(4)                         (5)                              (6)

mesylates, or halides with sodium selenophenolate (**5**, 275). The complete sequence involves oxidation to the selenoxide (2), deprotonation, reaction with a carbonyl compound, reduction to the hydroxy selenide (5), and, finally, reaction with methanesulfonyl chloride (3 eq.) and triethylamine. The olefin (6) is obtained by formal loss of the methanesulfonate of phenylselenenic acid, $C_6H_5SeOH$. The process is applicable even to tetrasubstituted alkenes. The stereochemistry of (6) is determined by that of the hydroxy selenide (5).

*Examples:*

[1]H. J. Reich and F. Chow, *J.C.S. Chem. Comm.*, 790 (1975).

**Sodium telluride**, $Na_2Te$. Mol. wt. 173.58. Supplier: Alfa.

*Benzocyclobutene.* The reaction of $\alpha,\alpha'$-dibromo-*o*-xylene (1) in dry DMF (20°, 16 hr.) with $Na_2Te$ gives 1,3-dihydrobenzo[*c*]tellurophene (2) in "moderate" yield. Pyrolysis of (2) at 500° under reduced pressure in a stream of He over quartz wool in a silica tube of a flow system gives benzocyclobutene (3) in 74% yield by loss of tellurium.[1]

(1)                    (2)                    (3)

The same procedure was used to prepare naphtho[*b*]cyclobutene (6) from (4) in unreported yield.

(4)                    (5)                    (6)

The same laboratory has reported[2] extrusion of tellurium from (8) when it is heated to 175° to give 3,4-homotropilidene (9, bicyclo[5.1.0]octa-2,5-diene) in quantitative yield. The sulfur analogue of (8) is stable at 200° for 16 hr.

(7)                    (8, 18%)        (9, 20%)

[1] E. Cuthbertson and D. D. MacNicol, *Tetrahedron Letters*, 1893 (1975).
[2] *Idem, J.C.S. Chem. Comm.*, 498 (1974).

## Sodium tetracarbonylferrate(+II) (Disodium tetracarbonylferrate), 3, 267– 268; 4, 461–465; 5, 624–625.

*Aldehydes from carboxylic acids.* Japanese chemists[1] have reported the synthesis of aldehydes or aldehydic acids from carboxylic anhydrides using this reagent. Yields are in the range of 30–90%.

*Examples:*

$$(C_6H_5CO)_2O \xrightarrow[73\%]{} 2\ C_6H_5CHO$$

An alternative route to aldehydes from carboxylic acids[2] involves conversion of the acid into the carboxylic ethylcarbonic anhydride (1) by reaction with ethyl chloroformate. These anhydrides react with sodium tetracarbonylferrate

$$RCOOH + ClCOOC_2H_5 \xrightarrow{N(C_2H_5)_3} \underset{(1)}{RC\overset{O}{\overset{\|}{}}{-}O\overset{O}{\overset{\|}{}}COC_2H_5} \xrightarrow[25^\circ]{Na_2Fe(CO)_4}$$

$$\left[ R\overset{O}{\overset{\|}{C}}Fe(CO)_4 \right] \xrightarrow[40-80\%]{HOAc} \underset{(2)}{RCHO}$$

(+II) to form acyl iron carbonyl complexes, which are hydrolyzed by acid to aldehydes (2).

*Cyclic ketones.* Reaction of $\alpha,\alpha'$-dibromo-*o*-xylene (1) with $Na_2Fe(CO)_4$ gives the complex (2) in 35% yield. Reaction of (2) in benzene with a slight excess of aluminum chloride at $25^\circ$ leads to 2-indanone (3), formally derived

by addition of carbon monoxide to the quinonedimethide. The complex (2) behaves as a normal aromatic compound; for example, it undergoes Friedel–Crafts acetylation to give (4), convertible into (5).

This reaction has some limitations. It apparently fails in the synthesis of strained cyclic ketones and it proceeds in low yield ($\sim$5%) in the case of complex of butadiene.[3]

*Ethyl ketones.*[4] The reactions of alkyl halides (and tosylates) with $Na_2Fe(CO)_4$ and ethylene in THF at 5 to $25^\circ$ gives ethyl ketones in high yields (80–95%). Intermediates (a)–(c) are probably involved. Actually, the hydrogen

$$RX + Na_2Fe(CO)_4 \rightarrow \underset{(a)}{\left[ Na^+RFe^-(CO)_4 \right.} \xrightarrow{C_2H_4} \underset{(b)}{Na^+RCOFe^-(C_2H_4)(CO)_3} \longrightarrow$$

$$\underset{(c)}{Na^+RCOCH_2CH_2Fe^-(C_2H_4)(CO)_3 \Big]} \xrightarrow{H^+} RCOCH_2CH_3$$

atom replacing iron in (c) does not come from the added proton source, but probably from ethylene.

This process is not useful for preparation of ketones from higher alkenes.

*Tricarbonyliron complexes of cyclobutadienes.* Roberts *et al.*[5] have reported a useful synthesis of (2) by reaction of (1) with sodium tetracarbonylferrate in THF. Use of diiron nonacarbonyl leads to (2) in lower yields.[6] The

(1)    (2)

diester can be converted into a number of other tricarbonyliron complexes of 1,2-disubstituted butadienes.

*Review.*[7] Collman cites a new practicable method of preparation of $Na_2Fe(CO)_4$ from $Fe(CO)_5$ and molten sodium in the presence of benzophenone ketyl in boiling dioxane.

[1]Y. Watanabe, M. Yamashita, T. Mitsudo, M. Tanaka, and Y. Takegami, *Tetrahedron Letters*, 3535 (1973).

[2]Y. Watanabe, M. Yamashita, T. Mitsudo, M. Igami, K. Tomi, and Y. Takegami, *ibid.*, 1063 (1975); Y. Watanabe, M. Yamashita, T. Mitsudo, M. Igami, and Y. Takegami, *Bul. Chem. Soc. Japan*, **48**, 2490 (1975).

[3]B. F. G. Johnson, J. Lewis, and D. J. Thompson, *Tetrahedron Letters*, 3789 (1974).

[4]M. P. Cooke, Jr., and R. M. Parlman, *Am. Soc.*, **97**, 6863 (1975).

[5]G. Berens, F. Kaplan, R. Rimerman, B. W. Roberts, and A. Wissner, *Am. Soc.*, **97**, 7076 (1975).

[6]The product (2) has been obtained by reaction of a dibromoanalog of (1) with $Fe_2(CO)_9$ in DMF: E. K. G. Schmidt, *Ber.*, **107**, 2440 (1974).

[7]J. P. Collman, *Accts. Chem. Res.*, **8**, 342 (1975).

**Sodium thiophenoxide**, $C_6H_5SNa$. Mol. wt. 132.16. Prepared by reaction of sodium and thiophenol in absolute ethanol.

*Protection of α-methylene-γ-butyrolactones.* Sodium thiophenoxide undergoes Michael addition to α-methylene-γ-butyrolactones in nearly quantitative yield. Deblocking involves oxidation of the sulfide to the sulfoxide and thermal elimination of benzenesulfenic acid. The yield in all steps is high.[1]

[1]P. A. Grieco and M. Miyashita, *J. Org.*, **40**, 1181 (1975).

**Sodium triacetoxyborohydride,** $NaBH(OAc)_3$. Mol. wt. 211.96.

*Preparation.* Sodium borohydride (4 eq.) suspended in benzene is treated with glacial acetic acid (3.25 eq.). A clear solution of $NaBH(OAc)_3$ is formed after refluxing for 15 min. $(N_2)$.

*Selective reduction of aldehydes.* Aldehydes, but not ketones, are reduced to alcohols in high yield by this reagent. The mild reducing property may be associated with the bulky nature and with the inductive electron-withdrawing effect of the three acetoxy groups.[1]

[1]G. W. Gribble and D. C. Ferguson, *J.C.S. Chem. Commun.*, 535 (1975).

**Sodium trichloroacetate, 1,** 1107–1118; **2,** 388–389.

*Dichlorocarbene.* Sepiol and Soulen[1] were able to double the yield of (1) by using about half the quantity of glyme used previously[2] for this reaction.

They were also able to improve the yield in the conversion of (2) into (3) by use of freshly sublimed antimony trifluoride.

[1]J. Sepiol and R. L. Soulen, *J. Org.*, **40,** 3791 (1975).
[2]S. W. Tobey and R. West, *Am. Soc.*, **88,** 2478 (1966).

**Sodium trithiocarbonate,** . Mol. wt. 154.20.

A solution of this salt is prepared from sodium sulfide in water and carbon disulfide (reflux).

*Hydrolysis of 2-haloethyl esters.* 2,2,2-Trichloroethyl esters have been used for protection of carboxyl groups because they are readily cleaved by zinc–acetic acid at 0° (**3,** 295–296). Ho has reported that 2-haloethyl esters can be cleaved by sodium trithiocarbonate in refluxing acetonitrile in 75–85% yield. The suggested mechanism is formulated.[1]

[1]T.-L. Ho, *Synthesis*, 715 (1974).

**Stannic chloride, 1,** 1111–1113; **3,** 269; **5,** 627–631.

*Enamines.* Recently, two laboratories[1,2] have reported that the classical method for preparation of enamines (pyrrolidine, *p*-TsOH catalysis, **1,** 972) failed in the case of 2-norbornanone (1), probably because of strain attending forma-

(1)                    (2)    $N(CH_3)_2$

tion of the double bond. One group[1] effected the desired transformation by a method introduced by Nelson and Pelter[3] for preparation of enamines of hindered ketones. The ketone (1) was heated with tris(dimethylamino)borane, $B[N(CH_3)_2]_3$, dimethylamine, and potassium carbonate in an autoclave at about $100°$ for 110 hr. The desired enamine (2) was obtained in 62% yield. The other group[2] treated (1) with dimethylamine and stannic chloride in pentane at room temperature (72 hr.) and obtained somewhat impure (2) in 58% yield. The actual reagent in this case may be $Sn[N(CH_3)_2]_4$.

*Cyclization of polyenes.*[4] Treatment of *trans*-geranylgeranic acid chloride (1) with 1 eq. of stannic chloride in methylene chloride at $-78°$ (1.5 hr.) effects cyclization to (2) in 71% yield. The reaction is of interest because it may be representative of the biogenesis of the 14-membered cembrene diterpenes.

(1)                    (2)

[1] K. G. R. Sundelin, R. A. Wiley, R. S. Givens, and D. R. Rademacher, *J. Med. Chem.*, **16**, 235 (1973).
[2] D. W. Boerth and F. A. Van-Catledge, *J. Org.*, **40**, 3319 (1975).
[3] P. Nelson and A. Pelter, *J. Chem. Soc.*, 5142 (1965).
[4] T. Kato, T. Kobayashi, and Y. Kitahara, *Tetrahedron Letters*, 3299 (1975).

**Stannous chloride, 1,** 1113–1116; **2,** 389; **5,** 631–632.

*Monomethylation of* cis-*glycols by diazomethane.*[1] Stannous chloride di-hydrate is a particularly active catalyst for the methylation of adenosine (1) by

(1)                    (2, 44%)                    (3, 56%)

diazomethane. 2′-O-Methyladenosine (2) and 3′-O-methyladenosine (3) are obtained quantitatively. Stannic chloride is equally effective, as are ferrous and ferric chloride. Boron trifluoride etherate even in DMF (1, 193) is ineffective in this case.

*Denitration of* vic-*dinitro compounds.* trans-Stilbene (2) and 9,9′-bifluorenylidene (4) can be obtained by denitration of (1) and (3), respectively, with anhydrous stannous chloride or the dihydrate (4 eq.). This method fails in the case of aliphatic vic-dinitro compounds.[2]

(1)                                    (2)

(3)                                    (4)

[1] M. J. Robins, A. S. K. Lee, and F. A. Norris, *Carbohydrate Res.*, 41, 304 (1975).
[2] K. Fukunaga, *Synthesis*, 442 (1975).

**Succinimidodimethylsulfonium tetrafluoroborate,** (1). Mol. wt. 233.03, m.p. 167–169°.

The salt is prepared by treatment of a suspension of the corresponding chloride (4, 87–90) in $CH_2Cl_2$ with a solution of $AgBF_4$ in $CH_3CN$ at −20°.

*Oxidation of hydroquinones.* Hydroquinones (o- or p-) are oxidized to o- or p-quinones in high yield by (1) or the corresponding chloride in $CH_2Cl_2$ or $CH_3CN$ at −20 to −50° followed by addition of triethylamine. The oxidation involves the intermediate (a), which is converted by base into either the ylide (b) or the phenoxide (c).[1]

[1] J. P. Marino and A. Schwartz, *J.C.S. Chem. Comm.*, 812 (1974).

**N-Sulfinylaniline**, $C_6H_5NSO$. Mol. wt. 139.18, b.p. 88–95°/17–20 mm.
*Preparation*[1]:

$$C_6H_5NH_2 \xrightarrow[91-94\%]{SOCl_2} C_6H_5N=S=O$$

*Degradation of α-amino acids.* α-Amino acids (preferably the sodium salts) are converted into carbonyl compounds by reaction with N-sulfinylaniline in DMSO (80°) (equation I). Esters are degraded to α-keto esters (rather low

$$(I) \quad \begin{matrix} R^1 \\ \diagdown \\ R^2 \diagup \end{matrix} \underset{NH_2}{\overset{|}{C}}COONa \ + \ C_6H_5NSO \xrightarrow[35-75\%]{DMSO} \begin{matrix} R^1 \\ \diagdown \\ R^2 \diagup \end{matrix} C=O$$

$$(II) \quad \underset{NH_2}{R\overset{|}{C}HCOOC_2H_5} \ + \ C_6H_5NSO \xrightarrow[\sim 30\%]{DMSO} RCOCOOC_2H_5$$

yield) (equation II). The paper by Taguchi *et al.* discusses several possible pathways.[2]

[1] P. Rajagopalan, B. G. Advani, and C. N. Talaty, *Org. Syn., Coll. Vol.*, **5**, 504 (1975).
[2] T. Taguchi, S. Morita, and Y. Kawazoe, *Chem. Pharm. Bull. Japan*, **23**, 2654 (1975).

**Sulfur**, **1**, 1118–1121; **3**, 273–275; **5**, 632–633.
    *2-Pyrones.* 2-Pyrones can be prepared by cyclization of δ-keto acids to give

(1)                    (2)                    (3)

3,4-dihydro-2-pyrones (2). Dehydrogenation of (2) to (3) with high-potential quinones or by bromination–dehydrobromination is not successful. Dehydrogenation can be carried out with sulfur at 185–195°. When R or $R' = C_6H_5$, palladium on charcoal can be used equally well.[1]

[1]M. Trolliet, R. Longeray, and J. Dreux, *Bull. soc.*, 1484 (1974).

### Sulfur–Hexamethylphosphoric triamide.

*Oxidation of aromatic compounds.* Lawesson *et al.* have reported some interesting oxidations with sulfur in HMPT at 190°. A variety of aromatic substrates are converted into N,N-dimethylthiocarboxamides.[1]

*Examples:*

$$C_6H_5CHO \longrightarrow C_6H_5\overset{\overset{S}{\parallel}}{C}N(CH_3)_2 \quad + \quad By\text{-}products$$
$$(58\%)$$

$$C_6H_5\overset{\overset{O}{\parallel}}{C}CH_3 \longrightarrow C_6H_5\overset{\overset{S}{\parallel}}{C}N(CH_3)_2 \quad + \quad C_6H_5CH_2\overset{\overset{S}{\parallel}}{C}N(CH_3)_2$$
$$(71\%) \qquad\qquad (10\%)$$

$$C_6H_5CH_2NH_2 \longrightarrow C_6H_5\overset{\overset{S}{\parallel}}{C}NHCH_2C_6H_5$$
$$(90\%)$$

$$C_6H_5CH_2NHC_2H_5 \longrightarrow C_6H_5\overset{\overset{S}{\parallel}}{C}NHC_2H_5 \quad + \quad C_6H_5\overset{\overset{S}{\parallel}}{C}N(CH_3)_2$$
$$(65\%) \qquad\qquad (7\%)$$

$$(53\%)$$

Carboxylic acid esters are also converted into N,N-dimethylthiocarboxamides under these conditions[2]:

$$RCOOR' \xrightarrow[40\text{-}90\%]{S,\,HMPT,\,185°} R\overset{\overset{S}{\parallel}}{C}N(CH_3)_2$$

*Reductions.* Unexpectedly, some substrates, (1)–(5), are reduced to E-stilbene (6) as the main product by sulfur in HMPT at 205°.[3]

$$\underset{(1)}{C_6H_5\overset{\displaystyle O}{\overset{\|}{C}}CH_2C_6H_5} \qquad \underset{(2)}{C_6H_5\overset{\displaystyle O}{\overset{\|}{C}}-\overset{\displaystyle OH}{\overset{|}{C}}HC_6H_5} \qquad \underset{\substack{(3,\ X=H,\ Y=OH) \\ (4,\ X=Y=OH)}}{C_6H_5\overset{\displaystyle X}{\overset{|}{C}}H\overset{\displaystyle Y}{\overset{|}{C}}HC_6H_5} \qquad \underset{(5)}{C_6H_5\overset{\displaystyle O}{\overset{\|}{C}}-\overset{\displaystyle O}{\overset{\|}{C}}C_6H_5}$$

$$\Big\downarrow\ S, HMPT, 205^\circ$$

$$\underset{(6)}{\overset{C_6H_5}{\underset{H}{}}C=C\overset{H}{\underset{C_6H_5}{}}}$$

[1] J. Perregaard, I. Thomsen, and S.-O. Lawesson, *Acta Chem. Scand.*, **B29**, 538 (1975).
[2] J. Perregaard and S.-O. Lawesson, *ibid.*, **B29**, 604 (1975).
[3] J. Perregaard, I. Thomsen, and S.-O. Lawesson, *ibid.*, **B29**, 599 (1975).

**Sulfur dioxide, 1**, 1122; **2**, 292; **4**, 469; **5**, 633.

*Cycloaddition.* Sulfur dioxide undergoes [4 + 2] cycloaddition with conjugated diallenes at room temperature. The reaction involves rotation of the two

$$\underset{R}{\overset{R^1}{}}C=C=C\underset{\underset{R^2}{}}{\overset{R^2}{}}\ \ C=C=C\underset{\underset{R^1}{}}{\overset{R}{}} \quad \xrightarrow[75-85\%]{SO_2} \quad \underset{R}{\overset{R^1}{}}C\overset{\overset{\displaystyle R^2\ \ R^2}{}}{=}\underset{\underset{O_2}{S}}{}\overset{}{=}C\underset{R}{\overset{R^1}{}}$$

most bulky groups away from each other.[1]

*Azetidine-2-ones.* Benzylideneaniline (1) reacts with haloketenes (2) in liquid $SO_2$ to give azetidine-2-ones (3) and/or 4-oxo-1,3-thiazolidine-1,1-dioxides (4). The latter products are formed when X or Y is Cl, Br, or SR.[2]

$$\underset{(1)}{C_6H_5N=CHC_6H_5}\ +\ \underset{(2)}{\overset{X}{\underset{Y}{}}C=C=O} \quad \xrightarrow{\underset{CH_2Cl_2}{SO_2}} \quad (3) \quad + \quad (4)$$

[1] K. Kleveland and L. Skattebøl, *Acta Chem. Scand.*, **29B**, 827 (1975).
[2] D. Belluš, *Helv.*, **58**, 2509 (1975).

**Sulfuric acid, 4**, 470–472; **5**, 633–639.

*2-Hydroxy-3-alkylcyclopent-2-ene-1-ones.* The substances (2) are available in 50–70% yield by treatment of 2,3-epoxy-2-alkylcyclopentanones (1, obtained by alkaline epoxidation of 2-alkylcyclopent-2-ene-1-ones) with acetic acid containing 2% of concentrated sulfuric acid at 55°. Use of boron trifluoride etherate, the classical reagent for rearrangement of cyclic 2,3-epoxy ketones (1,

72), gives the same product, but in low yield. A carbonium ion intermediate is postulated.[1]

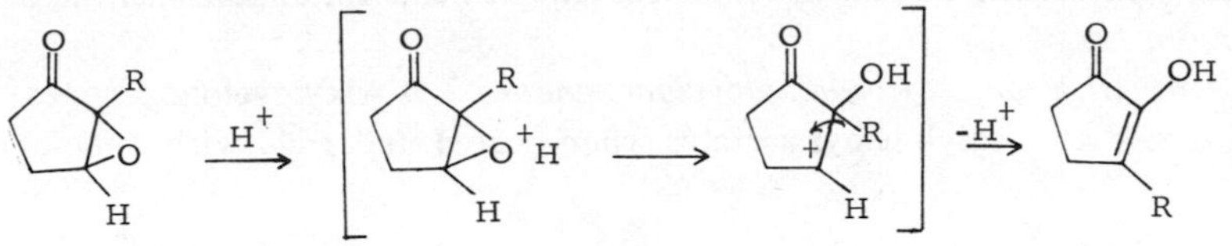

(1)                                        (2)

*6H-1,3,5-Oxathiazines.* These new heterocycles have been prepared by the reaction of aliphatic aldehydes with thioamides in acetic acid containing 100% sulfuric acid. In one case boron trifluoride etherate in chloroform was used.[2]

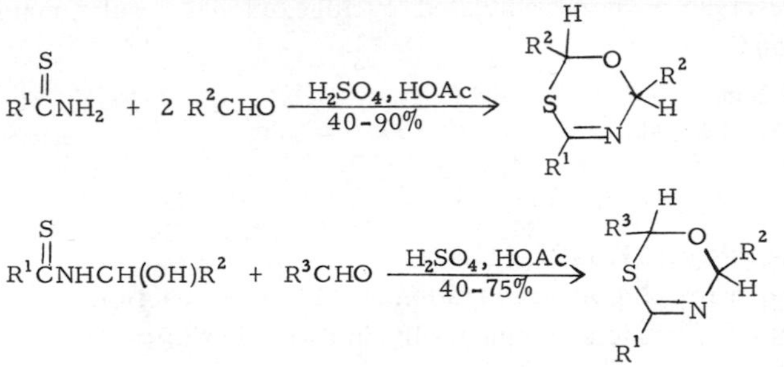

*Stereoselective allylic rearrangement.*[3] Irradiation of either chanoclavine-I (1) or isochanoclavine-I (2) (ergot alkaloids) in the presence of sulfuric acid

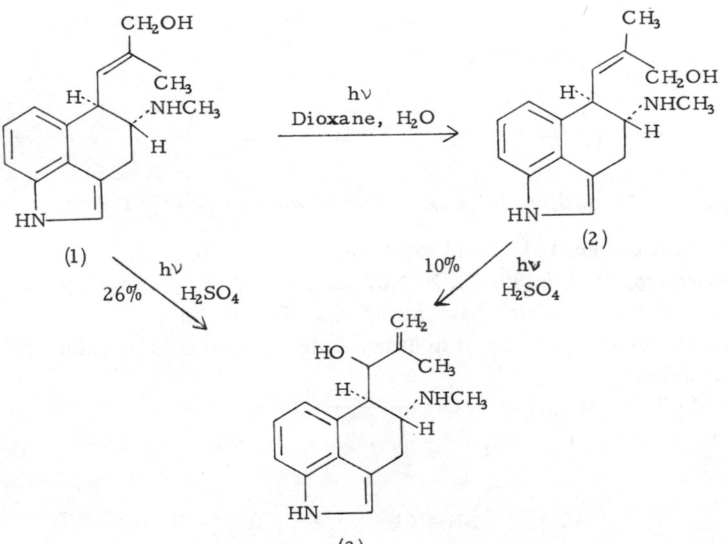

(1)

(2)

(3)

results in allylic rearrangement to paliclavine (3), a metabolite of a strain of *Claviceps paspali.*

Fehr and Stadler discuss the chemical requirements for and mechanism of this reaction.

*Aromatization   of   α-alkylcyclohexanones.*[4] α-Alkylcyclohexanones are aromatized to *o*-alkylphenyl acetates when heated under $N_2$ with 2 moles each

of concentrated sulfuric acid and acetic anhydride together with acetic acid. Similar treatment of non-α-alkylated cyclohexanones results mainly in aldol condensation.

[1] A. Barco, S. Benetti, G. P. Pollini, and R. Taddia, *Synthesis*, 104 (1975).
[2] C. Giordano and A. Belli, *Synthesis*, 789 (1975).
[3] T. Fehr and P. A. Stadler, *Helv.* **58**, 2484 (1975).
[4] M. S. Kablaoui, *J. Org.*, **39**, 2126 (1974).

**Sulfur monochloride, 1,** 1122–1123.

*Reaction with 2,4,6-tri-t-butylaniline* (1). The reaction of sulfur monochloride with this hindered amine results in formation of two tautomers, (2) and (3).[1]

[1] Y. Inagaki, R. Okazaki, and N. Inamoto, *Tetrahedron Letters*, 4575 (1975).

**Sulfur tetrafluoride, 1,** 1123–1125; 2, 392–393; 5, 640.

*Fluorodehydroxylation,* COH → CF. Ordinarily, $SF_4$ shows low reactivity to alcohols. Merck chemists[1] have found that $SF_4$ in liquid HF at $-78°$ readily replaces the hydroxyl group of hydroxy amino compounds by fluorine.

*Examples:*

$$\xrightarrow{32\%}$$

[1] J. Kollonitsch, S. Marburg, and L. M. Perkins, *J. Org.*, **40**, 3808 (1975).

**Sulfur trioxide–Dioxane, 1,** 1126–1127; **2,** 393; **5,** 643.

*Insertion reaction with silyl azides.* Sulfur trioxide (as the 1:1 complex with dioxane) undergoes insertion with silyl azides and with compounds having a tri-

$$(CH_3)_3Si-N_3 + SO_3 \xrightarrow[45\%]{} N_3SO_2-O-Si(CH_3)_3$$

$$(CH_3)_3Si-NH-Si(CH_3)_3 + SO_3 \xrightarrow[47\%]{} (CH_3)_3Si-NH-SO_2OSi(CH_3)_3$$

$$C_6H_5-\underset{\underset{CH_3}{|}}{N}-Si(CH_3)_3 + SO_3 \longrightarrow C_6H_5-\underset{\underset{CH_3}{|}}{N}-SO_2-OSi(CH_3)_3$$

methylsilyl group attached to nitrogen.[1]

[1] H. R. Kricheldorf and E. Leppert, *Synthesis*, 49 (1975).

**Sulfuryl chloride, 1,** 1128–1131; **2,** 394–395; **3,** 276; **4,** 474–475; **5,** 641.

*α,β-Unsaturated esters.* Japanese chemists[1] have extended the synthesis of olefins from β-hydroxy sulfoxides of Durst (**5**, 127–129) to a synthesis of α,β-unsaturated esters. A β-hydroxy sulfoxide (3) is prepared by reaction of an aldehyde or ketone with the Grignard reagent (2) derived from ethyl α-(t-butyl-sulfinyl)acetate (1). Reaction of (3) with sulfuryl chloride at room temperature

for 10 min. gives the α,β-unsaturated ester (4). Usually the ester obtained in the case of an aldehyde has the E-configuration. However, the E and Z-esters are obtained in a ratio of about 1:1 in the case of ethyl methyl ketone.

[1] J. Nokami, N. Kunieda, and M. Kinoshita, *Tetrahedron Letters*, 2179 (1975).

**Sulfuryl chlorofluoride**, $FSO_2Cl$. Mol. wt. 118.52, b.p. 7.1°, m.p. −124.7°. Supplier: ROC/RIC.

*Chlorination.*[1] This reagent reacts with cyclohexenones to form 2-chloro-3-oxocyclohexenes in 50–75% yield (equation I). The reaction of sulfuryl

(I)

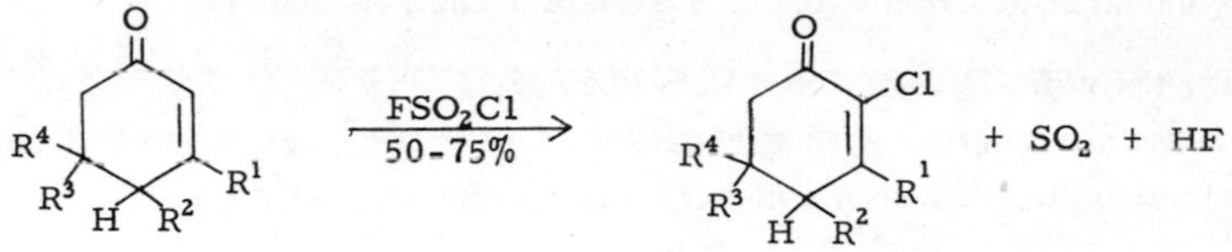

chloride with cyclohexenones leads to an array of products (equation II). The main product is formed by chlorination followed by addition to the double

(II)

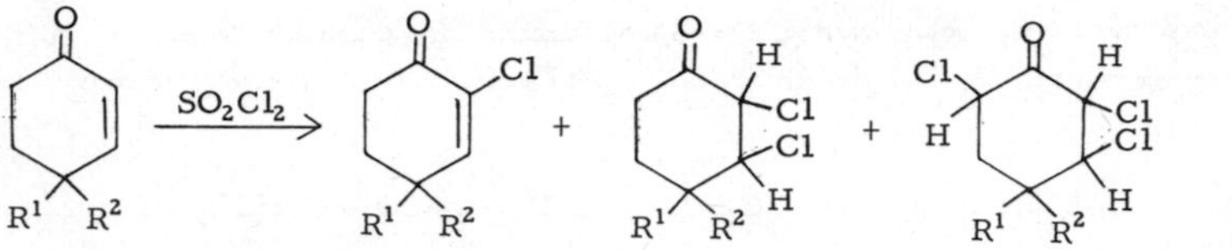

(usually predominates)

bond of the hydrogen chloride formed.

[1] M. F. Grenier-Loustalot, P. Iratcabal, F. Métras, and J. Petrissans, *Synthesis*, 33 (1976).

# T

1,3,4,6-Tetraacetylglycouril, $o$⟨structure⟩$o$ (1). Mol. wt. 310.21, m.p. 234–238° dec. Supplier: Aldrich.

This substance is prepared by acetylation of glycouril with acetic anhydride catalyzed with $HClO_4$.

*Acetylation of amines, phenols, and thiols.* This reagent readily transfers two acetyl groups to amines, phenols, and thiols under mild conditions (25°) with conversion to 1,4-diacetylglycouril. Aliphatic primary amines react most readily; in fact, primary amino groups can be selectively acetylated in the presence of a secondary amino group. Acetylation of aromatic amines is slower, but can be catalyzed by an acid (*e. g.*, acetic acid). Phenols are also acetylated, but aminophenols can be selectively N-acetylated. Thiophenols and alkyl thiols are acetylated readily, but alcohols react only slowly, even at elevated temperatures.[1,2]

[1] D. Kühling, *Ann.*, 263 (1973).
[2] C. Hase and D. Kühling, *ibid.*, 95 (1975).

## 2,4,4,6-Tetrabromocyclohexa-2,5-dienone, 4, 476–477; 5, 643–644.

*Oxidation of thiols to disulfides.* Sodium thiolates (2 eq.) are oxidized to disulfides by this reagent (1 eq.). The reaction is exothermic and rapid. Benzene is used as solvent because the by-products are insoluble in this medium.[1]

$$ RSNa + RSBr \xrightarrow[80-90\%]{} RSSR + NaBr $$

[1] T.-L. Ho, T. W. Hall, and C. M. Wong, *Synthesis*, 872 (1974).

## Tetra-*n*-butylammonium azide, $(C_4H_9)_4NN_3$. Mol. wt. 284.48, m.p. 80°.

This relatively safe azide has been prepared from hydrazoic acid and tetra-*n*-butylammonium hydroxide.[1] It is conveniently prepared in quantitative yield from a mixture of tetra-*n*-butylammonium hydroxide and sodium azide containing excess NaOH in water by extraction with methylene chloride, which extracts the tetra-*n*-butylammonium azide into the organic phase.[2]

*Acyl azides; isocyanates.*[2] The reagent undergoes metathesis readily with carboxylic acid chlorides in benzene or toluene. If the solution is heated to 50–90°, Curtius rearrangement leads to isocyanates in 60–90% yield.

[1] G. Opitz, A. Griesinger, and H. W. Schubert, *Ann.*, **665**, 91 (1963); V. Gutmann, G. Hampel, and O. Leitmann, *Monatsh.*, **95**, 1034 (1964).

[2] A. Brändström, B. Lamm, and I. Palmertz, *Acta Chem. Scand.*, **28B**, 699 (1974).

**Tetra-*n*-butylammonium borohydride,** $(C_4H_9)_4NBH_4$. Mol. wt. 257.31; soluble in $CH_2Cl_2$, insoluble in ether.

*Preparation.*[1] Tetraalkylammonium borohydrides can be prepared by addition of a slight excess of sodium borohydride to a solution or suspension of a tetraalkylammonium hydrogen sulfate in an aqueous solution of NaOH. The resulting tetraalkylammonium borohydride is extracted with methylene chloride. The solid salt can be obtained by evaporation of the methylene chloride and crystallization from ethyl acetate. These salts are mild reducing agents. They are converted into diborane and a tetraalkylammonium halide by treatment in methylene chloride with an alkyl halide (methyl iodide, ethyl bromide). The advantage of generation of diborane in this way is that anhydrous methylene chloride is easily obtained.

*Conversion of carboxylic acids to aldehydes.* Raber and Guida[2] have reported a general method for partial reduction of carboxylic acids to aldehydes. The acid is converted into the 2-methoxyethyl ester (1), which is then alkylated with triethyloxonium tetrafluoroborate in $CH_2Cl_2$. Crystalline 1,3-dioxolanium tetrafluoroborates (2) are obtained by initial alkylation of the methoxy oxygen

(a)

atom. The salts (2) are conveniently reduced by tetra-*n*-butylammonium borohydride in methylene chloride at $-78°$ under nitrogen. Overall yields of 1,3-dioxolanes (3) are in the range 75–85%.

This present procedure is an alternative to one described by Johnson and Rickborn[3] that also involves reduction of an acid to an acetal.

[1] A. Brändström, U. Junggren, and B. Lamm, *Tetrahedron Letters*, 3173 (1972).
[2] D. J. Raber and W. C. Guida, *Synthesis*, 808 (1974).
[3] M. R. Johnson and B. Rickborn, *Org. Syn.*, 51, 11 (1971).

**Tetra-*n*-butylammonium halides,** $[CH_3(CH_2)_3]_4 \overset{+}{N} X^-$.

*Alkyl iodides* $\rightleftharpoons$ *alkyl chlorides.*[1] These conversions can be effected in high yield by ion-pair extractions. The alkyl chloride (0.1 mole), sodium iodide (0.11 mole), and tetra-*n*-butylammonium iodide (0.014 mole) are refluxed in chlorobenzene–water for 90 min. Yields of the alkyl iodide are 90% or more. The reverse reaction can be effected as follows: The alkyl iodide (0.1 mole) and anhydrous tetra-*n*-butylammonium chloride (0.108 mole) are heated under reduced pressure at a temperature below $140°$. The alkyl chloride is formed in over 90% yield in a few minutes.

[1] A. Brändström and H. Kolind-Andersen, *Acta Chem. Scand.*, B29, 201 (1975).

**Tetra-*n*-butylammonium hydrogen sulfate,** $[CH_3(CH_2)_3]_4 \overset{+}{N} HSO_4^-$. Mol. wt. 339.54, m.p. $169–171°$. Suppliers: Aldrich, Fluka.

*Methylene diesters of carboxylic acids,* $(RCOO)_2 CH_2$.[1] These esters can be prepared in 60–90% yield by dissolving equimolar amounts of the acid and the salt in aqueous NaOH; the salt of the acid forms and is extracted into methylene

$$2\ RCOOH \longrightarrow 2\ RCOO^- \overset{+}{N}(\underline{n}\text{-}Bu)_4 + CH_2Cl_2 \xrightarrow[60-90\%]{\Delta} (RCOO)_2CH_2$$

chloride. The solution is dried and then refluxed for about 4 days. The method fails with dicarboxylic acids (succinic acid, phthalic acid).

*Ether synthesis.*[2] Ethers can be prepared in high yield by the reaction of primary or secondary alcohols with primary alkyl chlorides in a two-phase system composed of excess 50% aqueous sodium hydroxide and the alkyl chloride functioning as the organic solvent. Tetra-*n*-butylammonium hydrogen sulfate (3–5 mole %) is used as the phase-transfer catalyst. Primary alcohols are alkylated in 3–4 hr; secondary alcohols require longer reaction times or larger amounts of catalyst. Yields are unsatisfactory with secondary alkyl chlorides.

$$ROH + R'Cl \xrightarrow[\text{Cat.}]{50\%\ NaOH} ROR' + NaCl + H_2O$$

The counterion has a pronounced effect on the rate of reaction; $HSO_4^-$ is more effective than $I^-$, which in turn, is more effective than $ClO_4^-$.

*N-Alkylation of indole.* Indole is alkylated exclusively on nitrogen in a two-phase system with a phase-transfer catalyst.[3]

[1] K. Holmberg and B. Hansen, *Tetrahedron Letters*, 2303 (1975).
[2] H. H. Freedman and R. A. Dubois, *ibid.*, 3251 (1975).
[3] A. Barco, S. Benetti, G. P. Pollini, and P. G. Baraldi, *Synthesis*, 124 (1976).

**Tetra-*n*-butylammonium iodide,** 5, 646–647.

*Solid-liquid phase-transfer catalysis.* The final step in a recent synthesis of the β-lactam (2) involved ring closure of the amide chloride (1). This was accomplished with sodium hydride in methylene chloride containing this quaternary salt (0.1 eq.). This β-lactam synthesis is of interest since it is based on

a substituted L-cysteinyl-L-valine peptide precursor and may be related to the biosynthesis of penicillins.[1]

*Alkylation of tosylmethylisocyanide.*[2] Tosylmethylisocyanide (4, 514–516; 5, 684–685) can be monoalkylated in high yield (75–95%) by primary alkyl halides (including alkyl chlorides and benzyl bromide) under phase-transfer conditions with either this quaternary salt or the less reactive benzyltriethylammonium chloride. Yields are lower with secondary alkyl halides. With more reactive alkyl halides, the reaction is conducted at $0°$ to avoid dialkylation.

*Wittig phosphonate olefin synthesis.*[3] The phosphonate synthesis of α,β-unsaturated nitriles and esters can be carried out by addition of both substrates simultaneously to a two-phase system of methylene chloride and 50% aqueous

sodium hydroxide containing the phase-transfer catalyst, tetra-*n*-butylammonium iodide. The reaction is strongly exothermic and is complete within 15 min. The reaction is slow when $R^3$ = 2-pyridyl. The olefin is obtained as a mixture of Z- and E-isomers, in which the latter predominates.

[1] J. E. Baldwin, A. Au, M. Christie, S. B. Haber, and D. Hesson, *Am. Soc.*, 97, 5957 (1975).
[2] A. M. van Leusen, R. J. Bouma, and O. Possel, *Tetrahedron Letters*, 3487 (1975).
[3] C. Piechucki, *Synthesis*, 869 (1974).

**Tetra-*n*-butylammonium *p*-toluenesulfinate, (1). Mol. wt. 397.67.**

This salt can be prepared[1] by the ion-pair extraction method[2] from tetra-*n*-butylammonium bromide and sodium *p*-toluenesulfinate (equation I).

(I)   $(C_4H_9)_4\overset{+}{N}\ \overset{-}{Br}$  +  $NaSO_2$—⟨ ⟩—$CH_3$  $\xrightarrow[-NaBr]{H_2O-CH_2Cl_2}$  $(C_4H_9)_4\overset{+}{N}\ \overset{-}{O_2S}$—⟨ ⟩—$CH_3$
                                                                                          (1)

*Sulfone synthesis.* The salt (1) reacts with alkyl halides in THF at 20–40° (2–4 hr.) to give alkyl 4-methylphenyl sulfones, usually in yields of 60–90%. The analogous reaction with sodium *p*-toluenesulfinate itself proceeds in much lower yields, mainly because O-alkylation competes with S-alkylation.[1]

[1] G. E. Vennstra and B. Zwaneburg, *Synthesis*, 519 (1975).
[2] A. Brändström, B. Lamm, and I. Palmertz, *Acta Chem. Scand.*, 28B, 699 (1974).

**Tetracyanoethylene (TCNE), 1, 1133–1136; 2, 397; 5, 647–648.**

*Cycloaddition* (5, 647–648). The first step in a synthesis of (+)-2,3-dihydro-triquinacenone-2 (5) involves addition of TCNE to cyclooctatetraeneiron tricarbonyl (1) in a 1,3-bonding scheme. The yield is surprisingly high (96%). The iron atom is then extruded by oxidation with ceric ammonium nitrate (4, 72)

to form (3), again in high yield. Although TCNE adducts are usually recalcitrant to hydrolysis, (3) is converted into the lactone acid (4) by concd. HCl (130°, 11 hr.). The product is converted by standard reactions into (5). Ultimately, TCNE is the source of $C_2$ and $C_3$ of (5). The absolute configuration of

(5) has been established as 1S by chiroptical measurements of the $n \longrightarrow \pi^*$ transitions in the region 260–360 nm.[1]

TCNE undergoes an unusual cycloaddition reaction with dispiro[4.2.2.0] deca-7,9-diene (6) in THF at 0°. The structure of the adduct (7) has been elucidated by X-ray analysis.[2]

(6)

(7)

[1] L. A. Paquette, W. B. Farnham, and S. V. Ley, *Am. Soc.*, **97**, 7272 (1975).
[2] D. Kaufmann, A. de Meijere, B. Hingerty, and W. Saenger, *Angew. Chem. internat. Ed.*, **14**, 816 (1975).

**Tetraethylammonium bromide**, $(C_2H_5)_4N^+Br^-$. Mol. wt. 210.16, m.p. 285–287° dec. Suppliers: Aldrich, Eastman, Fisher, others.

*α-Linked disacharrides.* β-Glucopyranosides are fairly readily available by the Koenigs–Knorr reaction of an alcohol with the anomeric center of gluco-pyranosyl halides. The reason for this result is that the α-halide, even though axial, is the more stable anomer and reacts with inversion to form the β-glucoside. On the other hand, the β-halide is the more reactive anomer. Lemieux *et al.*[1] have shown that added halide ions promote anomerization of the α-halide to the β-halide by way of ion-pair intermediates and that α-glycosides can then be obtained preferentially. Thus the reaction of tetra-O-benzyl-α-D-glucopyranosyl bromide (1) with methanol in methylene chloride containing an equimolar amount of tetraethylammonium bromide (25°, 40 hr.) gives methyl tetra-O-benzyl-α-D-glucopyranoside (3) in 95% yield, formed with inversion from (2).

(1)                    (2)

(3)

Lemieux and co-workers[2] have used this halide-ion catalysis for the practical synthesis of several α-linked disacharrides as well as of several complex oligosach-arrides of biological importance.

[1] R. U. Lemieux, K. B. Hendriks, R. V. Stick, and K. James, *Am. Soc.*, 97, 4056 (1975).
[2] R. U. Lemieux and H. Driguez, *ibid.*, 97, 4063, 4069 (1975); R. U. Lemieux, D. R. Bundle, and D. A. Baker, *ibid.*, 97, 4076 (1975).

**Tetraethylammonium cyanide**, $(C_2H_5)_4N^+CN^-$. Mol. wt. 156.26. Sensitive to light, hygroscopic. Supplier: Fluka.

*Preparation.*[1]

*Synthesis of nitriles.*[2] The reactions of alkyl or acyl halides with this salt in a polar solvent ($CH_2Cl_2$, $CH_3CN$, or DMSO) at -15 to 50° gives the corresponding nitriles in yields generally in the range 50–85%. The alkyl halide can be primary, secondary, or tertiary, saturated or α,β-unsaturated. Benzyl halides also undergo this reaction. Work-up is simple. The solvent is removed; the nitrile

$$(C_2H_5)_4N^+CN^- + RX \longrightarrow RCN + (C_2H_5)_4N^+X^-$$

is dissolved in ether and filtered from the tetraethylammonium halide.

[1] J. Solodar, *Syn. Inorg. Metal-org. Chem.*, 1, 141 (1971).
[2] G. Simchen and H. Kobler, *Synthesis*, 605 (1975).

**Tetraethylthiuram disulfide**, $(C_2H_5)_2NC(S)SSC(S)N(C_2H_5)_2$ (1). Mol. wt. 296.54, m.p. 71–72°. Supplier: Aldrich.

*(E)-α,β-Unsaturated aldehydes.* The anion of 2-alkenyl N,N-diethyldithio-carbamates (2) reacts with (1) at the α-carbon atom to form a product (a) that undergoes facile [3,3] sigmatropic rearrangement to form (3). The products are hydrolyzed to aldehydes (4) by red mercuric oxide and boron trifluoride etherate (3, 136).[1]

[1] I. Hori, T. Hayashi, and H. Midorikawa, *Synthesis*, 727 (1975).

**Tetrahydrofurane, 1**, 1140–1141; **2**, 398; **3**, 398; **4**, 278; **5**, 649.

*Aryl—alkyl coupling.* The reaction of aryl bromides with alkyllithium compounds in ether results mainly in halogen—metal exchange. However, when THF is used as solvent aryl bromides react with primary alkyllithiums (1 hr., 25°) to give the cross-coupled products in 50–70% yields:

$$ArBr + RLi \xrightarrow[50-70\%]{THF, 25^0} ArR + LiBr$$

Cross-coupled products are formed in insignificant amounts if *sec*- or *tert*-alkyllithiums are employed. However, if 1-bromonaphthalene is treated first with *sec*-butyllithium and then with *n*-octyl bromide, 1-*n*-octylnaphthalene can be obtained in 74% yield.[1]

[1] R. E. Merrill and E. Negishi, *J. Org.*, **39**, 3452 (1974).

**Tetrahydro-2H-1,3-oxazine-2-one,**  (1). Mol. wt. 101.11, m.p. 80°.

This urethane can be prepared from 3-amino-1-propanol and ethylene carbonate (73% yield).[1]

*Halopropyl isocyanates.*[2] Reaction of (1) with pyridine and tosyl chloride in $CH_2Cl_2$ at 20–25° leads to a crystalline pyridinium salt (2). Reaction of (2) with pyridinium chloride, bromide, or iodide gives the halopropyl isocyanates (3) in >90% yield. The five-membered urethane corresponding to (1) reacts

more slowly but is converted into β-chloroethyl isocyanate (46% yield).

[1] E. Dyer and H. Scott, *Am. Soc.*, **79**, 672 (1957); H. Najer, P. Chabrier, R. Giudicelli, and J. Sette, *Bull. soc.*, 1609 (1959).
[2] A. Krantz and B. Hoppe, *Tetrahedron Letters*, 695 (1975).

**Tetrakis(triphenylphosphine)nickel(0),** $[(C_6H_5)_3P]_4Ni$. Mol. wt. 1107.89, reddish brown solid, m.p. 123–128°. The complex decomposes rapidly upon exposure to air as a solid or in solution.

*Preparation.*[1] The reagent is prepared in 55% yield by the reaction of bis-(2,4-pentanedionato)nickel (1 mole) with 6 moles of triphenylphosphine followed by treatment with triethylaluminum with use of standard techniques for manipulation of air-sensitive compounds.

[1] R. A. Schunn, *Inorg. Syn.*, **13**, 124 (1973).

**Tetrakis(triphenylphosphine)palladium(0),**   $Pd[P(C_6H_5)_3]_4$   (1).   Mol.   wt. 1155.80, m.p. 100–105° dec., yellow, unstable to air. Suppliers: ROC/RIC, Strem.

*Preparation.* The complex can be obtained in 90–94% yield by reduction of palladium dichloride by hydrazine hydrate in the presence of triphenylphosphine.[1] The complex can also be obtained by reaction of palladium bis(di-

$$2PdCl_2 + 8P(C_6H_5)_3 + 5NH_2NH_2 \cdot H_2O \rightarrow 2Pd[P(C_6H_5)_3]_4 +$$
$$4NH_2NH_2 \cdot HCl + N_2 + 5H_2O$$

benzylideneacetone) with triphenylphosphine (89% yield).[2]

The reagent melts with decomposition at temperatures of 105–115°. It is moderately unstable to air and should be stored under nitrogen. It is moderately soluble in benzene, $CH_2Cl_2$, and $CHCl_3$, and insoluble in saturated hydrocarbon solvents.

*Olefins from vinyl halides and alkyllithiums.*[2]  Vinyl halides react with this palladium(0) complex to give $\rho$-vinylpalladium complexes (2) with retention of configuration.[3] These complexes react with alkyllithium or Grignard reagents to

$$\begin{array}{c}R^1\\R^2\end{array}C=C\begin{array}{c}H\\X\end{array} + (1) \xrightarrow{C_6H_6} \begin{array}{c}R^1\\R^2\end{array}C=C\begin{array}{c}H\\PdX[P(C_6H_5)_3]_{4-n}\end{array} \xrightarrow[\substack{60-99\% \\ \text{overall}}]{\substack{RLi(RMgX^1) \\ \text{Ether}}}$$

(X=Br, Cl)                                        (2)

$$\begin{array}{c}R^1\\R^2\end{array}C=C\begin{array}{c}H\\R\end{array}$$

(3)

form alkenes in good to fairly high yields. More importantly the olefins are obtained in high isomeric purity (99–100%). Thus (E)-β-bromostyrene is converted into (E)-propenylbenzene in 98% yield with complete retention of stereochemistry. The reaction has been used for synthesis of di- and trisubstituted

$$\begin{array}{c}C_6H_5\\H\end{array}C=C\begin{array}{c}H\\Br\end{array} \xrightarrow[98\%]{\substack{1)\ (1) \\ 2)\ CH_3Li}} \begin{array}{c}C_6H_5\\H\end{array}C=C\begin{array}{c}H\\CH_3\end{array}$$

olefins. The Pd complex is essential for this coupling. The reaction of methyllithium alone with β-bromostyrene produces mainly phenylacetylene by dehydrobromination. If the Grignard reagent is fairly unreactive, the Pd complex can be used in catalytic amounts because of regeneration.

*Beckmann fragmentation of ketoximes.* Monoketoximes of aromatic ketones are converted into nitriles and aldehydes by treatment with this Pd(0)

(I)   $C_6H_5\diagdown C \diagup COC_6H_5$
      $\|$
      $N \diagdown$
         $OH$

$\xrightarrow[\text{2) O}_2]{\begin{array}{c}\text{1) Pd}[\ddot{\text{P}}(C_6H_5)_3]_4\\ \text{CH}_3\text{CN, }60^0\end{array}}$   $C_6H_5CN$  +  $C_6H_5CHO$

(85%)         (43%)

(II)  $C_6H_5\diagdown C \diagup CH_2C_6H_5$
      $\|$
      $N \diagdown$
         $OH$

$\longrightarrow$  $C_6H_5CN$  +  $C_6H_5CHO$

(58%)         (46%)

complex and then with oxygen. Yields are highest with monoketoximes of α-diketones (equation I). The reaction fails with aliphatic ketoximes.[5]

*Aryl- and vinyl-substituted acetylenes.*[6]  Acetylene or monosubstituted acetylenes react with aryl or vinyl halides in the presence of this palladium(0) complex as catalyst and of base (CH$_3$ONa) in DMF at 40–80° to give aryl- or vinylacetylenes:

$$C_6H_5I \; + \; C_6H_5C\equiv CH \xrightarrow[95\%]{\text{Pd}[\text{P}(C_6H_5)_3]_4} C_6H_5C\equiv CC_6H_5$$

$$CH_2=CHBr \; + \; C_6H_5C\equiv CH \xrightarrow[52\%]{} C_6H_5C\equiv CCH=CH_2$$

Nickel(0) complexes effect this reaction, but the product is an acetylenic nickel complex; hence the reaction is not catalytic.

*Dehydrobromination of α-bromo ketones.* This complex is useful for dehydrobromination of bromo ketones that can result in phenolic products, as in equation I. In other, more general, cases it does not seem to be useful (equation II).[7]

(I)

$\xrightarrow{\text{Pd}[\text{P}(C_6H_5)_3]_4}$

$\xrightarrow{94\%}$

(II)

$\longrightarrow$  +

(70%)         (22%)

*Ketone synthesis.* Ketones can be prepared in 65–85% yield by the reaction of aromatic or aliphatic acyl halides with diaryl- or dialkylmercury(II) compounds in HMPT with this palladium complex as catalyst.[8]

$$R^1\overset{\overset{O}{\|}}{C}Br \;+\; PdL_4 \;\xrightarrow{\;HMPT\;}\; \left[ \overset{\overset{O}{\|}}{\underset{Br}{R^1C}}PdL_4 \xrightarrow[-HgR^2Br]{R^2_2Hg} \overset{\overset{O}{\|}}{\underset{R^2}{R^1C}}PdL_4 \right] \;\longrightarrow\; R^1\overset{\overset{O}{\|}}{C}R^2 \;+\; PdL_4$$

*Extrusion of mercury from bis(propenyl)mercury.*[9] Both bis(*cis*-propenyl)mercury and bis(*trans*-propenyl)mercury (1) have been prepared by the following route[10] :

$$C_3H_5Li \xrightarrow{\;HgBr_2\;} C_3H_5HgBr \xrightarrow{\;Na_2SnO_3\;} (C_3H_5)_2Hg$$
$$(1)$$

The intermediate vinylmercury salts are now readily available by hydroboration–mercuration.[11] Treatment of either isomer of (1) with this palladium(0) complex in acetonitrile at 25° results in extrusion of mercury and stereoselective formation of 2,4-hexadiene in almost quantitative yields. The *cis*-isomer is converted almost exclusively into *cis, cis*-2,4-hexadiene; the *trans*-isomer is converted entirely into *trans,trans*-2,4-hexadiene.

A variety of palladium(II) complexes also induce this reaction, but yields are variable and stereoselectivity is generally low. Side products containing palladium are also formed.

[1] D. R. Coulson, *Inorg. Syn.*, **13**, 121 (1972).
[2] Procedure of Dr. Y. Takahashi, cited in reference[7].
[3] M. Yamamura, I. Moritani, and S.-I. Murahashi, *J. Organometal. Chem.*, **91**, C39 (1975).
[4] P. Fitton and J. E. McKeon, *Chem. Commun.*, 4 (1968).
[5] K. Maeda, I. Moritani, T. Hosokawa, and S.-I. Murahashi, *J.C.S. Chem. Comm.*, 689 (1975).
[6] L. Cassar, *J. Organometal. Chem.*, **93**, 253 (1975).
[7] J. M. Townsend, I. D. Reingold, M. C. R. Kendall, and T. A. Spencer, *J. Org.*, **40**, 2976 (1975).
[8] K. Takagi, T. Okamoto, Y. Sakakibara, A. Ohno, S. Oka, and N. Hayama, *Chem. Letters*, 951 (1975).
[9] E. Vedejs and P. D. Weeks, *Tetrahedron Letters*, 3207 (1974).
[10] D. Moy, M. Emerson, and J. P. Oliver, *Inorg. Chem.*, **2**, 1261 (1963).
[11] H. C. Brown and S. K. Gupta, *Am. Soc.*, **94**, 4371 (1972); R. C. Larock and H. C. Brown, *J. Organometal. Chem.*, **36**, 1 (1972).

**Tetramethylammonium dimethyl phosphate,** $(CH_3O)_2\overset{\overset{O}{\|}}{P}-\bar{O}\;\overset{+}{N}(CH_3)_4$ (1). Mol. wt. 199.18, m.p. 215°.

The salt is prepared in >90% yield by the reaction of excess trimethylamine with trimethyl phosphate.[1]

*Dehydrohalogenation.* French chemists[2] have utilized this salt for dehydrobrominations, particularly of 4-bromo- and 2,4-dibromo-3-keto steroids. Reactions were conducted in DMF or acetonitrile.

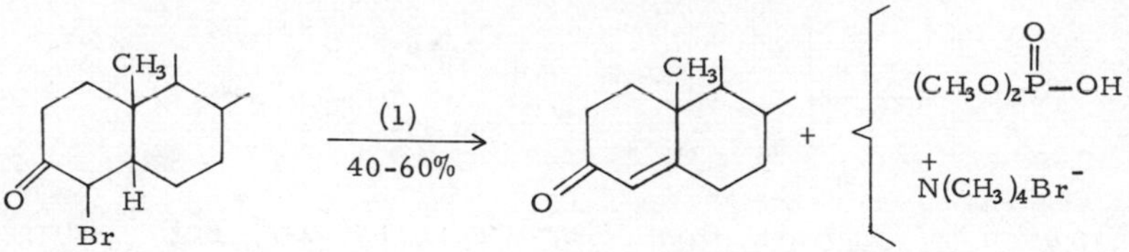

In a synthesis of $\Delta^{5,7}$-cholestadiene-1$\alpha$,3$\beta$-diol (1$\alpha$-hydroxyprovitamin D$_3$), Israeli chemists[3] effected dehydrobromination of the dibromide (1) at 130° (4 hr.) in HMPT containing about 10% of triethylmethylammonium dimethyl phosphate. A 1.3:1 mixture of the $\Delta^{5,7}$-diene (2) and the $\Delta^{4,6}$-diene (3) was obtained in almost quantitative yield. The mixture was separated on a silica column impregnated with silver nitrate.

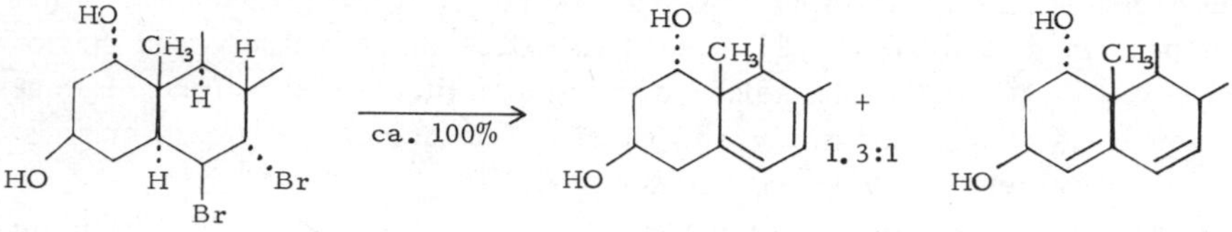

(1)                    (2)                    (3)

[1] A. Carayon-Gentil, T. N. Thanh, G. Gonzy, and P. Chabrier, *Bull. soc.*, 1616 (1967).
[2] G. Sturtz and A. Raphalen, *Tetrahedron Letters*, 1529 (1970); J. L. Kraus and G. Sturtz, *Bull. soc.*, 2551 (1971).
[3] D. Freeman, A. Acher, and Y. Mazur, *Tetrahedron Letters*, 261 (1975).

**Tetramethylcyclobutadiene–Aluminum chloride complex** (1). Mol. wt. 187.43.

The complex is prepared[1] by slow addition of 2-butyne (0.37 mol.) in CH$_2$Cl$_2$ to a stirred suspension of aluminum chloride in CH$_2$Cl$_2$ at 0–5°.

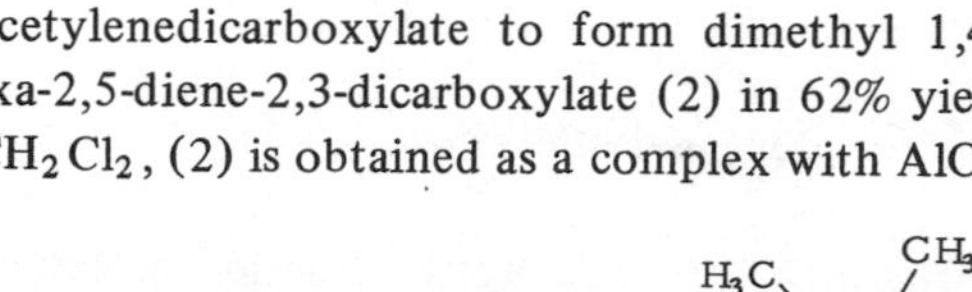

(1)

*Cycloadditions; derivatives of "Dewar" benzenes.* The complex (1) reacts with dimethyl acetylenedicarboxylate to form dimethyl 1,4,5,6-tetramethyl-bicyclo[2.2.0]hexa-2,5-diene-2,3-dicarboxylate (2) in 62% yield. If the reaction is conducted in CH$_2$Cl$_2$, (2) is obtained as a complex with AlCl$_3$. If the reaction

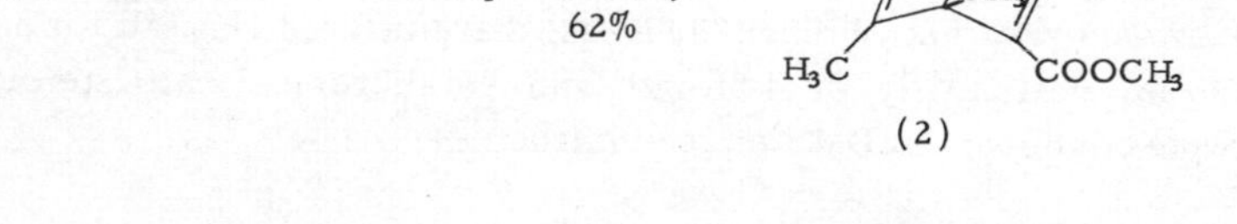

(2)

is conducted in DMSO, (2) is obtained directly. In either case the reaction involves formation of tetramethylcyclobutadiene *in situ*, followed by [2 + 2] cycloaddition.

The reaction of (1) with dimethyl maleate leads to (3) in 35% yield; reaction with dimethyl fumarate gives (4) in 30% yield.[1]

(3)                                      (4)

Wynberg *et al.*[2] have used this cycloaddition reaction for the first synthesis of a chiral "Dewar" benzene (6). Thus the reaction of (1) with methyl phenylpropiolate in DMSO at 0–5° gives the ester (5) in 75% yield (crude). The corresponding free acid was resolved with (+)-dehydroabietylamine (1, 183) to give optically active material, $[\alpha]_D$ −136°. This was esterified to give the optically

$$(1) + C_6H_5C\equiv CCOOCH_3 \xrightarrow[75\%]{DMSO}$$

(5, R=CH₃)
(6, R=H, $[\alpha]_D$ −136.7°)

(7)                                      (8)

active methyl ester (5a). Pyrolysis of (5a) gives the achiral diphenyl derivative (7); photolysis of (5a) gives the optically inactive prismanic derivative (8) in remarkably high yield.

[1] J. B. Koster, G. J. Timmermans, and H. van Bekkum, *Synthesis*, 139 (1971).
[2] J. H. Dopper, B. Greijdanus, and H. Wynberg, *Am. Soc.*, 97, 216 (1975).

**N,N,N′,N′-Tetramethyldiaminophosphorochloridite,** $ClP[N(CH_3)_2]_2$ (1). Mol. wt. 154.59, b.p. 29°/1 mm.

*Preparation*[1]:

$$2\ P[N(CH_3)_2]_3 + PCl_3 \xrightarrow[91.5\%]{60°} 3\ ClP[N(CH_3)_2]_2$$

(1)

*Diphenylacetylenes.* In a recent synthesis of diphenylacetylenes,[2] the phosphorochloridite (1) was allowed to react with a benzyl alcohol in the presence of triethylamine to form a phosphorodiamidite (2). Reaction of (2) with a benzotrichloride gave (3), which was converted into a diphenylacetylene (4) on dehydrochlorination. Diphenylacetylene itself was obtained in 62% overall yield;

$$(1) \; + \; X-C_6H_4CH_2OH \; \xrightarrow{N(C_2H_5)_3} \; X-C_6H_4CH_2OP[N(CH_3)_2]_2 \; \xrightarrow[-ClP(O)[N(CH_3)_2]_2]{C_6H_5CCl_3} \;$$

$$(2)$$

$$X-C_6H_4CH_2CCl_2C_6H_5 \; \xrightarrow[C_2H_5OH]{KOH} \; X-C_6H_4C{\equiv}CC_6H_5$$
$$(3) \hspace{5cm} (4)$$

lower yields were reported for the case of X = Cl and F. However, benzyl alcohols and benzotrichlorides are relatively easy to obtain.

[1] H. Noth and H. J. Vetter, *Ber.*, **94**, 1505 (1961).
[2] J. H. Hargis and W. D. Alley, *J.C.S. Chem. Comm.*, 612 (1975).

**Tetramethylethylenediamine (TMEDA)**, 2, 403; 3, 284–285; 4, 485–489; 5, 652.

*Enynes.* [1] Vinylic organic cuprates react with 1-bromo- or 1-iodo-1-alkynes in the presence of 1.2–2 eq. of TMEDA to give, after hydrolysis, enynes usually in yields of 75–85%.

$$\underset{C_2H_5}{\overset{R}{>}}C{=}C\underset{Cu}{\overset{H}{<}}\cdot MgX_2 \; + \; R'C{\equiv}CX \; \xrightarrow[\substack{2) \; H_3O^+ \\ 75-85\%}]{\substack{1) \; (C_2H_5)_2O, \; THF \\ TMEDA, \; -15^0}} \; \underset{C_2H_5}{\overset{R}{>}}C{=}C\underset{C{\equiv}CR'}{\overset{H}{<}}$$

$$X = Br, I$$

$$R = CH_3, \underline{n}\text{-}Bu$$

The reaction is particularly useful for synthesis of 4-alkene-2-yne-1-ols:

$$\underset{C_2H_5}{\overset{\underline{n}\text{-}Bu}{>}}C{=}C\underset{Cu}{\overset{H}{<}}\cdot MgBr_2 \; + \; THPOCH_2C{\equiv}CBr \; \xrightarrow{80\%} \; \underset{C_2H_5}{\overset{\underline{n}\text{-}Bu}{>}}C{=}C\underset{C{\equiv}CCH_2OH}{\overset{H}{<}}$$

This reaction was used for a four-step synthesis of bombykol (4), the sex pheromone of the female silkworm moth, as shown in the formulation.

$$CH_3CH_2CH_2MgBr \; \xrightarrow{\substack{1) \; CuBr \\ 2) \; HC{\equiv}CH \\ 3) \; BrC{\equiv}CCH_2OSi(CH_3)_3 \\ 4) \; H_3O^+}} \; \underset{CH_3(CH_2)_2}{\overset{H}{>}}C{=}C\underset{C{\equiv}CCH_2OH}{\overset{H}{<}} \; \xrightarrow[93\%]{\substack{1) \; LiAlH_4 \\ 2) \; Acetylation}}$$

1) Cu(I), THF, -30°

BrMg(CH$_2$)$_8$OCHOC$_2$H$_5$

$\underset{CH_3}{|}$

(3)

2) H$_3$O$^+$

$\xrightarrow{\hspace{1cm} 75\% \hspace{1cm}}$

(4)

[1] J. F. Normant, A. Commercon, and J. Villieras, *Tetrahedron Letters*, 1465 (1975).

**(+)   and   (−)-α-(2,4,5,7-Tetranitro-9-fluorenylideneaminoxy)propionic   acid (TAPA), 1, 1147.**

*Resolution of racemic amines.* This resolving agent is recommended for resolution of racemic weakly alkaline amines that do not form stable salts with the usual resolving acids, for highly insoluble amines that do not readily form soluble salts, and for amines that do not form crystalline salts with readily available resolving acids.[1]

[1] F. I. Carroll, B. Berrange, and C. P. Linn, *Chem. Ind.*, 477 (1975).

**Tetraphenylcyclopentadienone (Tetracyclone), 1, 1149–1150; 4, 490–491.**

*Scavenger for singlet oxygen.* This highly colored substance has been used as a scavenger for singlet oxygen.[1]

[1] R. W. Murray, W. C. Lumma, Jr., and J. W.-P. Lin, *Am. Soc.*, 92, 3205 (1970); M. E. Brennan, *Chem. Commun.*, 956 (1970); J. R. Sanderson and P. R. Story, *J. Org.*, 39, 3183 (1974).

**Tetraphenylethylene, (C$_6$H$_5$)$_2$C=C(C$_6$H$_5$)$_2$, 2, 404.**

*Recent synthesis.*[1]

*Diazoalkane decomposition.*[2] Surprisingly, tetraphenylethylene is almost as efficient as various copper catalysts for decomposition of diazoalkanes to carbenoids. For example, diazomethane and cyclohexene in the presence of this catalyst react to form norcarane in 15 ± 5% yield; with copper catalysis the yield of norcarane is 24%. Cyclopropanations have been observed with this hydrocarbon catalyst with a variety of diazo compounds: diazomethane, α-diazoacetophenone, and diazofluorene. Diphenyldiazomethane, however, is converted mainly into benzophenone azine, (C$_6$H$_5$)$_2$C=NN=C(C$_6$H$_5$)$_2$.

[1] C. Y. Meyers, W. S. Matthews, G. J. McCollum, and S. J. Branca, *Tetrahedron Letters*, 1105 (1974).

[2] C.-T. Ho, R. T. Conlin, and P. P. Gaspar, *Am. Soc.*, 96, 8109 (1974).

**Thallium (I) ethoxide, 2, 407–411; 4, 501–502; 5, 656.**

*Etherification.* Ethers can be prepared in good yield by alkylation of thallium (I) alkoxides in acetonitrile (equation I). The reaction has been used to date

(I)   ROH $\xrightarrow[\text{quant.}]{\substack{\text{TlOC}_2\text{H}_5 \\ \text{C}_6\text{H}_6, 20°}}$ ROTl $\xrightarrow[25-95\%]{\substack{\text{R'X, CH}_3\text{CN} \\ 20-60°}}$ ROR'

only with primary alkyl halides. The reaction is accelerated by an increase in solvent polarity: $C_6H_6$, $CCl_4 \ll CH_3CN < DMF$. The rate also increases in the order alkyl $< ROCH_2 < TlOCH_2 < COOR < CONR_2$ for X in $X CH_2 OTl$. The method is valuable in the case of long-chain alkyl halides.[1]

*N-Alkyl-N-triflylarylsulfonimides.* N-Alkylarylsulfonamides (1) react with thallium(I) ethoxide in benzene at $25°$ ($N_2$) to form the stable thallium(I) compounds (2) in high yield. Treatment of (2) with triflic anhydride leads to (3).

Alkyl N-tresylarylsulfonimides can be obtained by the reaction of (2) with tresyl chloride (2,2,2-trifluoroethanesulfonyl chloride).[2]

[1] H.-O. Kalinowski, D. Seebach, and G. Crass, *Angew. Chem. internat. Ed.*, **14**, 762 (1975).
[2] H.-L. Pan and T. L. Fletcher, *Synthesis*, 39 (1975).

**Thallium(I) hydroxide**, TlOH. Mol. wt. 221.38. Suppliers: Alfa, ROC/RIC.

*Thallium(I) salts of β-dicarbonyl compounds.*[1] The first step in a recent synthesis of methyltriacetic lactone (5) involved conversion of *t*-butyl acetoacetate (1) into the thallium salt (2). This underwent C-acylation with ethyl

methylmalonyl chloride (3) in ether to give (4), which was not purified, but was refluxed in benzene containing some *p*-toluenesulfonic acid for 12 hr. Decarboalkylation and ring closure gave the lactone (5) in 55% overall yield.

[1] E. Suzuki and S. Inoue, *Synthesis*, 259 (1975).

**Thallium(III) nitrate (TTN)**, 4, 492–497; 5, 656–657.

*Cyclothallation of myrcene.*[1] Reaction of myrcene (1) with thallium(III) nitrate trihydrate in methanol at $-60°$ followed immediately by neutralization with sodium carbonate, gives a 1:1 mixture of (2) and (3) in 50% yield. Intermediates (a) and (b) are suggested.

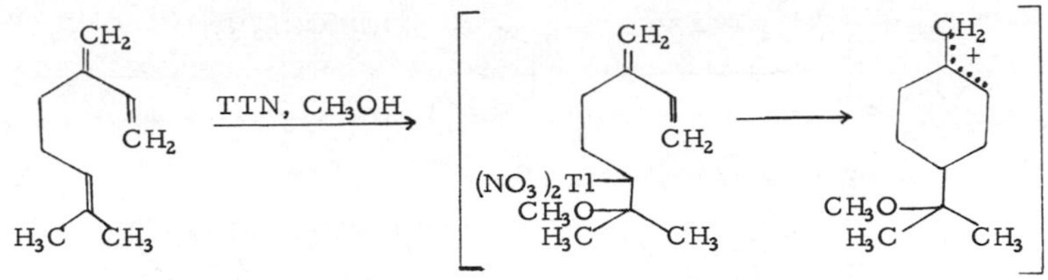

(1)                                    (a)                    (b)

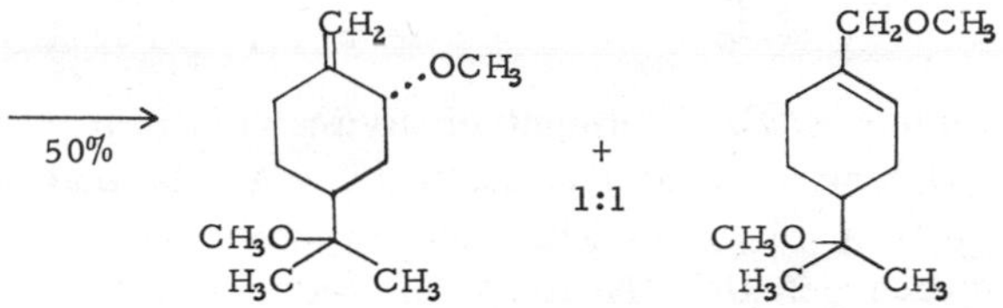

50%                    +
                      1:1

(2)                              (3)

[1] A. Anteunis and A. De Smet, *Synthesis*, 868 (1974).

**Thallium(III) trifluoroacetate (TTFA)**, 3, 286–289; 4, 498–500; 5, 658–659.

*Arylthallium(III) bistrifluoroacetates* (2). These substances can be prepared in high yield by replacement of a trimethylsilyl group (*cf.* 4, 282–283):

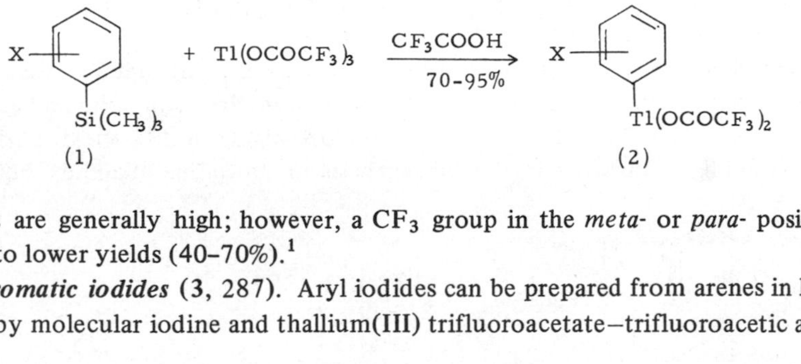

Yields are generally high; however, a $CF_3$ group in the *meta-* or *para-* position leads to lower yields (40–70%).[1]

*Aromatic iodides* (3, 287). Aryl iodides can be prepared from arenes in high yield by molecular iodine and thallium(III) trifluoroacetate–trifluoroacetic acid:

$$2 \ C_6H_6 + I_2 + Tl(CF_3CO_2)_3 \xrightarrow[100\%]{68°} 2 \ C_6H_5I + Tl(CF_3CO_2) + 2 \ CF_3COOH$$

Polyiodination of benzene and mesitylene can also be effected stepwise. Yields are lower when trifluoroacetic acid is replaced by $CCl_4$, $CHCl_3$, $CH_3CN$, or $CH_3OH$.[2]

*Nitroaryl iodides.* These substances can be prepared conveniently by the two-step procedure illustrated for *m*-xylene. Note that the iodine enters the position originally occupied by thallium.[3]

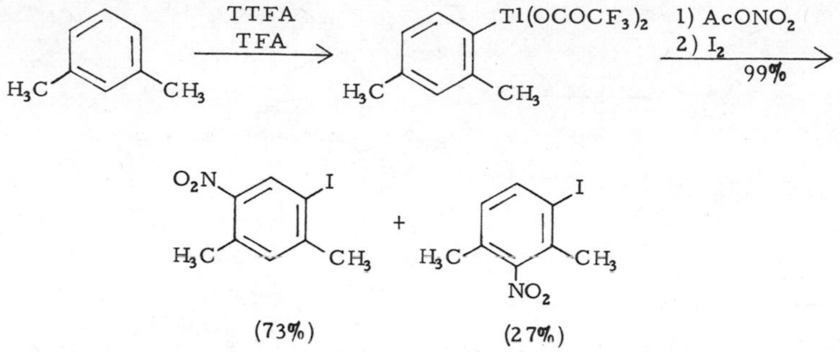

(73%)         (27%)

*Phenols; arylboronic acids.* Treatment of arylthallium bistrifluoroacetates with diborane in THF leads to an intermediate, possibly $ArBH_2$, that is converted into arylboronic acids on treatment with water or into a phenol on treatment with alkaline hydrogen peroxide. Treatment of the intermediates with silver nitrate fails to produce biaryls.[4]

*Oxidative phenol coupling* (**4**, 499–500). Schwartz and Mami[5] have effected oxidative *para-ortho* coupling of the benzyltetrahydroisoquinoline alkaloid (1) to (2) with TTFA in $CH_2Cl_2$ (–78 to 20°, overnight) in 23% yield. This oxidative coupling is involved in the biosynthesis of morphine alkaloids, but it has

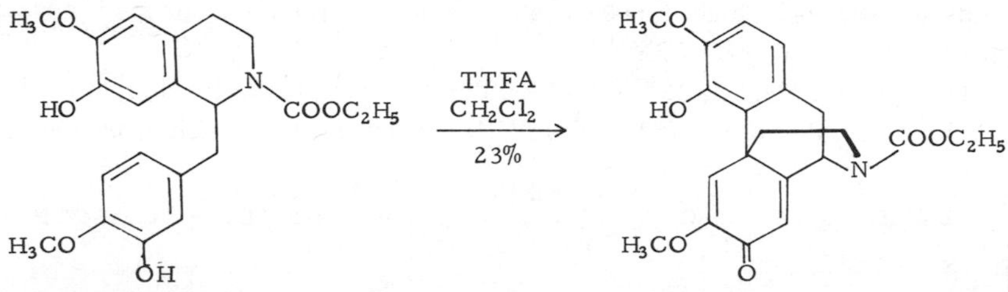

(1)                  (2)

only been effected in the laboratory in one instance by Barton *et al.*[6] in a yield of 0.03% by ferricyanide oxidation.

The product (2) was converted into (±)-thebaine (3) by reduction with LiAlH$_4$ and treatment with 1 $N$ HCl.

(3)

[1]H. C. Bell, J. R. Kalman, J. T. Pinhey, and S. Sternhell, *Tetrahedron Letters*, 3391 (1974).
[2]N. Ishikawa and A. Sekiya, *Bull. Chem. Soc. Japan*, 47, 1680 (1974).
[3]E. C. Taylor, H. W. Altland, and A. McKillop, *J. Org.*, 40, 3441 (1975).
[4]S. W. Breuer, G. M. Pickles, J. C. Podesta, and F. G. Thorpe, *J.C.S. Chem. Comm.*, 36 (1975).
[5]M. A. Schwartz and I. S. Mami, *Am. Soc.*, 97, 1239 (1975).
[6]D. H. R. Barton, D. S. Bhakuni, R. James, and G. W. Kirby, *J. Chem. Soc. (C)*, 128 (1967).

**Thallous 2-methylpropane-2-thiolate,** TlSC(CH$_3$)$_3$ (1). Mol. wt. 293.55, m.p. 170–175° dec., bright yellow.

The salt is obtained[1] in 95% yield by reaction of thallous ethoxide in benzene with 2-methylpropane-2-thiol (20°) for a few minutes.

*Synthesis of esters and lactones.*[1] Thallous 2-methylpropane-2-thiolate (1) reacts with acid chlorides to give S-$t$-butyl thioates (2) quantitatively. The $t$-butyl thioate group is relatively stable to mild acidic and alkaline reagents, but

is converted into the carboxylic acid RCOOH by wet organic solvents. Hence this group is useful for protection of carboxylic acids. S-$t$-Butyl thioates are converted into esters (3) by reaction with an alcohol in the presence of mercury(II) salts. In the case of secondary, tertiary, and hindered primary alcohols, mercuric trifluoroacetate is efficient; for methyl and ethyl esters, the combination of mercuric chloride and cadmium carbonate is recommended. Yields of esters are high, often quantitative.

Masamune has made use of this property for synthesis of medium ring lactones (macrolides), for example, zearalenone dimethyl ether ethylene ketal (5) from the hydroxy ester (4). On treatment with mercuric trifluoroacetate at 25°, (4) is converted within 5 min. into the lactone (5) in >90% yield (recrystallized).

$[4, R = SC(CH_3)_3]$

$$\xrightarrow[> 90\%]{\substack{Hg(OCOCF_3)_2 \\ CH_3CN, 25^\circ}}$$

(5)

This lactonization procedure was used by Masamune *et al.*[2] in a total synthesis of methynolide (6), the aglycone of the macrolide antibiotic methymycin.

(6)

[1] S. Masamune, S. Kamata, and W. Schilling, *Am. Soc.*, 97, 3515 (1975).
[2] S. Masamune, H. Yamamoto, S. Kamata, and A. Fukuzawa, *ibid.*, 97, 3513 (1975).

## Thioanisole, $C_6H_5SCH_3$.

*Polymeric thioanisole*, (P)—$C_6H_4SCH_3$. Crosby *et al.*[1] have prepared a polymeric thioanisole reagent from macroreticular polystyrene, (P)—$C_6H_5$, and have shown that it can be used in the Corey–Kim oxidation (4, 89) of primary and secondary alcohols. Yields are somewhat lower than those obtained with $Cl_2$ — $C_6H_5SCH_3$. The main advantages of the polymeric reagent are ease of work-up with recovery of the reagent for reuse and freedom from odor.

[1] G. A. Crosby, N. M. Weinshenker, and H.-S. Uh, *Am. Soc.*, 97, 2232 (1975).

**Thiobenzoyl chloride,** $C_6H_5\overset{\displaystyle S}{\overset{\|}{C}}Cl$ . Mol. wt. 156.63, violet-red, lachrymatory liquid, b.p. 88° (3.75 mm.). Unstable above 78°.

*Preparation.* Several methods of preparation have been reported, but the most convenient is probably the reaction of $CS_2$ with phenylmagnesium bromide or phenyllithium to give dithiobenzoic acid; this is converted into thiobenzoyl chloride with thionyl chloride.[1] Oxygen must be rigorously excluded or benzoyl chloride is formed. The reagent reacts with alcohols in dry pyridine to give thiobenzoic acid O-esters, $RO\overset{\displaystyle S}{\overset{\|}{C}}C_6H_5$, in 50–90% yields.[2] These esters can

also be prepared by reaction of sodium alkoxides with (thiobenzoylthio)acetic
acid, $C_6H_5\overset{\overset{S}{\|}}{C}-SCH_2COOH$.[2,3]

*Dehydration of homoallylic alcohols.* O-Cholesteryl thiobenzoate $(1, \lambda_{max}$ 256, 288, and 420 nm) on irradiation in cyclohexane with a medium-pressure mercury vapor lamp for 12 min. is converted into 3,5-cholestadiene (2) and thiobenzoic acid in quantitative yield. Similar irradiation of O-cholestanyl thiobenzoate results merely in rearrangement to the thiobenzoic acid S-ester; hence the elimination reaction appears to require some conjugation in the transition state of the newly formed bond.[4]

$$(1) \xrightarrow{h\nu} (2) + C_6H_5\overset{\overset{S}{\|}}{C}OH$$

Similarly, irradiation of O-phenethyl thiobenzoate (3) gives styrene (4) in 90% yield.[5] Thus this method is useful for dehydration under very mild conditions of

$$C_6H_5CH_2CH_2O\overset{\overset{S}{\|}}{C}C_6H_5 \xrightarrow{h\nu} C_6H_5CH=CH_2$$
$$(3) \qquad\qquad\qquad (4)$$

homoallylic alcohols. Note, however, that irradiation of (5) and (6) gives complex mixtures.

$$HC\equiv CCH_2CH_2O\overset{\overset{S}{\|}}{C}C_6H_5 \qquad \triangleright\!-CH_2CH_2O\overset{\overset{S}{\|}}{C}C_6H_5$$
$$(5) \qquad\qquad\qquad\qquad (6)$$

[1] R. Mayer and S. Scheithauer, *Ber.*, **98**, 829 (1965); E. J. Hedgley and H. G. Fletcher, Jr., *J. Org.*, **30**, 1282 (1965).
[2] D. H. R. Barton, C. Chavis, M. K. Kaloustian, P. D. Magnus, G. A. Poulton, and P. J. West, *J.C.S. Perkin I*, **1571 (1973)**.
[3] K. A. Jensen and C. Pedersen, *Acta Chem. Scand.*, **15**, 1087 (1961).
[4] S. Achmatowicz, D. H. R. Barton, P. D. Magnus, G. A. Poulton, and P. J. West, *J.C.S. Perkin I*, 1567 (1973).
[5] D. H. R. Barton, M. Bolton, P. D. Magnus, K. G. Marathe, G. A. Poulton, and P. J. West, *ibid.*, 1574 (1973).

**N,N'-Thiocarbonyldiimidazole, 1,** 1151–1152; **2,** 411–412; **5,** 661.

*7-Oxanorbornadiene* (2). This hydrocarbon has been obtained from (1) by the modified Corey–Winter synthesis (**5**, 661). Some interesting reactions of

(2) have been reported: [2 + 2 + 2] cycloaddition to give (3), [2 + 2] cycloaddition to (4), and peracid oxidation to the *exo,exo*-bisepoxide (5).[1]

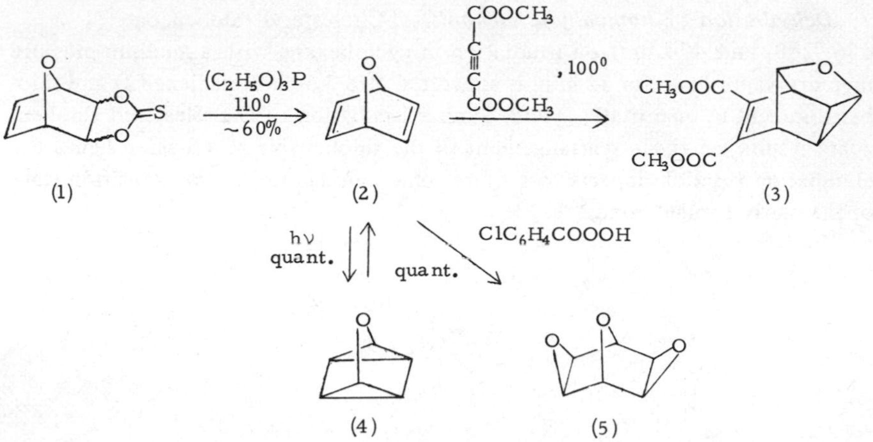

(1)                           (2)                                                    (3)

(4)                    (5)

[1] H. Prinzbach and H. Babsch, *Angew. Chem. internat. Ed.*, **14**, 753 (1975).

**2-Thiomethyl-4,4-dimethyl-2-oxazoline**, (1). Mol. wt. 145.23.

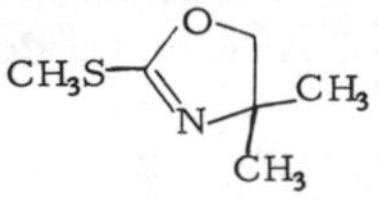

This heterocycle is prepared[1] by treatment of 4,4-dimethyloxazoline-2-thione[2] with NaH and $CH_3I$ in THF.

*Homologation of aldehydes and ketones to thiiranes (3) and 1-alkenes (4).*[1]
2,2-Disubstituted thiiranes and 1-alkenes can be prepared from the lithio salt of (1) as formulated. This synthesis is not general for 2,3-disubstituted thiiranes or

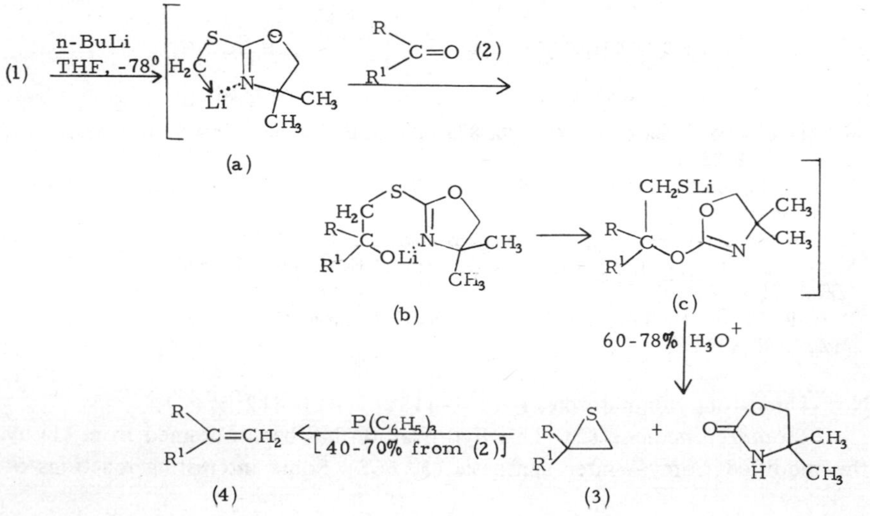

(a)

(b)                                      (c)

60-78% $\Big|$ $H_3O^+$

(4)                                                      (3)

internal alkenes because the lithio salt corresponding to (a) can be generated only when the S-alkyl group is methyl, benzyl, allyl, cyanomethyl, and so forth. Johnson et al.[3] have used the related heterocycle 2-thiomethyl-2-thiazoline (5) for the synthesis of 2-phenylthiirane (7). The reagent (5) suffers from the same

limitations as (1): It is not applicable when the methyl group on sulfur is replaced by a higher alkyl group.

[1] A. I. Meyers and M. E. Ford, *Tetrahedron Letters*, 2861 (1975).
[2] M. Skulski, D. L. Garmaise, and A. F. McKay, *Can. J. Chem.*, 34, 815 (1956).
[3] C. R. Johnson, A. Nakanishi, N. Nakanishi, and K. Tanaka, *Tetrahedron Letters*, 2865 (1975).

**Thionyl chloride, 1, 1158–1163; 2, 412; 3, 290; 4, 503–505; 5, 663–667.**

*Sulfonyl chlorides* (**1**, 1159). *Correction:* The reference cited (13) refers to the preparation of $p$-toluenesulfinyl chloride:

$$\underline{p}\text{-CH}_3\text{C}_6\text{H}_4\text{SO}_2\text{Na} \cdot 2\ \text{H}_2\text{O} + 3\ \text{SOCl}_2 \xrightarrow[86\text{-}92\%]{} \underline{p}\text{-CH}_3\text{C}_6\text{H}_4\text{SOCl} + \text{NaCl} + 3\ \text{SO}_2 + 4\ \text{HCl}$$

*Deoxygenation of aromatic sulfoxides.* Aromatic sulfoxides are deoxygenated by treatment with thionyl chloride at 70°. Chlorinated by-products are obtained unless cyclohexene is added to react with the chlorine formed.

$$\text{Ar}_2\text{SO} + \text{SOCl}_2 \longrightarrow \text{Ar}_2\text{S} + \text{Cl}_2 + \text{SO}_2$$

Under these conditions diaryl sulfides can be obtained from the sulfoxides in 85–90% yield.[1]

[1] I. Granoth, *J.C.S. Perkin I*, 2166 (1974).

**Thiophenol, 4, 505.**

*Isomerization of Z- to E-olefins.* Henrick et al.[1] report the isomerization of the (2Z,4E)-2,4-dienoic acid (1) to a mixture of the (2E,4E)-acid (2) and the (2Z,4E)-acid in the ratio 65:35 when it is heated with about 1% by weight of

thiophenol at 100° for 1–2 hr. The presence of azobisisobutyronitrile or light had no effect on the equilibration (*cf.* 4, 505). Several related reagents (thiobenzoic S-acid, diphenyl disulfide, thioacetic S-acid, thioglycolic acid) were found to be less effective than thiophenol. The isomerization conditions were shown to be effective for many olefins.

[1]C. A. Henrick, W. E. Willy, J. W. Baum, T. A. Baer, B. A. Garcia, T. A. Mastre, and S. M. Chang, *J. Org.*, **40**, 1 (1975).

**Thiourea, 1**, 1164–1167; **2**, 412–413; **3**, 290–291.

*Reduction of epidioxides.* Schenck and Dunlap[1] used thiourea for reduction of epidioxides, obtained by 1,4-addition of singlet oxygen to conjugated dienes at low temperatures, to the corresponding diols. Kaneko *et al.*[2] report that the diols can be obtained in one step by irradiation of the diene with singlet oxygen in the presence of thiourea. Under these conditions irradiation can be conducted conveniently at room temperature. This procedure was used to prepare *cis*-2-cyclopentene-1,4-diol and *cis*-2-cyclohexene-1,4-diol.

[1]G. O. Schenck and D. E. Dunlap, *Angew. Chem.*, **68**, 248 (1956).
[2]C. Kaneko, A. Sugimoto, and S. Tanaka, *Synthesis*, 876 (1974).

**Thiourea dioxide (Formamidinesulfinic acid), 4**, 506; **5**, 668–669.

*Reduction of disulfides and N-tosylsulfilimines.*[1] These sulfur compounds are reduced to thiols and sulfides, respectively, by thiourea dioxide under phase-transfer conditions (aqueous NaOH, hexadecyltributylphosphonium bromide).

*Reduction of steroidal ketones.*[2] Steroidal ketones have been reduced with this substance in the presence of potassium alkoxides. More recent studies indicate that the hydride originates from the alkoxide ion (Meerwein–Ponndorf reaction) and that thiourea dioxide plays only a minor role.

[1]G. Borgogno, S. Colonna, and R. Fornaiser, *Synthesis*, 529 (1975).
[2]R. Caputo, L. Mangoni, P. Monaco, G. Palumbo and L. Previtera, *Tetrahedron Letters*, 1041 (1975).

**Titanium(III) acetate, Ti(OCOCH$_3$)$_3$.** Mol. wt. 225.04.

*Preparation.* Sodium acetate is added to commercial 30% titanium(III) chloride solution; the precipitate of Ti(OAc)$_3$ is washed and dried.

*Reductive acetylation of oximes.* Boar, Barton, *et al.*[1] note that this reagent is a more powerful reductant than chromium(II) acetate. They used it for the preparation of 1-acetylaminocyclohexene (1) and other enamides.

[1]R. B. Boar, J. F. McGhie, M. Robinson, D. H. R. Barton, D. C. Horwell, and R. V. Stick, *J.C.S. Perkin I*, 1237 (1975).

## Titanium(III) chloride, 2, 415; 4, 506–508; 5, 669–671.

*Cleavage of 2,4-dinitrophenylhydrazones.*[1] Carbonyl compounds can be regenerated in high yield (80–95%) from the 2,4-DNP derivatives by treatment in DME with a 20% aqueous solution of $TiCl_3$[2] at reflux temperature ($N_2$). McMurry and Sylvestri consider that titanous ion reduces the nitro groups to amino groups and then cleaves the hydrazone N—N bond to generate an imine, which is then hydrolyzed readily to the carbonyl compound.

*Cleavage of tosylhydrazones.* Tosylhydrazones are cleaved to ketones in 85–95% yield by treatment with $TiCl_3$ in dioxane and sodium acetate in 50% aqueous acetic acid at room temperature.[3]

*Olefins from carbonyl compounds.* Polish chemists[4] have described the preparation of a lower valent titanium complex designated as [M] from titanium (III) chloride as indicated:

$$TiCl_3 + 3\ THF \xrightarrow[80\%]{Argon} \underset{(blue)}{TiCl_3 \cdot 3\ THF} \xrightarrow[40^0]{2.5\ Mg} \underset{(black)}{[M]}$$

This reagent [M] reacts with carbonyl compounds in the ratio $TiCl_3 \cdot 3THF/Mg/R_1R_2C{=}O = 1:2.5:2$ to form olefins or coupled *vic*-diols, often in high yield. Compare **Tungsten hexachloride, 4,** 569–570.

$$2\ (CH_3)_2C{=}O \xrightarrow[98\%]{[M]} (CH_3)_2C{=}C(CH_3)_2$$

$$2\ (C_6H_5)_2C{=}O \xrightarrow[67\%]{[M]} (C_6H_5)_2C{=}C(C_6H_5)_2$$

$$2 \quad \text{(cyclohexanone)} \xrightarrow[45\%]{[M]} \text{(dicyclohexyl pinacol, OH HO)}$$

$$2 \, C_6H_5COOCH_3 \xrightarrow{[M]} C_6H_5COCOC_6H_5 + C_6H_5(C_6H_5CO)C=C(C_6H_5CO)C_6H_5$$
$$(46\%) \qquad\qquad (20\%)$$

*Reduction of heterocyclic N-oxides.* N-Oxides can be reduced in 70–95% yield by 20% aqueous TiCl$_3$ in methanol or THF (five examples).[5]

*Review.*[6]

[1] J. E. McMurry and M. Silvestri, *J. Org.*, **40**, 1502 (1975).
[2] Matheson, Coleman, and Bell.
[3] B. P. Chandrasekhar, S. V. Sunthankar, and S. G. Telang, *Chem. Ind.*, 87 (1975).
[4] S. Tyrlik and I. Wolochowicz, *Bull. soc.*, 2147 (1973).
[5] J. M. McCall and R. E. TenBrink, *Synthesis*, 335 (1975).
[6] J. E. McMurry, *Accts. Chem. Res.*, **7**, 281 (1974).

**Titanium(III) chloride–Lithium aluminum hydride.**

The reduction of TiCl$_3$ with 4 eq. of LiAlH$_4$ in THF gives a fine black suspension of a titanium(II) species. A 4:1 ball-milled mixture of TiCl$_3$ and LiAlH$_4$ is available from Alfa Inorganics ("McMurry's Reagent"). Addition of this stable solid to THF leads to evolution of hydrogen and formation of titanium(II).

*Reduction of epoxides to olefins.* Epoxides are reduced to olefins by this reagent in 40–75% yields.

$$\text{(epoxide)} + \text{Ti(II)} \rightarrow \, \overset{\diagup}{\underset{\diagdown}{C}} = \overset{\diagup}{\underset{\diagdown}{C}} + \, O=Ti(IV)$$

The reaction is not stereospecific; thus reduction of either *cis-* or *trans*-5-decene oxide gives a 4:1 mixture of *trans-* and *cis*-decene.[1]

*Coupling of allylic and benzylic alcohols* (2, 415).[2] This 3:1 reagent (or the commercial mix) converts allylic and benzylic alcohols into the products of reductive coupling. The reaction is carried out in refluxing glyme (16 hr.). Products of primary–primary and primary–tertiary coupling are obtained from farnesol (last example).

*Examples:*

$$2 \, C_6H_5CH_2OH \xrightarrow[78\%]{TiCl_3-LiAlH_4} C_6H_5CH_2CH_2C_6H_5$$

$$2 \, C_6H_5\overset{\displaystyle CH_3}{\underset{\displaystyle CH_3}{C}}-OH \xrightarrow{95\%} C_6H_5\overset{\displaystyle CH_3}{\underset{\displaystyle CH_3}{C}}-\overset{\displaystyle CH_3}{\underset{\displaystyle CH_3}{C}}C_6H_5$$

$$2 \, \text{(cycloheptenyl-OH)} \xrightarrow{87\%} \text{(bicycloheptenyl)}$$

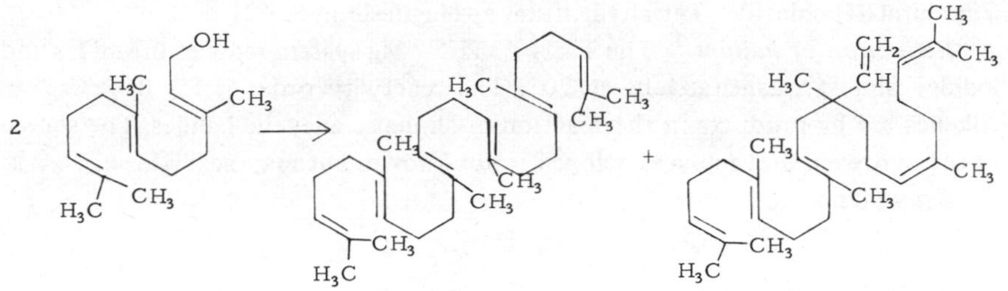

(33%)     (15%)

The reaction is considered to involve titanium(II) alkoxides.[3]

*Bromohydrins → alkenes.* Bromohydrins are reduced generally in satisfactory yields to alkenes by this low-valent titanium species. The reaction is not stereoselective.[4]

*Examples:*

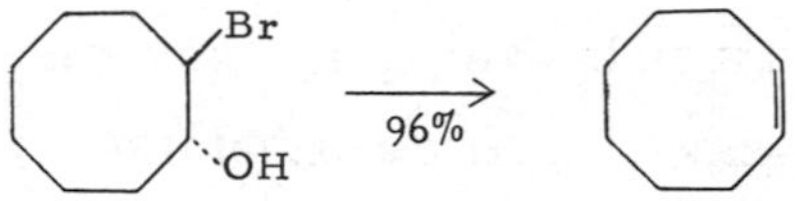

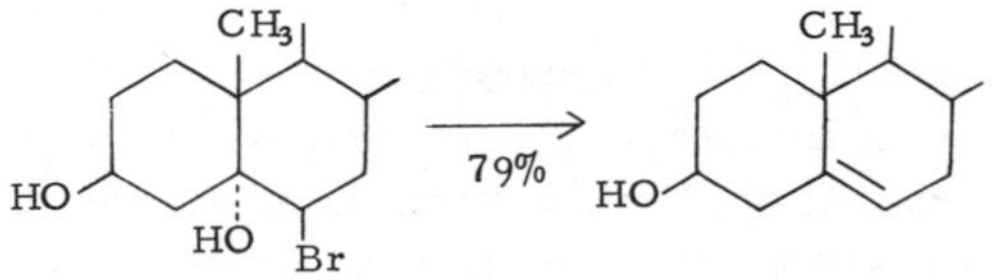

2-Bromo-1-decanol $\xrightarrow[74\%]{}$ 1-Decene

*McMurry–Fleming olefin synthesis.* Two laboratories[5][6] have reported the synthesis of the crowded alkene tetraisopropylethylene by reductive coupling of 2,4-dimethyl-3-pentanone; yields are 6% and 12%. The NMR indicates that the isopropyl groups of this alkene are not equivalent.

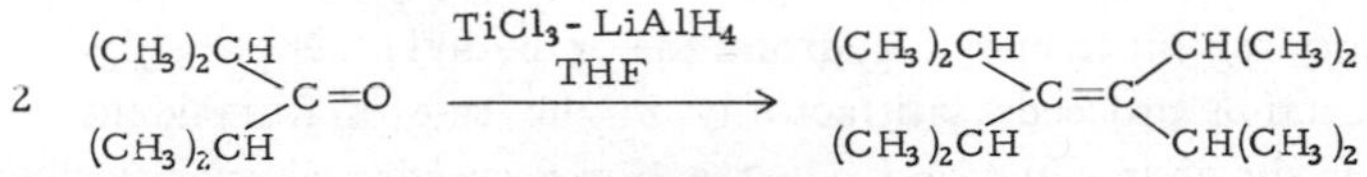

[1] J. E. McMurry and M. P. Fleming, *J. Org.*, **40**, 2555 (1975).
[2] J. E. McMurry and M. Silvestri, *ibid.*, **40**, 2687 (1975).
[3] E. E. van Tamelen, B. Akermark, and K. B. Sharpless, *Am. Soc.*, **91**, 1552 (1969).
[4] J. E. McMurry and T. Hoz, *J. Org.*, **40**, 3797 (1975).
[5] R. F. Langler and T. T. Tidwell, *Tetrahedron Letters*, 777 (1975).
[6] D. S. Bomse and T. H. Morton, *ibid.*, 781 (1975).

**Titanium(III) chloride–Tetrahydrofurane–Magnesium, 5, 671.**

*Reduction of halides.*[1] The $TiCl_3 \cdot 3THF \cdot Mg$ system reduces bromides and iodides almost quantitatively at $20°$. The reactivity order is $I > Br > Cl > F$. Alkenes are by-products in the reaction of aliphatic alicyclic halides. The source of the hydrogen that replaces halogen is not known, but may be THF.

*Examples:*

$$C_6H_5Br \rightarrow C_6H_6 \quad (89\%)$$

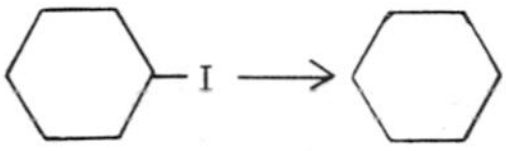

$(61\text{-}66\%)$

$$C_5H_{11}Br \rightarrow C_5H_{12} \quad + \quad C_3H_7CH=CH_2$$
$$(79\%) \qquad\qquad (12\%)$$

$$C_6H_5CH_2Cl \longrightarrow C_6H_5CH_3 \quad (96\text{-}100\%)$$

[1]S. Tyrlik and I. Wolochowicz, *J.C.S. Chem. Comm.*, 781 (1975).

**Titanium(IV) chloride, 1,** 1169–1171; **2,** 414–415; **3,** 291; **4,** 507–508; **5,** 671–672.

*Aldol condensation.* Japanese investigators[1] report that trimethylsilyl enol ethers of aldehydes or ketones react with carbonyl compounds in the presence of an equimolar amount of titanium tetrachloride to give the cross-aldol products in satisfactory yields. The success of the method probably depends on formation of an intermediate titanium chelate:

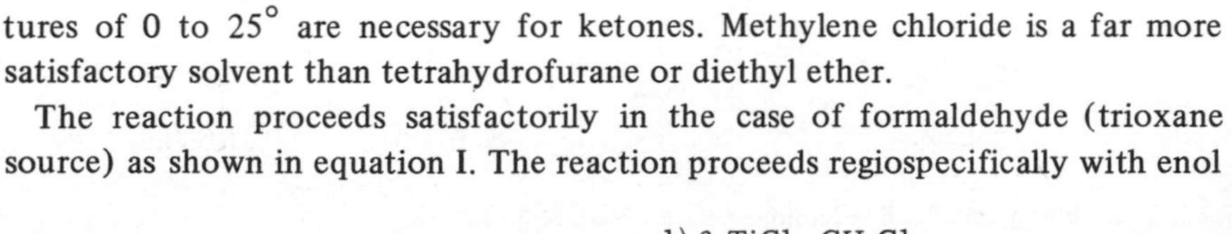

The reaction proceeds readily at $-78°$ in the case of aldehydes, but temperatures of 0 to $25°$ are necessary for ketones. Methylene chloride is a far more satisfactory solvent than tetrahydrofurane or diethyl ether.

The reaction proceeds satisfactorily in the case of formaldehyde (trioxane source) as shown in equation I. The reaction proceeds regiospecifically with enol

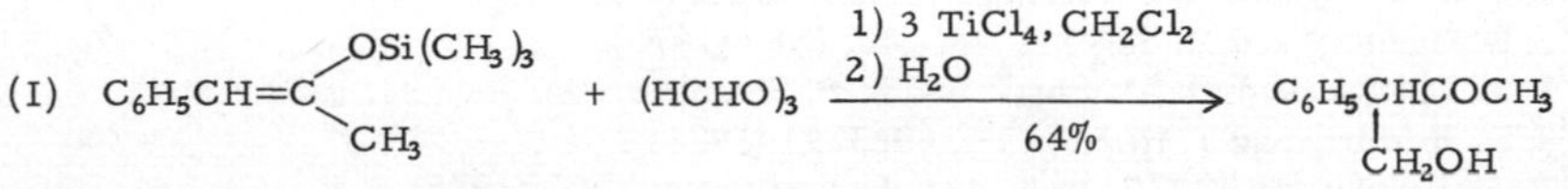

ethers of unsymmetrical ketones (equations II and III).

(II)   + $C_6H_5CHO$   $\xrightarrow[\text{58\%}]{\text{1) TiCl}_4,\ CH_2Cl_2\ \ 2)\ H_2O}$

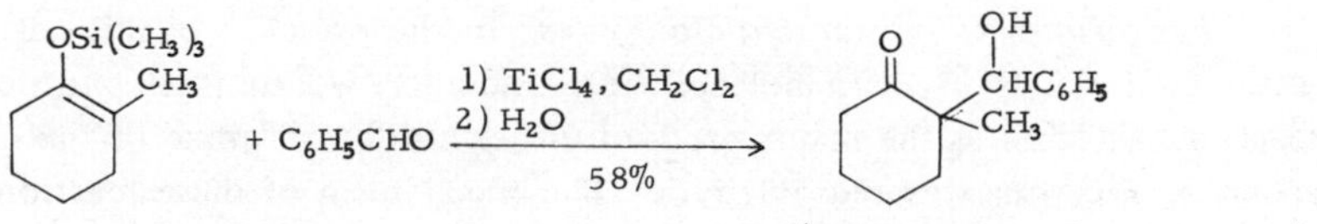

(threo- and erythro-isomers)

(III)   + $C_6H_5CHO$   $\xrightarrow{81\%}$

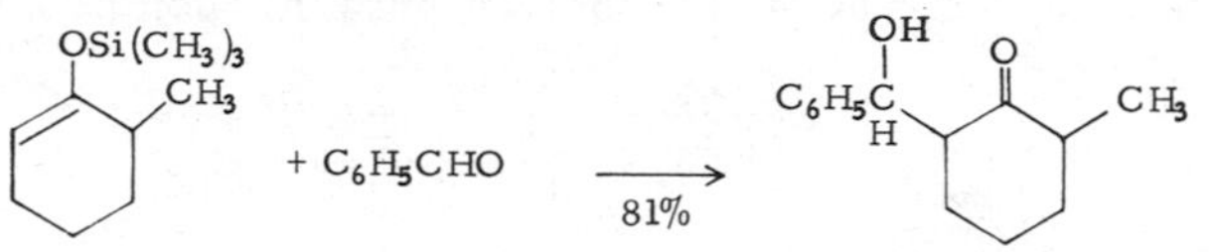

(4-isomers)

*Furanes.*[2] In a furane synthesis described recently the α-bromo acetal (1) and silyl enol ethers (2) react, in the presence of $TiCl_4$, to form β-alkoxy-γ-bromoketones (3). The products undergo dehydrohalogenation and loss of $CH_3OH$ when heated in toluene for 5–8 hr. to form furanes (4). The furanes (4) are sometimes accompanied by (5), formed by elimination of only $CH_3OH$.

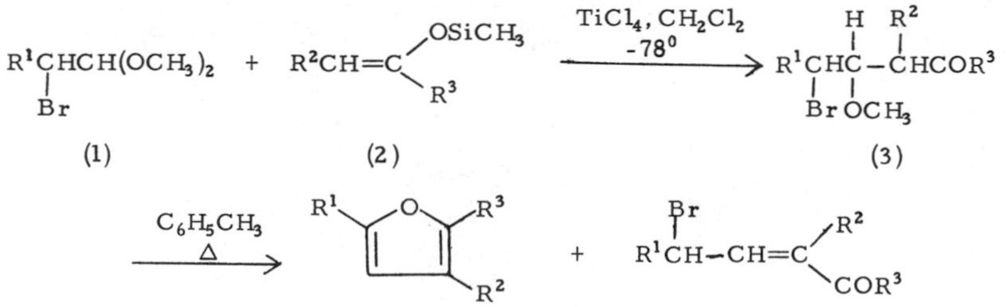

$R^1CHCH(OCH_3)_2$ (1) + $R^2CH=C$ (2) $\xrightarrow[-78^0]{\text{TiCl}_4,\ CH_2Cl_2}$ $R^1CHC-CHCOR^3$ (3)

$\xrightarrow[\Delta]{C_6H_5CH_3}$ (4, 32-82%) + (5, 0-21%)

The reaction can be extended to cyclic α-bromo ketals as shown in equation I.

(I)   + $H_2C=C$   $\xrightarrow{77\%}$

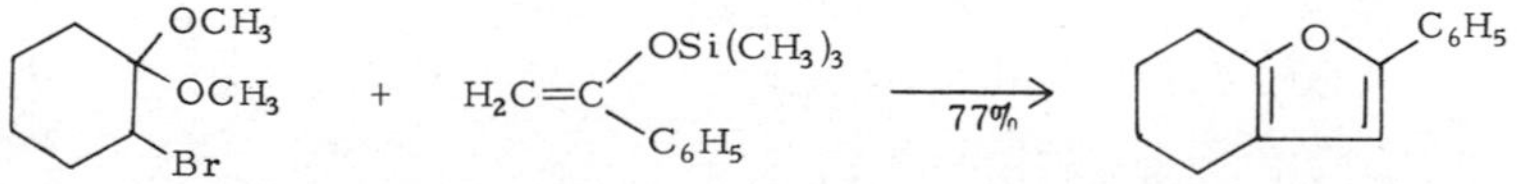

*Cyclopentenones.*[3] The synthesis of furanes described above has been modified to a synthesis of 2-(1-alkenyl)-2-cyclopentenones (equation II).

(II)   +   $\xrightarrow{\text{TiCl}_4}$

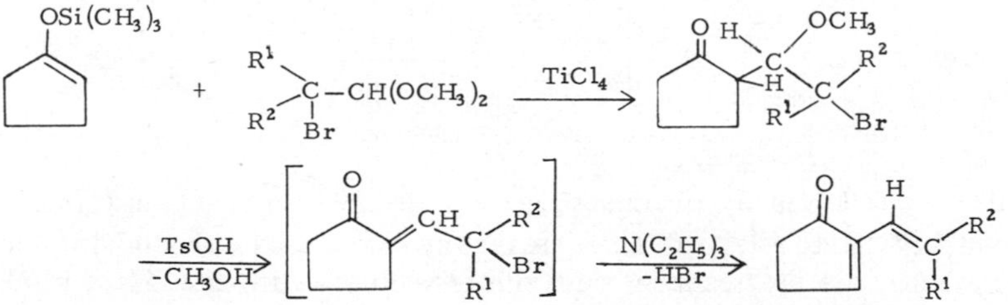

$\xrightarrow[-CH_3OH]{\text{TsOH}}$   $\xrightarrow[-HBr]{N(C_2H_5)_3}$ (45-90%)

*Dihydrofuranes; unsaturated aldehydes.* In the presence of this salt, α, β-unsaturated ketones in methanol solution undergo two unusual photochemical reactions to form as the major product either a dihydrofurane or the dimethyl acetal of a β,γ-unsaturated aldehyde. The mechanism of these reactions is not known, but in both reactions the oxygen of the original ketone is replaced by a new carbon atom. Four examples of this photochemical reaction have been reported (equations I–IV).[4]

(I)

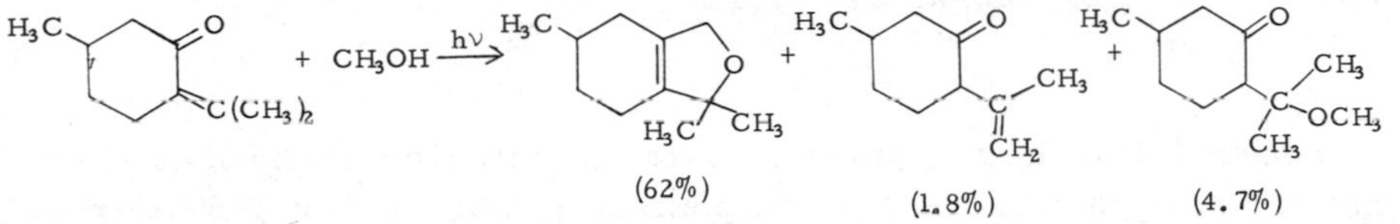

(62%)    (1. 8%)    (4. 7%)

(II)

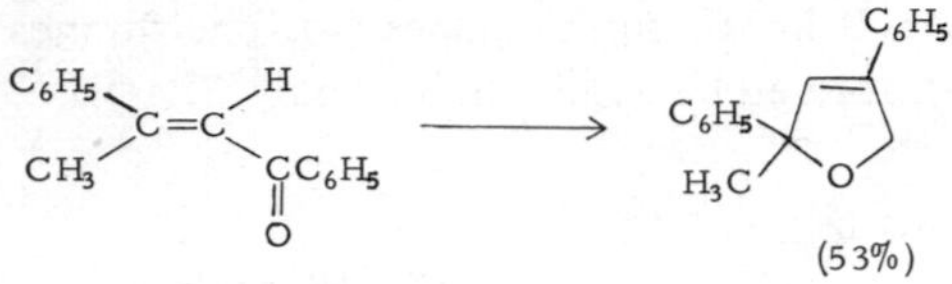

(53%)

(III)

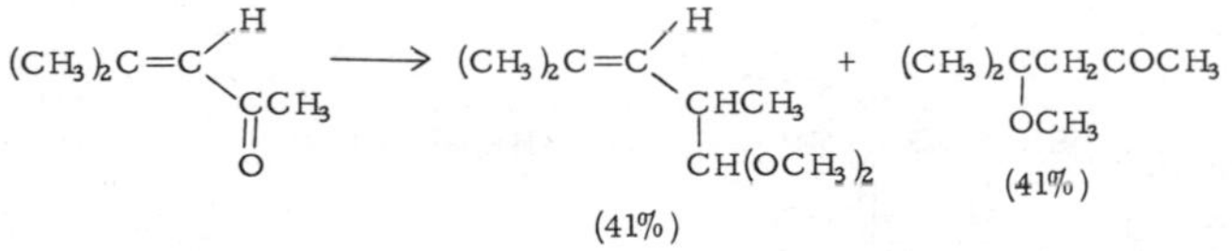

(41%)    (41%)

(IV)

(33%)

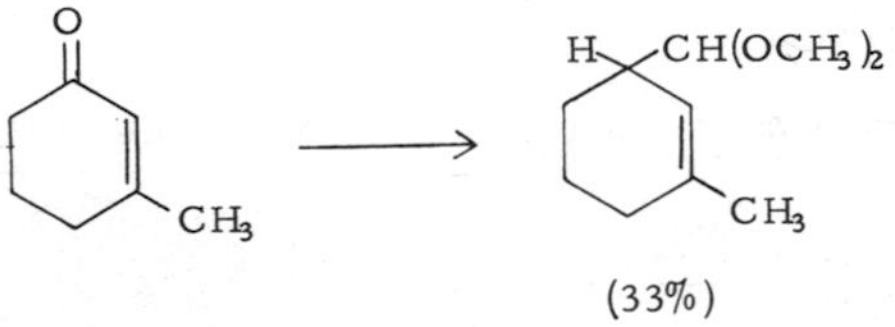

*δ-Hydroxy-β-keto esters.* Diketene reacts with acetals under catalysis with TiCl₄ to give δ-alkoxy-β-keto esters (1) in good yield.[5] Reaction with aldehydes

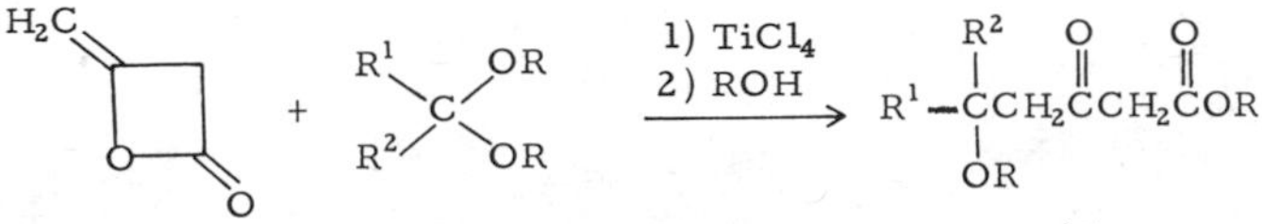

(1)

at −78° followed by treatment with an alcohol also results in formation of δ-hydroxy-β-keto esters (2). These are readily lactonized to 5,6-dihydro-4-hydroxy-2-pyrones (3) by treatment with dilute sodium hydroxide. These products are readily methylated by dimethyl sulfate.[6]

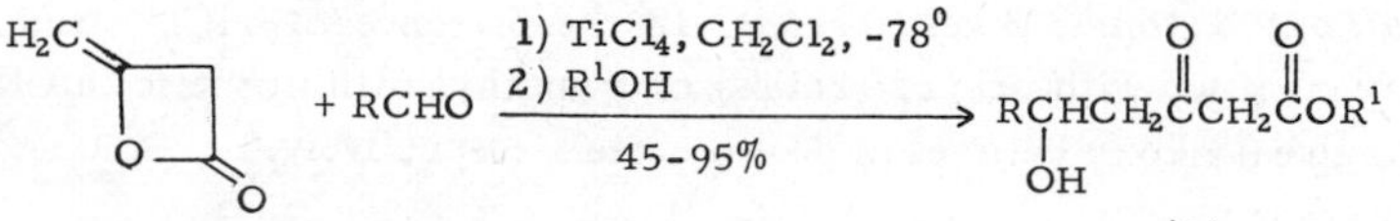

$$RCHCH_2\overset{O}{\overset{||}{C}}CH_2\overset{O}{\overset{||}{C}}OR^1$$
$$\underset{OH}{|}$$

(2)

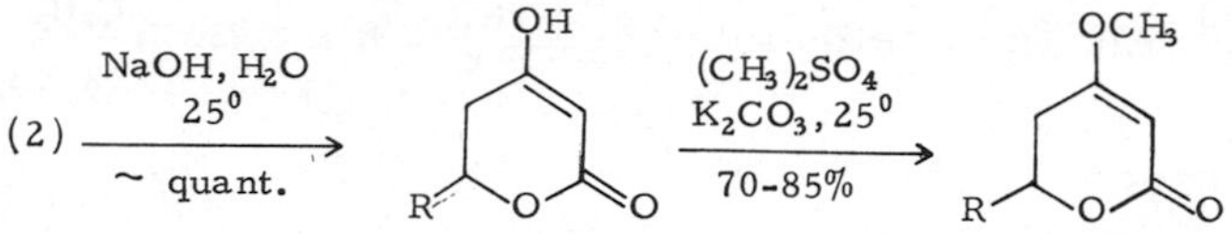

(3)                    (4)

*1,5-Diketones.* In the presence of $TiCl_4$, silyl enol ethers undergo conjugate addition to $\alpha,\beta$-unsaturated ketones and esters. This Michael reaction is carried out in methylene chloride at $-78°$ and is complete in less than 1 hr.[7]

*Examples:*

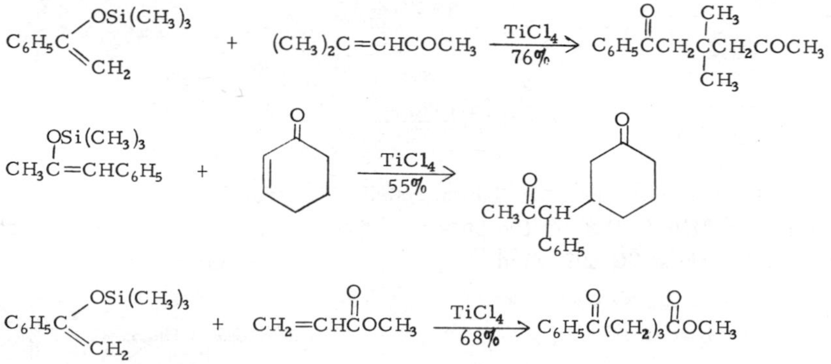

*Reaction with enol ethers of ethyl acetoacetate.* Reaction of the ethyl enol ether of ethyl acetoacetate (1) with $TiCl_4$ gives the derivative (2) of *p*-orsellinic acid; the phenyl enol ether (3) under the same conditions is converted into the derivative (4) of *o*-orsellinic acid.[8]

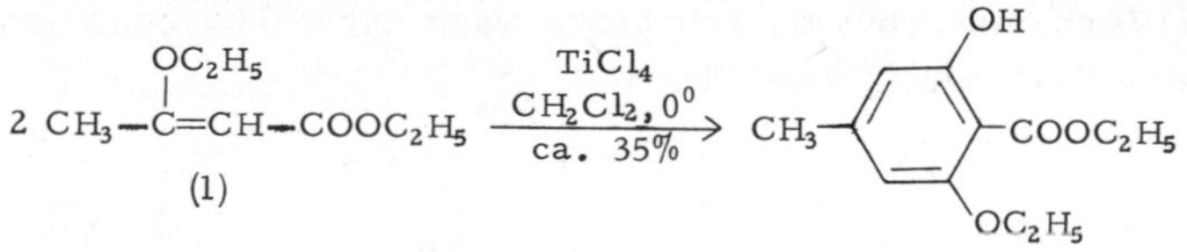

(1)                             (2)

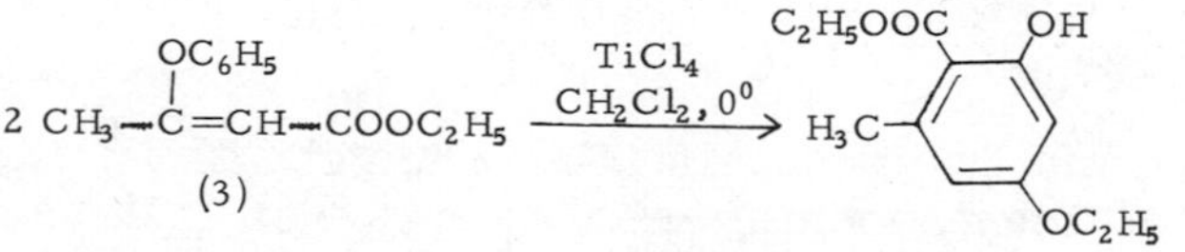

(3)                             (4)

*β-Alkoxy ketones; β-keto acetals.* In the presence of $TiCl_4$, trimethysilyl enol ethers react with acetals (ketals) or trimethyl orthoformate in $CH_2Cl_2$ at $-78°$ to give β-alkoxy ketones or β-keto acetals, respectively.[9]

$$\underset{OSi(CH_3)_3}{\overset{\displaystyle |}{CH_3C}}=CHC_6H_5 \quad + \quad C_6H_5CH(OC_2H_5)_2 \quad \xrightarrow[95\%]{TiCl_4} \quad \underset{OC_2H_5}{\overset{\displaystyle |}{C_6H_5CH}}-CH\overset{C_6H_5}{\underset{COCH_3}{\big\langle}}$$

$$\underset{OSi(CH_3)_3}{\overset{\displaystyle |}{CH_3C}}=CHC_6H_5 \quad + \quad HC(OCH_3)_3 \quad \xrightarrow[64\%]{TiCl_4} \quad (CH_3O)_2CHCH\overset{C_6H_5}{\underset{COCH_3}{\big\langle}}$$

$$\text{(cyclohexene-OSi(CH}_3)_3) \quad + \quad C_6H_5CH(OC_2H_5)_2 \quad \xrightarrow[95\%]{TiCl_4} \quad \text{(cyclohexanone-CH} \overset{OC_2H_5}{\underset{C_6H_5}{\big\langle}})$$

$$\text{(cyclohexene-OSi(CH}_3)_3) \quad + \quad HC(OCH_3)_3 \quad \xrightarrow[71\%]{TiCl_4} \quad \text{(cyclohexanone-CH(OCH}_3)_2)$$

*Allyl ethers.* Acetals of $\alpha,\beta$-unsaturated aldehydes (1) react with Grignard reagents in THF at $-78°$ in the presence of titanium tetrachloride to give allyl ethers (2) in about 70–80% yield.[10]

$$\underset{(1)}{R^1CH=CHCH(OCH_3)_2} \quad + \quad R^2MgX \quad \xrightarrow[70-80\%]{\overset{TiCl_4}{THF}} \quad R^1CH=CHCH\overset{OCH_3}{\underset{R^2}{\big\langle}} \quad (2)$$

*Cross aldol condensation.*[11] This aldol condensation has also been carried out with a variety of keto esters and silyl enol ethers.[12] The reaction is carried out in methylene chloride at $20°$ with equimolecular amounts of reactants and titanium(IV) chloride. Only the keto group reacts under these conditions and dehydration of the products is not observed.

*Examples:*

$$\underset{CH_3CH_2}{\overset{(CH_3)_3SiO}{\big\langle}}C=CHCH_3 \quad + \quad CH_3\overset{O}{\overset{\|}{C}}COOC_2H_5 \quad \xrightarrow[87\%]{TiCl_4} \quad CH_3CH_2\overset{O}{\overset{\|}{C}}\underset{CH_3}{\overset{\displaystyle |}{C}}H\underset{CH_3}{\overset{OH}{\underset{\displaystyle |}{C}}}COOC_2H_5$$

$$\text{(cyclohexene-OSi(CH}_3)_3) \quad + \quad CH_3\overset{O}{\overset{\|}{C}}COOC_2H_5 \quad \xrightarrow{88\%} \quad \text{(cyclohexanone-}\overset{OH}{\underset{CH_3}{\overset{\displaystyle |}{C}}}\overset{COOC_2H_5}{\big\langle})$$

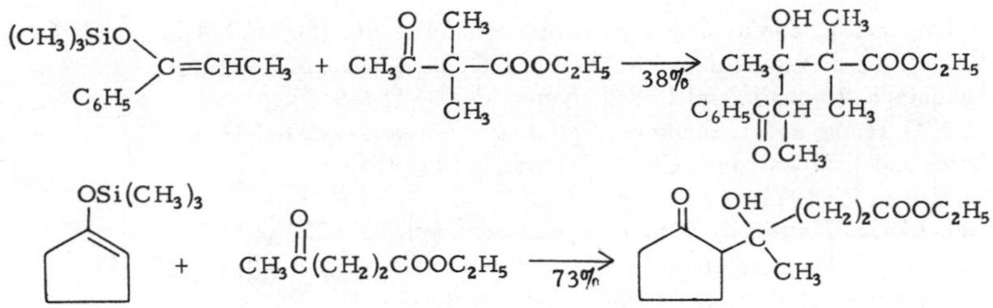

$$(CH_3)_3SiO \begin{matrix} \\ \\ C_6H_5 \end{matrix} C=CHCH_3 \; + \; CH_3\overset{O}{\underset{CH_3}{\overset{\|}{C}}}-\overset{CH_3}{\underset{CH_3}{C}}-COOC_2H_5 \xrightarrow{38\%} CH_3\overset{OH}{\overset{|}{C}}-\overset{CH_3}{\underset{|}{C}}-COOC_2H_5 \\ C_6H_5\underset{O\,CH_3}{\overset{|}{C}}CH \; CH_3$$

*1-Chlorobicyclooctanes.* [13]  The reaction of the bridgehead tosylates (1) with anhydrous $TiCl_4$ in ether gives the corresponding chlorides (2) in good yield. No reaction occurs in strongly solvating solvents (THF, dioxane, HMPT). $FeCl_3$ can be used, but yields are somewhat lower.

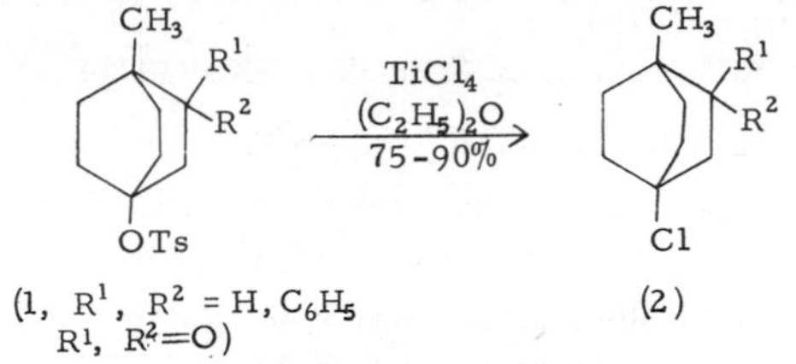

$$\text{(1, } R^1, \; R^2 = H, C_6H_5 \\ R^1, \; R^2 = O)$$

(2)

The tosylates (1) are converted into bromides by reaction with $MgBr_2$ [14]; iodides are obtained by reaction with $MgI_2$. [15]

*Knoevenagel condensation.* The condensation of aldehydes and sodio acetylacetonate in the presence of $TiCl_4$ leads to 3-alkylidene-2,4-pentanediones in 50–95% yield. The products exist as the chelates (a). [16]

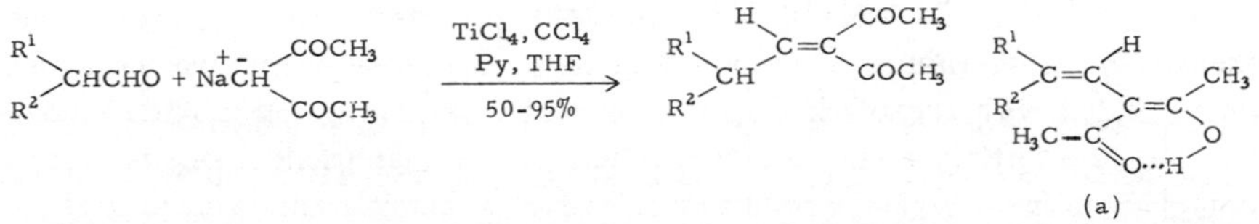

(a)

*Oxygen-transfer reactions.* Titanium tetrachloride is relatively inactive as a catalyst for photosensitized oxygenation of ergosteryl acetate (5, 483). However, if this Lewis acid is used in 1–2 molar eq., the chlorohydrin (2) is formed in yields as high as 75%. This reaction proceeds in the dark and involves triplet oxygen. The scope and mechanism of this interesting reaction await elucidation. [17]

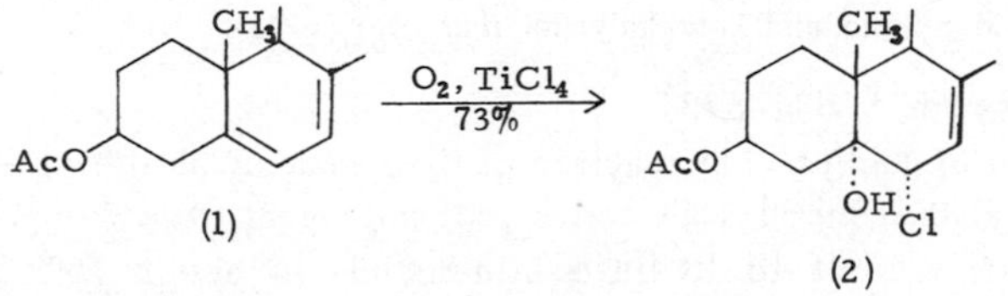

[1] T. Mukaiyama, K. Banno, and K. Narasaka, *Am. Soc.*, **96**, 7503 (1974).
[2] T. Mukaiyama, H. Ishihara, and K. Inomata, *Chem. Letters*, 527 (1975).
[3] H. Ishihara, K. Inomata, and T. Mukaiyama, *ibid.*, 531 (1975).
[4] T. Sato, G. Izumi, and T. Imamura, *Tetrahedron Letters*, 2191 (1975).
[5] T. Izawa and T. Mukaiyama, *Chem. Letters*, 1189 (1974).
[6] *Idem, ibid.*, 161 (1975).
[7] K. Narasaka, K. Soai, and T. Mukaiyama, *Chem. Letters*, 1223 (1974).
[8] G. Declercq, G. Moutardier, and P. Mastagli, *Compt. rend.*, **28** (C), 279 (1975).
[9] T. Mukaiyama, and M. Hayashi, *Chem. Letters*, 15 (1974).
[10] T. Mukaiyama and H. Ishikawa, *ibid.*, 1077 (1974).
[11] T. Mukaiyama, K. Banno, and K. Narasaka, *Am. Soc.*, **96**, 7503 (1974).
[12] K. Banno and T. Mukaiyama, *Chem. Letters*, 741 (1975).
[13] W. Kraus, and H.-D. Gräf, *Angew. Chem. internat. Ed.*, **14**, 824 (1975).
[14] D. Seyferth, H. P. Hofmann, R. Burton, and J. F. Helling, *Inorg. Chem.*, **1**, 227 (1962).
[15] K. Nützel, *Houben-Weyl-Müller*, Vol. 13/2a, Thieme, Stuttgart, 1973, p. 174.
[16] W. Lehnert, *Synthesis*, 667 (1974).
[17] D. H. R. Barton and R. K. Haynes, *J.C.S. Perkin I*, 2065 (1975).

**Titanium(IV) chloride–Lithium aluminum hydride.**

**gem-*Dichlorocyclopropanes*.** The low-valent titanium chloride obtained from $TiCl_4$ and $LiAlH_4$ effects the synthesis of *gem*-dichloropropanes from olefins and carbon tetrachloride. The reaction takes place at $0°$ in THF and has the advantage that a strong base is not required.[1]

**Methylenation.** 2-(Phenylthio)ethanols (2), prepared from ketones (1) and phenylthiomethyllithium, undergo reductive β-elimination to give 1-alkenes when treated with the black reagent prepared from $TiCl_4$ and $LiAlH_4$ in the presence of a tertiary amine [1,8-bis(dimethylamino)naphthalene or tri-*n*-butylamine]. Benzene–dioxane is used as solvent for the elimination (4 hr. reflux).[2]

(1)                          (2)                          (3)

[1] T. Mukaiyama, M. Shiono, K. Watanabe, and M. Onaka, *Chem. Letters*, 711 (1975).
[2] Y. Watanabe, M. Shiono, and T. Mukaiyama, *ibid.*, 871 (1975).

**Titanocene system, 5, 672–673.**

*Reduction of halides.* The reaction of titanocene dichloride with magnesium under argon at $0°$ generates an active reducing agent. Azo compounds are reduced by this reagent to hydrazo compounds in about 75% yield. Alkyl chlorides and bromides are reduced to the parent hydrocarbons in about 80–85% yield. In addition, α-halo ketones and esters are reduced. The reducing hydrogen has been shown to originate from the cyclopentadienyl group.[1]

*Examples:*

$\underline{n}\text{-}C_9H_{19}Br \longrightarrow \underline{n}\text{-}C_9H_{20}$   (84%)

2-Bromonaphthalene $\longrightarrow$ Naphthalene   (85%)

2-Bromocycloheptanone $\longrightarrow$ Cycloheptanone (68%) + Cycloheptanol (10%)

$BrCH(COOC_2H_5)_2 \longrightarrow CH_2(COOC_2H_5)_2$   (79%)

[1] T. R. Nelsen and J. J. Tufariello, *J. Org.*, **40**, 3159 (1975).

*p*-**Toluenesulfonic acid, 1,** 1172–1178; **4,** 508–510; **5,** 673–675.

*Selective cleavage of BOC groups.* Selective cleavage of the *t*-butyloxycarbonyl amino protecting group in the presence of benzyl esters has been reported with an acidic ion-exchange resin (one example)[1] and with 85% formic acid (two examples).[2] This reaction has recently been carried out (seven examples)[3] with 1 eq. of TsOH at room temperature (3 hr. or more) in about 90% yield. The resulting amino ester tosylates can be used directly in peptide synthesis. The BOC group can also be selectively cleaved under these conditions in the presence of a *p*-methoxybenzyl ester group.

[1] C. J. Gray and A. M. Khoujah, *Tetrahedron Letters*, 2647 (1969).
[2] H. Kinoshita and H. Kotake, *Chem. Letters*, 631 (1974).
[3] J. Goodacre, R. J. Ponsford, and I. Stirling, *Tetrahedron Letters*, 3609 (1975).

*p*-**Toluenesulfonyl azide (Tosyl azide), 1,** 1178–1179; **2,** 415–417; **3,** 291–292; **4,** 510; **5,** 675.

*α,β-Unsaturated diazo ketones.* Regitz *et al.*[1] reported that substituted vinyl methyl ketones can be converted into α, β-unsaturated diazo ketones by formylation followed by reaction with tosyl azide (**3,** 291–292). Harman *et al.*[2] report that improved yields are obtained when ethyl formate is replaced by diethyl oxalate. The method is formulated for preparation of 4-diazo-3-oxo-1-phenyl-1-butene.

$$C_6H_5CH=CHCOCH_3 + (H_5C_2OOC)_2 \xrightarrow[\ \ ]{\underset{C_2H_5OH}{Na}} C_6H_5CH=CHC\overset{O}{\overset{\|}{C}}\overset{Na^+}{\overset{}{\bar{C}}}HCOCOOC_2H_5$$

$$\xrightarrow[73\%\ \text{overall}]{\underset{C_2H_5OH}{TosN_3}} C_6H_5CH=CHC\overset{O}{\overset{\|}{C}}CHN_2$$

[1] M. Regitz, F. Menz, and A. Liedhegener, *Ann.*, **739,** 174 (1970).
[2] R. E. Harman, V. K. Sood, and S. K. Gupta, *Synthesis*, 577 (1974).

*p*-Toluenesulfonyl chloride, 1, 1179–1185; 3, 292; 4, 510–511; 5, 676–677.

*N-Methyl nitrones* → *N-methylamides* (4, 510–511). Details are now available for the rearrangement of ketonic nitrones.[1]

*Further examples:*

*Allylic chlorides* (3, 293). The preparation of geranyl chloride in 82–85% yield from the alcohol by the Stork procedure has been described in detail.[2]

Bunton *et al.*[3] have described a simple variant of Stork's procedure in which the alcohol is dissolved in dry pentane and treated with mesyl chloride at $-5°$. Pyridine is then added and the reaction mixture is allowed to come to room temperature. Geranyl chloride was obtained from the alcohol by this method in 79% yield.

[1] D. H. R. Barton, M. J. Day, R. H. Hesse, and M. M. Pechet, *J.C.S. Perkin I*, 1764 (1975).
[2] G. Stork, P. A. Grieco, and M. Gregson, *Org. Syn.*, **54**, 68 (1974).
[3] C. A. Bunton, D. L. Hachey, and J.-P. Leresche, *J. Org.*, **37**, 4036 (1972).

*p*-Toluenesulfonylhydrazine, 1, 1185–1187; 2, 417–423; 3, 293; 4, 511–512; 5, 678–681.

*Olefin synthesis by reaction of alkyllithiums with tosylhydrazones.* Review.[1] Two laboratories[2,3] have reported evidence for the mechanism formulated in equation I. If the reaction is carried out in TMEDA (an excellent solvent for the reaction), the vinyl anion intermediate can be trapped by electrophiles such as $D_2O$:

$$\xrightarrow[98\%]{\begin{array}{l}1)\ \underline{n}\text{-BuLi, TMEDA}\\2)\ D_2O\end{array}}$$

The postulated dianion has been trapped by addition of the electrophile to the reaction mixture at $-78°$ after 5 min.

High regioselectively has also been reported in an acyclic case: 2-Octanone is converted into essentially pure 1-octene.

*3H-1,2-Diazepines.* β-Alkyl-α,β;γ,δ-unsaturated ketones (1) react with tosylhydrazine to form 3,4-dihydro-2-tosyl-1,2-diazepines (2) rather than tosylhydrazones. The products eliminate *p*-toluenesulfinic acid when treated with base to form 3H-1,2-diazepines (3). Formation of (2) has been observed only

$$\xrightarrow[55\text{-}77\%]{\begin{array}{c}\text{TsNHNH}_2\\\text{H}^+\end{array}}$$

$$\xrightarrow[70\text{-}78\%]{\begin{array}{c}\text{NaOC}_2\text{H}_5\\\text{C}_6\text{H}_5\text{CH}_3\end{array}}$$

(1, R or $R^1$ = H, $CH_3$, $C_6H_5$)      (2)      (3)

when an alkyl substituent is at the β-position of (1); in the absence of this group only tosylhydrazones of (1) are formed. The diazepines (3) do not appear to exist as the bicyclic tautomers (as in the cycloheptatriene–norcaradiene system).[4]

*Transposition of an enone function.* Patel and Reusch[5] have used the Shapiro–Heath olefin synthesis (**2**, 418–419) in a method for transposition of a cyclic enone function (*cf.* **Wharton reaction, 1,** 419). The method is illustrated

$$\xrightarrow[75\%]{\begin{array}{c}\text{H}_2\text{O}_2\text{, NaOH}\\\text{CH}_3\text{OH}\end{array}}$$

$$\xrightarrow[80\%]{\begin{array}{c}\text{KOH}\\\text{CH}_3\text{OH}\end{array}}$$

$$\xrightarrow[85\%]{\begin{array}{c}\text{TsNHNH}_2\\\text{C}_2\text{H}_5\text{OH}\end{array}}$$

(1)      (2)      (3)

$$\xrightarrow[>95\%]{\text{CH}_3\text{Li}}$$

$$\xrightarrow[>95\%]{\text{HCl, THF}}$$

(4)      (5)      (6)

for conversion of isophorone (1) into the isomeric 4,4,6-trimethyl-2-cyclohexenone (6). The method is limited by the fact that it is not applicable to an $\alpha$-substituted $\alpha,\beta$-unsaturated ketone; in addition, a substituent at the alternate $\alpha'$-position can induce Favorskii-like rearrangement in the reaction of cyclic $\alpha,\beta$-epoxyketones with base.

[1] R. H. Shapiro, *Org. React.*, in press.
[2] R. H. Shapiro, M. F. Lipton, K. J. Kolonko, R. L. Buswell, and L. A. Capuano, *Tetrahedron Letters*, 1811 (1975).
[3] J. E. Stemke and F. T. Bond, *ibid.*, 1815 (1975).
[4] C. D. Anderson, J. T. Sharp, H. R. Sood, and R. S. Strathdee, *J.C.S. Chem. Comm.*, 613 (1975).
[5] K. M. Patel and W. Reusch, *Syn. Commun.*, 27 (1975).

**Tosylmethyl isocyanide (TosMIC), 4,** 514–516; **5,** 684–685.

*Preparation.* The preparation of the reagent from *p*-toluenesulfinic acid has been described.[1]

$$\underset{\text{90-95}^0}{\overset{\text{HCOOH}}{\xrightarrow{\hspace{1cm}}}}$$

$$\underline{p}\text{-CH}_3\text{C}_6\text{H}_4\text{SO}_2\text{H} + \text{CH}_2\text{O} + \text{H}_2\text{NCHO} \xrightarrow[\text{42-47\%}]{} \underline{p}\text{-CH}_3\text{C}_6\text{H}_4\text{SO}_2\text{CH}_2\text{NHCHO}$$

$$\xrightarrow[\text{76-84\%}]{\overset{\text{POCl}_3,\text{Glyme,}}{-5^0\rightarrow 0^0, \text{N}(\text{C}_2\text{H}_5)_3}} \underline{p}\text{-C}_6\text{H}_4\text{SO}_2\text{CH}_2\text{N}=\text{C}: + \text{HOPOCl}_2 + \text{HCl}$$

*Nitriles.* Details for the conversion of 2-adamantanone into 2-adamantanecarbonitrile (84–90% yield) are available.[2]

*17-Acetyl steroids.* 17-Keto steroids can be converted into 17-acetyl steroids by reaction with tosylmethyl isocyanide. The resulting nitriles are obtained as epimeric mixtures in which the 17$\beta$-epimer predominates. The mixtures can be separated by crystallization or chromatography. The 17-carbonitriles are converted into 17-acetyl steroids in satisfactory yield by brief treatment with methyllithium.[3]

$$\alpha:\beta\sim 2:3$$

[1] B. E. Hoogenboom, O. H. Oldenziel, and A. M. van Leusen, *Org. Syn.*, submitted (1975).
[2] O. H. Oldenziel, J. Wildeman, and A. M. van Leusen, *ibid.*, submitted (1975).
[3] J. R. Bull and A. Tuiman, *Tetrahedron*, 31, 2151 (1975); J. R. Bull, J. Floor, and A. Tuinman, *ibid.*, 31, 2157 (1975).

**Trialkynylalanes, $(\text{RC}\equiv\text{C})_3\text{Al}$.**

These substances are obtained by the reaction of aluminum chloride with an alkynyllithium.

*Alkynes.* These reagents undergo clean coupling with tertiary alkyl halides or secondary alkyl sulfonates.[1]

$$\text{(I)} \quad (RC{\equiv}C)_3Al \ + \ R'X \xrightarrow[0^0]{CH_2Cl_2} RC{\equiv}CR'$$

*Examples:*

$$(CH_3)_3CCl \ + \ (C_6H_5C{\equiv}C)_3Al \xrightarrow{90\%} (CH_3)_3CC{\equiv}CC_6H_5$$

[1] E. Negishi and S. Baba, *Am. Soc.*, **97**, 7385 (1975).

## Tri-*n*-butylamine, 1, 1189–1190.

*Dichlorocarbene.*[1] Dichloronorcarane can be prepared in about 75% yield from cyclohexene, chloroform, and an aqueous solution of sodium hydroxide in the presence of tri-*n*-butylamine or the hydrochloride as catalyst. Some other tertiary amines or quaternary ammonium salts are equally effective: tetra-*n*-butylammonium bromide, N-*n*-butylpiperidine, N,N-di-*n*-butylpiperidinium iodide. No primary or secondary amines were found to have this catalytic activity.

Makosza et al.[2] have confirmed the strong catalytic effect of trialkylamines in the generation of dichlorocarbene and also of dibromocarbene. Actually, dibromocarbene generated by catalysis with trialkylamines adds to 1-alkenes, a reaction that is not observed with dibromocarbene generated with tetraalkyl-ammonium halide catalysts. The authors have presented indirect evidence that dichlorocarbene reacts with a trialkylamine to form a basic salt that abstracts a proton from chloroform:

$$:CCl_2 \ + \ R_3N \overset{+ \ -}{\underset{}{\rightleftharpoons}} R_2NCCl_2 \xrightarrow[]{HCCl_3} \overset{+}{R_3}NCHCl_2CCl_3^- \longrightarrow \overset{+}{R_3}NCHCl_2Cl^- \ + \ :CCl_2$$

[1] K. Isagawa, Y. Kimura, and S. Kwon, *J. Org.*, **39**, 3171 (1974).
[2] M. Makosza, A. Kacprowicz, and M. Fedorynski, *Tetrahedron Letters*, 2119 (1975).

**Tri-*n*-butylphosphine–Dialkyl disulfides.**

*Preparation of sulfides.*[1] The reaction of the nucleoside adenosine (1.1 mmole) with dimethyl disulfide (10 mmole) and tri-*n*-butylphosphine (10 mmole) in DMF at 25° gives 5'-S-methylthio-5'-desoxyadenosine (2) in 73% yield. The reaction is general for desoxynucleosides and disulfides. This reaction

proceeds sluggishly when triphenylphosphine is used in place of tri-*n*-butylphosphine. Note that triphenylphosphine has been used as one component in related coupling reactions (*see* **Triphenylphosphine—Diethyl azodicarboxylate, 4,** 553–555 and **2,2'-Dipyridyl disulfide—Triphenylphosphine, 5,** 285–286).

[1] I. Nakagawa and T. Hata, *Tetrahedron Letters*, 1409 (1975).

*trans*-1-Tri-*n*-butylstannyl-1-propene-3-tetrahydropyranyl **ether** (1). Mol wt. 431.21, b.p. 140–142°/0.1 mm.

*Preparation.* This reagent is prepared in one step by the reaction of bis(tri-*n*-butyltin) oxide (Alfa), polymethylhydrosiloxane (reducing agent, **4,** 393–394), and propargyl tetrahydropyranyl ether at 80° initiated with azobisiso-butyronitrile. The *cis*-isomer is formed at temperatures below 80°.[1]

*trans-Allylic alcohols.*[1] This organotin reagent undergoes quantitative metal–metal exchange with *n*-butyllithium in THF at −78° to form the lithium compound (2), which reacts with alkyl halides to give *trans*-allylic alcohols in high yield. It has been used to prepare the *trans,trans*-allylic diol (3) by the reaction of (2) with 1,8-dibromooctane in about 85% isolated yield. Alcohols of this type are useful because the corresponding bromides can be coupled with

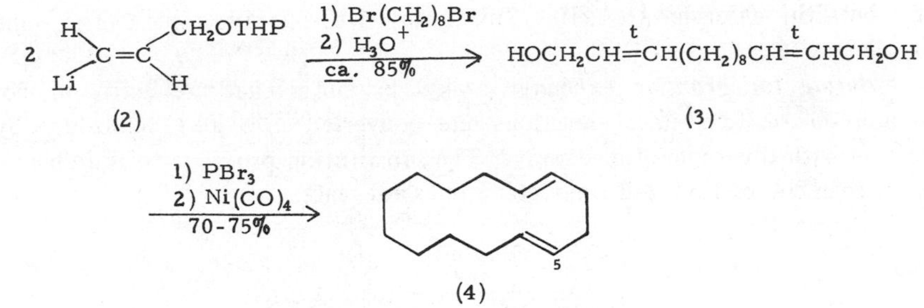
H
CH2OTHP
2 C=C
Li
H
1) Br(CH2)8Br
2) H3O+
ca. 85%
t
t
HOCH2CH=CH(CH2)8CH=CHCH2OH
(2)
(3)
1) PBr3
2) Ni(CO)4
70-75%
5
(4)

nickel carbonyl to form cyclic 1,5-dienes (4) (2, 291–293).

The lithium reagent (2) can also be used to obtain vinylated conjugate adducts. Thus it can be converted into the mixed cuprate (5) by reaction with 1-pentynylcopper in THF (−78°, 60 min., −50°, 10 min.). The cuprate undergoes

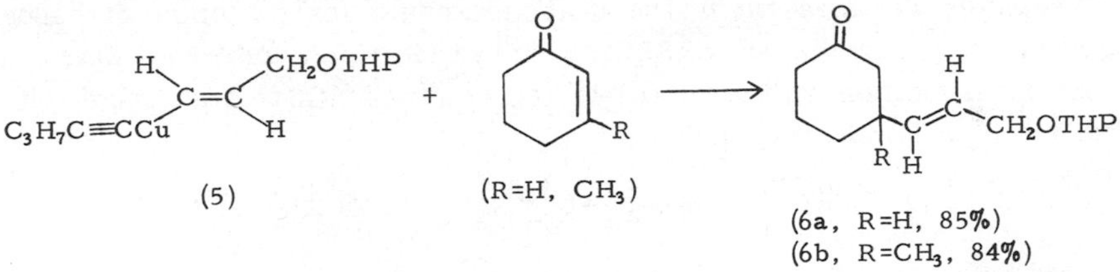
H
CH2OTHP
C=C
C3H7C≡CCu
H
+
O
R
(5)
(R=H, CH3)
O
H
C
C
CH2OTHP
R H
(6a, R=H, 85%)
(6b, R=CH3, 84%)

conjugate addition to cyclohexenone to give (6a) in high yield.[2] The corresponding adduct (7) from 2-cyclopentenone has been used to synthesize (12), which had been transformed earlier into 11-desoxyprostaglandin $E_2$ (13).

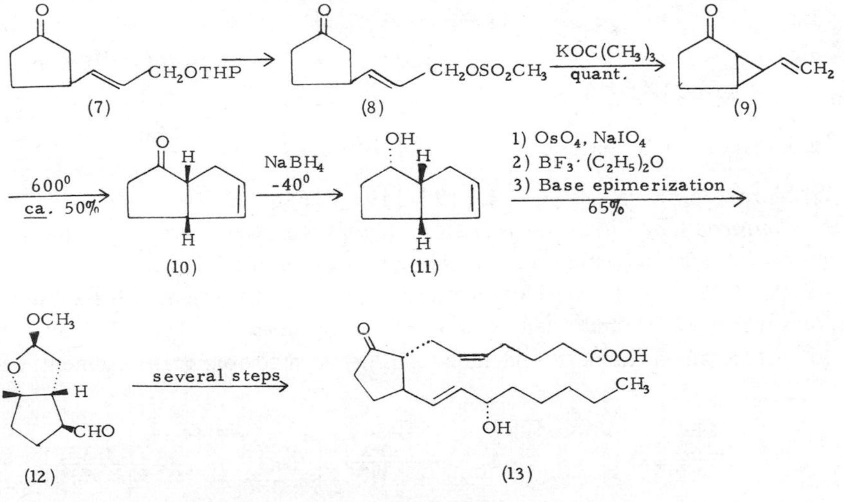
O
CH2OTHP
(7)
O
CH2OSO2CH3
(8)
KOC(CH3)3
quant.
O
CH2
(9)
600°
ca. 50%
O
H
H
(10)
NaBH4
-40°
OH
H
H
(11)
1) OsO4, NaIO4
2) BF3 · (C2H5)2O
3) Base epimerization
65%
OCH3
O
H
H
CHO
(12)
several steps
O
COOH
CH3
OH
(13)

[1]E. J. Corey and R. H. Wollenberg, *J. Org.*, **40**, 2265 (1975).

[2]Alkenyl groups are transferred in preference to alkynyl groups in mixed cuprates.

**Tri-*n*-butyltin chloride,** $(n\text{-}C_4H_9)_3SnCl$. Mol. wt. 325.49, b.p. 172°/5 mm. Supplier: Alfa.

*Chloride for bromide exchange.* Alkyl bromides that are active in carbonium-ion or free-radical reactions are converted into alkyl chlorides by reaction with this organotin reagent.[1] The substitution proceeds to equilibrium rather than completion. 1-Bromooctane does not react.

*Example:*

$$(C_6H_5)_2CHBr \xrightarrow[\quad\quad]{\substack{Bu_3SnCl \\ 50°}} (C_6H_5)_2CHCl$$

$$(30\%) \qquad\qquad\qquad (70\%)$$

[1] E. C. Friedrich, P. F. Vartanian, and R. L. Holmstead, *J. Organometal. Chem.*, **102**, 41 (1975).

**Tri-*n*-butyltin hydride,** 1, 1192–1193; 2, 424; 3, 294; 4, 518–520; 5, 685–686.

*Catalytic dehalogenation.* The usual procedure for reduction of halides utilizes a stoichiometric amount of the tin hydride. Corey and Suggs[1] have conducted this reaction with use of 0.1–0.3 eq. of a trialkyltin chloride and about 2 eq. of sodium borohydride in ethanol:

$$2\ R'_3SnX + 2\ NaBH_4 \longrightarrow 2\ R'_3SnH + 2\ NaCl + B_2H_6$$

$$RX + R'_3SnH \longrightarrow RH + R'_3SnX$$

The reduction is usually carried out under irradiation. Yields are high (86–100%).

The ethanolic solution of tri-*n*-butyltin hydride can be used for hydrostannation:

$$HC\equiv CCH_2OTHP + Bu_3SnH \xrightarrow[73.5\%]{h\nu} \underset{Bu_3Sn}{\overset{H}{\diagdown}}C=C\underset{H}{\overset{CH_2OTHP}{\diagup}}$$

[1] E. J. Corey and J. W. Suggs, *J. Org.*, **40**, 2554 (1975).

**Trichloroacetonitrile,** $CCl_3CN$, 1, 1194–1195; 5, 686.

*Conversion of allylic alcohols into rearranged allylic amines.*[1] This transposition can be accomplished by conversion of an allylic alcohol, for example geraniol (1), into the trichloroacetimidate (2) by treatment with sodium or potassium hydride and then with trichloroacetonitrile (ether, –10° and then 20°). This allylic imidate undergoes a [3.3] sigmatropic rearrangement when

(1)                    (2)                         (3)                   (4)

heated in toluene (8 hr.) to give the allylic trichloroacetamide (3), which on alkaline hydrolysis gives the amine (4). The reaction is general, although yields are low in the case of 3-alkyl-2-cyclohexene-1-ols.

[1] L. E. Overman, *Am. Soc.*, **96**, 597 (1974); *idem, Tetrahedron Letters*, 1149 (1975); L. A. Clizbe and L. E. Overman, *Org. Syn.*, submitted (1975).

## 2,2,2-Trichloroethanol, 3, 295–296; 4, 521–522.

*2,2,2-Trichloroethyl dihydrogen phosphate* (2). This substance can be prepared[1] as a crystalline solid (m.p. 120–121°) in good yield by the following procedure: The alcohol and phosphoryl chloride are refluxed with a catalyst (AlCl$_3$, KCl, or pyridine) for 4 hr. (HCl is evolved). The resulting phosphorodichloridate (1) is then heated with a small excess of water for 1 hr. at 80°.

$$CCl_3CH_2OH + POCl_3 \xrightarrow[76\%]{AlCl_3, \, \Delta} \underset{(1)}{CCl_3CH_2O\diagdown \underset{O\diagup}{P}\diagup^{Cl}_{\diagdown Cl}} \xrightarrow[85\%]{\underset{80°}{H_2O}} \underset{(2)}{CCl_3CH_2O\diagdown \underset{O\diagup}{P}\diagup^{OH}_{\diagdown OH}}$$

This method is a general route to aryl and alkyl dihydrogen phosphates. These products have been used in synthesis of oligonucleotides via phosphotriester intermediates.[2]

[1] G. R. Owen, C. B. Reese, C. J. Ransom, J. H. van Boom, and J. D. H. Herscheid, *Synthesis*, 704 (1974).

[2] N. J. Cusack, C. B. Reese, and J. H. van Boom, *Tetrahedron Letters*, 2209 (1973) and references cited therein; E. S. Werstiuk and T. Neilson, *Canad. J. Chem.*, **50**, 1283 (1972).

## Trichloroisocyanuric acid, 2, 426–427; 3, 297; 5, 687. Trade name: Chloreal.[1]

*Chlorination of phenyl allyl sulfides.* Cohen *et al.*[2] have reported that this inexpensive reagent is more effective than NCS for chlorination of phenyl allyl sulfides to give 1-thiophenoxy-3-chloroalkenes (equation I).

$$(I) \quad \underset{\text{(triazine ring with Cl, O, N substituents)}}{\boxed{}} + \underset{C_6H_5SCH-CH=CHR^2}{\overset{R^1}{|}} \xrightarrow[88-98\%]{5°, CCl_4} \underset{C_6H_5SC=CHCHR^2}{\overset{R^1}{|} \; \overset{Cl}{|}}$$

These products can be used as synthetic equivalents of $\alpha,\beta$-unsaturated aldehydes in Michael reactions. Thus $C_6H_5SCH=CHCH_2Cl$[3] can be used as the equivalent of acrolein, $CH_2=CHCHO$, which normally undergoes 1,2- rather than 1,4-addition with organometallic reagents. An interesting example is the preparation of an intermediate (1) to the pyridine annelation reagent of Danishefsky and Cain.[4]

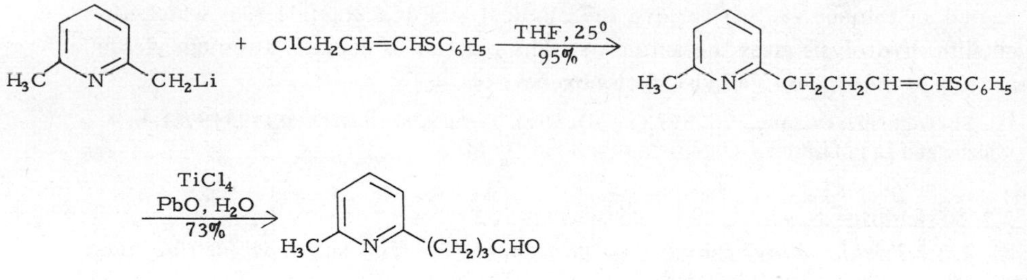

(1)

[1] *Merck Index*, **8**, 1069 (1968).
[2] A. J. Mura, Jr., D. A. Bernett, and T. Cohen, *Tetrahedron Letters*, 4433 (1975).
[3] Phenyl allyl sulfide is available from Columbia Organic Chemicals.
[4] S. Danishefsky and P. Cain, *J. Org.*, **39**, 2925 (1974); *idem, Am. Soc.*, **97**, 5282 (1975); see **2-Methyl-6-vinylpyridine**, this volume.

**Trichlorosilane,** 3, 298–299; 4, 525–526; 5, 687–688.

*Reductive deoxygenation of esters.* Under irradiation, esters of nonprimary aliphatic alcohols are reduced to ethers and hydrocarbons (equation I). The relative amounts of the two products are profoundly affected by the $R'$ group.[1]

$$(I) \quad R\overset{\overset{\displaystyle O}{\|}}{C}OR' \xrightarrow[h\nu]{Cl_3SiH} RCH_2OR' + R'H$$

*Examples:*

$$CH_3(CH_2)_{10}CH_2OCOCH_3 \longrightarrow \underset{98:2}{CH_3(CH_2)_{10}CH_2OCH_2CH_3} + CH_3(CH_2)_{10}CH_3$$

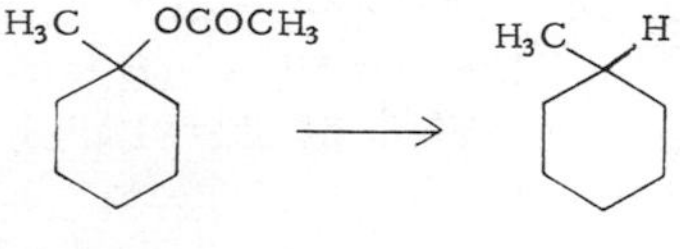

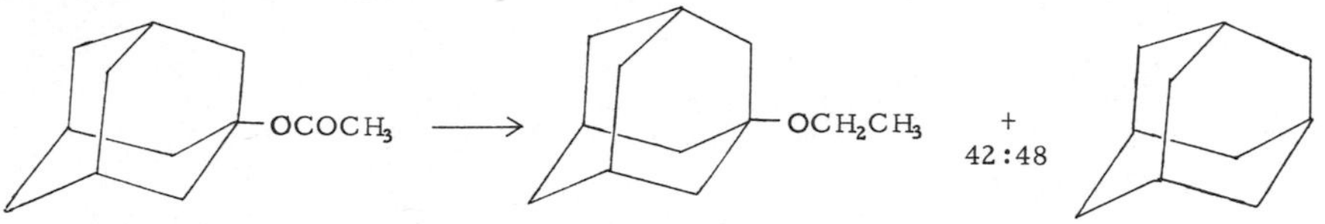

[1] S. W. Baldwin and S. A. Haut, *J. Org.*, **40**, 3885 (1975).

**1,1,1-Trichloro-3,3,3-trifluoroacetone,** $F_3C\overset{\overset{\displaystyle O}{\|}}{C}CCl_3$. Mol. wt. 215.40, b.p. 83.5–84.5°. Supplier: PCR.

The reagent trifluoroacetylates amino groups of amino acids in DMSO at 25–35° (22 hr.).[1]

$$F_3CCCCl_3 + H_2N- \longrightarrow CF_3CNH- + CHCl_3$$

[1] C. A. Panetta, *Org. Syn.*, submitted (1974).

**Triethylallyloxysilane,** $CH_2=CHCH_2OSi(C_2H_5)_3$ (1). Mol. wt. 172.34.

This allyl ether is prepared by treatment of the sodium or lithium salt of allyl alcohol with triethylchlorosilane in ether at room temperature.

*Synthesis of aldehydes.*[1,2] Treatment of (1) with *sec*-butyllithium in THF at −78° leads to the carbanion (a), which reacts primarily at the γ-position with alkyl iodides and bromides to give enol ethers of aldehydes (3). Exclusive formation of the *cis*-isomer (3) suggests that the reactive anionic species is (b). The

main limitation to this aldehyde synthesis is that (4) predominates over (3) in the case of highly hindered alkylating reagents.

Less favorable ratios of (3) to (4) are also obtained in the case of ordinary allylic esters, $CH_2=CHCH_2OR(Ar)$. *See also* **Allyl methyl ether,** this volume.

[1] W. C. Still and T. L. Macdonald, *Am. Soc.*, **96**, 5561 (1974).
[2] D. A. Evans, G. C. Andrews, and B. Buckwalter, *ibid.*, **96**, 5560 (1974).

**Triethylaminoaluminum hydride,** $(C_2H_5)_3NAlH_3$. Mol. wt. 127.16, m.p. 18–19°.

This reducing agent is prepared[1] from lithium aluminum hydride and triethyl-ammonium chloride (82% yield). It is more stable than aluminum hydride.

*Reduction of steroid ketones.*[2] 3-Ketosteroids are reduced by triethyl-aminoaluminum hydride in THF mainly to the 3β-alcohol (75–95% yield). α,β-Unsaturated 3-ketosteroids are reduced to the α,β-unsaturated alcohols.

[1] J. K. Ruff and M. F. Hawthorne, *Am. Soc.*, **82**, 2141 (1960).
[2] S. Cacchi, B. Giannoli, and D. Misiti, *Synthesis*, 728 (1974).

**Triethyl orthoacetate,** 3, 300–302; 4, 527.

*Trisubstituted C=C bonds.*[1] Claisen rearrangement of the β-hydroxy α-methylene acetals (1)[2] under the conditions of Johnson, Faulkner, *et al.* (3, 301)

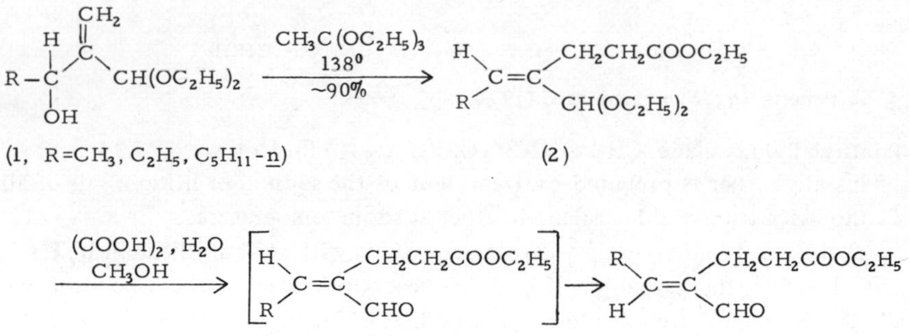

(1, $R = CH_3, C_2H_5, C_5H_{11}\text{-}\underline{n}$)     (2)

(a)     (3)

gives Z-trisubstituted olefins (2) in yields of ~90%. Deacetalization of the acetal function is accompanied by rearrangement to the more stable E-aldehyde (3).

The advantage of this approach is that the two functional groups of (2) or (3) are capable of various transformations to give either (E) or (Z) trisubstituted olefins. An example is the conversion of (3) into (6).

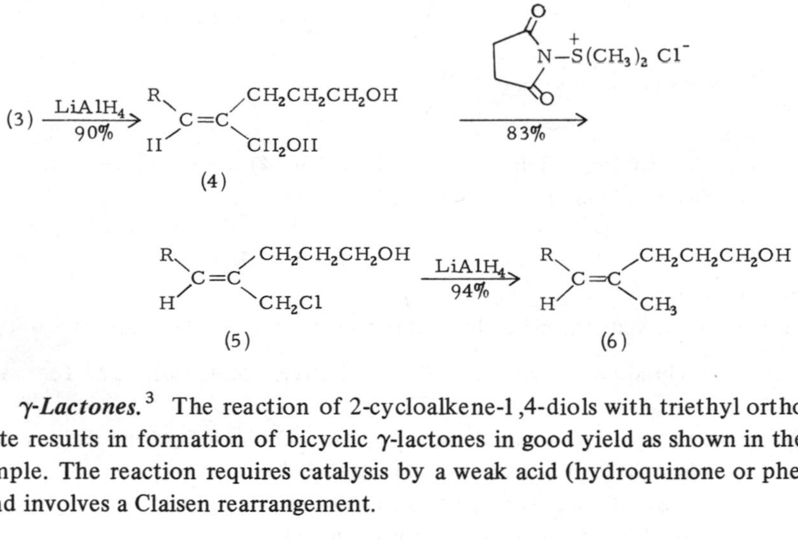

(4)

(5)     (6)

*γ-Lactones.*[3]  The reaction of 2-cycloalkene-1,4-diols with triethyl orthoacetate results in formation of bicyclic γ-lactones in good yield as shown in the example. The reaction requires catalysis by a weak acid (hydroquinone or phenol) and involves a Claisen rearrangement.

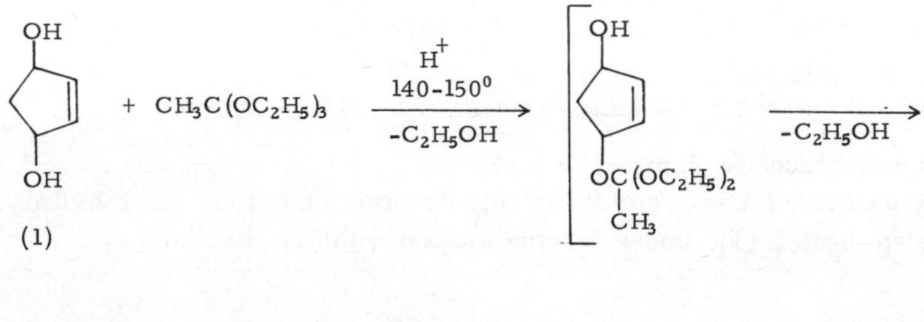

(1)

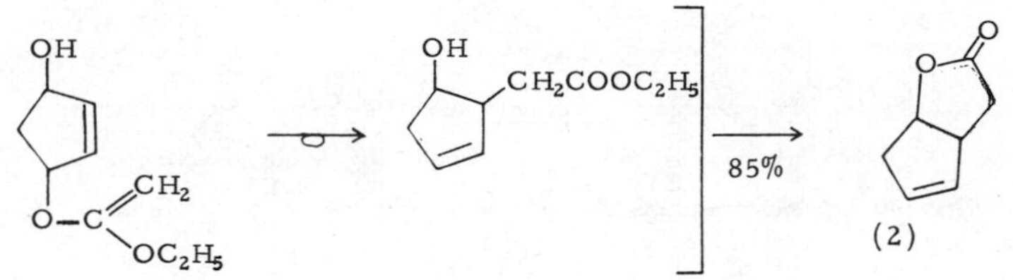

The lactol (3) corresponding to (2) can be prepared by an analogous rearrangement:

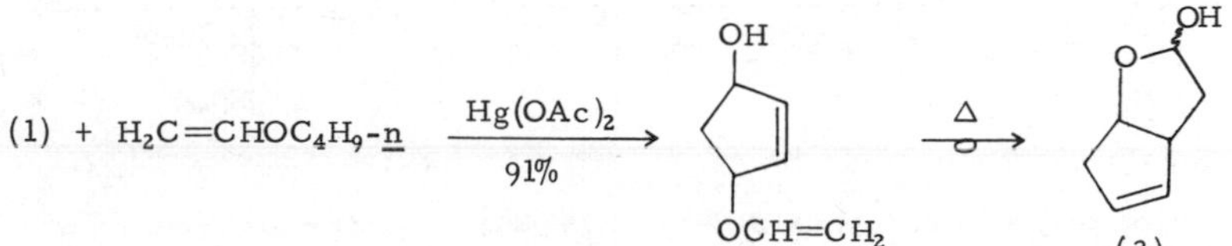

(1) + $H_2C=CHOC_4H_9$-$\underline{n}$  $\xrightarrow{\text{Hg(OAc)}_2}$  91%

*α, β-Unsaturated γ-lactones (butenolides).* Alkylated butenolides can be prepared in moderate yield by condensation of 2-alkyne-1,4-diols with this ortho ester followed by Claisen rearrangement.[4]

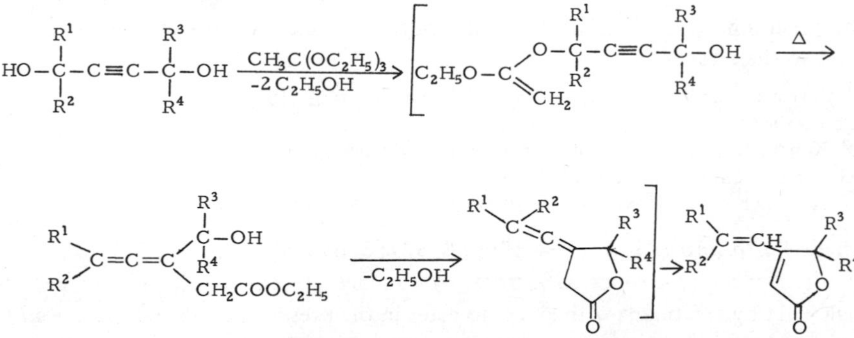

*Ortho ester rearrangement of allylic alkynyl alcohols.* The allylic alkynyl alcohols (1) and (4) undergo stereoselective ortho ester Claisen rearrangement when refluxed with triethyl orthoacetate (propionic acid catalysis). In both cases

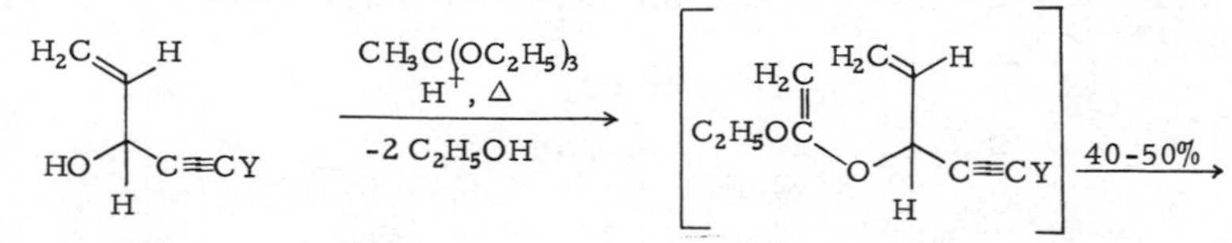

$[1, Y = H, Si(CH_3)_3]$

(a)

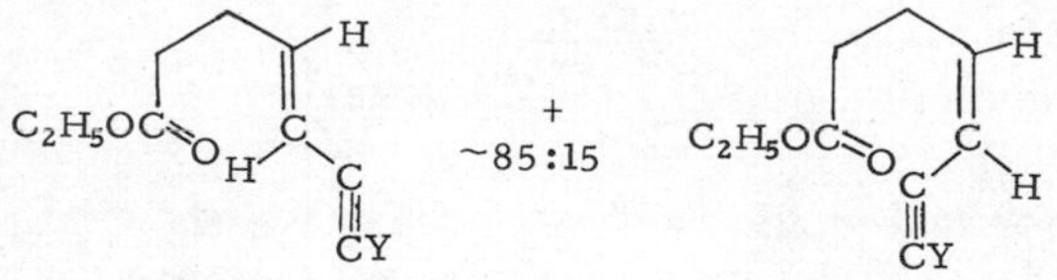

(2)     ~85:15     (3)

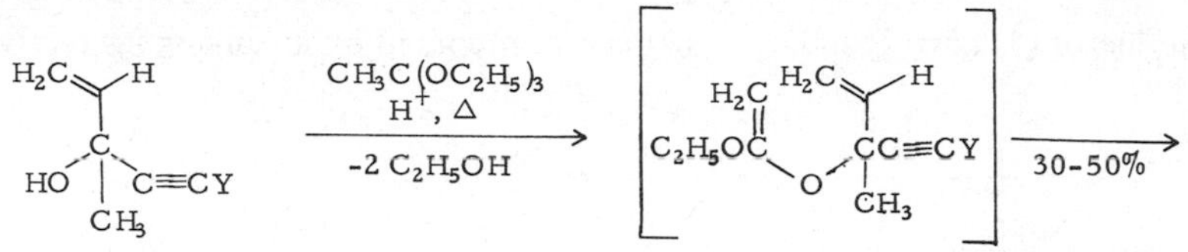

[ 4, Y = H, Si(CH₃)₃ ]     (b)     30–50%

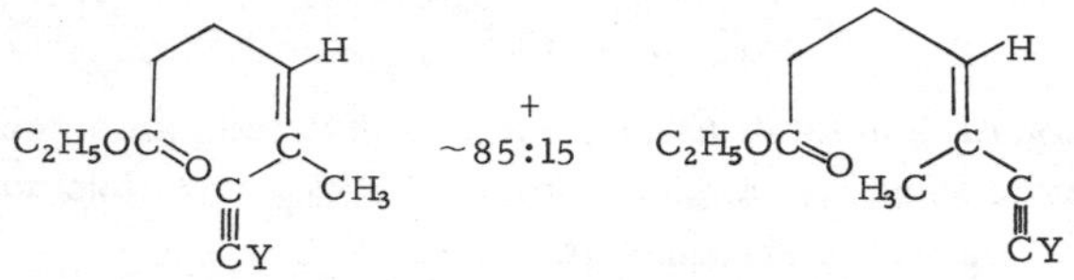

(5)     ~85:15     (6)

the predominant olefin formed is the one in which the bulkier substituent is *trans* to the ester side chain.[5]

[1] J. C. Depezay and Y. Le Merrer, *Tetrahedron Letters*, 3469 (1975).
[2] Preparation: *idem, ibid.*, 2751 (1974).
[3] K. Kondo, M. Matsumoto, and F. Mori, *Angew. Chem., internat. Ed.*, **14**, 103 (1975).
[4] G. Bennett, *Chem. Letters*, 939 (1975).
[5] K. A. Parker and R. W. Kosley, Jr., *Tetrahedron Letters*, 691 (1975).

**Triethyl orthoformate, 1**, 1204–1209; **4**, 527; **5**, 690–691.

*Conversion of oximes to nitriles.* Aldoximes are converted into nitriles in high yield by treatment with an ortho ester in the presence of a catalytic amount of an acid (methanesulfonic acid). The primary product is the oxime dialkyl ortho ester, which can be isolated in the absence of the acid catalyst.[1]

$$RCH{=}NOH + R'C(OC_2H_5)_3 \underset{-C_2H_5OH}{\rightleftharpoons} RCH{=}NOCR'(OC_2H_5)_2 \xrightarrow{H^+} RCN + R'COOC_2H_5 + C_2H_5OH$$

Ortho esters are also effective for Beckmann fragmentation of α-oximino ketones.

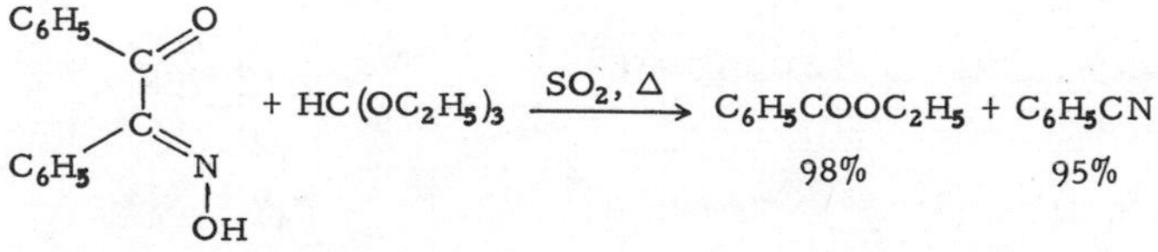

*Cyclization of an α, β-unsaturated ketoamide.*[2] In a recent practicable synthesis of (±)-perhydrohistrionicotoxin (4),[3] a key step involved cyclization of the α,β-unsaturated ketoamide (1), prepared in two steps from methyl 4-(chloroformyl)butyrate, to the more stable isomeric ketolactam (2). This reaction was accomplished by treatment with triethyl orthoformate in ethanol with

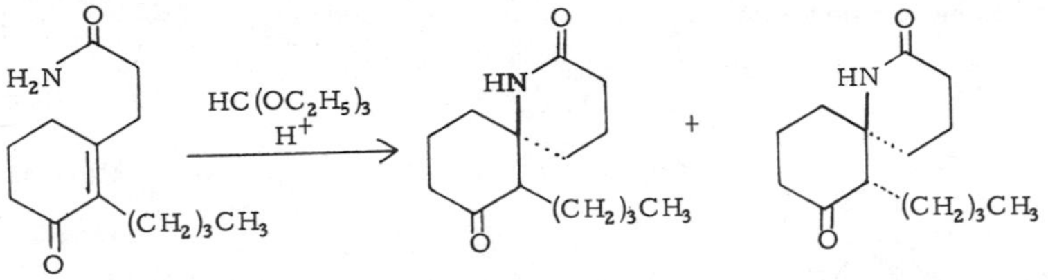

(1)                    (2)                    (3)

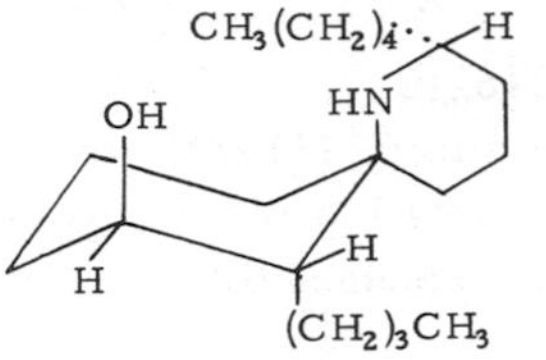

(4)

*d*-10-camphorsulfonic acid as catalyst. A mixture of (2) and (3) in the ratio of 2:1 was obtained in almost quantitative yield. Conventional reactions were used to convert the mixture into the toxin (4), which was obtained in almost 14% overall yield from the starting ester.

[1] M. M. Rogić, J. F. van Peppen, K. P. Klein, and T. R. Demmin, *J. Org.*, **39**, 3424 (1974).
[2] T. Fukuyama, L. V. Dunkerton, M. Aratani, and Y. Kishi, *ibid.*, **40**, 2011 (1975).
[3] Histrionicotoxins are toxic alkaloids from the venom of the frog *Dendrobates histrionicus;* they inhibit the cholinergic receptor.

**Triethyloxonium tetrafluoroborate, 1,** 1210–1212; **2,** 430–431; **3,** 303; **4,** 527–529; **5,** 691–693.

*Reaction with cyclic β-diketones.* Cyclohexane-1,3-dione (also dimedone) reacts with 1 eq. each of Meerwein's reagent and ethyldiisopropylamine to give the monoenol ether in quantitative yield. When excess reagent is used, 1,3-diethoxycyclohexadiene-1,3 is obtained in essentially quantitative yield. These

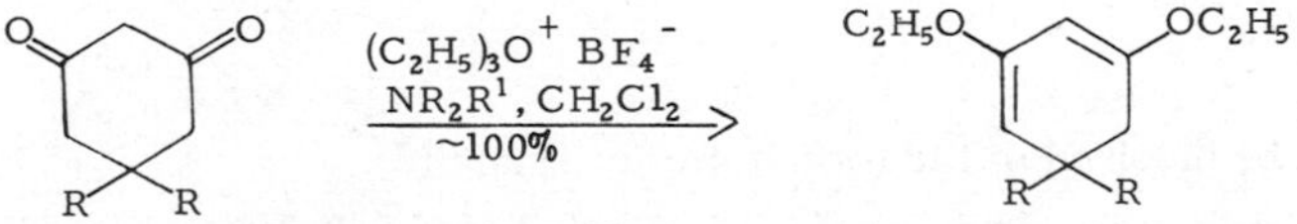

(R = H, CH₃)

1,3-diethoxy-1,3-dienes have high reactivity in Diels–Alder reactions. They had been obtained previously by Birch reduction of 1,3-diethoxybenzenes.[1]

[1]E. Rizzardo, *J.C.S. Chem. Comm.*, 644 (1975).

**Triethyl phosphite, 1,** 1212–1216; **2,** 432–433; **3,** 304; **4,** 529–530; **5,** 693.

*Desulfurization.* A recent, improved two-step synthesis of [2.2]paracyclophane has been reported.[1]

[1]M. Brink, *Synthesis*, 807 (1975).

**Triethyl phosphonoacetate, 1,** 1216–1217.

*Aziridines.* Treatment of the nitrone (1) with sodio triethyl phosphonoacetate in DME at 70° results in formation of the two aziridines (2) and (3).[1] Formation of two products can be explained by tautomerization of the nitrone.

Since Huisgen and Wulff[2] have shown that methylenetriphenylphosphorane reacts with nitrones to give oxazaphospholidines (4), an intermediate of this

type may be involved in the formation of (2) and (3).

[1]E. Breuer and I. Ronen-Braunstein, *J.C.S. Chem. Comm.*, 949 (1974).
[2]R. Huisgen and J. Wulff, *Ber.*, **102**, 746 (1969).

**Triethylsilyl thiophenoxide,** $(C_2H_5)_3SiSC_6H_5$. Mol. wt. 224.44, b.p. 77° (0.25 mm).

*Preparation.* The reagent is prepared in quantitative yield by the reaction of triethylsilane and thiophenol catalyzed by tris(triphenylphosphine)chlororhodium at 50° (15 min.).

$$(C_2H_5)_3SiH + C_6H_5SH \xrightarrow[\text{quant.}]{\text{cat.}} (C_2H_5)_3SiSC_6H_5 + H_2$$

(1)

*Silyl enol ethers.*[1] The reagent (1) reacts with enolizable ketones in refluxing xylene (7 hr.) to give triethylsilyl enol ethers (2):

$$R^1COCH_2R^2 + (1) \xrightarrow{100\text{-}140^0} R^1-\underset{\underset{OSi(C_2H_5)_3}{|}}{C}=CHR^2 + C_6H_5SH$$

(2)

The reaction can also be carried out with (1) generated *in situ:*

$$R^1COCH_2R^2 + (C_2H_5)_3SiH + C_6H_5SH \xrightarrow[-H_2]{\substack{[(C_6H_5)_3P]_3RhCl \\ C_6H_6, \ \Delta}} (2)$$

Nonenolizable ketones are hydrosilylated without dehydrogenation under these conditions.

Yields are comparable by both procedures.

[1] I. Ojima, M. Nihonyanagi, and Y. Nagai, *J. Organometal. Chem.*, **50**, C26 (1973).

**Triethyltin methoxide,** $(C_2H_5)_3SnOCH_3$.

*Conversion of alcohols to carbonyl compounds.*[1] Alcohols are converted into aldehydes or ketones by the following two-step procedure:

$$\underset{R^2}{\overset{R^1}{>}}\!C\!\underset{H}{\overset{OH}{<}} + (C_2H_5)_3SnOCH_3 \xrightarrow[78\text{-}92\%]{C_6H_5CH_3, \ \Delta} \underset{R^2}{\overset{R^1}{>}}\!C\!\underset{H}{\overset{OSn(C_2H_5)_3}{<}} \xrightarrow[91\text{-}96\%]{\substack{1) \ Br_2, THF \\ 2) \ Na_2S_2O_3, KOH}} \underset{R^2}{\overset{R^1}{>}}C=O$$

Geraniol has been oxidized to geranial in this way in 94% yield without formation of neral or geranic acid.

[1] K. Saigo, A. Morikawa, and T. Mukaiyama, *Chem. Letters*, 145 (1975).

**Trifluoroacetic acid, 1,** 1219–1221; **2,** 433–434; **3,** 305–308; **4,** 530–532; **5,** 695–700.

*Olefin cyclization* (**3,** 305–307; **4,** 531–532; **5,** 696–697). A key step in a novel synthesis of (±)-longifolene (6) from (1) involves the acid-catalyzed cyclization of (2) to (3) with TFA at 0°. The original paper suggests a mechanism for the cyclization of (2) mainly to the bicyclic product (3) rather than to a hydroazulene derivative.[1]

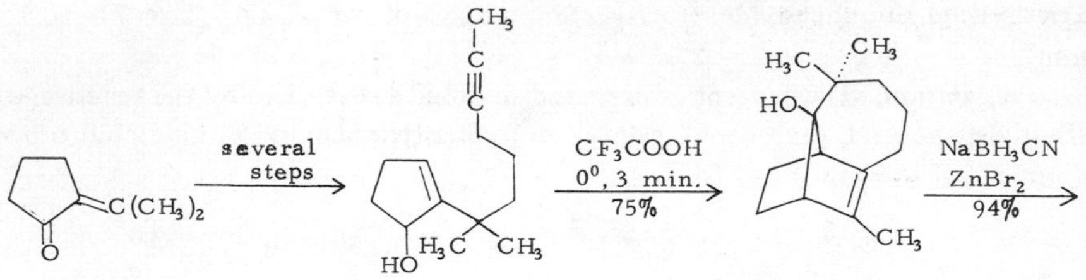

(1)       (2)       (3)

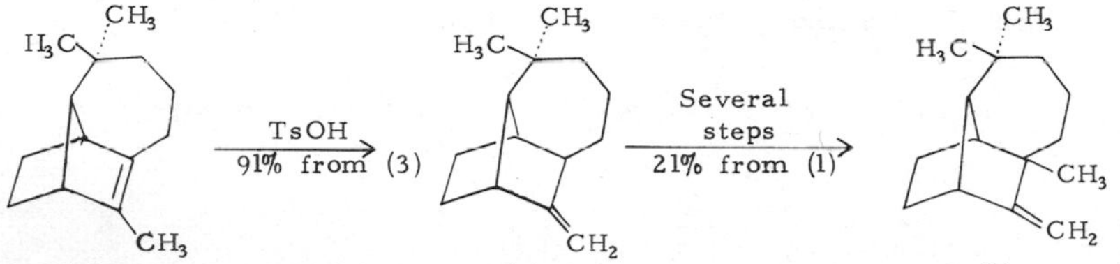

(4)       (5)       (6)

The Johnson synthesis of steroids and triterpenes has been extended to a synthesis of *dl*-serratenediol (7), a pentacyclic triterpene with a seven-membered C ring and nine asymmetric centers.[2]

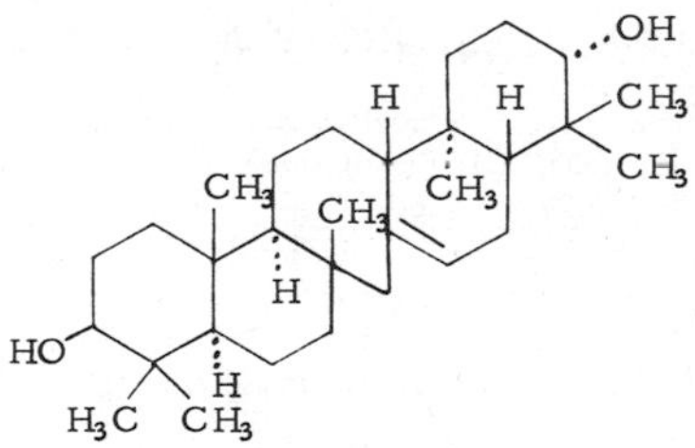

(7)

*Spirodienones from phenolic diazomethyl ketones.* Mander et al.[3] have reported examples of intramolecular cyclizations of phenolic diazoketones by use of an acid catalyst such as trifluoroacetic acid or boron trifluoride etherate. Nitromethane is used as solvent for the latter catalyst; trifluoroacetic acid serves both as catalyst and solvent. Thus treatment of the phenolic diazoketone (1) with $BF_3$ etherate in nitromethane leads to the spirodienone (2); this substance is extremely labile and is rapidly rearranged to the indanone (3), isolated as the

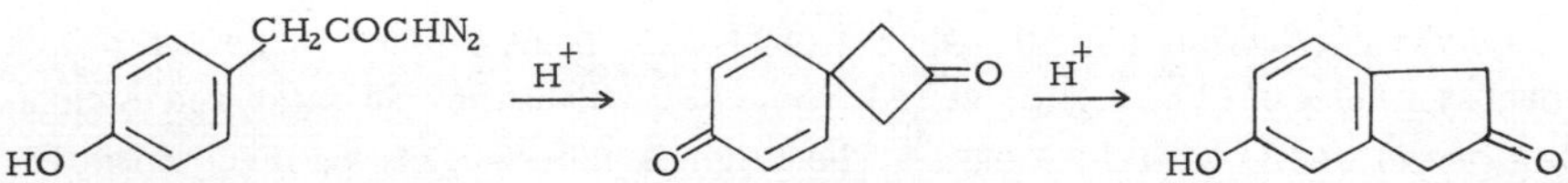

(1)       (2)       (3)

major product of the reaction. Application of this reaction to tetrahydronaph-thyl diazomethyl ketones such as (4) and (6) leads to spirocyclohexa-2,5-dienones in high yield. The products (5) and (7), unlike (2), are stable to TFA at 80° for several days. Application of the same reaction to (8) leads to the tricyclic cyclohexa-2,4-dienone (9).

$$ \text{TFA, } -20°$$
$$\sim 100\%$$

(4)          (5)

$$ \text{TFA, } -20°$$
$$48\%$$

(6)          (7)

$$ \text{TFA, } 0°$$
$$96\%$$

(8)          (9)

*Selective hydrogenation of quinolines and isoquinolines.*[4] Catalytic hydrogenation of quinolines and isoquinolines usually occurs preferentially in the pyridine ring. However, if the hydrogenation is conducted in trifluoroacetic acid, the reverse situation obtains and the benzene ring is reduced more rapidly. The same result can be obtained with mineral acids, but such hydrogenations are much slower. Both 2- and 4-phenylpyridine can also be reduced preferentially in the benzene ring. Platinum oxide or palladium or rhodium catalysts can be used. Further reduction of 5,6,7,8-tetrahydroquinolines with sodium and ethanol provides a convenient route to *trans*-decahydroquinolines.

[1] R. A. Volkmann, G. C. Andrews, and W. S. Johnson, *Am. Soc.*, 97, 4777 (1975).

[2] G. D. Prestwich and J. N. Labovitz, *Am. Soc.*, 96, 7103 (1974).

[3] D. J. Beames and L. N. Mander, *Australian J. Chem.*, 27, 1257 (1974); D. J. Beames, T. R. Klose, and L. N. Mander, *ibid.*, 27, 1269 (1974); D. W. Johnson and L. N. Mander, *ibid.*, 27, 1277 (1974).

[4] F. W. Vierhapper and E. L. Eliel, *J. Org.*, 40, 2729 (1975); *idem, ibid.*, 40, 2734 (1975).

**Trifluoroacetic acid—Alkylsilanes, 5, 695.**

*Ionic hydrogenation.*[1] Noncatalytic hydrogenation of $C=C$, $C=O$, $C=N$, C—OH, C—X groups can be effected with this combination. The acid supplies a proton and the silane a hydride ion:

$$\begin{matrix} \diagup \\ \diagdown \end{matrix}C{=}Y \xrightarrow{\ H^+\ } \begin{matrix} \diagup \\ \diagdown \end{matrix}\overset{+}{C}{-}YH \xrightarrow{\ H^-\ } \begin{matrix} \diagup \\ \diagdown \end{matrix}CH{-}YH$$

Trifluoroacetic acid is commonly used as the proton donor and a silane as the hydride donor. The donating ability of silanes shows the sequence:

$$(C_2H_5)_3SiH \ > \ (C_2H_5)_2SiH_2 \ > \ (C_6H_5)_2SiH_2 \ > \ (C_6H_5)_3SiH$$

This method of reduction has been developed mainly by Russian chemists and has now been reviewed.[1] One advantage is that this reduction proceeds by *trans*-addition to carbon—carbon double bonds and usually results in formation of the thermodynamically more stable isomer. Thus reduction of steroid $\Delta^{8(19)}$ and $\Delta^{9(11)}$ double bonds leads to the natural *trans,anti,trans* configuration. A further advantage is that sulfur-containing heterocycles such as thiophenes are hydrogenated in yields of 60–80%.

*Reduction of cyclohexanones.* Doyle and West[2] have conducted extensive investigations on the reduction of cyclohexanones by organosilanes in acidic media. Steric factors play a dominant role in the stereochemical outcome. However, inductive effects of alkyl substituents in the silane are pronounced. The geometry of transition states in this hydride transfer has been discussed.

[1] D. N. Kursanov, Z. N. Parnes, and N. M. Loim, *Synthesis*, 633 (1974).
[2] M. P. Doyle and C. T. West, *J. Org. Chem.*, **40**, 3821, 3829, 3835 (1975).

**Trifluoroacetic anhydride, 1,** 1221–1226; **3,** 308; **5,** 701.

*Dicarboxylic anhydrides.* Duckworth[1] prepared acid anhydrides in about 90% yield by reaction of 1 mole of trifluoroacetic anhydride with 1 mole of a dicarboxylic acid in ether. Treatment of the mono-mixed anhydride with 1 mole of pyridine then led to the anhydride and pyridinium trifluoroacetate, which was precipitated by addition of petroleum ether. Alternatively, the mixed anhydride can be cyclized by heat treatment in vacuum at about 70°. This latter method was used by Moore and Kelly[2] to prepare the anhydride (3) of tetra-hydrofurane-*cis*-2,5-dicarboxylic acid (1) in >90% yield.

(1)                          (2)                          (3)

*Pyrroles.* Enamines of an α-amino acid such as (1) react rapidly with trifluoroacetic anhydride at 0° to form pyrrole derivatives (2).[3]

$$
\underset{\substack{\text{C}_6\text{H}_5\text{CHCOONa} \\ | \\ \text{HNC}=\text{CHCOOC}_2\text{H}_5 \\ | \\ \text{CH}_3}}{\phantom{x}}
\quad \xrightarrow[\text{30\%}]{(\text{CF}_3\text{CO})_2\text{O}} \quad
$$

(1)

(2)

[1] A. C. Duckworth, *J. Org.*, **27**, 3146 (1962).
[2] J. A. Moore and J. E. Kelly, *Org. Prep. Proc. Int.*, **6**, 255 (1974).
[3] S. K. Gupta, *Synthesis*, 726 (1975).

## 2,2,2-Trifluoroethylamine, $CF_3CH_2NH_2$. Mol. wt. 99.06. The hydrochloride is available from Aldrich.

*Isomerization of β,γ-unsaturated ketones.* This primary amine has been found to be far more effective than acids or bases in the isomerization of (1) to (2). The intermediate protonated Schiff base (a) was identified as a transient intermediate. It is suggested that this process may represent a model for the corresponding enzymatic isomerizations.[1]

$$
\text{(1)} + CF_3CH_2NH_2 \xrightarrow{H^+} \left[ \text{(a)} \right] \xrightarrow{H_2O} \text{(2)} + CF_3CH_2NH_2
$$

(1)                                          (a)                                          (2)

[1] R. H. Kayser and R. M. Pollack, *Am. Soc.*, **97**, 952 (1975).

## Trifluoromethanesulfonic acid (Triflic acid, TFSA), 4, 533; 5, 701–702.

*Reaction of phenylhydroxyl amine with benzene.* The reaction of phenylhydroxylamine (1) with benzene catalyzed with this acid gives (2) and (3) as the major products.[1] When the catalyst is trifluoroacetic acid, (4) becomes the

$$
\text{(1)} + C_6H_6 \xrightarrow{\text{TFSA}} \text{(2, 48\%)} + \text{(3, 23\%)} + (C_6H_5)_2NH \quad (4, <1\%)
$$

(1)                                          (2, 48%)        (3, 23%)

$$
\text{(1)} + C_6H_6 \xrightarrow{\text{TFA}} \text{(4, 56\%)} + \text{(2, 9\%)} + \text{(3, 8\%)}.
$$

major product.[2] The paper by Okamoto *et al.*[1] should be consulted for possible reasons for this unusual effect.

*Cyclization.*[3] The azabicyclooctenones (1) are cyclized to (2) when treated with the acid for 15 min. at 20°. The reaction is comparable to the known

$$\xrightarrow[33-100\%]{CF_3SO_3H}$$

(1)                                    (2)

cyclization of cinnamoyl chloride by aluminum chloride to 3-phenylindanone.[4]

*Cleavage of amino acid protective groups* (5, 702). In a recent synthesis of neurotensin, a bovine hypothalamic principle, the last step involved cleavage of tosyl, benzyl, and benzyloxycarbonyl protective groups. This deblocking was carried out with triflic acid at 40° (60 min.).[5]

[1] T. Okamoto, K. Shudo, and T. Ohta, *Am. Soc.*, 97, 7184 (1975).
[2] K. Shudo and T. Okamoto, *Tetrahedron Letters*, 1839 (1973).
[3] N. Dennis, B. E. D. Ibrahim, and A. R. Katritzky, *Synthesis*, 105 (1976).
[4] K. M. Johnston and J. F. Jones, *J. Chem. Soc. (C)*, 814 (1969).
[5] H. Yajima, K. Kitagawa, T. Segawa, M. Nakano, and K. Kataoka, *Chem. Pharm. Bull. Japan*, 23, 3299 (1975).

**Trifluoromethanesulfonic-alkanesulfonic anhydrides,** $RSO_2 OSO_2 CF_3$.

These reagents are prepared by reaction of alkanesulfonyl bromides with silver trifluoromethanesulfonate. They are stable at room temperature when R is a primary alkyl group. When $R = CH(CH_3)_2$, the anhydride decomposes on attempted vacuum distillation.

The reagents ($R = CH_3$, $C_2H_5$) react with activated arenes without Friedel-Crafts catalysts to form alkyl aryl sulfones. Alkylation rather than sulfonylation is observed when $R = CH(CH_3)_2$. Sulfonylation involves the sulfonylium ion $RSO_2^+$; when R is a secondary alkyl group the ion readily loses $SO_2$ to give the carbonium ion.[1]

[1] K. Huthmacher, G. König, and F. Effenberger, *Ber.*, 108, 2947 (1975).

**Trifluoromethanesulfonic anhydride (Triflic anhydride),** 4, 533–534; 5, 702–705.

*Rearrangement of quadricyclyl-7-carbinyl triflate* (2).[1] The triflate ester (2), prepared as formulated, on solvolysis in the nonacidic, nonnucleophilic solvent trifluoroethanol[2] with 1 eq. of triethylamine (scavenger for TfOH) is converted into (3) as the major product. This solvolysis involves a Wagner–Meerwein rearrangement, but does not result in an olefinic product. However, the present

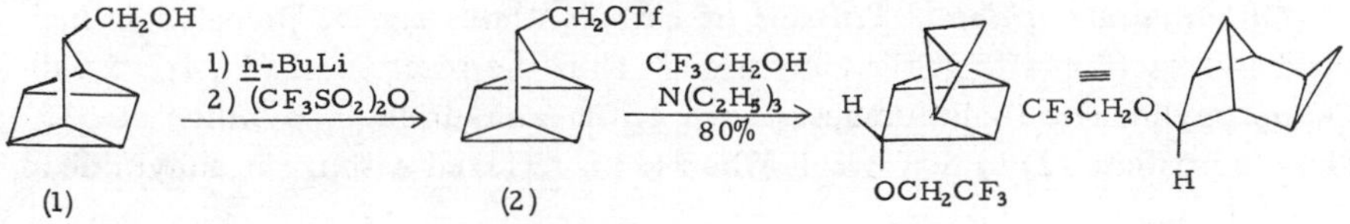

(1)    (2)    (3a)    (3b)

rearrangement does involve transformation of a primary carbonium ion into a secondary ion.

*Carbenes.*[3] Primary vinyl triflates also serve as precursors to unsaturated carbenes. Thus isopropylidenecarbene (2) has been generated by treatment of the triflate (1) with potassium *t*-butoxide at $-20°$. This unsaturated carbene adds stereospecifically to *cis*- and *trans*-2-butene, an indication that the carbene is a singlet.[4]

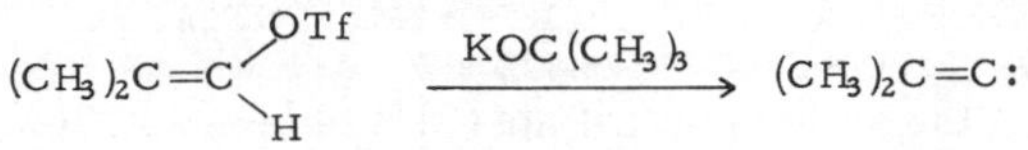

*Allenes.* Allenes can be prepared from ketones in a two-step procedure: conversion to the vinyl triflate and elimination of triflic acid by quinoline at $100°$. Unsymmetrical ketones give mixtures of triflates, but the same allene is

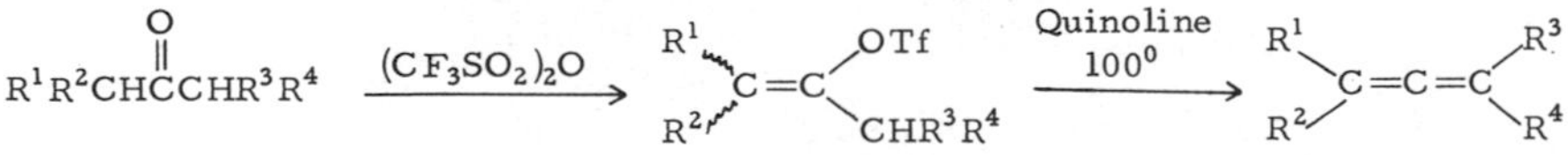

obtained from the mixture. The method can only be used for preparation of 1,1-di- or higher substituted allenes, since elimination from a triflate of type (a)

*Examples:*

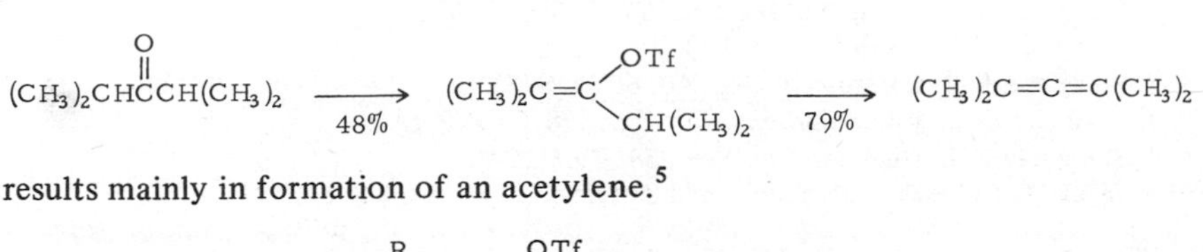

results mainly in formation of an acetylene.[5]

*Carbohydrate triflates.* Triflates of carbohydrates can be prepared in high yield by reaction with triflic anhydride.[6] Thus the reaction of (1), 1,2; 5,6-di-O-isopropylidene-α-D-allofuranose, with triflic anhydride in pyridine at $-15°$ gives the triflate (2) in 80% yield. When (1) is refluxed with triflic anhydride in

(1, R = OH)
(2, R = OTf)
(3, R = N+C5H5ŌTf)

pyridine for 25 hr., the pyridinium triflate (3) is obtained in 70% yield. The 3-O-tosyl derivative of (1) corresponding to (2) is unchanged when refluxed in pyridine for 24 hr. Thus triflates of carbohydrates are more reactive in $S_N2$ reactions than sulfonates.

*Vinyl triflates* (4, 532–533). The procedure for the preparation of vinyl triflates from ketones has been published.[7]

45%

*Synthesis of cyclobutanone* (4, 533). The synthesis of cyclobutanone from 3-butyn-1-yl trifluoromethanesulfonate has been published.[8]

[1] W. G. Dauben and J. W. Vinson, *J. Org.*, **40**, 3756 (1975).
[2] W. S. Trahanovsky and M. P. Doyle, *Tetrahedron Letters*, 2155 (1968).
[3] P. J. Stang and M. G. Mangum, *Am. Soc.*, **97**, 1459 (1975).
[4] G. L. Closs, *Topics Stereochem.*, **3**, 193 (1968).
[5] P. J. Stang and R. J. Hargrove, *J. Org.*, **40**, 657 (1975).
[6] L. D. Hall and D. C. Miller, *Carbohydrate Res.*, **40**, C1 (1975).
[7] P. J. Stang and T. E. Duebner, *Org. Syn.*, **54**, 79 (1974).
[8] M. Hanack, T. Dehesch, K. Hummel, and A. Nierth, *ibid.*, **54**, 84 (1974).

## Trifluoromethanesulfonylimidazole, 5, 705.

*Vinyl triflates.* Attempts to prepare the triflate (2) from 6-octyne-2-one (1) with triflic anhydride under usual conditions failed, but the preparation was achieved with trifluoromethanesulfonylimidazole with potassium hydride as base. The triflate (2) was obtained as a mixture of inseparable Z- and E-isomers and was accompanied by the isomeric triflate (3).[1]

$CH_3C \equiv C(CH_2)_3COCH_3$ + [imidazole with $SO_2CF_3$] $\xrightarrow{\text{KH, THF}}$ $CH_3C \equiv C(CH_2)_2CH = C$ [with $CH_3$ and $OTf$]

(1)                                                                              (2, E and Z)

+ $CH_3C \equiv C(CH_2)_3C$ [with $CH_2$ and $OTf$] + [imidazole $N \diagdown NH$]

(3)

The triflate (2) undergoes solvolysis very slowly in aqueous ethanol, mainly to (1). In a highly ionized solvent ($CF_3CH_2OH$), solvolysis leads to cyclic products as well as the starting ketone because of participation of the triple bond.

[1]M. J. Chandy and M. Hanack, *Tetrahedron Letters*, 4515 (1975).

**Trifluoromethylthiocopper, $CuSCF_3$ (1). Mol. wt. 164.61.**
*Preparation:*

$$AgSCF_3 + CuBr \xrightarrow[100\%]{CH_3CN} CuSCF_3 + AgBr$$

*Trifluoromethyl sulfides.* The reagent (1) reacts with aryl and heteroaryl iodides in N-methylpyrrolidone, quinoline, or DMF to form trifluoromethyl sulfides:

(1)

$R$ [benzene ring] $-I$ $\xrightarrow{150-165°}$ $R$ [benzene ring] $-SCF_3$

Yields of products are 70–75% with iodides containing electron-withdrawing substitutents; yields are decreased to 30–35% with iodides containing electron-donating groups.[1]

[1]L. M. Yagupolskii, N. V. Kondratenko, and V. P. Sambur, *Synthesis*, 721 (1975).

**Trifluoro(trichloromethyl)silane, $CCl_3SiF_3$. Mol. wt. 203.47, b.p. 43.5°.**
*Preparation:*

$$CH_3SiCl_3 \xrightarrow[90\%]{Cl_2} CCl_3SiCl_3 \xrightarrow[45\%]{SbF_3} CCl_3SiF_3$$

*Dichlorocarbene.* This silane decomposes in the vapor phase at 120–140° to liberate dichlorocarbene and $SiF_3Cl$. Added olefins are converted into *gem*-dichlorocyclopropanes in 85–95% yields. The reaction is stereospecific with *cis*- and *trans*-2-butene.[1]

[1]J. M. Birchall, G. N. Gilmore, and R. N. Haszeldine, *J.C.S. Perkin I*, 2530 (1974).

**Trifluorovinyllithium,** $F_2C=CFLi$. Mol. wt. 87.96.

   *Preparation:*

$$F_2C=C\overset{F}{\underset{Cl}{\big\langle}} + \underline{n}\text{-}C_4H_9Li \xrightarrow[\sim 100\%]{\text{THF, ether,}\\ \text{pentane, }-135^0} F_2C=C\overset{F}{\underset{Li}{\big\langle}}$$

   *Reactions.*[1]

$$F_2C=C\overset{F}{\underset{Li}{\big\langle}} + \overset{R^1}{\underset{R^2}{\big\rangle}}C=O \xrightarrow{\sim 85\%} R^2\overset{R^1}{\underset{OH}{\overset{|}{-}C}}\overset{|}{-}C=CF_2 \xrightarrow[75-80\%]{H_2SO_4\\ -10^0} \overset{R^1}{\underset{R^2}{\big\rangle}}C=CFCOF$$

$$\overset{R^1}{\underset{R^2}{\big\rangle}}C=CFCOF$$

$$\xrightarrow[70-90\%]{H_2O} \overset{R^1}{\underset{R^2}{\big\rangle}}C=CFCOOH$$

$$\xrightarrow[70-85\%]{ROH} \overset{R^1}{\underset{R^2}{\big\rangle}}C=CFCOOR$$

$$\xrightarrow[70-95\%]{HNR_2} \overset{R^1}{\underset{R^2}{\big\rangle}}C=CFCONR_2$$

$$\bigcirc{=}CFCOF + (CH_3)_2CuLi \xrightarrow[70\%]{} \bigcirc{=}CFCOCH_3$$

[1] J. F. Normant, J. P. Foulon, D. Masuri, R. Sauvêtre, and J. Villieras, *Synthesis*, 122 (1975).

**2,4,6-Triisopropylbenzenesulfonyl chloride** (TPS), **1,** 1228-1229; **3,** 308.

   *Polynucleotide synthesis.* Narang and collaborators[1] have reported the synthesis of a pentadecanucleotide of thymidine by a modified phosphotriester method.

[1] K. Itakura, N. Katagiri, C. P. Bahl, R. H. Wightman, and S. A. Narang, *Am. Soc.*, **97,** 7327 (1975).

**Trimethylaluminum,** **5,** 707-708.

   *Handling.* Alkylaluminum compounds are highly reactive to air and moisture. A full face shield and leather gloves are essential. Contact with flammable solvents should be avoided. Harney *et al.*[1] have described in detail suitable apparatus and techniques for handling this and other pyrophoric reagents.

   *C-Methylation of t-alcohols and arylalkylcarbinols.*[1] Alcohols of this type undergo C-methylation with trimethylaluminum at 120-130°. An excess of

$$R^1R^2R^3COH \xrightarrow[C_6H_6\ \text{or}\ C_6H_5CH_3]{(CH_3)_3Al} R^1R^2R^3CCH_3$$

reagent is advantageous, possibly necessary. The reaction is autocatalytic; addition of a catalyst increases the initial rate. Water, a carboxylic acid, and hydrogen sulfide can function as catalysts. Usually a trace of benzoic acid is used.

Methylation is most facile for arylcarbinols and tertiary alcohols. 1,1,1-Triphenylethane is readily obtained in high yield from triphenylmethanol, and 1-methyladamantane is formed from adamantane-1-ol. Yields are lower when dehydration is possible. Methylation is not stereospecific; the same mixture of epimeric products is obtained from the isomeric 4-t-butyl-1-ethylcyclohexanols.[2] These features are characteristic of carbonium ion reactions.

*C-Methylation of ketones.*[3] Various ketones (and some aldehydes) are *gem*-dimethylated by trimethylaluminum in a hydrocarbon solvent at 120–180°. Yields range from moderate to high. 3-Cholestanone can be converted into 3α-methylcholestane-3β-ol or exhaustively methylated to 3,3-dimethylcholestane. Benzaldehyde is converted quantitatively into isopropylbenzene. Anthrone is converted into 9-methylanthracene. As before, the main side reaction is dehydration to olefin by-products.

*C-Methylation of carboxylic acids.*[4] Carboxylic acids can be C-methylated to t-butyl compounds by excess trimethylaluminum in benzene at about 120°. For example, benzoic acid is converted into t-butylbenzene. Adamantane-1-carboxylic acid is converted into a mixture of 1-t-butyladamantane and 1-isopropenyladamantane.

*Bridgehead methylation.* Trimethylaluminum reacts rapidly with 1-bromoadamantane and 1-bromobicyclo[2.2.2]octane in $CH_2Cl_2$ at $0°$[5] to give 1-methyladamantane (1, 82% yield) and 1-methylbicyclo[2.2.2]octane (2, 98%

(1)          (2)          (3)

yield). 1-Bromobicyclo[2.2.1]heptane (3) does not react under these conditions or even at 100° in a bomb for 24 hr.[6]

*Conjugate addition to α,β-unsaturated ketones.* Nickel(II) acetylacetonate catalyzes the reaction of ketones with trimethylaluminum to give methylcarbinols; however, other unexpected products are formed, sometimes extensively. This organonickel compound also catalyzes the conjugate addition of trimethylaluminum to α,β-unsaturated ketones (equation I).[1] The yields of

(I)

β-methylated ketones range from 30 to 100%. 1,2-Addition resulting in allylic alcohols is the main side reaction. The reaction closely resembles conjugate addition reactions of dimethylcopperlithium; thus copper is not unique in promoting conjugate addition to α,β-unsaturated ketones.[7]

*Methyl ketones.* Nitriles (aliphatic and aromatic) are converted into methyl ketones by reaction with trimethylaluminum in the presence of various catalysts,

$$R-C\equiv N \;+\; Al(CH_3)_3 \xrightarrow[\substack{55-80\%}]{\substack{1)\ Ni(\text{acac})_2 \\ 2)\ H_2O^+}} RCOCH_3$$

particularly nickel(II) acetylacetonate.[8]

[1] D. W. Harney, A. Meisters, and T. Mole, *Australian J. Chem.*, **27**, 1639 (1974).
[2] R. G. Salomon and J. K. Kochi, *J. Org.*, **38**, 3715 (1973).
[3] A. Meisters and T. Mole, *Australian J. Chem.*, **27**, 1655 (1974).
[4] Idem, *ibid.*, **27**, 1665 (1974).
[5] Aluminum alkyls can react violently with alkyl halides.
[6] E. W. Della and T. K. Bradshaw, *J. Org.*, **40**, 1638 (1975).
[7] E. A. Jeffery, A. Meisters, and T. Mole, *Australian J. Chem.*, **27**, 2569 (1974); L. Bagnell, E. A. Jeffery, A. Meisters, and T. Mole, *ibid.*, **28**, 801, (1975); L. Bagnell, A. Meisters, and T. Mole, *ibid.*, **28**, 817, 821 (1975).
[8] L. Bagnell, E. A. Jeffery, A. Meisters, and T. Mole, *ibid.*, **27**, 2577 (1974).

**Trimethylamine N-oxide, 1,** 1230–1231; **2,** 434; **3,** 309–310.

*Oxidation of organoboranes* (**3,** 309–310). Kabalka and Hedgecock[1] have reported that organoboranes are oxidized efficiently by trimethylamine N-oxide dihydrate (Aldrich), which is soluble in both hydrocarbon and ethereal solvents. Diglyme is generally used as solvent, since a convenient oxidation rate is obtained at reflux. The oxide is more convenient and safer to handle than hydrogen peroxide, and the yields of alcohols are at least as high as, and in some cases higher than, those obtained with hydrogen peroxide. The reagent is compatible with a variety of functional groups:

$$-CHO, \; {>}C{=}O, \; -COOR, \; -C\equiv N.$$

*Liberation of organic ligands from iron carbonyl complexes.* Oxidizing reagents (CAN, ferric chloride, lead tetraacetate) have been used to free an organic ligand from complexes with $Fe(CO)_3$ and $Fe(CO)_4$. Israeli chemists[2] report that amine oxides, particularly trimethylamine oxide, are useful for this purpose. Relatively low temperatures (25–81°) can be used and the oxide is transformed into trimethylamine (volatile). Carbon dioxide is formed.

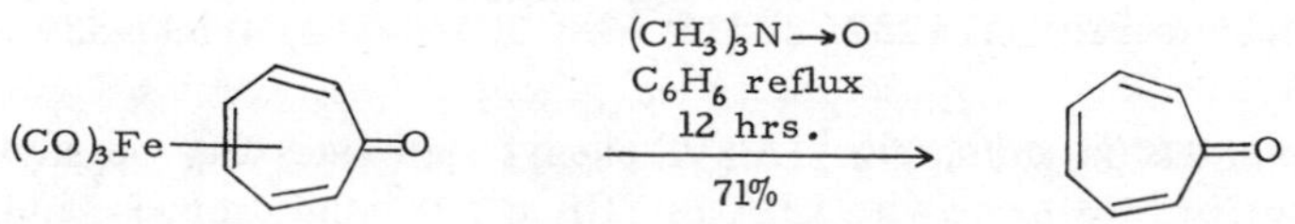

$(CH_3)_3N \rightarrow O$
$C_6H_6$ reflux
12 hrs.
71%

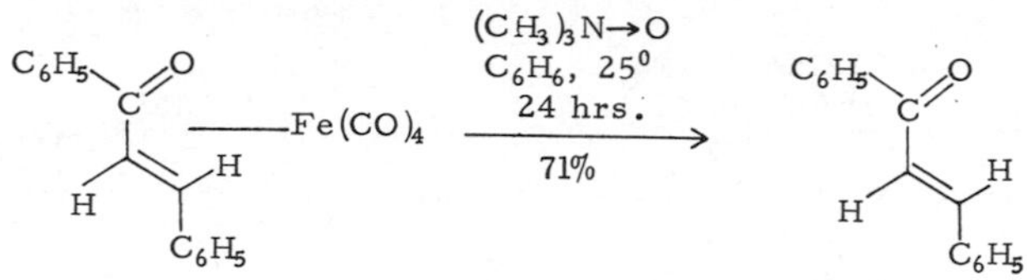

$(CH_3)_3N \rightarrow O$
$C_6H_6$, 25°
24 hrs.
71%

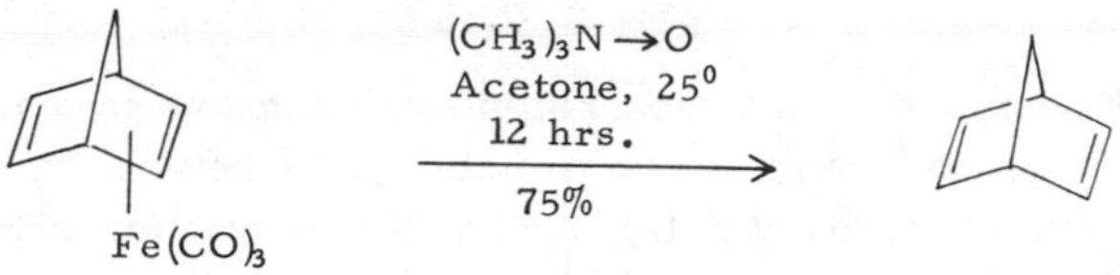

$(CH_3)_3N \rightarrow O$
Acetone, 25°
12 hrs.
75%

$Fe(CO)_3$

The reverse of the dissociation reaction cited above has been realized. Thus when $Fe(CO)_5$ is added to the amine oxide in benzene containing a diene at 0°, gas is evolved immediately. The diene–$Fe(CO)_3$ complex is isolated in satisfactory yield (usually 60–80%) after 1 hr. reflux.[3]

Diene + $Fe(CO)_5$ + $(CH_3)_3NO \rightarrow$ Diene-$Fe(CO)_3$ + $(CH_3)_3N$ + $CO_2$ + $CO$

The direct complexation of dienes with $Fe(CO)_5$ requires temperatures of about 120°.

[1] G. W. Kabalka and H. C. Hedgecock, Jr., *J. Org.*, **40**, 1776 (1975).
[2] Y. Shvo and E. Hazum, *J.C.S. Chem. Comm.*, 336 (1974).
[3] *Idem, ibid.*, 829 (1975).

## 2,4,6-Trimethylbenzenesulfonyl chloride (2-Mesitylenesulfonyl chloride), 1, 661.

*Esterification* (1, 1183–1184). Swiss chemists have found this hindered sulfonyl chloride superior to *p*-toluenesulfonyl chloride for intramolecular esterifications in the synthesis of the macrotetrolide antibiotic nonactin (1).[1]

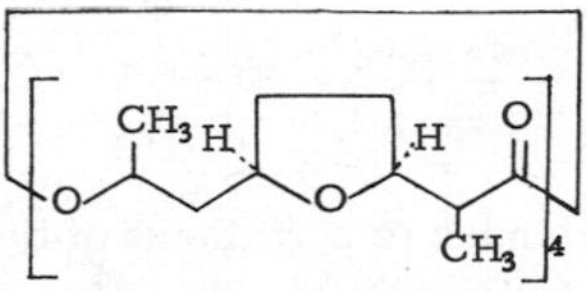

(1)

[1] H. Gerlach, K. Oertle, A. Thalmann, and S. Servi, *Helv.*, **58**, 2036 (1975).

**Trimethylchlorosilane, 1**, 1232; **2**, 435–438; **3**, 310–312; **4**, 537–539; **5**, 709–713.

*α-Trimethylsilyl sulfoxides.*[1] Alkyl phenyl sulfoxides can be silylated in good yields by treatment with LDA in THF at $-78°$ and dropwise addition of the anion to trimethylchlorosilane at $-78°$ (equation I). The products (1) are

(I)   $CH_3SOC_6H_5 \xrightarrow[\substack{95\%}]{\substack{1)\ LDA,\ THF \\ 2)\ (CH_3)_3SiCl}} (CH_3)_3SiCH_2SOC_6H_5 \xrightarrow{60°} \left[ CH_2{=}S{\substack{\diagup OSi(CH_3)_3 \\ \diagdown C_6H_5}} \right]$

(1)                                    (a)

$\xrightarrow{72\%} (CH_3)_3SiOCH_2SC_6H_5$

(2)

of interest because they undergo a Pummerer-like rearrangement under relatively mild conditions to hemithioacetal trimethylsilyl ethers (2) by migration of silicon from carbon to oxygen (a). Actually, this reaction occurs below room temperature on silylation of dialkyl sulfoxides (equation II). In this case substantial amounts of vinyl thioethers are also formed.

(II)   $(\underline{n}\text{-}C_4H_9)_2SO \xrightarrow[\substack{2)\ (CH_3)_3SiCl,\ -78°\rightarrow 20°}]{\substack{1)\ LDA,\ THF}} \underline{n}\text{-}C_4H_9SCH{\substack{\diagup OSi(CH_3)_3 \\ \diagdown C_3H_7\text{-}\underline{n}}}$

(45%)

$+\quad \underline{n}\text{-}C_4H_9SCH{=}CHC_2H_5 \quad + \quad (\underline{n}\text{-}C_4H_9S)_2CHC_3H_7\text{-}\underline{n}$

(30%)                         (13%)

*Synthesis of ketones from esters via acyloin condensation.* (**4**, 537). Wakamatsu *et al.*[2] have developed a convenient synthesis of ketones from esters as shown in formulation I. Although the synthesis involves several steps, yields are usually high.

(I)   $2\ RCOR' \rightarrow R{-}\underset{\substack{\| \\ }}{\overset{\substack{O \\ \|}}{C}}{-}OSi(CH_3)_3 \ R{-}\overset{\|}{C}{-}OSi(CH_3)_3 \xrightarrow[DME]{CH_3Li} R{-}\overset{\substack{\| \\ }}{C}{-}O^- \atop R{-}\overset{\|}{C}{-}O^-\ \ 2\ Li^+$

$\xrightarrow[65\text{-}100\%]{R'X} \underset{\substack{| \\ OH}}{R\overset{\substack{OR' \\ \| \ |}}{C}C{-}R} \xrightarrow[quant.]{NaBH_4} \underset{\substack{| \ | \\ OHOH}}{R\overset{\substack{H\ R' \\ | \ |}}{C}{-}CR} \xrightarrow[ether]{LTA} RCOR' + RCHO$
                                                                        90-100%

The method has been extended to a synthesis of the 1,4-diketone (2) from the ethylene ketal of methyl levulinate (1). The diketone was cyclized by base to dihydrojasmone (3). The overall yield of (3) from (1) was 49.5%. The process is shown in scheme II.[3]

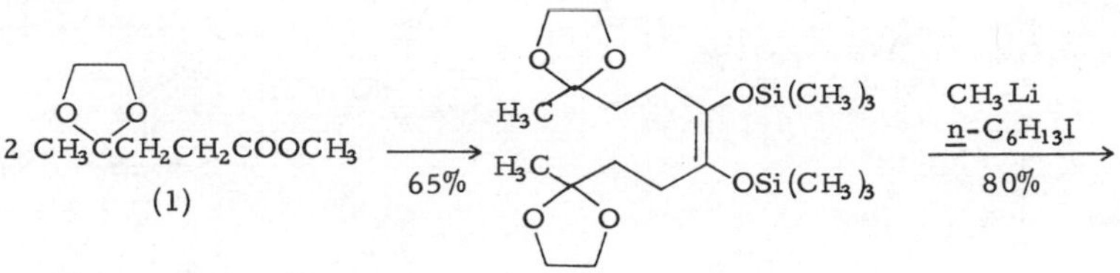

2 $CH_3CCH_2CH_2COOCH_3$
(1)

(II)

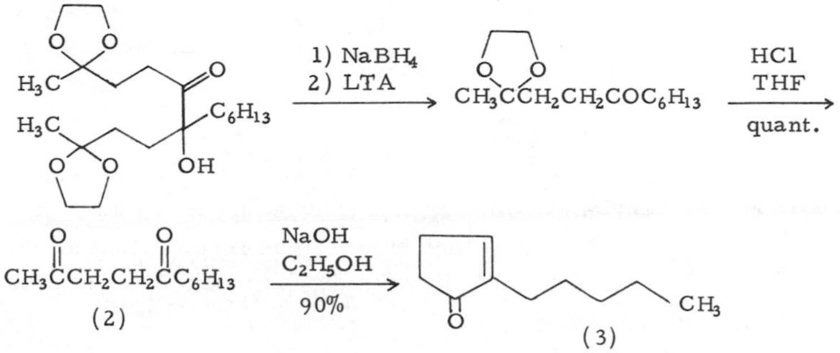

1) $NaBH_4$
2) LTA

$CH_3CCH_2CH_2COC_6H_{13}$

HCl
THF
quant.

$CH_3CCH_2CH_2CC_6H_{13}$
(2)

NaOH
$C_2H_5OH$
90%

$CH_3$
(3)

*Acyloin condensation of ethyl acetate and ethyl propionate.*[4] Acyloin condensation of ethyl acetate with sodium and trimethylchlorosilane (3, 311–312; 4, 537) leads mainly to Z-2,3-bis(trimethylsilyloxy)butene-2 (1); the more stable E-isomer (2) is formed in traces. Similarly, acyloin condensation of ethyl propionate gives the Z- and E-isomers of 3,4-bis(trimethylsilyloxy)hexene-3 in the ratio 88:12.

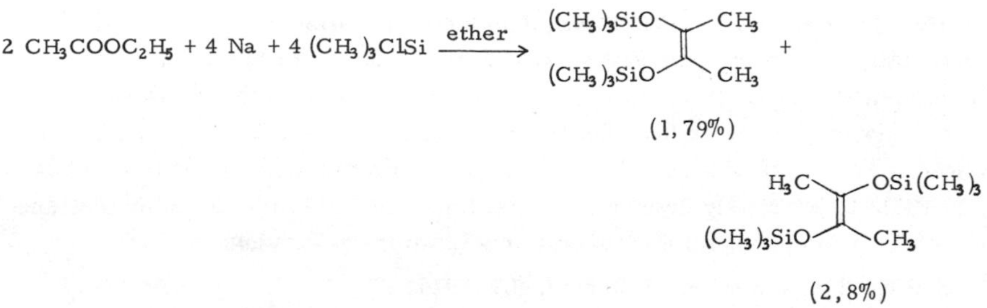

2 $CH_3COOC_2H_5$ + 4 Na + 4 $(CH_3)_3ClSi$  ether →

$(CH_3)_3SiO$   $CH_3$
$(CH_3)_3SiO$   $CH_3$

(1, 79%)

+

$H_3C$   $OSi(CH_3)_3$
$(CH_3)_3SiO$   $CH_3$

(2, 8%)

*2,6-Dienic carboxylic acids.* Fráter[5] has described a method for addition of an isoprene unit to an allylic alcohol. The method is illustrated for the conversion of the ester (1) of geraniol into farnesenic acid (3). The ester is treated with lithium N-isopropylcyclohexylamide (LiICA, 4, 306–309) and then with trimethylchlorosilane. The resulting trimethylsilylketene acetal (a) undergoes

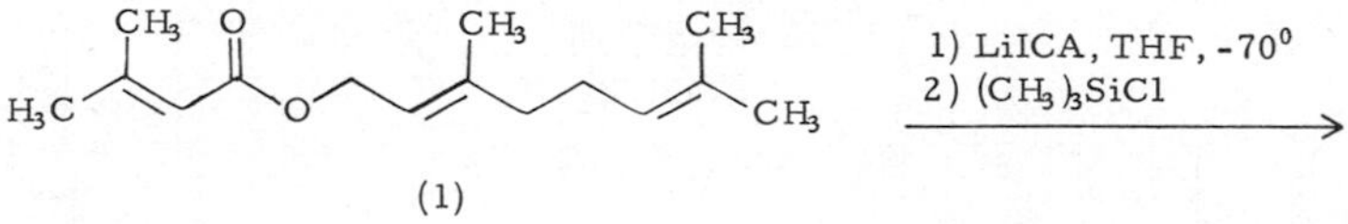

1) LiICA, THF, -70°
2) $(CH_3)_3SiCl$

(1)

(a)

(2)

(3, E and Z isomers)

allylic rearrangement at 20–40° to give, after aqueous work-up, the acid (2). This acid (or the methyl ester) undergoes Cope rearrangement at 156° to give a mixture of 2E,6E; 2Z,6Z; 2E,6Z; and 2Z,6E-isomers of farnesenic acid (3), in which the last two predominate.

[1] E. Vedejs and M. Mullins, *Tetrahedron Letters*, 2017 (1975).
[2] T. Wakamatsu, K. Akasaka, and Y. Ban, *Tetrahedron Letters*, 3879 (1974).
[3] *Idem, ibid.*, 3883 (1974).
[4] C. M. Cookson and G. H. Whitham, *J.C.S. Perkin I*, 806 (1975); G. A. Russell, D. F. Lawson, H. L. Malkus, R. D. Stephens, G. R. Underwood, T. Takano, and V. Malatesta, *Am. Soc.*, **96**, 5830 (1974).
[5] G. Fráter, *Helv.*, **58**, 442 (1975).

**Trimethylchlorosilane–Zinc, 5, 714.**

*Deoxygenation of ketosteroids.* The deoxygenation of ketones with this combination has been extended to ketosteroids. $\Delta^2$-Cholestene can be obtained in about 70% yield from 5α-cholestane-3-one with trimethylchlorosilane and zinc dust in THF at reflux. The reaction fails with 6-, 7-, 12-, 17-, and 20-ketones. This result suggests that a very bulky intermediate is involved. Thus it is possible to selectively deoxygenate diketones. For example, $\Delta^2$-5α-cholestene-7-one can be prepared from 5α-cholestane-3,7-dione (72% yield).[1]

[1] P. Hodge and M. N. Khan, *J.C.S. Perkin I*, 809 (1975).

**Trimethylene dithiotosylate, 4, 539–540.**

Details for the preparation of trimethylene dithiotosylate and of ethylene dithiotosylate have been published.[1] Use of the former reagent for preparation of dithianes from enamines[2] and use of the latter reagent for preparation of dithiolanes from the hydroxymethylene derivative of a ketone[3] have also been reported.

*1,2-Diketones.* 1,3-Diketones, for example acetylacetone, react with the

$$CH_3COCH_2COCH_3 \quad + \quad TsS(CH_2)_3STs \longrightarrow$$
(1)

reagent in methanol to give 2-acyl-1,3-dithianes (3) by cleavage of the intermediate diacyldithianes (2). Alkylation of the product (3) gives protected derivatives of 1,2-diketones.[4]

[1] R. B. Woodward, I. J. Pachter, and M. L. Scheinbaum, *Org. Syn.*, **54**, 33 (1974).
[2] *Idem, ibid.*, **54**, 39 (1974).
[3] *Idem, ibid.*, **54**, 37 (1974).
[4] R. J. Bryant and E. McDonald, *Tetrahedron Letters*, 3841 (1975).

**2,4,4-Trimethyl-2-oxazoline,** 3, 313–314; 5, 714–716.

(±)-*cis-2-Methylcyclopentanecarboxylic acid.* Meyers *et al.*[1] have reported a highly stereoselective synthesis of this less stable isomer of 2-methylcyclopentanecarboxylic acid (5) by way of the oxazoline (1).

[1] A. I. Meyers, E. D. Mihelich, and K. Kamata, *J.C.S. Chem. Comm.*, 768 (1974).

**N,4,4-Trimethyl-2-oxazolinium iodide, 4, 540–541.**

The preparation of this reagent and the use for synthesis of *o*-anisaldehyde have been published.[1]

[1] R. S. Brinkmeyer, E. W. Collington, and A. I. Meyers, *Org. Syn.*, **54**, 43 (1974).

**Trimethyloxonium tetrafluoroborate, 1,** 1232; **2,** 438; **3,** 314–315; **4,** 541; **5,** 716.

*Vinylenedisulfonium salts.*[1] The *cis*-dithioethers (1) react with this alkylating reagent (1 : 1) in $CH_2Cl_2$ (20°) to give the *cis*-monosulfonium salts (2) in 88% yield. This product undergoes photochemical isomerization to the *trans*-isomer (3). Both (2) and (3) can be alkylated further to two diastereomeric disulfonium salts (4)–(7).

These disulfonium salts undergo specific reactions with base, possibly through an ylidic intermediate (a) (equation I).

[1] H. Braun and A. Amann, *Angew Chem. internat Ed.*, **14**, 755 (1975).

**Trimethyl phosphite—Copper(I) iodide, 5,** 717–718.

*Decomposition of diethyl diazomalonate* (1). Decomposition of (1) at 50° in ethanol catalyzed by trimethyl phosphite—copper(I) iodide results in formation of diethyl ethoxymalonate (2) in 82% yield. This product is formed by insertion of dicarboethoxycarbene into the O—H bond of ethanol. The dimer of

$$N_2C(COOC_2H_5)_2 \xrightarrow[\begin{array}{c}P(OCH_3)_3/CuI\\C_2H_5OH\end{array}]{} C_2H_5O-CH(COOC_2H_5)_2 + (C_2H_5OOC)_2C=C(COOC_2H_5)_2$$

(1)     (2, 82%)     (**3**, 12%)

the carbene (3) is formed to a slight extent.[1] Photolysis of (1) in ethanol results in low yields of (2).[2]

[1] R. Pellicciari and P. Cogolli, *Synthesis*, 269 (1975).
[2] S. Julia, H. Ledon, and G. Linstrumelle, *Compt. Rend.*, **272** (C), 1898 (1971).

**Trimethylsilylacetic acid,** $(CH_3)_3SiCH_2COOH$. Mol. wt. 132.24, m.p. 40°.

*Preparation.*[1] The acid is obtained on carbonation of trimethylsilymethyl-magnesium chloride.

*α,β-Unsaturated acids; α-trimethylsilyl carboxylic acids; α-trimethylsilyl-γ-butyrolactones.*[2] This acid can be converted into the dianion (1) by 2.2 eq. of LDA in THF at 0°. This dianion reacts with aldehydes and ketones at low temperatures to form α,β-unsaturated acids (Peterson elimination reaction, **5**, 724).

*Examples:*

$$(1) + C_6H_5CHO \xrightarrow[88\%]{} C_6H_5CH=CHCOOH$$
$$(E/Z=1:1)$$

The dianion reacts with epoxides to form γ-hydroxy acids, which can be dehydrated to trimethylsilyl-γ-butyrolactones:

The butyrolactone can be converted into the α-ethylidene-γ-butyrolactone as shown.

Reaction of (1) with an alkyl bromide or iodide gives an α-trimethylsilyl carboxylic acid:

$$(1) + RX \xrightarrow[85-98\%]{20°} \underset{\underset{Si(CH_3)_3}{|}}{RCHCOOH}$$

Note that reaction of trimethylchlorosilane with enolates of ketones, esters, and butyrolactones gives products of O-silylation predominately.

[1] L. H. Sommer, J. R. Gold, G. M. Goldberg, and N. S. Marans, *Am. Soc.*, **71**, 1509 (1949).
[2] P. A. Grieco, C.-L. J. Wang, and S. D. Burke, *J.C.S. Chem. Comm.*, 537 (1975).

**Trimethylsilyl azide, 1**, 1236; **3**, 316; **4**, 542; **5**, 719–720.

*Oxazine-2,6-diones.*[1] The reaction of trimethylsilyl azide with maleic anhydrides neat or in $CHCl_3$ provides a superior route to these compounds (1). When

(1)                              (2)

R or X = H, 4-substituted derivatives of (1) are formed preferentially. N-Methylation of (1) is accomplished readily with dimethyl sulfate buffered with sodium bicarbonate to give (2).

[1] J. D. Warren, J. H. MacMillan, and S. S. Washburne, *J. Org.*, **40**, 743 (1975).

**Trimethylsilyl cyanide, 4**, 542–543; **5**, 720–722.

*Synthesis of ketones* (**5**, 721). Hünig[1] has published further examples of his synthesis of ketones from aromatic or heterocyclic aldehydes (equation I).

The alkylating agent can be a primary halide (chlorides react rather slowly) or a primary tosylate or dialkyl sulfate. In one example cyclohexyl bromide was used, but an extended reaction period was required. The paper also includes two mild conditions for the hydrolysis of (4) to the ketone (5). In one method the intermediates (4) are hydrolyzed by addition of 2 N hydrochloric acid (overnight, 25°) to the reaction medium. In the other, triethylamine hydrofluoride in THF is added to the reaction mixture; after addition of water, the organic products are extracted with methylene chloride. This solution is then treated with dilute NaOH. Triethylamine hydrofluoride presumably effects cleavage of (4) to a cyanohydrin (*cf.* **4**, 177, 477), which is then converted to the ketone (5) by base. This salt is prepared by treatment of triethylamine with 48% hydrofluoric acid with cooling. The solution is left standing overnight, and the heavier layer is concentrated in vacuum and dried by azeotropic distillation with benzene. It is obtained as a syrup. The exact structure is not known.

*Reaction with ketenes.* Trimethylsilyl cyanide adds to the carbonyl group of ketenes to give β-substituted α-trimethylsiloxyacrylonitriles (1) in 60–95% yield. The reaction can be carried out on the free ketenes or on ketenes generated *in situ* from acyl chlorides and tertiary amines. Generally, mixtures of (1)

$$R^1 \diagdown C=C=O \; + \; (CH_3)_3SiCN \; \xrightarrow[60-95\%]{} \; R^1 \diagdown C=C \diagup^{OSi(CH_3)_3}_{CN}$$
$$R^2 \diagup \hspace{5.5cm} R^2 \diagup$$

(1)

are obtained when $R^1 \neq R^2$. This reaction is one of the few known cases of addition to the C=O function of ketenes with retention of the C=C double bond.[2]

*Reaction with isocyanates.* The reagent reacts with *p*-tosyl isocyanate to give the 1:1 adduct (1). In general, aryl isocyanates form 1:2 adducts (2).[3]

$$(CH_3)_3SiCN + \underline{p}\text{-}CH_3C_6H_4SO_2N=C=O \rightarrow \underline{p}\text{-}CH_3C_6H_4SO_2N-\overset{\overset{\displaystyle Si(CH_3)_3}{|}}{C}CN$$
$$\underset{\displaystyle O}{\overset{\displaystyle \|}{\phantom{C}}}$$

(1)

$$(CH_3)_3SiCN + 2\ ArN=C=O \longrightarrow$$

(2)

*α-Aminonitriles.* Cyanosilylation of Schiff bases and of oximes with trimethylsilyl cyanide catalyzed by $AlCl_3$, $ZnI_2$, or $Al(OR)_3$ provides a useful route to aminonitriles.[4]

$$R^1 \diagdown C=NR^3 + (CH_3)_3SiCN \xrightarrow{AlCl_3} R^1 \diagdown \underset{CN}{\overset{\overset{\displaystyle Si(CH_3)_3}{|}}{C}}-NR^3 \xrightarrow[80-98\%]{H_2O} R^1 \diagdown \underset{CN}{\overset{|}{C}}NHR^3$$
$$R^2 \diagup \hspace{3cm} R^2 \diagup \hspace{3cm} R^2 \diagup$$

$$R^1 \diagdown C=NOH + (CH_3)_3SiCN \xrightarrow{ZnI_2} R^1 \diagdown \underset{CN}{\overset{|}{C}}NHOSi(CH_3)_3 \xrightarrow[65-80\%]{CH_3OH} R^1 \diagdown \underset{CN}{\overset{|}{C}}NHOH$$
$$R^2 \diagup \hspace{3cm} R^2 \diagup \hspace{3cm} R^2 \diagup$$

[1] S. Hünig and G. Wehner, *Synthesis*, 180 (1975).
[2] U. Hertenstein and S. Hünig, *Angew. Chem. internat. Ed.*, **14**, 179 (1975).
[3] I. Ojima, S. Inaba, and Y. Nagai, *J.C.S. Chem. Comm.*, 826 (1974).
[4] I. Ojima, S. Inaba, and K. Nakatsugawa, *Chem. Letters*, 331 (1975).

**Trimethylsilyldiethylamine, 3**, 317; **4**, 544–545.

*Amino acids from the hydrochlorides.* Amino acids can be liberated from the hydrochlorides or hydrobromides by treatment with trimethylsilyldiethylamine in DMF or with bis(trimethylsilyl)acetamide (**1**, 61; **2**, 30; **3**, 23–24) in THF. Average yields exceed 96%.

$$HCl \cdot H_2N-\overset{\overset{R}{|}}{C}H-COOH + 2 \ (CH_3)_3SiN(C_2H_5)_2 \longrightarrow HCl \cdot HN(C_2H_5)_2 \ +$$

$$(CH_3)_3SiNH\overset{\overset{R}{|}}{C}HCOOSi(CH_3)_3 \ \xrightarrow{H_2O} \ H_2N\overset{\overset{R}{|}}{C}HCOOH$$

Blomquist *et al.*[2] have used triethylamine for this purpose.

[1] S. V. Rogozhin, Y. A. Davidovich, and A. I. Yurtanov, *Synthesis*, 113 (1975).
[2] A. T. Blomquist, B. F. Hiscock, and D. N. Harpp, *Syn. Commun.*, **3**, 343 (1973).

**Trimethylsilyl isocyanate,** $(CH_3)_3SiNCO$. Mol. wt. 115.20, b.p. 91°.
*Preparation:*[1]

$$(CH_3)_3SiCl + AgNCO \longrightarrow (CH_3)_3SiNCO + AgCl$$

*Primary amides.* The reaction of a Grignard reagent with an isocyanate usually leads to a secondary amide:

$$RMgX + R'NCO \longrightarrow R\overset{\displaystyle O}{\underset{\displaystyle NHR'}{C}}$$

Primary amides can be obtained if trimethylsilyl isocyanate is used, since the trimethylsilyl group is removable with ammonium chloride[2]:

$$RMgBr \xrightarrow[\sim 50\%]{\begin{array}{l}1) \ (CH_3)_3SiNCO \\ 2) \ NH_4Cl, H_2O, dioxane\end{array}} R\overset{\displaystyle O}{\underset{\displaystyle NH_2}{C}}$$

Chloroacetyl isocyanate[3] can also be used; in this case the initial product is an imide, from which the chloroacetyl group can be removed by NaOH in methanol or by zinc dust in methanol.

$$RMgBr \xrightarrow[55-85\%]{\begin{array}{l}1) \ ClCH_2\overset{\overset{\displaystyle O}{\|}}{C}NCO \\ 2) \ H_2O\end{array}} R\overset{O}{\overset{\|}{C}}NH\overset{O}{\overset{\|}{C}}CH_2Cl \xrightarrow[50-90\%]{\begin{array}{c}NaOH \ or \\ Zn\end{array}} R\overset{\displaystyle O}{\underset{\displaystyle NH_2}{C}}$$

[1] G. S. Forbes and H. H. Anderson, *Am. Soc.*, **70**, 1222 (1948).
[2] K. A. Parker and E. G. Gibbons, *Tetrahedron Letters*, 981 (1975).
[3] A. J. Speziale and L. R. Smith, *Org. Syn.*, *Coll. Vol.*, **V**, 204 (1973).

**Trimethylsilylketene,** $(CH_3)_3Si$ $\overset{(CH_3)_3Si}{\underset{H}{>}}C=C=O$. Mol. wt. 114.22, b.p. 81–82°; stable under $N_2$ for many weeks at 25°.

*Preparation:*

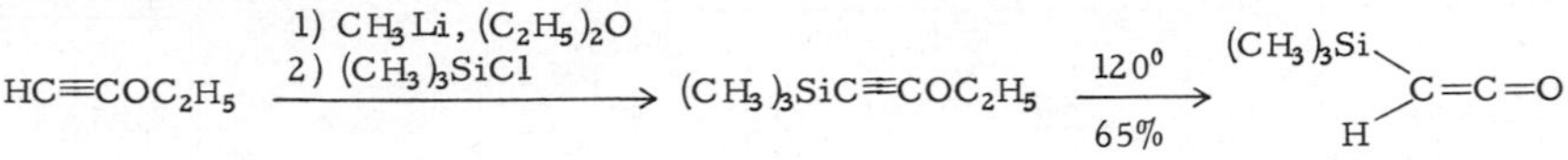

$$HC{\equiv}COC_2H_5 \xrightarrow[\quad 2)\ (CH_3)_3SiCl \quad]{1)\ CH_3Li,\ (C_2H_5)_2O} (CH_3)_3SiC{\equiv}COC_2H_5 \xrightarrow[65\%]{120°} \overset{(CH_3)_3Si}{\underset{H}{>}}C=C=O$$

(1)

*Reactions.*[1] This reactive ketene acylates hindered amines to produce amides in almost quantitative yield. Tertiary alcohols are converted into esters; this reaction is strongly catalyzed by boron trifluoride etherate.

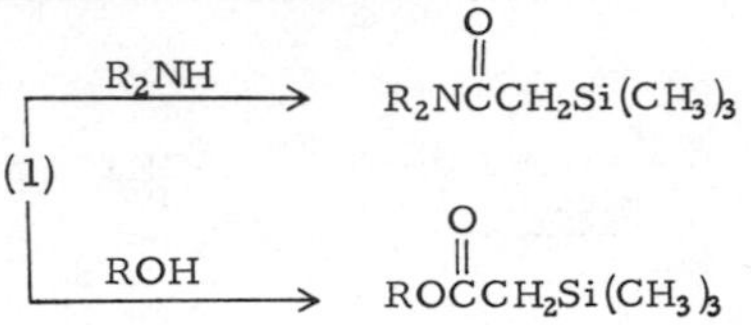

The ketene reacts with carboethoxymethylenetriphenylphosphorane in $CH_2Cl_2$ at −5° to give the allenic ester:

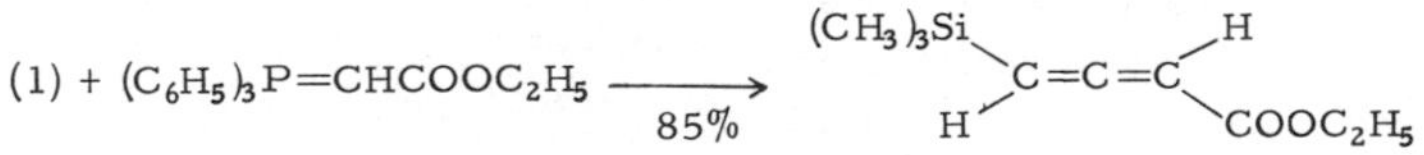

$$(1) + (C_6H_5)_3P{=}CHCOOC_2H_5 \xrightarrow{85\%} \overset{(CH_3)_3Si}{\underset{H}{>}}C=C=C\overset{H}{\underset{COOC_2H_5}{<}}$$

A complex mixture is obtained from the reaction of (1) with methylenetriphenylphosphorane.

[1] R. A. Ruden, *J. Org.*, **39**, 3607 (1974).

**Trimethylsilylmethyllithium,** $(CH_3)_3SiCH_2Li$ (1). Mol. wt. 94.16.

The reagent is prepared by reaction of lithium dispersion (excess) with chloromethyltrimethylsilane.

*β,γ-Unsaturated esters, amides, and nitriles.* Trimethylsilylmethyl ketones can be prepared by reaction of (1) with aldehydes or acids (equations I and II).[1] Ketones of this type can be converted into β,γ-unsaturated esters (III), amides

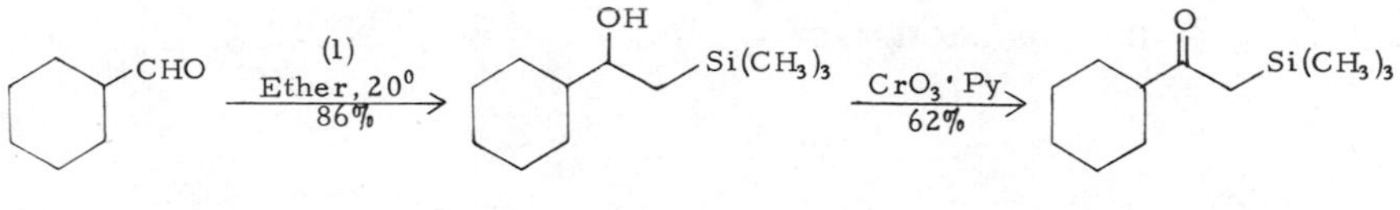

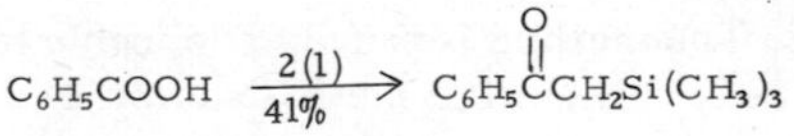

(IV), and nitriles (V) by reaction with *t*-butyl lithioacetate, (5, 371), lithio N,N-dimethylacetamide,[2] and lithio acetonitrile,[3] respectively, followed by elimination of trimethylsilanol (Peterson reaction).

(III)
$$RCCH_2Si(CH_3)_3 \xrightarrow[68-72\%]{LiCH_2COOC(CH_3)_3} RC-CH_2COOC(CH_3)_3 \xrightarrow[THF,\,0^0]{HClO_4}$$

$$\overset{CH_2}{RC}-CH_2COOC(CH_3)_3 \;+\; (CH_3)_3SiOH$$

(IV)

$$\xrightarrow[60\%]{1)\ LiCH_2CON(CH_3)_2 \quad 2)\ H_3O^+}$$

(V)   $CH_3COCH_2Si(CH_3)_3 \xrightarrow[75\%]{1)\ LiCH_2CN \quad 2)\ H_3O^+} CH_3CCH_2CN$

Organolithium compounds add to these ketones to give adducts that are convertible into olefins (equation VI).

(VI)

$$\xrightarrow[80\%]{1)\ C_6H_5Li \quad 2)\ H_2O^+}$$

[1] R. A. Ruden and B. L. Gaffney, *Syn. Commun.*, **5**, 15 (1975).
[2] D. Seebach and D. N. Crouse, *Ber.*, **101**, 3113 (1968).
[3] E. M. Kaiser and C. R. Hauser, *J. Org.*, **33**, 3402 (1968).

## Trimethylsilylmethylmagnesium chloride, 5, 724–725.

*Methylenation.* Chan and Chang[1] have published details of their procedure for effecting methylenation of ketones. The reaction is carried out in refluxing ether for 1–2 hr. to form, after hydrolysis, the adduct (2), which on treatment with thionyl chloride or acetyl chloride at 25° is converted into the alkene (3).

$$\underset{R^2}{\overset{R^1}{>}}C=O + (CH_3)_3SiCH_2MgCl \longrightarrow \underset{R^2}{\overset{R^1}{>}}C\underset{CH_2}{\overset{OH}{<}} \xrightarrow[ca.\,50-60\%]{CH_3COCl} \underset{R^2}{\overset{R^1}{>}}C=CH_2$$

(1)                                        (2)   Si(CH_3)_3                              (3)

Overall yields are in the range 50–60%. The method is usually applicable to α,β-unsaturated ketones, but the yield is only 20% in the methylenation of cyclohexene-2-one. Trimethylsilylmethyllithium, $(CH_3)_3SiCH_2Li$,[2] can also be used, but yields are somewhat lower.

*Related reagents.* The reaction of carbonyl compounds with trimethylsilyl-benzyllithium, $(CH_3)_3SiCHLiC_6H_5$, leads to benzylidenes, $>C=CHC_6H_5$. The reagent is prepared by the reaction of methyllithium with benzyltrimethylsilane.

Reagents of the type (1) can be prepared by the reaction of triphenylvinyl-silane with organolithium compounds. They react with carbonyl compounds to

$$(C_6H_5)_3SiCH=CH_2 + RLi \longrightarrow (C_6H_5)_3SiCHLiCH_2R \xrightarrow{\begin{array}{c} R^1 \\ \diagdown \\ R^2 \end{array} C=O} \begin{array}{c} R^1 \\ \diagdown \\ R^2 \end{array} C=CHCH_2R$$
$$(1) \hspace{6cm} (2)$$

form substituted ethylidenes (2); in the cases reported, E- and Z-isomers were obtained in equal amounts.

The silicon method was used to synthesize the sex pheromone of the gypsy moth, *cis*-7,8-epoxy-2-methyloctadecane, as illustrated:

$$(CH_3)_2CH(CH_2)_2CH_2Li + (C_6H_5)_3SiCH=CH_2 \longrightarrow (CH_3)_2CH(CH_2)_4C\overset{\diagup Si(C_6H_5)_3}{\underset{\diagdown Li}{H}} \xrightarrow[50\%]{CH_3(CH_2)_9CHO}$$

$$(CH_3)_2CH(CH_2)_4CH=CH(CH_2)_9CH_3 \xrightarrow[\text{quant.}]{m\text{-}ClC_6H_4COOOH} (CH_3)_2CH(CH_2)_4\underset{\diagdown O \diagup}{CH-CH}(CH_2)_9CH_3$$

(<u>cis</u> and <u>trans</u> 1:1)

[1] T. H. Chan and E. Chang, *J. Org.*, **39**, 3264 (1974).
[2] A. G. Brook, J. M. Duff, and D. G. Anderson, *Canad. J. Chem.*, **48**, 561 (1970).

**Trimethylsilylmethylpotassium,** $(CH_3)_3SiCH_2K$. Mol. wt. 126.32.

This organopotassium reagent is obtained as a grey suspension by reaction of bis(trimethylsilylmethyl)mercury (Fluka) with potassium/sodium alloy (5:1 by weight) in pentane for 30 min. A similar reaction with sodium metal is slower.[1]

$$[(CH_3)_3SiCH_2]_2Hg + 2K \longrightarrow 2(CH_3)_3SiCH_2K + Hg$$

*Metalation of allylic methylene and methine groups.*[2] This base selectively metalates the allylic position of cyclohexene. This hydrogen–metal exchange has been used in a high-yield synthesis of 2-(3-cyclohexenyl)ethanol (1). The reaction is carried out by stirring cyclohexene with a suspension of the base in

(1)

$$(CH_3)_2CHCH=CH_2 \xrightarrow[\text{THF, } -78°]{\text{Base}} \left[ \begin{array}{l} \rightarrow (CH_3)_2CCH=CH_2 \xrightarrow{\triangle, 25°} (CH_3)_2CCH=CH_2 \quad (2, 37\%) \\ \quad\quad | \quad\quad\quad\quad\quad\quad\quad\quad\quad\quad\quad | \\ \quad\quad K \quad\quad\quad\quad\quad\quad\quad\quad\quad CH_2CH_2OH \\ \\ \rightarrow (CH_3)_2C=CHCH_2K \xrightarrow{\triangle, 25°} (CH_3)_2C=CHCH_2CH_2CH_2OH \\ \quad\quad\quad\quad\quad\quad\quad\quad\quad\quad\quad\quad\quad\quad (3, 28\%) \end{array} \right.$$

pentane for 20 hr. at 25° followed by addition of ethylene oxide at 0°. Work-up after 30 min. at 25° gives (1) in 83% yield. This is the product of direct substitution ($S_E2$). Use of $n$-butyllithium activated by potassium $t$-butoxide gives (1) in only 46% yield.

Application of this sequence to 3-methyl-1-butene leads to formation of both (2) and (3), the products of direct substitution ($S_E2$) and of vinylogous attack ($S_E2'$), respectively.

The formation of C—C bonds by way of allylpotassium compounds has been reviewed.[3]

*Metalation of alkylbenzenes.*[4] The base converts alkylbenzenes into benzyl anions:

$$(CH_3)_3SiCH_2K \quad + \quad Ar\overset{\displaystyle R}{\underset{\displaystyle R}{C}}-H \longrightarrow Ar\overset{\displaystyle R}{\underset{\displaystyle R}{C}}{}^-K^+ \quad + \quad (CH_3)_4Si$$

[1] D. Seyferth and W. Freyer, *J. Org.*, **26**, 2604 (1961).
[2] J. Hartmann and M. Schlosser, *Synthesis*, 328 (1975).
[3] M. Schlosser, *Angew. Chem. internat. Ed.*, **13**, 701 (1974).
[4] A. J. Hart, D. H. O'Brien, and C. R. Russell, *J. Organometal. Chem.*, **72**, C19 (1974).

**1-Trimethylsilylpropynylcopper**, $(CH_3)_3SiC\equiv CCH_2Cu$ (1). Mol. wt. 174.78.

This organocopper reagent is prepared by addition of lithio-1-trimethylsilyl-propyne (2, 239–241) to an ether slurry of cuprous iodide (1.1 eq.) at $-78°$ ($N_2$); gradual warming to $-20$ to $-10°$ produces a dark brown heterogeneous solution of (1).

*1,5-Enynes and 1,4,5-trienes.*[1] The reagent undergoes 1,6-addition to $\Delta^{2,4}$-dienoic esters to give mixtures of 1,5-enynes and 1,4,5-trienes. An excess of reagent substantially increases the yields. The allene/acetylene ratio is strikingly sensitive to substitution on the $\delta$-carbon of the substrate.

*Examples:*

$$+ \quad (CH_3)_3SiC\equiv CCH_2CH_2\diagdown_{C=C}\diagup^H_{CH_2COOC_2H_5}$$
4:1

$$\underset{CH_3}{\overset{}{\diagup}}C=C\diagup^H \quad \diagup^H C=C\diagdown^H_{COOC_2H_5} + (1) \xrightarrow{70\%} H_2C=C=C\diagdown^{Si(CH_3)_3}_{CHCH_3}\diagdown_{C=C}\diagup^H_{CH_2COOC_2H_5}$$

$$+ \quad (CH_3)_3SiC\equiv CCH_2CHCH_3 \diagdown_{C=C}\diagup^H_{CH_2COOC_2H_5}$$
18:82

Similar results are obtained with 1-t-butyldimethylsilylpropynylcopper.

[1] B. Ganem, *Tetrahedron Letters*, 4467 (1974).

**Trimethylsilyl perchlorate**, $(CH_3)_3SiClO_4$. Mol. wt. 172.64, b.p. 35–38°/14 torr. *Caution:* explosive, fumes in the air.

*Preparation.*[1]

$$(CH_3)_3SiCl + AgClO_4 \xrightarrow{C_6H_6, N_2} (CH_3)_3SiClO_4 + AgCl$$

*Synthesis of nucleosides.*[2] Reaction of silylated heterocycles with a protected 1-O-acyl or 1-O-alkyl sugar in 1,2-dichloroethane with either trimethylsilyl perchlorate or trimethylsilyl trifluoromethanesulfonate, $(CH_3)_3SiOSO_2$-$CF_3$,[3] as catalyst gives nucleosides in high yield. The reaction is formulated for the reaction of silylated[4] uracil (1) with 1-O-acetyl-2,3,5-tri-O-benzoyl-β-D-ribofuranose (2) either at 22° for a prolonged period or for 4 hr. at reflux temperature. The function of the catalyst is to convert the sugar into the glycosyl

perchlorate, which then reacts with (1) to give 4-O-silylated uridine tri-O-benzoate (5). Since (3) is regenerated in the reaction, it is required in only catalytic amounts. Work-up with bicarbonate saponifies (5) to give uridine tri-O-benzoate. Typical yields of nucleosides obtained in this way are 65–90%.

This reaction has also been conducted with stannic chloride as the catalyst.[5]

*Selective cleavage of BOC groups.* BOC groups are selectively cleaved in the presence of CBO groups and benzyl esters by this reagent in $C_6H_6$–$CH_2Cl_2$ at $24°$ in a few minutes.[6]

[1] C. Eaborn, *J. Chem. Soc.*, 2517 (1955); U. Wannagat and W. Liehr, *Angew. Chem.*, **69**, 783 (1957).

[2] H. Vorbrüggen and K. Krolikiewicz, *Angew. Chem. Int. Ed.*, **14**, 421 (1975).

[3] H. C. Marsmann and H. G. Horn, *Z. Naturforsch.*, **27B**, 1448 (1972).

[4] Prepared by reaction of uracil with hexamethyldisilazane.

[5] U. Niedballa and H. Vorbrüggen, *J. Org.*, **39**, 3654, 3660, 3664, 3668, 3673 (1974).

[6] H. Vorbrüggen and K. Krolikiewicz, *Angew. Chem. internat. Ed.*, **14**, 818 (1975).

## Trimethylstannylethylidenetriphenylphosphorane,

$(C_6H_5)_3P{=}CHCH_2Sn(CH_3)_3$ (1). Mol. wt. 453.13; red.

The reagent is prepared by the reaction of trimethyltinlithium and vinyltriphenylphosphonium bromide (**4**, 572) in ether–THF at $-93°$:

$$(CH_3)_3SnLi \; + \; CH_2{=}CH\overset{+}{P}(C_6H_5)_3B\bar{r} \quad \xrightarrow{\sim 100\%} (1)$$

*Wittig reaction.*[1] The reagent reacts with cyclohexanone to give the tin compound (2) in almost quantitative yield. The product is cleaved by aqueous hydrochloric acid to vinylcyclohexane (3):

(2)

(3)

Similar products are obtained with acetone and acetaldehyde, but in somewhat lower yields.

[1] S. J. Hannon and T. G. Traylor, *J.C.S. Chem. Comm.*, 630 (1975).

## Trioctylpropylammonium chloride, $[CH_3(CH_2)_7]_3N^+(Cl^-)CH_2CH_2CH_3$. Mol. wt. 432.22. Supplier: Eastman.

*Conversion of bromides into acetates.*[1] Bromides can be converted into the corresponding acetates by treatment with potassium acetate in a two-phase system, water–chloroform, with this quaternary ammonium ion as a phase-transfer catalyst.

$$\underline{n}\text{-}C_8H_{17}Br + KOAc \quad \xrightarrow[68\%]{\overset{\text{Cat.}}{H_2O,\ HCCl_3}} \quad \underline{n}\text{-}C_8H_{17}OAc$$

The method was found to be particularly useful for the synthesis of *cis*-3,5-diacetoxycyclopentene, since acetolysis of *cis*-3,5-dibromocyclopentene with KOAc in HOAc–Ac$_2$O gives a mixture of *cis*- and *trans*-3,5-diacetoxycyclopentene and *cis*- and *trans*-3,4-diacetoxycyclopentene.

[1]T. Toru, S. Kurozumi, T. Tanaka, S. Miura, M. Kobayashi, and S. Ishimoto, *Synthesis*, 867 (1974).

## Trioxo(*t*-butylnitrido)osmium(VIII)  (N-*t*-Butyl  osmiamate),  $(CH_3)_3CNOsO_3$.

Mol. wt. 309.32, bright yellow needles decomposing at 112°.

*Preparation.* The reagent is prepared by the reaction of equimolar amounts of osmium tetroxide and *t*-butylamine in olefin-free pentane. In one method the reaction is carried out at 0° (yield about 60%).[1] Higher yields (>90%) can be

$$OsO_4 + (CH_3)_3CNH_2 \xrightarrow[\quad 0^0 \quad]{\text{Pentane}} O=\overset{\text{O}}{\underset{\text{O}}{\overset{\|}{\underset{\|}{Os}}}}=NC(CH_3)_3 + H_2O$$

(1)

obtained if the solvent is removed after 5 min. at reduced pressure and the mixture allowed to stand for 12 hr. in the dark before sublimation of the product.[2]

*Oxyamination of olefins.*[2] This reagent reacts with a variety of olefins to give, after reductive cleavage of intermediate osmate esters, *cis*-vicinal amino alcohols in fair to high yield. The reaction is regiospecific in that the new C–N bond is formed at the least substituted carbon atom of the olefin. The reagent reacts more rapidly with monosubstituted olefins than with di- and trisubstituted olefins. Diols are obtained as the main products from hindered olefins. Methylene chloride or THF can be used as solvents, but pyridine is the solvent of choice.

*Examples:*

$$C_6H_5CH{=}CH_2 \xrightarrow[92\%]{\begin{array}{l}1)\ (1),\ \text{Pyridine}\\ 2)\ \text{LiAlH}_4,\ (C_2H_5)_2O\end{array}} C_6H_5\underset{\underset{\text{OH}}{|}}{CH}{-}CH_2NHC(CH_3)_3$$

$$C_8H_{17}CH{=}CH_2 \xrightarrow[89\%]{} C_8H_{17}CHCH_2NHC(CH_3)_3$$
$$\underset{OH}{}$$

$$\xrightarrow{85\%}$$

(38%)          +          (45%)

Sharpless[2] has also prepared the reagent in which the *t*-butyl group of (1) is replaced by 1-adamantyl.

[1] A. F. Clifford and C. S. Kobayashi, *Inorg. Syn.*, **6**, 204 (1960).
[2] K. B. Sharpless, D. W. Patrick, L. K. Truesdale, and S. A. Biller, *Am. Soc.*, **97**, 2305 (1975).

**Triphenylmethyl isocyanide (Trityl isocyanide)**, $(C_6H_5)_3CNC$. Mol. wt. 269.33, m.p. 127–128°.

*Preparation.*[1]

*Nitriles; ketones.* In a further study of synthetic uses of lithium aldimines (**3**, 279–280), Periasamy and Walborsky[2] noted that *t*-butylnitrile is obtained in 88% yield by addition of *t*-butyllithium to trityl isocyanide. However, use of most lithium reagents results in formation of ketones (see below). The authors

$$(C_6H_5)_3CNC + (CH_3)_3CLi \longrightarrow \left[ \begin{array}{c} (C_6H_5)_3C \\ N{\equiv}C \\ \quad C(CH_3)_3 \end{array} \right] \xrightarrow{88\%} (CH_3)_3CC{\equiv}N + (C_6H_5)_3CLi$$

then found that secondary nitriles could be prepared in good yield by reaction of trityl isocyanide with secondary Grignard reagents as shown in the examples.

$$(C_6H_5)_3CNC + \underset{2:3}{} \underset{CH_3}{\overset{C_2H_5}{>}}CHMgBr \xrightarrow{98\%} \underset{CH_3}{\overset{C_2H_5}{>}}CHC{\equiv}N$$

$$(C_6H_5)_3CNC + C_6H_5CH_2MgBr \xrightarrow{78\%} C_6H_5CH_2C{\equiv}N$$
$$\underset{1:2}{}$$

$$(C_6H_5)_3CNC + C_6H_{11}MgBr \xrightarrow{78\%} C_6H_{11}C{\equiv}N$$
$$\underset{1:2}{}$$

Yields were low with primary Grignard reagents except for benzylmagnesium bromide; tertiary and aromatic Grignard reagents failed to react with the isocyanide to any appreciable extent.

The synthesis of symmetrical ketones noted above was developed into a practicable method by use of 2 eq. of the organolithium reagent for 1 eq. of the isocyanide. The reaction involves formation of the nitrile followed by reaction of this product with another equivalent of the lithium reagent:

$$\overset{\text{NLi}}{\underset{\|}{}} \qquad \overset{\text{O}}{\underset{\|}{}}$$
$$\text{RCN} + \text{RLi} \longrightarrow \text{RCR} \longrightarrow \text{RCR}$$

*Examples:*

$$(\text{C}_6\text{H}_5)_3\text{CNC} + 2\ \underline{\text{n}}\text{-BuLi} \xrightarrow[\ 59\%\ ]{} (\underline{\text{n}}\text{-Bu})_2\text{C}{=}\text{O}$$

$$(\text{C}_6\text{H}_5)_3\text{CNC} + 2\ \underline{\text{t}}\text{-BuLi} \xrightarrow[\ 75\%\ ]{} (\underline{\text{t}}\text{-Bu})_2\text{C}{=}\text{O}$$

$$(\text{C}_6\text{H}_5)_3\text{CNC} + 2\ \text{C}_6\text{H}_5\text{Li} \xrightarrow[\ 94\%\ ]{} (\text{C}_6\text{H}_5)_2\text{C}{=}\text{O}$$

[1]H. M. Walborsky and G. E. Niznik, *J. Org.*, **37**, 187 (1972).
[2]M. P. Periasamy and H. M. Walborsky, *ibid,*, **39**, 611 (1974).

**Triphenylphosphine, 1,** 1238–1247; **2,** 443–445; **3,** 317–320; **4,** 548–550.

*Olefin synthesis by double extrusion* (3, 319–320; 4, 550). Barton *et al.*[1] have prepared some very hindered olefins by reaction of hindered thioketones with diazo compounds to afford $\Delta^3$-1,3,4-thiadiazolines. When heated with triphenylphosphine or tri-*n*-butylphosphine, these afford olefins by extrusion of $\text{N}_2$ and S. The method is illustrated for the synthesis of (+)-2-diphenylmethylenefenchane (2).

(1)                                                    (a)

(2)

*Dequaternization of pyridinium salts.* Pyridinium salts, particularly those bearing an electron-withdrawing group on the ring, undergo dequaternization when heated in acetonitrile with triphenylphosphine.[2]

$$(R = CH_2C_6H_5, CH_3,$$
$$Y = CN, COCH_3, C_2H_5)$$

[1] D. H. R. Barton, F. S. Guziec, jun., and I. Shahak, *J.C.S. Perkin I*, 1974 (1974).
[2] J. P. Kutney and R. Greenhouse, *Syn. Commun.*, **5**, 119 (1975).

## Triphenylphosphine–Carbon tetrabromide–Lithium azide.

*Conversion of −OH to −N₃.* Japanese chemists[1] have converted the primary $5'$-hydroxyl groups of nucleosides into azided groups in 50–90% yield by reaction with triphenylphosphine, carbon tetrabromide, and lithium azide in DMF at 20° (24 hr.). The mechanism in equation I is suggested.

(I)      $(C_6H_5)_3P \ + \ CBr_4 \longrightarrow (C_6H_5)_3\overset{+}{P}-CBr_3 \ \ Br^- \xrightarrow{\ LiN_3\ } (C_6H_5)_3\overset{+}{P}CBr_3 \ N_3^-$

$\xrightarrow[-CHBr_3]{ROH} \ (C_6H_5)_3\overset{+}{P}-OR \ N_3^- \longrightarrow RN_3 \ + \ (C_6H_5)_3P=O$

[1] T. Hata, I. Yamamoto, and M. Sekine, *Chem. Letters.*, 977 (1975).

## Triphenylphosphine—Carbon tetrachloride, 1, 1274; 2, 445; 3, 320; 4, 551–552; 5, 727.

*β-Halovinyl ketones.*[1] 1,3-Diketones are converted into β-halo-α,β-unsaturated ketones by reaction with triphenylphosphine and carbon tetrachloride or tetrabromide. Only resinous products are formed in the case of β-keto aldehydes.

*Examples:*

*Alkyl chlorides.* Hungarian chemists[2] have isolated and identified the salts (1) and (2) in the reaction of triphenylphosphine and carbon tetrachloride with alcohols, enolizable ketones, and acid halides. They have suggested a new mechanism in which these salts participate.

$$[ (C_6H_5)_3PCH_2Cl]^+ Cl^- \qquad [ (C_6H_5)_3PCHCl_2]^+ Cl^-$$

$$(1) \qquad\qquad\qquad (2)$$

*Polymeric reagent.* Hodge and Richardson[3] have prepared a polymer-supported triphenylphosphine by bromination of a polystyrene cross-linked with divinylbenzene followed by reaction with lithium diphenylphosphide (4, 303). A similar reagent has been described by Regen and Lee.[4] Polymeric material is available from Strem.

The polymeric reagent has been used in conjunction with carbon tetrachloride to convert acids into acid chlorides and alcohols into alkyl chlorides. Yields are generally satisfactory and the work-up merely involves removal of the polymeric triphenylphosphine oxide by filtration.

*Allylic chlorides* (4, 552). The preparation of geranyl chloride from geraniol has been published.[5]

[1] L. Gruber, I. Tömösközi, and L. Radics, *Synthesis*, 708 (1975).
[2] I. Tömösközi, L. Gruber, and L. Radics, *Tetrahedron Letters*, 2473 (1975).
[3] P. Hodge and G. Richardson, *J.C.S. Chem. Comm.*, 622 (1975).
[4] S. L. Regen and D. P. Lee, *J. Org.*, **40**, 1669 (1975).
[5] J. G. Calzada and J. Hooz, *Org. Syn.*, **54**, 63 (1974).

**Triphenylphosphine–Diethyl azodicarboxylate, 4, 553–555, 5, 727–728.**

*Alkyl aryl ethers.* Two laboratories[1,2] have reported independently that alkyl aryl ethers can be prepared by reaction of an alcohol and a phenol with this combination of reagents. The reaction proceeds with inversion in the case of $3\beta$-cholestanol (but not cholesterol). The reaction proceeds at room temperature in THF.

*Transesterification.*[3] Ethyl and higher esters are converted into methyl esters in high yield under mild conditions by reaction with methanol (excess) and this complex. The reverse reaction proceeds slowly, if at all.

[1] S. Bittner and Y. Assaf, *Chem. Ind.*, 281 (1975).
[2] M. S. Manhas, W. H. Hoffman, B. Lal, and A. K. Bose, *J.C.S. Perkin I*, 461 (1975).
[3] S. Bittner, Z. Barneis, and S. Felix, *Tetrahedron Letters*, 3871 (1975).

**Triphenylphosphine dibromide, 1, 1247–1249; 2, 446; 3, 320–322; 4, 555; 729–731.**

*α,α′-Dibromoalkynes.* α,α′-Dibromoalkynes (2) can be obtained in high yield by reaction of alkynediols (1) with triphenylphosphine dibromide in acetonitrile. Use of phosphorus tribromide in this transformation results in low yields of (2) owing to propargylic rearrangements.[1]

$$HO-\underset{\underset{H}{|}}{\overset{\overset{H(CH_3)}{|}}{C}}-C\equiv C-\underset{\underset{H}{|}}{\overset{\overset{H(CH_3)}{|}}{C}}-OH \quad + \quad 2 \ (C_6H_5)_3PBr_2 \quad \xrightarrow[85-90\%]{\overset{CH_3CN}{40-50°}}$$

(1)

$$Br-\underset{\underset{H}{|}}{\overset{\overset{H(CH_3)}{|}}{C}}-C\equiv C-\underset{\underset{H}{|}}{\overset{\overset{H(CH_3)}{|}}{C}}-Br \quad + \quad 2 \ (C_6H_5)_3P=O \quad + \quad 2 \ HBr$$

(2)

*Cleavage of lactones.* Lactones are cleaved by the reagent, but yields are low.[2]

$$\xrightarrow[30\%]{\begin{array}{l}1) \ (C_6H_5)_3PBr_2, CH_3CN \\ 2) \ CH_3OH\end{array}} \quad Br(CH_2)_4COOCH_3$$

*Azetidines.*[3] Reaction of $\gamma$-amino alcohols with triphenylphosphine dibromide and then with triethylamine first at $0°$ and then at $25°$ results in formation of triphenylphosphine oxide and N-substituted azetidines. This method had been used previously to prepare aziridines from $\beta$-amino alcohols (**3**, 322).

$$RNHCH_2CH_2\overset{\overset{R^1}{|}}{C}HOH \quad \xrightarrow[45-65\%]{\begin{array}{c}C_6H_5PBr_2 \\ N(C_2H_5)_3\end{array}} \quad$$

*Cleavage of esters to acid chlorides* (**5**, 730). The definitive paper on cleavage of esters with $(C_6H_5)_3PCl_2$, $(C_6H_5)_3PBr_2$, and $(C_6H_5)_3\overset{+}{P}ClBF_3Cl^-$ has been published.[4]

[1] R. Machinek and W. Lüttke, *Synthesis*, 255 (1975).
[2] E. E. Smissman, H. N. Alkaysi, and M. W. Creese, *J. Org.*, **40**, 1640 (1975).
[3] J. P. Freeman and P. J. Mondron, *Synthesis*, 894 (1974).
[4] D. J. Burton and W. M. Koppes, *J. Org.*, **40**, 3026 (1975).

## Triphenylphosphine dichloride, 1, 1247–1249.

*cis-1,2-Dichlorides.* Triphenylphosphine dichloride reacts with cyclohexene oxide (1) with inversion at both asymmetric centers to form *cis*-1,2-dichlorocyclohexane (2). Reaction of (1) with triphenylphosphine dibromide gives a mixture of *cis*- and *trans*-1,2-dibromocyclohexane.[1]

(1)                (a)                (2)

*Polymer-supported forms.* Chemists at General Electric[2] have reported the preparation by three routes of two trisubstituted phosphine dichlorides (1) and (2) bound to cross-linked polystyrene beads. The final step in all cases involved

$$\text{(P)}-C_6H_4CH_2-\overset{\overset{\displaystyle Cl}{|}}{\underset{\underset{\displaystyle Cl}{|}}{P}}(C_6H_5)_2 \qquad \text{(P)}-C_6H_4-\overset{\overset{\displaystyle Cl}{|}}{\underset{\underset{\displaystyle Cl}{|}}{P}}(C_6H_5)_2$$

$$\text{(1)} \qquad\qquad\qquad \text{(2)}$$

the reaction of phosgene with the phosphine oxide reagents corresponding to (1) and (2).

These reagents can be used in the same way as triphenylphosphine dichloride (equations I–V). Each reaction leads to the polymer-bound corresponding

$$(\text{I}) \qquad C_6H_5CH_2COOH + (1) \xrightarrow[100\%]{CH_2Cl_2} C_6H_5CH_2COCl$$

$$(\text{II}) \qquad C_6H_5CONH_2 + (1) \xrightarrow[78\%]{CH_2Cl_2} C_6H_5C\equiv N$$

$$(\text{III}) \qquad C_6H_5CONHC_6H_5 + (2) \xrightarrow[93\%]{CH_3CN} \underset{C_6H_5}{\overset{Cl}{>}}C=NC_6H_5$$

$$(\text{IV}) \qquad C_6H_5CH_2OH + (2) \xrightarrow[88\%]{CH_3CN} C_6H_5CH_2Cl$$

$$(\text{V}) \qquad C_6H_5COCH_3 + (2) \xrightarrow[75\%]{CH_3CN} \underset{Cl}{\overset{C_6H_5}{>}}C=CH_2$$

phosphine oxide reagent, which can be reconverted to the dichloride reagent and reused.

[1] P. E. Sonner and J. E. Oliver, *J. Org.*, 000 (0000); J. E. Oliver and P. E. Sonnett, *Org. Syn.*, submitted (1976).

[2] H. M. Relles and R. W. Schluenz, *Am. Soc.*, **96**, 6469 (1974).

## Triphenylphosphine dihalides.

*β-Halo- α,β-unsaturated ketones.* These reagents react with cyclic β-diketones in the presence of triethylamine at room temperature to give β-halo-α, β-unsaturated ketones in high yield (90–97%), presumably by the mechanism formulated in equation I.[1]

Yields of products are low under these conditions with triphenylphosphine diiodide, but use of acetonitrile and a reaction time of about 4 days gives $\beta$-iodo-$\alpha$, $\beta$-unsaturated ketones; however, yields are only about 70%.

[1] E. Piers and I. Nagakura, *Syn. Commun.*, 5, 193 (1975).

**Triphenylphosphine ditriflate,** $(C_6H_5)_3\overset{+}{P}OSO_2CF_3\overset{-}{O}SO_2CF_3$ (1). Mol. wt. 560.43, m.p. 74–75°.

The reagent is prepared by the reaction of triphenylphosphine oxide with trifluoromethanesulfonic anhydride at 0°. It is unstable to air or moisture, but can be stored for some time under nitrogen.

*Reactions.*[1] The reagent reacts with secondary or tertiary alcohols at room temperature to form a complex, $RO\overset{+}{P}(C_6H_5)_3\overset{-}{O}SO_2CF_3$, which is converted into an olefin when heated. For example, cyclohexanol can be converted into cyclohexene (90% yield). However, an array of products is formed from tertiary alcohols.

The complex formed from (1) and an acid reacts with an amine to form an amide or with an alcohol to form an ester:

Amides can be dehydrated to nitriles in 50–80% yield by treatment with the reagent at room temperature for 24 hr.

[1] J. B. Hendrickson and S. M. Schwartzman, *Tetrahedron Letters*, 277 (1975).

**Triphenylphosphite methiodide (Methyl triphenoxyphosphonium iodide)**, **1**, 1249; **2**, 446; **4**, 448, 557–559.

*Dehydration* (**4**, 558–559). Labile dienes and trienes are prepared advantageously by reaction with this reagent in HMPT (50–60°).[1]

*Examples:*

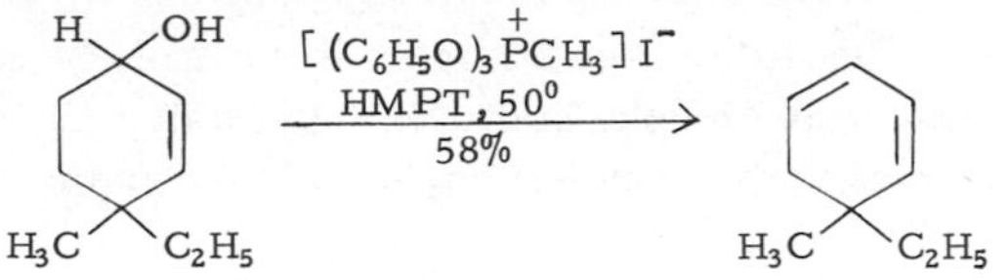

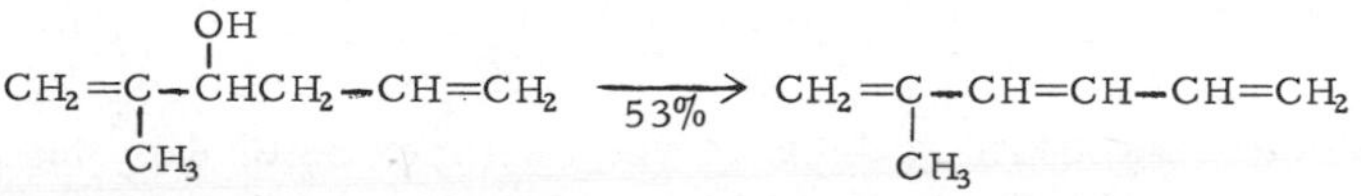

Dehydration of either *erythro*- or *threo*-1,2-diphenyl-1-propanol, (1) and (2), with triphenylphosphite methiodide in HMPT gives a mixture of α-methylstilbenes in which the less stable Z-olefin predominates. Dehydration of (1) and

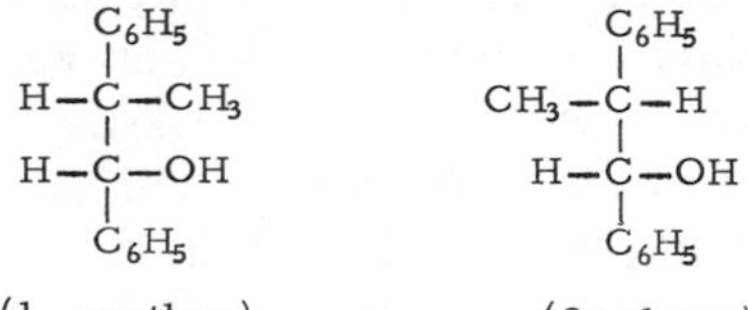

(1, <u>erythro</u>)                    (2, <u>threo</u>)

(2) with iodine or with *p*-toluenesulfonic acid in refluxing xylene gives E- and Z-α-methylstilbene in the ratio 72:28. This ratio is the equilibrium composition of α-methylstilbene.[2]

[1] C. W. Spangler and T. W. Hartford, *Synthesis*, 108 (1976).
[2] W. Reeve and R. M. Doherty, *J. Org.*, **40**, 1662 (1975).

**Triphenyltin hydride (Triphenylstannane)**, **1**, 1250–1251; **3**, 448; **4**, 559; **5**, 734.

$\Delta^{14}$-*Pregnene-20-ones* (**5**, 293–294). This monohydride reduces typical $\Delta^{14,16}$-pregnadiene-20-ones (1) to $\Delta^{14}$-pregnene-20-ones (2) in >80% yield. Yields with tri-*n*-butylstannane are ~35%.[1]

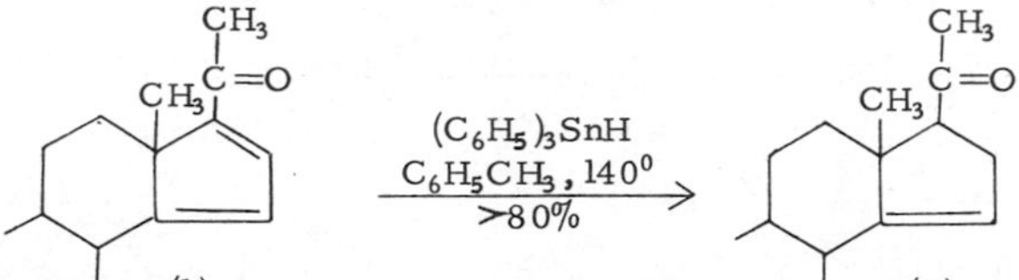

[1] E. Yoshii, T. Koizumi, H. Ikeshima, K. Ozaki, and I. Hayashi, *Chem. Pharm. Bull. Japan*, **23**, 2496 (1975).

## Tris(aquo)hexa-$\mu$-acetato-$\mu_3$-oxotriruthenium(III,III,III) acetate,

$[Ru_3O(OCOCH_3)_6(H_2O)_3]^+\bar{O}COCH_3$ (1). Mol. wt. 780.53. Green-black solid. Soluble in water, lower alcohols, DMF; insoluble in benzene, chloroform.

*Preparation.*[1] This substance is prepared by the interaction of ruthenium trichloride with acetic acid and sodium acetate in refluxing ethanol.

*α,β-Unsaturated ketones.* This ruthenium complex is particularly active as the catalyst for homogeneous intramolecular hydrogen transfer of 1-alkene-3-ols (equation I).[2] It is also the most active catalyst for transfer of hydrogen from

$$(I) \qquad H_2C=CHCHR \xrightarrow[>95\%]{Cat.} CH_3CH_2CR$$

$α,β$-unsaturated secondary alcohols to vinyl ketones, particularly methyl vinyl ketone (equation II).[3] The highest yields (82–90%) are obtained at 100° with a

$$(II) \qquad RCH=CHCHR' + H_2C=CHCOCH_3 \xrightarrow{Cat.} RCH=CHCR' + CH_3CH_2COCH_3$$

donor/acceptor molar ratio of 1:5. This intermolecular process thus provides a general route to $α,β$-unsaturated ketones. Chalcone, benzalacetone, and aceto-phenone are inactive as hydrogen acceptors in this reaction. And vinyl ketones are not reduced by other common alcoholic donors such as benzyl alcohols. If $RuCl_2[P(C_6H_5)_3]_3$ or $RuCl_3$ is used as catalyst, intramolecular (I) and inter-molecular (II) hydrogen-transfer reactions obtain.

[1] P. Legzdins, R. W. Mitchell, G. L. Rempel, J. P. Ruddick, and G. Wilkinson, *J. Chem. Soc. (A)*, 3322 (1970); A. Spencer, and G. Wilkinson, *J.C.S. Dalton*, 1570 (1972).
[2] Y. Sasson and G. L. Rempel, *Tetrahedron Letters*, 4133 (1974).
[3] *Idem, Canad. J. Chem.*, 52, 3825 (1974).

## Tris(phenylthio)methyllithium, $(C_6H_5S)_3C\bar{L}i^+$. Mol. wt. 346.45.

This lithium salt (1) is easily prepared[1] by treatment of triphenyl orthothio-formate[2] or of tetrakis(phenylthio)methane[3] with *n*-butyllithium in THF at −78° ($N_2$).

This unusually stable carbenoid (1) is apparently in equilibrium with a carbene:

$$(1) + CH_3I \xrightarrow{95\%} (C_6H_5S)_3CCH_3$$

$$(1) + C_6H_5CHO \xrightarrow{80\%} (C_6H_5S)_3CCHC_6H_5 \atop \qquad\qquad\qquad OH$$

$$(1) + H_2C=C(SC_6H_5)_2 \xrightarrow{78\%}$$

For example, it decomposes above 20° to give tetrakis(phenylthio)ethylene.[3]

*Typical reaction.*[3]

$$(C_6H_5S)_3CLi \rightleftharpoons (C_6H_5S)_2C: + C_6H_5SLi$$

(1)    $\downarrow 20°$

$$(C_6H_5S)_2=C(SC_6H_5)_2$$

*γ-Keto esters.*[1] α,β-Unsaturated ketones undergo conjugate addition with (1) to give γ-keto orthothioesters, usually in high yield. The reaction is illustrated for the case of cyclohexenone. The resulting γ-keto orthothioester (2) can

(2)    (3)

(4)

be hydrolyzed to the γ-keto ester (4) or converted into 3-methylcyclohexanone (3) by treatment with Raney nickel.

Hindered β,β-disubstituted enones such as 3-methylcyclohex-2-enone and $(CH_3)_2C=CHCOCH_3$ are not useful for this reaction. Also, the corresponding salt $(C_6H_5S)_3\bar{C}Na^+$ gives significantly lower yields of ortho thioesters. α,β-Unsaturated aldehydes react with (1) to give products of 1,2-addition.

[1] A.-R. B. Manas and R. A. Smith, *J.C.S. Chem. Comm.*, 216 (1975).
[2] A. Fröling and J. F. Arens, *Rec. trav.*, 81, 1009 (1962).
[3] D. Seebach, *Ber.*, 105, 487 (1972).

**N,N′,N″-Tris(tetramethylene)phosphoroamide,** $[\underline{CH_2(CH_2)_3N}]_3P=O$. Mol. wt. 257.31. Supplier: Fluka.

This polar aprotic solvent has a higher known electron-donating power than any known solvent (about 25% higher than HMPT).[1]

[1] Y. Ozari and J. Jagur-Grodzinski, *J.C.S. Chem. Comm.*, 295 (1974).

**Tris(trimethylsilyl)hydrazidocopper,** $[(CH_3)_3Si]_2N-N\overset{Si(CH_3)_3}{\underset{Cu}{\diagup}}$ (1). Mol. wt. 311.24.

The reagent is prepared from tris(trimethylsilyl)hydrazidolithium and cuprous iodide.

*Aryl hydrazines.*[1] The reagent reacts with aryl iodides at 135–140° (6 hr.) to form the protected hydrazine (2), which is hydrolyzed by acid to the hydrazine (3).

$$ArI \ + \ (1) \ \xrightarrow[\substack{135-140° \\ 45-55\%}]{\text{Quinoline}} \ [(CH_3)_3Si]_2N-N\overset{Si(CH_3)_3}{\underset{Ar}{\diagdown}} \ \xrightarrow[\substack{CH_3OH \\ 90\%}]{H_3O^+} \ ArNHNH_2$$

$$(2) \qquad\qquad (3)$$

[1] F. D. King and D. R. M. Walton, *Synthesis*, 738 (1975).

**Tris(triphenylphosphine)chlororhodium(I), 1,** 1252; **2,** 248–253; **3,** 325–329; **4,** 559–562; **5,** 736–740.

*Selective hydrogenation.* Butynediol can be selectively hydrogenated to *cis*-2-butene-1,4-diol in $CF_3CH_2OH$ at –20° or in $C_6H_6-CF_3CH_2OH$ at 0° with this catalyst. The product distribution is 94% of the enediol and 4% of the butanediol.[1] The same selective hydrogenation can be realized with $IrCl(CO)$-$[P(C_6H_5)_3]_2$ as catalyst in toluene–trifluoroethanol at 60° under weak UV irradiation.[2]

Mesityl oxide, $(CH_3)_2C{=}CHCOCH_3$, can be selectively hydrogenated to methyl isobutyl ketone with this catalyst.[3]

*Transfer hydrogenation.* Japanese chemists[4] report that some cyclic amines, particularly indoline and pyrrolidine, are much more reactive in catalyzed transfer hydrogenation than oxygenated and hydroaromatic compounds.

*Hydrosilylation of carbonyl compounds.*[5] This rhodium complex is an effective catalyst for hydrosilylation of aldehydes, ketones, $\alpha,\beta$-unsaturated aldehydes and ketones, and $\alpha$-diketones. Hydrosilylation followed by hydrolysis is equivalent to reduction of either the carbonyl group or the $\alpha,\beta$-unsaturation.

*Examples:*

$$C_6H_5CHO \ + \ HSi(C_2H_5)_3 \ \xrightarrow[95\%]{Cat.} \ C_6H_5CH_2OSi(C_2H_5)_3 \ \xrightarrow[\text{quant.}]{H_2O} \ C_6H_5CH_2OH$$

$$(CH_3)_2C=CH-\overset{\overset{\displaystyle CH_3}{|}}{C}=O \;+\; HSi(C_2H_5)_3 \xrightarrow[94\%]{Cat.} (CH_3)_2CHCH=\overset{\overset{\displaystyle CH_3}{|}}{C}-OSi(C_2H_5)_3 \xrightarrow[quant.]{H_2O} (CH_3)_2CHCH_2\overset{\overset{\displaystyle CH_3}{|}}{C}=O$$

$$+ \quad HSi(C_2H_5)_3 \xrightarrow[92\%]{Cat.}$$

$(C_2H_5)_3SiO$

$$+ \quad HSi(C_2H_5)_3 \xrightarrow[80\%]{Cat.}$$

(dl and meso)

$$CH_3COCH_2COCH_3 \;+\; HSi(C_2H_5)_3 \xrightarrow[92\%]{Cat.} CH_3COCH=C\overset{\diagup OSi(C_2H_5)_3}{\diagdown CH_3}$$

(E and Z isomers)

*Decarbonylation* (2, 451; 3, 327–329; 4, 559). 1,3-Diynes (2) can be prepared from diethynyl ketones (1) by treatment with 1 eq. of tris(triphenylphosphine)chlororhodium in refluxing benzene (15 hr.) or refluxing xylene (1–4 hr.). Yields are in the range of 50–95%.[6]

$$ArC\equiv C\overset{\overset{\displaystyle O}{\|}}{C}C\equiv CAr + [(C_6H_5)_3P]_3RhCl \longrightarrow ArC\equiv C-C\equiv CAr + RhCl(CO)[P(C_6H_5)_3]_2 + P(C_6H_5)_3$$

(1)                                                 (2)

α- and β-Diketones are decarbonylated to monoketones by this complex in refluxing toluene.[7]

$$R\overset{\overset{\displaystyle O}{\|}}{C}-(CH_2)_n-\overset{\overset{\displaystyle O}{\|}}{C}R' \quad \xrightarrow{RhCl[P(C_6H_5)_3]_3} \quad R\overset{\overset{\displaystyle O}{\|}}{C}-(CH_2)_n-R'$$

(n = 0, 1)

*Polymer bound.* This material is available from Strem.

[1] W. Strohmeier and K. Grünter, *J. Organometal. Chem.*, **90**, C45 (1975).
[2] *Idem, ibid.*, **90**, C48 (1975).
[3] W. Strohmeier and E. Hitzel, *J. Organometal. Chem.*, **91**, 373 (1975); *idem, ibid.*, **102**, C37 (1975).
[4] T. Nishiguchi, K. Tachi, and K. Fukuzumi, *J. Org.*, **40**, 237, 240 (1975).
[5] I. Ojima, M. Nihonyanagi, T. Kogure, M. Kumagai, S. Horiuchi, and K. Nakatsugawa, *J. Organometal. Chem.*, **94**, 449 (1975).
[6] E. Müller and A. Segnitz, *Ann.*, 1583 (1973).
[7] K. Kaneda, H. Azuma, M. Wayaku, and S. Teranishi, *Chem. Letters*, 215–216 (1974).

**Tris(triphenylphosphine)nickel(0) [Ni(TPP)₃].** Mol. wt. 845.55, red-brown.

This zerovalent nickel reagent can be generated[1] *in situ* by reduction of bis(triphenylphospine)nickel(II) chloride[2] with zinc in the presence of triphenylphosphine in DMF.

*Ullmann reaction.*[1] Semmelhack (**4**, 33) has shown that bis(1,5-cyclo-octadiene)nickel(0) and tetrakis(triphenylphospine)nickel(0) are useful reagents for coupling of aryl halides (Ullmann reaction). However, these zerovalent nickel complexes are not easy to prepare and are extremely sensitive to moisture. Kende reasoned that the coordinatively unsaturated Ni(TPP)$_3$ is actually the species involved and indeed found that aryl halides couple to biaryls in the presence of 1 eq. of this complex. Aryl halides with two *ortho*-substituents do not couple. Otherwise, yields of biaryls are in the range of 60–80%.

*Examples:*

$$2 \ C_6H_5Br \xrightarrow[73\%]{} C_6H_5-C_6H_5$$

In addition, vinylic and allylic halides can be coupled with this reagent (equations I and II).[3]

(I)     $2 \ C_6H_5CH=CHBr \xrightarrow[43\%]{} C_6H_5CH=CH-CH=CHC_6H_5$

(mainly <u>trans</u>, <u>trans</u>)

(II)     $2 \ C_6H_5CH=CHCH_2Br \xrightarrow[50\%]{} C_6H_5CH=CHCH_2CH_2CH=CHC_6H_5$

[1]A. S. Kende, L. S. Liebeskind, and D. M. Braitsch, *Tetrahedron Letters*, 33 75 (1975).
[2]K. Yamamoto, *Bull. Chem. Soc. Japan*, **27**, 501 (1954).
[3]C. A. Tolman *et al.*, *Am. Soc.*, **94**, 2669 (1972); *idem, ibid.*, **96**, 53 (1974).

## Tris(triphenylphosphine)ruthenium dichloride, 4, 564; 5, 740–741.

*Reduction of anhydrides to γ-lactones.*[1] Carboxylic acid anhydrides are reduced by homogeneous hydrogenation catalyzed by this ruthenium complex to γ-lactones. The corresponding dicarboxylic acids are formed as well by hydrolysis by water formed in the reaction. A number of other typical homogeneous catalysts are ineffective. The reduction is considered to involve cleavage of the C—O bond to form a hydroxy carboxylic acid, which then cyclizes.

*Hydrogenation of nitro groups.* Knifton[2] has reported the hydrogenation of nitroparaffins to secondary alkyl primary amines with this catalyst.

Aromatic nitro groups are selectively reduced to amino groups with the catalyst in the presence of halogen, ester, and nitrile groups. Nitrobenzene is reduced to aniline quantitatively. More interestingly, dinitroarenes can be selectively reduced to nitroanilines in good yield. Such selectivity is not possible with most heterogeneous catalysts.[3]

*Transfer hydrogenation* (4, 564). Sasson and Blum[4] have reported kinetic studies on the reduction of α,β-unsaturated ketones by catalyzed transfer of hydrogen from primary and secondary alcohols. They also report catalyzed transfer of hydrogen from secondary carbinols to saturated ketones:

$$(C_6H_5CH_2)_2C{=}O + CH_3\overset{\underset{\displaystyle C_6H_5}{|}}{C}HOH \xrightarrow[\text{180}^0,\ \text{4 hr.}]{[C_6H_5)_3P]_3RuCl_2} (C_6H_5CH_2)_2CHOH + \underset{C_6H_5}{\overset{CH_3}{\diagdown}}C{=}O$$

$$93\%$$

This reaction is successful when the two ketones involved have significantly different oxidation potentials.

*Addition of carbon tetrachloride to cyclohexene.*[5] Addition of carbon tetrachloride to cyclohexene catalyzed by this Ru(II) complex results in formation of *trans*-1-trichloromethyl-2-chlorocyclohexane (63% yield, pure) as the major product. Catalysis by dibenzoyl peroxide results in a 53:47 mixture of the *trans*- and *cis*-adducts (low yield).

[1] J. E. Lyons, *J.C.S. Chem. Comm.*, 412 (1975).
[2] J. F. Knifton, *J. Org.*, **40**, 519 (1975).
[3] *Idem*, *Tetrahedron Letters*, 2163 (1975).
[4] Y. Sasson and J. Blum, *J. Org.*, **40**, 1887 (1975).
[5] H. Matsumoto, T. Nikaido, and Y. Nagai, *Tetrahedron Letters*, 899 (1975).

**Tris(triphenylsilyl)vanadate**, $[(C_6H_5)_3SiO]_3VO$. Mol. wt. 893.14, stable.

*Preparation.*[1] The catalyst can also be prepared *in situ* from tri-*n*-propyl orthovanadate and triphenylsilanol (Arapahoe).[2]

*17(20)-Ene-21-al steroids.*[2] 17α-Ethynyl-17β-hydroxy steroids (1) are rearranged to this key system (2) for the synthesis of corticosteroids by tris-

(triphenylsilyl) vanadate–triphenylsilanol in refluxing xylene containing a trace of benzoic acid. An intermediate vanadate is postulated.

(1)                                                      (2, E and Z isomers)

[1] H. Pauling, *Chimia*, **27**, 383 (1973).
[2] G. L. Olson, K. D. Morgan and G. Saucy, *Synthesis*, 25 (1976).

**Tri(tetra-*n*-butylammonium)hexacyanoferrate(III),** $(Bu_4N)_3Fe(CN)_6$ (1). Mol. wt. 938. Soluble in $CHCl_3$, $CH_3OH$.

The reagent is prepared by ion-pair extraction from $K_3Fe(CN)_6$ and tetrabutylammonium hydroxide.[1]

*Oxidation of 2,6-di-t-butyl-substituted phenols.* The reagent oxidizes the phenol (1) to (2) and (3) in a ratio of 6–10:1.[1] The phenol is oxidized by the

(2)                          (3)                          (4)

more acidic hydrogen hexacyanoferrate(III) to these compounds also, but in reverse proportions.[2] The first oxidation may involve a radical mechanism involving quinone methide intermediates, whereas the second oxidation may involve an ionic mechanism.

[1] G. Brunow and S. Sumelius, *Acta. Chem. Scand.*, **B29**, 499 (1975).
[2] L. Taimr and J. Pospisil, *Tetrahedron Letters*, 2809 (1971).

**Trityl chloride, 1,** 1254–1256; **2,** 453–454; **4,** 565; **5,** 741.

*Polymeric reagent.*[1] A polymeric form (1) of trityl chloride can be prepared from a styrene-divinylbenzene polymer(P)as formulated. The reagent has been used to protect the primary hydroxyl groups of glucopyranosides. Use of the polymeric reagent does not improve the yield, but simplifies the work-up.

[1] J. M. J. Fréchet and K. E. Haque, *Tetrahedron Letters*, 3055 (1975).

### Trityl tetrafluoroborate, 1,1256–1258; 2, 454; 4, 565–567.

*Dehydrogenation* (**1**, 1257). Vogel and Ippen[1] effected dehydrogenation of the dibromoheptalene (1) by conversion into the tropylium salt (2) by treatment with trityl tetrafluoroborate in $CH_2Cl_2$ (25°, 45 min.). The salt was then deprotonated with HMPT (at 25°, 30 min.) to give (3). HMPT was superior to

trimethylamine, which has usually been used for this transformation. Dibromoheptalene is stable to oxygen at room temperature; heptalene itself has a pronounced tendency to polymerize.

Dehydrogenation of (4) via the tropylium salt was unsuccessful, but dehydrogenation was effected with DDQ in benzene at 160° (15 min.). The resulting heptalene is also stable to air at 25°.

*Photooxygenation of dienes* (**4**, 566). Complete details are now available for this catalyzed photooxygenation.[2]

[1] E. Vogel and J. Ippen, *Angew. Chem. internat. Ed.*, **13**, 734 (1974).
[2] D. H. R. Barton, R. K. Haynes, J. Leclerc, P. D. Magnus, and I. D. Menzies, *J.C.S. Perkin I*, 2055 (1975).

### Trityl trifluoroacetate, $(C_6H_5)_3 \overset{+}{C}OCOCF_3{}^-$. Mol. wt. 356.33.

The reagent is generated *in situ* from triphenylmethanol in trifluoroacetic acid.

*Dehydrogenation.*[1] The reagent is comparable to trityl perchlorate and trityl tetrafluoroborate (**1**, 1287) for dehydrogenation of hydroaromatic compounds to aromatic compounds. In a typical experiment the hydroaromatic compound is refluxed with triphenylmethanol in TFA for 1–20 hr. Yields, in general, are high, and the method is sometimes superior to dehydrogenation with *n*-butyl-lithium–TMEDA (**5**, 86).

[1] P. P. Fu and R. G. Harvey, *Tetrahedron Letters*, 3217 (1974).

**Tungsten hexacarbonyl,** $W(CO)_6$. Mol. wt. 351.98. Suppliers: Alfa, ROC/RIC, Strem.

*Olefin metathesis.* Irradiation of *trans*-2-pentene and tungsten hexacarbonyl in carbon tetrachloride gives 2-butene and 3-hexene (both mainly in the *trans*-form).[1] No metathesis is observed when $CCl_4$ is replaced by *n*-hexane.

Krausz and co-workers suggest that the first step in this photochemical reaction involves formation of $W(CO)_5Cl$ by reaction with the solvent.

[1] P. Krausz, F. Garnier, and J. E. DuBois, *Am. Soc.*, 97, 437 (1975).

**Tungsten hexachloride, 4,** 569–570; **5,** 742.

*Deoxygenation of epoxides* (**4**, 569–570). Detailed instructions for deoxygenation of *trans*-cyclododecene oxide (**1**) are available.[1]

[1] M. A. Umbreit and K. B. Sharpless, *Org. Syn.*, submitted (1976).

# U

**Urushibara catalysts, 4,** 571; **5,** 743.

*Review.* Shiota[1] has reviewed some recent applications of Urushibara catalysts. A new, related nickel catalyst known as U-Ni-N (neutral) has been obtained by refluxing precipitated nickel with isopropanol.[2] As shown in equations I and II this new catalyst shows high selectivity.

(I)    $(CH_3)_2C{=}CHCOCH_3 \xrightarrow{\text{U-Ni-N}} (CH_3)_2CHCH_2COCH_3$

(II)

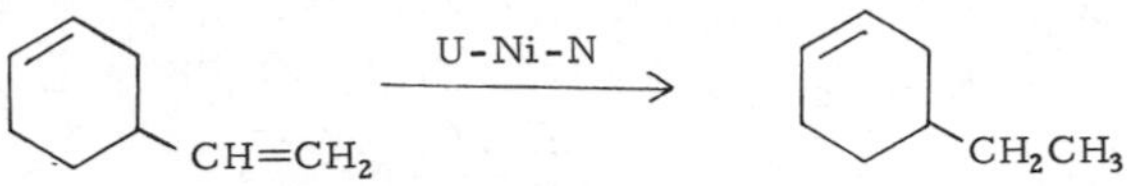

*Hydrogenation of 3-keto steroids.* Urushibara nickel A is the best catalyst for reduction of cholestane-3-one to the 3α-ol. The ratio of epicholestanol to

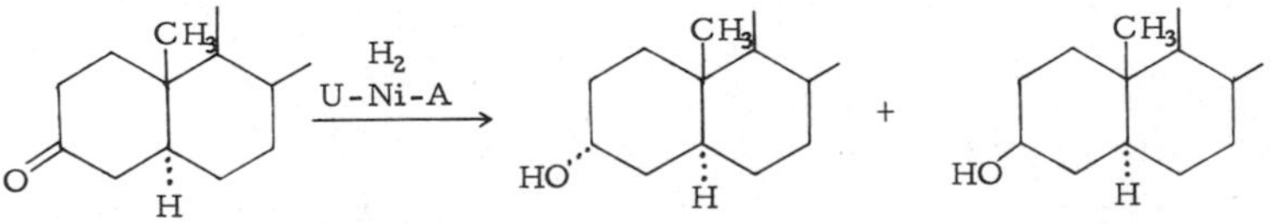

(60-87%)     (13-40%)

cholestanol is dependent on the solvent and, to a lesser extent, on the reaction temperature. This catalyst can also be used to obtain epicholesterol in high yield from $\Delta^5$-cholestene-3-one.[3]

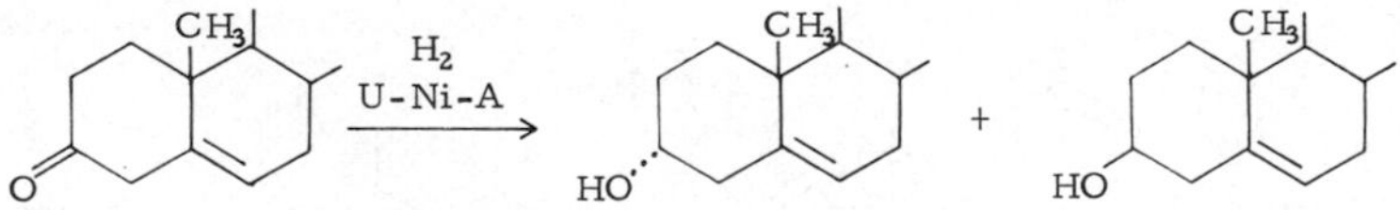

(77-94%)     (6-23%)

[1] M. Shiota, *Strem Chem.,* **III,** No. 1, 13 (1975).
[2] M. Kajitani, J. Okada, T. Ueda, A. Sugimori, and Y. Urushibara, *Chem. Letters,* 777 (1973).
[3] M. Ishige and M. Shiota, *Canad. J. Chem.,* **53,** 1700 (1975).

# V

**Vanadium oxytrifluoride, 5, 745–746.**

  *Nonphenolic oxidative coupling.* Kupchan *et al.*[1] have shown that $VOF_3-$ TFA oxidation of the N-formylbenzylisoquinoline derivative (1) to (2) proceeds through a morphinanedienone intermediate (a). Kupchan and Kim[2] have reported *in vitro* syntheses of dibenzazonine and aporphine alkaloids, such as (3)

(1)                    (a)                    (2)

and (4), from morphinanedienones, which may be *in vivo* precursors to such alkaloids as well.

(3)                    (4)

[1]S. M. Kupchan, V. Kameswaran, J. T. Lynn, D. K. Williams, and A. J. Liepa, *Am. Soc.*, **97**, 5622 (1975).
[2]S. M. Kupchan and C.-K. Kim, *ibid.*, **97**, 5623 (1975).

β-Vinylbutenolide, CH₂=CH [structure] (1). Mol. wt. 110.11. Relatively unstable neat.

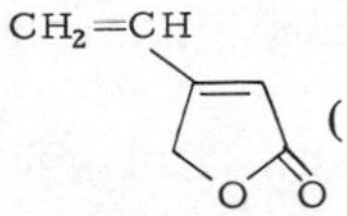

*Preparation from β-vinylbutyrolactone:*

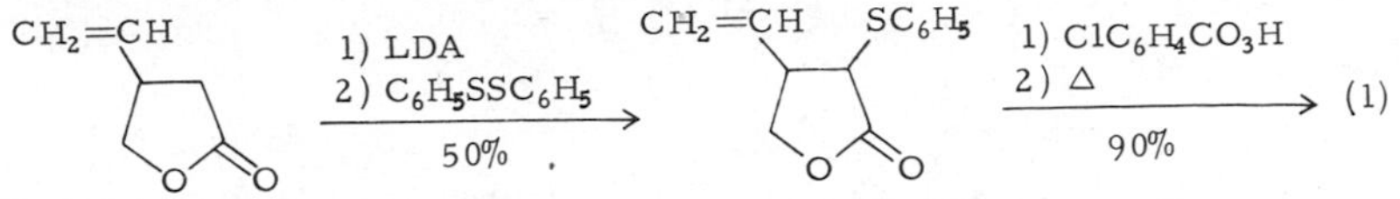

$$CH_2=CH \quad \xrightarrow[50\%]{\begin{array}{c}1)\ LDA \\ 2)\ C_6H_5SSC_6H_5\end{array}} \quad CH_2=CH \quad SC_6H_5 \quad \xrightarrow[90\%]{\begin{array}{c}1)\ ClC_6H_4CO_3H \\ 2)\ \triangle\end{array}} \quad (1)$$

*Lactone annelation.*[1]  The enolate of 2-methylcyclohexane-1,3-dione (excess) undergoes 1,6-conjugate addition to (1) followed by cyclization when refluxed in THF for 15 hr. to give (2) and (3) in the ratio of 11:1. Both are *cis*-fused

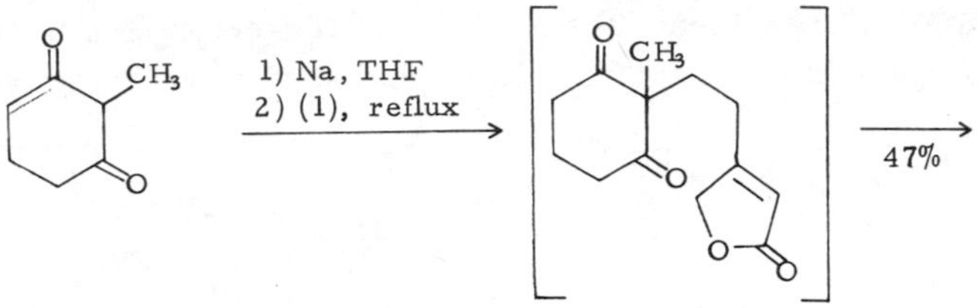

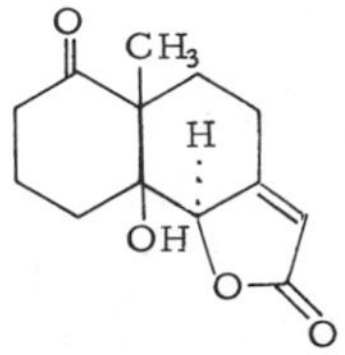

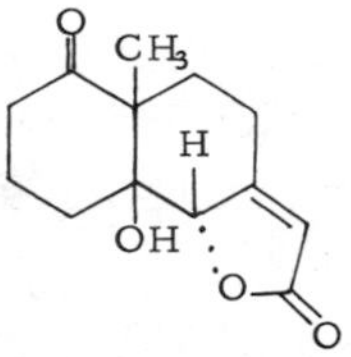

(2)          (3)

quant. | SOCl₂, Py          quant. | SOCl₂, Py

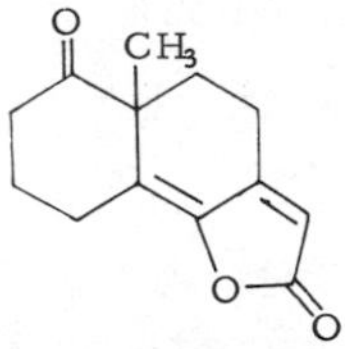

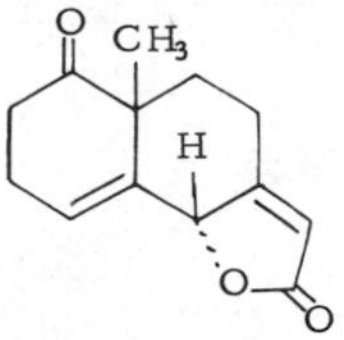

(4)          (5)

decalin lactones. On dehydration they are converted into (4) and (5), respectively. Either (2) or (3) on equilibration with base gives an equilibrium mixture of (2) and (3) in the ratio of 7:2.

Annelation of 2-ethoxycarbonylcyclohexanone with (1) gives (6) as the major product (40% yield).[2]

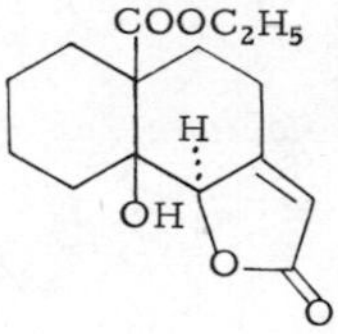

(6)

[1] F. Kido, T. Fujishita, K. Tsutsumi, and A. Yoshikoshi, *J.C.S. Chem. Comm.*, 337 (1975).
[2] K. Kondo and F. Mori, *Chem. Letters*, 741 (1974); K. Iwai, H. Kosugi, and H. Uda, *ibid.*, 1237 (1974).

## Vinyl carbomethoxymethyl ketone (Methyl 3-oxo-4-pentenoate), 1, 1272–1273; 5, 746–747.

*Annelation.*[1] The reagent (2) has been used for preparation of 3,4,6,7,8,9-hexahydro-2-quinolizones, a useful building block in alkaloid syntheses. Thus the

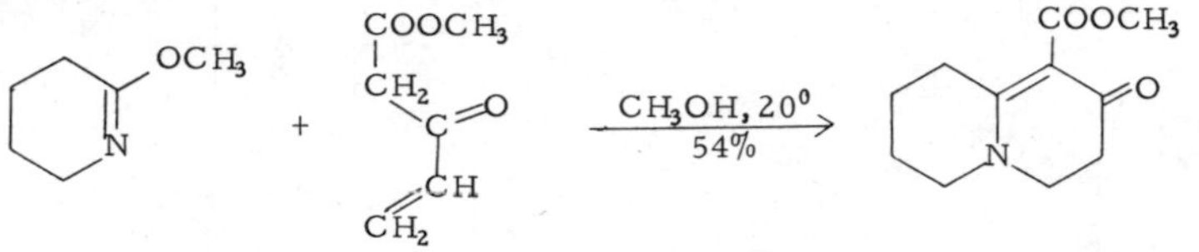

(1)                    (2)                    (3)

imino ether (1) reacts with (2) to form (3) in satisfactory yield.

[1] B. M. Trost and R. A. Kunz, *Am. Soc.*, 97, 7152 (1975).

**Vinylcopper reagents,** $\underset{R'}{\overset{R}{\diagdown}}C=C\underset{H}{\overset{Cu}{\diagup}}$ .

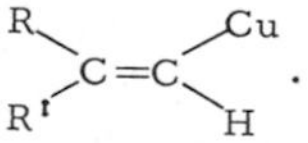

*Preparation.*[1,2] A Grignard reagent reacts with cuprous bromide in ether at $-40°$ ($N_2$) to form an alkylcopper reagent (1), which adds stereospecifically *cis* to 1-alkynes to form vinylcopper derivatives (2); these unstable reagents are not isolated, but are used in solution for various transformations.

$$RMgBr + CuBr \xrightarrow[\quad -40°\quad]{(C_2H_5)_2O} RCu \cdot MgBr_2$$

(1)

$$(1) + R'C\equiv CH \xrightarrow[\quad -15°\quad]{(C_2H_5)_2O} \underset{R'}{\overset{R}{\diagdown}}C=C\underset{H}{\overset{Cu}{\diagup}}$$

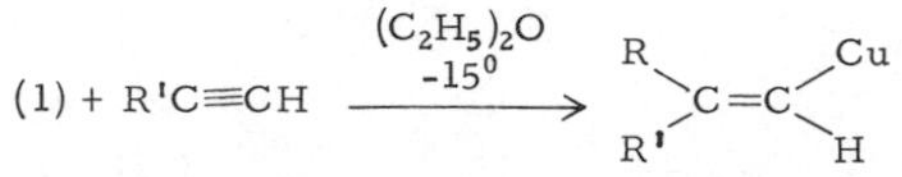

(2)

*Vinylic iodides.*[1,2] The reagents (2) are converted into 1-iodo-1-alkenes in 65–75% yield by treatment with iodine at $-30°$.

$$(2) + I_2 \xrightarrow[\substack{-30^0 \\ 65-75\%}]{(C_2H_5)_2O} \quad \underset{R'}{\overset{R}{>}}C=C\underset{H}{\overset{I}{<}} \;+\; CuI$$

*Vinylic bromides.*[3] Reaction of vinylcopper reagents (2) with bromine leads to products of oxidation (dienes). However, vinylic bromides can be prepared from (2) by a two-step procedure. For example, (3) undergoes copper–mercury exchange with retention of configuration on treatment with mercuric bromide (0.5 eq.) to give (4) in 74% yield. This compound is converted by bromine in

$$2\;\underset{CH_3}{\overset{C_2H_5}{>}}C=C\underset{H}{\overset{Cu}{<}} + HgBr_2 \xrightarrow[74\%]{(C_2H_5)_2O \\ -30^0} \left(\underset{CH_3}{\overset{C_2H_5}{>}}C=C\underset{H}{\overset{Hg}{<}}\right)_2 \xrightarrow[\substack{77\% \\ \text{overall}}]{Br_2,\,Py} 2\;\underset{CH_3}{\overset{C_2H_5}{>}}C=C\underset{H}{\overset{Br}{<}}$$

(3)                      (4)                      (5)

pyridine with retention of configuration into the vinylic bromide (5). Actually, it is not necessary to isolate the intermediate (4). Thus successive treatment of (3) with mercuric bromide and a solution of bromine in pyridine gives (5) in 77% overall yield from ethylmagnesium bromide.

*Conjugated dienes.*[2] Conjugated dienes are obtained stereospecifically by passing oxygen through a solution of (2) at $-15°$; copper is precipitated.

$$2\;\underset{R'}{\overset{R}{>}}C=C\underset{Cu}{\overset{H}{<}} \xrightarrow[60-75\%]{O_2} \underset{R'}{\overset{R}{>}}C=C\underset{H}{\overset{H}{<}}C=C\underset{R}{\overset{R'}{<}}$$

*Di- and trisubstituted olefins.*[2] The vinylcopper reagents are not stable at ambient temperatures, but in the presence of various ligands they can be alkylated at room temperature. Triethyl phosphite is usually used for this purpose; satisfactory yields also require HMPT as cosolvent. Under these conditions di- and trisubstituted olefins can be obtained readily. Yields depend on the nature

$$(2) + R^2X \rightarrow \underset{R'}{\overset{R}{>}}C=C\underset{H}{\overset{R^2}{<}}$$

of $R^2X$, increasing in the order $RI > RBr > RCl \gg ROTs$. However, the alkylation fails with propargylic or homopropargylic halides. This alkylation reaction was used for the synthesis of dihydromyrcene (4) from (3) in 68% overall yield.

(3)                                                   (4)

The alkylation reaction can be used for a synthesis of allylic alcohols as illustrated.

$\alpha,\beta$-*Unsaturated acids.*[4] $\alpha,\beta$-Unsaturated acids are obtained almost quantitatively from the vinylcopper compounds (2). Again HMPT is the cosolvent of choice, and triethyl phosphite is used as ligand. An example of the reaction is the synthesis of (Z)-3-methyl-2-pentenoic acid in 85% yield from ethylmagnesium bromide:

The acid can be reduced in high yield (90%) by lithium aluminum hydride to the corresponding allylic alcohol.

[1] J. F. Normant and M. Bourgain, *Tetrahedron Letters*, 2583 (1971).
[2] J. F. Normant, G. Cahiez, C. Chuit, and J. Villieras, *J. Organometal. Chem.*, **77**, 269 (1974).
[3] J. F. Normant, C. Chuit, G. Cahiez, and J. Villieras, *Synthesis*, 803 (1974).
[4] J. F. Normant, G. Cahiez, C. Chuit, and J. Villieras, *J. Organometal. Chem.*, **77**, 281 (1974).

**Vinyldiazomethane,** $CH_2=CH-CHN_2$. Mol. wt. 68.08. *Caution:* Vinyldiazomethane is potentially explosive.

This substance is unstable, but can be prepared in methanol–pentane solution in 82–85% yield by the reaction of sodium methoxide with ethyl allylnitrosocarbamate.[1,2] The reagent in solution decomposes slowly at 0°.

*Vinylcyclopropanes.*[2] The reaction of vinyldiazomethane with olefins with catalysis by copper salts, particularly cupric trifluoromethanesulfonate[3] or cupric hexafluoroacetylacetonate,[4] gives vinylcyclopropanes in moderate yield.

Vinyldiazomethane reacts thermally with strained olefins to give pyrazolines, which decompose to vinylcyclopropanes, as shown for the reaction with cyclo-pentene.[5]

[1] J. Hooz and H. Kono, *Org. Prep. Proc. Int.*, **3**, 47 (1971).
[2] R. G. Salomon, M. F. Salomon, and T. R. Heyne, *J. Org.*, **40**, 756 (1975).
[3] C. L. Jenkins and J. K. Kochi, *Am. Soc.*, **94**, 843 (1972).
[4] R. A. Zelonka and M. C. Baird, *J. Organometal. Chem.*, **33**, 267 (1971).
[5] M. Schneider and I. Merz, *Tetrahedron Letters*, 1995 (1974).

**1,2-Vinylenebis(triphenylphosphonium) dibromide,**

$(C_6H_5)_3\overset{+}{P}CH=CH\overset{+}{P}(C_6H_5)_3 \, 2Br^-$ (1). Mol. wt. 448.14.
   *Preparation:*[1]

$$2\,(C_6H_5)_3P + 2CH_3COBr \xrightarrow[\text{quant.}]{} (1)$$

*Vinyl ethers.*[2]   The reagent reacts with alcohols or phenols in the presence of triethylamine at room temperature to form the phosphonium salts (2), which are hydrolyzed by sodium hydroxide to vinyl ethers (3).

$$ROH + (1) + (C_2H_5)_3N \xrightarrow[\substack{-(C_6H_5)_3P \\ -(C_2H_5)_3N\cdot HBr}]{CHCl_3,\ 20^0} ROCH=CH\overset{+}{P}(C_6H_5)_3Br^- \xrightarrow[\substack{50\text{-}75\% \\ \text{overall}}]{NaOH.\ H_2O}$$

$$ROCH=CH_2 + (C_6H_5)_3P=O$$
$$(3)$$

[1] H. Christol, H. J. Cristau, and J. -P. Joubert, *Bull. Soc.*, 1421 (1974).
[2] H. Christol, H.-J. Cristau, and M. Soleiman, *Synthesis*, 736 (1975).

**Vinyllithium,** $CH_2=CHLi$. ROC/RIC supplies the reagent as a 2 *M* solution in THF.
   *Monosubstituted alkenes.*[1]   Monosubstituted alkenes can be prepared by alkylation of commercial vinyllithium in THF:

$$CH_2=CHLi + CH_3(CH_2)_6CH_2I \xrightarrow[92\%]{\substack{THF \\ -28^0\ to\ 10^0}} CH_2=CHCH_2(CH_2)_6CH_3$$

$$CH_2=CHLi + Br(CH_2)_4Br \xrightarrow[67\%]{} CH_2=CH(CH_2)_4Br$$

This reaction can be extended to di- and trisubstituted alkenes by use of substituted vinyllithium reagents. These can be prepared quantitatively from vinyl chlorides or bromides by reaction in ether with lithium powder (containing 2% sodium) or from vinyl bromides by halogen—metal exchange with an alkyllithium.

*Examples:*

[1] J. Millon, R. Lorne, and G. Linstrumelle, *Synthesis*, 434 (1975).

**Vinyl triphenylphosphonium bromide**, **1**, 1274–1275; **2**, 456–457; **3**, 333; **4**, 572; **5**, 750–751.

*Dihydrothiophenes; 1,3-dienes; thiophenes.* The salt reacts with α-mercapto ketones to give dehydrothiophenes (1) in moderate yields (about 50–75%).

The corresponding sulfones (2) are useful as a source of 1,3-dienes (**2**, 390). Thus when a solution of (3) is refluxed in xylene in the presence of dimethyl acetylenedicarboxylate, (4) is obtained in 78% yield.

By use of various substituted vinylphosphonium salts, $R^4_{\phantom{5}}\!\!\diagdown\!\!C\!=\!CH\overset{+}{P}(C_6H_5)_3X^-$,

it is possible to prepare dihydrothiophenes of the general type (5) in a regio-specific manner.

(5)

McIntosh and Khalil[2] report that 2,5-dihydrothiophenes can be dehydrogenated to thiophenes by chloranil (**1**, 125–127; **2**, 66–67; **3**, 46; **4**, 75–76) in either *t*-butanol or pyridine in 70–90% yield. Dehydrogenation is blocked by a fully substituted carbon atom. Thus thiophenes substituted at the 2-, 3-, and 4- positions by alkyl groups are obtainable in two steps in overall yields of 55–85%.

*Cyclopentenones.* Japanese chemists[3] have reported a new synthesis of substituted cyclopentenones involving a Wittig reaction of the anion of an α-diketone with vinyl triphenylphosphonium salts. A representative synthesis is formulated. Isopropenyl triphenylphosphonium bromide, $CH_2\!=\!C\overset{CH_3}{\underset{P(C_6H_5)_3}{\diagup}}$ $Br^-$ can

also be used, but yields are about 10% in the case of propenyl triphenylphos-

phonium bromide, $CH_3CH\!=\!CH\overset{+}{P}(C_6H_5)_3Br^-$. The dithiocyclopentenone derivatives are desulfurized in satisfactory yield by Raney nickel (W-2) in refluxing benzene.

[1] J. M. McIntosh, H. B. Goodbrand, and G. M. Masse, *J. Org.*, **39**, 202 (1974).
[2] J. M. McIntosh and H. Khalil, *Canad. J. Chem.*, **53**, 209 (1975).
[3] I. Kawamoto, S. Muramatsu, and Y. Yura, *Tetrahedron Letters*, 4223 (1974).

# W

**Wittig reaction, 4**, 573; **5**, 752.

    *Bis-Wittig reaction.* Synthesis of nonbenzenoid aromatic compounds by the bis-Wittig reaction has been reviewed.[1]

[1]K. P. C. Vollhardt, *Synthesis*, 765 (1975).

# X

**Xenon difluoride, XF$_2$.**

*Preparation.*[1]

*Fluorination of aromatic compounds.* Fluoroaromatic hydrocarbons can be obtained in satisfactory yield with this reagent, although biphenyls and even polyphenyls are often formed as by-products. The fluorination requires catalysis by hydrogen fluoride.[2]

Phenols and ethers react to give monosubstituted products in yields of 37–71%; in this case HF is not required as catalyst. Thus phenol is converted into o-fluorophenol, m-fluorophenol, and p-fluorophenol in the ratio 2:2:1 (overall yield 47%).[3]

*Other examples:*

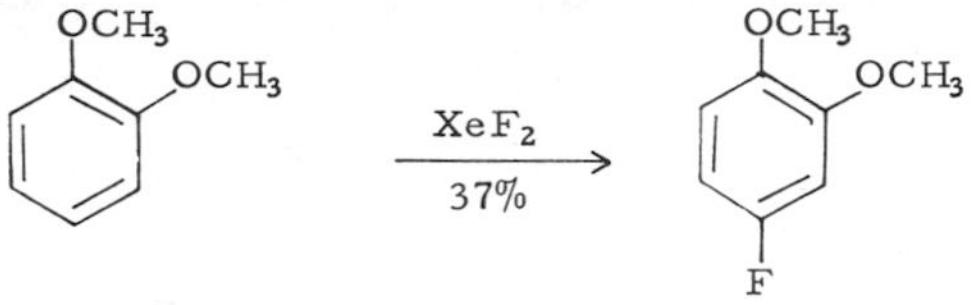

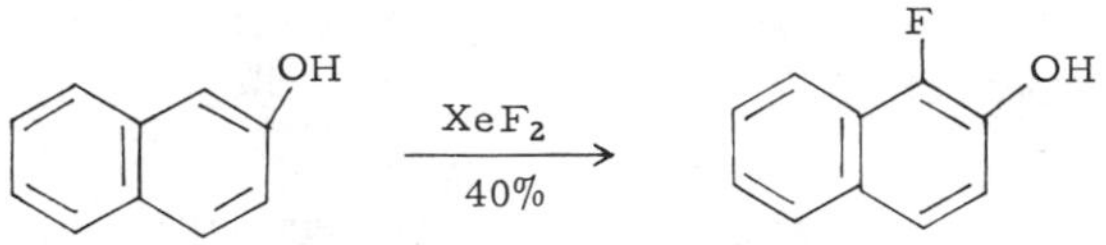

*Fluorination of phenanthrene and anthracene.*[4,5] The fluorination of phenanthrene with this reagent is more complex than chlorination or bromination.

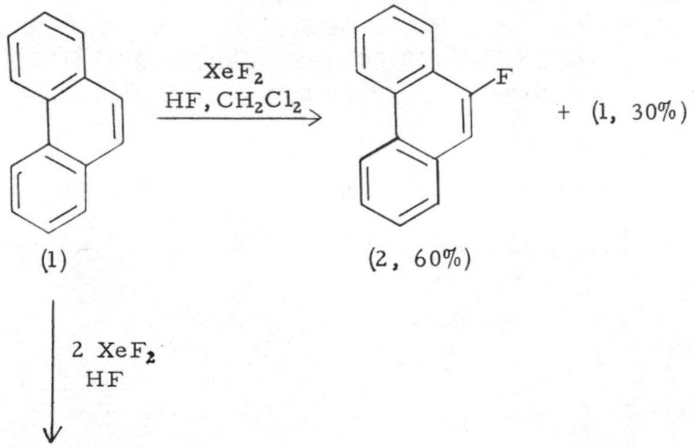

(2, 46%)    +    (3, 20%)    +    (4, 10%)    +    (5, 3%)

1-Fluoroanthracene (7) is the main product from fluorination of anthracene (6).

(6)    $\xrightarrow[\text{HF}]{\text{XeF}_2}$    (7, 45%)    +    (8, 26%)    +    (9, 9%)

*Fluorination of pyrene.*[6] Pyrene (1) reacts with xenon difluoride (HF catalysis necessary) to give mainly 1-fluoropyrene (2) and some 2-fluoropyrene

(1)    $\xrightarrow[\text{HF}]{\text{XeF}_2}$    (2, 16-22%)    +    (3, 11-14%)    +    4-Fluoropyrene (traces)

(3). The pattern of substitution is exceptional because $C_1$ is almost the exclusive site of electrophilic substitutions of pyrene.

[1] J. L. Weeks, C. L. Chernick, and M. S. Matheson, *Am. Soc.*, 84, 4612 (1962).

[2] M. J. Shaw, J. A. Weil, H. H. Hyman, and R. Filler, *Am. Soc.*, 95, 5096 (1970); M. J. Shaw, H. H. Hyman, and R. Filler, *ibid.*, 95, 5096 (1970); *idem, J. Org.*, 36, 2917 (1971).

[3] S. P. Anand, L. A. Quarterman, H. H. Hyman, K. G. Migliorese, R. Filler, *ibid.*, 40, 807 (1975).

[4] M. Zupan and A. Pollak, *ibid.*, 40, 3794 (1975).

[5] S. P. Anand, L. A. Quarterman, P. A. Christian, H. H. Hyman, and R. Filler, *ibid.*, 40, 3796 (1975).

[6] E. D. Bergmann, H. Selig, C.-H. Lin, M. Rabinovitz, and I. Agranat, *ibid.*, 40, 3793 (1975).

# Y

**Ytterbium(III) nitrate,** $Yb(NO_3)_3 \cdot 5H_2O$. Mol. wt. 449.15. Suppliers: Alfa, ROC/RIC.

*Oxidation of benzoins to benzils.*[1]   Benzoins (1) are oxidized to benzils (2) in high yield by catalytic amounts of this lanthanide nitrate and 1 eq. of HCl in aqueous glyme. Benzil can also be obtained from desoxybenzoin and diphenylacetylene under these conditions. This oxidation has recently been reported with

(1, R, R' = H, OCH₃)                                            (2)

thallium(III) nitrate (**4,** 494), but in this case the thallium salt does not function catalytically. Girard and Kagan consider that in this new reaction the actual oxidizing reagent is $NO_3^-$ and that ytterbium(III) functions as a catalyst for re-oxidation of the nitrite ion to nitrate ion by oxygen.

In contrast to thallium salts, ytterbium salts are not known to be toxic.

[1]P. Girard and H. B. Kagan, *Tetrahedron Letters*, 4513 (1975).

# Z

**Zinc, 1**, 1276–1284; **2**, 459–462; **3**, 344–337; **4**, 574–577; **5**, 753–756.

*3,4-Benzotropilidenes.* Swenton and Madigan[1] have developed a general route to 3,4-benzotropilidenes that employs mild, neutral conditions, which are essential to avoid isomerization to the more stable 1,2-isomer. The method is illustrated for the synthesis of 7-carbomethoxy-3,4-benzotropilidene (4). Copper sulfate catalyzed reaction of ethyl diazoacetate with 1,4-dihydronaphthalene (1) leads to the expected adduct, *exo*-7-carboethoxy-3,4-benzo[4.1.0]heptane, which is hydrolyzed and then esterified with methanol to give the corresponding methyl ester (2). This product is dibrominated (dibenzoyl peroxide initiation). The crude dibromide (3) is then refluxed in THF containing a trace of

HOAc with zinc dust for 0.5 hr. under $N_2$ (prolonged treatment results in polymerization). Dehydrogenation is accompanied by cleavage of the cyclopropane ring to give (4). This method of ring expansion is fairly general, but fails in preparation of the parent hydrocarbon and of the 7-cyano and 7,7-dicyano derivatives because of difficulties in the bromination step.[2]

Vogel and Hogrefe[3] have used this route in a synthesis of dimethyl 3,8-heptalenedicarboxylate (9). In contrast to dihalocarbenes, carboethoxycarbene preferentially attacks the outer bonds of isotetralin (5) to give, after hydrolysis and reesterification with diazomethane, (6). Tetrabromination (azobisisobutyronitrile) gives a mixture of tetrabromides (7). Dehydrobromination leads to (8). Dehydrogenation with DDQ then gives the 3,8-heptalene derivative (9).

672

Heterocyclic pseudoquinones.[4] Treatment of the known diketones (1) with bromine or phenyltrimethylammonium perbromide (**1**, 855–856) leads to the

$(1, Y = S, O, NCH_3)$        (2)        (a)

dibromo diketones (2). When these are treated with zinc that has been activated by treatment with $CuSO_4$,[5] intramolecular ring closure occurs, probably via a biradical (a), to give (3). These diketones can be dehydrogenated by DDQ or chloranil to heterocyclic quinones (4).

*Monoalkylketenes.* Methylketene and ethylketene can be obtained in 60–65% yield by reaction of 2-bromopropionyl bromide and 2-bromobutyryl bromide in THF with zinc activated by treatment with hydrochloric acid (**1**, 1276) under reduced pressure (100 mm). The ketenes codistill with THF and are free from starting materials and zinc salts. The present method is a modification of Standinger's ketene synthesis, which gave only low yields of these ketenes.[6]

*Stereoselective reduction of triple bonds.* Morris et al.[7] have reported that zinc powder in 50% aqueous *n*-propanol at reflux reduces triple bonds to *cis*-double bonds stereoselectively.

Application of the method by Näf et al.[8] to the Z-dienyne (1), however, leads to Z,Z,E-2,4,6-undecatriene (2), probably formed from the expected triene

(1)                                                    (2)

~75%  |  Zn, KCN
      |  $C_3H_7OH$, 25°

(3)

(3) by a thermal [1,7]-hydrogen shift. The Swiss chemists then found the desired reduction to (3) can be accomplished at room temperature without a prototropic shift with zinc activated by potassium cyanide. The method, however, is not always reproducible.

This reduction could not be carried out by hydrogenation using Lindlar's catalyst or Wilkinson's catalyst or with diimide because of lack of selectivity.

The paper suggests a possible mechanism for the *cis*-selectivity in the zinc reduction.

*Reformatsky reaction* (**1**, 1285–1286; **3**, 334–335; **5**, 753–754). More recent developments in the Reformatsky reaction have been reviewed.[9]

Rieke and Uhm[10] have published more details concerning the preparation of a highly reactive zinc by reduction of anhydrous zinc chloride with potassium metal in refluxing THF (**5**, 753). The Reformatsky reaction of ethyl α-bromoacetate with various aldehydes and ketones can be conducted in ether at 20° when this reactive form of zinc is used. Yields of β-hydroxy esters are usually greater than 95%.

**Trimethylsilyl enol ether of camphor.**[11] The usual methods for preparation of trimethylsilyl enol ethers are unsatisfactory in the case of camphor. However, 3-*endo*-bromocamphor (1)[12] is converted into the trimethylsilyl enol ether of camphor (2, >80% yield) by reaction with activated zinc in ether in the presence of trimethylchlorosilane.

(1)                                    (2)

[1] J. S. Swenton and D. M. Madigan, *Tetrahedron*, **28**, 2703 (1972).

[2] K. A. Burdett, F. L. Shenton, D. H. Yates, and J. S. Swenton, *Tetrahedron*, **30**, 2057 (1974).

[3] E. Vogel and F. Hogrefe, *Angew. Chem. internat. Ed.*, **13**, 735 (1974).

[4] E. Ghera, Y. Gaoni, and D. H. Perry, *J.C.S. Chem. Comm.*, 1034 (1974).

[5] J. B. Lambert, F. R. Koeng, and J. W. Hamersma, *J. Org.*, **36**, 2941 (1971).

[6] C. C. McCarney and R. S. Ward, *J.C.S. Perkin I.*, 1600 (1975).

[7] S. G. Morris, S. F. Herb, P. Magidman, and F. E. Luddy, *J. Am. Oil Chem. Soc.*, **49** (1972).

[8] F. Näf, R. Decorzant, W. Thommen, B. Wilhalm, and G. Ohloff, *Helv.*, **58**, 1016 (1975).

[9] M. W. Rathke, *Org. React.*, **22**, 423 (1975).

[10] R. D. Rieke and S. J. Uhm, *Synthesis*, 452 (1975).

[11] G. C. Joshi and L. M. Pande, *ibid.*, 450 (1975).

[12] K. Tsuda, E. Ohki, and S. Nozoe, *J. Org.*, **28**, 783 (1963).

## Zinc–Zinc chloride.

*Reductive cleavage of strained cyclopropanes and cyclobutanes.* Dekker *et al.*[1] have reported reductive cleavage of some strained ring systems by zinc and zinc chloride in protic solvents as shown in the examples. The carbonyl groups are essential for this cleavage. No cleavage is observed in aprotic solvents (*e.g.*, benzene).

[1] J. Dekker, F. J. C. Martins, J. A. Kruger, and A. J. Goosen, *Tetrahedron Letters*, 3721 (1974); J. Dekker, F. J. C. Martins and J. A. Kruger, *ibid.*, 2489 (1975).

**Zinc chloride, 1,** 1289–1292; **2,** 464; **3,** 338; **5,** 763–764.

*Directed aldol condensation.* House *et al.*[1] have published a detailed procedure for the aldol condensation of the lithium enolate of phenylacetone with *n*-
butyraldehyde in the presence of zinc chloride to give *threo*-4-hydroxy-3-phenyl-
2-heptanone. Ether or ether–dimethoxyethane mixtures are the most suitable

$$C_6H_5CH = \overset{\overset{\displaystyle OLi}{|}}{C}CH_3 + \underline{n}\text{-}C_3H_7CHO \xrightarrow[53-60\%]{ZnCl_2} CH_3CO\overset{\overset{\displaystyle H}{|}}{\underset{\underset{\displaystyle C_6H_5}{|}}{C}} - \overset{\overset{\displaystyle H}{|}}{\underset{\underset{\displaystyle OH}{|}}{C}}(CH_2)_2CH_3$$

solvents. Optimum temperatures are $-10$ to $+10°$ and optimum reaction periods are 2–5 min. By-products result from longer reaction periods and higher
reaction temperatures.

*ROH $\longrightarrow$ RCl.* In some instances zinc chloride has been found to be more
effective than pyridine, the usual catalyst (**1,** 1160–1161), in the reaction of
alcohols with thionyl chloride. The reaction proceeds mainly with inversion of
configuration.[2]

*Tertiary alkyl azides.* Azides can be prepared by reaction of tertiary alkyl,
allylic, and benzylic chlorides with sodium azide with catalysis by zinc chloride.

$$\underline{t}\text{-}RCl + NaN_3 \xrightarrow[CS_2, 20°]{ZnCl_2} \underline{t}\text{-}RN_3 + NaCl$$

The exchange is slow (10–100 hr.) but yields usually are $>80\%$.[3]

[1] R. A. Auerbach, D. S. Crumbine, D. L. Ellison, and H. O. House, *Org. Syn.*, **54,** 49 (1974).
[2] V. D. Grob, T. G. Squires, and J. R. Vercellotti, *Carbohydrate Res.*, **10,** 595 (1969); T. G.
Squires, W. W. Schmidt, and C. S. McCandlish, Jr., *J. Org.*, **40,** 134 (1975).
[3] J. A. Miller, *Tetrahedron Letters*, 2959 (1975).

# INDEX OF REAGENTS
# ACCORDING TO TYPES

ACETALIZATION: Aluminum chloride.
ACETOACETIC ESTER CONDENSATION: Methyltricaprylammonium chloride.
ACETOXYLATION: Lead tetracetate.
ACETYLATION: Acetyl bromide. 1,3,4,6-Tetraacetylglycouril.
ACYLOIN CONDENSATION: Trimethylchlorosilane.
ALDOL CONDENSATION: Dimethylcopperlithium. Lithium diisopropylamide. Titanium
    (IV) chloride. Zinc chloride.
ALKYLATION: Dimethylcopperlithium. Methyl fluorosulfonate. Palladium acetate.
    Sodium chloride.
C-ALKYLATION: Silver oxide.
N-ALKYLATION: Lithium aluminum hydride. Potassium hydridotetracarbonylferrate.
    Tetra-*n*-butylammonium hydrogen sulfate.
O-ALKYLATION: Sodium hydride.
ALKYLATION, π-ALLYLPALLADIUM COMPLEXES: Palladium(II) chloride.
ALKYLATION, GEMINAL: Diphenylsulfonium cyclopropylide.
ALKYLATION, ARENES: Chromium hexacarbonyl.
ALKYLATIVE ELIMINATION: Methyl 2-phenylsulfinylacetate.
ALKYL HALIDE EXCHANGE: Tetra-*ṅ*-butylammonium halides.
ALLYLIC OXIDATION: Selenium dioxide.
ALLYLIC REARRANGEMENT: Isoprene epoxide. Sulfuric acid.
π-ALLYLPALLADIUM CHLORIDE COMPLEXES: Bis(benzonitrile)palladium(II)
    chloride.
ANNELATION: *t*-Butyl-γ-iodotiglate. 1-Chloro-3-pentanone. 1-Fluorovinyl methyl
    ketone. 1-Lithiocyclopropyl phenyl sulfide. Methyl α-trimethylsilylvinyl ketone.
    Methyl vinyl ketone. 6-Methyl-2-vinylpyridine. Vinyl carbomethoxymethyl ketone.
AROMATIZATION: Acetic anhydride-Phosphoric acid. Hydrobromic acid. Sulfuric
    acid.
ASYMMETRIC HYDROGENATION: (–)-and (+)-2,3-O-Isopropylidene-2,3-dihydroxy-
    1,4-bis(diphenylphosphino)butane. Neomethyldiphenylphosphine.
ASYMMETRIC SYNTHESIS:
    (–)-N-Benzyl-N-methylephedrinium bromide. Diisopinocamphylborane. *trans*-2,4-
    Dimethoxymethyl-5-phenyloxazoline. (–)-N,N-Dimethylephedrinium bromide.
    Dimethylsulfonium methylide. (–)-Menthoxyacetyl chloride. 2-Methyl-*t*-methoxy-
    methyl-5-phenyl-2-oxazoline. L-Proline. Quinine, chinchonine.

BAEYER–VILLIGER OXIDATIONS: Hydrogen peroxide–Acetic acid.
BECKMANN FRAGMENTATION: S-Methyl *p*-toluenethiosulfonate. Tetrakis(triphenyl-
    phosphine)palladium. Triethyl orthoformate.
BENZOYLATION: S-Benzoic O,O-diethyl phosphorodithioic anhydride. Benzoyl cyanide.
    Dibenzoyl peroxide.
BIRCH REDUCTIONS: Sodium borohydride.
BROMINATION: 2-Bromo-2-cyano-N,N-dimethylacetamide. 2-Carboxyethyltriphenyl-
    phosphonium perbromide. Chlorine bromide. Iodine bromide.

CARBENES, GENERATION: Benzyl-*n*-butyldimethylammonium chloride. Benzyltriethylammonium chloride. Crown ethers. Dibenzo-18-crown-6. Dicyclohexyl-18-crown-6. Lithium triethylmethoxide. *see* Phase-transfer catalysts. Tri-*n*-butylamine. Trimethyl phosphite–Copper(I) iodide.

CARBENE PRECURSORS: 2,2-Dimethylvinyl triflate. Fluorodiiodomethane. Phenyl diazomethyl sulfoxide. Trifluoro(trichloromethyl)silane.

CARBONYLATIONS: Bis(triphenylphosphine)palladium dichloride. Chlorodicarbonyl-rhodium(I) dimer. Carbon dioxide. Chloromethyl methyl ether. Magnesium methyl carbonate.

CARBYNES: Diethyl mercurybisdiazoacetate.

CHELATION: Cobalt(II) chloride.

CHLORINATION: Antimony(V) chloride. Iodobenzene dichloride. Molybdenum(V) chloride. Sodium hypochlorite. Sulfuryl chlorofluoride. Trichloroisocyanuric acid.

CLAISEN CONDENSATIONS: Sodium bis(trimethylsilyl)amide.

CLAISEN REARRANGEMENT: *t*-Butyldimethylchlorosilane. Hexamethylphosphoric triamide. Triethyl orthoacetate.

CLEAVAGE OF:

AMINO PROTECTIVE GROUPS: Palladium block. *p*-Toluenesulfonic acid. Trifluoro-methanesulfonic acid.

CYCLOALKANONES: Lead tetracetate.

2,4-DINITROPHENYLHYDRAZONES: Titanium(III) chloride.

EPOXIDES: Alumina.

ESTERS: Triphenylphosphine dibromide. Diisobutylaluminum hydride. Ferric chloride-Acetic anhydride. Magnesium bromide–Acetic anhydride.

ETHYLENE DITHIOKETALS: N-Chlorosuccinimide.

2-HALOETHYL ESTERS: Sodium ethanedithiolate.

HYDRAZONES, OXIMES: Acetone.

β-KETO ESTERS: 3-Quinuclidinolol.

LACTONES: Triphenylphosphine dibromide.

METHYL ETHERS: Pyridine hydrochloride.

PHTHALOYL GROUPS: Hydrazine.

SECONDARY AMIDES: Diphenyldi(1,1,1,3,3,3-hexafluoro-2-phenyl-2-propoxy)sulfurane.

TOSYLHYDRAZONES: Titanium(III) chloride.

CONJUGATE ADDITIONS: Bis(methylthio)(trimethylsilyl)methyllithium. Diethylalum-inum cyanide. Di(*a*-methoxyvinyl)copperlithium. Ethyl diethoxyacetate. Ethyl methylsulfinylacetate. Lithium *a*-carboethoxy vinyl(1-hexynyl)cuprate. Potassium fluoride. Quinine, chinchonine. Titanium tetrachloride. Trimethylaluminum. Tris-(phenylthio)methyllithium.

CONJUGATE REDUCTION:

Cuprous bromide–Sodium bis(2-methoxyethoxy)aluminum hydride. Lithium aluminum hydride–Cuprous iodide. Potassium tri-*sec*-butylborohydride.

COREY–WINTER OLEFIN SYNTHESIS: Methyl iodide. N,N'-Thiocarbonyldiimidazole.

COUPLING, ALLYLIC AND BENZYLIC ALCOHOLS: Titanium(III) chloride–Lithium aluminum hydride.

CROSS–COUPLING: Dichloro[1,3-bis(diphenylphosphino)propane] nickel(II). Tetra-hydrofurane.

CYANOBORATION: Sodium cyanide.

CYANOSILATION: *t*-Butyldimethylsilyl cyanide.

CYCLIZATION: *n*-Butyllithium. Hydrobromic acid. Hydrogen fluoride. Ion-exchange resins. 2-Lithio-2-trimethylsilyl-1,3-dithiane. Lithium diisopropylamide. Polyphos-phoric acid. Tetra-*n*-butylammonium iodide. Stannic chloride. Trifluoroacetic acid. Trifluoromethanesulfonic acid.

CYDOADDITIONS: Antimony(V) fluoride. Carbon disulfide. 2,3-Dichloro-5,6-dicyano-1,4-benzoquinone. Dimethyl acetylenedicarboxylate. Dimethylketene. Ethylaluminum dichloride. 2-Methoxyallyl bromide. Ninhydrin. Sodium cyclopentadienyldicarbonylferrate. Sodium iodide–Copper. Tetracyanoethylene. Tetramethylcyclobutadiene–Aluminum chloride complex.

CYCLOBUTYL ANNELATION: 1-Lithiocyclopropylphenyl sulfide.

CYCLODEHYDRATION: Magnesium methyl carbonate. Pyridine hydrochloride.

CYCLODEHYDROGENATION: Sodium aluminum chloride.

CYCLOPROPANATION: Copper–Isonitrile complexes. Cupric chloride. Diethylzinc–Bromoform–Oxygen. Palladium acetate. Titanium(IV) chloride–Lithium aluminum hydride.

DARZENS CONDENSATION: Chloroacetonitrile. Dibenzo-18-crowns-6. Sodium hydride–Dimethylformamide.

DEALKYLATION: Cuprous chloride.

DEALKYLATION, AMINES: Phenyl isocyanide dichloride. Silver nitrite.

DEAMINATION: m-Chloroperbenzoic acid. Sodium borohydride.

DEBENZYLATION: n-Propylthiolithium.

DECARBOALKYLATION: Hexamethylphosphoric triamide. Hydrochloric acid. Sodium cyanide–Hexamethylphosphoric triamide.

DECARBOXYLATION: Tris(triphenylphosphine)chlororhodium. t-Butyl hydroperoxide. Ferrous perchlorate. Lead tetraacetate–N-Chlorosuccinimide.

DECARBOXYLATIVE DEHYDRATION: Dimethylformamide dineopentyl acetal.

DEHALOGENATION: 1,8-Bis(dimethylamino)napthalene. Lithium diphenylphosphide. Potassium hydridotetracarbonylferrate. Sodium–t-Butanol–Tetrahydrofurane. Sodium borohydride. Sodium iodide. Tri-n-butyltin hydride.

DEHYDRATION: Diethoxyaluminum chloride. Lithium n-propylmercaptide. N-Methyl-N,N'-di-t-butylcarbodiimidium tetrafluoroborate. o-Nitrobenzenesulfonyl chloride. Potassium acetate. Thiobenzoyl chloride. Triphenylphosphine ditriflate. Triphenylphosphine methiodide.

DEHYDRATIVE DECARBOXYLATION: N,N-dimethylformamide dimethyl acetal.

DEHYDROBROMINATION: Aluminum chloride. Tetramethylammonium dimethyl phosphate.

DEHYDROGENATION: t-Butyl hypochlorite. 2,3-Dichloro-5,6-dicyanobenzoquinone. Pyridine. Pyridinium hydrobromide. Sodium bis-(2-methoxyethoxy)aluminum hydride. Sulfur. Trityl tetrafluoroborate. Trityl trifluoroacetate.

DEHYDROHALOGENATION: Potassium t-butoxide. Potassium hydroxide. Sodium amide. Tetrakis(triphenylphosphine)palladium(O).

DEMETHYLATION: Lithium triethylborohydride.

DEOXIMATION: Chromic acid.

DEOXYGENATION, EPOXIDES: Ferric chloride–n-Butyllithium. Potassium selenocyanate. Sodium cyclopentadienyldicarbonylferrate. Tungsten hexachloride.

ESTERS: Trichlorosilane.

N-OXIDES: Boron trifluoride etherate.

endo-PEROXIDES: Naphthalene–Lithium.

NITRORARENES: Hexamethylditin.

SULFOXIDES: Iodine–Pyridine–Sulfur dioxide.

DESULFURIZATION: Hexamethylphosphorous triamide. Potassium hydridotetracarbonylferrate. Triethyl phosphite.

DEUTERATION: Rhodium trichloride hydrate.
DIELS–ALDER CATALYSTS: Aluminum chloride. Boron trifluoride etherate.
DIELS–ALDER REACTIONS: Benzocyclopropene. 1,2-Bis($\beta$-tosylethoxycarbonyl)-diazene. Cyclobutadiene iron tricarbonyl. 1,2-Dicyanocyclobutene. Diethyl ketomalonate. 1,3,4,6-Heptatetraene. 2-Hydroxy-5-oxo-5,6-dihydro-2$H$-pyrone. 3-Hydroxy-2-pyrone. Isopropylidene isopropylidenemalonate. Lithium tetramethylpiperidide. 4-Methoxy-5-acetoxymethyl-o-benzoquinone. 4-Methoxy-5-methyl-o-benzoquinone. trans-1-Methoxy-3-trimethyl-silyloxy-1,3-butadiene. Perfluorotetramethylcyclopentadiene. 4-Phenyl-1,2,4-triazoline-3,5-dione. Potassium t-butoxide.
DIMERIZATION, BUTADIENE: Bis(triphenylphosphine)nickel dibromide.
DIMERIZATION, VINYL ARENES: Ion-exchange resins.

ENE REACTIONS: Oxygen, singlet.
EPIMERIZATION, sec-ALCOHOLS: Potassium superoxide.
EPOXIDATION: N-Benzoylperoxycarbamic acid. O-Benzylmonoperoxycarbonic acid. Perbenzoic acid. Potassium hypochlorite. Silver oxide.
EPOXIDES, REARRANGEMENT: Sulfuric acid.
ESCHENMOSER $\alpha,\beta$-epoxy ketone cleavage: 2,4-Dinitrobenzenesulfonylhydrazine.
ESTER CONDENSATION: Polyphosphoric acid. Potassium hydride.
ESTERIFICATION: 6-Chloro-1-p-chlorobenzenesulfonyloxybenzotriazole. Graphite bisulfate. Hexamethylphosphoric triamide. 1-Methyl-2-bromopyridinium iodide. Triphenylphosphine ditriflate. Hexamethylphosphoric triamide.

FETIZON'S REAGENT: Silver carbonate–Celite.
FISCHER INDOLE SYNTHESIS: Phenylhydrazine.
FLUORINATION: Diethylaminosulfur trifluoride. Diethyl(2-chloro-1,1,2-trifluoroethyl)-amine. Phenyliodine(III) difluoride. Xenon difluoride.
FLUORODEHYDROXYLATION: Sulfur tetrafluoride.
FORMYLATION: 1,3-Dithiane. Hexamethylphosphorous triamide.
o-FORMYLATION, PHENOLS: N-Chloroccinimide–1,3-Dithiane complex.
FRIEDEL–CRAFTS ACYLATION: Polyphosphoric acid.
FRIES REARRANGEMENT: $\beta$-Cyclodextrin.

GABRIEL SYNTHESIS: N-Benzyltriflamide.

HALOGEN EXCHANGE: Molybdenum pentachloride.
HALOMETHYLATION: 1-Chloro-4-bromomethoxybutane. 1-Chloro-4-chloromethoxy-butane.
HOMOLOGATION, ORGANOBORANES: Methylthiomethyllithium.
HYDRATION, ACETYLENES: Mercuric oxide.
HYDROACYLATION: Potassium hydridotetracarbonylferrate.
HYDROBORATION: 1,3,2-Benzodioxaborole. 9-Borabicyclo[3.3.1]nonane. Diborane. Diisopinocamphenyl borane. Dimethyl sulfide–Borane.
HYDROGENATION, ASYMMETRIC: (–)-and (+)-2,3-O-Isopropylidene-2,3-dihydroxy-1,4-bis(diphenylphosphine)butane. Neomenthyldiphenylphosphine.
HYDROGENATION CATALYSTS: trihapto–Allyltris(trimethylphosphite)cobalt(I). Lindlar catalyst. Palladium catalysts. Palladium(II) chloride. Rhodium-on-carbon. Tris(triphenylphosphine)chlororhodium. Tris(triphenylphosphine)ruthenium dichloride. Urushibara catalysts.

HYDROGENATION, IONIC: Trifluoroacetic acid–Alkylsilanes.
HYDROGENATION, TRANSFER: Hexafluoroantimonic acid. Palladium black. Tris-(aquo)hexa-$\mu$-acetate-$\mu_3$-oxotriruthenium(III,III,III) acetate. Tris(triphenylphosphine)chlororhodium. Tris(triphenylphosphine)ruthenium(II) dichloride.
HYDROLYSIS, ESTERS: Quinoline–Acetic acid.
HYDROPEROXIDATION: Oxygen, singlet.
HYDROSILYLATION: Tris(triphenylphosphine)chlororhodium.
HYDROXYLATION: Di-$t$-butyl diperoxycarbonate. Dichlorodicyanobenzoquinone. Hydrogen peroxide. Oxygen.
cis-HYDROXYLATION: Sodium nitrite–Acetic acid.
HYDROZIRCONATION: Di($\eta^5$-cyclopentadienyl)(chloro)hydridozirconium(IV).

INTERCHANGE, THIOACETAL–HEMITHIOACETAL–ACETAL: Methyl fluorosulfonate.
IODOCYCLOPROPANATION: Diethylzinc–Iodoform.
ION PAIR EXTRACTION: Benzyl tri-$n$-butylammonium chloride.
ISOMERIZATION: ALKENES: Potassium selenocyanate. Siloxene. Thiophenol.
  ALKYNES: Potassium 3-amino propylamide.
  QUADRICYCLANE: Bis(acrylonitrile)nickel(0).
  $\beta$, $\gamma$-UNSATURATED KETONES: 2,2,2-Trifluoroethylamine.
ISOPROPENYLATION: Isopropenyllithium. Dimethylcopperlithium.

JONES REAGENT: see Chromic acid in Subject Index.

KNOEVENAGEL CONDENSATION: $\beta$-Alanine. 1,5-Diazabicyclo[5.4.0]undecene-5. Titanium(IV) chloride.
KNORR REACTION: Dimethyl sulfoxide.
KOENIGS–KNORR REACTION: Mercuric oxide–Mercuric bromide. Silver carbonate.
KOENIGS–KNORR SYNTHESIS: Silver carbonate–Celite.

LACTONE ANNELATION: $\beta$-Vinylbutenolide.
LACTONIZATION: 2,2'-Dipyridyl disulfide–Triphenylphosphine.
LEMIEUX–RUDLOFF OXIDATION: Periodates.

MALONIC ESTER SYNTHESIS: Ethyl malonate.
MESYLATION: Diethylmethyl(methylsulfonyl)ammonium fluorosulfonate.
METHYLATION: Dimethyl sulfoxide. Methyl fluorosulfonate. Methyl trifluoromethanesulfonate. Simmons-Smith reagent.
C-METHYLATION: Trimethylaluminum.
N-METHYLATION, AMIDES: Chloromethyl methyl sulfide.
O-METHYLATION: Sodium hydride.
METHYLENATION: N-Methylphenylsulfonimidoylmethyl lithium. Phenylthioacetic acid. Titanium(IV) chloride–Lithium aluminum hydride. Trimethylsilylmethylmagnesium chloride.

NEF REACTION: Sodium methoxide.
NITRATION: Nitronium trifluoromethanesulfonate.
NITROSATIVE CYCLIZATION: Sodium nitrite.

OXIDATION, CATALYSTS: Salcomine.
OXIDATION, REAGENTS: Alumina. N–Bromosuccinimide. Ceric ammonium nitrate.

Cerium(IV) oxide–Hydrogen peroxide. *m*-Chloroperbenzoic acid. N-Chlorosuccinimide–Triethylamine. Chromic acid. Chromyl chloride. Cupric sulfate. Dimethyl
sulfide ditriflate. Dimethyl sulfoxide. Dimethyl sulfoxide–Iodine. Dimethyl sulfoxide-
Sulfur trioxide. Diphenylseleninic anhydride. Iodobenzene ditrifluoroacetate. Lead
tetraacetate. Jones reagent. Manganese dioxide. Oxygen. Ozone. Periodates. Potassium permanganate–Acetic anhydride. Potassium superoxide. Pyridinium chlorochromate. Ruthenium tetroxide. Silver carbonate–Celite. Succinimidodimethylsulfonium fluoroborate. Sulfur–Hexamethylphosphoric triamide. Trimethylamine N-oxide.
Tri(tetra-*n*-butylammonium)hexacyanoferrate(III). Ytterbium(III) nitrate.

OXIDATIVE ACETOXYLATION: Bis-(2,2-dipyridyl)-silver(II) peroxydisulfate.

OXIDATIVE CLEAVAGE, ACID HYDRAZIDES: Oxygen.

OXIDATIVE COUPLING: Cupric acetate. Potassium ferricyanide. Silver oxide. Thallium-
(III) trifluoroacetate. Vanadium oxytrifluoride.

OXIDATIVE CYCLIZATION: Mercuric acetate.

OXIDATIVE DECARBOXYLATION: N-Chlorosuccinimide. Cobalt(III) acetate. Lead
tetraacetate. Oxygen.

OXIDATIVE DECYANATION: Oxygen.

OXIDATIVE REARRANGEMENT, AMIDES: Lead tetraacetate.

OXYAMINATION: Trioxo(*t*-butylnitrido)osmium(VIII).

OXYGEN, SINGLET: Cerium(IV) oxide–Hydrogen peroxide. Phthaloyl peroxide.
Potassium perchromate.

OXYGENATION: Potassium *t*-butoxide.

OXYMERCURATION: Mercuric acetate.

OXYSELENATION: Dimethyl selenoxide.


PENTANNELATION: 2,2-Dimethyl-6-(*p*-chlorothiophenylmethylene)cyclo-
hexanone.

PEPTIDE SYNTHESIS: μ-Oxo-bis[tris(dimethylamino)phosphonium]bistetrafluoro-
borate.

PETERSON ELIMINATION: Trimethylsilylacetic acid.

PFITZNER–MOFFATT OXIDATION: Dimethyl sulfoxide.

PHASE–TRANSFER CATALYSTS: Adogen 464. (–)-N-Benzyl-N-methylephedrinium
bromide. Benzyltriethylamine. Benzyl triethylammonium chloride. 18-Crown-6.
Dibenzo-18-crown-6. Diethyl phenylsulfinylmethylphosphonate. Hexadecyltributylphosphonium bromide. Methyltricaprylylammonium chloride. Tetra-*n*-butylammo-
nium hydrogen sulfate. Tetra-*n*-butylammonium iodide.

PHENOLS, REDUCTION: Raney nickel.

PHENYLATION: π-Benzenechromium tricarbonyl. π-(Chlorobenzene)chromium tricarbonyl.

PHOSPHONATE REACTION: Benzyltriethylammonium chloride.

PHOSPHORYLATION: Cyclohexyl isocyanide. Di(2-*t*-butyl phenyl)phosphorochloridate.
2-(N,N-Dimethylamino)-4-nitrophenyl phosphate. N-(1,2-Dimethylethylenedioxy-
phosphoryl)imidazole.

PHOSPHORYLATION, AMINES: Benzyltriethylamine.

PHOTOOXYGENATION: Trityl tetrafluoroborate.

PINACOL REDUCTION: Magnesium amalgam.

PROTECTION OF: ALDEHYDES: Methyl fluorosulfonate.

ALKENES: (Cyclopentadienyl)(isobutenyl)iron dicarbonyl tetrafluoroborate.

AMINO GROUPS: 5-Benzisoxazolemethylylene chloroformate. *t*-Butyl azidoformate.

AMINO GROUPS: 2-*t*-Butyloxycarbonyloxyimino-2-phenylacetonitrile. *p*-Dihydroxy-

borylbenzyloxycarbonyl chloride. Diphenyldi(1,1,1,3,3-hexafluoro-2-phenyl-2-propoxy)sulfurane.
CARBONYL GROUPS: N-Bromosuccinimide. Chromic acid. 2-Methylene-1,3-propanedid.
CARBOXYL GROUPS: Hydrazine hydrate. Sodium ethanedithiolate. Sodium trithiocarbonate.
1,3-DIOLS: Benzaldehyde.
HYDROXYL GROUPS: *t*-Butyldimethylsilane. *t*-Butyldiphenylsilyl chloride. Chloromethyl methyl sulfide. Iodomethyl methyl sulfide.
  Levulinic anhydride. Methanesulfonyl chloride. 4-Methoxy-5,6-dihydro-2H-pyrane. Methyal. α-Naphthyldiphenylmethyl chloride. Trityl chloride.
KETONES: Sodium hypochlorite.
α-METHYLENE-γ-BUTYROLACTONES: Sodium thiophenoxide.
THIOL GROUPS: 2-Picolyl chloride 1-oxide hydrochloride. β-Nitro-styrene.
PUMMERER REARRANGEMENT: Acetic anhydride–Methanesulfonic acid. Acetic anhydride–Sodium acetate.

REARRANGEMENTS, RAMBERG–BACKLUND: Sodium 2-methyl-2-butoxide.
REDUCTION, ASYMMETRIC: (+)-(2S,3R)-4-Dimethylamino-3-methyl-1,2-diphenyl-2-3-methyl-1,2-diphenyl-2-butanol.
REDUCTION, CATALYST: D-(-)-N-Dodecyl-N-methylephedrinium bromide.
REDUCTION, REAGENTS: Aluminum hydride. Aluminum tricyclohexoxide. 1-Benzyl-1,4-dihydronicotinamide. Bis(4-methyl-1-piperazinyl)aluminum hydride. 9-Borabicyclo[3.3.1]nonane. 9-Borabicyclo[3.3.1]nonane ate complexes. Borane–Dimethyl sulfide. *n*-Butylphenyltin dihydride, polymeric. Catecholborane. Cuprous bromide–Sodium bis-(2-methoxyethoxy)-aluminum hydride. Diborane. $\eta^5$-Dicyclopentadienyl-titanium dichloride. N,N-Diethylhydroxylamine. Diimide. Diisobutylaluminum hydride. Hydrazine. Hydriodic acid. Isobutylaluminum dichloride. Lithium aluminum hydride. Lithium aluminum hydride–Cuprous iodide. Lithium cyanoborohydride. Lithium methoxyaluminum hydride. Lithium tri-*sec*-butylborohydride. Lithium triethylborohydride. Magnesium. Phosphorus–Iodine. Potassium borohydride. Potassium tetracarbonylhydridoferrate. Potassium tri-*sec*-butylborohydride. Potassium tri-*sec*-butylborohydride–Cuprous iodide. Raney nickel. Sodium bis-(2-methoxyethoxy)aluminum hydride. Sodium cyanoborohydride. Sodium hydrogen selenide. Sodium triacetoxyborohydride. Sulfur–Hexamethylphosphoric triamide. Thiourea dioxide. Titanium(III) chloride. Titanium(III) chlꞓide–Lithium aluminum hydride. Titanium(III) chloride–Magnesium. Titanocene system. Triethylaminoaluminum hydride. Triethylsilane–Trifluoroacetic acid. Trimethylchlorosilane–Zinc. Triphenyltin hydride. Zinc.
REDUCTIVE ALKYLATION: Lithium–Ammonia
REDUCTIVE CLEAVAGE: Lithium–Hexamethylphosphoric triamide.
REDUCTIVE ALKYLATION: Sodium cyanoborohydride.
REDUCTIVE METHYLATION: Sodium bis-(2-methoxyethoxy)aluminum hydride.
REFORMATSKY REACTION: Indium. Zinc.
RESOLUTION: 3β-Acetoxy-17β-chloroformylandrostene-5. R-(-)-1-(1-Naphthyl)ethyl isocyanate. α-(2,4,5,7-Tetranitro-9-fluorenylideneaminoxy)propionic acid.
RING EXPANSION: Dibromomethyllithium. Ethyl diazoacetate. Iron pentacarbonyl.

SCAVENGER, ACID: Dimethyl sulfoxide. Molecular sieves.
SINGLET OXYGEN: Tetraphenylcyclopentadienone.

SHAPIRO–HEATH OLEFIN SYNTHESIS: p-Toluenesulfonylhydrazine.
SPIROANNELATION: Carboethoxycyclopropyltriphenylphosphonium tetrafluoroborate.
STEVENS REARRANGEMENT: Sodium methylsulfinylmethide.
SURFACTANTS: Brij 35.
SYNTHESIS OF:
ACID FLUORIDES: Dialkylaminosulfur trifluorides.
ACYL AZIDES: Tetra-n-butylammonium azide.
ACYLOINS: m-Chloroperbenzoic acid. 5-(2'-Hydroxyethyl)-4-methyl-3-benzylthia-
zolium chloride.
ALCOHOLS: Benzalsodium. Di($\eta^5$-cyclopentadienyl)(chloro)hydridozirconium(IV).
ALDEHYDES: Bis(4-methylpiperazinyl)aluminum hydride. Di($\eta^5$-cyclopentadienyl)-
(chloro)hydridozirconium(IV). Dihalobis(triphenylphosphine)palladium(II). 7,8-
Dimethyl-1,5-dihydro-2,4-benzodithiepin. Grignard reagents. Lithium bis(ethylene-
dioxyboryl)methide. 3-Methyl-1-phenyl-2-phospholene. 2-Methyl-2-thiazoline.
Methylthioacetic acid. 3-Methylthio-1,4-diphenyl-s-triazium iodide. Sodium meth-
oxide. Methylthiomethyl N,N-dimethyldithiocarbamate. Sodium tetracarbonylferrate(II).
Tetra-n-butylammonium borohydride. Triethylallyloxysilane. N,4,4-Trimethyl-2-
oxazolinium iodide.
ALKANECARBOXYLATES: Ozone.
ALKENES: Allyl dimethyldithiocarbamate. Bis($\eta^5$-cyclopentadienyl)niobium trihydride.
Cyanogen bromide. Di-n-butylcopperlithium. $\alpha,\alpha$-Dichloromethyl methyl ether. 2,3-
Dimethyl-2-butylborane. N,N-Dimethyl dichlorophosphoramide. Diphenyl diselenide.
Di-n-propylcopperlithium. Ferric chloride. Grignard reagents. Iodine. Lithium
phenylethynolate. Lithium 2,2,6,6-tetramethylpiperidide. Methyl iodide. o-Nitro-
phenyl selenocyanate. Propargyl bromide. trans-1-Propenyllithium. Selenium.
Tetrakis(triphenylphosphine)palladium. Titanium(III) chloride. Titanium trichloride–
Lithium aluminum hydride. p-Toluenesulfonylhydrazine. Triphenylphosphine. Vinyl-
copper reagents. Vinyllithium. Zinc.
1-ALKENES: Grignard reagents.
$\beta$-ALKOXY KETONES: Titanium tetrachloride.
3-ALKYLALKANOIC ACIDS: (4S,5S)-2-Methyl-4-methoxymethyl-5-phenyl-2-oxazoline.
3-ALKYL-1-ALKYNES: n-Butyllithium.
ALKYL ARYL KETONES: $\alpha$-Azidostyrene.
ALKYL ARYL SULFONES: Trifluoromethanesulfonic-alkenesulfonic anhydrides.
ALKYL AZIDES: Triphenylphosphine–Carbon tetrabromide–Lithium azide. Zinc
chloride.
ALKYL CLORIDES: Antimony pentachloride. Ferric chloride. Dimethylformanide–
Phosphorus pentachloride. Hexamethylphosphoric triamide. Tri-n-butyltin chloride.
ALKYL FLUOROFORMATES: Polyhydrogen fluoride–Pyridine.
ALKYL HALIDES: t-Butyl hydroperoxides. Polyhydrogen fluoride–Pyridine. Zinc
chloride.
ALKYL IODIDES: Iodine.
t-ALKYL IODIDES: Ferric chloride.
ALKYLKETENES: Zinc.
ALKYNES: Diacetatobis(triphenylphosphine)palladium(II). Dichlorobis(triphenyl-
phosphine)palladium(II). Diethyl lithiodichloromethylphosphonate. 1,3-Dilithio-
propyne. Iodine. Potassium hydroxide. Tetrakis(triphenylphosphine)palladium(0).
Trialkynylalanes.
1-ALKYNES: Cuprous chloride.
ALLENEDIYNES: Cuprous bromide.

ALLENES: Carbon dioxide. Cuprous bromide. Trifluoromethanesulfonic anhydride.

α-ALLENIC ALCOHOLS: Grignard reagents.

ALLYL ETHERS: Titanium tetrachloride.

ALLYLIC ALCOHOLS: Benzeneselenol. 9-Borabicyclo[3.3.1]nonane. *n*-Butyllithium. Cuprous iodide–Grignard reagents. Diethylaluminum 2,2,6,6-tetramethylpiperidide. Fluorodimethoxyborane. *trans*-1-Tri-*n*-butylstannyl-propene-3-tetrahydropyranyl ether.

ALLYLIC CHLORIDES: 2,4-Dinitrofluorobenzene. *p*-Toluenesulfonyl chloride. Triphenyl phosphine–Carbon tetrachloride.

AMIDES: Boron tribromide. Boron trifluoride etherate. 6-Chloro-1-*p*-chlorobenzene-sulfonyloxybenzotriazole. Diethylphosphoryl cyanide. Dihalobis(triphenylphosphine)-palladium(II). Dihalobis(triphenylphosphine)palladium(II) complexes. Palladium(II) chloride. Sodium amide. Trimethylsilyl isocyanate. Triphenylphosphine ditriflate.

AMINES: N-Benzyltriflamide. Lindlar catalyst.

*sec* or *t*-AMINES: Iron pentacarbonyl. Palladium black.

AMINO ACIDS: L-Proline. Ruthenium tetroxide.

α-AMINO ALDEHYDES: Diisobutylaluminum hydride.

α-AMINO ALCOHOLS: Rhodium on alumina.

α-AMINO-γ-KETO ACIDS: α-Hydroxyhippuric acid.

α-AMINONITRILES: Trimethylsilyl cyanide.

ANTHRACENES: N-Lithium 2,2,6,6-tetramethylpiperide. Polyphosphoric acid.

ANTHRAQUINONES: Potassium hydroxide.

ARENE OXIDES: (–)-Menthoxyacetyl chloride.

O-ARENESULFONYLHYDROXYLAMINES: O,N-Bis(trimethylsilyl)hydroxylamine.

ARYLALKANES: Tetrahydrofurane.

ARYLAMINES: Bis(trimethylsilyl)amidocopper.

ARYL HALIDES: Hydrogen fluoride–Pyridine.

ARYL HYDRAZINES: Tris(trimethylsilyl)hydrazidocopper.

ARYL IODIDES: Thallium(III) trifluoroacetate.

ARYL KETONES: *n*-Butyllithium.

5-ARYL-γ-LACTONES: Silver carbonate–Celite.

ARYLOXYSILANES: Hexamethyldisilazane.

N-ARYLSULFIMIDES: N-Chlorosuccinimide–Dialkyl sulfides.

AZETIDINE-2-ONES: Sulfur dioxide.

AZETIDINES: Triphenylphosphine dibromide.

α-AZIDOKETONES: Iodobenzene diacetate–Trimethylsilyl azide.

AZIRIDINES: N-Chlorosuccinimide. Triethyl phosphonoacetate.

1-AZIRINES: Diazabicyclo[2.2.2]octane.

AZOXY COMPOUNDS: Cuprous chloride.

BENZALDEHYDES: Mercurous nitrate.

BENZALS: Benzaldehyde ethylene dithioacetal.

BENZAMIDINES: Iron pentacarbonyl.

BENZCYCLOBUTENES: $\eta^5$-Cyclopentadivinylcobalt dicarbonyl.

BENZOFURANES: N-Chlorosuccinimide–Dimethyl sulfide.

3,4-BENZOTROPILIDENES: Zinc.

BENZYL ALCOHOLS: Sodium borohydride.

BENZYL ESTERS: Benzyldimethylanilinium hydroxide.

BIARYLS: Bis(triphenylphosphine)nickel(II) chloride. Palladium(II) acetate.

α-BROMOACETALS: Bromine.

2-BROMO-1-ALKENES: Hydrogen bromide.
β-BROMO KETONES: Bromine.
1,2,3-BUTATRIENES: Diphosphorus tetraiodide.
BUTENOLIDES: Phenyl selenenyl chloride. Phenylthioacetic acid.
γ-BUTYROLACTONES: Ammonium persulfate. Chromic acid. Ion-exchangeresins.
   (4S,5S)-2-Methyl-4-methoxymethyl-5-phenyl-2-oxazoline. Naphthalene–Lithium.
   Perchloric acid. Potassium tri-n-butylborohydride.
CARBAMATES: Selenium.
CARBODIIMIDES: Palladium(II) chloride.
CARBONATES: N,N′-Carbonyldiimidazole. Hexamethylphosphorus triamide–Dialkyl
   azodicarboxylates.
CARBONYL COMPOUNDS: Mercuric chloride. Triethyltin methoxide.
CARBOTHIOATE S-ESTERS: Ethyl chlorothiolformate.
CARBOXYLIC ACIDS: Carbon dioxide. Chloroacetonitride. Di($\eta^5$-Cyclopentadivinyl)
   (chloro)hydridozirconium. Dihalobis(triphenylphosphine)palladium(II) complexes.
CARDENOLIDES: S-Methyl p-toluene thiosulfonate.
CAROTINOIDS: Sodium benzenesulfinate.
α-CHLORO ACID CHLORIDES: N-Chlorosuccinimide.
α-CHLOROCYCLOALKANONES: Lithium piperidide.
CYCLIC CARBONATES: N,N′-Carbonyldiimidazole.
CYCLIC KETONES: Sodium bistrimethylsilylamide.
CYCLOALKANONES: Methyl methylthiomethyl sulfoxide. Diborane. Silver per-
   chlorate. Sodium cyclopentadivinyldicarbonylferrate.
CYCLOBUTANONES: Methyl methylthiomethyl sulfoxide. Simmons-Smith reagent.
   Trifluoromethanesulfonic anhydride.
CYCLOBUTENONES: Alumina. 1-Chloro-N,N,2-trimethylpropenylamine. Methyl
   fluorosulfonate.
1,3-CYCLOHEXADIENES: trans-1-Butadienyltriphenylphosphonium bromide.
CYCLOHEXA-2,5-DIENONES: Antimony(V) chloride.
CYCLOHEXANONES: Magnesium.
CYCLOHEXYLIDENEACETALDEHYDES: Manganese dioxide.
CYCLOPENTANONES: Magnesium. 1-Phenylthiocyclopropyltriphenylphosphonium
   fluoroborate. Simmons Smith reagent. Sodium tetracarbonylferrate(-II).
CYCLOPENTENONES: 3-Benzyl-5-(2-hydroxyethyl)-4-methyl-1,3-thiazolium chloride.
   Boron trifluoride etherate. Lithium dimethoxyphosphinylmethylide. Silver tetra-
   fluoroborate. Titanium tetrachloride. Triethyl phosphite. Vinyl triphenylphos-
   phonium bromide.
CYCLOPROPANES: Cuprous t-butoxide.
CYCLOPROPANONES: Potassium fluoride.
CYCLOPROPENES: Cesium fluoride.
CYCLOPROPYL KETONES: Sodium chloride–Dimethyl sulfoxide.
DEPSIPEPTIDES: 1-Hydroxybenzotriazole.
DESOXYBENZOINS: Phosphorus–Iodine.
DIALKYLACETIC ACIDS: 2-Methyl-4-methoxymethyl-5-phenyl-2-oxazoline.
trans-2,3-DIALKYLCYCLOPENTANONES: Lithium methyl(vinyl) cuprate.
α, α-DIALKYL KETONES: Diphenyl disulfide.
DIALKYLMALONIC ESTERS: n-Butyllithium.
DIALKYL PEROXIDES: Potassium superoxide.
DIAMANTANES: Chlorosulfonic acid.
3H-1,2-DIAZEPINES: p-Toluenesulfonylhydrazine.

α,α-DIBROMOALKYNES: Triphenylphosphine dibromide.
1,4-DICARBONYL COMPOUNDS: Ketene dimethylthioacetal monoxide.
DICARBOXYLIC ANHYDRIDES: Trifluoroacetic anhydride.
*vic*-DICHLORIDES: Molybdenum pentachloride.
1,1-DICHLOROALKENES: Diethyl lithiodichloromethanephosphonate.
α,α-DICHLOROCARBOXYLIC ACIDS: Lithium diethylamide.
*gem*-DICHLOROCYCLOPROPANES: Titanium(IV) chloride–Lithium aluminum hydride.
DIENE–Fe(CO)$_3$ COMPLEXES: Dimethylamine oxide.
1,3-DIENES: Diethylaluminum 2,2,6,6-tetramethylpiperidide. Lithium aluminum hydride. Lithium diphenylphosphide. Methylcopper.
1,4-DIENES: Diisobutylaluminum hydride. Diphosphorus tetraiodide.
1,5-DIENES: Lithium bis(dialkylamino) cuprates.
2,6-DIENIC CARBOXYLIC ACIDS: Trimethylchlorosilane.
DIHYDROFURANES: Manganic acetate.
DIARYL SULFONES: Silver trifluoromethanesulfonate.
N,N-DIFLUOROAMIDES: Fluorooxytrifluoromethane.
*gem*-DIFLUORO COMPOUNDS: Molybdenium hexafluoride.
2,3-DIHYDROFURANES: Carboethoxycyclopropyltriphenylphosphonium tetrafluoroborate.
1,2- AND 1,4-DIHYDROPYRIDINES: Methyl chloroformate.
DIHYDROTHIOPHENES: Vinyl triphenylphosphonium bromide.
1,2-DIKETONES: Dimethyl sulfoxide. *m*-Chloroperbenzoic acid. Magnesium iodide. Manganese acetate. Oxygen, singlet. Trimethylenethiotosylate.
1,3-DIKETONES: Di-*n*-butylcopperlithium.
1,4-DIKETONES: 3-Benzyl-5-(2-hydroxyethyl)-4-methyl-1,3-thiazolium chloride. Bis(methylthio)(trimethylsilyl)methyllithium. Cupric chloride. Di(α-methoxyvinyl)copperlithium. Lithium diethylamide–Hexamethylphosphoric triamide. Manganic acetate. S-(2-Methoxyalkyl)N,N-dimethyldithiocarbamate. 1,3-Propanedithiol. Silver oxide.
1,5-DIKETONES: Aniline. Titanium tetrachloride.
1,2-DIOLS: Dimethyl sulfoxide.
*cis*-1,2-DIOLS: Iodine–Potassium iodate.
1,4-DIOLS: (4S,5S)-2-Methyl-4-methoxymethyl-5-phenyl-2-oxazoline.
1,2-DIOXETANES: Hydrogen peroxide–Dibromantin.
DIPHENIC ACIDS: 2-(*o*-Methoxyphenyl)-4,4-dimethyl-2-oxazoline.
DIPHENYLACETYLENES: N,N,N′,N′-Tetramethyldiaminophosphorochloridite.
DISULFIDES: 2,4-Dinitrobenzenesulfenyl chloride.
DITHIANES: Trimethylene dithiotosylate.
DITHIOACETALS: Methyltricapryylammonium chloride.
DIYNES: Iodine.
ENAMINES: Stannic chloride.
ENAMIDES: Acetic anhydride–Pyridine. Titanium(III) acetate.
1,4-ENEDIONES: Pyruvaldehyde 2-phenylhydrazone.
ENIMIDES: Acetic anhydride–Pyridine.
ENOL ACETATES: Cuprous acetate.
ENOL ETHERS: Grignard reagents.
ENYNES: Diacetatobis(triphenylphosphine)palladium(II). Tetrakis(triphenylphosphine)-palladium(0). Tetramethylethylenediamine. 1-Trimethylsilylpropynylcopper.
EPISULFONIUM SALTS: Methyl(bismethylthio)sulfonium hexachloroantimonate.
EPOXIDES: Bromomethyllithium. Diphenyl selenide. Methaneselenol.

ESTERS: Cryptates. Dihalobis(triphenylphosphine)palladium(II). Ion-exchange resins. Peracetic acid. Silver tetrafluoroborate. Thallous 2-methylpropane-2-thiolate. 2,4,6-Trimethylbenzenesulfonyl chloride.

ETHERS: Antimony(V) fluoride. Hexamethylphosphorus triamide. Tetra-$n$-butyl hydrogen sulfate. Thallium(I) ethoxide. Triphenylphosphine–Diethyl azodicarboxylate.

ETHYL KETONES: Sodium tetracarbonylferrate(II).

ETHYNYLCARBINOLS: Lithium acetylide.

α-FLUOROCARBOXYLIC ACIDS: Polyhydrogen fluoride–Pyridine.

α-FLUORO KETONES: Polyhydrogen fluoride–Pyridine.

2,3-FURANEDIONES: Oxalyl chloride.

FURANES: Diisobutylaluminum hydride. Phenylselenenyl chloride. Titanium tetrachloride.

GLYCOSIDES: Tetraethylammonium bromide. Silver carbonate–Celite.

α-HALO ACYL HALIDES: N-Bromosuccinimide.

β-HALO-α,β-UNSATURATED KETONES: Triphenylphosphine dihalides.

β-HALOVINYL KETONES: Triphenylphosphine–Carbon tetrachloride.

HETEROCYCLES: Carbon suboxide. Diethyl isocyanomethylphosphonate. Ethoxycarbonyl isothiocyanate. Propargyltriphenylphosphonium bromide. Selenium.

HOMOALLYLIC ALCOHOLS: π-2-Methylallylnickel bromide. 2-Methyl-2-thiazoline.

HOMOALLYLIC HALIDES: Magnesium iodide.

HOMOPHTHALIC ACIDS: Cuprous bromide.

HYDRAZONES: N-Benzoyl-N'-trifylhydrazine.

HYDROAZULENES: 2-Methylcyclopentenone. 3-Dimethylsulfoxonium methylide.

HYDROQUINONES: N,N-Diethylhydroxylamine.

α-HYDROXY ACIDS: Acetic anhydride–Sodium acetate. $m$-Chloroperbenzoic acid.

β-HYDROXY ALDEHYDES: 2-Methyl-2-thiazoline.

α-HYDROXYAMINO ACIDS: Lithium cyanoborohydride.

2-HYDROXY-3-ALKYLCYCLOPENT-2-ENE-1-ONES: Sulfuric acid.

α-HYDROXY CARBONYL COMPOUNDS: $m$-Chloroperbenzoic acid.

α-HYDROXYCARBOXYLIC ACIDS: $t$-Butyldimethylsilyl cyanide.

4-HYDROXYCYCLOPENTENONES: Chloral.

β-HYDROXY ESTERS: Alkyl orthotitanates. Ethyl α-phenylsulfinylacetate.

δ-HYDROXY-β-KETO ESTERS: Titanium(IV) chloride.

α-HYDROXYMETHYL KETONES: Formaldehyde.

β-HYDROXY SULFIDES: Phenylthioacetic acid.

β-HYDROXY THIOL ESTERS: Lithium diisopropylamide.

γ-HYDROXY-α,β-UNSATURATED ALDEHYDES: 1,3-Bis(methylthio)allyllithium.

IMINES: Sodium hydride.

INDANES: π-Cyclopentadienylcobalt dicarbonyl.

INDENES: Hydrobromic acid.

INDENONES: Bis(benzonitrile)palladium dichloride.

INDOLES: Phenylhydrazine. Sodium amide–Sodium $t$-butoxide.

ISOCYANATES: Tetra-$n$-butylammonium azide.

ISOPRENOIDS: 1-Cyclobutenylmethyllithium.

ISOTHIOCYANATES: Carbon disulfide.

KETENE THIOACETALS: Bis(dimethylaluminum)1,3-propanedithiolate. Bis(methylthio)(trimethylsilyl)methyllithium.

β-KETO ACETATES: Titanium tetrachloride.

β-KETO ACIDS: $t$-Butyl α-lithioisobutyrate.

α-KETO ACID CHLORIDES: Dichloromethyl methyl ether.

β-KETOALDEHYDES: S-Methyl *p*-toluenethiosulfonate. 1-Methylthio-3-methyoxypropane.

γ-KETOALKYNES: Propargyl bromide.

α-KETO ESTERS: Diphenyl disulfide. Ethyl methylsulfinylacetate. Magnesium iodide.

β-KETO ESTERS: Ethyl α-phenylsulfinylacetate.

γ-KETO ESTERS: Di(α-methoxyvinyl)copperlithium. Ethyl acrylate. Tris(phenylthio)-methyllithium.

KETONES: *t*-Butyl α-lithioisobutyrate. Benzoin. Chlorocarbonylbis(triphenylphosphine)-rhodium(I). *m*-Chloroperbenzoic acid. Chromic acid. Dimethylcopperlithium. Diphenyl disulfide. Formaldehyde diphenyl thioacetal. Methylthioacetic acid. Sodium cyanide. Tetrakis(triphenylphosphine)palladium(0). Trimethylchlorosilane. Triphenylmethyl isocyanide. Trimethylsilyl cyanide.

γ-KETONITRILES: Sodium cyanide.

β-KETO PHENYL SULFOXIDES: Methyl benzenesulfinate.

20-KETO STEROIDS: Methoxymethylenetriphenylphosphorane.

β-KETO SULFOXIDES: Ethyl phenyl sulfoxide.

β-LACTAMS: Tetra-*n*-butylammonium iodide.

δ-LACTOLS: β-(Phenylsulfonyl)propionaldehyde ethylene acetal.

LACTONES: 2,2′-Dipyridyl disulfide. Manganic acetate. Potassium carbonate–Dimethyl sulfoxide. Silver carbonate–Celite. Silver fluoroborate. Sodium hydride. Thallous 2-methylpropane-2-thiolate.

β-LACTONES: Lithium phenylethynolate.

γ-LACTONES: Triethyl orthoacetate.

δ-LACTONES: Ozone.

LITHIUM ORGANOCUPRATES: Dimethylsulfide–Cuprous bromide.

METACYCLOPHANES: Bis(1,3-diphenylphosphinopropane)nickel(II) chloride.

2-METHOXYALKANOIC ACIDS: *trans*-2,4-Dimethoxymethyl-5-phenyloxazoline.

β-METHOXY KETONES: Isopropenyl acetate.

N-METHYLAMIDES: *p*-Toluenesulfonyl chloride.

α-METHYLENE-γ-BUTYROLACTONES: π-(2-Carboethoxy)nickel bromide. 1,5-Diazabicyclo[4.3.0]nonene-5. Lithium α-carboethoxyvinyl(1-hexynyl)cuprate. *p*-Methoxybenzyl itaconate. Methoxyphenylthiomethyllithium.

METHYLENECYCLOALKENES: Nickel carbonyl.

METHYLENECYCLOPROPANES: Sodium bis(trimethylsilyl)amide.

α-METHYLENE-γ-LACTONES: Formaldehyde. 2-Phenylthiopropionic acid. Sodium cyanoborohydride.

METHYL KETONES: Chromic acid. Ethyl α-phenylsulfinylacetate. Lithium acetylide. Lithium diethylamide–Hexamethylphophoric triamide. S-(2-Methoxyallyl)-N,N-dimethyldithiocarbamate. α-Methoxyvinyllithium. Phenylthioacetic acid. Potassium tetracarbonylhydridoferrate. Silver(II) oxide. Trimethylaluminium.

α-METHYL-α,β-UNSATURATED KETONES: Methyllithium.

MONOBROMOCYCLOPROPANES: Diethylzine–Chloroform-Oxygen.

1,5-NAPHTHOQUINONES: 2,3-Dichloro-5,6-dicyano-1,4-benzoquinone.

NITRILES: 18-Crown-6. Dichloromethylenedimethylammonium chloride. O-2,4-Dinitrophenylhydroxylamine. N-Ethylacetonitrilium tetrafluoroborate. Hydroxylamine-O-sulfonic acid. Methyltricarylylammonium chloride. Phenyl chlorsulfite. Phosphonitrilic chloride. Tetraethylammonium cyanide. Tosylmethyl isocyanide. Triethyl orthoformate. Triphenylmethylisocyanide.

NITROALKANES: Hexafluoroantimonic acid.

NITROARYL IODIDES: Thallium(III) trifluoroacetate.

NUCLEOSIDES: Trimethylsilyl perchlorate.

NUCLEOTIDES: 1-(Mesitylene)-1,2,4-triazole.

cis-Δ¹-3-OCTALONES: trans-1-Methoxy-3-trimethylsilyloxy-1,3-butadiene.

OXAZIRIDINES: (R)-(+)-α-Phenylethylamine.

OXETANE: Bis(tri-n-butyl)tin oxide.

OXIRANES: Phenylthioacetic acid.

PEPTIDES: Azidotris(dimethylamino)phosphonium hexafluorophosphate. Benzotriazole-N-hydroxytris(dimethylamino)phophonium hexafluorophosphate. Cyanogen bromide. Dicyclohexylcarbodiimide. Diethyl phophorocyanidate. 2-Ethyl-7-hydroxybenzis-oxazolium tetrafluoroborate. 1-Hydroxybenzotriazole. N-Hydroxy-S-norbornene-endo-2,3-dicarboximide. Ion-exchange resins. Isobutyl chloroformate.

PHENOLS: Methyl vinyl ketone. Thallium(III) trifluoroacetate.

2-PHENYLINDOLES: Phenacylidinedimethylsulfurane.

PHOSPHODIESTERS: Acetoinenediolcyclopyrophosphate.

PIPERIDINES: Palladium nitrate.

POLYCYCLIC ARENES: Pyridinium hydrobromide perbromide.

POLYENALS: Molecular sieves.

POLYENES: Allylidenetriphenylphophorane.

POLYNUCLEOTIDES: o-Chlorophenyl phosphorodichloridite, 2,4,6-Triisopropylben-zenesulfonyl chloride.

POTASSIUM ENOLATES: Potassium hydride.

PURINES: Diethyl azodicarboxylate.

2-PYRIDONES: Potassium t-butoxide.

PYRIDINES: Cobaltocene.

PYROPHOSPHATES: Acetoinenediolcyclopyrophosphate.

PYRROLES: Trifluoroacetic anhydride.

QUINOL ACETATES: Lead tetracetate.

QUINOLINES: Hexamethylphophoric triamide.

p-QUINOLS: Oxygen.

QUINONES: Salcomine.

SELENOKETONES: Selenium.

SILYL ENOL ETHERS: Triethylsilyl thiophenoxide.

SPIRODIENONES: Trifluoroacetic acid.

SPIROLACTONES: Dimethylcopperlithium.

SULFIDES: Hexadecyltributylphosphonium bromide. N,N-Methylphenylaminotriphenyl-phosphonium iodide. Methyltricarylylammonium chloride. Tri-n-butylphosphine–Di-alkyl disulfides.

SULFONES: Tetra-n-butylammonium p-toluenesulfinate.

TERPENOIDS: Bis(cyclopentadienyl)titanium dichloride. Palladium chloride.

1,2,4,5-TETRAENES: Cuprous chloride.

TETRAHYDROPYRANES: Hexamethylphosphoric triamide.

TETRALINS: π-Cyclopentadienylcobalt dicarbonyl.

THEXYLMONOALKYLBORANES: 2,3-Dimethyl-2-butylborane.

THIIRANES: 3-Methylbenzothiazole-2-thione. 2-Thiomethyl-4,4-dimethyl-2-oxazoline.

THIIRENIUM SALTS: Methyl(bismethylthio)sulfonium hexachloroantimonate.

THIOACYL CHLORIDES: Hydrogen chloride.

THIOL ESTERS: Diethyl phosphorochloridate.

THIOLS: Sodium borohydride, sulfurated.

α,p-TOLUENESULFINYL KETONES: Methyl p-toluenesulfinate.

1,2,3-TRICARBONYL COMPOUNDS: Dimethyl sulfoxide.
1,4,5-TRIENES: 1-Trimethylsilylpropynylcopper.
TRIFLATES: Trifluoromethanesulfonic anhydride.
TRIFLUOROMETHYL SULFIDES: Trifluoromethylthiocopper.
α-TRIMETHYLSILYL ACIDS: Trimethylsilylacetic acid.
α-TRIMETHYLSILYL-γ-BUTYROLACTONES: Trimethylsilylacetic acid.
TRIMETHYLSILYLENOL ETHERS: Bis(trimethylsilyl)formamide. Zinc.
TRIMETHYLSILYLETHYNYL KETONES: Cuprous trimethylsilylacetylide.
TROPONES: Diiron nonacarbonyl.
TROPONOIDS: Diiron nonacarbonyl.
α,β-UNSATURATED ACIDS: Trimethylsilylacetic acid. Vinyl copper reagents.
γ,δ-UNSATURATED ALCOHOLS: Diisobutylaluminum hydride.
α,β-UNSATURATED ALDEHYDES: S-Alkyl N,N-dimethylaminodithiocarbamate. Bis-(dimethylaluminum)1,3-propanedithiolate. β-(Phenylsulfonyl)propionaldehyde ethylene acetal. Sodium N,N-dimethyldithiocarbamate. Tetraethylthiuram disulfide.
β,γ-UNSATURATED ALDEHYDES: 2-Methoxycyclopropyllithium.
β,γ-UNSATURATED AMIDES: Trimethylsilylmethyllithium.
α,β-UNSATURATED ESTERS: Diethoxyaluminum chloride. Dimethyl disulfide. Ethyl methylsulfinylacetate. Methyl 2-phenylsulfinylacetate. Sulfuryl chloride.
β,γ-UNSATURATED ESTERS: t-Butyl lithioacetate. Trimethylsilylmethyllithium.
α,β-UNSATURATED KETONES: Bis(dimethylaluminum)1,3-propanedithiolate. Lithio-2-trimethylsilyl-1,3-dithiane. Phenylselenenyl bromide and chloride. Sodium borohydride. Tris(aquo)hexa-μ-acetato-μ_3-oxotriruthenium(III,III,III) acetate.
γ,δ-UNSATURATED KETONES: Dimethylcopperlithium.
α,β-UNSATURATED γ-LACTONES: Triethyl orthoacetate.
β,γ-UNSATURATED NITRILES: Trimethylsilylmethyllithium.
α,β-UNSATURATED SULFONES: Benzyltriethylammonium chloride.
α,β-UNSATURATED SULFOXIDES: Diethyl phosphorylmethyl methyl sulfoxide.
β,γ-UNSATURATED-δ-VALEROLACTONES: Diethyl ketomalonate.
VINYL ACETATES: Cuprous acetate.
VINYLCYCLOPROPANES: Copper–t-Butyl isonitrile. Vinyldiazomethane.
VINYL ETHERS: 1,2-Vinylenebis(triphenylphosphonium)dibromide.
VINYL FLUORIDES: Fluormethylenetriphenylphosphorane.
VINYL HALIDES: Di(η^5-cyclopentadienyl)(chloro)hydridozirconium(IV). Vinylcopper reagents.
VINYL ISOCYANIDES: Diethyl isocyanomethylphosphonate.
VINYL TRIFLATES: Trifluoromethanesulfonic anhydride. Trifluoromethanesulfinyl imidazole.
XANTHONES: Cupric bromide.

THIOKETALIZATION: Methylthiotrimethylsilane.
TRANSACETALIZATION: Diethylene orthocarbonate.
TRANSESTERIFICATION: Boron tribromide. Triphenylphosphine–Diethyl azodicarboxylate.
TRIPHASE CATALYSIS: Benzyl-n-butyldimethylammonium chloride.

ULLMANN DIPHENYL ETHER SYNTHESIS: Pentafluorocopper.
ULLMANN REACTION: Copper(I) trifluoromethanesulfonate. Diethylamino(trimethyl)silane. Tris(triphenylphosphine)nickel(0).

WILLGERODT REACTION:  Ammonium polysulfide.
WITTIG-HORNER REACTION:  Diethyl isocyanomethylphophonate.  Diethyl lithiodichloromethanephosphonate.  Diethyl phenylsulfinylmethylphosphonate.  Diethyl phenylthiomethanephosphonate.  Diethyl phosphorylmethyl methyl sulfoxide.  Lithium diphenylphosphide.
WITTIG REACTION:  Allylidenetriphenylphosphorane.  18-Crown-6.  Fluoromethylenetriphenylphosphorane.  Hexamethylphosphoric triamide.  Methoxymethylenetriphnylphosphorane.  Sodium methylsulfinylmethylide.  Trimethylstannylethylidenetriphnylphosphorane.
WITTIG REARRANGEMENT:  *n*-Butyllithium.

# AUTHOR INDEX

Aalbersberg, W. G. L., 154
Abbas, S. A., 35
Abiko, 149
Abou-Donia, M. B., 456
Abramovitch, A., 91
Abramson, N. L., 19
Acharya, S. P., 468
Acher, A., 574
Achmatowicz, O., Jr., 291
Achmatowicz, S., 583
Adam, R., 500
Adamek, J. P., 534
Adams, R., 465
Adamson, J. R., 453
Adcock, W., 454
Adinolfi, M., 295
Adkins, T. J., 91
Advani, B. G., 556
Afzali, A., 452
Agawa, T., 158
Agranat, L., 475, 670
Aguiar, A. M., 341, 416
Ahlfaenger, B., 201
Ahmed, F. U., 20
Aigami, K., 526
Ainsworth, C., 114
Aithie, G. C. M., 308
Aizawa, T., 96
Akabori, S., 138
Akasaka, K., 628
Akashi, K., 127
Akermark, B., 589
Akhtar, M. N., 358
Akiyama, S., 138
Akutagawa, S., 49
Albertson, N. F., 283
Alcock, N. W., 301
Alhaique, F., 541
Ali, S. M., 157
Alkaysi, H. N., 646
Allan, A. R., 385
Alley, W. D., 576
Allred, A. L., 399
Alonso, C., 162
Alper, H., 198, 485
Atland, H. W., 581

Altman, J., 290
Alward, S. J., 181
Amada, T., 500
Amann, A., 630
Amaro, A., 170, 171
Amel, R. T., 456
Amice, P., 523
Amick, D. R., 118, 119, 120
Ammons, A. S., 19
Anand, N., 284
Anand, S. P., 670
Andersen, N. H., 19, 321, 322
Andersen, R. J., 316
Anderson, C. D., 600
Anderson, D. G., 637
Anderson, G. C., 30, 32
Anderson, H. H., 634
Anderson, J. E., 81
Anderson, R. J., 165
Anderson, R. W., 416
Ando, W., 436
Andrews, G. C., 32, 607, 613
Anet, F. A. L., 15
Angerer, J., 359
Anteunis, A., 579
ApSimon, J., 288
Arakawa, M., 95
Aranda, G., 501
Arase, A., 259, 510
Aratani, M., 611
Araújo, H. C., 124, 441
Archer, S., 283
Arens, J. F., 651
Arentzen, R., 366
Arganat, I., 475
Armstrong, V. W., 97
Arnarp, J., 406
Arnason, B., 381
Arora, S., 530
Arz, A. A., 17
Asaoka, M., 279
Ashby, E. C., 327
Askam, V., 285
Assaf, Y., 645
Atherton, F. R., 43
Atlani, P., 441

Attanasi, O., 162
Au, A., 567
Auerbach, R. A., 676
Aumann, R., 306
Autrey, R. L., 400, 403
Averbeck, H., 304
Avram, M., 279
Aya, T., 73
Ayres, D. C., 506
Ayrey, G., 224, 241
Ayyar, K. S., 19
Azuma, H., 653

Baardman, F., 252
Baba, S., 601
Baba, Y., 201
Babiarz, J. E., 102
Babler, J. H., 1, 338
Babsch, H., 584
Baccolini, G., 534
Bachmann, J.-P., 215
Back, T. G., 509
Bacquet, C., 345
Baer, T. A., 224, 586
Bagnell, L., 624
Bahl, C. P., 622
Baine, O., 40
Baird, M. C., 665
Baird, M. S., 385
Baker, D. A., 569
Baker, J. D., Jr., 98
Bal, K., 74
Balanson, R. D., 319
Baldwin, D., 283, 606
Baldwin, J. E., 205, 373, 567
Baldwin, S. W., 606
Baldwin, W. A., 517
Balenović, K., 486
Balsamini, C., 357
Ban, Y., 628
Bánko, K., 43
Banno, K., 596
Banquer, C., 254
Banthorpe, D. V., 263
Bao, L. Q., 76, 118
Barager, H. J., III, 457
Baraldi, P. G., 566
Baran, J. S., 348
Barbaro, G., 306
Barbee, T. G., Jr., 546

Barborak, J. C., 198
Barco, A., 497, 560, 566
Barkovich, A. J., 154
Barnard, D., 224, 241
Barneis, Z., 645
Barnier, J. P., 523
Barone, G., 295
Barreau, M., 141
Barreiro, E., 215
Bartholomew, J. T., 102
Bartoletti, I., 61
Barton, D. H. R., 5, 99, 100, 109, 124,
   187, 226, 229, 241, 264, 508, 509, 580,
   581, 583, 587, 596, 598, 643, 644, 657
Barton, D. L., 354
Bashall, A. P., 10
Basu, N. K., 124
Basus, V. J., 15
Bates, A. J., 425
Bates, G. S., 192
Bateson, J. H., 186
Báttioni, P., 165
Baudouy, R., 342
Bauer, D. P., 229
Bauer, H., 422
Bauer, L., 291
Baum, J. W., 224, 586
Baumgarten, H. E., 317
Baumgarten, R. J., 531, 534
Baxter, G. J., 96
Bayet, P., 29, 40, 240
Beak, P., 348, 383
Beal, D. A., 105
Beames, D. J., 615
Beasley, G. H., 419
Beck, A. K., 29, 90
Becker, M., 74
Beckett, A. H., 538
Beevor, P. S., 547
Behan, J. M., 488
Bekowies, P. J., 487
Bell, H. C., 581
Belli, A., 560
Belluco, U., 450
Bellůs, D., 172, 558
Belzecki, C., 457
Benetti, S., 497, 560, 566
Ben-Ishai, D., 290
Bennett, G., 610
Bensoam, J., 412

Berchtold, G. A., 436
Berens, G., 552
Berger, G., 407
Bergeron, R., 44
Bergmann, E. D., 216, 670
Berler, Z., 290
Berlin, K. D., 475
Berliner, E., 475
Bernardi, L., 110, 515
Bernassau, J.-M., 455
Bernstein, Z., 254
Berrange, B., 577
Bershas, J. P., 15
Bertelo, C. A., 179
Bertin, J., 23, 269
Bertini, F., 74
Bertrand, M., 223
Bessière-Chrétien, Y., 498
Besters, J. S. M. M., 514
Bestmann, H. J., 95, 275, 279, 359, 381
Beyl, V., 273
Bhakuni, D. S., 581
Bhandari, K. S., 515
Bhanu, S., 203
Bhatnagar, P. K., 255
Bhatt, M. V., 283
Bickel, H., 186
Biggs, J., 57
Biller, S. A., 642
Billups, W. E., 477, 478
Bindra, R., 165
Binger, P., 530
Birch, A. J., 323, 475
Birchall, J. M., 621
Bird, J. G., 283
Birkofer, L., 390
Bissinger, W. E., 456
Bither, T. A., 81
Bittner, S., 645
Black, C. J., 76, 118
Black, D. St. C., 114
Blackburn, T. F., 179
Blackman, N. A., 114
Blake, M. R., 504
Blackesley, C. N., 546
Blaney, F., 121
Blankenhorn, G., 547
Blatt, A. H., 240
Blomquist, A. T., 634
Blossey, E. C., 19

Blout, E. R., 137
Blucher, W. G., 420
Blum, D. M., 407
Blum, J., 108, 542, 655
Boar, R. B., 5, 9, 587
Boch, M., 141
Bock, M. G., 110, 302
Boden, R. M., 135, 137, 436
Bodrikov, I. V., 232
Boeckelheide, V., 85, 90
Boeckman, R. K., Jr., 205, 407
Boerth, D. W., 554
Bogdanowicz, M. J., 155, 243
Boggio, R. J., 534
Boigegrain, R., 280
Boldrini, G. P., 306
Boldt, P., 170
Bolduc, P. R., 436
Bolton, M., 583
Bomse, D. S., 589
Bond, F. T., 600
Bonjoulkian, R., 187
Borch, R. F., 542
Borgman, R. J., 9
Borgogno, G., 586
Bos, H., 143
Boschi, T., 450
Boschung, A. F., 434, 436
Bose, A. K., 645
Bosisio, G., 515
Bost, M., 141
Boudjouk, P., 58
Boullier, P. A., 263
Boulos, A. L., 503
Bouma, R. J., 567
Bourelle-Wargnier, F., 124
Bourgain, M., 270, 331, 664
Boxler, D., 422
Bowers, C. W., 137
Boya, M., 417
Boyd, D. R., 358
Bozzato, G., 215
Bradshaw, J. S., 137
Bradshaw, T. K., 624
Braitsch, D. M., 654
Branca, S. J., 70, 577
Branchaud, B., 110
Brandsma, L., 144, 398
Brändström, A., 41, 564, 565, 567
Brase, D., 430

Braun, H., 630
Bredereck, H., 243
Brehme, R., 43
Brennan, M. E., 577
Brenner, G. S., 412
Breslow, R., 289, 299, 300
Breuer, E., 612
Breyer, S. W., 581
Bridges, A. J., 395
Brindle, J.-R., 534
Brink, M., 357, 612
Brinkmeyer, R. S., 630
Britten-Kelly, M. R., 509
Britton, R. W., 530
Broadbent, H. S., 493
Broadhurst, M. D., 419
Brockington, R., 381
Brocksom, T. J., 339
Brodowski, W., 263
Brook, A. G., 637
Brooke, G. S., 234
Brossmer, R., 511
Broussard, J., 479
Brown, C. A., 476, 482, 483, 492
Brown, E. V., 21
Brown, F. C., 515
Brown, H. C., 33, 34, 63, 64, 71, 73, 172,
    208, 259, 284, 295, 325, 342, 350, 573
Brown, R. F. C., 96
Brown, T. M., 414
Bruggink, A., 144
Brunelle, D. J., 343, 465
Brunow, G., 655
Bruza, K. J., 205
Bryant, R. J., 629
Brynjolffssen, J., 466
Bryson, T. A., 354
Buch, V., 33
Buchan, G. M., 170
Büchi, G., 77, 285, 514
Buchner, E., 190
Buckwalter, B., 607
Buddrus, J., 263
Buisson, J.-P., 498
Bukowski, P., 291
Bull, J. R., 600
Buncel, E., 158
Bundle, D. R., 569
Bunnett, J. F., 192
Bunton, C. A., 249, 598

Burdett, K. A., 675
Burke, S., 460
Burke, S. D., 632
Burnett, R. E., 416
Burns, W., 473
Bürstinghaus, R., 55, 118
Burton, D. J., 261, 262, 646
Burton, R., 596
Buswell, R. L., 600
Butterworth, R., 261
Bycroft, B. W., 492, 493
Byrne, M. P., 137
Bywood, R., 453

Cacchi, S., 470, 534, 607
Cadogan, J. I. G., 419
Caglioti, L., 162
Cahiez, G., 270, 664
Cahnmann, H. J., 432, 436
Cain, E. N., 76
Cain, P., 410, 411, 606
Caine, D., 492
Cainelli, G., 74, 304, 486
Calamai, E. C., 545
Calas, R., 323
Caló, V., 71, 73, 375
Calzada, J. G., 645
Campbell, M. M., 115
Campbell, R. G., 353
Campbell, R. H., 187
Cann, K., 198
Canonica, L., 518
Caplar, V., 474
Capozzi, G., 375
Capuano, L. A., 600
Caputo, R., 586
Carayon-Gentil, A., 574
Cardillo, G., 43
Carey, F. A., 35
Carlson, B. A., 172, 350
Carlson, R. G., 546
Carlson, R. M., 366, 395, 496
Carmody, M. J., 279
Carpenter, W., 300
Carrié, R., 472
Carroll, F. I., 577
Carstens, E., 547
Caruso, T. C., 137
Casey, C. P., 495
Caspi, E., 702

Cassar, L., 573
Castro, B., 25, 35, 280
Caton, M. P. L., 284
Caubere, P., 70, 526
Cava, M. P., 451, 452
Cella, J. A., 111, 114
Cerný, M., 529
Chabrier, P., 570, 574
Chan, A. S. K., 198
Chan, H.-F., 385
Chan, T. H., 100, 323, 482, 636, 637
Chandrasekhar, B. P., 588
Chandy, M. J., 621
Chang, E., 636, 637
Chang, S. M., 142, 224, 386
Chanon, F., 228
Chanon, M., 228
Chanot, J. J., 70
Chapman, D. D., 409
Chapman, P. H., 234
Chattopadhyaya, J. B., 296
Chau, L. V., 175, 511
Chavdarian, C. G., 205, 441
Chavis, C., 583
Chaykovsky, M., 342
Chebaane, L., 503
Chen, C. H., 476
Chen, R. H. K., 95
Chen, W.-Y., 193
Chênevert, R., 441
Chern, C.-I., 490
Chernick, C. L., 670
Chidgey, R., 544
Chiou, B. L., 399
Chishti, N. H., 97
Chiusoli, G. P., 451
Cho, H., 316
Chorev, M., 288
Chow, F., 316, 548, 549
Chow, W. Y., 477, 479
Christensen, J. J., 137
Christian, P. A., 670
Christie, J. J., 146
Christie, M., 567
Christol, H., 665
Christy, K. J., 79
Chu, C.-Y., 225, 288
Chuit, C., 270, 664
Chujo, Y., 130, 436
Chung, S.-K., 435, 436

Ciabattoni, J., 114, 479
Cinquini, M., 191
Cistone, F., 534
Claesson, A., 148, 325
Clark, F. R. S., 443
Clark, G., 28
Clark, R. D., 215, 339
Clarke, J. A., 73
Clarke, J. E., 307
Clauss, K., 122
Clemans, G. B., 500
Clerici, A., 21
Cleve, G., 443
Clifford, A. F., 642
Clizbe, L. A., 605
Closs, G. L., 620
Coates, G. E., 362
Coates, R. M., 40, 238, 400, 462
Coffee, E. C. J., 284
Cogolli, P., 631
Cohen, T., 133, 359, 606
Cohen, Z., 441
Cole, C.-A., 17
Cole, L. L., 279
Collington, E. W., 630
Collins, J. F., 10
Collman, J. P., 552
Colomer, E., 179
Colon, I., 244
Colonna, S., 586
Commercon, A., 143, 270, 577
Comninellis, C., 284
Concannon, P. W., 114
Confalone, P. N., 254
Conia, J. M., 521, 523
Conlin, R. T., 577
Constantinides, D., 390
Conway, W. P., 395
Cook, F. L., 137
Cooke, M. P., Jr., 349, 552
Cookson, C. M., 628
Cookson, R. C., 19
Coon, C. L., 420
Corbin, V. L., 165
Corcoran, R. J., 300
Cordes, E. H., 70
Corey, E. J., 11, 50, 56, 75, 76, 80, 81,
    95, 110, 133, 215, 232, 233, 246, 247,
    248, 267, 292, 302, 319, 322, 339, 352,
    367, 380, 381, 383, 417–419, 489, 490,

493, 499, 603, 604
Corriu, R. J. P., 179
Cortez, C., 230, 326
Cotterrell, G., 165
Coulson, D. R., 573
Counts, K. M., 360
Coustard, J.-M., 273
Coutrot, P., 189, 191, 354
Cox, G. R., 165
Crabbé, P., 214, 215, 536
Craig, L. C., 2
Cram, D. J., 134, 135, 137
Cram, J. M., 137
Cramer, F., 152
Crandell, J. K., 78, 165
Crass, G., 135, 137, 578
Crawford, H. T., 91
Creary, X., 192
Creese, M. W., 646
Cresp, T. M., 138
Criegee, R., 441, 548
Crison, C., 498
Cristau, H.-J., 665
Cristea, I., 133
Crociani, B., 450
Crosby, G. A., 92, 582
Cross, B. E., 186
Crouse, D. N., 81, 84, 636
Crumrine, D. S., 676
Curci, R., 113, 114, 487
Curtin, D. Y., 516
Curtis, A. B., 489, 490
Curtis, W. D., 137
Cusack, N. J., 605
Cuthbertson, E., 550
Cuvigny, T., 279, 323, 333, 352
Czaja, R. F., 19
Czernecki, S., 233, 280

Dabral, V., 284
Dagli, D. J., 252
Dagli, J., 542
Dahn, H., 229
Daigle, D., 341
Dalavoy, V. S., 475
Dalton, J. C., 385
D'Amico, J. J., 187
Damon, R. E., 253
Damps, K., 9
Danen, W. C., 419

d'Angelo, J., 264, 267, 530
Daniels, C. K., 146
Daniels, J., 64
Daniewski, A. R., 10
Danishefsky, S., 43, 217, 230, 354, 364,
    368, 372, 409–411, 606
Dansette, P., 477
Dao, H. L., 229
Datta, M. C., 45
Dauben, W. G., 94, 309, 310, 419, 620
Dave, V., 284, 458
Davidovich, Y. A., 639
Davidson, A. H., 340, 341
Dawes, C. C., 187
Dawson, A. D., 230
Day, M. J., 124, 598
Dayer, F., 229
Debal, A., 352
Deber, C. M., 126, 137
Debon, A., 300
De Buyck, L., 118
de Castiglione, R., 110
Dechatre, J. P., 514
DeChristopher, P. J., 534
Declercq, G., 596
Decorzant, F., 215, 675
de Graaf, C., 270
Dehesch, T., 620
Dehm, D., 137
de Jonge, C. R. H. I., 507
De Kimpe, N., 118
Dekker, J., 675
de Koning, A. J., 503
de Lattre, J., 369
Delaumeny, M., 270
Del Cima, F., 27
Della, E. W., 624
De Lucchi, O., 375
De Lue, N. R., 295
Demailly, G., 162
De Man, J. H. M., 172
De Meijere, A., 568
Demmin, T. R., 611
Denivelle, L., 514
Dennis, N., 618
Denniston, A. D., 322
Denzel, T., 95
Depezay, J.-C., 331, 520, 610
DePuy, C. H., 73
de Roch, I. S., 428, 430

DeRosa, M., 543
Dershowitz, S., 381
Deslongchamps, P., 438, 441
De Smet, A., 579
de Souza, J. P., 3, 374
Dessau, R. M., 127, 356, 357
Dev, S., 475
Devaprabhakara, D., 523
De Voe, S. V., 494
de Vos, D., 458
Dey, A. S., 477
Diakur, J., 192
Dick, K. F., 124
Dickinson, R. A., 70
Dickson, R. S., 454
Dieck, H. A., 61, 157
Dietsche, T. J., 395
Dietz, G., 547
Dietz, S. E., 534
DiFuria, F., 114, 487
Dilbeck, G. A., 475
Dineshkumar, 252
Ding, J. Y., 288
Dinizo, S. E., 113, 114
Dinulescu, I. G., 279
Disnar, J. R., 331
Dittmer, D. C., 102
Dixit, V. M., 326
Djarmari, Z., 201
Dodd, G. H., 79
Doering, W. v. E., 137
Doherty, R. M., 649
Doleschall, G., 396
DoMinh, T., 190
Dondoni, A., 306
Donndelinger, P., 146
Dopper, J. H., 525, 575
Dormoy, J. R., 25, 35
Douglas, G. H., 258
Douteau, M.-H., 273
Downie, I. M., 280
Doyle, M. P., 616, 620
Dreux, J., 557
Dreux, M., 189, 557
Driguez, H., 569
Dubois, J.-E., 165, 215
Dubois, R. A., 566
Dubuis, R., 319
Ducep, J.-B., 344
Duckworth, A. C., 617

Dueber, T. E., 620
Duff, J. M., 637
Duffaut, N., 323
Duggan, A. J., 323
du Manoir, J. R., 190
Dumont, W., 29, 90, 240, 362
Dunbar, B. I., 137
Duncan, D. M., 446
Duncan, G. R., 542
Duncan, W. G., 78, 304
Dunham, M., 381
Dunkelblum, E., 43
Dunkerton, L. V., 611
Dunlap, D. E., 586
Dunogues, J., 323
Durandetta, J. L., 403, 404
Duranti, E., 357
Durst, H. D., 121, 137, 406
Durst, T., 118
Dutta, P. K., 297
Dyer, E., 570
Dykman, E., 470

Eaborn, C., 640
Eastlick, D. T., 453
Eastwood, F. W., 96
Eaton, J. T., 534
Eberstein, K., 545
Eder, U., 411
Edwards, D. A., 143
Edwards, O. E., 327
Edwards, W. B., III, 339
Effenberger, F., 420, 521, 618
Egli, Ch., 409
Ehrenfreund, J., 297
Eisenbraun, E. J., 9, 20, 304
Eizember, R. F., 19
Ekouya, A., 323
Elberling, J. A., 174
Eliel, E. L., 615
Ellison, D. L., 676
Ellison, R. A., 255
Ellison, R. H., 454
Elwood, J. K., 409
Emerson, M., 573
Emke, A., 467
Enders, D., 339
Endo, T., 232
Entwistle, I. D., 446
Erbland, M. L., 244

Erdtman, H., 516
Erickson, B. W., 53, 76
Eschenfelder, V., 511
Eschenmoser, A., 91, 222
Esmail, R., 250
Evans, D., 105
Evans, D. A., 30, 32, 399, 607
Evans, G., 39
Evans, S. M., 455
Evin, G., 35

Falck, J. R., 133
Falling, S. N., 254
Fanta, P. E., 133
Farina, J. S., 331
Farnham, W. B., 352, 579
Fatiadi, A. J., 357
Faulkner, D. J., 316
Feder, H. M., 172
Fedorynski, M., 601
Feher, F., 288
Fehr, T., 560
Feinberg, R. S., 304
Feiring, A. E., 479
Felix, S., 645
Felkin, H., 270
Ferguson, D. C., 553
Ferles, M., 1, 2
Ferreira, G. A. L., 124, 441
Ferretti, A., 465
Ferris, A. F., 423
Fétizon, M., 430, 455, 501
Fiaud, J.-C., 40
Field, L., 374
Fierz, G., 544
Fieser, L. F., 5, 525
Fieser, M., 5, 525
Filby, J. E., 288
Filler, R., 6, 228, 670
Finch, N., 226, 228, 229
Findlay, J. W. A., 170
Finnan, J. L., 115, 415
Fischer, W. F., Jr., 203, 215
Fischli, A., 526, 527
Fisher, R. P., 195, 293, 295, 403
Fitt, J. J., 228, 229
Fitton, P., 573
Fitzpatrick, J. D., 150
Fizet, C., 290
Fleming, M. P., 348, 589

Fletcher, H. G., Jr., 583
Fletcher, T. L., 17, 578
Floor, J., 600
Floyd, D. M., 331
Fontana, A., 419
Foote, C. S., 241, 436
Forbes, C. P., 226, 229
Forbes, G. S., 634
Ford, M. E., 360, 585
Fordham, W. D., 263
Foreman, G. M., 195
Fornasier, R., 586
Forte, P. A., 456
Fortunato, J. M., 348
Foster, C. H., 436
Fouad, H., 422
Foulon, J. P., 622
Fournier, P., 165
Fowler, F. W., 375
Fox, D. P., 231
Franck, R. W., 415
Franck-Neumann, M., 407
Franz, J. A., 239
Fráter, G., 627, 628
Frazier, H. X., 229, 252
Fréchet, J. M. J., 26, 657
Freedman, H. H., 566
Freeman, D., 574
Freeman, J. P., 128, 646
Fréhel, D., 438, 441
Freyer, W., 638
Friedman, L., 203
Friedman, N., 166, 203
Friedrich, E. C., 254, 604
Frimer, A., 490
Fritz, H.-G., 317
Fröhlich, K., 306
Fröling, A., 651
Fu, P. P., 658
Fuchs, P. L., 77, 234
Fuchs, R., 279
Füldner, H. U., 375
Fuji, K., 374
Fujii, S., 233
Fujikura, Y., 359, 526
Fujimoto, T. T., 32
Fujinami, T., 96
Fujino, M., 291
Fujisawa, T., 260
Fujishita, T., 662

Fujita, E., 374
Fujita, S., 187
Fujita, T., 415
Fukuda, T., 291
Fukui, K., 436
Fukumoto, K., 283
Fukunaga, K., 555
Fukuyama, T., 611
Fukuzawa, A., 582
Fukuzumi, K., 653
Fuller, G. B., 496
Funamizu, M., 130
Fung, M. K., 469
Funk, M. O., 167
Funk, R. L., 152
Furui, S., 224
Furukawa, J., 451
Furukawa, N., 369
Furukawa, S., 392

Gaffney, B. L., 84, 636
Galante, J. J., 534
Gallagher, G., 453
Galli, C., 479
Galpin, I. J., 425
Ganem, B., 111, 114, 260, 348, 407, 455, 492, 639
Gannon, J. J., 128
Ganschow, S., 35, 36
Gaoni, Y., 675
Garbarino, J. A., 109
Garcia, B. A., 224, 586
Garcia, G. A., 101, 339
Garmaise, D. L., 585
Garneau, F. X., 190
Garnett, J. L., 504
Garratt, P. J., 470
Gaspar, P. P., 577
Gasparrini, F., 162
Gassman, P. G., 91, 118, 119, 120, 248
Gasteiger, J., 24
Geiger, R., 288
Geke, J., 420
Gemal, A. L., 3
Georgoulis, C., 233, 280
Geraghty, M. B., 530
Gerkin, R. M., 103
Gerlach, H., 107, 247, 520, 625
Germain, A., 103
Ghatak, U. R., 70

Ghera, E., 675
Ghosez, L., 123
Giannoli, B., 607
Gibbons, E. G., 634
Giering, W. P., 155, 539
Gil, G., 223
Gillard, B. K., 25
Gillespie, J. S., Jr., 341, 468
Gilman, H., 40
Gilmore, G. N., 621
Gimbarzevsky, B. P., 118
Ginsburg, D., 107
Giordano, C., 560
Girard, C., 521, 523
Girard, P., 671
Girodeau, J. M., 137
Giudicelli, R., 570
Givens, R. S., 554
Gleason, J. G., 76, 118
Glemser, O., 186
Gless, R. D., 2
Godefroi, E. F., 514
Goe, G. L., 436
Goering, H. L., 142
Goh, S. H., 230, 326
Gokhale, P. D., 475
Gold, H., 229
Gold, J., 84
Gold, J. R., 632
Goldberg, G. M., 632
Goldberg, N. N., 339
Golding, B. T., 79
Goldner, H., 547
Goldsmith, B., 534
Goldsmith, D. J., 353
Golfier, M., 514
Gonzy, G., 574
Goodacre, J., 597
Goodbrand, H. B., 667
Goodman, I. J., 9
Goosen, A. J., 675
Gordinier, A., 486
Gore, J., 223, 342
Gorski, R. A., 252
Gould, K. J., 208, 494
Gräf, H.-D., 596
Graf, E., 137, 138
Grammaticakis, M. P., 526
Gramstad, T., 521
Granoth, I., 585

Gras, J. R., 223
Graselli, P., 74
Graves, J. M. H., 258
Gray, C. J., 597
Green, J. W., 360
Green, M. L. H., 48
Greene, A., 536
Greenhorn, J. D., 344
Greenhouse, J., 496
Greenhouse, R., 644
Greenlimb, P. E., 262
Gregor, I. K., 504
Gregson, M., 598
Greijdanus, B., 575
Grenier-Loustalot, M. F., 562
Gribble, G. W., 534, 553
Grieco, C., 327
Grieco, P. A., 236, 238, 266, 267, 359,
    402, 403, 422, 460, 464, 552, 598, 632
Griesbaum, K., 441
Grieshaber, P., 243
Griesinger, A., 564
Griffin, C. E., 244
Griffin, T. S., 543
Grimm, D., 122
Grimm, E. G., 399
Grisdale, E. E., 135, 137
Grob, C. A., 403
Grob, V. D., 676
Gröbel, B.-T., 55, 118, 322
Groen, M. B., 525
Grohman, K., 170, 171
Gromelski, S., Jr., 546
Grosser, J., 171
Groves, J. T., 261, 534
Gruber, L., 645
Grünter, K., 653
Grundon, M. F., 359
Grynkiewicz, G., 280
Grzejszczak, S., 147, 190, 193
Grzonka, Z., 78
Guerrieri, F., 451
Guida, W. C., 564, 565
Gund, T. M., 17
Gundermann, K.-D., 470
Gupta, S. K., 33, 34, 573, 617
Gurria, G. M., 17
Gutmann, V., 564
Guyot, M., 503
Guziec, F. S., Jr., 509, 644

Haak, P., 230
Haber, S. B., 567
Hachey, D. L., 598
Hänssle, P., 373
Haffner, H. E., 534
Hagedorn, I., 268
Hageman, H. J., 507
Hagihara, N., 61
Hagiwara, D., 106
Hagiwara, K., 11, 91
Hahnfeld, J. L., 261
Haigh, F. C., 360
Haines, A. H., 35
Hajeck, M., 518
Hajos, Z. G., 267, 411, 542
Halasz, S. P. v., 186
Hales, H. J., 524
Hall, D. R., 547
Hall, H. K., Jr., 480
Hall, H. T., 21, 104
Hall, L. D., 620
Hall, S. S., 323
Hall, T. W., 563
Hallett, A., 425
Halperin, G., 297
Halpern, B., 38
Halpern, J., 172
Halpern, Y., 24
Halvey, N., 407
Hamajima, R., 510
Hamaoka, T., 71, 73
Hamersma, J. W., 675
Hamlin, K. E., 526
Hamming, M. C., 20
Hamon, D. P. G., 524
Hampel, C. A., 152
Hampel, G., 564
Hampton, K. G., 146
Hanack, M., 479, 620, 621
Hanafusa, T., 244
Hancock, W. S., 149
Hands, D., 467
Hanessian, S., 81
Hannan, W., 504
Hannon, S. J., 640
Hansen, B., 566
Hansen, J. F., 128
Hansen, L. D., 137
Hanson, J. R., 283
Haque, K. E., 657

Hara, H., 316
Hara, S., 222, 403
Hara, Y., 377
Hardegger, E., 408, 409
Harder, R. J., 517
Harding, K. E., 124
Hargis, J. H., 576
Hargrove, R. J., 620
Harmon, A. D., 538
Harmon, R. E., 597
Harney, D. W., 622, 624
Harpp, D. N., 76, 118, 280, 634
Harrington, K. J., 96
Harris, L. S., 283
Harrison, C. R., 208, 453, 494
Harrison, M. J., 316
Hart, A. J., 638
Hart, D. J., 94
Hart, D. W., 179
Hartenstein, J., 514
Hartford, T. W., 649
Hartke, K., 379
Hartley, D., 258
Hartley, F. R., 61
Hartmann, J., 638
Hartwig, I., 390
Haruki, E., 95
Harvey, R. G., 229, 230, 316, 326, 657
Hase, C., 563
Hase, T., 381, 383
Hasegawa, K., 424
Hasenhuettl, G., 492
Hashimoto, H., 194, 216
Haslanger, M. F., 215
Hassner, A., 114, 122, 319
Haszeldine, R. N., 521, 621
Hata, K., 602, 644
Hata, N., 70
Hata, T., 602, 644
Hatcher, A. S., 398
Hathaway, B. J., 133
Hauptmann, H., 82
Hauser, C. R., 84, 636
Haut, S. A., 606
Haveaux, B., 123
Hawley, G. G., 152
Hawthorne, M. F., 607
Hayakawa, S., 430
Hayakawa, Y., 198, 201
Hayama, N., 573

Hayashi, I., 403, 649
Hayashi, M., 596
Hayashi, N., 170
Hayashi, T., 12, 14, 398, 569
Hayes, N. F., 502
Haynes, R. K., 596, 657
Hazum, E., 306, 525
Heaney, H., 279, 280, 524
Heathcock, C. H., 205, 215, 339, 441
Heavner, G. A., 115
Hecht, S. M., 159
Heck, R. F., 61, 157
Hedayatullah, M., 514
Hedgecock, H. C., Jr., 624, 625
Hedgley, E. J., 583
Hegedus, L. S., 93, 105, 361, 418, 419
Heiba, E. I., 127, 356, 357
Heibl, C., 285
Heine, H.-G., 34
Helali, S. E., 409
Helder, R., 501
Helferich, B., 511
Helling, J. F., 596
Helquist, P. M., 361, 417, 419
Henderson, W. A., Jr., 137
Hendrickson, J. B., 36, 43, 44, 225, 648
Hendriks, K. B., 569
Heng, K. K., 215
Hennig, G. R., 269
Henrick, C. A., 163, 165, 224, 586
Hentschel, P., 263
Hepburn, D. R., 220
Herb, S. F., 675
Herman, G., 133
Hernández, E., 162
Herold, A., 22, 23
Herr, R. W., 245
Herriott, A. W., 404, 406
Herrmann, J. L., 312
Herron, D. K., 383
Herscheid, J. D. H., 605
Herten, B. W., 99
Hertenstein, U., 633
Hes, J., 168
Hesbain-Frisque, A. M., 123
Heseltine, D. W., 409
Hess, H. M., 409
Hesse, R. H., 124, 264
Hesson, D., 567
Heuman, P., 534

Heusel, G., 422
Heusser, H., 91
Heyne, T. R., 665
Higgins, R., 436
Higuchi, M., 185
Higuchi, S., 158
Hill, A. E., 364
Hill, K. A., 83
Hill, M. E., 420
Hillard, R. L., III, 154
Hilscher, J.-C., 529
Hine, J., 261
Hingerty, B., 568
Hino, K., 399, 445
Hirama, M., 170
Hirao, K., 453
Hirao, T., 450
Hiroi, K., 238, 266, 267, 313, 316, 403, 464
Hirsekorn, F. J., 15
Hiscock, B. F., 634
Hitzel, E., 653
Hiyama, T., 91, 220, 385
Ho, C.-T., 577
Ho, T.-L., 50, 229, 250, 251, 470, 497, 529, 540, 543, 553, 563
Hoa, K., 504
Hodge, P., 453, 628, 645
Hodgson, K. O., 35
Hoekstra, M. S., 416
Höfle, G. A., 373
Höft, V. E., 35, 36
Högberg, S. A. G., 134, 137
Hoentjen, G., 507
Hoffman, J. A., 307
Hoffman, W., 441
Hoffman, W. H., 645
Hoffmann, H., 243
Hoffmann, H. M. R., 364, 544
Hofmann, A., 2
Hofmann, H. P., 596
Hogeveen, H., 24
Hogrefe, F., 675
Hohler, I., 268
Hoinowski, A. M., 412
Hojo, M., 510
Holah, D. G., 133
Holland, H. L., 201
Hollinshead, J. H., 524
Holmberg, G. H., 415

Holmberg, K., 566
Holmstead, R. L., 604
Holschneider, F., 32
Hoogenboom, B. E., 600
Hoornaert, C., 123
Hooten, K. A., 399
Hooz, J., 645, 665
Hopf, H., 271
Hoppe, B., 570
Hoppe, I., 343
Hopps, H. B., 64
Hori, I., 398, 569
Horiuchi, S., 653
Horiike, M., 145
Horn, H. G., 640
Horn, P., 243
Horrocks, W. D., Jr., 51
Hortmann, A. G., 25
Horwell, D. C., 5, 587
Horwitz, J. P., 221
Hosaka, K., 11
Hosokawa, T., 573
House, H. O., 203, 215, 225, 676
Howarth, B. D., 147
Hoyer, G.-A., 443
Hoyng, C. F., 32
Hoz, T., 589
Hsia, S.-L., 254
Hsu, I. H. S., 228, 229
Hsu, R. Y., 102
Huang, B.-S., 501
Huang, C. T., 248
Huang, S.-P., 283
Hudrik, P. F., 201, 245
Hudson, B., 426
Hudson, D., 425
Hudson, H. R., 220
Hünig, S., 24, 633
Hughes, G. A., 258
Huisgen, R., 17, 24, 612
Hummel, K., 620
Hung, W. M., 4
Hunter, D. H., 160
Huselton, J. K., 137
Husson, H.-P., 361
Hutchings, M. G., 536
Hutchins, R. O., 531, 538
Hutchinson, C. R., 538
Huthmacher, K., 521, 618
Hutmacher, H.-M., 317

Huttemann, T. J., 469
Hyman, H. H., 670

Ibrahim, B. E. D., 618
Ichijima, S., 158
Ichikizaki, I., 174
Igami, M., 552
Ignatiadou-Ragoussis, V., 430
Iguchi, K., 11
Ihara, M., 283
Ikeda, H., 526
Ikeda, K., 492
Ikeda, M., 233
Ikeshima, H., 317, 403, 649
Ikota, N., 193
Ila, H., 284
Imai, S., 244
Imamura, T., 596
Imanaka, T., 450
Imoto, E., 95
Imuta, M., 436
Inaba, S., 633
Inagaki, S., 436
Inagaki, Y., 560
Inamoto, N., 560
Inamoto, Y., 359, 526
Inomata, K., 596
Inoue, I., 70
Inoue, S., 510, 578
Inouye, Y., 145
Ioannou, P. V., 79
Ippen, J., 33, 478, 479, 657
Iratcabal, P., 562
Ireland, R. E., 277, 279
Iriuchijima, S., 6
Isaac, R., 167
Isagawa, K., 601
Ishida, A., 412
Ishigama, T., 232
Ishige, M., 659
Ishihara, H., 596
Ishii, Y., 187
Ishikawa, H., 596
Ishikawa, N., 581
Ishimoto, S., 165, 641
Isobe, M., 265, 267
Itahara, T., 430, 479
Itakura, K., 361, 622
Ito, H., 174
Ito, K., 73

Ito, S., 90, 170
Ito, T., 95
Ito, Y., 129, 141, 450, 517
Itoh, I., 295, 308, 352, 496
Itoh, M., 91, 106, 149, 215
Itohi, K., 187
Ittah, Y., 542
Iwai, K., 464, 662
Iyoda, M., 138
Izatt, R. M., 137
Izawa, T., 308, 596
Izydore, R. A., 253

Jackson, A. H., 419
Jackson, W. R., 450
Jacobs, W. A., 2
Jacobson, R. M., 545
Jacobsson, U., 24
Jacto-Guillarmod, A., 11
Jacquesy, J.-C., 273
Jacquesy, R., 273
Jagur-Grodzinski, J., 651
Jain, T. C., 304
James, K., 569
James, R., 581
Jaouen, G., 126
Jarboe, C. H., 292
Javet, P., 284
Jeanne-Carlier, R., 124
Jeffery, E. A., 624
Jefford, C. W., 434, 436, 455
Jenkins, C. L., 133, 665
Jensen, F. R., 82
Jensen, K. A., 583
Jerina, D. M., 358, 477
Jesthi, P. K., 329
Jetuah, F. K., 5
Johnson, A. W., 456
Johnson, B. F. G., 39, 552
Johnson, C. R., 245, 395, 585
Johnson, D., 195
Johnson, D. W., 615
Johnson, F., 541
Johnson, F. P., 91
Johnson, G., 115
Johnson, G. S., 423
Johnson, J. L., 534
Johnson, M. R., 565
Johnson, R. A., 490
Johnson, S. E., 195

Johnson, W. S., 613
Johnston, D. E., 121
Johnston, D. N., 419
Johnston, K. M., 618
Johnstone, R. A. W., 446, 488
Jonczyk, A., 43
Jones, D. N., 47
Jones, E. R. H., 304, 451, 511, 514
Jones, G., 291
Jones, G. H., 137
Jones, J. F., 618
Jones, N. R., 493
Jones, R. W., 398
Jones, S. R., 317
Jonkers, F. L., 381
Jorgenson, M. J., 368
Jorritsma, H., 24
Joshi, G. C., 675
Joshua, C. P., 34
Joubert, J.-P., 665
Joyce, M. A., 137
Julia, M., 141
Jung, G., 422
Junggren, U., 565
Junjappa, H., 456
Jurczak, J., 280
Jurd, L., 518

Kabalka, G. W., 98, 624, 625
Kablaoui, M. S., 560
Kacher, M., 538
Kacprowicz, A., 601
Kader, A. T., 56
Kagami, M., 37
Kagan, H. B., 23, 269, 671
Kagi, D. A., 19
Kaiser, E. M., 84, 339, 525, 526, 636
Kajfež, F., 474
Kajimoto, T., 430
Kajitani, M., 659
Kakis, F. J., 430
Kalinowski, H.-O., 578
Kalman, J. R., 581
Kaloustian, M. K., 583
Kamata, K., 389, 629
Kamata, S., 192, 582
Kameswaran, V., 660
Kametani, T., 283
Kamiya, T., 91
Kammeyer, C. W., 254

Kanamaru, H., 392
Kanazawa, R., 529
Kandasamy, D., 538
Kaneda, K., 450, 653
Kaneko, C., 586
Kaneko, T., 378
Kanematsu, K., 43, 297
Kano, S., 547
Kantlehner, W., 58, 243
Kaplan, F., 552
Kapoor, V. M., 170
Kappe, T., 96
Kar, S. K., 548
Karady, S., 412
Kariyone, K., 65
Karrer, P., 37
Karstashov, V. R., 232
Kasai, Y., 193
Kashdan, D. S., 91
Kashima, C., 534
Kashitani, T., 138
Kasina, S., 223
Katagiri, N., 361, 622
Kataoka, K., 618
Kato, M., 170, 482
Kato, T., 554
Katritzky, A. R., 618
Katz, J.-J., 172, 208
Katzenellenbogen, J. A., 79, 180
Katzin, M. I., 316
Kaufmann, D., 568
Kautsky, H., 510
Kautzner, B., 179
Kawai, M., 464, 466
Kawamoto, I., 667
Kawano, Y., 534
Kawasaki, M., 70
Kawashima, T., 500
Kawauchi, H., 45
Kawazoe, Y., 556
Kayser, R. H., 617
Keeley, D. E., 319, 320
Keen, G. W., 20, 304
Kees, K., 469
Keii, T., 504, 510
Keinan, E., 441
Keiser, J. E., 9
Kelley, J. A., 114
Kellog, M. S., 455
Kelly, J. E., 617

Kelly, R. B., 181
Kelly, T. R., 383
Kemp, D. S., 32, 194, 195, 254
Kemp, G., 279, 280
Kendall, M. C. R., 573
Kendall, P. M., 105, 204
Kende, A. S., 390, 654
Kenehan, E. F., 114
Kenne, L., 406
Kennedy, E., 353
Kennepohl, G. J. A., 190
Kenner, G. W., 425
Kerber, R. C., 419
Kergomard, A., 358
Kester, M., 472
Keul, H., 441
Keziere, R. T., 530
Khalil, H., 667
Khan, E. A., 90
Kahn, M. N., 628
Khan, N., 514
Khan, S. A., 288
Khanna, J. M., 326
Khanna, V. K., 546
Kharasch, M. S., 232, 450
Khor, T. C., 454
Khoujah, A. M., 597
Kido, F., 530, 662
Kieboom, A. P. G., 149
Kieczykowski, G. R., 312
Kiehlmann, E., 521
Kiessel, M., 24
Kiji, J., 451
Kikuchi, K., 534
Kikugawa, Y., 534
Kim, B., 190
Kim, C.-K., 660
Kim, C. S., 494
Kimoto, S., 69, 70
Kimura, M., 304
Kimura, Y., 601
King, F. D., 57, 58, 185, 652
King, J. F., 190
King, R. W., 469
Kinnick, M. D., 468
Kinoshita, H., 257, 597
Kinoshita, M., 561
Kinzig, C. M., 17
Kirby, G. W., 581
Kirchhoff, R. A., 395

Kirkpatrick, D., 208
Kirsanov, A. V., 186
Kishi, Y., 611
Kishida, Y., 399
Kiso, Y., 51, 59
Kitagawa, K., 618
Kitagawa, Y., 328
Kitahara, T., 372
Kitahara, Y., 170, 554
Kitai, M., 295
Kitatani, K., 91, 385
Kitatsuji, E., 403
Kiyohara, Y., 187
Klamann, D., 263
Klausner, Y. S., 288
Klayman, D. L., 543
Klein, H., 441
Klein, K. P., 611
Klein, S. A., 534
Klein, U., 304
Kleiner, H.-J., 375
Kleveland, K., 558
Klose, T. R., 264, 615
Klumpp, G. W., 143
Knaus, G., 204, 387, 389
Knifton, J. F., 655
Knights, E. F., 64
Knoevenagel, E., 260
Knorr, L., 228
Knoth, W. H., 81
Knox, S. D., 47
Kobayashi, C. S., 642
Kobayashi, J., 152
Kobayashi, M., 165, 641
Kobayashi, S., 291
Kobayashi, T., 554
Kobler, H., 569
Kobylecki, R. J., 147
Koch, P., 509
Kochi, J. K., 57, 133, 259, 624, 665
Kocienski, P. J., 478, 479
Kocór, M., 10
Kodama, M., 90
Kodama, S., 51
Koebernick, W., 545
Köbrich, G., 171
Koehl, W. I., Jr., 356
Koeng, F. R., 675
König, G., 618
Koenig, K. E., 35

König, W., 288
Koga, K., 307
Kogure, T., 653
Koizumi, T., 317, 403, 649
Kolb, M., 55, 322
Kolbah, D., 474
Kolind-Andersen, H., 565
Kollonitsch, J., 561
Kolonko, K. J., 600
Komatsu, M., 158
Komin, J. B., 78
Komiyama, E., 547
Kondo, A., 297
Kondo, K., 436, 462, 509, 610, 662
Kondo, T., 44
Kondratenko, N. V., 621
Kono, H., 665
Konoike, T., 141, 517
Kopecky, K. R., 287, 288
Kopp, R., 479
Koppel, G. A., 468
Koppes, W. M., 646
Kopple, K. D., 307
Koreeda, M., 202
Kornblum, N., 229, 419
Korth, T., 141
Koseley, R. W., Jr., 610
Koster, J. B., 575
Kosugi, H., 464, 466, 662
Kotake, H., 597
Kotsuki, H., 529
Kottwitz, J., 443
Kouwenhoven, A. P., 252
Kovacic, P., 146
Kovitch, G. H., 190
Kowalski, C., 101
Kozar, L. G., 339
Kozarich, J. W., 159
Kozikowski, A. P., 50, 292, 310
Kramer, G. W., 63, 64
Kranz, A., 570
Krapcho, A. P., 91
Kratzl, K., 503
Kraus, G. A., 339
Kraus, J. L., 574
Kraus, W., 596
Krause, J. G., 456
Kricheldorf, H. R., 561
Krief, A., 29, 90, 240, 361, 362, 488
Krieger, J. K., 495
Krishnamurthy, S., 63, 64, 348

Krishna Rao, G. S., 498
Krolikiewicz, K., 640
Krow, G. R., 383
Krüger, W., 511
Kruger, J. A., 675
Ku, A. T., 24, 230
Kucherov, V. F., 520
Kühle, E., 32
Kühling, D., 563
Kugel, W., 58
Kuhlmann, H., 39, 289
Kuhn, D., 133
Kukhar, V. P., 170
Kulik, S., 419
Kumada, M., 51, 59
Kumagai, M., 653
Kunerth, D. C., 274, 279
Kung, F. E., 456
Kunieda, N., 254, 561
Kunz, R. A., 662
Kuo, Y.-N., 114
Kupchan, S. M., 660
Kuroda, S., 170
Kuroda, Y., 187
Kuromizu, K., 445
Kurosawa, T., 259
Kurozumi, S., 165, 238, 464, 641
Kursanov, D. N., 616
Kurtz, D. W., 409
Kurzer, F., 32, 250
Kutney, J. P., 97, 496, 644
Kuwajima, I., 44
Kwan, T., 141
Kwant, P. W., 24
Kwiatkowski, G. T., 339
Kwon, S., 601


Laasch, P., 250
Labaw, C. S., 494
Labinger, J. A., 48, 179
Labovitz, J. N., 436, 615
Lahav, M., 166
Lai, R., 172
Laidler, D. A., 137
Lajšić, S. P., 201
Lal, B., 645
Lallemand, J.-Y., 141
Lalloz, L., 526
LaMattina, J. L., 446
Lambert, J. B., 91, 675

Lamby, E. J., 486
Lamm, B., 41, 564, 565, 567
Lammek, B., 78
Lammert, S. R., 281
Lamon, R. W., 250
Landick, R. C., 466
Landini, D., 175, 272, 405, 406, 543
Landor, P. D., 144
Landor, S. R., 144, 147
Lane, C. F., 64, 208, 259, 349, 538
Langford, P. B., 261
Langford, R. B., 232
Langhals, H., 81, 82
Langler, R. F., 589
Lantos, I., 363
Larchevêque, M., 279, 323, 333, 352
Larock, R. C., 573
Larson, G. L., 162
Larson, W. D., 15
Lasch, I., 375
Laub, R. J., 208
Laurenco, C., 191
Lavallee, P., 81
Lawesson, S.-O., 279, 558
Lawson, D. F., 628
Lecher, H., 32
Leclerc, G., 458
Leclerc, J., 657
Leclercq, D., 279
Ledon, H., 631
Lee, A. S. K., 555
Lee, D. G., 506
Lee, D. L., 2
Lee, D. P., 38, 645
Lee, G. R., 492, 493
Lee, J.-K., 383
Lee, L.-F., 546
Lee, M. K. T., 24
Lee, S. J., 390
Lee, T. B. K., 412
Lee, V., 360, 503
Lee, W., 160
Leffler, A. J., 414
LeGoff, E., 198
Legris, C., 354
Legzdins, P., 650
Lehn, J.-M., 137, 138
Lehnert, W., 596
Leichter, L. M., 243
Leighton, P., 144

Leiserowitz, L., 166
Leitmann, O., 564
Lemal, D. M., 454
Le Merrer, Y., 610
Lemieux, R. U., 44, 568, 569
Lentz, C. M., 343
Leong, B. K. J., 133
Leovey, E. M. K., 19
L'Eplattenier, F. A., 417
Leppert, E., 561
Leresche, J.-P., 598
Lester, R., 547
Letsinger, R. L., 115, 254, 415
Leung, H. W., 158
Leung, K. K., 466
Lever, O. W., Jr., 205, 373
Levine, R., 339
Levy, A. B., 325
Lewellyn, M. E., 215
Lewis, J., 39, 552
Ley, S. V., 568
Liaaen-Jensen, S., 542
Liard, J.-L., 534
Liberatore, F., 534
Liebeskind, L., 390, 406, 654
Liedhegener, A., 597
Liehr, W., 640
Liepa, A. J., 660
Lin, C.-H., 670
Lin, H. C., 24, 103, 273
Lin, J. J., 327
Lin, J. W.-P., 577
Lindberg, B., 406
Lindlar, H., 319
Lindsey, R. V., Jr., 81
Linn, C. P., 577
Linstrumelle, G., 495, 631, 666
Lion, C., 165, 215
Liotta, C. L., 136, 137
Lipshutz, B. H., 428, 430
Lipton, M. F., 600
Liso, G., 534
Liu, H. J., 253
Liu, K. H., 55
Lo, S. M., 105
Locke, J. M., 374
Lockwood, P. A., 288
Loeber, D. E., 514
Loev, B., 363
Logue, M. W., 84, 148

Loim, N. M., 616
Long, M. A., 504
Longeray, R., 557
Lönngren, J., 406
Lonsky, W., 503
Loomis, G. L., 215
Loozen, H. J. J., 514
Lopez, L., 71, 73, 375
Lord, P. D., 534
Lorne, R., 666
Loubinoux, B., 70
Loudon, G. M., 233
Louis, J.-M., 514
Louw, R., 256
Lucas, C. R., 48
Lucchini, V., 375
Luche, J. L., 23, 215, 269
Luczak, J., 229, 295
Luddy, F. E., 675
Ludwikow, M., 160
Lüttke, W., 646
Luk, K.-C., 286
Lukas, J. H., 252
Lumma, W. C., 412
Lumma, W. C., Jr., 577
Lunsford, W. B., 115
Lusch, M. J., 377
Lusinchi, X., 73
Lyle, R. E., 446
Lynd, R. A., 198, 201
Lyon, G. D., 534
Lyons, D. E., 254
Lyons, J. E., 655
Lythgoe, B., 341

Ma, K. W., 534
Macaione, D. P., 142
McCall, J. M., 161, 588
McCandlish, C. S., Jr., 676
McCann, E. L., III, 414
McCarney, C. C., 675
McCloskey, J. E., 304
McCollum, G. J., 577
McCormack, W. B., 393
McCormick, J. P., 354
McCurry, P., 364
McDermott, J. X., 124
McDonald, E., 481, 629
Macdonald, T. L., 386, 607
McGhie, J. F., 5, 587

McGrath, J. P., 111, 114
Machida, Y., 319, 490
Machinek, R., 121, 646
McIntosh, J. M., 667
Mack, M. P., 121
McKay, A. F., 585
McKeon, J. E., 573
McKervey, M. A., 121, 473
McKillop, A., 144, 147, 205, 360, 581
MacLean, D. B., 201
McLean, S., 253
McLoughlin, B. J., 258
MacMillan, J. H., 632
Mc Murry, J. E., 255, 447, 450, 587, 588, 589
MacNicol, D. D., 550
McNutt, R. W., 383
Macomber, R. S., 229
McPhillips, J. J., 9
Madhavarao, M., 539
Madigan, D. M., 672, 675
Maeda, K., 573
Maestrone, T., 110
Maggerramov, A. M., 232
Magidman, P., 675
Magnus, P. D., 100, 109, 187, 241, 583, 657
Mah, T., 348
Mahajan, J. R., 124, 441, 542
Mai, R. S., 86, 90
Maitte, M. P., 514
Majetich, G., 359
Majumdar, S. P., 153
Maki, Y., 534
Makino, S., 198
Mąkosza, M., 41–43, 160, 601
Malatesta, V., 628
Malaval, A., 441
Maldonado, L., 530
Malek, J., 518, 529
Maletina, I. I., 301
Mali, R. S., 86, 90
Mallon, C. B., 135, 137
Mami, I. S., 580, 581
Manas, A.-R. B., 651
Mander, L. N., 614, 615
Mandolini, L., 479
Mane, R. B., 498
Manescalchi, F., 304
Mangoni, L., 296, 586

Mangum, M. G., 231, 620
Manhas, M. S., 645
Maniwa, K., 6
Manners, G. D., 518
Manning, R. A., 500
Manske, R. F. H., 201
Marans, N. S., 632
Marathe, K. G., 583
Marburg, S., 561
Marchese, L., 375
Marchini, P., 534
Marchiori, F., 419
Marcuzzi, F., 114
Marecek, J. F., 8, 219
Marica, E., 279
Marino, J. P., 121, 209, 331, 378, 466,
  556
Markovski, L. N., 184, 186
Marmor, R. S., 189
Marrero, R., 114
Marschall, H., 73
Marshall, G. R., 149
Marshall, J. A., 215, 454
Marsmann, H. C., 640
Martin, J. C., 239
Martin, M. M., 166
Martinez, J., 470
Martins, F. J. C., 675
Marx, J. L., 102
Masaki, Y., 267, 422
Masamune, S., 56, 192, 582
Masler, W. F., 416
Masse, G. M., 667
Massiot, G., 361
Masson, S., 300
Massuda, D., 100
Mastagli, P., 596
Mastre, T. A., 224, 586
Masuda, A., 55
Masuda, R., 510
Masuda, T., 369
Masuda, Y., 259, 510
Masure, D., 622
Mateescu, G., 279
Matheson, M. S., 588, 670
Mathey, F., 393, 412
Matsuda, H., 399
Matsuda, M., 377
Matsuda, T., 59
Matsugo, S., 436

Matsuki, Y., 90
Matsumoto, H., 655
Matsumoto, K., 158
Matsumoto, M., 432, 436, 610
Matsumoto, S., 185
Matsumoto, Y., 194
Matsumura, N., 95
Matsuura, T., 430, 436, 479
Matteson, D. S., 329
Matthews, W. S., 577
Matuszko, A., 58
Maurin, R., 223
Mauz, O, 202
Mauzerall, D., 37
May, K. D., 546
May, L. M., 124
Mayer, H., 526, 527
Mayer, R., 583
Maynez, S. R., 9
Mayo, F. R., 450
Mayr, H., 17
Mazur, Y., 441, 574
Mazza, S., 364
Meakins, G. D., 304, 514
Medici, A., 470
Meerwein, H., 250
Mehta, A. M., 170
Mehta, G., 287, 297
Meienhofer, J., 445
Meijer, J., 144, 146, 270
Meinwald, J., 150
Meisters, A., 624
Mélin, J., 22, 23
Mellor, J. M., 317
Melvin, L. S., Jr., 215, 247
Mende, U., 443
Menger, F. M., 70
Menz, F., 597
Menzies, I. D., 657
Mercer, G. D., 504
Merkel, C., 267
Merlini, L., 517
Merrer, Y. L., 331
Merrifield, R. B., 303, 304, 307
Merrill, R. E., 91, 570
Mersch, R., 250
Mertes, M. P., 168
Merz, A., 43, 665
Mesbergen, W. B., 209
Meth-Cohn, O., 73

Métras, F., 562
Metzger, J., 228
Meunier, B., 179
Meyer, A., 126
Meyer, W. L., 500
Meyers, A. I., 20, 204, 368, 387, 389, 403,
    404, 585, 629, 630
Meyers, C. Y., 577
Miarka, S. V., 291
Michel, R. E., 419
Michener, E., 383
Micović, I. V., 201
Middleton, W. J., 184
Midgley, J. M., 467
Midland, M. M., 284, 325
Midorikawa, H., 12, 14, 398, 569
Midura, W., 191
Migliorese, K. G., 670
Mihelich, E. D., 20, 368, 387, 389, 629
Mijs, W. J., 507
Mikolajczyk, M., 191, 193, 229, 295
Miles, D. H., 501
Miller, A. D., 339
Miller, C. H., 369
Miller, D. C., 620
Miller, J. A., 259, 308, 676
Miller, L. L., 418, 419
Miller, M. J., 233
Miller, R. B., 490, 492
Millon, J., 666
Mimoun, H., 428, 430
Minamikawa, J., 233
Minato, A., 59
Minisci, F., 21
Mioduski, J., 150
Mishima, T., 220, 385
Misiti, D., 162, 607
Misumi, S., 500
Mitchell, A. R., 307
Mitchell, R. H., 90, 235
Mitchell, R. W., 650
Mitsudo, T., 485, 486, 552
Miura, C., 201
Miura, S., 165, 641
Miwa, T., 403
Miyake, K., 279
Miyake, N., 51, 59
Miyano, S., 194
Miyashita, M., 552
Miyaura, N., 149, 215, 295, 308

Miyoshi, M., 158, 224
Mizoguchi, T., 307
Mizuno, K., 534
Mizuno, Y., 152, 472
Modena, G., 375
Moder, T. I., 82
Mödhammer, U., 271
Moffatt, J. G., 229
Mole, T., 624
Molines, H., 263
Molko, D., 503
Mollan, R. C., 254
Monaco, P., 586
Mondron, P. J., 646
Montanari, F., 175, 191
Monteiro, H. J., 3, 374
Moody, R. J., 329
Moore, G. L., 148
Moore, H. W., 78
Moore, J. A., 467, 617
Moorthy, S. N., 523
Moracci, F. M., 534
Moran, T. A., 341
Moreau, C., 438, 441
Moreno-Mañas, M., 417
Morgan, K. D., 656
Morgan, P. H., 538
Mori, F., 610, 662
Mori, K., 216
Moriarty, R. M., 477
Morikawa, A., 613
Morita, S., 556
Moritani, I., 64, 377, 392, 443, 445, 573
Morrell, D. G., 57
Morris, D. G., 515
Morris, D. L., 475
Morris, S. G., 675
Morrison, J. D., 416
Morrow, C. J., 2, 141, 229, 416
Morton, T. H., 589
Moss, R. A., 135, 137
Mossman, A. B., 36
Mostowicz, D., 457
Moulines, J., 279
Moutardier, G., 596
Moy, D., 573
Müller, B., 186, 419
Müller, E., 653
Müller, N., 441
Müller, P., 19

Müller, P. M., 286
Müller, W. E., 304, 514
Müller, W. M., 511
Münzenmaier, W., 47
Muetterties, E. L., 15
Mukaiyama, T., 53, 118, 308, 376, 412, 596, 613
Mullins, M., 628
Mumford, C., 288
Munavu, R., 404
Muneyuki, R., 502
Mura, A. J., Jr., 133, 359, 606
Murahashi, S.-I., 64, 392, 443–445, 573
Murai, S., 73, 224, 424
Muraki, M., 53
Muramatsu, S., 667
Murray, A. M., 515
Murray, R. W., 577
Murray, W. P., 494
Mushika, Y., 207
Musser, J. H., 255
Musso, H., 317
Muth, R., 467
Myatt, H. L., 64
Mychajlowskij, W., 482

Näf, F., 215, 478, 479, 675
Nagai, Y., 633, 655
Nagakura, I., 344, 648
Nagamatsu, T., 547
Nagasampagi, B. A., 3
Nagasawa, H. T., 174
Nagasawa, K., 368
Nagata, T., 296
Nagel, A., 411
Nair, P. G., 34
Najer, H., 570
Nakagawa, I., 602
Nakagawa, M., 138
Nakai, T., 14, 365, 540
Nakamura, E., 44
Nakamura, H., 65
Nakamura, N., 56
Nakanishi, A., 585
Nakanishi, K., 202
Nakanishi, N., 585
Nakano, M., 618
Nakano, S., 374
Nakano, T., 158
Nakatsugawa, K., 633, 653

Nakatsuka, T., 51
Nakayama, J., 96
Nakayama, K., 129
Nambudiry, M. E. N., 341
Naoi, Y., 158
Narang, S. A., 361, 622
Narasaka, K., 596
Narasimhan, N. S., 86, 90
Narayanan, C. R., 3
Naruse, M., 215, 295
Nash, R. D., 490, 492
Nasutavicus, W. A., 541
Nathan, E. C., 479
Natsukawa, K., 509
Nayak, U. R., 475
Neckers, D. C., 19
Negishi, E., 87, 91, 208, 399, 492, 570, 601
Neilson, T., 605
Neiman, M., 114
Nelke, J. M., 141
Nelson, T. R., 101
Nelsen, P., 554
Nelson, V., 146
Nematollahi, J., 223
Nenitzescu, C. D., 279
Neoh, S. B., 470
Nerdel, F., 263
Nesbitt, B. F., 547
Neumann, S. M., 259
Neumeyer, J. L., 477
Newkome, G. R., 244, 479
Newman, M. S., 4, 360, 406, 503, 545, 546
Nicholas, K. M., 155
Nicholls, B., 104, 126
Nickel, D. L., 40
Nicolaou, K. C., 246, 247, 319, 490
Nidy, E. G., 490
Niedballa, U., 640
Niederprüm, H., 273
Nielsen, S. F., 137
Nierth, A., 620
Nieves, I., 162
Nihonyanagi, M., 653
Nikaido, T., 655
Nikishin, G. I., 316
Niles, G. P., 98
Nilsen, B. P., 517
Nilsson, A., 23

Nishi, A., 158
Nishiguchi, T., 653
Nishimura, O., 291
Nishinaga, A., 430, 479
Nishio, T., 161
Niznik, G. E., 643
Noguez, J. A., 267, 403
Nojima, H., 106
Nojima, M., 296
Nokami, J., 257, 561
Noland, W. E., 545
Nomura, M., 17
Norman, R. O. C., 70, 316, 443
Normant, H., 279, 333
Normant, J., 143
Normant, J. F., 143, 189, 270, 331, 577,
    622, 664
Norris, F. A., 555
Norton, J. R., 450
Norton, S. T., 472
Notani, J., 106
Noth, H., 576
Noyori, R., 45, 170, 198, 201
Nozaki, H., 91, 163, 183, 220, 222, 295,
    328, 385
Nozoe, S., 675
Nützel, K., 596
Nunes, B. J., 441
Nunn, M. J., 259
Nyberg, K., 52
Nybratten, G., 542

Oae, S., 369
Obayashi, M., 291
O'Brien, D. H., 638
O'Connell, E. J., Jr., 244
Oertle, K., 625
Ogawa, M., 316
Ogibin, Y. N., 316
Ogura, K., 391, 392, 446
Ohara, M., 152
Ohkata, K., 244
Ohki, E., 675
Ohloff, G., 478, 675
Ohnishi, Y., 37
Ohno, A., 37, 573
Ohno, M., 145, 215
Ohnuma, T., 312
Ohoi, F., 144
Ohshiro, Y., 158

Ohta, H., 260
Ohta, S., 70
Ohtomi, M., 138
Oikawa, Y., 258
Oine, T., 70
Oishi, T., 312
Ojima, I., 633, 653
Oka, K., 403, 573
Oka, S., 69
Okada, J., 659
Okamoto, H., 534
Okamoto, T., 573, 617, 618
Okamura, Y., 43, 500
Okano, M., 23, 141
Okawara, M., 14, 34, 365, 540
Okazaki, H., 219
Okazaki, R., 8, 560
Okita, T., 198
Oklobdžija, M., 474
Okogun, J. I., 139
Oku, M., 24
Okura, I., 504
Okwute, K. S., 139
Olah, G. A., 24, 103, 105, 109, 286
Olah, J. A., 105, 273, 474
Oldenziel, O. H., 600
Oliva, G., 430
Oliver, J. E., 573
Ollinger, J., 238
Olsen, D. O., 1
Olsen, R. K., 193
Olson, G. L., 656
Olsson, L.-I., 148
Omote, Y., 161
Onaka, M., 596
Ono, I., 23, 70
Ono, Y., 510
Onoe, A., 141
Openshaw, H. T., 43
Opitz, G., 564
Orda, V. V., 301
Osawa, E., 17
Osborn, J. A., 105
Oshima, A., 430
Oshima, K., 328
Osuch, C., 339
Otsubo, R., 90
Otsuji, Y., 95
Otsuka, S., 49
Ottenheijm, H. C., 172

Oudman, D., 525
Ourisson, G., 407
Overman, L. E., 605
Owen, G. R., 605
Owens, R. M., 406
Oyler, A. R., 366, 496
Ozaki, K., 317, 403, 649
Ozari, Y., 351

Pac, C., 534
Pacevitz, H. A., 40
Pachter, I. J., 629
Padwa, A., 137
Pajetta, P., 419
Palmer, W. J., 52, 564
Palmertz, I., 564, 567
Palmisano, G., 203
Palumbo, G., 586
Pan, H.-L., 17, 578
Pande, L. M., 675
Pandey, P. N., 287, 297
Panetta, C. A., 607
Panunzio, M., 306, 486
Papadakis, I., 514
Pappas, J. J., 103
Paquette, L. A., 24, 68, 70, 279, 352, 518, 519, 568
Paraskewas, S., 450
Parham, W. E., 283
Parish, E. J., 501
Parker, K. A., 610, 634
Parker, T., 284
Parker, V. D., 23
Parlman, R. M., 349, 552
Parnes, Z. N., 616
Parrilli, M., 296
Parrish, D. R., 267, 411
Parshall, G. W., 48
Pashinnik, V. E., 184, 186
Pasternak, V. I., 170
Patchornik, A., 319
Patel, K. M., 600
Patrick, D. W., 642
Patrick, T. B., 190
Patronik, V. A., 252
Pauling, H., 307, 655
Paulsen, H., 545
Pavey, D. F., 147
Pawlak, M., 77
Pechet, M. M., 124, 263, 598

Pedersen, C., 583
Pedersen, E. B., 279
Peet, J. H. J., 454
Pelegrina, D. R., 114
Pellé, G., 26
Pellegata, R., 203
Pelletier, S. W., 200, 201
Pellicciari, R., 631
Pelter, A., 208, 295, 494, 536, 554
Penque, R., 414
Peppard, D. J., 419
Periasamy, M. P., 642, 643
Perkins, L. M., 561
Perregaard, J., 558
Perriot, P., 189, 331
Perrotti, E., 509
Perry, D. H., 675
Peruzzotti, G., 201
Pesaro, M., 215
Pesce, G., 375
Pesnelle, P., 419
Peter, H., 186
Peters, J. W., 241, 487
Peterson, D., 201, 245
Peterson, J. R., 496
Peterson, P. E., 381
Petragnani, N., 339
Petrissans, J., 562
Petrova, J., 189, 191
Petrzilka, M., 76
Pettee, J. M., 234
Pettit, R., 149, 150
Petty, J. D., 339
Pfleger, H., 510
Pfleiderer, W., 547
Philip, G., 19
Philips, K. D., 221
Phillips, C., 416
Phillips, L. R., 151
Picard, J. P., 323
Picard, P., 279
Picker, D., 404−406
Picker, D. H., 19
Pickles, G. M., 581
Picot, A., 73
Piechucki, C., 567
Pierce, J. B., 91
Piers, E., 344, 376, 530, 648
Pierson, A. K., 283
Pietra, F., 27

Pigott, H. D., 238, 400
Pike, D., 141
Pilkiewicz, F. G., 137
Pincock, R. E. 515
Pines, S. H., 19
Pinhey, J. T., 581
Pinnick, H. W., 114
Pirisi, F. M., 175
Pirkle, W. H., 415
Pittman, C. U., Jr., 59
Pitts, J. N., Jr., 487
Pitzele, B. S., 348
Pizey, J. S., 357
Plattner, E., 284
Plattner, E., 284
Plattner, J. T., 2
Plenchette, A., 279
Pobiner, H., 430
Podesta, J. C., 581
Pogonowski, C. S., 236, 238, 460
Poje, M., 486
Pollack, R. M., 617
Pollak, A., 300, 670
Pollini, G. P., 497, 560, 566
Polovsky, S. B., 415
Ponsford, R. J., 597
Popovitz-Biro, R., 166
Poppi, R. G., 547
Porta, O., 21
Porter, N. A., 166, 167
Posner, G. H., 16, 17, 211, 215, 343, 465
Pospišil, J., 656
Possel, O., 567
Postlethwaite, J. D., 133
Potier, P., 361
Poulton, G. A., 99, 583
Povall, T. J., 446
Powell, D. L., 24
Powell, J. R., 32
Prager, R. H., 350
Pragnell, J., 304, 514
Prakasa Rao, A. S. C., 475
Prakash, S. R., 3
Prasad, K. S. N., 411
Preckel, M., 243
Prestwich, G. D., 613
Previtera, L., 586
Price, P., 201
Prinzbach, H., 584
Prior, M., 477

Profitt, J. A., 352
Prokipcak, J. M., 456
Pross, A., 254
Prossel, G., 122
Pugin, A., 417
Pummerer, R., 308
Puthenpurayil, J., 354

Qazi, T. U., 285
Quarterman, L. A., 670
Quici, S., 406, 543
Quillinan, A. J., 90, 424
Quinney, J. C., 100

Raber, D. J., 564, 565
Rabinovitz, M., 670
Radau, M., 379
Rademacher, D. R., 554
Radhakrishnan, J., 546
Radics, L., 645
Radlick, P., 91
Radüchel, B., 443
Radunz, H., 141
Rajagopalan, P., 556
Rakowski, M. C., 15
Ramage, R., 97, 425
Rama Rao, A. V., 296
Ramirez, F., 8, 218, 219, 281
Ranganathan, S., 548
Rangarajan, T., 20
Ransom, C. J., 229, 605
Rao, Y. S., 6, 228, 283
Raper, R., 411
Raphalen, A., 574
Rapoport, H., 1, 2, 141, 227, 229
Rasmussen, J. K., 122
Ratcliffe, A. H., 97
Rathke, M. W., 259, 675
Rauchfuss, T. B., 504
Rauchschwalbe, G., 262
Rauckman, E. J., 456
Raudenbusch, W., 403
Raulins, N. R., 127
Rausch, M. D., 27–28
Rautenstrauch, V., 323
Reap, J. J., 403
Rebek, J., Jr., 36, 254
Rebsdat, S., 243
Reeder, R. A., 137
Reese, C. B., 229, 367, 519, 605

Reeve, W., 649
Regen, S. L., 38, 645
Regitz, M., 597
Reho, A., 534
Reich, H. J., 235, 335, 338, 339, 459, 460, 548, 549
Reich, I. L., 235
Reifschneider, W., 465
Reindel, W., 308
Reingold, I. D., 573
Relles, H. M., 647
Rempel, G. L., 426, 650
René, L., 498
Renga, J. M., 235, 460
Rengaraju, S., 283
Renge, T., 73
Renick, K. J., 307
Rens, M., 123
Reusch, W., 600
Reuss, R. H., 114
Rhee, H. K., 70
Rhee, J. U., 70
Rhoads, S. J., 127
Riccieri, F. M., 541
Richards, R., 143
Richardson, G., 645
Rickborn, B., 103, 565
Rieke, R. D., 675
Rigassi, N., 409
Rimerman, R., 552
Rindone, B., 355, 518
Ring, H., 306
Rivetti, F., 375
Rizzardo, E., 612
Robbins, M. D., 548
Roberts, B. W., 552
Roberts, D. C., 194, 195
Roberts, F. E., 412
Roberts, S. M., 157
Robertson, D. A., 25
Robey, R. L., 546
Robins, M. J., 555
Robinson, L., 249
Robinson, M., 5, 587
Rocchi, R., 419
Rockett, B. W., 454
Rodé-Gowal, H., 229
Rodewald, P. G., 356
Rodrigo, R. G. A., 201
Rodrigues, R., 339

Röhle, G., 511
Roessler, P., 228
Rogers, D. Z., 17
Rogers, H. R., 124
Rogić, M. M., 259, 611
Rogozhin, S. V., 634
Roitburd, G. V., 520
Rolla, F., 272, 406, 543
Romanet, R. F., 312
Romano, L. J., 414
Rona, R. J., 245
Ronald, R. C., 109
Ronen-Braunstein, I., 612
Ronlán, A., 23
Rooney, J. J., 121, 473
Root, W. G., 198
Rosan, A., 539
Rosen, G. M., 456
Rosen, P., 430
Rosenblum, M., 155, 539
Rosenfeld, M. N., 241
Rosenthal, I., 490
Rosini, G., 76, 470, 534
Ross, J. A., 454
Rossbach, F., 379
Roundhill, D. M., 504
Rouot, B., 458
Roussel, A., 455
Rowe, K., 536
Royer, R., 498
Rozantzev, E. G., 114
Rua, L., 538
Rubenstein, M., 319
Rubottom, G. M., 114
Ruden, R. A., 84, 187, 322, 635, 636
Ruddick, J. P., 650
Rücchardt, C., 81, 82
Rueppel, M. L., 1, 2
Rüttimann, A., 107, 222
Ruff, J. K., 607
Runquist, A. W., 17
Ruppert, W., 187
Russell, C. R., 638
Russell, G. A., 628
Russell, K. E., 419, 470
Russell, T. W., 446
Ruston, S., 341
Ruzicka, L., 87, 91
Ryu, 1, 73

Sachdev, H. S., 232, 233
Sacks, C. E., 234
Sadovaja, N. K., 232
Saegusa, T., 129, 130, 137, 141, 145, 451, 517
Saenger, W., 568
Saha, C. R., 45
Saidi, M. R., 539
Saigo, K., 118, 308, 376, 613
Saika, D., 115
Saito, E., 460
Saito, I., 241, 432, 436
Saito, T., 453
Sakai, I., 430
Sakai, K., 158
Sakai, M., 198
Sakai, S., 96, 187
Sakakibara, Y., 573
Sakata, Y., 500
Sakurai, H., 534
Salbaum, H., 95
Salomon, R. G., 201, 624, 665
Salomon, M. F., 665
Salzmann, T. N., 218, 238
Sambur, V. P., 621
Sanderson, J. R., 577
San Filippo, J., Jr., 413, 414, 490
Sankaran, V., 546
Sankarappa, S. K., 546
Sano, K., 187
Sano, S., 158
Santamello, E., 518
Santosusso, T. M., 228
Santucci, E., 541
Sanyal, B., 70
Sasaki, T., 41, 43, 215, 297
Sasson, I., 436
Sasson, Y., 426, 655
Sato, T., 500, 596
Satzinger, G., 514
Saucy, G., 656
Sauer, G., 411
Sauer, J. D., 244
Sauter, H., 773
Sauvage, J.-P., 137
Sauvêtre, R., 622
Savignac, P., 189, 191, 279
Savoia, D., 43
Sawada, H., 220
Saward, C. J., 154

Sawaya, H. S., 215
Sayed, Y. A., 283
Sayer, T. S. B., 206
Scarponi, U., 110
Schaap, A. P., 469
Schadenberg, H., 525
Schamp, N., 118
Scharf, H.-D., 172
Schaub, F., 165
Schegolev, A. A., 520
Scheinbaum, M. L., 629
Scheinmann, F., 90, 203, 424
Scheithauer, S., 583
Schenck, G. O., 586
Schill, G., 267
Schilling, W., 582
Schlessinger, R. H., 73, 253, 311, 312
Schleyer, P. v. R., 17
Schlosser, M., 175, 262, 511, 638
Schluenz, R. W., 647
Schmand, H. L. D., 170
Schmidt, E. K. G., 552
Schmitt, G., 201
Schmitz, R. F., 143
Schneider, P., 186, 665
Schnur, R. C., 378, 379
Schöllkopf, U., 187, 343, 373
Scholl, R., 525
Schranzer, G., 124
Schreckenberg, M., 536
Schreurs, H., 270
Schriesheim, A., 430
Schröder, G., 548
Schroeder, L. R., 360
Schröder, R., 187
Schubert, H. W., 564
Schuun, R. A., 570
Schwartz, A., 121, 556
Schwartz, J., 48, 179, 451
Schwartz, M. A., 580, 581
Schwartzman, S. M., 225, 648
Schweizer, E. E., 494
Scolastico, C., 355, 518
Scott, A. I., 435, 436
Scott, H., 570
Scouten, C. G., 64
Scullard, P. W., 400, 403
Seebach, D., 29, 55, 84, 90, 118, 133, 238, 248, 267, 322, 339, 493, 578, 636, 651
Seer, C., 525

Segawa, T., 618
Segnitz, A., 653
Seki, K., 312
Seki, Y., 73
Sekine, M., 644
Sekiya, A., 581
Sekiya, M., 73
Selig, H., 670
Selikson, S. J., 430
Selve, C., 35, 280
Semmelhack, M. F., 28, 104, 361
Sen, D., 45
Sendoda, Y., 510
Sepiol, J., 553
Servi, S., 625
Seto, S., 534
Setoyama, O., 453
Sette, J., 570
Setton, R., 23, 269
Severin, T., 500
Seybold, G., 285
Seyferth, D., 89, 91, 189, 596,
    638
Seyler, R. C., 450
Shah, S. K., 335, 338
Shahak, I., 216, 542, 644
Shaikh, S., 456
Shamblee, D. A., 468
Shani, A., 203
Shanklin, J. R., 395, 464
Shannon, P. V. B., 419
Shapiro, R. H., 600
Sharkey, W. H., 81
Sharma, A. K., 230
Sharma, G. K., 453
Sharp, J. T., 600
Sharpless, K. B., 127, 235, 420, 487, 589,
    642
Sharts, C. M., 146
Shaw, A., 519
Shaw, J. E., 274, 279
Shaw, M. J., 670
Sheats, J. R., 105
Sheehan, J. C., 504
Shen, T. Y., 451
Shenton, F. L., 675
Shenton, K. E., 451
Shepard, K. L., 348
Sheppard, R. C., 425
Shibasaki, M., 490
Shibuya, S., 547

Shih, Y.-S., 475
Shim, S. C., 485, 486
Shimada, E., 376
Shimamura, T., 445
Shimazu, H., 432, 436
Shimizu, Y., 102, 401, 403
Shinagawa, S., 291
Shiner, C. S., 490
Shioiri, T., 193
Shiono, H., 14, 365, 540
Shiono, M., 596
Shiota, M., 659
Shiue, C.-Y., 288
Shoaf, C. J., 342
Shoenberg, A., 61
Shoji, Y., 170
Shu, J. S., 504
Shudo, K., 618
Shutt, R., 422
Shvo, Y., 306, 625
Sidani, A. R., 123
Siddall, J. B., 165, 258
Siegel, S., 195
Sih, C. J., 201
Siirala-Hansen, K., 361
Šilhavý, P., 518
Silveira, A., Jr., 399
Silvestri, M., 587, 588, 589
Sim, S. K., 160
Simchen, G., 243, 569
Simonneaux, G., 126
Singer, B., 43
Singh, G., 217, 392
Singh, R. K., 43, 230
Sinoway, L., 465
Sivanandaiah, K. M., 288
Skattebøl, L., 558
Skell, P. S., 190
Skotnicki, J., 534
Skuballa, W., 443
Skulski, M., 585
Sletzinger, M., 412
Sliam, E., 279
Small, V. R., Jr., 260
Smillie, R. D., 530
Smissman, E. E., 646
Smit, W. A., 520
Smith, A. B., III, 67, 70, 142
Smith, D. G., 393
Smith, D. J. H., 393
Smith, H., 258

Smith, H. L., 317
Smith, J. G., 416
Smith, J. R. L., 70
Smith, K., 295, 536
Smith, L. R., 36, 59, 634
Smith, R. A., 76, 280, 650
Smith, R. A. J., 215
Smith, R. V., 9
Smith, W. T., Jr., 40
Snider, B. B., 300
Soai, K., 596
Soleiman, M., 665
Solladié, G., 162
Solodar, J., 569
Sommer, L. H., 84, 632
Sondheimer, F., 138
Sonnet, P. E., 279, 647
Sonoda, N., 73, 224, 424, 509
Sonogashira, K., 61
Sood, A., 70
Sood, H. R., 600
Sood, R., 201
Sood, V. K., 597
Sorace, R., 467
Soucy, M., 376
Soulen, R. L., 553
Southard, G. L., 234
Sowa, J. R., 545
Sowerby, R. L., 463
Sowinski, A. F., 414
Spahic, B., 175
Spangler, C. W., 649
Sparrow, J. T., 304
Speck, D. H., 137
Spencer, A., 426, 650
Spencer, T. A., 573
Speziale, A. J., 36, 252, 634
Spille, J., 250
Spitzer, U. A., 506
Springer, J. M., 546
Spyroudis, S., 301
Squires, T. G., 676
Stadler, P. A., 560
Stafforst, D., 187
Staires, S. K., 479
Staklis, A., 317
Stam, M. F., 249
Stang, P. J., 231, 620
Stanton, J. L., 32
Stare, F. J., 37

Starks, C. M., 405, 406
Starratt, A. N., 124
Staunton, J., 9
Stein, R. G., 400
Steiner, K., 409
Steinfatt, M., 471
Steinman, D. H., 348
Steinmaus, H., 375
Steinmetz, A., 441
Stemke, J. E., 600
Sterling, C. J. M., 56
Sterling, J. J., 343
Stern, E. W., 124
Sternbach, D. D., 36, 44
Sternhell, S., 254, 581
Stetter, H., 39, 289, 536
Stevens, K. L., 518
Stewart, D., 359
Stick, R. V., 5, 569, 587
Still, W. C., 386, 607
Stille, J. K., 59
Stillings, M. R., 70
Stirling, I., 597
Stitzel, R. E., 9
Stobbe, J., 523
Stockis, A., 306
Stoddart, J. F., 137
Stojanac, N., 70
Stojanac, Z., 70
Stokes, B. G., 385
Stoll, A., 2, 309
Stork, G., 101, 103, 264, 265, 267, 339,
   341, 530, 598
Story, P. R., 577
Stotter, P. L., 83
Strand, G., 319
Stransky, W., 279
Strathdee, R. S., 600
Straub, H., 47
Strauss, J. U. G., 450
Strausz, O. P., 190
Strawson, C. J., 517
Strege, P. E., 395, 448, 450
Streith, J., 290
Streitwieser, A., Jr., 521
Strohmeier, W., 28, 104, 653
Strube, R. E., 255
Sturtz, G., 574
Subrahmanyam, C., 208
Subramanyam, G., 254

Sucrow, W., 304
Sudweeks, W. B., 493
Süess, R., 309
Suga, K., 412
Suggs, J. W., 380, 499, 604
Sugihara, Y., 192
Sugimori, A., 659
Sugimoto, A., 586
Sugimoto, K., 260
Sugiyama, A., 534
Suksamrarn, A., 481
Sullivan, P. T., 472
Sumelius, S., 656
Sumitani, K., 51
Sumoto, K., 233
Sundelin, K. G. R., 554
Šunjić, V., 474
Sunthankar, S. V., 19, 588
Suter, C., 419
Suzuki, A., 25, 149, 215, 259, 293, 295,
    308
Suzuki, E., 510, 578
Suzuki, J., 73
Suzuki, K., 73
Suzuki, M., 158, 392
Suzuki, Y., 244
Svoboda, J. J., 24, 109
Swenton, J. S., 672, 675
Swern, D., 115, 228, 230
Swierczewski, G., 270
Szechner, B., 291
Szilagyi, S., 454
Szmant, H. H., 27

Tabata, M., 25, 295
Tachi, K., 653
Taddia, R., 497, 560
Tadros, W., 503
Tämnefors, I., 148
Tagami, H., 308
Taguchi, H., 163, 222
Taguchi, T., 556
Taguchi, Y., 207
Taimr, L., 656
Takagi, K., 573
Takahagi, Y., 547
Takahashi, K., 170
Takahashi, N., 504
Takahashi, R., 201
Takai, K., 106

Takaishi, N., 526
Takamatsu, N., 174
Takami, Y., 95
Takamura, N., 307
Takaya, H., 45
Takayanagi, H., 430
Takegami, Y., 485, 486, 552
Takehira, Y., 146
Takei, H., 279
Takei, S., 534
Takenake, Y., 216
Takeshita, H., 170
Takigawa, T., 216
Talaty, C. N., 556
Tamao, K., 51, 59
Tamaru, Y., 396
Tamura, Y., 116, 118, 233
Tanaka, K., 65, 585
Tanaka, M., 552
Tanaka, S., 183, 586
Tanaka, T., 165, 641
Tanba, Y., 443
Tancrede, J., 539
Tang, C., 2
Tang, C.-P., 166
Tang, C. S. F., 141, 229
Tangari, N., 74, 506
Tanguy, L., 172
Tani, J., 70
Tanida, H., 502
Tanigawa, Y., 392
Tarchini, C., 175
Tardivat, J.-C., 359
Tasai, H., 232
Taylor, A. R., 304
Taylor, E. C., 581
Taylor, G. F., 524
Taylor, R., 89, 91
Taylor, R. T., 68, 70
Tebbe, F. N., 48
Teixeira, H. L. S., 339
Temple, D. L., 20
TenBrink, R. E., 161, 588
Teranishi, S., 450, 653
Terao, S., 436
Terasawa, I., 215
Terasawa, M., 450
Teschner, M., 238, 450
Texier, F., 472
Thalmann, A., 247, 520, 625

Thanh, T. N., 574
Thayer, A. L., 469
Thielke, D., 2
Thomas, C. B., 443
Thommen, W., 215, 675
Thompson, D. J., 552
Thompson, M. D., 137
Thomsen, I., 558
Thorpe, F. G., 581
Thuan, S. L. T., 514
Thuillier, A., 300
Tideswell, T., 341
Tidwell, T. T., 589
Tilak, M. A., 307
Timmermans, G. J., 575
Tirpak, J. G., 133
Tobey, S. W., 553
Toda, F., 146
Todd, A. R., 42, 43
Toder, B. H., 70
Tömösközi, I., 645
Tohda, Y., 61
Tohma, M., 304
Toi, H., 64
Tokita, S., 547
Tokoroyama, T., 529
Tokura, N., 296
Tolkmith, H., 115
Tolman, C. A., 654
Tomi, K., 552
Tomita, H., 451
Tomita, T., 304, 451
Tonnis, J. A., 146
Tordeux, M., 70
Tortorella, V., 506
Toru, T., 165, 247, 641
Toube, T. P., 514
Touster, O., 423
Townsend, J. M., 48, 573
Toy, A. D. F., 215
Toye, J., 123
Trahanovsky, W. S., 620
Traitler, H., 503
Traylor, T. G., 640
Trippett, S., 268
Trolliet, M., 557
Trost, B. M., 32, 116, 118, 155, 218, 238,
    243, 313, 316, 319, 320, 369, 393, 395,
    396, 447, 448, 450, 464, 466, 662
Troxler, F., 2

Truce, W. E., 377
Truesdale, L. K., 399, 642
Tsuchihashi, G., 6, 392, 446
Tsuda, K., 675
Tsuda, T., 130, 145, 534
Tsui, F. P., 273
Tsuji, J., 430
Tsujihara, K., 369
Tsukanaka, M., 220
Tsunoda, M., 118
Tufariello, J. J., 597
Tuiman, A., 600
Tulis, R. W., 504, 506
Tunemoto, D., 462
Turner, A. B., 170
Turner, L. M., 19
Turner, M., 1
Turro, N. J., 482
Tsutsumi, K., 662
Tyman, J. H. P., 517
Tyrlik, S., 588, 590

Ucciani, E., 172
Uda, H., 464, 466, 662
ud Din, Z., 503, 546
Ueda, M., 25, 76, 659
Ueda, T., 659
Ueda, Y., 322
Uemura, S., 23, 141
Ueno, Y., 34
Ugi, I., 8, 91
Uh, H.-S., 582
Uhm, S. J., 675
Uijttewaal, A. P., 381
Ulrich, P., 367
Umani-Ronchi, A., 43, 74, 306, 486
Umeda, I., 45
Upton, C. J., 348
Urushibara, Y., 659
Usui, M., 376
Utawanit, T., 179
Utimoto, K., 295

Vaidyanathaswamy, R., 523
Valcho, J. J., 91
Valenta, Z., 70
Valentine, J. S., 489, 490
Valenty, S., 190
van Asten, J. J. A., 256
van Beelen, D. C:, 458

van Bekkum, H., 149, 575
van Boom, J. H., 605
Van Brussel, W., 201
Van-Catledge, F. A., 554
van der Gen, A., 381
van der Kooi, H. O., 458
Van Der Puy, M., 261
Vandewalle, M., 201
Van Ende, D., 362, 488
Van Hecke, G. R., 51
van Leusen, A. M., 567, 600
van Muijlwijk, A. W., 149
van Peppen, J. F., 611
Van Strien, R. E., 84
van Tamelen, E. E., 378, 379, 589
Varkony, T. H., 441
Varma, R. K., 202
Varvoglis, A., 301
Vartanian, P. F., 604
Vaughn, H. L., 548
Vazquez, M. A., 114
Vedejs, E., 15, 384, 573, 628
Venier, C. G., 457
Vennstra, G. E., 567
Veracini, C. A., 27
Vercellotti, J. R., 676
Verhé, R., 118
Verkade, J. G., 469
Vermeer, P., 144, 146, 270
Vetter, H. J., 576
Vierhapper, F. W., 615
Vietti, D. E., 128
Vigneaud, V. du, 445
Villieras, J., 143, 189, 270, 331, 345, 354,
    577, 622, 664
Vinokur, E., 323
Vinson, J. W., 620
Viriot-Villaume, M. L., 70
Vogel, E., 33, 478, 479, 657, 675
Volkmann, R. A., 615
Vollhardt, K. P. C., 154, 668
von Bredow, K., 773
Vorbrüggen, H., 442, 443, 640
Voss, P., 273
Vostrowsky, O., 279
Vrielink, J. J., 143
Vuillerme, J.-P., 359
Vuitel, L., 11

Wachter, M., 511

Wackerle, L., 91
Waddington, T. C., 301
Wade, L. G., Jr., 497
Wagner, S. D., 93, 361
Wagner, W. J., 400
Wahl, R., 243
Wailes, P. C., 179
Waitkins, G. R., 422
Wakabayashi, M., 500
Wakamatsu, T., 626, 628
Wakatsuki, Y., 128
Wakselman, C., 70, 263
Walborsky, H. M., 91, 145, 642, 643
Wald, K., 76
Waldschmidt-Leitz, E., 445
Walker, B. J., 341
Walker, D., 234, 453
Walker, D. M., 268
Wallace, T. J., 430
Wallach, O., 357
Walsh, E. N., 215
Walton, D. R. M., 57, 58, 185, 652
Wamhoff, H., 76
Wang, C.-L. J., 464, 632
Wang, N., 368
Wang, S.-W., 254
Wannagut, U., 640
Wanzlick, H. J., 375
Ward, M. A., 457
Ward, R. S., 675
Warner, C. D., 525, 526
Warnhoff, E. W., 284
Warren, J. D., 632
Warren, S., 340, 341
Warwel, S., 201
Washburne, S. S., 632
Wasserman, H. H., 241, 428, 430, 436
Watanabe, K., 152, 596
Watanabe, S., 415
Watanabe, Y., 485, 552
Waterman, E. L., 93, 361
Watkins, G. L., 284
Watt, D. S., 113, 114, 352, 430
Watts, L., 150
Watts, W. E., 359
Wayaku, M., 653
Weber, L., 447, 450
Weber, W. P., 35
Webster, D. E., 426
Weedon, B. C. L., 511, 514

Weeks, J. L., 670
Weeks, P. D., 573
Wegener, J., 2
Wehlacz, J. T., 165
Wehner, G., 633
Weigold, H., 179
Weil, J. A., 670
Weinshenker, N. M., 92, 582
Weinstock, L. M., 412
Weinstock, J., 83
Weis, C. D., 173
Weiss, R. H., 192
Weissberger, E., 306
Welch, J., 286, 474
Welling, L. L., 76
Wells, A. G., 35
Wells, D., 32
Wells, P. B., 426
Wemple, J., 252, 358, 542
Wentworth, S. E., 142
Wepplo, P. J., 312
Wermuth, C. G., 458
Werstiuk, E. S., 605
West, C. T., 616
West, P. J., 583
West, R., 24, 553
Westheimer, F. H., 37
Weston, A. W., 526
Westphal, D., 318
Wetter, H., 107
Wexler, S., 436
Weyerstahl, P., 79, 263
Weyler, W., Jr., 78
Whalley, W. B., 467
Whatley, L. S., 24
White, D. A., 475
White, D. R., 103
White, J. D., 497
White, P. D., 338
White, W. N., 341
Whitesell, J. K., 338
Whitesides, G. M., 124, 343, 495
Whitham, G. H., 628
Whiting, M. C., 126, 451
Whitlock, B. J., 354
Whitlock, H. W., 354
Whitmore, F. C., 84
Whitten, C. E., 343, 389
Wicha, J., 74
Wick, A., 222

Widiger, G. N., 11
Wiechert, R., 411
Wieland, D. M., 245
Wightman, R. H., 622
Wildeman, J., 600
Wiley, R. A., 554
Wiley, R. H., 292
Wilhalm, B., 675
Wilkins, A. L., 304, 514
Wilkins, C. L., 521
Wilkins, J. M., 225
Wilkinson, G., 105, 426, 454, 650
Willard, A. K., 277, 279
Williams, D. J., 536
Williams, D. K., 660
Williams, D. R., 487
Williams, J. W., 40
Williams, K., 38
Williams, V. Z., Jr., 17
Willstätter, R., 445
Willy, W. E., 224, 586
Wilson, E. M., 453
Wilson, S. R., 151
Wilt, J. W., 400
Winer, A. M., 487
Winterfeldt, E., 2, 141, 198, 201
Winternitz, F., 470
Wissner, A., 552
Wistrand, L.-G., 52
Wold, S. F., 36
Wolfe, S., 345
Wollenberg, R. H., 56, 603
Wolman, Y., 78
Wolochowicz, I., 588, 590
Wolters, J., 458
Wong, C. M., 50, 229, 470, 497, 529, 543, 563
Wong, J. Y., 92
Wong, W. B., 447, 450
Wong, P. K., 59
Woodbridge, D. T., 224, 241
Woods, T. S., 543
Woodward, R. B., 629
Worrall, W. S., 103
Woznow, R. J., 70
Wright, G. F., 416
Wright, M. J., 488
Wright, P. W., 341
Wrobel, S. J., Jr., 254
Wu, D. K., 103

Wu, E. S. C., 384
Wudl, F., 121
Wüest, H., 77, 514
Wulff, G., 511
Wulff, J., 612
Wynberg, H., 501, 524, 525, 575

Yagen, B., 202
Yagi, H., 358
Yagupol'skii, L. M., 301, 611
Yajima, H., 618
Yamada, K., 295
Yamada, S., 193, 307
Yamada, S. I., 193
Yamada, Y., 11
Yamamoto, G., 509
Yamamoto, H., 51, 141, 162, 182, 183,
    222, 328, 582
Yamamoto, I., 644
Yamamoto, K., 451, 654
Yamamoto, Y., 64, 322, 377, 534
Yamamura, M., 443, 573
Yamane, M., 432, 436
Yamashita, A., 476
Yamashita, M., 392, 446, 485, 486, 552
Yamazaki, H., 128
Yamazaki, Y., 55
Yanami, T., 482
Yan Kui, Y. T., 367
Yang, D. T. C., 201
Yano, T., 445
Yanuka, Y., 297
Yaroslavsky, C., 510
Yasuda, A., 183
Yasuda, M., 21
Yatagai, H., 377
Yazawa, H., 65
Ykman, P., 480
Yokomatsu, T., 547
Yokoyama, T., 141
Yokoyama, Y., 193
Yoneda, F., 185, 547
Yonemitsu, O., 258, 453

Yonezawa, K., 129
Yoshii, E., 317, 403, 649
Yoshida, T., 399, 492
Yoshifuji, M., 28
Yoshikoshi, A., 482, 662
Yoshimura, N., 445
Young, M. W., 235, 420, 422
Young, R. N., 109, 524
Yu, S. H., 105
Yura, Y., 667
Yurtanov, A. I., 634
Yuyama, M., 161

Zamecnik, J., 171
Zamojski, A., 280, 291
Zanarotti, A., 517
Zapp, J. A., Jr., 278
Zaretzkii, Z., 166
Zatorski, A., 191, 193
Zbiral, 297, 318
Zefirov, N. S., 232
Zehavi, U., 420
Zelonka, R. A., 665
Zera, R. T., 174
Zergenyi, J., 403
Ziegler, E., 96
Zimmer, H., 392
Zimmerman, W. T., 310
Zlotogorski, C., 108
Zoller, U., 290
Zon, G., 273
Zoran, A., 108
Zoretic, P. A., 110
Zubiani, A., 74
Zubrick, J. W., 137
Zupan, M., 300, 670
Zur, Z., 470
Zwaneburg, B., 567
Zweifel, G., 198, 201, 293, 295
Zweig, J., 215
Zwierzak, A., 42, 43
Zweiezchowska, Z., 291

# SUBJECT INDEX

Page numbers referring to reagents are indicated in **boldface**.

Acetal interchange, 381-382
Acetalization, 19, 379-380
Acetals, 437
Acetamidobenzenesulfonyl chlorides, 497
Acetanilide, 276, 532
Acetic acid—Acetic anhydride, **1, 4**
Acetic anhydride, **2-3, 4**
Acetic anhydride—Boron trifluoride
   etherate, **3**
3$\beta$-Acetoxy-17$\beta$-chloroformylandrostene-5,
   9
2$\alpha$-Acetoxy-5$\alpha$-cholestane-3-one, 4
Acetic anhydride—Methanesulfonic acid, **3**
Acetic anhydride—Phosphoric acid, **3-4**
Acetic anhydride—Pyridine, **4-5**
Acetic anhydride—Sodium acetate, **5-6**
Acetic anhydride—Triethylamine, **6**
Acetoacetic acid, 21
Acetoacetic ester condensation, 405
Acetoacetic esters, 202
Acetoinenediolcyclopyrophosphate, **6-8**
Acetone, **9**
Acetonitrile, 130, 135, 136, 140, 142, 160
Acetonylcyclohexane, 518
3-Acetonylcyclohexane-1-one, 21
2-Acetonylfurane, 106, 107
Acetophenone, 125, 139, 187, 240
$\alpha$-Acetoxyacrylonitriles, 103
5-Acetoxyaldonic acid methyl esters, 437,
   438
2-Acetoxyalkyl methyl selenides, 224
3$\beta$-Acetoxyandrostene-5-one-17, 5, 74
$\alpha$-Acetoxy ketones, 206
Acetoxylation, 316
17$\beta$-Acetoxy-5,6-epoxyandrostane-7-ols,
   282
17$\beta$-Acetoxy-4-methylestratriene-1,3,5(10),
   282
3-Acetoxy-2-methyl-1-nonene, 79
1-Acetoxy-3-oxobutane, 255
9$\alpha$-Acetoxypinene, 497, 498
3$\beta$-Acetoxypregna-5,14-diene-20-one, 316
Acetylacetone, 290, 629
1-Acetylaminocyclohexene, 587

1-Acetylaspidoalbidine, 312
Acetyl bromide, **9**
1-Acetylcyclohexanol, 360
2-Acetyl-3,4-dihydro-6-methoxynaph-
   thalene, 169
Acetylene, 341, 450
Acetylenes, 61
17-Acetyl steroids, 600
Acetyl *p*-toluenesulfonate, **10**
1-O-Acetyl-2,3,5-tri-O-benzoyl-$\beta$-D-
   ribofuranose, 639
Acid fluorides, 184
Acid hydrazides, 429
Acrolein, 461, 605
Acrylonitrile, 45
2-Acylamino-2-alkene-3-ones, 533
Acyl azides, 564
$\alpha$-Acyl-$\gamma$-butyrolactones, 534
2-Acyl-1,3-dithianes, 76, 629
Acyl hydrazides, 161
N-Acylmorpholines, 52
Acyloin condensation, 626-627
Acyloins, 289
Acyl tetrafluoroborates, 520
2-Adamantanecarbonitrile, 600
Adamantane-1-carboxylic acid, 623
1-Adamantanol, 440, 623
Adamantanone, 286, 287, 440, 600
1-Adamantyl hexafluoroantimonate, 272
Adenosine, 288, 554, 602
Adenosine 5'-phosphate, 207
Adipic acid, 504
Adogen 464, **10**
Aeroplysinin-I, 314-315
Aflatoxin B$_1$, 286
$\beta$-Alanine, **10-11**
L-Alanine, 493
Aldehydes, 60, 216
Aldol condensation, 32, 213, 336, 590-
   591, 676
Aldols, 336, 365
Aldoximes, 4, 229
Aliquat, 336, **404-406**, 431
Alkali metal iron carbonylates, 305

Alkaneboronic acids, 34
Alkanesulfonyl fluorides, 284
Alkeneboronic acids, 34
1-Alkenes, 270, 463, 584-585
4-Alkene-2-yne-1-ols, 76
3-Alkenols, 199
Alkenylboranes, 324, 325
Alkenylation, 135, 259
Alkenylboronic acids, 72
1-Alkenyl bromides, 72
2-(1-Alkenyl)-2-cyclopentenones, 591
2-Alkenyl N,N-diethyldithiocarbamates, 569
1-Alkenyl iodides, 72
Alkenyloxysilanes, 424
Alkenylsilanes, 281
2-Alkoxy-1,3-benzodithioles, 96
β-Alkoxy-γ-bromoketones, 591
3-Alkoxycyclobutanones, 16
ω-Alkoxycarbonylalkanoic acids, 313
2-Alkoxy-3,4-dihydro-2H-pyranes, 322
δ-Alkoxy-β-keto esters, 592, 593
β-Alkoxy ketones, 594
3-Alkoxy-2-oxanorcaranes, 322
2-Alkoxy-3-oximinocycloalkenes, 422, 423
Alkoxysilanes, 179
Alkoxysulfonium salts, 377
Alkoxytris(dimethylamino)phosphonium
    chlorides, 280
3-Alkylalkanoic acids, 388
Alkyl ω-alkenoates, 313
3-Alkyl-1-alkynes, 85
6-N-Alkylaminouracils, 185
Alkyl aryl ethers, 645
Alkyl aryl ketones, 24
N-Alkylarylsulfonamides, 578
o-Alkylbenzaldehydes, 443
Alkyl benzoates, 347
o-Alkylbenzoic acids, 368
2-Alkyl-1-t-butylaziridines, 118
Alkyl t-butylperoxyglyoxalates, 82
Alkyl chlorides, 22, 259
6-Alkyl-Δ²-cyclohexenones, 236
5-Alkyl-Δ²-cyclopentenones, 236
Alkyl α,α-dichloroalkanoates, 331
Alkyldiphenylselenonium tetrafluoro-
    borates, 240
Alkyl fluoroformates, 473
2-Alkylfuranes, 32
2-Alkyl-4-hydroxycyclopentenones, 100-
    101

Alkylidenecyclobutanones, 223
2-Alkylidene-1-oxocyclopentanes, 284
3-Alkylidene-2,4-pentanediones, 595
Alkylidenetriphenylphosphoranes, 95
Alkyl iodides, 259, 294
Alkyl isocyanoacetates, 158
Alkyl 4-methylphenyl sulfones, 567
β-Alkyl-α-methylacroleins, 14
Alkyl orthotitanates, 11
2-Alkyl-1-oxo-2-cyclopentenes, 284
o-Alkylphenyl acetates, 560
Alkyl thiocyanates, 259
Alkyl N-tresylarylsulfonimides, 578
N-Alkyl-N-triflylarylsulfonimides, 578
β-Alkyl-α,β-unsaturated ketones, 344
2-Alkyne-1,4-diols, 609
1-Alkynes, 146, 189, 270
Alkynyloxiranes, 269
Allenediynes, 143
Allene oxide-cyclopropanone systems, 482
Allenes, 95, 143-144, 223, 523, 619
Allenic alcohols, 147-148, 269, 325
Allenic esters, 635
Allyl alcohol, O,2-dilithio derivative, 11
Allylamine, 445
Allyl bromide, 198
Allyl chloride, 11, 539
S-Allyl N,N-dimethyldithiocarbamate,
    11-13
Allyl ethers, 594
Allylic alcohols, 1, 17, 28-29, 30-31, 85-86,
    182, 261, 318, 335, 338, 425, 444, 588,
    602-603, 604, 664
Allylic amines, 604-605
Allylic chlorides, 598, 645
Allylic dibromides, 418
Allylic ethers, 269
Allylic halides, 327, 329
Allylic hydroperoxides, 287
Allylic phosphine oxides, 340
Allylic rearrangement, 11-12
Allylic trifluoroacetoxylation, 360-361
Allylidenetriphenylphosphorane, 14-15
π-Allylpalladium chloride complexes, 45-
    47
Allyl phenyl selenides, 338
Allylpotassium compounds, 638
Allyl selenides, 338
Allyltriphenylphosphonium bromide, 15
Allyltriphenylphosphorane, 77

Allyltris(trimethylphosphite)cobalt(I), 15
Alumina, 3, 4, 5, 16-17
Aluminum bromide, 17
Aluminum chloride, 17-19
Aluminum ethoxide, 180
Aluminum hydride, 19, 346
Aluminum isopropoxide, 19
Aluminum tricyclohexoxide, 20
Amberlite IRA, 302, 303
Amberlyst-15 sulfonic acid resin, 302
Amidation, 67
Amides, 60, 106, 193, 548, 635, 636, 648
Amidoalkylation, 290
Amino acids, 492-493, 505-506
α-Amino alcohols, 503
α-Amino aldehydes, 201
Aminoalkoxynaphthpyridines, 540
1-Amino-4-arylaminoanthraquinones, 18-19
Aminobenzenesulfonic acids, 497
3-endo-Aminobornane-2-one, 307
3-endo-Aminoborneol, 307
3-endo-Aminoisoborneol, 307
4-Aminocyclohexanecarboxylic acid, 193
Δ⁵-3β-Amino-4β-hydroxysteroids, 360-361
Δ⁴-3β-Amino-6β-hydroxysteroids, 361
α-Amino-γ-keto acids, 290
2-Amino-2-methyl-1-propanol, 20, 368
α-Aminonitriles, 633
2-Aminopyridine, 248
3-Aminopropylamine, 476
β-Aminopropenyltriphenylphosphonium
    bromides, 494
3-Amino-1-propanol, 570
Δ⁵-3β-Aminosteroids, 360
Ammonium tetrafluoroborate, 132
Ammonium persulfate, 20-21
Ammonium peroxydisulfate, 20-21
Ammonium polysulfide, 21
5α-Androstanediols, 511
5β-Androstane-17β-ol-3-one, 401
Androstenolone, 368
Androsterone, 299
Aniline, 21, 45
Aniline hydrobromide, 229
Anilines, 513
o-Anisaldehyde, 630
Annelation, 77, 83, 110, 243, 263, 264-265,
    407-409, 466, 661-662
Annulenes, 138
Anthracene, 347-348, 474-475, 533, 669-670

Anthracene 1,2-oxide, 358
Anthranilic acid, 96
Anthraquinone, 284
Anthrone, 120, 121, 623
Antimony(III) fluoride, 553
Antimony(V) fluoride, 23-24, 103
Antimony(V) chloride, 22-23, 108
Apinol tetramethyl ether, 424
Apocarotinoids, 527
Apomorphine, 9
Aporphines, 660
Aquovitamin B₁₂, 139
O-Arenecarbonylhydroxylamines, 58
Arene imines, 541
Arene oxides, 357-358, 435-436
O-Arenesulfonylhydroxylamines, 58
Argentic oxide, 518
Arndt-Eistert reaction, 141
Aromantization, 3-4, 169, 560
Aryl-alkyl coupling, 570
Aryl alkyl ethers, 280
Arylamines, 57
α-Aryl-γ-benzylidene-Δ^{α,β}-butenolides, 6
Arylboronic acids, 580
Aryldiazoalkanes, 453
Aryl iodides, 579
Aryl ketones, 87
5-Aryl-γ-lactones, 512-513
Aryloxysilanes, 273
Arylphosphonic acids, 192
N-Aryl sulfimides, 118
Arylsulfonyltriazoles, 361
Arylthallium(III) bistrifluoroacetates, 579
4-Aryl-1,2,4-triazoline-3,5-diones, 75
4-Arylurazoles, 75
Asymmetric alkylation, 39-40
Asymmetric epoxide synthesis, 219
Aurones, 358, 359
Azabicyclooctenones, 618
Azasulfonium salts, 118, 119, 120
Azetidine-2-ones, 558
Azetidines, 646
Azides, 319, 644, 676
α-Azidostyrene, 24-25, 297
Azidotris(dimethylamino)phosphonium
    hexafluorophosphate, 25
Aziridines, 612
Azelaic acid monomethyl ester, 217-218
1-Azirines, 157
Azobenzenes, 513

Azobisisobutyonitrile, 72, 585, 602, 672
Azo compounds, 273
Azoxy compounds, 146, 273

Baeyer-Villiger oxidation, 110, 286-287
Basketene, 41
Bates' Reagent, 425
9–BBN, **62-64**
Beckmann fragmentation, 400-402, 571-
    572, 610
Beckmann rearrangement, 170
Benzalacetone oxime, 127, 128
Benzaldehyde, **26,** 53, 446
Benzaldehyde ethylene dithioacetal, 26-27
Benzaldehydes, 360
Benzamide, 109, 525
Benzamidine, **27**
Benzamidines, 304
Benzanilide, 532
Benz[a]anthracene, 20
π-Benzenechromium tricarbonyl, **27-28**
Benzenediazonium-2-carboxylate, 96
Benzene oxide, 435
Benzeneselenol, **28-29**
Benzenesulfenic acid, 258
Benzenesulfenyl chloride, **30-32**
Benzene trioxide, 435
Benzil, 37, 225, 671
Benzil dimethyl ketal, 241
Benzils, 142, 671
5-Benzisoxazolemethylene chloroformate,
    **32**
Benzocyclopropenes, 19
Benzocyclopropene, 33, 88, 477
Benzocyclobutenes, 153
1,3,2-Benzodioxaborole, **33-34,** 98
1.3-Benzodithiole-2-carbene, 96
Benzofuranes, 118-119
Benzoic acid, 428
S-Benzoic O,O-diethyl phosphorodithioic
    anhydride, **34**
Benzoin, **34,** 37, 225, 408
Benzoin benzoate, 347
Benzoins, 142, 470, 671
Benzonitrile, 239
Benzophenone, 187
Benzophenone azine, 577
Benzophenone hydrazone, 120, 121, 452
Benzo[a]pyrene-4,5-quinone, 229
1,4-Benzoquinones, 187, 424, 507

Benzothiazole, 206
Benzotriazolyl-N-hydroxytris(dimethyl-
    amino)phosphonium hexafluorophosphate,
    34-35
Benzotrichloride, 576
Benzotropilidenes, 244, 672-673
Benzoylation, 34, 35
o-Benzoylbenzoates, 347
Benzoyl chloride, 171
Benzoyl cyanide, **35**
Benzoylhydrazine, 36
N-Benzoylimidazole, 35
Benzoyl isocyanate, 35
Benzoyl-L-leucine, 193
Benzoyl-L-leucylglycine ethyl ester, 193
Benzoyloxylation, 160
N-Benzoylperoxycarbamic acid, **35-36**
N-Benzoyl-N'-triflylhydrazine, **36**
Benzpyrene-1-ol, 316
Benzpyrene, 316
1-Benzoyl-2-vinylcyclopropane, 212
Benzobarrelene, 523, 524
Benzocyclobutene, 550
Benzyl acetates, 127
Benzyl alcohol, 428, 444
Benzyl alkyl ketones, 71
Benzylamine, 230, 239
6-N-Benzylaminouracil, 185
Benzylation, 44, 221
Benzyl benzoate, 347
Benzyl bromides, 360
Benzyl-n-butyldimethylammonium chloride,
    38
Benzyl chloride, 38
1-Benzyl-1,4-dihydronictinamide, 36-37
Benzyldimethylanilinium hydroxide, **37**
Benzyl esters, 37
Benzyl fluoride, 473
Benzyl-5-(2-hydroxyethyl)-4-methyl-1,3-
    thiazolium chloride, 38-39
Benzylidene acetals, 438-439
Benzylideneacetone, 39
Benzylideneacetone(tricarbonyl)iron, 39
Benzylideneaniline, 558
Benzylidenes, 637
(–)-N-Benzyl-N-methylephredrinium
    bromide, **39-40**
Benzylmonoperoxycarbonic acid, **40**
Benzyl nitriles, 135
p-Benzyloxybenzaldehyde, 158

Benzylsodium, 40
Benzyl phenyl selenides, 235
*N*-Benzyloxycarbonylcarpamic acid, 247
Benzyl tri-*n*-butylammonium chloride, 41
Benzyltriethylammonium chloride, 41-43,
  405
N-Benzyltriflamide, 43-44
Benzyl triflate, 44
Benzyl trifluoromethanesulfonate, 44
Benzyltrimethylammonium fluoride, 44
Benzyltrimethylsilane, 637
4-Benzylurazole, 75
$N^3$-Benzyluridine, 221
Benzyne, 96, 346
Bianthrone, 120, 121
Biaryls, 59, 443
Bicyclo[2.1.1.]hexene, 546
Bicyclo[3.2.2]nona-6,8-diene-3-one, 364
Bicyclo[5.1.0]octa-2,5-diene, 550
Bicyclo[3.3.0]octane-1-ol, 63
Bicyclo[3.3.0]octanes, 23
Bicyclo[4.2.0]octene-2, 196
Bicyclo[4.2.0]octene-3, 197
Bicyclo[4.2.0]octene-7, 196
9,9'-Bifluorenylidene, 555
2,2'-Biimidazole, 357
2,2'-Bi(2-imidazoline), 357
Bile acids, 426
Biphenyls, 313
Birch aromatization reaction, 475
Birch reduction, 88, 169, 409
Bis(acetylacetonate)palladium(II), 45
Bis(acrylonitrile)nickel(0), 45
Bis-annelations, 409-410
Bis(benzonitrile)palladium(II), 45-47
Bisbenzylisoquinolines, 452
Bis(1,5-cyclooctadiene)nickel(O), 45, 654
Bis($\eta^5$-cyclopentadienyl)niolium tri-
  hydride, 47-48
Bis(cyclopentadienyl)titanium dichloride,
  48-49
Bis(dimethylaluminum)1,2-ethanedithio-
  late, 50
Bis(dimethylaluminum)1,3-propanedithio-
  late, 49-50
1,8-Bis(dimethylamino)naphthalene, 50,
  59, 596
1,2-Bis(dimethylphosphino)ethane, 59
1,2-Bis(diphenylphosphino)ethane, 447
Bis(diphenylphosphino)propane, 59

[1,3-Bis(diphenylphosphino)propane]-
  nickel(II) chloride, 50-51
Bis(salicylidene)ethylenediiminocobalt(II),
  507
1,2-Bis($\beta$-tosylethoxycarbonyl)diazene,
  55-56
*trans*-1,2-Bis(tri-*n*-butylstannyl)ethylene,
  56
Bis(tributyltin)ethylene glycolate, 186
Bis(tri-*n*-butyl)tin oxide, 56-57, 602
Bis(triethylphosphine)nickel(II) bromide,
  57
Bis(trimethylsilyl)acetals, 113
Bis(trimethylsilyl)acetamide, 634
Bis(trimethylsilyl)acetylene, 153
Bis(2,2-dipyridyl)silver(II) peroxydisulfate,
  51-52
1,4-Bis(halomethoxy)butanes, 105
*trans*-Bishomobenzene, 68-69
1,3-Bishomocubane, 108
1,8-Bishomocubanes, 519
Bis(4-methylpiperazinyl)aluminum hydride,
  52-53
1,3-Bis(methylthio)allyllithium, 53
Bis(methylthio)(trimethylsilyl)methane, 53
Bis(methylthio)(trimethylsilyl)methyl-
  lithium, 53-54
Bis(methylthio)(trimethylstannyl)methyl-
  lithium, 54
Bismuth trioxide–Stannic oxide, 55
Bisnor–S, 17
Bis(2,4-pentanedionato)nickel, 570
Bis(phenylthio)methane, 267
Bis(propenyl)mercury, 573
Bis(trimethylsilyl)amidocopper, 57
Bis(trimethylsilyl)formamide, 58
O,N-Bis(trimethylsilyl)hydroxylamine, 58
Bis(trimethylsilylmethyl)mercury, 637
Bis(triphenylphosphine)nickel(II) bromide,
  58-59
Bis(trimethylphosphine)nickel(II) chloride,
  59, 653
Bis(triphenylphosphine)palladium(II)
  chloride, 59
Bis(triphenylphosphine)palladium(II)
  halides, 60-61
Bis(triphenylphosphine)rhodium carbonyl
  chloride, 105
Bis-Wittig reaction, 668
Bivalvane, 352

BMS, 64
Bombykol, 576, 577
9-Borabicyclo[3.3.1]nonane, **62-64**
Borane–Dimethyl sulfide, 64
Boric acid, 79
Borneol, 111
Boronic esters, 329
Boron tribromide, **64-65**
Boron trichloride, **65**
Boron trifluoride etherate, 3, **67-70**, 186, 200, 245, 248, 252, 261, 304, 316, 412, 467, 559, 614, 635
Bredt's rule, 526
Brij 35, 70
Bromination, 17, 71, 97
Bromine, **70-73**, 75
α-Bromo acetals, 72
ω-Bromoacetates, 352
Bromoacetyl fluoride, 103
Bromacetylium hexafluoroantimonate, **103**
1-Bromoadamantane, 271, 483, 514, 623
2-Bromo-1-alkenes, 284-285
1-Bromo-1-alkynes, 85
2-Bromoallyl alcohol, 11
*p*-Bromoaniline, 229
2-Bromobenzoic acids, 144
6-Bromobenzpyrene, 316
*o*-Bromobenzyl methyl ether, 83
1-Bromobicyclo[2.2.1]heptane, 623
1-Bromobicyclo[2.2.2]octane, 623
2-Bromobutyryl bromide, 674
Bromocamphor, 484, 675
ω-Bromocarboxylic acids, 479
*p*-Bromochlorobenzene, 179
2-Bromo-2-cyano-N,N-dimethylacetamide, **73**
Bromocyclobutanones, 157
3-Bromocyclohexene, 205
Bromocyclopropanes, 194
α-Bromocyclopropyl trifluoroacetates, 532
2-(2-bromoethyl)dioxolane-1,3, 512
Bromohydrins, 589
β-Bromo ketones, 70-71
*anti*-7-Bromo-*syn*-7-lithionorcarane, 89
1-Bromo-3-methyl-2-butene, 346
Bromomethyllithium, **74**
2-Bromo-1-methylnaphthalenes, 468
3-Bromo-1-methylnaphthalene, 468
1-Bromonaphthalene, 570

*o*-Bromonitrobenzene, 130
7-Bromonorcarane, 533
1-Bromooctane, 38, 135, 604
5-Bromopentanone-2 ethylene ketal, 74
2-Bromopropionyl bromide, 674
1-Bromopropyne-3-ols, 146-147
β-Bromostyrene, 571
N-Bromosuccinimide, **74-76**, 169
11-Bromoundecanoic acid, 479
β-Bromo-α,β-unsaturated ketones, 344
1,3-Butadiene, 58
*trans*-1-Butadienyltriphenylphosphonium bromide, **76-77**
Butadiynyl(trimethyl)silane, 143
1,4-Butanediol, 379
1,4-Butanediols, 387-388
*t*-Butanol, 24, 126
2-Butanone, 426
Buta-1,2,3-trienes, 243-244
1-Butene, 504
2-Butene, 504, 510
2-Butene-1,4-diol, 652
$\Delta^{\alpha,\beta}$-Butenolides, 464, 609
*t*-Butoxybis(dimethylamino)methane, 242
*t*-Butyl acetoacetate, 578
*t*-Butylacetylene, 225, 478
1-*t*-Butyladamantane, 623
*n*-Butylamine, 199
*t*-Butylamine, 117
*t*-Butyl azidoformate, **77-78**
B-*n*-butyl-9-BBN, 62-63
N-*n*-Butylbenzanilide, 234
*t*-Butylbenzene, 623
*t*-Butyl α-benzoylisobutyrates, 84
*t*-Butyl bromide, 17
*t*-Butyl chloride, 24
*t*-Butyl chromate, 126
*t*-Butylcyanoketenes, **78**
4-*t*-Butylcyclohexanone, 380, 386
*t*-Butyldimethylchlorosilane, **78-79**
*t*-Butyldimethylsiloxyvinyl ethers, 79
*t*-Butyldimethylsilyl cyanide, **80-81**
*t*-Butyldimethylsilyl ethers, 26
1-*t*-Butyldimethylsilylpropynylcopper, 639
*t*-Butyldiphenylsilyl chloride, **81**
*t*-Butyldiphenylsilyl ethers, 81, 183
4-*t*-Butyl-1-ethylcyclohexanols, 623
*t*-Butyl hydroperoxide, 81-82
*t*-Butyl hypochlorite, 75, **82**, 118
*t*-Butyl γ-iodotiglate, **82-83**

*t*-Butyl isonitrile, 129
*t*-Butyl lithioacetate, **84**, 636
*t*-Butyl α-lithioisobutyrate, 84
*n*-Butyllithium, 28, 30, 31, **85-91**, 95-96, 235, 276
*sec*-Butyllithium, 607
*t*-Butyllithium, 81
*n*-Butyllithium–TMEDA, 151, 202
4-*t*-Butyl-1-methylcyclonexanol, 386
*t*-Butyl nitrate, 272
*t*-Butyl nitrile, 642
*n*-Butyl nitrite, 127
N-*t*-Butyl osmiamate, **641-642**
*t*-Butyloxycarbonylation, 91
*t*-Butyloxycarbonyl azide, 77-78
*t*-Butyloxycarbonyl fluoride, **91**
1-*t*-Butyloxycarbonyl-3-formylindole, 78, 91
2-*t*-Butyloxycarbonyloxyimino-2-phenylacetonitrile, 91
*n*-Butylphenyltin dihydride, 92
N-*n*-Butylpiperidine, 601
S-*t*-Butyl thioates, 581
γ-Butyrolactones, 124, 302, 387-388, 415, 453-454, 490-491
1-Butyne, 178
Butynediol, 652
Butyne-1-ol, 451
2-Butynoic acid, 345, 346
3-Butyn-1-yl trifluoromethanesulfonate, 620

Caglioti reduction, 532
Calcium carbonate, 110
Calcium silicide, 510
Camphor, 97, 111, 675
*d*-10-Camphorsulfonic acid, 611
Camptothecin, 2, 80-81, 140, 227, 228
CAN, 99
Cannizzaro reaction, 379
Carbamates, 416
Carbenes, 135-136
Carboalkylation, 59, 60
Carbodiimides, 450
2-Carboethoxyallyl bromide, 93
π-(2-Carboethoxyallyl)nickel bromide, **93**, 361
7-Carboethoxy-3,4-benzo[4.1.0]heptane, 672
Carboethoxycyclopropyltriphenylphosphonium tetrafluoroborate, **93-94**

Carboethoxymethyne, 190
1-Carbomethoxy-4-alkyl-1,4-dihydropyridines, 376
7-Carbomethoxy-3,4-benzotropilidene, 672
2-Carbomethoxycyclohexanone, 217
Carboethoxymethylenetriphenylphosphorane, 635
Carbonates, 97, 280
Carbon dioxide, **94-95**, 188
Carbon dioxide fixation, 130
Carbon disulfide, **95-96**
Carbon monoxide, 59, 60, 177
Carbonylation, 108
Carbonyl chloride fluoride, 91
Carbonylcyclopropane, 96
N,N'-Carbonyldiimidazole, 97
1,1-Carbonyldi-1,2,4-triazole, 35
Carbon suboxide, 96
Carbothioate S-esters, 252
4-Carboxybenzaldehyde, 263
2-Carboxyethyltriphenylphosphonium perbromide, **97**
Carboxylation, 94-95, 354
Carboxylic acids, 95, 102-103
Carboxymethylenetriphenylphosphorane, 389
Carboxypeptidase A, 149
Carbynes, 190
Cardenolide glycosides, 514
Cardenolides, 403
Caro's acid, 97-98
Carotinoids, 526-527, 541
Carpaine, 247
Carvone, 330, 331, 491
Cataline, 315
Catechol, 429
Catecholborane, 33-34, 98
Catechols, 10
$C_{18}$-Cecropia juvenile hormone, 163, 164
Cembrene diterpenes, 554
Cembrene-A, 87
Cembrene, 419
Cephalosporins, 186, 280, 383
Cephams, 115
Ceric ammonium nitrate, 99, 149, 197
Cerium(IV) oxide–Hydrogen peroxide, **99-100**
Cesium fluoride, 100
Cetyltrimethylammonium bromide, 70
Chanoclavine-I, 559

Chloral, 100-101
o-Chloranil, 168, 667, 674
p-Chloranil, 168
Chloreal, 605
Chlorination, 23, 562
Chlorine, 101-102
Chlorine bromide, 102
Chloroacetaldehyde, 267
Chloroacetonitrile, 102-103
Chloroacetyl fluoride, 103
Chloroacetyl isocyanate, 634
Chloroacetylium hexafluorantimonate,
    103
α-Chloroacrylonitrile, 142
α-Chloroacyl chlorides, 117
2-Chloroalkanoic acids, 204
1-Chloro-1-alkynes, 189
Chlorobenzene, 179
π-(Chlorobenzene)chromium tricarbonyl,
    103-104
p-Chlorobenzenesulfonyl chloride, 106
2-Chlorobicyclo[2.2.1]heptene-1, 90
7-Chlorobicyclo[4.1.0]heptene-6, 100
1-Chlorobicyclooctanes, 595
α-Chloroboronic esters, 171, 172
1-Chloro-4-bromomethoxybutane, 104-105
4-Chloro-1-butanol, 104
Chloro(carbonyl)bis(triphenylphosphine)-
    rhodium(I), 105
α-Chlorocarboxylic acid esters, 118
6-Chloro-1-p-chlorobenzenesulfonyloxy-
    benzotriazole, 106
1-Chloro-4-chloromethoxybutane, 104-105
α-Chlorocycloalkanones, 344-345
α-Chlorocyloheptanone, 344, 345
1-Chlorocyclopropene, 100
α-Chloro-N-cyclohexylacetaldonitrone, 107
α-Chloro-N-cyclohexylpropanaldonitrone,
    106-107
Chlorodecarbomethoxylation, 480
Chlorodicarbonylrhodium(I) dimer, 108
Chlorodiphenylmethylium hexachloro-
    antimonate, 108-109
α-Chloroepoxy ketones, 354-355
Chlorofluorocyclopropanes, 511
Chlorofluorocarbene, 175, 511
β-Chloroethyl isocyanate, 570
Chloroformates, 226
α-Chloroglycidic esters, 353
6-Chloro-N-hydroxybenzotriazole, 106

3-Chloro-4-hydroxybenzyl cyanide, 105
3-Chloroindole, 543
N-Chloroindole, 543
Chloroiodination, 140
Chloromethylation, 18
Chloromethylcarbene, 530
4-Chloro-4-methylcyclohexa-2,5-dienone,
    22
1-Chloro-1-methylcyclopropanes, 530
Chloromethyl ethyl ether, 304
2-Chloromethyl-4-methoxymethyl-5-
    phenyloxazoline, 204
Chloromethyl methyl ether, 18, 109
Chloromethyl methyl ethers, 404
Chloromethyl methyl sulfide, 109-110, 301
2-Chloromethyl-4-nitroanisole, 363
Chloromethylphosphonyl dichloride, 188
Chloromethylthiocarbene, 136
Chloromethyltrimethylsilane, 635
2-Chloro-3-oxocyclohexenes, 562
1-Chloro-3-pentanone, 110
m-Chloroperbenzoic acid, 46, 110-114,
    197, 441, 455-456, 457, 459, 487, 508
o-Chlorophenyl phosphorodichloridite,
    114-115
3-Chloropropionaldehyde, 529
N-Chloro-N-sodiourethane, 115
Chlorospiro[2.4]heptadienes, 170, 171
N-Chlorosuccinimide, 115-118
N-Chlorosuccinimide—Dialkyl sulfides,
    118
N-Chlorosuccinimide—Dimethyl sulfide,
    118-119
N-Chlorosuccinimide—1,3-Dithiane complex,
    119-120
N-Chlorosuccinimide—Triethylamine, 120-
    121
Chlorosulfonyl isocyanate, 122
Chlorosulfuric acid, 121
1-Chloro-N,N,2-trimethylpropenylamine,
    122-123
3-Chloro-2-triphenylsilyl-1-phenylpropene,
    482
Δ$^{3,5}$-Cholestadiene, 361, 583
Δ$^{4,6}$-Cholestadiene, 361
Δ$^{5,7}$-Cholestadiene-1α-,3β-diol, 574
Cholestane-2,3-dione, 433
5α-Cholestane-3,7-dione, 628
5α-Cholestane-3-one, 628
Δ$^{5,25}$-Cholestadiene-3β-ol, 202

α-Cholestanol, 659
β-Cholestanol, 299, 659
3α-Cholestanyl acetate, 298
O-Cholestanyl thiobenzoate, 583
$\Delta^2$-Cholestene, 628
$\Delta^4$-Cholestene, 46
$\Delta^5$-Cholestene, 46, 361
$\Delta^4$-Cholestene-3α-ol, 304
$\Delta^4$-Cholestene-3β-ol, 304
$\Delta^{9\,(11)}$-Cholestene-3α-ol, 298
$\Delta^5$-Cholestene-4α-ol, 47
$\Delta^5$-Cholestene-4β-ol, 47
$\Delta^4$-Cholestene-6α-ol, 47
$\Delta^4$-Cholestene-6β-ol, 47
$\Delta^4$-Cholestenone-3, hydrazone, 537
$\Delta^5$-Cholestene-3-one, 659
$\Delta^5$-Cholestene-4-one, 47
$\Delta^4$-Cholestene-6-one, 47
$\Delta^2$-5α-Cholestene-7-one, 628
$\Delta^{17(20)}$-3-Cholestenol, 299
$\Delta^{14}$-3α-Cholestenyl acetate, 298
$\Delta^{16}$-3α-Cholestenyl acetate, 299
Cholesterol, 109
Cholesterol β-epoxide, 304
O-Cholesteryl thiobenzoate, 583
Chromic acid, **123-124,** 225, 437
Chromium(II) acetate, 4, 5, 587
Chromium hexacarbonyl, 27, 103, **125-126**
Chromyl chloride, **126-127**
Cinchonine, **501**
Cinnamoyl chloride, 618
Cinnamyl alcohol, 116
Cinnanyl aldehyde, 116
Citronellal, 499
Citronellol, 499
Claisen condensation, 285, 348, 530
Claisen rearrangement, 79, 127, 276-277, 607-608, 609
Claisen's alkali, **127**
Claisen self-condensation, 331
Clorox, 435, 506
Cobalt(III) acetate, **127**
Cobalt(II) chloride, **127-128,** 145
Cobaltocene, **128**
2,4,6-Collidine, 7
Collins reagent, 126, 499
Conjugate addition, 53-54, 163-164, 204-205, 212-213, 253, 256, 289, 311, 330-331, 378, 399, 465, 481, 538,
593, 623-624, 638, 661
Conjugate cyanation, 180-181
Conjugate reduction, 144, 172, 326, 348, 491-492
Cope rearrangement, 23, 390, 628
Copper—Isonitrile complexes, 128-129
Copper(I) phenylacetylide—Tri-n-butylphosphine, **130**
Copper(I) trifluoromethanesulfonate, **130-133**
Corey-Kim oxidation, 582
Corey-Winter olefin synthesis, 384-583
Coronene, 524
Cortexolone, 298
Corticosteriods, 4-5
Cortisone, 298
Corynantheine, 400
Costunolide, 303
p-Cresol, 22
p-Cresols, 477
Cross aldol condensation, 594
Cross-coupling, 50-51, 203, 259
Crotylmagnesium chloride, 48
12-Crown-4, 134
15-Crown-5, 134
18-Crown-6, 134, 135, 136, 137, 431, 488
Crown ethers, **133-137,** 159-160
Cryptate 222, 138
Cryptates, **137-138**
Cupric acetate, **138,** 140
Cupric bromide, **138-139,** 226
Cupric chloride, **139-141,** 259
Cupric hexafluoroacetylacetonate, 664
Cupric sulfate, **141-142**
Cupric tetrafluoroborate, **142**
Cupric triflate, 130, 664
Cuprous acetate, 127, **142-143**
Cuprous bromide, 18, **143-144,** 183
Cuprous bromide—Sodium bis-(2-methoxyethoxy)aluminum hydride, 144
Cuprous t-butoxide, 144-145
Cuprous chloride, 145-146, 198, 251, 429
Cuprous cyanide, 146-147
Cuprous iodide, **147,** 199, 204, 205, 269, 327
Cuprous iodide—Grignard reagents, 147-148
Cuprous triflate, **130-133**
Cuprous trimethylsilylacetylide, 148
Curtius degradation, 188, 316

Curtius rearrangement, 564
2-Cyanobenzyl cyanide, 541
Cyanoboration, 535-536
1-Cyanocyclopropanecarboxylic acid, 42
Cyanogen bromide, 148-149
1-Cyanooctane, 38
Cyanosilylation, 80-81
Cyanuric chloride, 149
Cyclanones, 351
Cyclic amidines, 78
Cyclic ketoimides, 441
Cyclization, 10
Cycloadditions, 23, 96, 122, 123, 168,
    208-209, 223, 251-252, 364, 544, 567,
    579-575, 584
Cycloalkanones, 391-392
2-Cycloalkene-1,4-diols, 608-609
Cycloalkenes, 162, 417-418
Cyclobutadiene, 55, 149
Cyclobutadiene iron tricarbonyl, 55, 149-
    150
1,3-Cyclobutanedicarboxylic acid, 546
Cyclobutanes, 334-335
Cyclobutanone, 391, 620
Cyclobutanone cyanohydrin, 530
Cyclobutanone dimethyl dithioacetal
    S-oxide, 390, 391
Cyclobutanones, 390-391, 522
2-Cyclobutenones, 16, 123, 381-382
Cyclobutenylideneammonium salts, 123
1-Cyclobutenylmethyllithium, 151
Cyclobutyl annelation, 319-320
β-Cyclocitral, 440
Cyclocostunolides, 303
Cyclodehydrogenation, 354, 474, 498,
    524-525
β-Cyclodextrin, 151-152
Cyclododecanone, 162
Cyclododecene, 658
Cyclododecene oxide, 658
α-Cyclogeraniol, 263
Cycloheptaamylose, 151-152
Cyclohepta-3,6-diene-1,2,5-trione, 168
Cyclohepta-4,6-diene-1,2,3-trione, 169
Cycloheptanone, 131, 132
1,3-Cyclohexadienes, 77, 222
Cyclohexane-1,2-dione, 433
Cyclohexane-1,3-dione, 611
Cyclohexane-1,4-diones, 283
Cyclohexanol, 116, 260, 648

Cyclohexanone, 116, 124, 340, 344, 504
4-(2-Cyclohexanone)butyric acid, 217
Cyclohexanones, 351
Cyclohexene, 124, 136, 140, 445, 518,
    648
2-Cyclohexene-1,4-diol, 586
Cyclohexene epoxide, 71, 333, 466, 488
Cyclohexene-3-ol, 428
Cyclohexenone, 21
Cyclohexenones, 371
2-(3-Cyclohexenyl)ethanol, 637
N-Cyclohexenylmorpholine, 433
Cyclohexylamine, 230
Cyclohexylideneacetaldehydes, 357
Cyclohexylidenecyclohexane, 171
Cyclohexylidenethyl bromide, 327, 328
Cyclohexyl isocyanide, 129, 152
1-Cyclohexyl-3-(2-morpholinoethyl)-carbodi-
    imide, metho-p-toluensulfonate, 227
Cyclooctadiene-1,5, 23, 62, 445
Cyclooctane-1,2-diol, 384
Cyclooctane-1,5-diol, 64
Cyclooctanol, 183
Cyclooctatetraene, 23
Cyclooctatetraene epoxide, 306, 519
Cyclooctatetraeneiron tricarbonyl, 567
Cyclooctene, 23, 183, 445
Cyclooctene oxide, 334
Cyclooctyl fluoride, 183
Cyclopentadiene, 123, 291
Cyclopentadienone, 454
π-Cyclopentadienylcobalt dicarbonyl,
    153-154
π-(Cyclopentadienyl)(isobutenyl)iron
    dicarbonyl tetrafluoroborate, 154-155
Cyclopentanecarboxylates, 129
Cyclopentane-1,3-dicarboxaldehyde, 434
Cyclopentanedicarboxylates, 129
Cyclopentanone, 131
Cyclopentanones, 351, 466, 522
2-Cyclopentene-1,4-diol, 586
Cyclopentenol, 64
2-Cyclopentenone, 64
3-Cyclopentenone, 391, 392
Cyclopentenone annelation, 67-68
Cyclopentenones, 67, 339, 520, 591-592,
    667
Cyclo(L-Pro-Gly)$_n$peptides, 135
1,2-Cyclopropa-4,5-cyclobutabenzene,
    153-154

Cyclopropane-1,1-dicarboxylic acid, 42-216

Cyclopropanes, 128, 129, 140, 141, 144-145, 219-220, 442-443, 521-523

Cyclopropanols, 71

Cyclopropanone cyanohydrin, 529-530

Cyclopropenes, 100

Cyclopropylcarbinols, 353

1-Cyclopropylcyclopropanols, 523

Cyclopropyldiphenylsulfonium fluoroborate, **155**

Cyclopropylidenemethenone, 96

Cyclopropyl ketones, 523, 534

Cyclopropyl lactones, 212

3α,5-Cyclosteroids, 3

Cyclothallation, 578, 579

Cyclotridecanone, 163

DAD, 185

β-Damascenone, 17

β-Damascone, 17

δ-Damascone, 310

Darzens condensations, 103, 159, 353, 542

DBN, 157

DBU, **158**

DDQ, **168-170**

Dealkylation, 145, 355, 458

Deamination, 113-114

Debenzylation, 496

Debromination, 50, 341

Decahydroquinolines, 615

γ-Decalactone, 460

9β,10α-Decalin-2β,3β-diol, 438

Decalols, 502

Decarboalkylation, 279, 447, 472, 536

Decarboethoxylation, 284

Decarbomethoxylation, 472

Decarbonylation, 653

Decarboxylation, 81-82, 160, 224, 260, 292, 317, 343, 652

Decarboxylative Michael reaction, 21

Decene-1, 140

Dechlorocarbomethoxylation, 480

Decinine, 362

Decyanation, 351

5-Decyne, 294

Dediazotiation, 285-286

Deformylation, 213

Dehalogenation, 483-484, 497, 523-524, 532, 543

Dehydration, 180, 281, 378, 475-476, 583, 648, 649

Dehydroabietylamine, 575

Dehydroboration, 207

Dehydrobromination, 17-18, 545, 572

15,15'-Dehydro-β-carotene, 526, 527

Dehydrocordrastine, 200

Dehydrodimerization, 55

14-Dehydroequilenin, 10

Dehydrogenation, 82, 169, 499, 500, 556-557, 657, 658, 667, 672, 672

Dehydrohalogenation, 477-478, 479, 486, 525, 573-574

6,7-Dehydrotropinones, 544

Dehydroxylation, 243, 244, 304

Demethylation, 348

Denitration, 555

Deoximation, 124, 304, 305

Deoxygenation, 67, 260, 415, 419, 446, 470, 539, 585, 588, 606, 628, 658

Dephthaloylation, 280-281

Depsipeptides, 288

Desilylation, 78-79

8-Desmethylfervenulin, 547

Desoxybenzoin, 383, 671

Desoxybenzoins, 470

Desoxycholic acid, 166

Desoxyketoses, 544-545

11-Desoxyprostaglandin E$_2$, 603

Desoxyvernolepin, 266, 267

3-Desoxyvitamin D$_2$, 340

Desulfurization, 278, 296, 400, 401, 484, 612

Deuterium exchange, 504

Dewar benzenes, 149-150, 518-519, 574, 575

Diacetatobis(triphenylphosphine)-palladium(II), 156-157

3,5-Diacetoxycyclopentene, 641

Diacetylapomorphine, 9

1,4-Diacetylglycouril, 563

Diadamantane, 17

Dialkenylchloroboranes, 377

Dialkylacetic acids, 387

Dialkylacetoinyl phosphates, 7

Dialkylalkenylboranes, 293

Dialkylcyanothexylborates, 535

N,N-Dialkylhydroxylamines, 538

DABCO, 87, **157-158**, 202, 278, 525

2,3-Dialkylindoles, 457-458

α,α'-Dialkyl ketones, 235-236
Dialkylmalonic acids, 88
Dialkyl peroxides, 488
Dialkyl phosphates, 8
Dialkylphosphorohalidates, 42
Dialkyl selenides, 510
Diallenes, 558
4,9-Diamantanol, 121
Diaminoaluminum hydrides, 52
2,4-Diamino-7,8-dihydropteridines, 342
2,3-Dialkylcyclopentanones, 342-343
Diaryl sulfones, 521
1,3-Diazaazulenes, 27
2,3-Diazabicyclo[2.2.0] hexene-5, 55
1,5-Diazabicyclo[2.2.2] nonene-5, 157, 411
Diazabicyclo[2.2.2] octane, 157-158, 525
1,5-Diazabicyclo[5.4.0] undecene-5, 94,
    95, 158, 411
2,3-Diaza-Dewar benzene, 56
3H-1,2-Diazepines, 599
Diaziridines, 253
α-Diazoacetophenone, 577
Diazoalkanes, 514, 577
3-Diazocamphor, 515
Diazodiphenylmethanes, 113
α-Diazo esters, 307
Diazofluorene, 577
α-Diazo ketones, 113, 141-142
Diazomethane, 159, 443, 577
Diazomethyl ketones, 67, 68, 614-615
4-Diazo-3-oxo-1-phenyl-1-butene, 597
Dibenzazecine, 481
Dibenzazonine alkaloids, 660
Dibenzo-15-crown-5, 174
Dibenzo-18-crown-6, 159-160, 174
Dibenzoylacetylene, 150
2,3-Dibenzoylbicyclo[2.2.0] hex-2,5-diene,
    150
1,2-Dibenzoylcyclopropane, 244
Dibenzoyl peroxide, 160-161, 655, 672
1,6-Dibenzoyltetracyclo[4.4.0²,⁵ 0⁷,¹⁰] deca-
    3,8-diene, 150
Dibenzyl peroxydicarbonate, 40
Diborane, 161-162, 564
α,α'-Dibromoacetone, 195
1,1-Dibromo-1-alkenes, 189
α,α'-Dibromoalkynes, 645-646
1,4-Dibromo-2-butene, 76
Dibromocarbene, 68, 136, 229, 468, 601
1,2-Dibromocyclohexane, 646

3,5-Dibromocyclopentene, 641
gem-Dibromocyclopropanes, 89
1,3-Dibromo-5,5-dimethylhydantoin, 287
1,2-Dibromoethane, 42
1,2-Dibromoethylene, 341
Dibromoheptalene, 657
Dibromomethane, 274
Dibromomethylenation, 211
Dibromomethylenetriphenylphosphorane,
    211
Dibromomethyllithium, 162-163
7,7-Dibromomonorcarane, 89, 533
Dibromostilbene, 545
α,α'-Dibromo-o-xylene, 550, 551
Di-t-butylcarbodiimide, 378
Di-n-butylcopperlithium, 163-165
Di-t-butyl diperoxycarbonate, 166
Di-t-butyl ketone, 508
2,6-Di-t-butyl-4-methylpyridine, 406
Di-t-butylperoxyoxalate, 166-167
2,6-Di-t-butylphenols, 656
Di(2-t-butylphenyl)phosphorochloridate,
    167-168
N,N-Di-n-butylpiperidinium iodide, 601
2,6-Di-t-butylpyridine, 406
Dicarboethoxycarbene, 631
Dicarboxylic anhydrides, 616-617
vic-Dichlorides, 413
1,1-Dichloro-1-alkenes, 189
1,2-Dichloralkenes, 140-141
o-Dichlorobenzene, 136
2,6-Dichlorobenzoates, 340, 341
2,6-Dichloro-p-benzoquinone, 101, 102
7,7-Dichlorobicyclo[4.1.0] heptane, 228
2,8-Dichlorobicyclo[3.2.1] octane, 23
Dichlorocarbene, 38, 41, 100, 136, 159,
    171, 219-220, 349-350, 553, 601, 621
3,4-Dichlorocyclobutene, 274
1,2-Dichlorocyclohexane, 646
1,4-Dichlorocyclooctane, 23
5,6-Dichlorocyclooctenes, 23
gem-Dichlorocyclopropanes, 350, 596
4,9-Dichlorodiamantane, 121
2,3-Dichloro-5,6-dicyano-1,4-benzoquinone,
    168-170, 511, 657, 672, 674
4,4'-Dichlorodiphenyl diselenide, 420, 421
Dichlorodiphenylmethane, 108
Dichlorodiphenylsilane, 81
1,1-Dichloroethane, 530
1,2-Dichloroethane, 17

1,2-Dichloroethylene, 294, 341
Dichlorofluoromethane, 175
N,N-Dichlorohexylamine, 145
N,N-Dichloro-1-methylcyclohexylamine, 145
Dichloromethylenecyclohexane, 345
Dichloromethylenedimethylammonium chloride, 170
Dichloromethyllithium, 170-171, 344, 345
α,α-Dichloromethyl methyl ether, 171-172
2,6-Dichloropyridine, 51
Dichlorovinylene carbonate, 172
Dicobalt octacarbonyl, 172
Dicyanobenzenes, 533
2,3-Dicyano-1,3-butadiene, 173
Dicyanocyclobutane, 172
1,2-Dicyanocyclobutene, 172-173
1,2-Dicyanopropane, 136
Dicyclohexylborane, 62
Dicyclohexylcarbodiimide, 174, 318
Dicyclohexyl-18-crown-6, 174-175, 480, 489
Dicyclohexyl-18-crown-6 – Potassium cyanide, 80
Dicyclopentadiene, 108, 445
Di($\eta^5$-cyclopentadienyl)(chloro)hydrido-zirconium(IV), 175-179
$\eta^5$-Dicyclopentadienyltitanium dichloride, 179
Dicyclopentadienylzirconium dichloride, 176
Dicyclopropylidenemethane, 479
Di-(4,5-dimethyl-2-oxo-1,3,2-dioxa-phospholenyl)oxide, 6-8
Dieckmann reaction, 542
Diels-Alder catalysts, 17-18
Diels-Alder reaction, 33, 55-56, 65-66, 142, 149-150, 173, 188, 231, 271, 291-292, 309-310, 346-347, 363-364, 370-372, 432, 433, 454, 467, 478, 612
1,3-Dienes, 182, 325, 340, 377, 431, 666
1,4-Dienes, 244
1,5-Dienes, 327-328
2,6-Dienic carboxylic acids, 627-628
Diepoxides, 74
Diethoxyaluminum chloride, 180-181
1,3-Diethoxybenzenes, 612
1,3-Diethoxycyclohexadiene-1,3, 611
3,3-Diethoxy-1-methylthiopropyne, 398
Di(α-ethoxyvinyl)copperlithium, 205

Diethylaluminum chloride, 181
Diethylaluminum cyanide, 180-181
Diethylaluminum dialkylamides, 181
Diethylaluminum 2,2,6,6-tetramethyl-piperidide, 181-183
Diethylaminosulfur trifluoride, 183-184
N,N-Diethylaminotrimethylsilane, 183, 184-185
N,N-Diethylaniline, 157
Diethyl arylphosphonates, 192
Diethyl azodicarboxylate, 185
Diethyl(2-chloro-1,1,2-trifluoroethyl)-amine, 186
Diethyl diazomalonate, 630
Diethylene glycol, 378, 379
Diethylene orthocarbonate, 186-187
Diethyl ethoxymalonate, 631
N,N-Diethylhydroxylamine, 187
Diethyl isocyanomethylphosphonate, 187
Diethyl ketomalonate, 188
Diethyl lithiodichloromethylphosphonate, 188-189
Diethyl malonate, 42, 255
Diethyl mercurybisdiazoacetate, 190
Diethylmethyl(methylsulfonyl)ammonium fluorosulfonate, 190
Diethyl oxalate, 402, 597
Diethyl phenylsulfinylmethylphosphonate, 191
Diethyl phosphorocyanidate, 192-193
Diethyl phosphorochloridate, 192
Diethyl phosphonate, 192
Diethylphosphoryl cyanide, 192-193
Diethyl phosphorylmethyl methyl sulfoxide, 193
Diethyl sulfoxide, 257
Diethyl trichloromethylphosphonate, 189
Diethylzinc, 194, 515
Diethylzinc–Bromine–Oxygen, 194
Diethylzinc–Iodoform, 194
gem-Difluorides, 300
N,N-Difluoroamines, 263-264
gem-Difluoro compounds, 412
α,α-Dihalo aldehydes, 117
7,7-Dihalobenzocyclopropenes, 19
1,1-Dihalospiropentanes, 42
1,2-Dihydrazones, 429
9,10-Dihydroanthracene, 533
1,3-Dihydrobenzo[c]tellurophene, 550
2,3-Dihydro-1,4-benzoxazine, 434

Dihydrocarvone, 77
Dihydrocortisone, 298, 299
Dihydrofuranes, 94, 356, 592
2,5-Dihydro-2-furoic acid, 323
Dihydroheptalenes, 497
5,6-Dihydro-4-hydroxy-2-pyrones, 592, 593
Dihydrojasmone, 333, 626, 627
Dihydromyrcene, 663
3,4-Dihydro-2-pyrones, 556
Dihydrothiophenes, 666, 667
3,4-Dihydro-2-tosyl-1,2-diazepines, 599
2,3-Dihydrotriquinacenone-2, 567, 568
3α,12α-Dihydroxy-5β-androstane, 166
p-Dihydroxyborylbenzyloxycarbonyl chloride, 194-195
1,5-Dihydroxy-3,7-di-t-butylnaphthalene, 170
5,7-Dihydroxy-4-methylphthalide, 517
Δ⁴-3β,6β-Dihydroxy steroids, 360
Diimide, 195
Dihydroxytetramethylcyclobutene, 548
1,3-Diiodo-5,5-dimethylhydantoin, 287
1,3-Diiodopropane, 129
Diiron nonacarbonyl, 39, 195-198, 552
Diisobutylaluminum hydride, 64, 198-201, 460, 528, 538
Diisopinocampheylborane, 202
Diisopropylethylamine, 24
1,2;5,6-Di-O-isopropylidene-α-D-allofuranose, 620
Diketene, 202, 592
3,17-Diketo-5α-androstanes, 302
2,4-Diketo(16-crown-5), 133
2,4-Diketo(19-crown-6), 133
1,2-Diketones, 113, 226-227, 353, 354, 361, 373, 433, 629
1,3-Diketones, 145, 164, 290, 356, 533, 542, 629
1,4-Diketones, 39, 53, 139, 204-205, 311-312, 332-333, 356, 365, 493, 515-516, 626, 627
1,5-Diketones, 593-594
β,ε-Diketophosphonates, 339
1,3-Dilithiopropyne, 202-203
Dilithium acetylide, 324
Dilithium ethynylbis(trialkylborates), 148
Dilithium tris(pent-1-ynyl)cuprate, 203
Dimedone, 611
Dimerization, 145

2,9-Dimethoxy-1,8-dihydroheptalene, 497
Dimethoxymethane, 374
2,4-Dimethoxymethyl-5-phenyloxazoline, 204, 205, 244
2,5-Dimethoxy-4-nitroaniline, 445
21-Dimethoxypregnene-14-one-20, 402
Di(α-methoxyvinyl)copperlithium, 204-205
2,7-Dimethoxy-1,4,5,8-tetrahydronaphthalene, 497
Dimethoxystilbene, 241
Dimethyl acetylenedicarboxylate, 150, 206, 370, 574, 666
Dimethyl alkanephosphonates, 532
3-Dimethylallyl alcohol, 1
Dimethylamine, 140, 554
N,N-Dimethylaminoacetonitrile, 190
4-Dimethylaminobenzaldehyde, 502
(2S,3R)-4-Dimethylamino-3-methyl-1,2-diphenyl-2-butanol, 206
2-Dimethylamino-4-methylquinoline, 276
2-(N,N-Dimethylamino)-4-nitrophenyl phosphate, 207
2-Dimethylaminoquinolines, 276
Dimethylaniline, 502
3,7-Dimethylanthrarufin, 486
2,6-Dimethylbenzoquinone, 65
Dimethylbromoformiminium bromide, 220
2,3-Dimethyl-2-butylborane, 207-208
Dimethyl 2-chloroethylene-1,1-dicarboxylate, 480
Dimethylchloroformiminium chloride, 220
4,5-Dimethyl-2-chloro-2-oxo-1,3,2-dioxaphosphole, 7-8
2,2-Dimethyl-6-(p-chlorophenylthiomethylene)cyclohexanone, 208-209
3,3-Dimethylcholestane, 623
Dimethylcopperlithium, 209-215, 386
N,N-Dimethylcyanoacetamide, 73
4,4-Dimethylcyclobutenone, 382-383
2,3-Dimethylcyclohexanone, 407
2,2-Dimethylcyclohexyl chloride, 82
5,6-Dimethyl-Δ¹(¹⁰)-decalone-2, 407
2,2-Dimethyl-1,3-dioxane-4,6-dione-5-spiropropane, 96
7,8-Dimethyl-1,4-dihydro-2,4-benzodithiepin, 216
N,N-Dimethyl dichlorophosphoramide, 215
Dimethyl 1-diazoalkanephosphonates, 532
6,6-Dimethyl-5,7-dioxaspiro[2,5]octane-4,8-dione, 216

Dimethyl diphenate, 132
Dimethyl diselenide, 361
Dimethyl disulfide, 13, **217-218**
N,N-Dimethylephedrinium bromide, 40, **219-220**
N-(1,2-Dimethylethenylenedioxyphosphoryl)imidazole, **218-219**
Di(1,2-dimethylethenylene)pyrophosphate, 218
2,3-Dimethyl-6-ethylphenol, 409
Dimethylformanide, 4, 17, 26, 122, 139, 146, 280
Dimethylformamide—Phosphorus pentachloride, **220**
6,6-Dimethylfulvene, 170, 171
N,N-Dimethylformamide dineopentyl acetal, 222
N,N-Dimethylformamide dimethyl acetal, 221-222
N,N-Dimethylformanide dibenzyl acetal, 221
Dimethyl 4-hydroxy-o-phthalate, 370
Dimethyl 3-hydroxyhomophthalate, 158
N,N-Dimethylhydrazine, **223**
Di-π-cyclopentadienylcobalt, 128
Dimethyl 3,8-heptalenedicarboxylate, 672, 673
Dimethyl fumarate, 451, 576
1,2-Dihydropyridines, 376
1,4-Dihydropyridines, 376
5,6-Dihydro-2H-pyrane-2-ones, 495
3,6-Dihydrophthalic anhydride, 317
9,10-Dihydrophenanthrene, 533
1,4-Dihydronaphthalene, 533-534, 672
Dimethyl 4-hydroxy-o-phthalate, 370
3,5-Dimethylisoxazole, 533
Dimethylketene, **223**
Dimethyl β-ketoadipate, 354
Dimethyl maleate, 450, 451, 575
Dimethyl malonate, 140
Dimethyl muconate, 451
3,3-Dimethyloxaziridines, 114
4,4-Dimethyl-Δ²-oxazolines, 20
Dimethyloxosulfonium methylide, 41
2,4-Dimethyl-3-pentanone, 589
4,4-Dimethyl-2-pentene, 251
2,4-Dimethylpyridine, 224
Dimethylselenonio methylide, 240
Dimethyl selenoxide, **224**
Dimethyl sulfate, 592

Dimethyl sulfide, 204, 248
Dimethyl sulfide—Cuprous bromide, **225**
Dimethyl sulfide ditriflate, **225**
Dimethylsulfonium methylide, 219
Dimethyl sulfoxide, 29, 160, **225-229**, 377, 467, 535
Dimethyl sulfoxide—Bromine, **229**
Dimethyl sulfoxide—Chlorine, **229**
Dimethyl sulfoxide—Sulfur trioxide, **229**
Dimethyl sulfoxide—Trifluoroacetic anhydride, **230**
Dimethyl 1,4,5,6-tetramethylbicyclo-[2.2.0]hexa-2,5-diene-2,3-dicarboxylate, 574
3,3-Dimethylthiobutyryl chloride, 285
N,N-Dimethylthiocarboxamides, 557
6,6-Dimethyl-2-vinyl-5,7-dioxaspiro[2,5]-octane-4,8-dione, **230**
2,2-Dimethylvinyltriflate , **230-231**
Dimsyllithium, 255, **342**
2,2'-Dinitrobiphenyl, 130
2,4-Dinitrobenzenesulfonylhydrazine, 232-233
2,4-Dinitrobenzenesulfenyl chloride, **231-232**
m-Dinitrobenzene, 419
2,6-Dinitroanilines, 446
Dimsylsodium, 6, **546-547**
vic-Dinitro compounds, 555
2,4-Dinitrofluorobenzene, **233**
Dinitrogen pentoxide, 420
Dinitrogen tetroxide, 284
2,4-Dinitrophenylaldoximes, 233
2,4-Dinitrophenylhydrazones, 587
0-2,4-Dinitrophenylhydroxylamine, 233
Dinucleotides, 379
1,2-Diols, 97, 206
p-Dioxane, 379
1,2-Dioxetanes, 241, 287-288, 433, 434
3,5-Dioxocyclopentenes, 201
1,3-Dioxolanium tetrafluoroborates, 564
Dipeptides, 379
Diphenic acids, 368
Diphenylacetonitrile, 159
Diphenylacetylene, 231, 545, 576, 671
Diphenylacetylenes, 576
9,10-Diphenylanthracene, 471
Diphenylcopperlithium, **234**
Diphenyldiazomethane, 120, 121, **234**, 452, 508, 577

1,4-Diphenyl-1,4-di(2-pyridyl)butatrienes, 243-244
Diphenyl-2,2'-dicarboxylic acid, 35
Diphenyldi(1,1,1,3,3,3-hexafluoro-2-phenyl-2-propoxy)sulfurane, **239**
Diphenyl diselenide, 235
Diphenyl disulfide, 31, **235-236**, 313
1,1-Diphenylethylene, 300
1,3-Diphenylisobenzofurane, 100
2-Diphenylmethylenefenchone, 643
Diphenylmethyl esters, 452
3,4-Diphenylphenol, 408
Diphenyl phosphoroazidate, 193
1,2-Diphenyl-1-propanol, 649
Diphenyl selenide, **240**
Diphenylseleninic anhydride, **240-241**
Diphenyl sulfide, **241**
S,S-Diphenylsulfilimines, 239
Diphenyl sulfoxide, 241
Diphenyl thioacetals, 130
Diphenyl thioketals, 130
1,4-Diphenylthiosemicarbazide, 396
1,4-Diphenyl-s-triazoline-3-thione, 396
Diphos, 447
Diphosphopyridine nucleotide, 36
Diphosphorus tetraiodide, 243-244
Di-3-pinanylborane, 161
Di-n-propylcopperlithium, **245**
2,2'-Dipyridine, 52
2,2'-Dipyridyl disulfide, **246-247**
α-Disaccharides, 569
Disiamylborane, 62
Disodium tetracarbonylferrate, 550-552
Disodium tetrachloropalladate, 45
Dispiro[2.0.2.4]deca-7,9-diene, 568
Disulfides, 231-232, 563, 586
N,N-Disulfonimides, 531
Dithallous ethylene glycolate, 186
Dithia[3.3]metacyclophanes, 85
1,3-Dithiane, **248**, 406
Dithianes, 628-629
Dithiobenzoic acid, 582
Dithiolanes, 629
Divinylallene, **271**
4,4'-Divinylbenzil, 142
4,4'-Divinylbenzoin, 142
Divinylcyclopropanes, 378
Divinylcuprates, 266
Diynediols, 147
Diynes, 294, 653

Dodecacarbonyliron, 39
N-Dodecyl-N-methylephedrinium bromide, **249**
Drierite, 360

"Easy Mesyl," 190
Enamides, 4, 587
Enamines, 82, 251, 409, 433, 553-554
1,4-Endoperoxides, 431
Ene-allenes, 341-342
17(20)-Ene-21-ol steroids, 655-656
3-Ene-1,2-diols, 182, 183
1,4-Enediones, 500
Ene reactions, 381, 432, 433
Enimides, 4
Enol ethers, 270
Enynes, 576
1,5-Enynes, 638-639
Eosin-Y, 431
ψ-Ephedrine, 219
Epicholestanol, 659
Epicholesterol, 659
Epidioxides, 586
2,7-Epi-(±)-perhydrohistrionicotoxin, 322
Episulfides, 375
Episulfonium salts, 375
Epi-β-vetivone, 213
Epoxidation, 35-36, 40, 431, 453, 486-487
Epoxide cleavage, 16, 29, 226, 245
Epoxides, 29, 41, 74, 171, 182, 183, 219, 240, 362-363, 375, 464, 588
2,3-Epoxy-2-alkylcyclopentanones, 558
1,4-Epoxy-1,4-dihydronaphthalenes, 405
2,3-Epoxyindanone, 516-517
α,β-Epoxy ketoximes, 214
7,8-Epoxy-2-methyloctadecane, 637
1,2-Epoxypropane, 226
α,β-Epoxysilanes, 245
Equilenins, 528
Equilins, 528
$\Delta^{7,22}$-Ergostadiene-3β-ol, 502
$\Delta^5$-5α-Ergostene-3β-ol, 502
Ergosterol, 502
Ergosteryl acetate, 100, 595
Ergosteryl acetate tricarbonyliron, 39
Erythronolide B, 215
Eschenmoser cleavage, 232-233
Eschenmoser's reagent, 106
Esterification, 106, 138, 246, 269, 274,

303, 375-376, 452, 519-520, 581-582, 625, 648
Estranes, 10
Estrone, 257-258, 410-411
Estrone methyl ether, 32, 237, 313, 314
Ethanedithiol, 532
Ethanolamine, 64
Ether cleavage, 198, 260, 352-353
Ethers, 565, 577-578
2-Ethoxycarbonylcyclohexanone, 662
Ethoxycarbonyl isothiocyanate, 250
1-Ethoxycyclododecene, 385
1-Ethoxycyclotetradecene, 385
α-Ethoxyvinyllithium, 205
Ethyl acetate, 627
Ethyl acetoacetate, 264, 593
N-Ethylacetonitrilium tetrafluoroborate, 250
Ethyl acrylate, 251
Ethyl allylnitrosocarbamate, 664
Ethylaluminium dichloride, 251-252
Ethyl benzoylformate, 37
Ethyl α-bromoacetate, 293, 674
Ethyl α-bromoacrylate, 329
Ethyl α-(t-butylsulfinyl)acetate, 561
S-Ethyl carbonochloridothioate, 252
Ethyl chloroformate, 336
Ethyl chlorothiolformate, 252
Ethyl crotonate, 449
Ethyl cyanoacetate, 42, 140
7-Ethylcyclohepta-1,3,5-trienes, 194
Ethyl diazoacetate, 252-253, 672
Ethyl diethoxyacetate, 253
Ethyldiisopropylamine, 611
Ethyl 3,3-dimethylacrylate, 67
Ethylene dithioketals, 76
2,2'-Ethylenebis(m-dithiane), 493
Ethylene carbonate, 570
Ethylenediamine, 324, 417
Ethylene dibromide, 203
Ethylene dithiotosylate, 628-629
Ethylene oxide, 199, 638
Ethyl formate, 597
2-Ethylhexahydropyrimidine, 445
2-Ethyl-7-hydroxybenzisoxazolium fluoro-borate, 253-254
Ethylidene iodide, 254
N-Ethylindoline, 531
Ethyl isocyanate, 232
Ethylketene, 674

Ethyl ketones, 551-552
Ethyl lithioacetate, 255
Ethyl lithiotrimethylsilylacetate, 255
Ethylmagnesium bromide, 115
Ethyl malonate, 255
Ethyl (R)-(−)-mandelate, 37
Ethyl mesitoate, 274
N,N-Ethylmethylephedrinium bromide, 219-220
Ethyl methylmalonyl chloride, 578
Ethyl α-methylsulfinylacetate, 255-256
Ethyl phenylacetate, 81
Ethyl α-phenylsulfinylacetate, 256
Ethyl phenyl sulfoxide, 257-258
Ethyl propionate, 627
Ethyl pyruvate, 10, 11
Ethyl 1,4,5,6-tetrahydronicotinate, 99
Ethyl vinyl ether, 530
Ethyl vinyl ketone, 31, 110
Ethynylcarbinols, 324, 541
1-Ethynylcyclohexanol, 360
17α-Ethynyl-17β-hydroxy steroids, 655
Etianic acids, 368
Eugenol, 155
Eusiderin, 516

Farnesenic acid, 627, 628
Farnesol, 447, 517, 588, 589
Farnesyl bromide, 328, 517
Favorskii rearrangement, 600
Fenchyl chloroacetates, 497, 498
Fenchylidenefenchane, 508, 509
Ferric chloride, 259
Ferric chloride−Acetic anhydride, 260
Ferric chloride−n-Butyllithium, 260
Ferric ethoxide, 95
Ferric thiocyanate, 259
Ferrous chloride, 428
Ferrous perchlorate, 260-261
Ferrous sulfate, 82
Fervenulin, 547
Fetizon reagent, 511-514
Finkelstein reaction, 174
Fischer esterification, 37
Fischer indole synthesis, 457
9-Fluorenol, 16
Fluorination, 183-184, 186, 669-670
1-Fluoroadamantane, 514
Fluoro allyl alcohols, 511
Fluoroamine reagents, 66

1-Fluoroanthracene, 670
Fluoroantimonic acid, 24
π-(Fluorobenzene)chromium tricarbonyl, 104
Fluorocarbene, 261
α-Fluorocarboxylic acids, 473
Fluorocyclopropanes, 261
Fluorodehydroxylation, 560
Fluorodienes, 175
Fluorodiiodomethane, **261-262**
Fluorodimethoxyborane, **261-262**
6-Fluoro-4,4-dimethylcyclohexene-2-one-1, 263
Fluoroiodomethyltriphenylphosphonium iodide, 262
3-Fluoroisoprene, 173
α-Fluoro ketones, 473
2-Fluoro-3-methyl-1,3-butadiene, 175
Fluoromethylenetriphenylphosphorane, **262**
m-Fluorophenol, 669
o-Fluorophenol, 669
p-Fluorophenol, 669
1-Fluoropyrene, 670
2-Fluoropyrene, 270
Fluorosulfuric acid, 262-263
1-Fluorovinyl methyl ketone, **263**
Fluoroxytrifluoromethane, **263-264**
Formaldehyde, 18, **264-267**, 379, 590
Formaldehyde diphenyl thioacetal, **267**
Formamidinesulfinic acid, **586**
Formic acid, 597
Formylation, 119, 120, 248, 280
N-Formylbenzylisoquinolines, 660
2-Formylcyclohexanone, 93
Formyl esters, 280
3-Formylindole, 78, 91
α-Formyllactones, 537
Formylmethylenetriphenylphosphorane, **267-268**
o-Formylphenols, 119-120
Fragranol, 320
Freon 113, 103
Friedel-Crafts reaction, 105
Fries rearrangements, 151-152
Fulvenes, 170
Funtaphyllamine, 162
2,3-Furanediones, 424
Furanes, 69, 200, 460, 591
2(5H)-Furanones, 200

3-Furanones, 397, 398
2-Furoic acid, 323

Gabriel synthesis, 43
Galvinoxyl, 510
Geranial, 613
Geraniol, 182, 183, 262, 263, 613
Geranylacetone, 49
Geranyl chloride, 598, 645
Geranylgeranic acid chloride, 554
Geranylgeraniol, 448
trans,trans-Geranyllinalool, 87
β-Glucopyranosides, 568
β-Glucosides, 511
Glutathione, 422
Glycidic thiol esters, 542
Glycidonitriles, 103
Glycouril, 563
Glyoxylic acid, 289, 485
Grandisol, 319, 320, 334, 335
Graphite bisulfate, **269**
Grifolin, 517
Grignard reagents, **269-270**
Guanosine, 221
Guaiacylacetone, 169

Haller-Bauer reaction, 525
α-Halo acyl chlorides, 75
2-Haloethyl esters, 553
γ-Halo ketones, 524
Halomethyl aryl ketones, 103
Halomethyl aryl sulfones, 41
Halomethylation, 105
ω-Halo-1-phenyl-1-alkynes, 165
Halopropyl isocyanates, 570
2-Halopyridines, 479
β-Halo-α,β-unsaturated ketones, 647-648
β-Halovinyl ketones, 644
Haworth succinic anhydride synthesis, 474
Hemithioacetal interchange, 381-382
1,6-Heptadiyne, 153
Heptalene, 657
3,8-Heptalenes, 672
n-Heptanal, 359, 381
1,3,4,6-Heptatetraene, **271**
Heterohelicenes, 524-525
Hexachloroethane, 413
Hexadecyltributylphosphonium bromide, 249, **271-272**, 543, 586
2,4-Hexadiene, 573

1,5-Hexadiyne, 153
Hexafluoroantimonic acid, **272-273**
3,4,6,7,8,9-Hexahydro-2-quinolizones, 662
Hexamethyl-Dewar benzene, 23, 443
Hexamethyldisilazane, 57, **273**, 640
Hexamethyldisiloxane, 226
Hexamethylditin, **273**
Hexamethylphosphoric triamide, 94, 104,
   116, 142, 158, 161, 164, 203, 204, 255,
   **273-279**, 331, 334, 383, 385, 399, 415,
   419, 425, 475, 496, 538, 657, 663
Hexamethylphosphorus triamide, 279-280,
   448
Hexamethylphosphorous triamide–Dialkyl
   azodicarboxylates, **280**
1,6-Hexanediol, 379
Hexanenitrile, 145
1-Hexene-3-ol, 425
1-Hexene-3-one, 425
3-Hexenyl phenyl ketone, 211
2-Hexyl-4-benzylfurane, 460
Hinokitiol, 196
Hinsberg test, 531
Histrionicotoxins, 611
HMP, 448
Hofmann elimination, 85
Hofmann rearrangement, 316
Homoallylic alcohols, 318, 361, 403, 404,
   583
Homoallylic halides, 353
Homoconjugate reactions, 216-217
Homoestrone, 410
Homogentisic acid, 432
4-Homoisotwistane, 525
D-Homo-19-nortestosterone, 265
Homophthalic acids, 144
2,3-Homotropilidene, 68, 69, 518
3,4-Homotropilidene, 550
Hydrangenol, 158
Hydration, 360
Hydrazine, **280-281**
Hydrazinolysis, 280
Hydrazobenzene, 428
Hydrazodicarboxylic acid, 513
Hydrazoic acid, 563
Hydrazones, 9, 36, 514
Hydridochlorotris(triphenylphosphine)-
   ruthenium(II), 425
Hydriodic acid, **281**, 470
Hydroacylation, 484-485

Hydroazulenes, 378
Hydroboration, 33-34, 62, 162
Hydrobromic acid, **282**
Hydrocarboxylation, 450-451
Hydrochloric acid, **283**
Hydrocyanation, 136
Hydrocyanosilylation, 633
Hydrofluoric acid, **284**
Hydroformylation, 177
Hydrogenation, 45
Hydrogen bromide, 282-284
Hydrogen chloride, **285**
Hydrogen fluoride, 285, 669, 670
Hydrogen fluoride–Pyridine, **285-286**
Hydrogen hexacyanoferrate(III), 656
Hydrogen peroxide, 197, **286**, 459
Hydrogen peroxide–Acetic acid, 286-287
Hydrogen peroxide–Dibromantin, 287-
   288
Hydrogen selenide, **288**
Hydroperoxidation, 435
$\alpha$-Hydroperoxides, 427
Hydroperoxyhexadienones, 99
3-Hydroperoxyindoline, 434
$\alpha$-Hydroperoxynitriles, 430
Hydrosilylation, 652-653
Hydrostannation, 604
p-Hydroxyacetophenone, 152
$\delta$-Hydroxyacetylenic acids, 495-496
$\alpha$-Hydroxy acids, 6, 37, 113
$\beta$-Hydroxy alcohols, 403, 404
$\alpha$-Hydroxy aldehydes, 112
$\beta$-Hydroxy aldehydes, 462
2-Hydroxy-3-alkylcyclopent-2-ene-1-ones,
   558-559
$\beta$-Hydroxyalkylsilanes, 199-200
p-Hydroxybenzaldehyde, 158
1-Hydroxybenzotriazole, 35, **288**, 425
m-Hydroxybenzyl alcohol, 126
4-Hydroxybutynenitriles, 147
$\alpha$-Hydroxycarboxamides, 427
$\beta$-Hydroxycarboxylic acids, 221-222
2'-Hydroxychalcones, 358, 359
4$\alpha$-Hydroxy-$\Delta^5$-cholestene, 46
6$\beta$-Hydroxy-$\Delta^4$-cholestene, 46
7$\alpha$-Hydroxy-$\Delta^5$-cholestene, 45, 47
25-Hydroxycholesterol, 74
26-Hydroxycholesterol, 202
p-Hydroxycinnamic acid, 102
3-Hydroxy-N,N-dimethyl-1-carboxamides, 302

β-Hydroxy esters, 11, 180, 256, 674, 675
5-(2'Hydroxyethyl)-4-methyl-3-benzylthiazolium chloride, **289**
5-(2'-Hydroxethyl)-4-methylthiazole, 38, **289**
α-Hydroxyhippuric acid, **289-290**
δ-Hydroxy-β-keto esters, 592-593
α-Hydroxy ketones, 112
α-Hydroxy ketoximes, 469
Hydroxylamine, 533
Hydroxylamine-O-sulfonic acid, **290**
Hydroxylation, 140, 166, 240, 286, 296, 427-428, 440
2-Hydroxy-4,10-methano[11]annulenone, 33
5-Hydroxymethylbenzisoxazole, 32
3-Hydroxy-2-methylcyclohexanone, 214
α-Hydroxymethyl ketones, 264-266
p-Hydroxymethylphenylboronic acid, 194
N-Hydroxy-5-norbornene-endo-2,3-dicarboximide, 290-291
2-Hydroxy-5-oxo-5,6-dihydro-2H-pyrane, **291**
15-Hydroxypentadecanoic acid, 246, 520
4-Hydroxy-3-phenyl-2-heptanone, 676
p-Hydroxyphenylpyruvate hydroxylase, 432
p-Hydroxyphenylpyruvic acid, 432
1α-Hydroxyprovitamin D₃, 574
1-Hydroxypyrazole 2-oxides, 127-128
3-Hydroxy-2-pyrone, 291-292
4-Hydroxy-2-pyrones, 424
4-Hydroxyquinolines, 503
β-Hydroxyselenides, 28, 29, 86, 549
β-Hydroxyselenocyanates, 487
N-Hydroxysuccinimide, 290, 425
β-Hydroxy sulfides, 363
β-Hydroxy sulfoxides, 5, 561
β-Hydroxy sulfoximines, 395
β-Hydroxy thiol esters, 334
2-Hydroxy-2,4,5-triphenylfuranone-3, 489, 490
3-Hydroxytropolone, 168
5-Hydroxytropolone, 168
γ-Hydroxy-α,β-unsaturated aldehydes, 53
4-Hydroxyvaleric acid, 511
Hydrozirconation, 176-179

Imidazole, 81, 97, 218, 219
20-Iminopregnanes, 161

Iminosulfuranes, 230
Indane 8,9-oxide, 435, 436
Indanes, 153
2-Indanone, 551
Indene dibromide, 50
Indenes, 282, 283
Indenones, 47
Indium, **293**
Indole, 531, 565, 566
Indoles, 526
Indoline, 652
Ing-Manske reaction, 280
Intramolecular ester condensation, 475
Iodine, 72, **293-295**, 333, 452, 579
Iodine–Dimethyl sulfoxide, 295
Iodine–Potassium iodate, **296**
Iodine–Pyridine N-oxide, **296**
Iodine–Pyridine–Sulfur dioxide, **296**
Iodine azide, **297**
Iodine bromide, **297**
Iodoacetonitrile, 494
α-Iodo acyl chlorides, 116
o-Iodoanisole, 130
Iodobenzene, 130, 443
π-(Iodobenzene)chromium tricarbonyl, 104
Iodobenzene diacetate–Trimethylsilyl azide, **297**
Iodobenzene dichloride, **298-300**
Iodobenzene difluoride, **300**
Iodobenzene ditrifluoroacetate, **301**
Iodocyclopropanation, 194
α-Iododiketones, 354
o-Iodofluorobenzene, 130
α-Iodo ketones, 470
Iodomethyl methyl sulfide, 301-302
o-Iodonitrobenzene, 130
p-Iodonitrobenzene, 130
α-Iodopyruvates, 354
Ion-exchange resins, **302-304**
Ionic hydrogenation, 616
β-Ionol, 526, 527
β-Ionone, 440
Ion-pair extraction, 41
β-Ionyl phenyl sulfone, 526, 527
Iron(III) acetylacetonate, **304**
Iron carbonyl complexes, 624
Iron pentacarbonyl, 39, **304-306**
Iron(II) perchlorate, **260-261**
Isoalantolactone, 490, 491

Isoamyl nitrite, 307, 547
4-Isoavenaciolide, 253
Isobenzothiophenes, 197
Isobutylaluminum dichloride, 307
Isobutyl chloroformate, 307
Isobutyronitrile, 104
Isobutyrophenone, 240
Isochanoclavine-I, 559
Isocoumarin, 86
Isocyanates, 69, 316, 564
Isomerization, olefins, 585
Isophorone, 599, 600
Isoprene, 48
Isoprene epoxide, 307-308
Isoprenoids, 151, 346
Isopropenyl acetate, 308
1-Isopropenyladamantane, 623
Isopropenylation, 308
Isopropenyllithium, 308
Isopropenyltriphenylphosphonium bromide, 667
N-Isopropylallenimine, 78
Isopropylbenzene, 623
Isopropyl α-chloroalkanoates, 331
2-Isopropylfurane, 196
3-Isopropylfurane, 196
Isopropylidenation, 211
Isopropylidenecarbene, 230-231, 619
Isopropylidenecyclopropenes, 231
2,3-Isopropylidene-2,3-dihydroxy-1,4-bis(diphenyl phosphino)butane, 309
Isopropylidene isopropylidenemalonate, 309-310
N-Isopropyl-β-lactamimide, 78
Isopropylmagnesium bromide, 179
Isoquinolinium perchlorate, 408
Isotetralin, 107, 672, 673
Isothiocyanates, 95-96
Itaconic acid, 366
Itaconic anhydride, 365

Jasmine lactone, 462
cis-Jasmone, 39, 67, 339
Jones reagent, 123-124, 228, 333
Juglone, 99

Ketalization, 302, 379, 380
Ketene, 11
Ketene S,S-dimethyl acetals, 54-55
Ketene dimethyl thioacetal monoxide, 311-312
Ketenes, 674
Ketene thioacetals, 49-50, 53, 118, 320, 321
Ketimines, 332
β-Keto acetals, 594
α-Keto acid chlorides, 172
α-Keto acids, 103
β-Keto acids, 84, 466
δ-Keto acids, 556
γ-Keto acids, 204, 205
β-Keto aldehydes, 397-398
Ketocarbenoids, 515
α-Keto esters, 238, 353-359, 550
β-Keto esters, 144, 212, 256, 356, 501
γ-Keto esters, 139, 204, 205, 251, 651
3-Keto-12α-hydroxycholanic acid, 166
1,2-Ketols, 408
1,4-Ketols, 333
α-Ketonitriles, 103
γ-Ketonitriles, 535
β-Keto phenyl sulfoxides, 374
11-Ketoprogesterone, 226
17-Ketosteroids, 368
20-Ketosteroids, 368
β-Keto sulfoxides, 5, 6, 257
Ketoximes, 4
Kindler reaction, 21
Koenigs-Knorr synthesis, 360, 511, 514, 568
Knoevenagel condensation, 10-11, 158, 484, 515
Knorr reaction, 225
Kornblum oxidation, 227
K Selectride, 490
Kuhn methylation, 541

Lactams, 504, 505
β-Lactams, 71, 122, 566
Lactols, 528-529
δ-Lactols, 462
Lactone annelation, 661-662
β-Lactones, 343
γ-Lactones, 355-356, 608, 654-655
δ-Lactones, 440
Lactonization, 246-247, 519-520, 581-582
Lanosterol, 426, 455
LDA, 334-339
Lead tetraacetate, 4, 313-317
Lead tetraacetate—N-Chlorosuccinimide, 317

Lead tetraacetate—Trifluoroacetic acid, 317
Lead tetrakis(trifluoroacetate), 318
Lemieux-Rudlolf oxidation, 455
Levulinic acid, 354
Levulinic anhydride, 318-319
Levulinic esters, 318-319
Limonene, 535
Lindlar catalyst, 67, 319
Lithioacetonitrile, 34, 636
1-Lithiocyclopropyl bromides, 89
1-Lithiocyclopropyl phenyl sulfide,
   319-320
2-Lithio-3,3-diethoxypropene, 330
Lithio-N,N-dimethylacetamide, 84, 636
2-Lithio-1,3-dithianes, 248
Lithiodithiophenoxymethane, 131
1-Lithio-3-methoxypropyne, 397
2-Lithiomethoxy-1,3-thiazoline, 399
α-Lithio selenides, 335
α-Lithio selenoxides, 335
2-Lithio-2-trimethylsilyl-1,3-dithiane,
   320-321
Lithio-1-trimethylsilylpropyne, 638
Lithium, 274, 351
Lithium—Alkylamines, 322
Lithium—Ammonia, 322-323, 463, 502
Lithium—Ethylamine, 12
Lithium—Hexamethylphosphoric triamide,
   323
Lithium—Trimethylchlorosilane, 323
Lithium acetylide, 324-325
Lithium alkynides, 189
Lithium aluminum hydride, 46, 118,
   325-326
Lithium aluminum hydride—Cuprous
   iodide, 326
Lithium azide, 327
Lithium bis(dialkylamino) cuprates, 327-
   328
Lithium bis(ethylenedioxylboryl)methidy,
   328-329
Lithium bis(trimethylsilyl)amide, 542
Lithium t-butoxide, 115, 116, 118
Lithium t-butylmercaptide, 343, 344
Lithium α-carboethoxyvinyl(1-hexenyl)-
   cuprate, 329-330
Lithium α-carbomethoxy(1-hexynyl)-
   vinylcuprate, 330
Lithium carbonate, 17, 18
Lithium chloroacetylide, 293

Lithium α-cyanohydroperoxides, 430
Lithium dicyclohexylamide, 162
Lithium diethylamide, 181, 331
Lithium diethylamide—Hexamethylphos-
   phoric triamide, 332-333
Lithium diisopropylamide, 11, 13, 15, 28,
   30, 49, 84, 88, 104, 116, 170, 235, 236,
   255, 334-339, 365, 366, 389, 393, 427,
   430, 495, 549
Lithium dimethoxyphosphinylmethylide,
   339
Lithium di(α-methoxyvinyl)cuprate, 204-
   205
Lithium dimethyl cuprate, 209-215
Lithium diphenylarsenide, 341
Lithium diphenyl cuprate, 234
Lithium diphenylphosphide, 340-341, 645
Lithium diphenylphosphinate, 340
Lithium di-n-propyl cuprate, 245
Lithium ethynyltrialkylborates, 324-325
Lithium fluoride, 17
Lithium N-isopropylcyclohexylamide, 79,
   238, 255, 394, 627
Lithium isopropylmercaptide, 343
Lithium methoxyaluminum hydride, 341-
   342
Lithium methylbromocuprate-diisobutyl-
   amine complex, 213
Lithium 4-methyl-2,6-di-t-butylphenoxide,
   95
Lithium methylsulfinylmethylide, 342
Lithium methyl(vinyl)cuprate, 342-343
Lithium naphthalenide, 415
Lithium organocuprates, 225
Lithium perchlorate, 103
Lithium phenylethynolate, 343
Lithium phenylthio(alkyl)cuprates, 344,
   465
Lithium phosphate, 334
Lithium piperidide, 345-346
Lithium n-propylmercaptide, 345-348
Lithium pyrrolidide, 327
Lithium 2,2,6,6-tetramethylpiperidide,
   181, 345-346
Lithium trialkylalkenylborates, 208, 283,
   493-494
Lithium tri-sec-butylborohydride, 348, 490
Lithium triethylborohydride, 348-349
Lithium triethylmethoxide, 349-350
Lithium trifluoroacetate, 103

Lithium trimethoxyaluminum hydride, 341
Longifolene, 613
2,4-Lutidine, 224
2,6-Lutidine, 130, 409
Lycoricidine, 69
Lysergic acid, 1

Macrolides, 479, 581-582
Magnesium, 351-352
Magnesium bromide, 270, 353
Magnesium bromide−Acetic anhydride,
    352-353
Magnesium iodide, 270, 353-354
Magnesium methyl carbonate, 354
Maleic anhydride, 271, 370
Malonic ester synthesis, 255
Malononitrile, 10, 11
Mandelic acid, 5, 6, 43
Manganese(III) acetate, 355-356
Manganese dioxide, 123, 357, 511
Mannitol, 134
McMurry-Fleming olefin synthesis, 589
McMurry's reagent, 588
Meerwein-Ponndorf reaction, 586
Meerwein's reagent, 611
Meisenheimer complexes, 157-158
Menthoxyacetyl chloride, 357-358
Menthyl chloride, 416
Mercuric acetate, 123, 198, 358-359
Mercuric chloride, 31, 53, 110, 359
Mercuric oxide, 248, 360
Mercuric oxide−Mercuric bromide,
    360
Mercury(I) nitrate, 360
Mercury(II) trifluoroacetate, 360-361, 581
Merrifield peptide synthesis, 148-149, 193
Mesityl bromide, 59
O-Mesitylenesulfonyl chloride, 625
O-Mesitylenesulfonylhydroxylamine, 320
1-(Mesitylenesulfonyl)-1,2,4-triazole, 361
Mesityl oxide, 408, 499, 652
Mesylates, 362-363
Mesylation, 190
[2.2]Metacyclophane, 499
Metacyclophanes, 51, 85
Methacrolein, 319, 371
Methacrylonitrile, 136, 144
Methallyl bromide, 361, 539
Methallyl chloride, 12, 175
S-Methallyl N,N-dimethyldithio-

carbamate, 14
π-2-Methallylnickel bromide, 361
Methaneselenol, 361-362
Methanesulfenyl chloride, 375
Methanesulfonamides, 190
Methanesulfonic acid, 3, 109
Methanesulfonyl chloride, 362-363, 598
1-Methanesulfonyloxybenzotriazole, 106
Methionine peptides, 149
4-Methoxy-5-acetoxymethyl-o-benzoqui-
    none, 363-364
2-Methoxyalkanoic acids, 204
Methoxyallene, 146
2-Methoxyallyl bromide, 364
(2-Methoxyallyl)-N,N-dimethyldithio-
    carbamate, 364-365
p-Methoxybenzaldehyde, 510
o-Methoxybenzoic acid, 368
p-Methoxybenzyl alcohol, 365
p-Methoxybenzyl esters, 597
p-Methoxybenzyl itaconate, 365-366
4-Methoxybutene-2-one, 370
2-(6'-Methoxycarbonylhexyl)cyclopentene-
    2-one-1, 475
2-Methoxycyclopropyllithium, 366-367
4-Methoxy-5,6-dihydro-2H-pyrane, 367
Methoxy-2,5-dinitroanisole, 445
2-Methoxyethyl esters, 564
Methoxyformylbenzofuranes, 498
β-Methoxy ketones, 308
Methoxymethyl aryl ethers, 109
4-Methoxy-5-methyl-o-benzoquinone,
    363-364
Methoxymethylenetriphenylphosphorane,
    368
Methoxymethyl ethers, 374
Methoxymethyl phenyl thioether, 369
2-(o-Methoxyphenyl)-4,4-dimethyl-2-
    oxazoline, 368
2-(4-methoxy-5-phenyl-3-thienyl)acrylic
    acid, 309
2-(4-methoxy-5-phenyl-3-thienyl)prop-
    ionic acid, 309
Methoxyphenylthiomethyllithium, 369
α-Methoxystilbene, 383
1-Methoxy-3-trimethylsilyloxy-1,3-butadiene,
    370-371
αMethoxyvinyllithium, 205, 372-373
Methyl acetoacetate, 290
Methyl acrylate, 371, 372

1-Methyladamantane, 623
2′-O-Methyladenosine, 554, 555
3′-O-Methyladenosine, 554, 555
Methylal, 374
2-Methylalkanoic acids, 387
N-Methylamides, 109, 598
9-Methylanthracene, 623
Methyl anthranilate, 130
N-Methylbenzanilide, 53
Methyl benzenesulfinate, 374
Methyl benzoate, 132
2-Methylbenzofurane, 119
3-Methylbenzothiazole-2-thione, 374-375
α-Methylbenzylamine, 457
Methyl benzylpenicillate, 344
1-Methylbicyclo[2.2.2]octane, 623
Methyl(bismethylthio)sulfonium hexa-
    chloroantimonate, 375
syn-7-Methyl-anti-7-bromonorcarane, 89
Methylbromoketene, 157
1-Methyl-2-bromopyridinium iodide,
    375-376
2-Methylbutane, 428
3-Methyl-1-butene, 638
2-Methylbutene-2-ol, 1
Methyl 2-butynoate, 163
Methyl chloroformate, 376
Methyl 4-(chloroformyl)butyrate, 611
Methyl-3-chloropropenoate, 14
Methyl α-chloropropionate, 144
Methyl chlorosulfinate—Dimethyl sulf-
    oxide, 377
Methyl cholanate, 426
3α-Methylcholestane-3β-ol, 623
Methyl cinnamate, 348
β-Methylcinnamic acid, 416
Methylcobalamin, 139
Methylcopper, 377
2-Methylcyclohexane-1,3-dione, 661
2-Methylcyclohexanone, 162, 482
3-Methylcyclohexanone, 651
2-Methylcyclohexenone, 372, 407
3-Methylcyclohexenone, 651
1-Methylcyclononene, 523
1-Methylcyclooctene, 523
Methylcyclopentadienes, 142
2-Methylcyclopentanecarboxylic acid, 629
2-Methylcyclopentenone-3-dimethyl-
    sulfoxonium methylide, 378
3-Methylcyclopentenones, 39

Methyl desoxypodocarpate, 243
Methyl 3,7-diacetylcholate, 475, 476
Methyl $\Delta^{11}$-3,7-diacetylcholenate, 475, 476
2-Methyl-1,10-dibromodecane, 51
N-Methyl-N,N′-di-t-butylcarbodiimium
    tetrafluoroborate, 378-379
1-Methyl-1,6-dihydrocarvone, 492
Methyl diphenylphosphinite, 265
Methylenation, 10, 395, 419, 463, 595,
    636-637
α-Methylene-γ-butyrolactones, 93, 330,
    366, 451, 552
Methylenecycloalkanes, 418
Methylenecyclobutane, 151
Methylenecyclobutanes, 251
Methylenecyclobutanones, 223
Methylenecyclohexane, 421
Methylenecyclopropanes, 42, 530
Methylene diesters, 565
α-Methylenelactam rearrangement, 2-3
α-Methylene-γ-lactones, 157, 460, 466,
    537
α-Methylene-δ-lactones, 369
2-Methylene-1,3-propanediol, 379-380
Methylenetriphenylphosphorane, 380-381,
    463, 612, 635
β-Methyl-α,β-ethylenic sulfoxides, 377
Methyl farnesoate, 448
Methyl fluoride—Antimony penta-
    fluoride—Sulfur dioxide, 381
Methyl fluorosulfonate, 26, 48, 381-382
Methyl geraniate, 447
Methyl gibberellate, 186
Methyl α-D-glucopyranoside, 26
β-Methyl glycopyranosides, 437, 438
Methyl 18α-glycyrrhetate, 501
4-Methyl-1,6-heptadiene-4-ol, 512
(S)-(+)-4-Methyl-3-heptanone, 105
1-Methyl-4a,5,6,7,8,8a-trans-hexahydro-
    naphthalene, 432
5-Methyl-2-hexyne, 178
Methylhydrazine, 281
Methyl hypobromite, 72
2-Methylimidazole, 357
2-Methyl-2-imidazoline, 357
1-Methylindenes, 468
Methyl iodide, 384
Methyl o-iodobenzoate, 130, 132
Methyl isobutyl ketone, 652
Methyl itaconate, 366

Methyl ketones, 123, 139, 216, 256, 257, 324-325, 332, 336, 365, 372, 380, 384, 402, 464, 484, 518, 624
Methyl levulinate, 626, 627
Methyl α-lithiodichloroacetate, 331
Methyllithium, 68, 90, 294, **384**
Methyllithium—Dimethylcopperlithium, **386**
Methyl lithocholate, 426
Methylketene, 674
Methyl mesityl ketoxime, 304, 305
Methyl methoxymagnesium carbonate, **354**
2-Methyl-4-methoxymethyl-5-phenyl-2-oxazoline, **386-388**
1-Methyl-3-methylene-2-piperidone, 1
Methyl γ-methylthiocrotonate, **389-390**
Methyl methylthiomethyl sulfoxide, **390-392**, 446
N-Methylmorpholine, 106, 425, 422
α-Methylnaphthalene, 222
1-Methylnipecotic acid, 2
N-Methyl nitrones, 598
1-Methyloxalylpiperidine, 504, 505
1-Methyloxalylpyrrolidine, 504, 505
Methyl 3-oxo-5β-cholanate, 297
Methyl 3-oxo-4-pentenoate, **662**
3-Methyl-2-pentenoic acid, 664
Methyl phenylacetate, 125
N,N-Methylphenylaminotriphenylphosphonium iodide, 392
Methyl phenyl phosphate, 152
3-Methyl-1-phenyl-2-phospholene, **392-393**
Methyl phenylpropiolate, 575
Methyl phenyl selenide, 29
Methyl 2-phenylsulfinylacetate, **393-395**
N-Methylphenylsulfonimidoylmethyllithium, **395**
2-Methylpropane-2-thiol, 581
Methylpyridines, 450
N-Methylpyrrolidone, 142, 621
Methylquinolines, 450
Methyl salicylate, 132
α-Methylstilbenes, 649
α-Methylstyrene, 38, 171, 302
2α-Methyltestosterone, 521, 522
4-Methyltestosterone, 521, 522
Methyl tetra-O-benzyl-α-D-gluco-pyranoside, 568
N-Methyltetrahydroprotoberberinium

iodide, 546, 547
2-Methyl-2-thiazoline, **403-404**
Methylthioacetic acid, **395-396**
S-γ-Methylthioallyl dithiocarbamate, 13
S-γ-Methylthioallyl dithiocarbamates, 539, 540
o-Methylthiobenzyl chloride, 18
p-Methylthiobenzyl chloride, 18
S-Methyl thiocarboxylates, 54-55
5'-S-Methylthio-5'-desoxyandenosine, 602
3-Methylthio-1,4-diphenyl-s-triazolium iodide, **396-397**
S-γ-Methylthiomethallyl dithiocarbamates, 539, 540
1-Methylthio-3-methoxypropyne, **397-398**
2-Methylthio-7-methoxytetralone-1, 400
Methylthiomethyl N,N-dimethyldithio-carbamate, 398
Methylthiomethyl ethers, 109, 225, 302
Methylthiomethyllithium, **398-399**
o-Methylthiomethyl phenol, 120
1-Methylthio-2-propanone, 118
Methyl thiotosylate, **400-403**
Methylthiotrimethylsilane, 399
Methyl p-toluenesulfinate, 374, **400**
S-Methyl p-toluenethiosulfonate, **400-403**
Methyltriacetic lactone, 578
Methyltricaprylylammonium chloride, **404-406**
Methyl trifluoromethanesulfonate, **406**
Methyl α-trimethylsilylvinyl ketone, **406-407**
Methyl triphenoxyphonium iodide, **649**
N-Methyltryptophol, 434
Methyl 10-undecenoate, 475
β-Methyl-α,β-unsaturated ketones, 385
Methyl vinyl ether, 372
Methyl vinyl ketone, 371, 390, **407-409**, 425, 462, 538, 650
Methyl vinyl ketones, 597
2-Methyl-6-vinylpyridine, **409-411**, 605, 606
Methylvitamin $B_{12}$, 139-140
Methymycin, 582
Methynolide, 582
Mevalonolactone, 255, 512
Michael addition, 14
MMC, 354
Moenocinol, 421-422
Molecular sieves, **411-412**

Molybdenum hexafluoride, 412
Molybdenum pentachloride, 412-413
Monoalkylboranes, 207
N-Monoalkylhydroxylamines, 538
Morphinanedienones, 660
Morpholine, 346
3-Morpholino-3-cholestene-2-one, 433
Mucic acid, 291
Muconic acid, 429
Muconitrile, 429
Muscone, 385
Muscopyridine, 51
Mycophenolic acid, 517
Myrcene, 151, 183, 578, 579
Myrinol, 358

Naphthalene, 506, 533
Naphthalene–Lithium, 415
Naphthalene 1,2-oxide, 358
Naphtho[b] cyclobutene, 550
Naphtho[b]-cyclopropene, 477, 478
Naphthols, 503
Naphtho[1,8-bc]-pyrane, 86
1,4-Naphthoquinones, 187
1,5-Naphthoquinones, 170
N-(1-Naphthyl)acetamides, 3
α-Naphthyldiphenylmethyl chloride, 415
α-Naphthyldiphenylmethyl ethers, 415
1-(1-Naphthyl)ethylamine, 416
1-(1-Naphthyl)ethyl isocyanate, 416
Nef reaction, 545
Neomenthyldiphenylphosphine, 416
Neomenthyldiphenylphosphine oxide, 416
Neopentyl alcohol, 222, 279
Nephrosterinic acid, 366
Nephthenol, 87
Nerol, 182, 183, 262, 263, 346
Neurotensin, 618
Nezukone, 196
Nickel(II) acetate, 417
Nickel(II) acetylacetonate, 417, 623, 624
Nickel carbonyl, 93, 417-419
Nickel tetracarbonyl, 417-419, 451
Nicotinamide, 36
Nicotine, 72
Nimbiol, 499
Nitration, 284, 420
Nitric acid, 284
Nitriles, 170, 229, 233, 250, 279, 290,
    449, 456, 569, 600, 610, 636, 648

1-Nitroadamantane, 272
Nitroalkanes, 272
o-Nitroaniline, 130
Nitroarenes, 273, 281
Nitroaryl iodides, 579-580
Nitrobenzene, 45
2-Nitrobenzenesulfenyl chloride, 419
o-Nitrobenzenesulfonyl chloride, 419-420
2-Nitrofluorene-9-ol, 16
3-Nitrofluorene-9-ol, 16
3-Nitrofluorene-9-one, 16
Nitrogen dioxide, 284
Nitrones, 538
Nitronium tetrafluoroborate, 420
Nitronium trifluoromethanesulfonate, 420
1-Nitrooctane, 135
p-Nitrophenyl(1,2-dimethylethenylene-
    dioxy)phosphate, 8
Nitrophenylenediamines, 446
o-Nitrophenyl selenocyanate, 420-422
2-Nitrophenylsulfenyl chloride, 422
Nitroamines, 336-337, 515
Nitrosation, 127
Nitrosative cyclization, 547
Nitroso compounds, 97-98
Nitrosolysis, 422
N-Nitroso-N-methylurea, 159
β-Nitrostyrene, 422
Nitrosyl chloride, 422-423
Nitrosyl tetrafluoroborate, 226
α-Nitrotriphenylmethane, 272
Nitroxides, 455-456
Nitroxyl radicals, 110
Nonactic acid, 106, 107
Nonactin, 106, 625
2-Norbornanone, 554
Norbornene, 232, 434
5-Norbornene-2-carboxaldehyde, 379
5-Norbornene-endo-2,3-dicarboxylic
    anhydride, 290
Norbornene epoxide, 435
Norbornyne, 90
Norcarane, 577
3-Norcarene, 68, 518
21-Norconanine, 72
Norpinene, 455
19-Norsteroids, 411
Nortricyclene, 232
Nuciferal, 461
Nucleosides, 167, 379, 639-640

Nucleotides, 167-168, 379

β-Ocimene, 183
1,7-Octadiyne, 153
Δ¹-3-Octalones, 371-372
Octanal, 116
Octanol-1, 116
Octanol-2, 116, 279
Octanone-2, 116, 124, 599
Octanone-3, 124
1,3,6-Octatriene, 58, 59
1-Octene, 599
2-Octene, 124
4-Octene, 199
1-Octene-4-yne, 155
1-n-Octylnaphthalene, 570
Oligonucleotides, 361
Oppenauer oxidation, 126
Orcinol, 517
o-Orsellinic acid, 593
p-Orsellinic acid, 593
Ortho ester rearrangement, 609
Osmium tetroxide, 230, 641
Osmium tetroxide−Sodium chlorate, 424,
    436
9-Oxabicyclo[4.2.1]nona-2,4,7-triene, 306
9-Oxabicyclo[6.1.0]nona-2,4,6-triene, 306
8-Oxabicyclo[3.2.1]oct-6-ene-3-one, 195
Oxalic acid, 511
Oxalyl chloride, 424
Oxalyldiacetone, 486
7-Oxanorbornadiene, 583-584
Oxathiazine 2,2-dioxide, 122
6H-1,3,5-Oxathiazines, 559
6-Oxatricyclo[3.2.1.0²'⁷]octane, 385
Oxazaphospholidines, 612
Oxazinediones, 122, 632
Oxaziridines, 35, 457
Oxazolidine-2-thiones, 225
Oxazolines, 69, 187
Oxepane, 379
Oxetanes, 56-57
Oxidative acetoxylation, 51-52
Oxidative coupling, 138, 516
Oxidative decarboxylation, 116-117, 313
    396, 427
Oxidative decyanation, 430
Oxidative phenol coupling, 480-481, 580,
    660
Oxidative ring cleavage, 313-314

N-Oxides, 35
Oximes, 9, 538
ω-Oximinocaproic acid, 423
α-Oximino ketones, 422, 423
μ-Oxo-bis[tris(dimethylamino)phosphonium]
    bistetrafluoroborate, 425
ζ-Oxo-α,β-enones, 213
α-Oxoketenes, 424
12-Oxo-11-oxa[4.4.3]propella-3,8-diene,
    107
7-Oxoprostaglandins, 164
Oxo reaction, 172
4-Oxo-1,3-thiazolidine-1,1-dioxides, 558
Oxotriruthenium acetate complex, 425-
    426
10-Oxoundecanal, 17
10-Oxoundecanol, 17
Oxyamination, 641-642
Oxygen, 142, 178, 194, 426-430, 595,663
Oxygen, singlet, 100, 241, 287, 431-436,
    469, 487, 577, 586
Oxygenation, 251, 477
Oxygen transfer, 595
Oxymercuration-demercuration, 359
Oxyselenation, 224
Ozone, 436-441

Paliclavine, 559-560
Palladium(II) acetate, 60, 61, 156, 442-443
Palladium acetate−Sodium chloride, 443
Palladium bis(dibenzlideneacetone), 571
Palladium black, 443-445
Palladium catalysts, 445-446
Palladium(II) chloride, 447-450
Palladium(II) chloride−Thiourea, 450-451
Palladium hydroxide, 81
Palladium nitrate, 451
[2.2]Paracyclophane, 612
Paraformaldehyde, 104
Patchouli alcohol, 478
Payne's reagent, 455-456
Penams, 115
Penicillins, 280, 566
1,3-Pentadiene, 65
Pentafluorophenylcopper, 451-452
2,4-Pentanedione, 484, 533
Pentannelation, 208-209
Pentazocine, 282
2-Pentenal, 546
1-Pentynylcopper, 603

Peptide synthesis, 25, 35, 106, 174, 194-195, 234, 254, 288, 290, 303-304, 307, 425
Peracetic acid, 452-453
Perbenzoic acid, 35, 111, 453
Perchloric acid, 231, 281, 453-454
Perchloryl fluoride, 454
Perfluorotetramethylcyclopentadienone, 454
Perhydrohistrionicotoxin, 75, 611
Perhydrophenanthrene-9-one, 535
Periodates, 454
Periodic acid, 124
Peropyrene, 499
Peroxy acids, 260
Peroxybenzimidic acid, 455-456
Peroxycyclohexanecarboylic acid, 260
Persulfoxides, 241
Peterson reaction, 631, 636
Pfitzner-Moffatt oxidation, 227-228, 467
Phase-transfer catalysts, 10, 136, 159, 174-175, 191, 219, 249, 271, 404-406, 452, 565, 566, 586, 640-641
Phenacyldimethylsulfonium bromide, 456
Phenacylidinedimethylsulfurane, 456
Phenanthrene, 35, 533, 669-670
9,10-Phenanthrenequinones, 476-477
Phenazines, 513
O-Phenethyl thiobenzoate, 583
Phenols, 580
o-Phenoxybenzaldehyde, 138, 139
6'-Phenoxylaudanosine, 452
Phenylacetaldehyde, 446
Phenyl acetate, 151, 152
Phenylalanine, 411, 493
Phenyl alkyl ketones, 34
Phenyl allyl sulfide, 605
1-Phenyl-2-amino-1,3-propanediol, 386
Phenylation, 28, 104
Phenylazomalondialdehyde, 417
3-Phenylbutanoic acid, 416
γ-Phenylbutyric acid, 20, 21, 127
δ-Phenyl-γ-butyrolactone, 20-21, 127
Phenyl chlorosulfite, 456
Phenyl chlorothiolformate, 252
Phenylcopper, 234
Phenyldiazomethane, 473
Phenyl diazomethyl sulfoxide, 457
o-Phenylenediamine, 428, 429
α-Phenylethylamine, 162, 457

α-Phenylethylammonium hexafluorophosphate, 134
Phenylfluorocarbene, 136
Phenylhydrazine, 457-458
2-Phenylhydrazonopropanal, 500
2-Phenylhydrazonopropylidenes, 500
Phenylhydroxyl amine, 617
3-Phenylindanone, 618
2-Phenylindole, 161
2-Phenylindoles, 456
2-Phenylindoxyl O-benzoate, 161
Phenyliodine dichloride, 298-300
Phenyliodine(III) difluoride, 300
Phenyliodine(III) ditrifluoroacetate, 301
Phenyl isocyanide dichloride, 458
α-Phenyl ketones, 234
Phenylmercuric acetate, 458
2-Phenylmesitylene, 59
Phenyl neopentyl sulfide, 272
Phenyl oxime-O-sulfonates, 456
2-Phenyloxirane, 219
Phenylpropargyl aldehydes, 6
Phenylselenenyl bromide, 459-460
Phenylselenenyl chloride, 459-460
Phenylselenenyl trifluoroacetate, 459
Phenylseleninyl chloride, 459-460
α-Phenylseleno ketones, 459
Phenylselenol, 235, 362
Phenylsulfinylcarbene, 457
α-Phenylsulfinylcyclohexanone, 236
Phenylsulfinylcyclopropanes, 457
Phenylsulfinyldiazomethane, 457
2-Phenylsulfonyl ketones, 3
β-(Phenylsulfonyl)propionaldehyde, ethylene acetal, 461-462
8-Phenyltheophylline, 185
2-Phenylthiirane, 585
Phenylthioacetic acid, 463-464
Phenylthiocopper, 465
1-Phenylthiocyclopropyltriphenylphosphonium tetrafluoroborate, 465-466
2-(Phenylthio)ethanols, 596
Phenylthiomethyllithium, 463, 596
2-Phenylthiopropionic acid, 466
Phenylthiotrimethylsilane, 399
cis-1-Phenyl-2-(p-tolylsulfonyl)ethene, 486
4-Phenyl-1,2,4-triazoline-3,5-dione, 253, 271, 299, 433, 467
Phenyl(tribromomethyl)mercury, 468
4-Phenylurazole, 467

5-Phenylvaleraldehyde, 403, 404
Phenyl vinyl sulfoxide, **468**
Phloroglucinol, 147
Phosgene, 166, 172
1-Phospha-2,8,9-trioxaadamantane, 469
1-Phospha-2,8,9-trioxaadamantane ozonide,
  469
Phosphodiesters, 7
Phosphonitrilic chloride, **469-470**
Phosphoric acid, 3-4
Phosphorodiamidites, 576
Phosphorus—Iodine, **470**
Phosphorus pentachloride, 172, **470-471**
Phosphorus pentasulfide, **470-471**
Phosphorus pentoxide, 374
Phosphorus tribromide, 183, 645
Phosphorus trichloride, 114, 172
Phosphorylation, 42, 152, 167-168, 207
Phosphotriester synthesis, 114-115, 605,
  622
Photo-Birch reduction, 533-534
Photooxygenation, 657
Phthalamic acids, 280, 281
Phthalans, 282, 283
Phthalic acid, 506
Phthalisoimides, 281
o-Phthaloyl dichloride, 471
N-Phthaloyl-L-glutamic anhydride, 174
Phthaloyl peroxide, 471
Phthalylhydrazide, 281
Phytoenes, 511
2-Picoline, 27
3-Picoline, 336
2-Picolyl chloride 1-oxide hydrochloride,
  **472**
Pinacolone, 139
Pinacol reduction, 352
α-Pinene, 306, 358
β-Pinene, 306, 358, 455
α-Pinenes, 498
Pinocarveol, 358
Piperidine, **472**
Piperidines, 451, 455
Piperidinosulfur trifluoride, 186
Piperonal, 70
Piperonylic acid, 70
Platinum—Silica catalyst, **472-473**
Polyenals, 411-412
Polyenes, 14-15
Polyhydrogen fluoride—Pyridine, **473-474**

Polyketides, 495
2,3-Polymethylenebenzofuranes, 436-437
Polymethylhydrosiloxane, 393, 602
Polynucleotides, 114-115, 622
Polyphosphate ester, **474**
Polyphosphoric acid, 302, **474-475**
Polystyrene—Aluminum chloride, 19
Porcine motilin, 193
Potassium, 274, 275, 293
Potassium acetate, **475-476**
Potassium 3-aminopropylamide, 476
Potassium bis(trimethylsilyl)amide, 482
Potassium borohydride, **476-477**
Potassium t-butoxide, 24, 29, 90, 228, 230,
  237, 334, **477-479**
Potassium t-butoxide—Dimethyl sulfoxide,
  479
Potassium carbonate—Dimethyl sulfoxide,
  479
Potassium chloride, **480**
Potassium cyanide, 135, 136
Potassium cyanide—18-Crown-6, 399
Potassium ferricyanide, **480-481**
Potassium fluoride, **480, 481-482**
Potassium hexamethyldisilizane, 28
Potassium hydride, 28, 199, 245, **482-483**
Potassium hydridotetracarbonylferrate,
  **483-486**
Potassium hydrogen sulfide, 103
Potassium hydroxide, **486**
Potassium hypochlorite, **486-487**
Potassium iodide, 227
Potassium N-methylanilide, 545
Potassium nitrite, 135
Potassium perchromate, 487
Potassium permanganate—Acetic anhydride,
  487
Potassium peroxydisulfate, 52
Potassium selenocyanate, **487-488**
Potassium sulfide, 10
Potassium superoxide, **488-490**
Potassium thiocyanate, 487
Potassium trialkylborohydrides, 482-483
Potassium tri-sec-butylborohydride, **490-**
  **492**
Potassium tri-sec-butylborohydride—Cuprous
  iodide, **492**
Prednisolone acetate, 298, 299
$\Delta^{14,16}$-Pregnadiene-20-ones, 649
$\Delta^{14}$-Pregnene-20-ones, 649

Pregnenolone, 368
Prenylation, 307-308
Prenyl bromide, 151
Princeton sludge catalyst, 17
Prismanes, 575
Proline, 411, **492-493**
1,3-Propanediamine, 445
1,3-Propanedithiol, 49, **493**
*n*-Propanol, 674
2-Propanol, 17
Propargyl bromide, **493-494**
Propargylic halides, 329
Propargyl tetradropyranyl ether, 602
Propargyltriphenylphosphonium bromide, 494
Propellane lactones, 107
Propenylbenzene, 571
1-Propenyllithium, 495
Propenyltriphenylphosphonium bromide, 667
Propiolic acid, **495-496**
Propionic acid, 149
Propionoin, 409
Propiophenone, 240
*n*-Propylthiolithium, **496**
Propylure, 293
Prostaglandin E$_2$, 101
Prostaglandin F$_{29}$, 489
Prostaglandins, 100, 101, 117, 167, 201, 203, 265-266, 342, 475, 536
Protolichesterinic acid, 366
Pseudoquinones, 673-674
Pummerer rearrangement, 3, 5-6, 209, 369
Purines, 185
Pyrane-2,4-diones, 424
Pyrazolines, 665
Pyrene, 499, 670
Pyridine, 4-5, 126, 190, 470, **497**, 616, 620
Pyridineacrylaldehydes, 268
Pyridinecarboxylic acids, 450
Pyridine hydrochloride, **497-498**
Pyridines, 128
Pyridinium acetamidobenzenesulfonates, 497
Pyridinium chlorochromate, **498-499**
Pyridinium hydrobromide perbromide, **499-500**
Pyridinium polydeuterium fluoride, 286
Pyridinium polyhydrogen fluoride, **285-286**

Pyridinium triflates, 620
(2,6)-Pyridinophanes, 51
2-Pyridones, 479
Pyridoxal pyrophosphate, 114
Pyrocatechase, 429
2-Pyrones, 556-557
Pyrophosphates, 7, 379
Pyrrole-2,4-dicarboxylates, 158
Pyrroles, 617
Pyrrolidine, 652
4-Pyrrolidinopyridine, 112
Pyrrolidone-2-hydrotribromide, 97
Pyruvaldehyde 2-phenylhydrazone, **500**
Pyruvic acid, 493

Quadricyclane, 45
Quadricyclyl-7-carbinyl triflate, 618-619
Quassin, 66
Queen butterfly pheromone, 79
Queen's substance, 217
Quinine, 80, **501**
Quinol acetates, 314-315
Quinoline, 383, 498, 619, 621
Quinoline—Acetic acid, 501
Quinolinecarboxylic acids, 450
*p*-Quinols, 428-429
Quinomycin, 193
Quinones, 120, 121, 555, 556
Quinonimines, 513
3-Quinuclidinol, **501**

Ramberg-Bäcklund rearrangement, 546
Raney cobalt, 172
Raney nickel, 13, 54, 76, 109, 119, 172, 337, 401, **502-503**, 651, 667
Reductive acetylation, 4
Reductive methylation, 528
Reformatsky reaction, 293, 674-675
Retro Diels-Alder reaction, 379
Retro Michael cleavage, 284
Rhodium on alumina, **503**
Rhodium on carbon, **503**
Rhodium carbonyl, **504**
Rhodium trichloride—Silica, 504
Rhodium trichloride hydrate, **504**
Ring expansion, 162-163, 194, 252-253, 314, 468, 672-674
Rose Bengal, 431
Rosenmund reaction, 393
Ruthenium dioxide–Sodium metaperiodate, 504

Ruthenium tetroxide, 504-506

Salcomine, 507
Samandarine, 401
Sarett reagent, 228
β-Schardinger dextrin, 151-152
Schiff bases, 117
Schmidt rearrangement, 316
Scholl reaction, 524
Schotten-Baumann reaction, 106, 252, 531
Schwartz reagent, 175
Seleniranes, 487
Selenium, 507-509
Selenium dioxide, 227, 509-510
Selenoacetals, 28
Selenofenchone, 508, 509
Selenoketones, 508
Selenoxide fragmentation, 28-29, 235, 335, 338, 420-422, 459-460
6-Selenoxo nucleosides, 288
Selinadienes, 211
Serratenediol, 614
Sesterterpenes, 421
Shapiro-Heath olefin synthesis, 598-599
Silica gel, 440, 510
Silicic acid, 510
Siloxene, 510
1-Siloxy-1-vinylcyclopropanes, 522-523
Silver(I) acetate, 52, 287, 511
Silver carbonate, 24, 511
Silver carbonate—Celite, 511-514
Silver fluoride, 514-515
Silver heptafluorobutanoate, 515
Silver nitrate, 171, 227, 426
Silver nitrite, 271, 515
Silver(I) oxide, 286, 515-518
Silver(II) oxide, 518
Silver perchlorate, 69, 246, 518-519
Silver tetrafluoroborate, 29, 90, 123, 240, 246, 519-520
Silver trifluoromethanesulfonate, 520-521, 618
Silyl azides, 561
Simmons-Smith reagent, 70, 323, 521-523
Snoutene, 41, 519
Sodio acetylacetonate, 595
Sodio triethyl phosphonoacetate, 612
Sodium, 215
Sodium—Ammonia, 12, 523
Sodium—t-Butanol—Tetrahydro-

furane, 523-524
Sodium—Potassium alloy, 524
Sodium aluminum chloride, 524-525
Sodium amalgam, 274, 462
Sodium amide, 428, 525-526
Sodium amide—Sodium t-butoxide, 526
Sodium anthracenide, 415
Sodium benzenesulfinate, 526-527
Sodium benzoate, 322
Sodium bis-(2-methoxyethoxy)aluminum hydride, 47, 528-529
Sodium bistrimethylsilylamide, 529-530, 545
Sodium borohydride, 199, 249, 446, 490, 530-534
Sodium borohydride, sulfurated, 534
Sodium chloride—Dimethyl sulfoxide, 534
Sodium cyanide, 38, 535-536
Sodium cyanide—Hexamethylphosphoric triamide, 536
Sodium cyanoborohydride, 537-538
Sodium cyclopentadienyldicarbonylferrate, 538-539
Sodium dichromate, 124, 225
Sodium N,N-dimethyldithiocarbamate, 539-540
Sodium diphenylphosphine, 416
Sodium diselenide, 542
Sodium ethanedithiolate, 540
Sodium ethoxide, 540-541
Sodium 3-(fluoren-9-ylidene)-2-phenyl-acrylate, 160
Sodium glycolate, 186
Sodium hydride, 109, 144, 339, 374, 394, 541-542
Sodium hydride—Dimethylformamide, 542
Sodium hydridotetracarbonylferrate, 485
Sodium hydrogen selenide, 542-543
Sodium hypochlorite, 543
Sodium iodide, 543
Sodium iodide—Copper, 544
Sodium iodide—Sodium acetate, 544-545
Sodium metaperiodate, 34, 396, 459
Sodium methoxide, 294, 545
Sodium N-methylanilide, 545-546
Sodium 2-methyl-2-butoxide, 103, 546
Sodium methylsulfinylmethylide, 546-547
Sodium naphthalenide, 221
Sodium nitrite, 547
Sodium nitrite—Acetic acid, 548

Sodium peroxide, 471, **548**
Sodium selenophenolate, 29, **548-549**
Sodium telluride, **550**
Sodium tetracarbonylferrate(+II), **550-552**
Sodium thiophenoxide, **552**
Sodium thiosulfate, 543
Sodium *p*-toluenesulfinate, 567
Sodium triacetoxyborohydride, **553**
Sodium triallkylcyanoborates, 535
Sodium trichloroacetate, **553**
Sodium trithiocarbonate, **553**
Sodium tungstate, 455
Solid-phase peptide synthesis, 148-149, 193
Solid-liquid phase-transfer catalysis, 566
Spiroannelation, 93-94
Spirobenzylisoquinolines, 200
Spirocyclohexa-2,5-dienonones, 615
Spirodienones, 614-615
1-Spiroisoquinolines, 546, 547
Spirolactones, 212
Spirovetivones, 93-94
Squalene, 328
Squalene-2,3-diol, 9
Stannic chloride, 105, **553-554**, 555, 639
Stannous chloride, **554-555**
13-*epi*-Steroids, 5
Stevens rearrangement, 85, 391, 546-547
Stilbene, 55
*trans*-Stilbene, 555, 557-558
Stilbene episulfide, 375
Stilbene oxide, 375
Styrene, 140, 583
Succindialdehyde, 493
Succinimidodimethylsulfonium tetra-
    fluoroborate, **555-556**
Sugiol, 499
Sulfenes, 190
Sulfenylation, 13, 31, 116, 217-218, 235,
    236, 313
Sulfides, 271-272
Sulfilimines, 230, 239
N-Sulfinylaniline, **556**
Sulfolane, 420, 480, 537
Sulfomonoperacid, **97-98**
Sulfonamides, 323
Sulfonates, 274
Sulfones, 567
Sulfonylation, 618
Sulfonyl chlorides, 497, 585
Sulfoxide fragmentation, 236, 320, 459

Sulfoxides, 529, 585
Sulfur, **556-557**
Sulfur–Hexamethylphosphoric triamide,
    **557-558**
Sulfur dioxide, 23, **558**
Sulfuric acid, **558-560**
Sulfur monochloride, 470-471, **560**
Sulfur tetrafluoride, **560-561**
Sulfur trioxide–Dioxane, 561
Sulfuryl chloride, 118, 298, 510, **561**
Sulfuryl chlorofluoride, 24, **562**
$\beta$-Sultines, 117
Surfactant catalysis, 70

Tartaric acid, 134
Terpenoids, 48-49
Testosterone, 521, 522
1,3,4,6-Tetraacetylglycouril, **563**
Tetraasterane, 317
Tetra-O-benzyl-$\alpha$-D-glucopyranosyl bromide,
    568
Tetrabromoacetone, 195, 196
1,3,5,7-Tetrabromoadamantane, 514
2,4,4,6-Tetrabromocyclohexa-2,5-dienone,
    **563**
1,1,3,3-Tetrabromo-4-methylpentane-2-one,
    196
Tetra-*n*-butylammonium azide, 563-564
Tetra-*n*-butylammonium borohydride,
    **564-565**
Tetra-*n*-butylammonium bromide, 601
Tetra-*n*-butylammonium fluoride, 78, 81
Tetra-*n*-butylammonium halides, **565**
Tetra-*n*-butylammonium hydrogen sulfate,
    **565-566**
Tetra-*n*-butylammonium hydroxide, 563
Tetra-*n*-butylammonium iodide, **566-567**
Tetra-*n*-butylammonium *p*-toluenesulfinate,
    **567**
Tetrachlorobenzyne, 523, 524
Tetrachloroethylene, 413
Tetracyanoethylene, 271, 567-568
Tetracyclines, 109
Tetracyclone, 489, 490, 548, **577**
9,11-Tetradecadien-1-yl acetate, 546
E-7-Tetradecene-1-ol acetate, 12
1,2,4,5-Tetraenes, 146
Tetraethylammonium bromide, 568-569
Tetraethylammonium chloride, 480
Tetraethylammonium cyanide, 569

Tetraethylammonium fluoride, 482
Tetraethylthiuram disulfide, **569**
Tetraethynylethanes, 82
Tetraethynylethylenes, 82
1,3,5,7-Tetrafluoroadamantane, 515
Tetrahydro-Bisnor-S, 121
Tetrahydrofurane, 347, 379, **570**
Tetrahydrofurane-*cis*-2,5-dicarboxylic acid, 616
Tetrahydrofuranes, 352
Tetrahydrofuyranyl ethers, 437
Tetrahydroisophosphinolinium salts, 475
Tetrahydronaphthyl diazomethyl ketones, 615
Tetrahydro-2*H*-1,3-oxazine-2-one, **570**
Tetrahydro-4-phenyltriazole-3, 467
Tetrahydrophosphinolinium salts, 475
Tetrahydropyranes, 276
Tetrahydropyranyl ethers, 437
Tetrahydropyrene, 499
Tetrahydro-γ-pyrone, 391, 392
Tetraisopropylethylene, 589
Tetrakisacetonitrilecopper(I) perchlorate, 130
Tetrakis(phenylthio)ethylene, **651-652**
Tetrakis(phenylthio)methane, 650
Tetrakis(triphenylphosphine)nickel(0), **570, 654**
Tetrakis(triphenylphosphine)palladium(0), **571-573**
Tetralins, 153
1-Tetralone, 242
Tetralones, 503
1,2,3,4-Tetramethoxybenzene, 424
Tetramethylammonium dimethyl phosphate, **573-574**
Tetramethylcyclobutadiene, 575
Tetramethylcyclobutadiene—Aluminum chloride complex, **574-575**
N,N,N′,N′-Tetramethyldiaminophosphorochloridite, **575-576**
Tetramethyl-1,2-dioxetane, 287
Tetramethylethylene, 176
Tetramethylethylenediamine, 163, 202, 203, 278, 385, 398, **576-577**, 598
1,1,3,3-Tetramethylguanidine, 254, 452
2,2,6,6-Tetramethylpiperidine, 110
2,2,6,6-Tetramethylpiperidine-1-oxyl, 110, 111
α-(2,4,5,7-Tetranitro-9-fluoroenlidene-

aminoxy)propionic acid, 577
Tetraphenylcyclopentadienone, 577
Tetraphenylethylene, **577**
Tetra-3-pinanyldiborane, **202**
Thaliporphine, 315
Thallium(I) alkoxides, 577
Thallium(I) ethoxide, 286, **577-578**
Thallium(I) hydroxide, **578**
Thallium(III) nitrate, **578-579**
Thallium(III) trifluoroacetate, **579-580**
Thallous 2-methylpropane-2-thiolate, **581-582**
Thebaine, 581
Thexylborane, **207-208**
Thexyldialkylboranes, 207
Thexylmonoalkylboranes, 207
9-Thiabarbaralane 9,9-dioxide, 23
3-Thiabicyclo[3.2.0]hepta-1,4-dienes, 470
Δ³-1,3,4-Thiadiazolines, 643
Thiiranes, 375, 488, **584-585**
Thiiranium salts, 375
Thiirenium salts, 375
Thioacetal interchange, 381-382
Thioacetals, 406
Thioacyl chlorides, 285
Thioanisole, 18, **582**
Thiobenzoic acid O-esters, 582
Thiobenzophenones, 197
Thiobenzoyl chloride, **582-583**
(Thiobenzoylthio)acetic acid, 583
Thiocarbamates, 509
N,N′-Thiocarbonyldiimidazole, **583-584**
Thio-Claisen rearrangements, 540
Thiofenchone, 508, 509
Thioketals, 296, 399
Thioketenes, 285
Thiolactones, 197
Thiol esters, 192-193
2-Thiomethyl-4,4-dimethyl-2-oxazoline, **584-585**
2-Thiomethyl-2-thiazoline, 585
Thionyl chloride, 75, **585**
Thiophenes, 667
Thiophenol, **585-586**
1-Thiophenoxy-3-chloroalkenes, 605
1-Thiophenoxycycloheptanecarboxaldehyde, 131, 132
Thiourea, **586**
Thiourea dioxide, **586**
α-Thujaplicin, 196

Titanium(III) acetate, 4, 586-587
Titanium(III) chloride, 327, 587-588
Titanium(III) chloride–Lithium aluminum hydride, 588-589
Titanium(III) chloride–Tetrahydrofurane–Magnesium, 590
Titanium(IV) chloride, 590-596
Titanium(IV) chloride–Lithium aluminum hydride, 596
Titanocene dichloride, 48-49, 179, 496
Titanocene system, 596-597
Toluene, 55
p-Toluenesulfenyl chloride, 30, 510
p-Toluenesulfinyl chloride, 585
α-p-Toluenesulfinyl ketones, 400
p-Toluenesulfonic acid, 1, 186, 597
p-Toluenesulfonyl azide, 597
p-Toluensulfonylazocyclohexene-1, 234
p-Toluenesulfonyl chloride, 598
p-Toluenesulfonylhydrazine, 232, 598-600
p-Toluenesulfonylhydrazones, 74-75, 98, 161, 543, 587
p-Toluenesulfonyl isocyanate, 467
N-Tosylsulfilimines, 586
2-p-Tolylsulfonylethyl chloroformate, 55
Tosyl azide, 597
β-Tosylethoxycarbonyl chloride, 55
p-Tosyl isocyanate, 633
Tosylmethyl isocyanide, 566, 600
Toxoflavin, 547
Transacetalization, 186-187
Transesterification, 64, 645
Transfer hydrogenation, 650, 652, 655
Tresyl chloride, 578
Triacetic lactone methyl ether, 509
1,4,6-Triacetoxynaphthalene, 370
Trialkylboranes, 24, 87, 212, 259, 293-294, 324, 398
Trialkyl borohydrides, 349
Trialkynylalanes, 600-601
Triamantane, 472-473
2,4,6-Triaminopyridines, 160
2,4,6-Tri-t-butylaniline, 560
1,3,5-Tribromobenzene, 147
Tri-n-butylamine, 596, 601
Tri-sec-butylborane, 490
Tri-t-butylcyclopropenyl fluoroborate, 478
Tributylethoxytin(III), 56, 57
2,4,6-Tri-t-butylphenol, 314
Tri-n-butylphosphine, 144, 643

Tri-n-butylphosphine–Dialkyl disulfides, 602
Tri-n-butylstannane, 649
1-Tri-n-butylstannyl-1-propene-3-tetra-hydropyranyl ether, 602-603
Tri-n-butyltin chloride, 604
Tri-n-butyltin hydride, 604
1,2,3-Tricarbonyl compounds, 227
Trichloroacetonitrile, 604-605
2,2,2-Trichloroethanol, 605
2,2,2-Trichloroethyl dihydrogen phosphate, 606
2,2,2-Trichloroethyl esters, 553
Trichlorofluoromethane, 9
Trichloroisocyanuric acid, 605-606
1-Trichloromethyl-2-chlorocyclohexane, 655
Trichlorosilane, 606
α,α,α-Trichlorotoluene, 70
1,1,1-Trichloro-3,3,3-trifluoroacetone, 606-607
Tricyclanone, 515
Tricyclo[5.2.1.0²,⁶]decane-3-one, 286, 287
Tricyclo[3.1.1.0³,⁶]heptanes, 150
Tricyclo[4.2.0.0²,⁵]octa-3,7-diene, 274
trans-Tricyclo[5.1.0.0²,⁴]octene-5, 68
Tricyclo[5.2.2.0²,⁶]undeca-3,8-diene, 359
1,4,5-Trienes, 638-639
Triethylallyloxysilane, 607
Triethylamine, 25, 32, 67, 91, 158, 172, 192, 193, 202, 218, 219, 379, 393, 425, 443, 509, 634, 665
Triethylamine hydrofluoride, 632
Triethylaminoaluminum hydride, 607
Triethylchlorosilane, 607
Triethylmethylammonium dimethyl phosphate, 574
Triethyl orthoacetate, 607-610
Triethyl orthoformate, 610-611
Triethyloxonium tetrafluoroborate, 250, 564, 611-612
Triethyl phosphite, 612, 663
Triethyl phosphonoacetate, 612
Triethylsilyl thiophenoxide, 613
Triethyltin methoxide, 613
Triflic acid, 617-618
Triflic anhydride, 36, 578, 618-620
Trifluoroacetic acid, 54, 84, 109, 124, 613-616, 657, 658
Trifluoroacetic acid–Alkylsilanes, 616

Trifluoroacetic anhydride, 616-617
Trifluoroacetylation, 606-607
2,2,2-Trifluoroethanesulfonyl chloride, 578
Trifluoroethanol, 618
2,2,2-Trifluorethylamine, 617
Trifluoromethanesulfonic acid, 617-618
Trifluoromethanesulfonic-alkanesulfonic anhydrides, 618
Trifluoromethanesulfonic anhydride, 44, 225, 618-620
Trifluoromethanesulfonylimidazole, 620-621
Trifluoromethyl sulfides, 621
Trifluoromethylthiocopper, 621
Trifluoro(trichloromethyl)silane, 621
Trifluorovinyllithium, 622
3α,5β,12α-Trihydroxycholanic acid, 166
Triisobutylaluminum, 198
2,4,6-Triisopropylbenzenesulfonyl chloride, 622
Trimethylaluminum, 49, 622-623
Trimethylamine N-oxide, 624-625
2,4,6-Trimethylbenzenesulfonyl chloride, 625
Trimethyl borate, 79, 261
Trimethylcarbenium fluoroantimonate, 24
Trimethylchlorosilane, 352, 626-628
Trimethylchlorosilane–Zinc, 628
4,4,6-Trimethyl-2-cyclohexenone, 599, 600
Trimethyl-1,2-dioxetane, 287
Trimethylenediamine, 476
Trimethylene dithioacetals, 382
α-Trimethylenedithiocyclobutanones, 242
Trimethylene dithiotosylate, 242, 628-629
2,4,4-Trimethyl-2-oxazoline, 629
N,4,4-Trimethyl-2-oxazolinium iodide, 630
Trimethyloxonium tetrafluoroborate, 378, 630
Trimethyloxosulfonium chloride, 71
Trimethyloxosulfonium iodide, 41
Trimethyl phosphite, 30
Trimethyl phosphite–Copper(I) iodide, 630-631
Trimethylsilanol, 84
α-Trimethylsiloxyacrylonitriles, 633
Trimethylsilylacetic acid, 631-632
Trimethylsilylacetone, 84
1-Trimethylsilyl-1-alkynes, 284
Trimethylsilyl azide, 632

Trimethylsilylbenzyllithium, 637
3-Trimethylsilyl-3-butene-2-one, 406-407
α-Trimethylsilyl-γ-butyrolactones, 631
α-Trimethylsilyl carboxylic acids, 631
Trimethylsilyl cyanide, 80, 135, 632-633
Trimethylsilyl cyclopropyl ethers, 70, 71
Trimethylsilyldiethylamine, 634
Trimethylsilyl enol ethers, 44, 70, 112, 162, 264
Trimethylsilyl ethers, 58, 515
Trimethylsilylethynyl ketones, 148
Trimethylsilyl isocyanate, 634
Trimethylsilylketene, 635
Trimethylsilylmethyl ketones, 635
Trimethylsilylmethyllithium, 635-636, 637
Trimethylsilylmethylmagnesium chloride, 636-637
Trimethylsilylmethylpotassium, 637-638
Trimethylsilyl perchlorate, 639
1-Trimethylsilylpropynylcopper, 638-639
α-Trimethylsilyl sulfoxides, 626
Trimethylsilyltrifluoroacetamide, 412
Trimethylsilyl trifluoromethanesulfonate, 639
Trimethylsilyl urethane, 412
Trimethylstannylethylidenetriphenylphosphorane, 640
Trimethylsulfonium iodide, 219
Trimethyltinlithium, 640
Trimethylvinylsilane, 100
1,3,5-Trinitrobenzene, 157-158
Trioctylpropylammonium chloride, 452, 640-641
Trioxane, 590
Trioxo(t-butylnitrido)osmium(VIII), 641-642
Triphase catalysis, 38
1,1,1-Triphenylethane, 623
Triphenylethylene, 42
Triphenylmethanol, 623, 647
Triphenylmethyl isocyanide, 642-643
Triphenylmethyllithium, 334
Triphenyl orthothioformate, 650
Triphenylphosphazines, 508
Triphenylphosphine, 59, 60, 61, 108, 125, 246, 443, 451, 643-647
Triphenylphosphine–Carbon tetrabromide–Lithium azide, 644
Triphenylphosphine–Carbon tetrachloride, 644-645

Triphenylphosphine–Diethyl azodi-
carboxylate, 280, **645**
Triphenylphosphine dibromide, 645-646
Triphenylphosphine dichloride, 646-647
Triphenylphosphine dihalides, **647-648**
Triphenylphosphine diiodide, 648
Triphenylphosphine ditriflate, 648
Triphenylphosphite methiodide, **649**
Triphenyl phosphite ozonide, 435
Triphenyl(prop-2-ynyl)phosphonium
bromide, **494**
Triphenylsilanol, 655, 656
Triphenylstannane, **649**
Triphenyltin hydride, **649**
Triphenylvinylsilane, 637
Triphosphopyridine nucleotide, 36-37
Tri-*n*-propyl orthovanadate, 655
Tris-annelation, 410
Tris(aquo)hexa-μ-acetato-μ$_3$-oxotri-
ruthenium acetate, **425-426, 650**
Tris(dimethoxyboryl)methane, 328, 329
Tris(dimethylamino)borane, 554
Trishomocubanone, 108
Tris(neomenthyldiphenylphosphine)-
chlororhodium, 416
Tris(phenylthio)methyllithium, 650-651
Tris(tetra-*n*-butylammonium)hexacyano-
ferrate(III), **656**
N,N′,N″-Tris(tetramethylene)phosphor-
amide, 651
Tris(trimethylsilyl)hydrazidolithium, 651
Tris(triphenylphosphine)chlororhodium,
613, **652-653**
Tris(triphenylphosphine)nickel(O), **653-654**
Tris(triphenylphosphine)ruthenium di-
chloride, **654-655**
Tris(triphenylsilyl)vanadate, **655-656**
*s*-Trithianes, 296
Tritiation, 89
Tri-*o*-tolylphosphine, 448
Triton B, 44, 446, 461
Trityl chloride, 656-657
Trityl isocyanide, **642-643**
Trityl tetrafluoroborate, **657**
Trityl trifluoroacetate, **657-658**
Tropine, 200
Tropinone, 200
α-Tropolone, 33
α-Tropolone methyl ether, 536
Troponoids, 195-196

*p*-Tropoquinone, 168
Tungsten hexacarbonyl, **658**
Tungsten hexachloride, 587
Tyrosine, 101

Ullmann diphenyl ether synthesis, 451-452
Ullmann reaction, 130, 132, 184, 363, 654
1,4-Undecadiene, 198
11-Undecanolide, 479
2,2,E-2,4,6-Undecatriene, 674
2-Undecene, 495
U-Ni-N, 659
α,β-Unsaturated acetals, 270
α,β-Unsaturated acetylenic esters, 163
α,β-Unsaturated acids, 631
β,γ-Unsaturated acids, 384-385
α,β-Unsaturated alcohols, 19
α,β-Unsaturated aldehydes, 13-14, 50, 461,
532, 538, 539-540, 594, 605, 651
β,γ-Unsaturated aldehydes, 366-367, 592
β,γ-Unsaturated amides, 84
α,β-Unsaturated butenolides, 460
α,β-Unsaturated γ-butyrolactones, 137
α,β-Unsaturated diazo ketones, 597
α,β-Unsaturated epoxides, 305
α,β-Unsaturated esters, 129, 144, 180, 217-
218, 393-394, 383-385, 561
β,γ-Unsaturated esters, 84, 635, 636
γ,δ-Unsaturated esters, 141
α,β-Unsaturated ketones, 49, 144, 320-321,
449, 459-460, 502, 522, 533, 537-538,
592, 623, 650, 651
γ,δ-Unsaturated ketones, 211-212
α,β-Unsaturated γ-lactones, 200
β,γ-Unsaturated methyl sulfinates, 381
α,β-Unsaturated nitriles, 351, 394, 535
β,γ-Unsaturated nitriles, 84
α,β-Unsaturated sulfones, 42-43, 436
α,β-Unsaturated sulfoxides, 193, 394
α,β-Unsaturated tosylhydrazones, 537
β,γ-Unsaturated-δ-valerolactones, 188
Uracil, 639, 640
Urethanes, 322
Uridine, 221
Uridine tri-O-benzoate, 639
Urushibara catalysts, **659**
Urushibara nickel A, 659

δ-Valerolactones, 402
Vanadium acetyl acetonate, 183

Vanadium oxytrifluoride, **660**
Vermiculine, 247
Vertaline, 246, 247
α-Vetispirene, 93-94
β-Vetivone, 93, 94, 213
Victor Meyer reaction, 272
Vilsmeier reagents, 66, 220
Vinyl acetates, 142-143
Vinylalanes, 198
Vinylallenes, 341
Vinyl azides, 157
Vinylboranes, 207
Vinyl bromides, 142, 663
β-Vinylbutenolide, **661-662**
β-Vinylbutyrolactone, 661
Vinylcarbinols, 1
Vinyl carbomethoxymethyl ketone, **662**
Vinylcopper reagents, **662-663**
Vinylcyclohexane, 640
4-Vinylcyclohexene, 155
Vinylcyclopropanes, 129, 378, 664-665
1-Vinylcyclopropanols, 522
Vinyldiazomethane, **664-665**
1,2-Vinylenebis(triphenylphosphonium)
    dibromide, **665**
Vinylenedisulfonium salts, 630
Vinyl ethers, 665
Vinyl fluorides, 262
Vinyl formamides, 187
Vinyl halides, 60, 178
Vinyl iodides, 142, 143, 662-663
Vinylketene, 291
Vinyl ketones, 650
Vinyllithium, 325, 342, **665-666**
Vinylmagnesium halides, 1
β-Vinylnaphthalene, 303
Vinyl oxiranes, 305
Vinyl phenyl sulfides, 130-131
Vinylsilanes, 482
Vinyl sulfides, 359
Vinyl triflates, 619, 620-621
Vinyltriphenylphosphonium bromide, 640,

666-667
Vitamin A, 411, 412
Vitamin A alcohol, 527

Wacker oxidation, 124
Wagner-Meerwein rearrangement, 618
Wharton reaction, 599
Wieland-Miescher ketone, 410
Wilkinson complexes, 416
Willgerodt reaction, 21
Williamson synthesis, 24
Wittig-Horner reactions, 190, 191, 193, 339,
    340, 566
Wittig reaction, 94, 135, 268, 274-275, 339,
    366-367, 368, 378, 389, 404, 509-510,
    546, 640, 667, 668
Wittig rearrangement, 85

Xanthones, 138-139
Xenon difluoride, 669-670

Yangonin, 509-510
Ylango sesquiterpenes, 530
Yneallenes, 341-342
Yohimban-17-one, 400
Yomogin, 491
Young test, 28
Ytterbium(III) nitrate, 671

Zearalenone dimethyl ether diethylene
    ketal, 581, 582
Ziegler-Thorpe cyclization, 545
Zinc, 183, **672-675**
Zinc–Acetic acid, 18, **500**
Zinc–Zinc chloride, **675**
Zinc acetate, 3
Zinc borohydride, 222
Zinc chloride, 123, 266, 370, 674,
    **676**
Zinc-copper couple, 195, 262
Zinc fluoride, 514
Zirconocene dichloride, 176

# ERRATA FOR VOLUMES 4-6

**4**, 99. The work on oxygenation of alkyltoluenes was carried out by chem-
of Gulf Oil.

**4**, 447. *For* M. G. Hutchings *read* M. G. Hutchins.

**4**, 522.    Trichloroisocyanuric acid and cyanuric chloride have different structures. The formula [(1), **4**, 522] corresponds to the latter reagent. The correct formula for the former reagent is shown in **2**, p. 426.

**4**, 572.    The reference for the preparation of vinylmagnesium chloride should
*read* H. E. Ramsden, J. R. Leebrick, S. D. Rosenberg, E. H. Miller, J. J.
Walburn, A. E. Balint, and R. C. Serr, *J. Org.*, **22**, 1602 (1957).

**4**, 600.  *For* L. F. Feiser *read* L. F. Fieser. *For* J. L. Garnatt *read* J. L. Garnett.

**4**, 603, *delete* M. G. Hutchings. *For* M. G. Hutchins *read* 447, 559.

**4**, 651, *for* Phenacy chloride *read* Phenacyl chloride.

**5**, 245. Change heading to **Dimethyl diazomalonate.**

**5**, 743. Read **Urushibara hydrogenation catalyst.**

**5**, 824. Read Aliquat 336 (trade name).

**6**, 70.    Ref.[11], change M. C. Stillings to M. R. Stillings.

**6**, 141. Ref.[5], change M. Bock to M. Boch.

**6**, 172. Ref.[2], bottom of page, change L. Taoguy to L. Tanguy.

**6**, 201. Ref.[10], change R. G. Saloman to R. G. Salomon.

**6**, 232. Ref.[2], change T. Ihigami to T. Ishigami.

**6**, 428, 430. Ref.[8], change I. S. De Roch to I. S. de Roch.